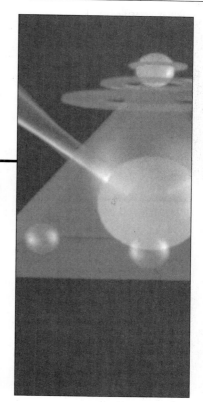

ANALYTICAL CHEMISTRY

Sixth Edition

Gary D. Christian
University of Washington

JOHN WILEY & SONS, INC.

EXECUTIVE EDITOR   Deborah Brennan
MARKETING MANAGER   Bob Smith
SENIOR PRODUCTION EDITOR   Elizabeth Swain
SENIOR DESIGNER   Kevin Murphy
ILLUSTRATION EDITOR   Sandra Rigby

Cover/chapter opening photo: Gregory Manicol/Photo Researchers

This book was set in 10.5 Times Roman by Matrix Publishing Services and printed
and bound by Courier Westford. The cover was printed by Lehigh Press.

This book is printed on acid-free paper. ∞

Library of Congress Cataloging in Publication Data
Christian, Gary D.
    Analytical chemistry / Gary D. Christian — 6th ed.
        p. cm.
    Includes index.
    ISBN 0-471-21472-8 (cloth)
    Wiley International Edition ISBN 0-471-45162-2
        1. Chemistry, Analytic—Quantitative. I. Title.

    QD101.2. C57 2003
    545—dc21   2002038040

Printed in the United States of America

10 9 8 7 6 5

To

Sue—for decades of joy and for her amazing grandmothering
Tanya and Taffy—for the sunshine they bring in our lives

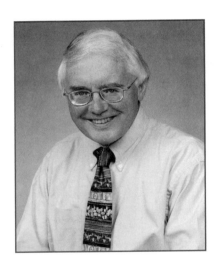

Gary Christian grew up in Oregon and has had a lifelong interest in teaching, inspired by a great high school science teacher in a rural school, Harald Platou. He received his B.S. degree from the University of Oregon and Ph.D. degree from the University of Maryland, where he gained an appreciation of the excitement of research from his mentor, Bill Purdy. He began his career as a research analytical chemist at Walter Reed Army Institute of Research, where he developed an interest in clinical and bioanalytical chemistry. He joined the University of Kentucky in 1967 and in 1972 moved to the University of Washington. He was Divisional Dean of Sciences during 1993–2001.

Gary wrote the first edition of this book in 1971. He is the author of over 300 research papers and has authored five other books, including *Instrumental Analysis*. His research interests include electroanalytical chemistry, atomic spectroscopy, process analysis, and flow injection analysis.

He was recipient of the American Chemical Society (ACS) Division of Analytical Chemistry Award for Excellence in Teaching and the ACS Fisher Award in Analytical Chemistry, a top recognition for contributions in analytical chemistry. He was a Fulbright Scholar and received the Talanta Gold Medal, Université Libre de Bruxelles Medal of Honor, Charles University Commemorative Medal, and University of Ghent Honorary Certificate of Research. He is joint editor-in-chief of *Talanta*, an international journal of analytical chemistry, and serves on the editorial boards of numerous other journals. He served as Chairman of the ACS Division of Analytical Chemistry.

Gary chaired preparation of the ACS Examination on Analytical Chemistry and was a preparer for the Chemistry Test for the GRE. He was a member of the team that prepared the exam for the International Chemistry Olympiad when it was held in the United States. He is a member of the American Chemical Society, Society for Applied Spectroscopy, Spectroscopy Society of Canada, and Society for Electroanalytical Chemistry.

Gary lives in Medina, Washington, with his wife of 41 years, Sue, and their wonderful granddaughters, Tanya and Taffy.

Preface

"Teachers open the door, but it is up to you to enter" —Anonymous

Analytical chemistry is concerned with the chemical characterization of matter, both qualitative and quantitative. It is important in nearly every aspect of our lives because chemicals make up everything we use. The late Charles N. Reilley said "analytical chemistry is what analytical chemists do." You will learn in this text what they do.

This text is designed for college students majoring in chemistry and fields related to chemistry. It deals with the principles and techniques of quantitative analysis, that is, how to determine how much of a specific substance is contained in a sample. You will learn how to design an analytical method, based on what information is needed, how to obtain a laboratory sample that is representative of the whole, how to prepare it for analysis, what measurement tools are available, and the statistical significance of the analysis. Chapters 24–26 illustrate applications of techniques you have learned to the fields of clinical chemistry, genomics and proteomics, and environmental sampling and analysis.

Examples of the use of analytical chemistry techniques are drawn from such areas as life sciences, clinical chemistry, air and water pollution, and industrial analyses. Analytical chemistry becomes meaningful when you realize that an incorrect blood analysis may endanger a patient's life, or that an error in quality control analysis may result in serious financial loss for a manufacturer. Millions of dollars are saved in the chemical industry by performing on-line automated analyses of chemical processes, to assure maximum efficiency in chemical production.

WHO SHOULD USE THIS TEXT?

The text is written for an undergraduate quantitative analysis course. It necessarily contains more material than normally can be covered in a one-semester or one-quarter course, so that your instructor can select those topics deemed most important. Some of the remaining sections may serve as supplemental material. Depending on how a quantitative analysis and instrumental analysis sequence is designed, it could serve for both courses. In any event, I hope you will take time to read some sections that look interesting to you that are not formally covered. They can certainly serve as a reference in the future.

WHAT IS THE SAME?

This sixth edition of *Analytical Chemistry* is extensively revised and updated (more on that below), but many features from previous editions remain. Each chapter is introduced with a summary paragraph that lists the topics to be covered, giving you a broad overview of each subject. **Boldface type** is used for key terms, and important equations are boxed to aid in review. *Margin notes* are generously used to further emphasize important concepts and aid in review.

Dimensional analysis is emphasized throughout to give you a better feel for the proper setting up of problems. SI units or symbols (e.g., L, mL, mol, and s) are used throughout. The concept of normality and equivalents is introduced, but emphasis remains on the use of molarity and moles. The presentation of normality is done in a way that allows it to be ignored if your instructor chooses not to assign it.

Problems and Recommended References are grouped by topic, for ease in assignment. References have been extensively updated, and numerous new problems added. There are 673 questions and problems for you to practice answering (there is Solutions Manual for these—see below).

A number of new topics were introduced in the prior edition, and most remain as important features of the text. Some are:

- Statistics of small sets of data
- Statistics of sampling
- Systematic approach to equilibrium calculations (mass and charge balance)
- Heterogeneous equilibria
- Logarithmic diagrams for describing multiple equilibria species (the preparation of these using spreadsheets is now introduced)
- Diode array spectrometers
- Fourier transform infrared spectroscopy
- Near-IR spectroscopy
- Fiber-optic sensors
- Gas chromatography–mass spectrometry

WHAT IS CHANGED?

Some chapters have been reordered to better tie together related topics. The chapter on basic tools and operations in analytical chemistry has been moved to the front of the text (Chapter 2), at the suggestion of a number of users. Different sections of it can be assigned by your instructor as needed for the laboratory. The chapter on gravimetric analysis and precipitation equilibria (Chapter 10) has been moved to just before the one on precipitation reactions and titrations, which better aligns these related topics.

Chromatography, a major analytical tool for analyzing mixtures of analytes, has undergone significant growth in recent years, with improved capabilities, and so the coverage of chromatographic techniques has been expanded and updated, comprising three chapters on principles of chromatography, gas chromatography, and liquid chromatography (Chapters 19–21). Older techniques such as paper chromatography have been deleted.

WHAT IS NEW?

This revision follows publication of the fifth edition of *Analytical Chemistry* by nearly a decade, a result of being a dean too long! Much has changed in that time. This sixth edition presents a number of new topics, new chapters, and changes in presentation. Major additions include:

- **Color** added to the text, for a more pleasing layout
- Spreadsheets (using Excel) introduced and used throughout the text for performing computations, statistical analysis, and graphing.

Many titration curves are derived using spreadsheets, as are the calculations of α-values and plots of α-pH curves, and of logarithm concentration diagrams. The spreadsheet presentations are given in a "user-friendly" fashion to make it easier for you to follow how they are set up. The way in which spreadsheet calculations are performed is by entering formulas in specific cells, e.g., cell B11 may contain a formula to calculate the ratio of numbers entered in cells A2 and A3, and the formula is (=A2/A3). The answer appears in cell B11 where the formula is entered. Cells that have formulas entered in them are **boldfaced**. Usually, you have to "reverse engineer" the cell entries to determine the equations from which they were derived, for example, the equation dividing one number by another to obtain the fraction. This is very awkward for complicated equations. To avoid this, the actual equations used for deriving the cell formulas are given in the documentation section below the spreadsheet setups; the formula to be entered in the identified cell number is given just below the equation. You will better understand and appreciate this when you begin studying the use of spreadsheets.

- New chapter on Good Laboratory Practice: Quality Assurance of Analytical Measurements

This aspect of the practice of analytical chemistry has become increasingly important as government agencies have established rather complex and stringent guidelines to assure that analytical measurements are accurate, for establishing or enforcing policy. Any budding analytical chemist will find that knowledge of these guidelines will be viewed favorably by an employer. This chapter serves as an introduction and reference to current practices, and includes:

- Validation of analytical methods
- Quality assurance
- Laboratory accreditation
- Electronic records and electronic signatures (the new 21 CRF, Part 11 regulations)

- New chapter on Genomics and Proteomics

Analytical chemistry played a key role in the completion of the historic Human Genome Project. You should know what that is. The technologies have become routine for DNA sequencing of complex organisms, for forensic science, and so forth. We now move into the realm of protein profiling (proteomics), an even more challenging analytical endeavor, and an introduction to this emerging field is given. This chapter discusses:

- The Human Genome Project
- How genes are sequenced
- The polymerase chain reaction (PCR)
- DNA chips
- 2-D PAGE and MALDI-TOF for protein profiling

A number of new topics throughout the text include:

- Calibration of glassware
- Accelerated and microwave extraction and digestion
- Solid-state ISFET electrodes
- Spectral databases—web-based: commercial and free
- Solid-phase extraction (SPE) expanded; solid-phase microextraction (SPME)
- Chromatography nomenclature: IUPAC-recommended terms and symbols used
- Theory of chromatography column efficiency expanded
- Chromatography simulation software for method development
- Capillary gas chromatography (GC) columns: updated and expanded
- Headspace, thermal desorption, and purge and trap GC analysis
- Fast gas and liquid chromatography
- High performance liquid chromatography-mass spectrometry (HPLC-MS)
- Mass analyzers for GC-MS and LC-MS
- HPLC stationary phases updated; narrowbore columns for high sensitivity
- Capillary electrophoresis expanded; capillary electrochromatography

EXPERIMENTS

There are 40 experiments illustrating most of the measurement techniques presented in your text. Experiments are grouped by topic. Each contains a description of the principles and chemical reactions involved, so you gain an overview of what is being analyzed and how. Solutions and reagents to prepare in advance of the experiment are listed, so experiments can be performed efficiently; your instructor will have prepared many of these. Experiments are designed where possible to avoid the use of asbestos, chloroform, carbon tetrachloride, or benzene, consistent with occupational health and safety. All experiments, particularly the volumetric ones, have been designed to minimize chemical waste by preparing the minimum volumes of reagents, like titrants, required to complete the experiment.

Deleted.   Three experiments that use mercury have been deleted. The paper chromatography experiment is deleted because thin-layer chromatography is predominantly used in its place today. For space reasons, the presentation of catalytic methods and the corresponding experiment are deleted from the text in favor of enzymatic kinetic methods. Also, the anion chromatography separation of cobalt and nickel is omitted.

New.   A new microscale titration experiment is included, provided by Professor John Richardson from Shippensburg State University, for the analysis of hard-water samples (Experiment 18). The tools and techniques used for that experiment could be used to design similar experiments for other titrations if desired. (If your instructor tries this with you, I may include your experiment in the next edition!) Two **team experiments** are added (Experiments 39 and 40) to illustrate the principles presented in Chapter 4 on statistical validation. One is on method validation and quality control, in which different members of teams perform different parts of the validation for a chosen experiment. The other is on proficiency testing, in which you calculate the z-values for all the student results of one or more class experiments and you compare your z-value to see how well you have performed.

Spreadsheets. You are encouraged (actually instructed) to use spreadsheets in your experiments to prepare calibration curves and to perform statistical analysis on your experimental results.

CD-ROM

Your textbook includes a CD-ROM that contains useful supplemental material to complement the text. When opening the CD-ROM, click first on "Read me" for details of the contents. There are files on chapter auxiliary data, chapter spreadsheet figures, chapter text spreadsheets, laboratory apparatus, spreadsheet problem solutions, and website URLs. There are useful hints for using the spreadsheets.

A WORD ABOUT WEBSITES

There are over 100 websites given throughout the text for access to useful supplemental material. I have heavy fingers when typing URLs, and the Web is unforgiving of typos. To efficiently access the websites, a list of all the URLs is posted on the Wiley website for your text, by chapter in order of appearance, including margin notes. You can use this list to access the websites without typing the URLs.

Click on the URL to open the site. If this does not work, copy it to your clipboard and then paste into your browser **Address** and then click on **Go** (or **Enter** on your keyboard). Sometimes a linked site won't open, but if you input just the homepage portion of the URL, you can open this and then link to the final URL. For company homepages, you may need to go the the Product link. Or try the site's search engine. Some URLs will change or be deleted with time. If you have trouble getting a website to open, or if it is changed, try going to a search engine, such as Google, and search for the company or topic.

TEXT WEBSITE

John Wiley & Sons, Inc. maintains a website for *Analytical Chemistry* that contains additional supplemental material, which may be updated or added to from time to time. Any text errors that are noted will be posted on this site. Materials on the website include supplemental materials for different chapters that expand on abbreviated presentations in the text. Chapters from *The Encyclopedia of Analytical Chemistry* on "Literature Searching Methodology" and "Analytical Problem Solving: Selection of Analytical Methods" are included. The website URLs in the text are also listed on this site and may be updated. **All figures and tables in the text are posted on the website and can be downloaded for preparation of transparencies.** You may access the website at: www.wiley.com/college/christian.

THANKS

The production of your text involved the assistance and expertise of numerous people. Special thanks go first to the users of the text who have contributed comments and suggestions for changes and improvements; these are always welcome. A number of colleagues served as reviewers of the text and manuscript and have aided immeasurably in providing specific suggestions for revision. They, naturally, express opposing views sometimes on a subject or placement of a chapter or section, but collectively have assured a near optimum outcome that I hope you find easy and enjoyable to read and study. Special mention goes to Professors Dennis Anjo (California State University at Long Beach), Kevin Chambliss (Baylor University),

Michael DeGrandpre (University of Montana), Jinmo Huang (The College of New Jersey), Ira Krull (Northeastern University), Gary Long and Harold McNair (Virginia Tech), Jody Redpenning (University of Nebraska), John Richardson and Thomas Schroeder (Shippensburg State University), Benjamin Rusiloski (Delaware State University), James Rybarczyk (Ball State University), Cheryl Klein Stevens (Xavier University of Louisiana), and Phillip Voegel (Midwestern State University). Professor Norman Dovichi (University of Washington) provided valuable input on Chapter 25 on genomics and proteomics. And thanks to Mack Carter for his computer wizardry, and Sheila Parker for helping keep my head above water.

The professionals at John Wiley & Sons have been responsible for producing a high-quality book. David Harris and Deborah Brennan, Acquisitions Editors, shepherded the process from beginning to end. Their assistant, Cathy Donovan, handled the whole review process and attended to many details. Elizabeth Swain was the production editor, arranging copyediting to printing. Sandra Rigby was the illustration editor responsible for the artwork in your text. Ernestine Franco at Pern Editorial Services was a real professional in copyediting the manuscript for a smooth printing. It has been a real pleasure working with this team and others in a long but rewarding process.

My wife and best friend, Sue, has been my strongest supporter during this two-year exercise. She kept the process on schedule by her efficient translation and typing of my scribbled handwritten pages. Thanks, thanks, thanks!

SOLUTIONS MANUAL

A comprehensive solutions manual is available for use by instructors and students in which all problems are completely worked out and all questions are answered. Answers for spreadsheet problems, which include the spreadsheets, are given in your CD-ROM. Answers to even-numbered problems are given in Appendix F.

GARY D. CHRISTIAN
January, 2003
Seattle

"To teach is to learn twice." —Joseph Joubert

Contents

Chapter 4
Good Laboratory Practice: Quality Assurance of Analytical Measurements   124

Chapter 5
Stoichiometric Calculations: The Workhorse of the Analyst   141

Chapter 6
General Concepts of Chemical Equilibrium   189

Chapter 7
Acid—Base Equilibria   219

Chapter 14
Redox and Potentiometric Titrations   414

Chapter 15
Voltammetry and Electrochemical Sensors   446

Chapter 16
Spectrochemical Methods   457

Chapter 17
Atomic Spectrometric Methods   522

Chapter 18
Sample Preparation:
Solvent and Solid-Phase Extraction   541

EXPERIMENTS   727

List of Spreadsheets Used throughout the Text

The use of spreadsheets for plotting curves and performing calculations is introduced in different chapters. Following is a list of the various applications of Microsoft Excel, by category, for easy reference for different uses. All spreadsheets are given on the CD. The Problem spreadsheets are only on the CD; others are in the text but are also on the CD. You should always practice preparing assigned spreadsheets before referring to the CD. You can save the spreadsheets on your CD to your desktop for use.

Use of Spreadsheets (Section 3.8)

Filling the Cell Contents, 78

Saving the Spreadsheet, 80

Printing the Spreadsheet, 80

Relative versus Absolute Cell References, 80

Use of Excel Statistical Functions (Paste functions), 81

Useful Syntaxes: LOG10, PRODUCT; POWER; SQRT; AVERAGE; MEDIAN; STDEV; TTEST; VAR, 82

Statistics Calculations

Standard Deviation: Chapter 3, Problems 14, 15, 16, 20, 21

Pooled Standard Deviation: Chapter 3, Problem 27

F-Test: Chapter 3, Problems 25, 26, 28

t-Test: Chapter 3, Problems 29, 30

t-Test, multiple samples: Chapter 3, Problem 42

Propagation of Error: Chapter 3, Problems 17 (add/subtract), 18 (multiply/divide)

Using Spreadsheets for Plotting Calibration Curves (Section 3.18, Figure 3.9)

Chart Wizard

Trendline

Least squares equation; R^2

Slope, Intercept, and Coefficient of Determination (without a plot) (Section 3.19, Chapter 3, Problems 37, 40, 41)

LINEST for Additional Statistics (Section 3.20, Figure 3.10, Examples 3.21, 3.22); ten functions: slope, std. devn., R^2, F, sum sq. regr., intercept, std. devn., std. error of estimate, d.f., sum sq. resid.

Calculating Unknown Concentration from Calibration Curve, 481

   Chapter 3, Problem 38

Standard Deviation of Unknown from Calibration Curve, 482

Multiple Standard Additions, Plus Standard Deviation, 534

Internal Standard Calibration, Plus Standard Deviation, 589

   Chapter 20, Problem 18

Plotting α vs. pH Curves (Figure 7.1, H_3PO_4), 245

Plotting log C vs. pH Curves (HOAc), 259

Chapter 7, Problem 61 (HOAc)

Plotting log C vs. pH Curves Using Alpha Values, 260

Chapter 7, Problem 64 (Malic acid, H_2A)

Chapter 7, Problem 68 (H_3PO_4, H_3A) (Figure 7.4)

Plotting Titration Curves

HCl vs. NaOH (Figure 8.1), 269

HOAc vs. NaOH, 277

H_2CrO_4 vs. NaOH (diprotic acid titration): CD, auxiliary data, Chapter 8, 284

NH_3 vs. HCl (Figure 8.8): Chapter 8, Problem 35

Ca^{2+} vs. EDTA: Chapter 9, Problem 21

Chapter One

ANALYTICAL OBJECTIVES, OR: WHAT ANALYTICAL CHEMISTS DO

"Unless our knowledge is measured and expressed in numbers,
it does not amount to much."
—Lord Kelvin

Analytical chemistry is concerned with the chemical characterization of matter and the answer to two important questions: what is it (qualitative) and how much is it (quantitative). Chemicals make up everything we use or consume, and knowledge of the chemical composition of many substances is important in our daily lives. Analytical chemistry plays an important role in nearly all aspects of chemistry, for example, agricultural, clinical, environmental, forensic, manufacturing, metallurgical, and pharmaceutical chemistry. The nitrogen content of a fertilizer determines its value. Foods must be analyzed for contaminants (e.g., pesticide residues) and for essential nutrients (e.g., vitamin content). The air in cities must be analyzed for carbon monoxide. Blood glucose must be monitored in diabetics (and, in fact, most diseases are diagnosed by chemical analysis). The presence of trace elements from gun powder on a murder defendant's hand will prove a gun was fired. The quality of manufactured products often depends on proper chemical proportions, and measurement of the constituents is a necessary part of **quality control**. The carbon content of steel will determine its quality. The purity of drugs will determine their efficacy.

Everything is made of chemicals. Analytical chemists determine what and how much.

1.1 What Is Analytical Science?

The above description of analytical chemistry provides an overview of the discipline of analytical chemistry. There have been various attempts to more specifically define the discipline. The late Charles N. Reilley said: "Analytical chemistry is what analytical chemists do" (Ref. 2). The discipline has expanded beyond the bounds of just chemistry, and many have advocated using the name *analytical science* to describe the field. This term is used in a National Science Foundation report from workshops on "Curricular Developments in the Analytical Sciences." Even this term falls short of recognition of the role of instrumentation development and application. One suggestion is that we use the term *analytical science and technology* (Ref. 3).

The Federation of European Chemical Societies held a contest in 1992 to define analytical chemistry, and the following suggestion by K. Cammann was selected [*Fresenius' J. Anal. Chem.*, **343** (1992):812–813.]

> **Analytical Chemistry provides the methods and tools needed for insight into our material world . . . for answering four basic questions about a material sample:**
> - **What?**
> - **Where?**
> - **How much?**
> - **What arrangement, structure or form?**

The Division of Analytical Chemistry of the American Chemical Society provides a comprehensive definition of analytical chemistry, which may be found on their website (www.acs-analytical.duq.edu/whatisanltycalchem.html). It is reproduced, in most part, here:

> **Analytical Chemistry seeks ever improved means of measuring the chemical composition of natural and artificial materials. The techniques of this science are used to identify the substances which may be present in a material and to determine the exact amounts of the identified substance.**
>
> **Analytical chemists work to improve the reliability of existing techniques to meet the demands for better chemical measurements which arise constantly in our society. They adapt proven methodologies to new kinds of materials or to answer new questions about their composition and their reactivity mechanisms. They carry out research to discover completely new principles of measurement and are at the forefront of the utilization or major discoveries, such as lasers and microchip devices for practical purposes. Their efforts serve the needs of many fields:**
> - **In *medicine*, analytical chemistry is the basis for clinical laboratory tests which help physicians diagnose disease and chart progress in recovery.**
> - **In *industry*, analytical chemistry provides the means of testing raw materials and for assuring the quality of finished products whose chemical composition is critical. Many household products, fuels, paints, pharmaceuticals, etc. are analyzed by the procedures developed by analytical chemists before being sold to the consumer.**
> - ***Environmental quality* is often evaluated by testing for suspected contaminants using the techniques of analytical chemistry.**
> - **The nutritional value of *food* is determined by chemical analysis for major components such as protein and carbohydrates and trace components such as vitamins and minerals. Indeed, even the calories in a food are often calculated from its chemical analysis.**
>
> **Analytical chemists also make important contributions to fields as diverse as forensics, archaeology, and space science.**

1.2 Qualitative and Quantitative Analysis: What Does Each Tell Us?

Qualitative analysis tells us what chemicals are present. Quantitative analysis tells us how much.

The discipline of analytical chemistry consists of **qualitative analysis** and **quantitative analysis**. The former deals with the identification of elements, ions, or

How Did Analytical Chemistry Originate?

That is a very good question. Actually, the tools and basic chemical measurements date back to the earliest recorded history. Fire assays for gold are referred to in Zechariah 13:9, and the King of Babylon complained to the Egyptian Pharoah Ammenophis the Fourth (1375–1350 BC) that gold he had received from the pharaoh was "less than its weight" after putting it in a furnace. The perceived value of gold, in fact, was probably a major incentive for acquiring analytical knowledge. Archimedes (287–212 BC) did nondestructive testing of the golden wreath of King Hieron. He placed lumps of gold and silver equal in weight to the wreath in jar full of water and measured the amount of water displaced by all three. The wreath displaced an amount between the gold and silver, proving it was not pure gold!

The balance is of such early origin that it was ascribed to the gods in the earliest documents found. The Babylonians created standard weights in 2600 BC and considered them so important that their use was supervised by the priests.

The alchemists accumulated the chemical knowledge that formed the basis for quantitative analysis as we know it today. Robert Boyle coined the term *analyst* in his 1661 book, *The Sceptical Chymist*. Antoine Lavoisier has been considered the "father of analytical chemistry" because of the careful quantitative experiments he performed on conservation of mass (using the analytical balance). (Lavoisier was actually a tax collector and dabbled in science on the side. He was guillotined on May 8, 1793, during the French Revolution because of his activities as a tax collector.)

Gravimetry was developed in the seventeenth century, and titrimetry in the eighteenth and nineteenth centuries. Guy-Lussac, in 1829, assayed silver by titration with 0.05% relative accuracy and precision!

Textbooks of analytical chemistry began appearing in the 1800s. Karl Fresenius published *Anleitung zur Quantitaven Chemischen Analyse* in Germany in 1845. Wilhelm Ostwald published an influential text on the scientific fundamentals of analytical chemistry in 1894 entitled *Die wissenschaflichen Grundagen der analytischen Chemie*, and this book introduced theoretical explanations of analytical phenomena using equilibrium constants (thank him for Chapter 6 and applications in other chapters).

The twentieth century saw the evolution of instrumental techniques. Steven Popoff's second edition of *Quantitative Analysis* in 1927 included electroanalysis, conductimetric titrations, and colorimetric methods. Today, of course, analytical technology has progressed to include sophisticated and powerful computer-controlled instrumentation and the ability to perform highly complex analyses and measurements at extremely low concentrations.

This text will teach you the fundamentals and give you the tools to tackle most analytical problems. Happy journey. For more on the evolution of the field, see Ref. 7.

compounds present in a sample (we may be interested in whether only a given substance is present), while the latter deals with the determination of how much of one or more constituents is present. The sample may be solid, liquid, gas, or a mixture. The presence of gunpowder residue on a hand generally requires only qualitative knowledge, not of how much is there, but the price of coal will be determined by the percent of sulfur impurity present.

Quantitative analysis   Qualitative analysis

(Courtesy of Merck KGaA. Reproduced by permission.)

Qualitative tests may be performed by selective chemical reactions or with the use of instrumentation. The formation of a white precipitate when adding a solution of silver nitrate to a dissolved sample indicates the presence of chloride. Certain chemical reactions will produce colors to indicate the presence of classes of organic compounds, for example, ketones. Infrared spectra will give "fingerprints" of organic compounds or their functional groups.

A clear distinction should be made between the terms **selective** and **specific**:

- A *selective* reaction or test is one that can occur with other substances but exhibits a degree of preference for the substance of interest.
- A *specific* reaction or test is one that occurs only with the substance of interest.

Few analyses are specific. Selectivity is achieved through proper preparation and measurement.

Unfortunately, few reactions are specific but many exhibit selectivity. Selectivity may be achieved by a number of strategies. Some examples are:

- Sample preparation (e.g., extractions, precipitation)
- Instrumentation (selective detectors)
- Target analyte derivatization (e.g., derivatize specific functional groups with detecting reagents)
- Chromatography, which provides powerful separation

For quantitative analysis, a history of the sample composition will often be known (it is known that blood contains glucose), or else the analyst will have performed a qualitative test prior to performing the more difficult quantitative analysis. Modern chemical measurement systems often exhibit sufficient selectivity that a quantitative measurement can also serve as a qualitative measurement. However, simple qualitative tests are usually more rapid than quantitative procedures. Qualitative analysis is composed of two fields: inorganic and organic. The former is usually covered in introductory chemistry courses, whereas the latter is best left until after the student has had a course in organic chemistry.

In comparing qualitative versus quantitative analysis, consider, for example, the sequence of analytical procedures followed in testing for banned substances at the Olympic Games. The list of prohibited substances includes about 500 different active constituents: stimulants, steroids, beta-blockers, diuretics, narcotics, analgesics, local anesthetics, and sedatives. Some are detectable only as their metabolites. Many athletes must be tested rapidly, and it is not practical to perform a detailed quantitative analysis on each. There are three phases in the analysis: the fast-screening phase, the identification phase, and possible quantification. In the fast-screening phase, urine samples are rapidly tested for the presence of classes of compounds that will differentiate them from "normal" samples. Various techniques include immunoassays, gas chromatography, and liquid chromatography. About 5% of the samples may indicate the presence of unknown compounds that may or may not be prohibited but need to be identified. Samples showing a suspicious profile during the screening undergo a new preparation cycle (possible hydrolysis, extraction, derivatization), depending on the nature of the compounds that have been detected. The compounds are then identified using the highly selective combination of gas chromatography/mass spectrometry (GC/MS). In this technique, complex mixtures are separated by gas chromatography, and they are then detected by mass spectrometry, which provides molecular structural data on the compounds. The MS data, combined with the time of elution from the gas chromatograph, provide a high probability of the presence of a given detected compound. GC/MS is expensive and time consuming, and so it is used only when necessary. Following the identification phase, some compounds must be precisely quantified since they may normally be present at low levels, for example, from food, pharmaceutical preparations, or endogenous steroids, and elevated levels must be confirmed. This is done using quantitative techniques such as spectrophotometry or gas chromatography.

This text deals principally with quantitative analysis. In the consideration of applications of different techniques, examples are drawn from the life sciences, clinical chemistry, environmental chemistry, occupational health and safety applications, and industrial analysis.

We describe briefly in this chapter the analytical process. More details are provided in subsequent chapters.

1.3 Getting Started: The Analytical Process

The general analytical process is shown in Figure 1.1. The analytical chemist should be involved in every step. The analyst is really a problem solver, a critical part of the team deciding what, why, and how. The unit operations of analytical chemistry that are common to most types of analyses are considered in more detail below.

DEFINING THE PROBLEM—WHAT DO WE REALLY NEED TO KNOW? (NOT NECESSARILY EVERYTHING)

Before the analyst can design an analysis procedure, he or she must know what information is needed, by whom, for what purpose, and what type of sample is to be analyzed. As the analyst, you must have good communication with the client. This stage of an analysis is perhaps the most critical. The client may be the Environmental Protection Agency (EPA), an industrial client, an engineer, or your grandmother—each of which will have different criteria or needs, and each having a different understanding of what a chemical analysis involves or means. It is

The way an analysis is performed depends on the information needed.

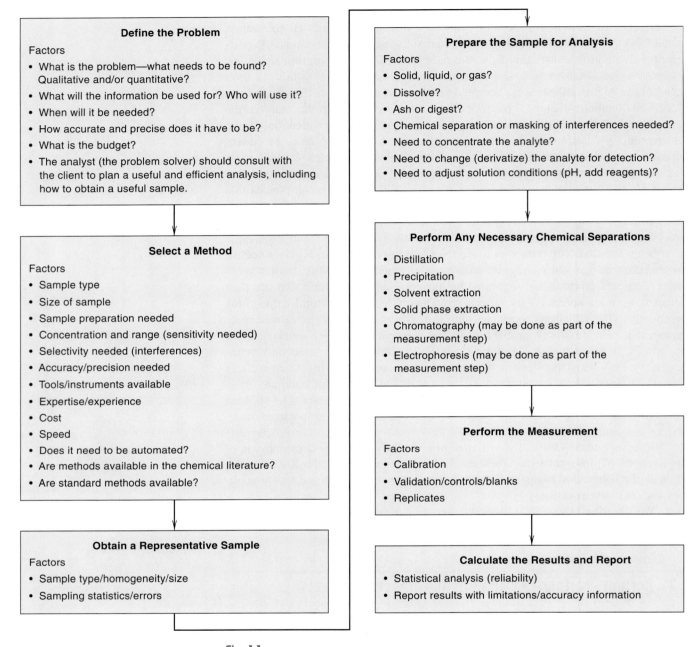

Define the Problem

Factors

- What is the problem—what needs to be found? Qualitative and/or quantitative?
- What will the information be used for? Who will use it?
- When will it be needed?
- How accurate and precise does it have to be?
- What is the budget?
- The analyst (the problem solver) should consult with the client to plan a useful and efficient analysis, including how to obtain a useful sample.

Select a Method

Factors

- Sample type
- Size of sample
- Sample preparation needed
- Concentration and range (sensitivity needed)
- Selectivity needed (interferences)
- Accuracy/precision needed
- Tools/instruments available
- Expertise/experience
- Cost
- Speed
- Does it need to be automated?
- Are methods available in the chemical literature?
- Are standard methods available?

Obtain a Representative Sample

Factors

- Sample type/homogeneity/size
- Sampling statistics/errors

Prepare the Sample for Analysis

Factors

- Solid, liquid, or gas?
- Dissolve?
- Ash or digest?
- Chemical separation or masking of interferences needed?
- Need to concentrate the analyte?
- Need to change (derivatize) the analyte for detection?
- Need to adjust solution conditions (pH, add reagents)?

Perform Any Necessary Chemical Separations

- Distillation
- Precipitation
- Solvent extraction
- Solid phase extraction
- Chromatography (may be done as part of the measurement step)
- Electrophoresis (may be done as part of the measurement step)

Perform the Measurement

Factors

- Calibration
- Validation/controls/blanks
- Replicates

Calculate the Results and Report

- Statistical analysis (reliability)
- Report results with limitations/accuracy information

Fig. 1.1. Steps in an analysis.

important to communicate in language that is understandable by both sides. If someone puts a bottle on your desk and asks, "What is in here?" or "Is this safe?", you may have to explain that there are 10 million known compounds and substances. A client who says, "I want to know what elements are in here" needs to understand that at perhaps $20 per analysis for 85 elements it will cost $1700 to test for them all, when perhaps only a few elements are of interest.

I have often had laypersons come to me with cosmetics they wish to "reverse engineer" so they can market them and make a fortune. When they realize it may cost a small fortune to determine the ingredients, requiring a number of sophisticated analyses, they always rethink their goals.

The concept of "safe" or "zero/nothing" is one that is hard to define or understand by many. Telling someone their water is safe is not for the analyst to say. All you can do is present the analytical data (and give an indication of its range of accuracy). The client must decide whether it is safe to drink, perhaps relying on other experts. Also, never report an answer as "zero," but as less than the detection limit, which is based on the measurement device/instrument. We are limited by our methodology and equipment, and that is all that can be reported. Some modern instruments, though, can measure ridiculously small amounts or concentrations, for example, parts per trillion. This presents a dilemma in making policy (often political in nature). A law may be passed that there should be zero concentration of a chemical effluent in water. In practice, the acceptable level is defined by how low a concentration can be detected; and the very low detectability may be far below the natural occurrence of the chemical or below the levels to which it can be reasonably reduced. We analysts and chemists need to be effective communicators of what our measurements represent.

Once the problem is defined this will dictate how the sample is to be obtained, how much is needed, how sensitive the method must be, how accurate and precise[1] it must be, and what separations may be required to eliminate interferences. The determination of trace constituents will generally not have to be as precise as for major constituents, but greater care will be required to eliminate trace contamination during the analysis.

Once the required measurement is known, the analytical method to be used will depend on a number of factors, including the analyst's skills and training in different techniques and instruments; the facilities, equipment, and instruments available; the sensitivity and precision required; the cost and the budget available; and the time for analysis and how soon results are needed. There are often one or more standard procedures available in reference books for the determination of an **analyte** (constituent to be determined) in a given **sample type**. This does not mean that the method will necessarily be applicable to other sample types. For example, a standard EPA method for groundwater samples may yield erroneous results when applied to the analysis of sewage water. The chemical literature (journals) contains many specific descriptions of analyses. *Chemical Abstracts* (http://info.cas.org), published by the American Chemical Society, is a good place to begin a literature search. It contains abstracts of all articles appearing in the major chemical journals of the world. Yearly and cumulative indices are available, and many libraries have computer search facilities. The major analytical chemistry journals may be consulted separately. Some of these are: *Analytica Chimica Acta, Analytical Chemistry, Analytical Letters, Analyst, Applied Spectroscopy, Clinica Chimica Acta, Clinical Chemistry, Journal of the Association of Official Analytical Chemists, Journal of Chromatography, Spectrochimica Acta,* and *Talanta.* While the specific analysis of interest may not be described, the analyst can often use literature information on a given analyte to devise an appropriate analysis scheme. Finally, the analyst may have to rely upon experience and knowledge to develop an analytical method for a given sample. The literature references in Appendix A describe various procedures for the analysis of different substances.

Examples of the manner in which the analysis of particular types of samples are made are given in Chapters 24 to 26. These chapters describe commonly performed clinical, biochemical, and environmental analyses. The various techniques described in this text are utilized for the specific analyses. Hence, it will be useful

The way you perform an analysis will depend on your experience, the equipment available, the cost, and the time involved.

The *analyte* is the substance *analyzed* for. Its concentration is *determined*.

Chemical Abstracts is a good source of literature.

---

[1]Accuracy is the degree of agreement between a measured value and a true value. Precision is the degree of agreement between replicate measurements of the same quantity and does not necessarily imply accuracy. These terms are discussed in more detail in Chapter 3.

for you to read through these applications chapters both now and after completing the majority of this course to gain an appreciation of what goes into analyzing real samples and why the analyses are made.

Once the problem has been defined, the following steps can be started.

OBTAINING A REPRESENTATIVE SAMPLE—WE CAN'T ANALYZE THE WHOLE THING

A chemical analysis is usually performed on only a small portion of the material to be characterized. If the amount of material is very small and it is not needed for future use, then the entire sample may be used for analysis. The gunshot residue on a hand may be an example. More often, though, the characterized material is of value and must be altered as little as possible in sampling.

The material to be sampled may be solid, liquid, or gas. It may be homogeneous or heterogeneous in composition. In the former case, a simple "grab sample" taken at random will suffice for the analysis. In the latter, we may be interested in the variation throughout the sample, in which case several individual samples will be required. If the gross composition is needed, then special sampling techniques will be required to obtain a representative sample. For example, in analyzing for the average protein content of a shipment of grain, a small sample may be taken from each bag, or tenth bag for a large shipment, and combined to obtain a **gross sample**. Sampling is best done when the material is being moved, if it is large, in order to gain access. The larger the particle size, the larger should be the gross sample. The gross sample must be reduced in size to obtain a **laboratory sample** of several grams, from which a few grams to milligrams will be taken to be analyzed (**analysis sample**). The size reduction may require taking portions (e.g., two quarters) and mixing, in several steps, as well as crushing and sieving to obtain a uniform powder for analysis. Methods of sampling solids, liquids, and gases are discussed in Chapter 2.

In the case of biological fluids, the conditions under which the sample is collected can be important, for example, whether a patient has just eaten. The composition of blood varies considerably before and after meals, and for many analyses a sample is collected after the patient has fasted for a number of hours. Preservatives such as sodium fluoride for glucose preservation and anticoagulants may be added to blood samples when they are collected; these may affect a particular analysis.

Blood samples may be analyzed as whole blood, or they may be separated to yield plasma or serum according to the requirements of the particular analysis. Most commonly, the concentration of the substance external to the red cells (the extracellular concentration) will be a significant indication of physiological condition, and so serum or plasma is taken for analysis.

If whole blood is collected and allowed to stand for several minutes, the soluble protein **fibrinogen** will be converted by a complex series of chemical reactions (involving calcium ion) into the insoluble protein **fibrin**, which forms the basis of a gel, or **clot**. The red and white cells of the blood become caught in the meshes of the fibrin network and contribute to the solidarity of the clot, although they are not necessary for the clotting process. After the clot forms, it shrinks and squeezes out a straw-colored fluid, **serum**, which does not clot but remains fluid indefinitely. The clotting process can be prevented by adding a small amount of an **anticoagulant**, such as heparin or a citrate salt (i.e., a calcium complexor). An aliquot of the unclotted whole blood can be taken for analysis, or the red cells can be centrifuged to the bottom, and the light pinkish-colored **plasma** remaining can be analyzed. Plasma and serum are essentially identical in chemical composition, the chief difference being that fibrinogen has been removed from the latter.

The *gross sample* consists of several portions of the material to be tested. The *laboratory sample* is a small portion of this, made homogeneous. The *analysis sample* is that actually analyzed. See Chapter 2 for methods of sampling.

Serum is the fluid separated from clotted blood. *Plasma* is the fluid separated from unclotted blood. It is the same as serum, but contains fibrinogen, the clotting protein.

Details of sampling other materials are available in reference books on specific areas of analysis. See the references at the end of the chapter for some citations.

Certain precautions should be taken in **handling and storing samples** to prevent or minimize contamination, loss, decomposition, or matrix change. In general, one must prevent contamination or alteration of the sample by (1) the container, (2) the atmosphere, or (3) light. Also, a chain of custody should be established and will certainly be required for any analysis that may be involved in legal proceedings. In the O.J. Simpson case, there were television news clips of people handling samples, purportedly without proper custody, placing them in the hot trunk of a car, for example. While this may not have affected the actual analyses and correctness of samples analyzed, it provided fodder for the defense to discredit analyses.

Care must be taken not to alter or contaminate the sample.

The sample may have to be protected from the atmosphere or from light. It may be an alkaline substance, for example, which will react with carbon dioxide in the air. Blood samples to be analyzed for CO_2 should be protected from the atmosphere.

The stability of the sample must be considered. Glucose, for example, is unstable, and a preservative such as sodium fluoride is added to blood samples. The preservation must not, of course, interfere in the analysis. Proteins and enzymes tend to denature on standing and should be analyzed without delay. Trace constituents may be lost during storage by adsorption onto the container walls.

Urine samples are unstable, and calcium phosphate precipitates out, entrapping metal ions or other substances of interest. Precipitation can be prevented by keeping the urine acidic (pH 4.5), usually by adding 1 or 2 mL glacial acetic acid per 100-mL sample. Store under refrigeration. Urine, as well as whole blood, serum, plasma, and tissue samples, can also be frozen for prolonged storage. Deproteinized blood samples are more stable than untreated samples.

Corrosive gas samples will often react with the container. Sulfur dioxide, for example, is troublesome. In automobile exhaust, SO_2 is also lost by dissolving in condensed water vapor from the exhaust. In such cases, it is best to analyze the gas by a stream process.

PREPARING THE SAMPLE FOR ANALYSIS—IT PROBABLY NEEDS TO BE ALTERED

The first step in analyzing a sample is to measure the amount being analyzed (e.g., volume or weight of sample). This will be needed to calculate the percent composition from the amount of analyte found. The analytical sample size must be measured to the degree of precision and accuracy required for the analysis. An analytical balance sensitive to 0.1 mg is usually used for weight measurements. Solid samples are often analyzed on a dry basis and must be dried in an oven at 110 to 120°C for 1 to 2 h and cooled in a dessicator before weighing, if the sample is stable at the drying temperatures. Some samples may require higher temperatures and longer heating time (e.g., overnight) because of the great affinity of moisture for their sample surface. The amount of sample taken will depend on the concentration of the analyte and how much is needed for isolation and measurement. Determination of a major constituent may require only a couple hundred milligrams of sample, while a trace constituent may require several grams. Usually **replicate samples** are taken for analysis, in order to obtain statistical data on the precision of the analysis and provide more reliable results.

The first thing you must do is measure the size of sample to be analyzed.

Analyses may be nondestructive in nature, for example, in the measurement of lead in paint by X-ray fluorescence in which the sample is bombarded with an X-ray beam and the characteristic reemitted X radiation is measured. More often, the sample must be in solution form for measurement, and solids must be dissolved. Inorganic materials may be dissolved in various acids, redox, or complexing

Solid samples usually must be put into solution.

media. Acid-resistant material may require fusion with an acidic or basic flux in the molten state to render it soluble in dilute acid or water. Fusion with sodium carbonate, for example, forms acid-soluble carbonates.

Ashing is the burning of organic matter. *Digestion* is the wet oxidation of organic matter.

Organic materials that are to be analyzed for inorganic constituents, for example, trace metals, may be destroyed by **dry ashing**. The sample is slowly combusted in a furnace at 400 to 700°C, leaving behind an inorganic residue that is soluble in dilute acid. Alternately, the organic matter may be destroyed by **wet digestion** by heating with oxidizing acids. A mixture of nitric and sulfuric acids is common. Biological fluids may sometimes be analyzed directly. Often, however, proteins interfere and must be removed. Dry ashing and wet digestion accomplish such removal. Or proteins may be precipitated with various reagents and filtered or centrifuged away, to give a **protein-free filtrate** (PFF).

If the analyte is organic in nature, these oxidizing methods cannot be used. Rather, the analyte may be extracted away from the sample or dialyzed, or the sample dissolved in an appropriate solvent. It may be possible to measure the analyte nondestructively. An example is the direct determination of protein in feeds by near-infrared spectrometry.

The pH of the sample solution will usually have to be adjusted.

Once a sample is in solution, the solution conditions must be adjusted for the next stage of the analysis (separation or measurement step). For example, the pH may have to be adjusted, or a reagent added to react with and "mask" interference from other constituents. The analyte may have to be reacted with a reagent to convert it to a form suitable for measurement or separation. For example, a colored product may be formed that will be measured by spectrometry. Or the analyte will be converted to a form that can be volatilized for measurement by gas chromatography. The gravimetric analysis of iron as Fe_2O_3 requires that all the iron be present as iron(III), its usual form. A volumetric determination by reaction with dichromate ion, on the other hand, requires that all the iron be converted to iron(II) before reaction, and the reduction step will have to be included in the sample preparation.

Always run a blank!

The solvents and reagents used for dissolution and preparation of the solution should be of high purity (reagent grade). Even so, they may contain trace impurities of the analyte. Hence, it is important to prepare and analyze replicate **blanks**, particularly for trace analyses. A blank theoretically consists of all chemicals in the unknown and used in an analysis in the same amounts (including water), run through the entire analytical procedure. The blank result is subtracted from the analytical sample result to arrive at a net analyte concentration in the sample solution. If the blank is appreciable, it may invalidate the analysis. Oftentimes, it is impossible to make a perfect blank for an analysis.

PERFORMING NECESSARY CHEMICAL SEPARATIONS

In order to eliminate interferences, to provide suitable selectivity in the measurement, or to preconcentrate the analyte for more sensitive or accurate measurement, the analyst must often perform one or more separation steps. It is preferable to separate the analyte away from the sample matrix, in order to minimize losses of the analyte. Separation steps may include precipitation, extraction into an immiscible solvent, chromatography, dialysis, and distillation.

PERFORMING THE MEASUREMENT—YOU DECIDE THE METHOD

The method employed for the actual quantitative measurement of the analyte will depend on a number of factors, not the least important being the amount of analyte present and the accuracy and precision required. Many available techniques possess varying degrees of selectivity, sensitivity, accuracy and precision, cost, and rapidity. **Gravimetric analysis** usually involves the selective separation of the analyte by precipitation, followed by the very nonselective measurement of mass (of the precipitate). In **volumetric**, or **titrimetric**, **analysis**, the analyte reacts with a measured volume of reagent of known concentration, in a process called **titration**. A change in some physical or chemical property signals the completion of the reaction. Gravimetric and volumetric analyses can provide results accurate and precise to a few parts per thousand (tenth of 1 percent) or better. But they require relatively large (millimole or milligram) quantities of analyte and are well suited for the measurement of major constituents. Volumetric analysis is more rapid than gravimetric analysis and so is preferred when applicable.

Instrumental techniques are used for many analyses and constitute the discipline of **instrumental analysis**. They are based on the measurement of a physical property of the sample, for example, an electrical property or the absorption of electromagnetic radiation. Examples are spectrophotometry (ultraviolet, visible, or infrared), fluorimetry, atomic spectroscopy (absorption, emission), mass spectrometry, nuclear magnetic resonance spectrometry (NMR), X-ray spectroscopy (absorption, fluorescence), electroanalytical chemistry (potentiometric, voltammetric, electrolytic), chromatography (gas, liquid), and radiochemistry. Instrumental techniques are generally more sensitive and selective than the classical techniques but are less precise, on the order of 1 to 5% or so. These techniques are usually much more expensive, capitalwise. But depending on the numbers of analyses, they may be less expensive when one factors in personnel costs. They are usually more rapid, may be automated, and may be capable of measuring more than one analyte at a time. Chromatography techniques are particularly powerful for analyzing complex mixtures. They perform the separation and measurement step simultaneously. Constituents are separated as they are washed down (eluted from) a column of appropriate material that interacts with the analytes to varying degrees, and the analytes are sensed with an appropriate detector as they emerge from the column, to give a transient peak signal, in proportion to the amount of analyte.

> Instruments are more selective and sensitive than volumetric and gravimetric methods. But they may be less precise.

Table 1.1 compares various analytical methods to be described in this text with respect to sensitivity, precision, selectivity, speed, and cost. The numbers given may be exceeded in specific applications, and the methods may be applied to other uses, but these are representative of typical applications. The lower concentrations determined by titrimetry require the use of an instrumental technique for measuring the completion of the titration. The selection of a technique, when more than one is applicable, will depend, of course, on the availability of equipment and personal experience and preference of the analyst. As examples, you might use sprectrophotometry to determine the concentration of nitrate in river water at the 1 ppm (1.6×10^{-5} M) level, using a diazotization reaction with nitrate to produce a color. Fluoride in toothpaste may be determined potentiometrically using a fluoride ion-selective electrode. A complex mixture of hydrocarbons in gasoline can be separated and determined using the technique of gas chromatography. Glucose in blood can be determined kinetically by the rate of the enzymatic reaction between glucose and oxygen, catalyzed by the enzyme glucose oxidase, with measurement of the rate of oxygen depletion or the rate of hydrogen peroxide production. The purity of a silver bar can be determined gravimetrically by dissolving a small sample in nitric acid and precipitating with chloride and weighing the purified precipitate.

The various methods of determining an analyte can be classified as either **absolute** or **relative**. Absolute methods rely upon accurately known fundamental constants for calculating the amount of analyte, for example, atomic weights. In gravimetric analysis, for example, an insoluble derivative of the analyte of known chemical composition is prepared and weighed, as in the formation of AgCl for

> Most methods require calibration with a standard.

Table 1.1
Comparison of Different Analytical Methods

Method	Approx. Range (mol/L)	Approx. Precision (%)	Selectivity	Speed	Cost	Principal Uses
Gravimetry	10^{-1}–10^{-2}	0.1	Poor–moderate	Slow	Low	Inorg.
Titrimetry	10^{-1}–10^{-4}	0.1–1	Poor–moderate	Moderate	Low	Inorg., org.
Potentiometry	10^{-1}–10^{-6}	2	Good	Fast	Low	Inorg.
Electrogravimetry, coulometry	10^{-1}–10^{-4}	0.01–2	Moderate	Slow–moderate	Moderate	Inorg., org.
Voltammetry	10^{-3}–10^{-10}	2–5	Good	Moderate	Moderate	Inorg., org.
Spectrophotometry	10^{-3}–10^{-6}	2	Good–moderate	Fast–moderate	Low–moderate	Inorg., org.
Fluorometry	10^{-6}–10^{-9}	2–5	Moderate	Moderate	Moderate	Org.
Atomic spectroscopy	10^{-3}–10^{-9}	2–10	Good	Fast	Moderate–high	Inorg, multielement
Chromatography	10^{-3}–10^{-9}	2–5	Good	Fast–moderate	Moderate–high	Org, multicomponent
Kinetic methods	10^{-2}–10^{-10}	2–10	Good–moderate	Fast–moderate	Moderate	Inorg., org., enzymes

chloride determination. The precipitate contains a known fraction of the analyte, in this case, fraction of Cl = at wt Cl/f wt AgCl = 35.453/143.32 = 0.24737.[2] Hence, it is a simple matter to obtain the amount of Cl contained in the weighed precipitate. Most methods, however, are relative in that they require comparison against some solution of known concentration. In titrimetric analysis, for example, the analyte is reacted with the solution of a reagent in a known stoichiometric ratio. Hydrochloric acid, for example, reacts with sodium hydroxide in a 1:1 ratio:

$$HCl + NaOH \rightarrow NaCl + H_2O$$

The volume of sodium hydroxide solution required to just completely react with the hydrochloric acid sample is measured. If we know the concentration of the sodium hydroxide solution in moles per liter, then the number of moles of NaOH added can be calculated (volume × molarity), and so we know the number of moles of HCl in the sample. Therefore, in this relative method, it is necessary to prepare a reacting solution (sodium hydroxide) of accurately known concentration.

Most instrumental methods of analysis are relative. Instruments register a signal due to some physical property of the solution. Spectrophotometers, for example, measure the fraction of electromagnetic radiation from a light source that is absorbed by the sample. This fraction must be related to the analyte concentration by comparison against the fraction absorbed by a known concentration of the analyte. In other words, the instrumentation must be **calibrated**.

Instrument response may be linearly or nonlinearly related to the analyte concentration. Calibration is accomplished by preparing a series of standard solutions of the analyte at known concentrations and measuring the instrument response to each of these (usually after treating them in the same manner as the samples) to prepare an **analytical calibration curve** of response versus concentration. The concentration of an unknown is then determined from the response, using the calibration curve. With modern computer-controlled instruments, this is often done electronically or digitally, and direct readout of concentration is obtained.

> A calibration curve is the instrument response as a function of concentration.

The sample matrix may affect the instrument response to the analyte. In such cases, calibration may be accomplished by the **method of standard additions**. A portion of the sample is spiked with a known amount of standard, and the increase in signal is due to the standard. In this manner, the standard is subjected to the same environment as the analyte. These calibration techniques are discussed in more detail when describing the use of specific instruments.

> Standard additions calibration is used to overcome sample matrix effects.

CALCULATING THE RESULTS AND REPORTING THE DATA—THIS IS THE WHOLE OBJECT

Once the concentration of analyte in the prepared sample solution has been determined, the results are used to calculate the amount of analyte in the original sample. Either an *absolute* or a *relative* amount may be reported. Usually, a relative composition is given, for example, percent or parts per million, along with the mean value for expressing accuracy. Replicate analyses can be performed (three or more), and a precision of the analysis may be reported, for example, standard deviation. A knowledge of the precision is important because it gives the degree of uncertainty in the result (see Chapter 3). The analyst should critically evaluate whether the results are reasonable and relate to the analytical problem as originally stated. Remember that the customer often does not have a scientific background so will

> The analyst must provide expert advice on the significance of a result.

---

[2] at wt = atomic weight; f wt = formula weight.

take a number as gospel. Only you, as analyst, can put that number in perspective, and it is important that you have good communication and interaction with the "customer" about what the analysis represents.

1.4 Validation of a Method—You Have to Prove It Works!

Great care must be taken that accurate results are obtained in an analysis. Two types of error may occur: *random* and *systematic*. Every measurement has some imprecision associated with it, which results in random distribution of results, for example, a Gaussian distribution. The experiment can be designed to narrow the range of this, but it cannot be eliminated. A systematic error is one that biases a result consistently in one direction. The sample matrix may suppress the instrument signal. A weight of an analytical balance may be in error, skewed either high or low. A sample may not be sufficiently dried.

> The best way to validate a method is to analyze a standard reference material of known composition.

Proper calibration of an instrument is only the first step in assuring accuracy. In developing a method, samples should be spiked with known amounts of the analyte (above and beyond what is already in the sample). The amounts determined (recovered) by the analysis procedure (after subtraction of the amount apparently present in the sample as determined by the same procedure) should be within the accuracy required in the analysis. A new method can be validated by comparison of sample results with those obtained with another accepted method. There are various sources of certified standards or reference materials that may be analyzed to assure accuracy by the method in use. For example, environmental quality control standards for pesticides in water or priority pollutants in soil are commercially available. The National Institute of Standards and Technology (NIST) prepares standard reference materials (SRMs) of different matrix compositions (e.g., steel, ground leaves) that have been certified for the content of specific analytes, by careful measurement by at least two independent techniques. Values are assigned with statistical ranges. Different agencies and commercial concerns can provide samples for round-robin or blind tests in which control samples are submitted to participating laboratories for analysis at random. The laboratory does not know the control value prior to analysis.

Standards should be run intermittently with samples. A *control sample* should also be run at least daily and the results plotted as a function of time to prepare a *quality control chart*, which is compared with the known standard deviation of the method. The measured quantity is assumed to be constant with time, with a Gaussian distribution, and there is a 1 in 20 chance that values will fall outside two standard deviations from the known value, and a 1 in 100 chance it will be 2.5 standard deviations away. Numbers exceeding these suggest uncompensated errors, such as instrument malfunction, reagent deterioration, or improper calibration.

> Good laboratory practice (validation) is required to assure accuracy of analyses.

Government regulations require careful established protocol and validation of methods and analyses when used for official or legal purposes. Guidelines of *good laboratory practice* (GLP) have been established to assure validation of analyses. They, of course, ideally apply to all analyses. These are discussed in detail in Chapter 4.

1.5 Range—What Size Sample?

Analytical methods are often classed according to size of sample. Such classification is arbitrary and there is no sharp dividing line. The analysis may be classed

Table 1.2
Classification of Analytical Methods According to Size of Sample

Method	Sample Weight (mg)	Sample Volume (μL)[a]
Meso	>100	>100
Semimicro	10–100	50–100
Micro	1–10	<50
Ultramicro	<1	

[a]μL = Microliter. Sometimes the symbol λ (lambda) is used in place of μL.

as **meso**, **semimicro**, **micro**, or **ultramicro**; the last two categories are often classified as "trace analysis." Table 1.2 gives approximate classifications according to sample weight or volume. The volume classifications are those employed in clinical laboratories. Special handling techniques and microbalances for weighing are required for micro and ultramicro operations.

ANALYZE VERSUS DETERMINE—THEY ARE DIFFERENT

The terms *analyze* and *determine* have two different meanings. We say a sample is **analyzed** for part or all of its constituents. The substances measured are called the **analytes**. The process of measuring the analyte is called a **determination**. Hence, in analyzing blood for its chloride content, we determine the chloride concentration.

You *analyze* a sample to *determine* the amount of analyte.

The constituents in the sample may be classified as **major** (>1%), **minor** (0.1 to 1%), or **trace** (<0.1%). A few parts per million or less of a constituent might be classed as **ultratrace**.

An analysis may be **complete** or **partial**; that is, either all constituents or only selected constituents may be determined. Most often, the analyst is requested to report on a specified chemical or perhaps a class of chemicals.

1.6 Some Useful Websites

In addition to the various literature and book sources we have mentioned, and those listed in Appendix A (Literature of Analytical Chemistry), there are a number of websites that are useful for supplementary resources for analytical chemists. These, of course, often change and new ones become available. But the following are good starting points for obtaining much useful information.

Chemistry in General

1. www.acs.org: The American Chemical Society home page. Information on journals, meetings, chemistry in the news, search databases (including *Chemical Abstracts*), and much more.

2. www.chemweb.com: This is a virtual club for chemists. The site contains databases and lists relating to chemistry and incorporates discussion groups that focus on specific areas such as analytical chemistry. You must join, but it is free.

3. www.chemsoc.org: This is a chemistry societies network, maintained by the Royal Society of Chemistry in Britain, with information provided by about 30 national chemistry societies. It has a virtual periodic table, and you can download a periodic table screensaver.

4. http://micro.magnet.fsu.edu/primer/java/scienceopticsu/powersof10/index.html: Check out this Powers of Ten visual scene, from protons to viewing the Milky Way 10 million light years from Earth.

Analytical Chemistry

1. www.acs-analytical.duq.edu: This is the home page of the Division of Analytical Chemistry of the American Chemical Society. There are a number of links and resources throughout this site that will take you all over the Internet to sites involving analytical chemistry.
2. www.chem.uni-potsdam.de/linkcenter/analchem.html: This is an analytical chemistry link center that connects to university and other sites all over the world.
3. www.cstl.nist.gov/nist839: The Analytical Division site of the National Institute of Standards and Technology (NIST). Provides information on each of its five divisions: Spectrochemical Methods, Organic Analytical Methods, Gas Metrology and Classical Methods, Molecular Spectrometry and Microfluidic Methods, and Nuclear Analytical Methods.
4. www.rsc.org/lap/rsccom/dab/analdiv.htm: The Analytical Division site of the Royal Society of Chemistry in Britain. The British equivalent of the ACS Division of Analytical Chemistry.
5. http://chemweb.chem.uconn.edu/Microchem/Index.html: American Microchemical Society site. Has links to analytical chemistry societies, conferences, and professors.
6. http://analytical.chemweb.com/home: The analytical chemistry forum of the Chemweb website above.
7. www.asdlib.org: Analytical Sciences Digital Library (ASDL). In its infancy, but will become the site of choice for listings of peer-reviewed websites dealing with pedagogy and techniques. See your text website for additional information about this site.

Learning Objectives

WHAT ARE SOME OF THE KEY THINGS WE LEARNED FROM THIS CHAPTER?

- Analytical science deals with the chemical characterization of matter—what, how much?, p. 1
- The analyst must know what information is really needed, and obtain a representative sample, pp. 5, 8
- Few measurements are specific, so operations are performed to achieve high selectivity, p. 9
- You must select the appropriate method for measurement, p. 10
- Validation is important, p. 14
- There are many useful websites dealing with analytical chemistry, p. 15

Questions

1. What is analytical chemistry?
2. Distinguish between qualitative analysis and quantitative analysis.
3. Outline the steps commonly employed in an analytical procedure. Briefly describe each step.
4. Distinguish between analyze, determine, sample, and analyte.
5. What is a blank?
6. List some of the common measuring techniques employed in analytical chemistry.
7. List some separation procedures employed in analytical chemistry.
8. Define instrumental analysis.
9. What is a calibration curve?
10. Distinguish between a specific reaction and a selective reaction.
11. Suggest a method from Table 1.1 to accomplish the following analyses: (a) the purity of NaCl in table salt, (b) the acetic acid content of vinegar, (c) the pH of swimming pool water.

Recommended References

GENERAL

1. J. Tyson, *Analysis. What Analytical Chemists Do.* London: Royal Society of Chemistry, 1988. A brief book that very succinctly discusses what analytical chemists do and how they do it.
2. R. W. Murray, "Analytical Chemistry Is What Analytical Chemists Do," Editorial, *Anal. Chem.*, **66** (1994) 682A.
3. G. M. Hieftje. "But Is It Analytical Chemistry? *Am. Lab.*, October (1993) 53.
4. T. Kuwana, Chair, *Curricular Developments in the Analytical Sciences*, A Report from Workshops held in 1996 and 1997, Sponsored by the National Science Foundation. (kuwana@sunflower.com)
5. R. A. DePalma and A. H. Ullman, "Professional Analytical Chemistry in Industry. A Short Course to Encourage Students to Attend Graduate School," *J. Chem. Ed.*, **68** (1991) 383. This is a description of an industrial analytical chemistry short course for teachers and students that Proctor & Gamble Company scientists deliver at universities and colleges at invitation, to explain what analytical chemists do in industry. The emphasis is "The analytical chemist as a problem solver." (parry.d@pg.com)
6. C. A. Lucy, "How to Succeed in Analytical Chemistry: A Bibliography of Resources from the Literature," *Talanta*, **51** (2000) 1125. Surveys the literature for advice on how to purchase equipment, how to write a manuscript, and how to get a job in analytical chemistry.
7. G. D. Christian, "Evolution and Revolution in Quantitative Analysis," *Anal. Chem.*, **67** (1995) 532A. Traces the history of analytical chemistry from the beginning of humans.

ENCYCLOPEDIAS AND HANDBOOKS

8. R. A. Meyers, editor-in-chief, *Encyclopedia of Analytical Chemistry.* Chichester: Wiley, 2000. A 15-volume set, with 10 volumes on applications and 5 on theory and instrumentation.

9. A. Townsend, editor-in-chief, *Encyclopedia of Analytical Science*. London: Academic, 1995. A 10-volume set with comprehensive coverage of the practice of analytical science. Covers all techniques that determine specific elements, compounds, and groups of compounds in any physical or biological matrix.

10. J. A. Dean, *Analytical Chemistry Handbook*. New York: McGraw Hill, 1995. Detailed coverage of conventional wet and instrumental techniques, preliminary operations of analysis, preliminary separation methods, and statistics in chemical analysis.

SAMPLING

11. F. F. Pitard, *Pierre Gy's Sampling Theory and Sampling Practice*. Vol. 1: *Heterogeneity and Sampling*. Vol. II: *Sampling Correctness and Sampling Practice*. Boca Raton, FL: CRC Press, 1989.

12. P. M. Gy and A. G. Boyle, *Sampling for Analytical Purposes*. Chichester: Wiley, 1998. An abridged guide by Pierre Gy to his formula originally developed for the sampling of solid materials but equally valid for sampling liquids and multiphase media.

13. P. M. Gy, *Sampling of Heterogeneous and Dynamic Material Systems*. Amsterdam: Elsevier, 1992.

14. G. E. Baiulescu, P. Dumitrescu, and P. Gh. Zugravescu, *Sampling*. New York: Ellis Horwood, 1991.

15. N. T. Crosby, ed., *General Principles of Good Sampling Practice*. London: Royal Society of Chemistry, 1995.

16. P. M. Gy, "Introduction to Theory of Sampling I. Heterogeneity of a Population of Uncorrelated Units," *Trends Anal. Chem.*, **14**(2) (1995) 67.

17. P. M. Gy, "Tutorial: Sampling or Gambling?" *Process Control and Quality*, 6 (1994) 97.

18. B. Kratochvil and J. K. Taylor, "Sampling for Chemical Analysis," *Anal. Chem.*, **53** (1981) 924A.

19. B. Kratochvil, "Sampling for Microanalysis: Theory and Strategies," *Fresenius' J. Anal. Chem.*, **337** (1990) 808.

20. J. F. Vicard and D. Fraisse, "Sampling Issues," *Fresenius' J. Anal. Chem.*, **348** (1994) 101.

21. J. M. Hungerford and G. D. Christian, "Statistical Sampling Errors as Intrinsic Limits on Detection in Dilute Solutions," *Anal. Chem.*, **58** (1986) 2567.

INTERFERENCES

22. W. E. Van der Linden, "Definition and Classification of Interference in Analytical Procedures," *Pure & Appl. Chem.*, **61** (1989) 91.

STANDARD SOLUTIONS

23. B. W. Smith and M. L. Parsons, "Preparation and Standard Solutions. Critically Selected Compounds," *J. Chem. Ed.*, **50** (1973) 679. Describes preparation of standard solutions of 72 inorganic metals and nonmetals.

24. M. H. Gabb and W. E. Latchem, *A Handbook of Laboratory Solutions*. New York: Chemical Publishing, 1968.

25. http://pubs.acs.org/reagents/committee.html, American Chemical Society Committee on Analytical Reagents.

CALIBRATION

26. H. Mark, *Principles and Practice of Spectroscopic Calibration.* New York: Wiley, 1991.

STANDARD REFERENCE MATERIALS

27. H. Klich and R. Walker, "COMAR—The International Database for Certified Reference Materials," *Fresenius' J. Anal. Chem.*, **345** (1993) 104.
28. Office of Standard Reference Materials, Room B311, Chemistry Building, National Institute of Standards and Technology, Gaithersburg, Maryland 20899.
29. National Research Council of Canada, Division of Chemistry, Ottawa K1A OR6, Canada.
30. R. Alverez, S. D. Rasberry, and G. A. Uriano, "NBS Standard Reference Materials: Update 1982," *Anal. Chem.*, **54** (1982) 1239A.
31. R. W. Seward and R. Mavrodineanu, "Standard Reference Materials: Summary of the Clinical Laboratory Standards Issued by the National Bureau of Standards," *NBS* (NIST) *Special Publications* 260–271, Washington, DC, 1981.
32. *Standard Coal Samples.* Available from U.S. Department of Energy, Pittsburgh Mining Technology Center, P.O. Box 10940, Pittsburgh, PA 15236. Samples have been characterized for 14 properties, including major and trace elements.

Chapter Two

BASIC TOOLS AND OPERATIONS OF ANALYTICAL CHEMISTRY

"Get your facts first, and then you can distort them as much as you please."
—Mark Twain

Read this chapter before performing experiments.

Analytical chemistry requires measurements in order to get the facts. Throughout the text, specific analytical equipment and instrumentation available to the analyst are discussed as they pertain to specific measuring techniques. Several standard items, however, are common to most analyses and will be required when performing the experiments. These are described in this chapter. They include the analytical balance and volumetric glassware and items such as drying ovens and filters. Detailed explanation of the physical manipulation and use of this equipment is best done by your laboratory instructor, when you can see and practice with the actual equipment, particularly since the type and operation of equipment will vary from one laboratory to another. Some of the general procedures of good laboratory technique will be mentioned as we go along

See your CD for pictures of commonly used glassware and apparatus in the analytical laboratory.

2.1 The Laboratory Notebook—Your Critical Record

A well-kept laboratory record will help assure reliable analyses.

You should first realize that in the analytical laboratory, more than anywhere else, cleanliness and neatness are of the utmost importance. This also applies to the keeping of an orderly notebook. All data should be recorded permanently in ink *when they are collected.* When you go into the analytical laboratory, you will find that this orderliness is to your advantage. First of all, there is a saving of time in not having to reorganize and rewrite the data. There is probably an additional saving of time since you will be more organized in carrying out the operations of the analysis if you have trained yourself to put the data down in an orderly fashion. Chances for a mistake are reduced. Second, if you make an immediate record, you will be able to detect possible errors in measurements or calculations. And data will not be lost or transferred incorrectly if they are recorded directly in a notebook instead of collected on scraps of paper.

For practicing analytical chemists and on-the-job applications, it is especially important to use the lab notebooks for entering observations and measurements directly. Complete documentation is essential for forensic or industrial laboratories for legal or patent considerations. In industrial research labs, the notebook must generally be signed (witnessed) and dated by another person familiar with the work to assure legal patent priority if applicable.

An example of a well-kept notebook with properly recorded data is illustrated below for the volumetric analysis of a soda ash unknown in your laboratory:

Date: 7 September, 2003

Analysis of soda ash unknown

Principle: The soda ash is dissolved in water and titrated to a bromcresol green end point with standard hydrochloric acid. The hydrochloric acid is standardized against primary standard sodium carbonate. Weigh sodium carbonate and soda ash unknown by difference.

Reference: Experiment 7

Titration Reaction $CO_3^{2-} + 2H^+ = H_2CO_3$

Standardization

$$M \text{ (HCl)} = \frac{\text{mg Na}_2\text{CO}_3}{\text{f wt Na}_2\text{CO}_3 \text{ (mg/mmol)} \times \frac{1}{2} \text{(mmol Na}_2\text{CO}_3/\text{mmol HCl)} \times \text{mL HCl}}$$

$$= \frac{\text{mg Na}_2\text{CO}_3}{105.99 \text{ (mg/mmol)} \times \frac{1}{2} \times \text{mL HCl}}$$

	#1	#2	#3
Bottle + sample	24.2689 g	24.0522 g	23.8597 g
Less sample	24.0522 g	23.8597 g	23.6269 g
g Na$_2$CO$_3$	0.2167 g	0.1925 g	0.2328 g
mg Na$_2$CO$_3$	216.7 mg	192.5 mg	232.8 mg
Buret reading	40.26 mL	35.68 mL	43.29 mL
Initial reading	0.03 mL	0.00 mL	0.02 mL
Net volume	40.23 mL	35.68 mL	43.27 mL
Molarity:	$0.1016_4 M$	$0.1018_0 M$	$0.1015_2 M$

Mean: 0.1016_5

Std. devn: 1.6 ppt

Range: 2.8 ppt

Soda Ash

$$\% \text{ Na}_2\text{CO}_3 = \frac{M \times \text{mL} \times \text{f wt Na}_2\text{CO}_3 \times \frac{1}{2} \text{ (mmol Na}_2\text{CO}_3/\text{mmol HCl)}}{\text{mg sample}} \times 100$$

$$= \frac{0.1016_5 \text{ (mmol/mL)} \times \text{mL} \times 105.99 \text{ (mg/mmol)} \times \frac{1}{2}}{\text{mg sample}} \times 100$$

Bottle + sample	25.6728 g	25.4673 g	25.2371 g
Less sample	25.4673 g	25.2371 g	25.0027 g
g sample	0.2055 g	0.2302 g	0.2344 g
mg sample	205.5 mg	230.2 mg	234.4 mg
Buret reading	35.67 mL	40.00 mL	40.70 mL
Initial reading	0.00 mL	0.01 mL	0.05 mL
Net volume	35.67 mL	39.99 mL	40.65 mL
% Na_2CO_3	93.50%	93.58%	93.42%

Mean: 93.50
Std. devn: 0.9 ppt
Range: 1.7 ppt

The above example is an abbreviated version in which actual calculation or numerical setups are omitted. For complete record keeping, you should include the computational setups in your notebook so an error can be tracked down later, if necessary.

Rather than fill all the space in the laboratory notebook, it is recommended you leave alternate pages for scratch pages (e.g., the left page, leaving the right page for summarizing data). It is also important that you record your data to the proper number of significant figures. Significant figures are discussed in Chapter 3, and you should review this material before beginning in the laboratory.

The correct number of significant figures in measurements and calculations is critical in giving the proper significance to an analysis. See Chapter 3.

Laboratory Notebook Documentation

The laboratory notebook is a record of your job as an analytical chemist. It documents everything you do. It is the source for reports, publications, and regulatory submissions. The success or failure of a company's product or service may depend on how well you do that documentation. The notebook becomes a legal document for patent issues, government regulation issues (validation, inspections, legal actions), and the like. Remember, "if it isn't written down, it wasn't done." The notebook is where you record your original ideas that may form the basis of a patent, and so it is important to record what went into those ideas and when.

What are the features of a well-maintained notebook? They will vary with individual preferences. but here are some good rules:

- Use a hardcover notebook (no loose leafs).
- Number pages consecutively.
- Record only in ink.
- Never tear out pages. If not used, put a line through the page.
- Date each page, sign it, and have it signed and dated (maybe shortly later) by someone else, stating "Read and Understood by"
- Record the name of the project, why it is being done, and any literature references.
- Record all data on the day you obtain it.

A word about *electronic notebooks.* Modern instrument software allows the analyst to collect, store, and process data directly from the instrument signal, based on appropriate calibration. It is important that the software and calibration be validated, as for the remainder of the analysis, as a part of good laboratory practice, discussed in Chapter 4.

2.2 Laboratory Materials and Reagents

Table 2.1 lists the properties of materials used in the manufacture of common laboratory apparatus. Borosilicate glass (brand names: Pyrex, Kimax) is the most commonly used material for laboratory apparatus such as beakers, flasks, pipets, and burets. It is stable to hot solutions and to rapid changes in temperature. For more

Table 2.1
Properties of Laboratory Materials

Material	Max. Working Temperature (°C)	Sensitivity to Thermal Shock	Chemical Inertness	Notes
Borosilicate glass	200	150°C change OK	Attacked somewhat by alkali solutions on heating	Trademarks: Pyrex (Corning Glass Works); Kimax (Owens-Illinois)
Soft glass		Poor	Attacked by alkali solutions	Boron-free. Trademark: Corning
Alkali-resistant glass		More sensitive than borosilicate		
Fused quartz	1050	Excellent	Resistant to most acids, halogens	Quartz crucibles used for fusions
High-silica glass	1000	Excellent	More resistant to alkalis than borosilicate	Similar to fused quartz Trademark: Vycor (Corning)
Porcelain	1100 (glazed) 1400 (unglazed)	Good	Excellent	
Platinum	ca. 1500		Resistant to most acids, molten salts. Attacks by aqua regia, fused nitrates, cyanides, chlorides at $>1000°C$. Alloys with gold, silver, and other metals	Usually alloyed with iridium or rhodium to increase hardness. Platinum crucibles for fusions and treatment with HF
Nickel and iron			Fused samples contaminated with the metal	Ni and Fe crucibles used for peroxide fusions
Stainless steel	400–500	Excellent	Not attacked by alkalis and acids except conc. HCl, dil. H_2SO_4, and boiling conc. HNO_3	
Polyethylene	115		Not attacked by alkali solutions or HF. Attacked by many organic solvents (acetone, ethanol OK)	Flexible plastic
Polystyrene	70		Not attacked by HF. Attacked by many organic solvents	Somewhat brittle
Teflon	250		Inert to most chemicals	Useful for storage of solutions and reagents for trace metal analysis.

specific applications, there are several other materials employed that may possess advantage with respect to chemical resistance, thermal stability, and so forth.

The different grades of chemicals are listed on the *back cover* of the text. In general, only *American Chemical Society (ACS) reagent-grade* or *primary standard* chemicals should be used in the analytical laboratory.

The American Chemical Society publishes a compendium of tests for evaluating the purity and quality of basic laboratory chemicals. Reagent chemicals that do not reference the ACS meet the manufacturer's own reagent specifications, which vary among suppliers.

The reagent-grade chemicals, besides meeting minimum requirements of purity, may be supplied with a report of analysis of the impurities (printed on the label). Primary standard chemicals are generally at least 99.95% pure. They are analyzed and the results are printed on the label. They are more expensive than reagent-grade chemicals and are used only for the preparation of standard solutions or for the standardization of a solution by reaction (titration) with it. Not all chemicals are available in primary standard grade. There are special grades of solvents for special purposes, for example, spectral grades or chromatographic grades. These are specifically purified to remove impurities that might interfere in the particular application. Likewise, there are trace metal analyzed acids that are specially refined and tested in greater detail for trace elemental impurities, typically in the parts per billion range.

In addition to commercial producers, the National Institute of Standards and Technology supplies primary standard chemicals. *NIST Special Publication 260* catalogs standard reference materials. (See http://ts.nist.gov/ts/htdocs/230/232/232.htm for information on the SRM program and lists of reference standards.) Reference standards are complex materials, such as alloys that have been carefully analyzed for the ingredients and are used to check or calibrate an analytical procedure.

The concentrations of commercially available acids and bases are listed on the inside of the *back cover*.

Reagent-grade chemicals are almost always used in analyses. Primary standards are used for preparing volumetric standard solutions.

2.3 The Analytical Balance—The Indispensible Tool

Weighing is a required part of almost any analysis, both for measuring the sample and for preparing standard solutions. In analytical chemistry we deal with rather small weights, on the order of a few grams to a few milligrams or less. Standard laboratory weighings are typically made to three or four significant figures, and so the weighing device must be both accurate and sensitive. There are various sophisticated ways of achieving this, but the most useful and versatile device used is the **analytical balance**.

Most analytical balances used today are electronic balances. The mechanical single-pan balance is still used, though, and so we will describe its operation. Both types are based on comparison of one weight against another (the electronic one for calibration) and have in common factors such as zero-point drift and air buoyancy. We really deal with masses rather than weights. The **weight** of an object is the force exerted on it by the gravitational attraction. This force will differ at different locations on Earth. **Mass**, on the other hand, is the quantity of matter of which the object is composed and is invariant.

The balance measures mass.

ELECTRONIC BALANCES—THE MOST CONVENIENT

Modern electronic balances offer convenience in weighing and are subject to fewer errors or mechanical failures than are mechanical balances. The operation of dialing

Electronic balances are more convenient to use.

Fig. 2.1. Electronic analytical balance. (Courtesy of Denver Instrument Co. Denver Instrument Company owns all images.)

weights, turning and reading micrometers, and beam and pan arrest of mechanical balances are eliminated, greatly speeding the measurement. A digital-display electronic balance is shown in Figure 2.1, and the operating principle of an electronic balance is illustrated in Figure 2.2. There are no weights or knife edges as with mechanical balances. The pan sits on the arm of a movable hanger (2), and this movable system is compensated by a constant electromagnetic force. The position of the hanger is monitored by an electrical position scanner (1), which brings the weighing system back to the zero position. The compensation current is proportional to the weight placed on the pan. This is sent in digital form to a microprocessor that converts it into the corresponding weight value, which appears as a digital display. The weight of the container can be automatically subtracted.

These balances use the principle of electromagnetic force compensation first described by Angstrom in 1895. But they still use the principle of comparing one weight with another. The balance is "zeroed," or calibrated, with a known weight. When the sample is placed on the pan, its weight is electronically compared with the known. This is a form of self-calibration. Modern balances may have such features as compensating for wandering from true zero and averaging variations due to building vibrations.

A single control bar is used to switch the balance on and off, to set the display to zero, and to tare a container automatically on the pan. Since results are

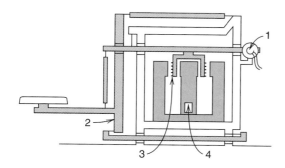

Fig. 2.2. Operating principle of electronic balance: 1, position scanner; 2, hanger; 3, coil; 4, temperature sensor. (From K. M. Lang, *American Laboratory*, March, 1983, p. 72. Reproduced by permission of American Laboratory, Inc.)

available as an electrical signal, they can be readily processed by a personal computer and stored. Weighing statistics can be automatically calculated.

Electronic analytical balances can be purchased with different weighing ranges and readabilities. A macrobalance will have a range on the order of 160 g, readable to 0.1 mg, and a semimicrobalance will have a range of about 30 g, readable to 0.01 mg. Microbalances weigh to 1 μg, and ultramicrobalances are available that are sensitive to 0.1 μg or less.

Electrochemical quartz balances are available with 100-μg range that can detect 1 ng (10^{-9} g) changes! The balance utilizes a thin quartz crystal disk oscillating at, for example, 10 MHz. The frequency of oscillation changes with any change in mass, and the frequency change measured by the instrument is converted to mass units. A film of gold is evaporated on the quartz, and the gold substrate can be coated with the material of interest. Mass changes as small as a few percent of a monolayer coverage of atoms or molecules on the gold surface can be measured. Mass changes with time can be recorded.

SINGLE-PAN MECHANICAL BALANCE

Mechanical balances are being replaced by electronic balances, but many are still in use (and are excellent balances), so they are described here.

The mechanical analytical balance is a first-class lever that compares two masses. Figure 2.3 illustrates such a balance. The fulcrum A lies between the points of application of forces B and C. The term M_1 represents the unknown mass and M_2 represents a known mass. The principle of operation is based on the fact that at balance, $M_1 L_1 = M_2 L_2$. If L_1 and L_2 are constructed to be as nearly equal as possible, then, at balance, $M_1 = M_2$. A pointer is placed on the *beam* of the balance to indicate on a scale at the end of the pointer when a state of balance is achieved. The operator adjusts the value of M_2 until the pointer returns to its original position on the scale when the balance is unloaded. Although mass is determined, the ratio of masses is the same as the ratio of weights when an equal-arm balance is used. it is customary, then, to use the term *weight* instead of *mass* and to speak of the operator as *weighing*. The known masses are called **standard weights**.

Most analytical weighings using mechanical balances are made using a single-pan balance. A schematic diagram of a typical mechanical single-pan balance is shown in Figure 2.4 (see below for a description of modern electronic balances).

Burns' Hog Weighing Method: "1) Get a perfectly straight plank and balance it across a sawhorse. 2) Put the hog on one end of the plank. 3) Pile rocks on the other end until the plank is again perfectly balanced. 4) Carefully guess the weight of the rocks."—Robert Burns

A first-class (unsymmetrical) lever is pivoted on a knife edge, and a pan is at one end in which the object is placed. However, there is no pan at the other end for placing weights. When the balance is not in use, a series of weights totaling 160 to 200 g are on the pan end of the beam. These are counterbalanced by a single weight on the other end of the beam, which also acts as part of a damping piston. When an object is placed on the pan, individual weights are *removed* from this end of the beam to restore it to equilibrium. This is accomplished by means of knobs on the front of the balance that lift weights or combinations of weights from the beam. Thus, the weights are never handled. These weights will be equal to the weight of the object on the pan.

Actually, the beam is not brought completely to balance, but weights are removed only to the nearest whole gram or 0.1 g, depending on the balance. The imbalance of the beam is registered optically and automatically on an illuminated

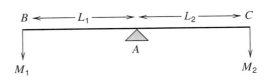

Fig. 2.3. Principle of analytical balance.

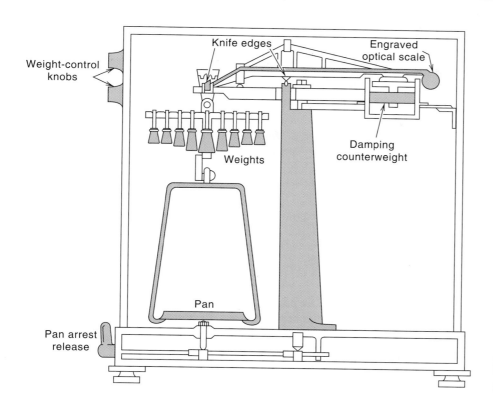

Fig. 2.4. Schematic diagram of a typical single-pan balance.

vernier scale by a light ray reflected from an engraved optical scale on the beam. The last digits (nearest 0.1 mg) are read from this scale. Alternatively, the imbalance may be read on a digital counter.

The original no-load reading or position is called the **zero point**, and the position under load is called the **rest point**. In operation, the rest point is made to coincide with the zero point. The zero point is generally adjusted to read zero by adjusting the vernier by means of a knob.

Single-pan balances are under constant load of 160 or 200 g, a required feature since they are not brought back to a state of balance. The sensitivity of a balance varies with the load because it is governed by the center of gravity of the beam; the beam bends slightly under load, causing a change in the center of gravity and the sensitivity. Calibration of the vernier or digital readout of a single-pan balance to read the amount of imbalance is done at a given sensitivity, that is, at a given load. Therefore, the load must remain constant.

All weights of a single-pan balance are concealed and are removed by means of control knobs on the front of the balance: one for tens (e.g., 10 to 90 g), one for units (1 to 9 g), and, if applicable, one for 0.1 units (0.1 to 0.9 g). The weights removed are registered on a counter on the front of the balance. The beam is brought to rest rapidly by means of an air piston damper.

Care must be taken not to damage the knife edges while the balance is not in operation and while objects are being placed or removed from the pan. A three-position beam-arrest knob is used to protect the knife edges and beam. The center position arrests the pan and beam; a second position partially releases the pan for use while finding the approximate weight of the object on the pan; and a third position completely releases the pan to allow the balance to come to rest.

A typical single-pan balance is shown in Figure 2.5. Weighings can be made in less than a minute with these balances.

Always check the zero point before making weighings.

Fig. 2.5. Typical single-pan balance. (Courtesy of Arthur H. Thomas Company.)

SEMIMICRO- AND MICROBALANCES

The discussion thus far has been limited to conventional macro or analytical balances. These perform weighings to the nearest 0.1 mg, and loads of up to 100 or 200 g can be handled. These are satisfactory for most routine analytical weighings. All of the above classes of balances can be made more sensitive by changing the parameters affecting the sensitivity, such as decreasing the mass of the beam and pans, increasing the length of the beam, and changing the center of gravity of the beam. Lighter material can be used for the beam since it need not be as sturdy as the beam of a conventional balance.

The *semimicrobalance* is sensitive to about 0.01 mg, and the *microbalance* is sensitive to about 0.001 mg (1 μg). The load limits of these balances are correspondingly less than the conventional balance, and greater care must be taken in their use.

ZERO-POINT DRIFT

The zero point of a balance is not a constant that can be determined or set and forgotten. It will drift for a number of reasons, including temperature changes, humidity, and static electricity. The zero point should therefore be checked at least once every half-hour during the period of using the balance.

WEIGHT IN A VACUUM—THIS IS THE MOST ACCURATE

The weighings that are made on a balance will, of course, give the weight in air. When an object displaces its volume in air, it will be buoyed up by the weight of air displaced (**Archimedes' principle**—see the box in Chapter 1 on how analytical chemistry originated). The density of air is 0.0012 g (1.2 mg) per milliliter. If the density of the weights and the density of the object being weighed are the same, then they will be buoyed up by the same amount, and the recorded weight will be equal to the weight in a vacuum, where there is no buoyancy. If the densities are markedly different, the differences in the buoyancies will lead to a small error in the weighing: One will be buoyed up more than the other, and an unbalance will result. Such a situation arises in the weighing of very dense objects [e.g., platinum vessels (density = 21.4) or mercury (density = 13.6)] or light, bulky objects [e.g., water (density ≈ 1)]; and in very careful work, a correction should be made for this error. For comparison, the density of weights used in balances is about 8. See Ref. 14 for air buoyancy corrections with a single-pan balance. (Reference 10 describes the calibration of the weights in a single-pan balance.)

Actually, in most cases, a correction is not necessary because the error resulting from the buoyancy will cancel out in percent composition calculations. The same error will occur in the numerator (as the concentration of a standard solution or weight of a gravimetric precipitate) and in the denominator (as the weight of the sample). Of course, all weighings must be made with the materials in the same type of container (same density) to keep the error constant.

An example where correction in vacuum is used is in the calibration of glassware. The mass of water or mercury delivered or contained by the glassware is measured. From a knowledge of the density of the liquid at the specified temperature, its volume can be calculated from the mass. Even in these cases, the buoyancy correction is only about one part per thousand. For most objects weighed, buoyancy errors can be neglected. Handbooks contain tables for converting weight of water or mercury in air to volume at different temperatures, using brass weights.

We make most quantitative weighings to 0.1 mg.

An object of 1-mL volume will be buoyed up by 1.2 mg!

The buoyancy of the weighing vessel is ignored, since it is subtracted.

Buoyancy corrections are usually significant in glassware calibration.

Weights of objects in air can be corrected to the weight in vacuum by

$$W_{vac} = W_{air} + W_{air}\left(\frac{0.0012}{D_o} - \frac{0.0012}{D_w}\right)$$ (2.1)

where
W_{vac} = weight in vacuum, g
W_{air} = observed weight in air, g
D_o = density of object
D_w = density of weights
0.0012 = density of air

The density of brass weights is 8.4 and that of stainless steel weights is 7.8. A calculation with water as the object will convince you that even here the correction will amount to only about one part per thousand.

Example 2.1

A convenient way to calibrate pipets is to weigh water delivered from them. From the exact density of water at the given temperature, the volume delivered can then be calculated. Suppose a 20-mL pipet is to be calibrated. A stoppered flask when empty weighs 29.278 g. Water is delivered into it from the pipet, and it now weighs 49.272 g. If brass weights are used, what is the weight of water delivered, corrected to weight in vacuum?

The same buoyancy corrections apply for mechanical or electronic balances (which are calibrated with weights of known density).

Solution

The increase in weight is the weight of water in air:

$$49.272 - 29.278 = 19.994 \text{ g}$$

The density of water is 1.0 g/mL (to 2 significant figures from 10 to 30°C—see Table 2.4). Therefore,

$$W_{vac} = 19.994 + 19.994\left(\frac{0.0012}{1.0} - \frac{0.0012}{8.4}\right) = 20.015 \text{ g}$$

Example 2.2

Recalculate the weight of the water delivered by the pipet in Example 2.1, using stainless steel weights at density 7.8 g/cm³.

Solution

Do not round off until the end of the calculation. Then the same value results:

$$W_{vac} = 19.994 + 19.994\left(\frac{0.0012}{1.0} - \frac{0.0012}{7.8}\right) = 20.015 \text{ g}$$

This illustrates that the buoyancy corrections in Table 2.4 are valid for either type of weight.

SOURCES OF ERROR IN WEIGHING

Several possible sources of error have been mentioned, including zero-point drift, the weights, and buoyancy. Change in ambient temperature or temperature of the object being weighed is probably the biggest source of error, causing a drift in the zero or rest point due to air current convections. Hot or cold objects must be brought to ambient temperature before being weighed. Hygroscopic samples may pick up moisture, particularly high-humidity atmosphere and exposure to air prior to and during weighing must be minimized.

GENERAL RULES FOR WEIGHING

We have seen that there are several types of balances, and the operation of these will differ according to the manufacturer. The specific operation of your particular balance will be explained by your instructor. The main objectives are to (1) protect the knife edges, (2) protect all parts from dust and corrosion, (3) avoid contamination or change in load (of sample or container), and (4) avoid draft (air convection) errors. Some general rules you should familiarize yourself with before weighing with any type of analytical balance are:

Learn these rules!

---

1. Never handle objects to be weighed with the fingers. A piece of clean paper or tongs should be used.
2. Weigh at room temperature, and thereby avoid air convection currents.
3. Never place chemicals directly on the pan, but weigh them in a vessel (weighing bottle, weighing dish) or on powder paper. Always brush spilled chemicals off immediately with a soft brush.
4. Always close the balance case door before making the weighing. Air currents will cause the balance to be unsteady.
5. When using a mechanical balance, never place objects or weights on the pans or remove them without securing the beam arrest and the pan arrest.

WEIGHING OF SOLIDS

Weighing by difference is required for hygroscopic samples.

Solid chemical (nonmetal) materials are usually weighed and dried in a **weighing bottle**. Some of these are shown in Figure 2.6. They have standard tapered ground-glass joints, and hygroscopic samples (which take on water from the air) can be weighed with the bottle kept tightly capped. Replicate weighings are most conveniently carried out by **difference**, at least with slower mechanical balances. With electronic balances, not much time is saved. The sample in the weighing bottle is weighed, and then a portion is removed (e.g., by tapping) and quantitatively transferred to a vessel appropriate for dissolving of the sample. The weighing bottle and sample are reweighed, and from the difference in weight, the weight of sample is calculated. The next sample is removed and the weight is repeated to get its weight by difference, and so on. This is illustrated under The Laboratory Notebook for the soda ash experiment.

It is apparent that by this technique an average of only one weighing for each sample, plus one additional weighing for the first sample, is required. However, each weight represents the difference between two weighings, so that the total

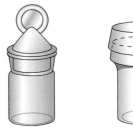

Fig. 2.6. Weighing bottles.

experimental error is given by the combined error of both weighings. Weighing by difference *with the bottle capped* must be employed if the sample is hygroscopic or cannot otherwise be exposed to the atmosphere before weighing. If there is no danger from atmospheric exposure, the bottles need not be capped.

For **direct weighing**, a **weighing dish** is used (Figure 2.7). Or, weighing paper or a plastic weighing boat is used. The dish, paper, or boat is weighed empty and then with the added sample. This requires two weighings for each sample. The weighed sample is transferred with the aid of a camel-hair brush after tapping. Direct weighing is satisfactory only if the sample is nonhygroscopic.

When making very careful weighings (e.g., to a few tenths of a milligram). you must take care not to contaminate the weighing vessel with extraneous material that may affect its weight. Special care should be taken not to get perspiration from the hands on the vessel because this can be quite significant. It is best to handle the vessel with a piece of paper. Solid samples must frequently be dried to a constant weight. Samples must always be allowed to attain the temperature of the balance room before weighing.

Fig. 2.7. Weighing dish.

WEIGHING OF LIQUIDS

Weighing of liquids is usually done by direct weighing. The liquid is transferred to a weighed vessel (e.g., a weighing bottle), which is capped to prevent evaporation during weighing, and is then weighed. If a liquid sample is weighed by difference by pipetting out an aliquot from the weighing bottle, the inside of the pipet must be rinsed several times after transferring. Care should be taken not to lose any sample from the tip of the pipet during transfer.

TYPES OF WEIGHING—WHAT ACCURACY DO YOU NEED?

There are two types of weighing done in analytical chemistry, **rough** and **accurate**. Rough weighings to two or three significant figures are normally used when the amount of substance to be weighed need only be known to within a few percent. Examples are reagents to be dissolved and standardized later against a known standard, or the apportioning of reagents that are to be dried and then later weighed accurately, or simply added as is, as for adjusting solution conditions. That is, only rough weighings are needed when the weight is not involved in the computation of the analytical result. Rough weighings need not be done on analytical balances but may be completed on triple-beam, top-loading, or torsion balances.

Accurate weighings are reserved for obtaining the weight of a sample to be analyzed, the weight of the dried product in gravimetric procedures, or the weight of a dried reagent being used as a standard in a determination, all of which must generally be known to four significant figures to be used in calculating the analytical result. **These are performed only on an analytical balance, usually to the nearest 0.1 mg**. An exact predetermined amount of reagent is rarely weighed (e.g., 0.5000 g), but rather an approximate amount (about 0.5 g) is weighed accurately (e.g., to give 0.5129 g). Some chemicals are never weighed on an analytical balance. Sodium hydroxide pellets, for example, are so hygroscopic that they continually, absorb moisture. The weight of a given amount of sodium hydroxide is not reproducible (and its purity is not known). To obtain a solution of known sodium hydroxide concentration, the sodium hydroxide is weighed on a rough balance and dissolved, and the solution is standardized against a standard acid solution.

Only some weighings have to be done on an analytical balance, those involved in the quantitative calculations.

2.4 Volumetric Glassware—Also Indispensible

Although accurate volume measurements of solutions can be avoided in gravimetric methods of analysis, they are required for almost any other type of analysis involving solutions.

VOLUMETRIC FLASKS

Volumetric flasks contain an accurate volume.

Volumetric flasks are used in the diluting of a sample or solution to a certain volume. They come in a variety of sizes, from 2 L or more to 1 mL. A typical flask is shown in Figure 2.8. These flasks are designed **to contain** an accurate volume at the specified temperature (20 or 25°C) when the bottom of the meniscus (the concave curvature of the upper surface of water in a column caused by capillary action—see Figure 2.14) just touches the etched "fill" line across the neck of the glass. The coefficient of expansion of glass is small, and for ambient temperature fluctuations the volume can be considered constant. These flasks are marked with "**TC**" to indicate "to contain." Other, less accurate containers, such as graduated cylinders, are also marked "TC." Many of these are directly marked on the face by the manufacturer as to the uncertainty of the container measurement; for example, a 250-mL volumetric flask is "± 0.24 mL," or roughly a 0.1% error.

Initially, a small amount of diluent (usually distilled water) is added to the empty flask. Reagent chemicals should never be added directly to a dry glass surface, as glass is highly absorbant. When using a volumetric flask, a solution should be prepared stepwise. The desired reagent chemical (either solid or liquid) to be diluted is added to the flask, and then diluent is added to fill the flask about two-thirds (taking care to rinse down any reagent on the ground glass lip). It helps to swirl the solution before diluent is added to the neck of the flask to obtain most of the mixing (or dissolving in the case of a solid). Finally, diluent is added so that the bottom of the meniscus is even with the middle of the calibration mark (at eye level). If there are any droplets of water on the neck of the flask above the meniscus, take a piece of tissue and blot these out. Also, dry the ground-glass stopper joint.

The solution is finally thoroughly mixed as follows. Keeping the stopper on securely by using the thumb or palm of the hand, invert the flask and swirl or shake it *vigorously* for 5 to 10 s. Turn right side up and allow the solution to drain from the neck of the flask. Repeat at least 10 times.

Note. Should the volume of liquid go over the calibration mark, it is still possible to save the solution as follows. Paste against the neck of the flask a thin strip of paper and mark on it with a sharp pencil the position of the meniscus, avoiding parallax error. After removing the thoroughly mixed solution from the flask, fill the flask with water to the calibration mark. Then by means of a buret or small volume graduated pipet, add water to the flask until the meniscus is raised to the mark on the strip of paper. Note and record the volume so added and use it to mathematically correct the concentration calculation.

Fig. 2.8. Volumetric flask.

PIPETS

Volumetric pipets deliver an accurate volume.

The pipet is used to transfer a particular volume of solution. As such, it is often used to deliver a certain fraction (**aliquot**) of a solution. To ascertain the fraction, the original volume of solution from which the aliquot is taken must be known, but it need not all be present, so long as it has not evaporated or been diluted. There are two common types of pipets, the **volumetric**, or **transfer**, **pipet** and the

measuring or **graduated pipet** (see Figures 2.9 and 2.10). Variations of the latter are also called **clinical**, or **serological**, **pipets**.

Pipets are designed **to deliver** a specified volume at a given temperature, and they are marked "**TD**." Again, the volume can be considered to be constant with small changes in temperature. Pipets are calibrated to account for the drainage film remaining on the glass walls. This drainage film will vary somewhat with the time taken to deliver, and usually the solution is allowed to drain under the force of gravity and the pipet is removed shortly after the solution is delivered. A uniform drainage time should be adopted.

The volumetric pipet is used for accurate measurements since it is designed to deliver only one volume and is calibrated at that volume. Accuracy to four significant figures is generally achieved, although with proper calibration, five figures may be obtained if necessary. See the table on the *back cover* for tolerances of class A transfer pipets. Measuring pipets are straight-bore pipets that are marked at different volume intervals. These are not as accurate because nonuniformity of the internal diameter of the device will have a relatively larger effect on total volume than is the case for pipets with a bulb shape. Also, the drainage film will vary with the volume delivered. At best, accuracy to three significant figures can be expected from these pipets, unless you make the effort to calibrate the pipet for a given volume delivered.

Most volumetric pipets are calibrated to deliver with a certain small volume remaining in the tip. This should not be shaken or blown out. In delivering, the pipet is held vertically and the tip is touched on the side of the vessel to allow smooth delivery without splashing and so that the proper volume will be left in the tip. The forces of attraction of the liquid on the wall of the vessel will draw out a part of this.

Some pipets are **blowout** types (including measuring pipets calibrated to the entire tip volume). The final volume of solution must be blown out from the tip to deliver the calibrated amount. These pipets are easy to identify, as they will always have one or two **ground bands or rings** around the top. (These are not to be confused with a colored ring that is used only as a color coding for the volume of the pipet.) The solution is not blown out until it has been completely drained by gravity. Blowing to increase the rate of delivery will change the volume of the drainage film.

Volumetric pipets are available in sizes of 100 to 0.5 mL or less. Measuring and serological pipets range from a total capacity of 25 to 0.1 mL. Measuring pipets can be used for accurate measurements, especially for small volumes, if they are calibrated at the particular volume wanted. The larger measuring pipets usually deliver too quickly to allow drainage as fast as the delivery, and they have too large a bore for accurate reading.

In using a pipet, one should always wipe the outside of the tip dry after filling. If a solvent other than water is used, or if the solution is viscous, pipets must be recalibrated for the new solvent or solution to account for difference in drainage rate.

Pipets are filled by suction, using a rubber pipet bulb, a pipet pump, or other such pipetting device. Before using a pipet, practice filling it with water. No solution should be pipetted by mouth.

SYRINGE PIPETS

These can be used for both macro and micro volume measurements. The calibration marks on the syringes may not be very accurate, but the reproducibility can be excellent if an automatic deliverer is used, such as a spring-load device that

Fig. 2.9. Transfer or volumetric pipets.

Fig. 2.10. Measuring pipets.

Fig. 2.11. Hamilton microliter syringe.

Syringe pipets are useful for delivering microliter volumes.

draws the plunger up to the same preset level each time. The volume delivered in this manner is free from drainage errors because the solution is forced out by the plunger. The volume delivered can be accurately calibrated. Microliter syringe pipets are used for introduction of samples into gas chromatographs. A typical syringe is illustrated in Figure 2.11. They are fitted with a needle tip, and the tolerances are as good as those found for other micropipets. In addition, any desired volume throughout the range of the syringe can be delivered.

The above syringe pipets are useful for accurate delivery of viscous solutions or volatile solvents; with these materials the drainage film would be a problem in conventional pipets. Syringe pipets are well suited to rapid delivery and also for thorough mixing of the delivered solution with another solution as a result of the rapid delivery.

A second type of syringe pipet is that shown in Figure 2.12. This type is convenient for rapid, one-hand dispensing of fixed volumes in routine procedures and is widely used in the clinical chemistry laboratory. It contains a disposable non-wetting plastic tip (e.g., polypropylene) to reduce both film error and contamination. A thumb button operates a spring-loaded plunger, which stops at an intake or a discharge stop; the latter stop is beyond the former to ensure complete delivery. The sample never contacts the plunger and is contained entirely in the plastic tip. These pipets are available in volumes of 1 to 1000 μL and are reproducible to 1 to 2% better, depending on the volume.

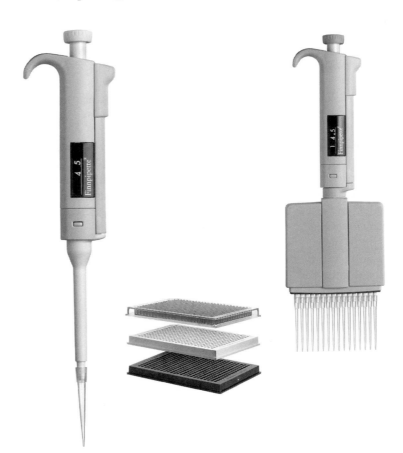

Fig. 2.12. Single-channel and multichannel digital displacement pipets and microwell plates. (Courtesy of Thermo Labsystems.)

The actual volume delivered by these and other micropipets frequently does not need to be known because they are used in relative measurements. For example, the same pipet may be used to deliver a sample and an equal volume of a standard solution for calibrating the instrument used for the measurement. Precision in delivery is usually more important than the absolute volume delivered. The standard for pipet calibration use in Europe (where many of these pipets are manufactured) is the German DIN 126650 (or a similar international standard ISO 8655). Calibrations are based on gravimetric testing (weighing of water). The DIN standard does not give separate limits for accuracy and precision, but rather uses a combined error limit equal to percent accuracy plus 2 times the standard deviation, that is, it gives a range in which we are 95% confident the delivery will fall (see Chapter 3 for a discussion of standard deviation and confidence limits). Table 2.2 lists the DIN error limits for single-channel displacement pipets. Table 2.3 lists accuracies and precisions for a typical model single-channel pipet.

> The volume may not be accurately known, but it is reproducible.

Besides the manually operated syringes, there are electronically controlled and variable-volume motor-driven syringes available for automated repetitive deliveries. Also, you may purchase pipets with multiple syringes for simultaneous delivery, with for example, 12 or 16 channels. These are useful for delivering solutions into microwell plates used in biotechnology or clinical chemistry laboratories that process thousands of samples (Figure 2.12). You may find more information on displacement pipets from representative manufacturers, for example, www.finnpipette.com and www.eppendorf.com.

BURETS

A buret is used for the accurate delivery of a variable amount of solution. Its principal use is in **titrations**, where a standard solution is added to the sample solution until the **end point** (the detection of the completion of the reaction) is reached. The conventional buret for macrotitrations is marked in 0.1-mL increments from 0 to 50 mL; one is illustrated in Figure 2.13. The volume delivered can be read to the nearest 0.01 mL by interpolation (good to about ±0.02 or ±0.03 mL). Burets

Table 2.2
DIN 12650 Error Limits for Single Channel Air Displacement Pipets[a]

Nominal Volume (μL)	Maximum Error (μL)	Relative Error (%)
1	±0.15	±15.0
2	±0.20	±10.0
5	±0.30	±6.0
10	±0.30	±3.0
20	±0.40	±2.0
50	±0.80	±1.6
100	±1.50	±1.5
200	±2.00	±1.0
500	±5.00	±1.0
1000	±10.00	±1.0
2000	±20.00	±1.0
5000	±50.00	±1.0
10000	±100.00	±1.0

[a]These limits apply to manufacturers with a controlled environment. If the tests are performed by a user in a normal laboratory environment, the limits in the table may be doubled.
Courtesy of Thermo Labsystems Oy, Finland.

Table 2.3
Accuracy and Precision for Single Channel Digital Finnpipettes

Range (μL)	Increment (μL)	Volume (μL)	Accuracy		Precision[a]	
			(μL)	(%)	s.d. (μL)	CV (%)
0.2–2	0.01	2	±0.050	±2.5	0.040	2.0
		0.2	±0.024	±12.0	0.020	10.0
0.5–10	0.1	10	±0.100	±1.0	0.050	0.5
		1	±0.025	±2.5	0.020	2.0
0.5–10	0.1	10	±0.100	±1.0	0.080	0.8
		1	±0.035	±3.5	0.030	3.0
2–20	0.1	20	±0.200	±1.0	0.080	0.4
		2	±0.060	±3.0	0.030	1.5
5–40	0.5	40	±0.240	±0.6	0.120	0.3
		5	±0.100	±2.0	0.100	2.0
10–100	1.0	100	±0.80	±0.8	0.20	0.2
		10	±0.30	±3.0	0.10	1.0
20–200	1.0	200	±1.20	±0.6	0.40	0.2
		20	±0.36	±1.8	0.14	0.7
200–1000	5.0	1000	±6.00	±0.6	2.00	0.2
		200	±1.80	±0.9	0.60	0.3
100–1000	5.0	1000	±6.00	±0.6	2.00	0.2
		100	±1.00	±1.0	0.60	0.6

[a] s.d. = standard deviation, CV = coefficient of variation.
Courtesy of Thermo Labsystems Oy, Finland.

are also obtainable in 10-, 25-, and 100-ml capacities, and microburets are available in capacities of down to 2 mL, where the volume is marked in 0.01-mL increments and can be estimated to the nearest 0.001 mL. Ultramicroburets of 0.1-mL capacity graduated in 0.001-mL (1-μL) intervals are used for microliter titrations.

Drainage film is a factor with conventional burets, as with pipets, and this can be a variable if the delivery rate is not constant. The usual practice is to deliver at a fairly slow rate, about 15 to 20 mL per minute, and then to wait several seconds after delivery to allow the drainage to "catch up." In actual practice, the rate of delivery is only a few drops per minute near the end point, and there will be no time lag between the flow rate and the drainage rate. As the end point is approached, fractions of a drop are delivered by just opening, or "cracking," the stopcock and then touching the tip of the buret to the wall of the titration vessel. The fraction of the drop is then washed down into the solution with distilled water.

CARE AND USE OF VOLUMETRIC GLASSWARE

We have mentioned a few precautions in the use of volumetric flasks, pipets, and burets. Your laboratory instructor will supply you with detailed instructions in the use of each of these tools. A discussion of some general precautions and good laboratory technique follows.

Cleanliness of glassware is of the utmost importance. If films of dirt or grease are present, liquids will not drain uniformly and will leave water breaks or droplets on the walls. Under such conditions the calibration will be in error. Initial cleaning should be by repeated rinses with water. Then try cleaning with dilute nitric

Fig. 2.13. Typical buret.

acid and rinse with more water. Use of a buret or test tube brush aids the cleaning of burets and necks of volumetric flasks—but be careful of scratching the interior walls. Pipets should be rotated to coat the entire surface with detergent. There are commercial cleaning solutions available that are very effective.

Pipets and burets should be rinsed at least twice with the solution with which they are to be filled. If they are wet, they should be rinsed first with water, if they have not been already, and then a minimum of *three* times with the solution to be used; about one-fifth the volume of the pipet or buret is adequate for each rinsing. A volumetric flask, if it is wet from a previously contained solution, is rinsed with three portions of water only since later it will be filled to the mark with water. It need not be dry.

> Rinse pipets and burets with the solution to be measured.

Note that analytical glassware should not be subjected to the common practice employed in organic chemistry laboratories of drying either in an oven (this can affect the volume of calibrated glassware) or by drying with a towel or by rinsing with a volatile organic solvent such as acetone (which can cause contamination). The glassware usually does not have to be dried. The preferred procedure is to rinse it with the solution that will fill it.

Care in reading the volume will avoid parallax error, that is, error due to incorrect alignment of the observer's eye, the meniscus, and the scale. This applies in the reading of any scale, such as the pointer scale of an analytical balance. Correct position is with your eye at the same level as the menicus. If the eye level is above the meniscus, the volume read will be smaller than that taken; the opposite will be true if the eye level is too low.

> Avoid parallax error in reading buret or pipet volumes.

After glassware is used, it can usually be cleaned sufficiently by immediate rinsing with water. If the glassware has been allowed to dry, it should be cleaned with detergent. Volumetric flasks should be stored with the stopper inserted, and preferably filled with distilled water. Burets should be filled with distilled water and stoppered with a rubber stopper when not in use.

GENERAL TIPS FOR ACCURATE AND PRECISE TITRATING

Your buret probably has a Teflon stopcock, and this will not require lubrication. Make sure it is secured tightly enough to prevent leakage, but not so tight as to make rotation hard. If your buret has a ground-glass stopcock, you may have to grease the stopcock. A thin layer of stopcock grease (not silicone lubricant) is applied uniformly to the stopcock, using very little near the hole and taking care not to get any grease in the hole. The stopcock is inserted and rotated. There should be a uniform and transparent layer of grease, and the stopcock should not leak. If there is too much lubricant, it will be forced into the barrel or may work into the buret tip and clog it. Grease can be removed from the buret tip and the hole of the stopcock by using a fine wire. If the buret contains a Teflon stopcock, it does not require lubrication. The buret is filled above the zero mark and the stopcock is opened to fill the tip. Check the tip for air bubbles. If any are present, they may work out of the tip during the titration, causing an error in reading. Work air bubbles out by rapid opening and closing of the stopcock to squirt the titrant through the tip or tapping the tip while solution is flowing. No bubbles should be in the barrel of the buret. If there are, the buret is probably dirty.

The initial reading of the buret is taken by allowing it to drain slowly to the zero mark. Wait a few seconds to make certain the drainage film has caught up to the meniscus. Read the buret to the nearest 0.02 mL (for a 50-mL buret). The initial reading may be 0.00 mL or greater. The reading is best taken by placing your finger just in back of the meniscus or by using a meniscus illuminator (Figure 2.14). The meniscus illuminator has a white and a black field, and the black field is

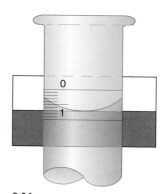

Fig. 2.14. Meniscus illuminator for buret.

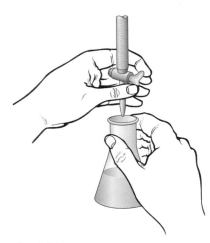

Fig. 2.15. Proper technique for titration.

Subsequent titrations can be speeded up by using the first to *approximate* the end-point volumes.

Class A glassware is accurate enough for most analyses. It can be calibrated to NIST specifications.

The variances or the uncertainties in each reading are additive. See propagation of error, Chapter 3.

positioned just below the meniscus. Avoid parallax error by making the reading at eye level.

The titration is performed with the sample solution in an Erlenmeyer flask. The flask is placed on a white background, and the buret tip is positioned within the neck of the flask. The flask is swirled with the right hand while the stopcock is manipulated with the left hand (Figure 2.15), or whatever is comfortable. This grip on the buret maintains a slight inward pressure on the stopcock to ensure that leakage will not occur. The solution can be more efficiently stirred by means of a magnetic stirrer and stirring bar.

As the titration proceeds, the indicator changes color in the vicinity where the titrant is added, owing to local excesses; but it rapidly reverts to the original color as the titrant is dispersed through the solution to react with the sample. As the end point is approached, the return to the original color occurs more slowly, since the dilute solution must be mixed more thoroughly to consume all the titrant. At this point, the titration should be stopped and the sides of the flask washed down with distilled water from the wash bottle. A drop from the buret is about 0.02 to 0.05 mL, and the volume is read to the nearest 0.02 mL. It is therefore necessary to split drops near the end point. This can be done by slowly turning the stopcock until a fraction of a drop emerges from the buret tip and then closing it. The fraction of drop is touched off onto the wall of the flask and is washed into the flask with the wash bottle, or it is transferred with a glass stirring rod. There will be a sudden and "permanent" (lasting at least 30 s) change in the color at the end point when a fraction of a drop is added.

The titration is usually performed in triplicate. After performing the first titration, you can calculate the approximate volume for the replicate titrations from the weights of the samples and the molarity of the titrant. This will save time in the titrations. The volume should not be calculated to nearer than 0.1 mL in order to avoid bias in the reading.

After a titration is complete, unused titrant should never be returned to the original bottle but should be discarded.

If a physical property of the solution, such as potential, is measured to detect the end point, the titration is performed in a beaker with magnetic stirring so electrodes can be placed in the solution.

TOLERANCES AND PRECISION OF GLASSWARE

The National Institute of Standards and Technology (NIST) has prescribed certain *tolerances*, or absolute errors, for different volumetric glassware, and some of these are listed on the *back cover* of the text. For volumes of greater than about 25 mL, the tolerance is within 1 part per thousand (ppt) relative, but it is larger for smaller volumes. The letter "A" stamped on the side of a volumetric flask, buret, or pipet indicates that it complies with class A tolerances. This says nothing about the precision of delivery. Volumetric glassware that meets NIST specifications or that is certified by NIST can be purchased, but at a significantly higher price than uncertified glassware. Less expensive glassware may have tolerances double those specified by NIST. It is a simple matter, however, to calibrate this glassware to an accuracy as good as or exceeding the NIST specifications (see Experiment 2).

The precision of reading a 50-mL buret is about ±0.02 mL. Since a buret is always read twice, the total absolute uncertainty may be as much as ±0.04 mL. The relative uncertainty will vary inversely with the total volume delivered. It becomes apparent that a titration with a 50-mL buret should involve about 40 mL titrant to achieve a precision of 1 ppt. Smaller burets can be used for increased precision at smaller volumes. Pipets will also have a certain precision of reading, but only one reading is required for volumetric pipets.

CALIBRATION OF GLASSWARE—FOR THE ULTIMATE ACCURACY

Example 2.1 illustrated how Equation 2.1 may be used in the calibration of glassware, to correct for buoyancy of the water used for calibration, that is, to correct to weight in vacuum. Dividing the weight of the water in vacuum by its density at the given temperature will convert it to volume.

Table 2.4 lists the calculated volumes for a gram of water in air at atmospheric pressure for different temperatures, corrected for buoyancy with stainless steel weights of density 7.8 g/cm³. These are used to give the volume of the glassware

Table 2.4
Glassware Calibration

	A	B	C	D	E	F	G	H
1	Table 2.4 Glassware Calibration							
2	Weight in vacuum assuming stainless steel weights, density 7.8 g/mL.							
3	Glass expansion for borosilicate glass, 0.000025 mL/mL/ °C.							
4								
5	Save this spreadsheet contained in your CD, Chapter 2, to your desktop, and use it to calculate							
6	calibrated volumes of glassware. Substitute into the appropriate Cell B, at temperature T, the							
7	weight of water in air, obtained from the glassware at the temperature of the measurement (Cell A).							
8	The calibration volume at the temperature, T (Cell D), and at 20°C (Cell F), is calculated.							
9	Round the calculated values to the appropriate number of significant figures, usually							
10	four or five.							
11								
12	T, °C	Wt. H₂O	Wt. in	Vol. at T,	Glass expansion,	Vol. at 20°,	Density,	
13		in air, g	vacuum, g	mL	at 20°C, mL	mL	g/mL	
14	10	1.0000	1.0010	1.0013	-0.000250	1.0016	0.9997026	
15	11	1.0000	1.0010	1.0014	-0.000225	1.0017	0.9996081	
16	12	1.0000	1.0010	1.0015	-0.000200	1.0017	0.9995004	
17	13	1.0000	1.0010	1.0017	-0.000175	1.0018	0.9993801	
18	14	1.0000	1.0010	1.0018	-0.000150	1.0020	0.9992474	
19	15	1.0000	1.0010	1.0019	-0.000125	1.0021	0.9991026	
20	16	1.0000	1.0010	1.0021	-0.000100	1.0022	0.9989460	
21	17	1.0000	1.0010	1.0023	-0.000075	1.0023	0.9987779	
22	18	1.0000	1.0010	1.0025	-0.000050	1.0025	0.9985896	
23	19	1.0000	1.0010	1.0026	-0.000025	1.0027	0.9984082	
24	20	1.0000	1.0010	1.0028	0.000000	1.0028	0.9982071	
25	21	1.0000	1.0010	1.0031	0.000025	1.0030	0.9979955	
26	22	1.0000	1.0010	1.0033	0.000050	1.0032	0.9977735	
27	23	1.0000	1.0010	1.0035	0.000075	1.0034	0.9975415	
28	24	1.0000	1.0010	1.0038	0.000100	1.0037	0.9972995	
29	25	1.0000	1.0010	1.0040	0.000125	1.0039	0.9970479	
30	26	1.0000	1.0010	1.0043	0.000150	1.0041	0.9967867	
31	27	1.0000	1.0010	1.0045	0.000175	1.0044	0.9965162	
32	28	1.0000	1.0010	1.0048	0.000200	1.0046	0.9962365	
33	29	1.0000	1.0010	1.0051	0.000225	1.0049	0.9959478	
34	30	1.0000	1.0010	1.0054	0.000250	1.0052	0.9956502	
35								
36	Formulas are entered into the boldface cells above as indicated below. They are							
37	copied down for all temperatures. See Chapter 3 for setting up a spreadsheet.							
38	Cell C14: $W_{vac} = W_{air} + W_{air}(0.0012/D_o - 0.0012/D_w) = W_{air}(0.0012/1.0 + 0.0012/7.8)$							
39		=	B14+B14*(0.0012/1.0-0.0012/7.8)		Copy down			
40	Cell D14: $V_T (mL) = W_{vac,T}(g)/D_T(g/mL)$							
41		=	C14/G14		Copy down			
42	Cell E14: Glass expans. = (T - 20) (deg) x 0.000025 (mL/mL/deg) x W_{air} (g)							
43		=	(A14-20)*0.000025*B14		Copy down			
44	Cell F14: $V_{20o} = V_T - Glass_{exp} =$		D14-E14		Copy down			

being calibrated, from the weight of water contained or delivered by the glassware. (The values are not significantly different for brass weights of 8.4 g/cm³ density. See Example 2.2.) The glass volumes calculated for the standard temperature of 20°C include slight adjustments for borosilicate glass (Pyrex or Kimax) container expansion or contraction with temperature changes (volumetric glassware has a cubical coefficient of expansion of about 0.000025 per degree centigrade, resulting in changes of about 0.0025% per degree; for 1 mL, this is 0.000025 mL per degree.). Water expands about 0.02% per degree around 20°C. Volume (concentration) corrections may be made using the water density data in Table 2.4, taking the ratios of the relative densities.

Your CD has the Table 2.4 spreadsheet, with formulas as indicated in the table. You can substitute specific weights of water in air, obtained from a flask, pipet, or buret, in cell B at the temperature of measurement to obtain the calculated calibration volume at temperature, T, and for 20°C. We describe the use of spreadsheets in Chapter 3. The CD also has a table and figure of the percent error for weight in vacuum as a function of sample density.

Example 2.3

(a) Use Table 2.4 to calculate the volume of the 20-mL pipet in Example 2.1, from its weight in air. Assume the temperature is 23°C. (b) Give the corresponding volume at 20°C as a result of glass contraction. (c) Compare with the volume calculated using the weight in air with that calculated using the weight in vacuum and the density in water (Example 2.1).

Solution

(a) From Table 2.4, the volume per gram in air is 1.0035 mL at 23°C:

$$19.994 \text{ g} \times 1.0035 \text{ mL/g} = 20.064 \text{ mL}$$

(b) The glass contraction at 20°C relative to 23°C is 0.0015 mL (0.000025 mL/ mL/°C × 20 mL × 3°C), so the pipet volume at 20°C is 20.062 mL.

(c) The density of water at 23°C is 0.99754 g/mL, so from the weight in vacuum:

$$20.015 \text{ g}/0.99754 \text{ g/mL} = 20.064 \text{ mL}$$

The same value is obtained.

Example 2.4

You prepared a solution of hydrochloric acid and standardized it by titration of primary standard sodium carbonate. The temperature during the standardization was 23°C, and the concentration was determined to be 0.1127_2 M. The heating system in the laboratory malfunctioned when you used the acid to titrate an unknown, and the temperature of the solution was 18°C. What was the concentration of the titrant?

Solution

$$M_{18°} = M_{23°} \times (D_{18°}/D_{23°})$$
$$= 0.1127_2 \times (0.99859/0.99754)$$
$$= 0.1128_4 \; M$$

(See Chapter 3 for significant figures and the meaning of the subscript numbers.)

• •

TECHNIQUES FOR CALIBRATING GLASSWARE

You generally calibrate glassware to five significant figures, the maximum precision you are likely to attain in filling or delivering solutions. Hence, your net weight of water needs to be five figures. If the glassware exceeds 10 mL, this means weighing to 1 mg is all that is needed. This can be readily and conveniently accomplished using a top-loading balance, rather than a more sensitive analytical balance. (Note: If the volume number is large without regard to the decimal, e.g., 99, then four figures will suffice—see Chapter 3 discussion on significant figures. A 10-mL pipet, e.g., may calibrate to 9.997 mL, which is as accurate as 10.003 mL, i.e., the last figure is 1 part in 10,000.)

1. Volumetric Flask Calibration. To calibrate a volumetric flask, first weigh the clean, dry flask and stopper. Then fill it to the mark with distilled water. There should be no droplets on the neck. If there are, blot them with tissue paper. The flask and water should be equilibrated to room temperature. Weigh the filled flask, and then record the temperature of the water to 0.1°C. The increase in weight represents the weight in air of the water contained by the flask.

2. Pipet Calibration. To calibrate a pipet, weigh a dry Erlenmeyer flask with a rubber stopper or a weighing bottle with a glass stopper or cap, depending on the volume of water to be weighed. Fill the pipet with distilled water (whose temperature you have recorded) and deliver the water into the flask or bottle, using proper pipetting technique, and quickly stopper the container to avoid evaporation loss. Reweigh to obtain the weight in air of water delivered by the pipet.

3. Buret Calibration. Calibrating a buret is similar to the procedure for a pipet, except that several volumes will be delivered. The internal bore of the buret is not perfectly cylindrical, and it will be a bit "wavy," so the actual volume delivered will vary both plus and minus from the nominal volumes marked on the buret, as increased volumes are delivered. You will ascertain the volume at 20% full-volume increments (e.g., each 10 mL, for a 50-mL buret) by filling the buret each time and then delivering the nominal volume into a dry flask. (The buret is filled each time to minimize evaporation errors. You may also make successive deliveries into the same flask, i.e., fill the buret only once. Make rapid deliveries.) Since the delivered volume does not have to be exact, but close to the nominal volume, you can make fairly fast deliveries, but wait about 10 to 20 s for film drainage. Prepare a plot of volume correction versus nominal volume and draw straight lines between each point. Interpolation is made at intermediate volumes from the lines. Typical volume corrections for a 50-mL buret may range up to ca. 0.05 mL, plus or minus.

Example 2.5

You calibrate a 50-mL buret at 10-mL increments, filling the buret each time and delivering the nominal volume, with the following results:

Buret Reading (mL)	Weight H_2O Delivered (g)
10.02	10.03
20.08	20.03
29.99	29.85
40.06	39.90
49.98	49.86

Construct a plot of volume correction versus volume delivered. The temperature of the water is 20°C and stainless steel weights are used.

Solution

From Table 2.4 (or use the CD Table 2.4 for automatic calculation of volumes):

$$W_{vac} = 10.03 + 10.03(0.00105) = 10.03 + 0.01 = 10.04 \text{ g}$$
$$\text{Vol.} = 10.04 \text{ g}/0.9982 \text{ g/mL} = 10.06 \text{ mL}$$

Likewise, for the others, we construct the table:

Nominal Volume (mL)	Actual Volume (mL)	Correction (mL)
10.02	10.06	+0.04
20.08	20.09	+0.01
29.99	29.93	−0.04
40.06	40.01	−0.05
49.98	50.00	0.00

Prepare a graph of nominal volume (y axis) versus correction volume. Use 10, 20, 30, 40, and 50 mL as the nominal volumes.

SELECTION OF GLASSWARE—
HOW ACCURATE DOES IT HAVE TO BE?

Only certain volumes need to be measured accurately, those involved in the quantitative calculations.

As in weighing operations, there will be situations where you need to accurately know volumes of reagents or samples measured or transferred (accurate measurements), and others in which only approximate measurements are required (rough measurements). If you wish to prepare a standard solution of 0.1 M hydrochloric acid, it can't be done by measuring an accurate volume of concentrated acid and diluting to a known volume because the concentration of the commercial acid is not known adequately. Hence, an approximate solution is prepared that is then standardized. We see in the table on the inside *back cover* that the commercial acid is about 12.4 M. To prepare 1 L of a 0.1 M solution, about 8.1 mL needs to be taken and diluted. It would be a waste of time to measure this (or the water used for dilution) accurately. A 10-mL graduated cylinder or 10-mL measuring pipet will suffice, and the acid can be diluted in an ungraduated 1-L bottle. If, on the other hand, you wish to dilute a stock standard solution accurately, then a transfer pipet must be used and the dilution must be done in a volumetric flask. Any volumetric measurement that is a part of the actual analytical measurement must be done with the accuracy required of the analytical measurement. This generally means four-significant-figure accuracy, and transfer pipets and volumetric flasks are required. This includes taking an accurate portion of a sample, preparation of a standard solution from an accurately weighed reagent, and accurate dilutions. Burets are used for accurate measurement of variable volumes, as in a titration. Preparation of reagents that are to be used in an analysis just to provide proper solution conditions (e.g., buffers for pH control) need not be prepared highly accurately, and less accurate glassware can be used, for example, graduated cylinders.

2.5 Preparation of Standard Base Solutions

Sodium hydroxide usually used as the titrant when a base is required. It contains significant amounts of water and sodium carbonate, and so it cannot be used as a primary standard. For accurate work, the sodium carbonate must be removed from the NaOH because it reacts to form a buffer that decreases the sharpness of the end point. In addition, an error will result if the NaOH is standardized using a phenolphthalein end point (in which case the CO_3^{2-} is titrated only to HCO_3^-), and then a methyl orange end point is used in the titration of a sample (in which case the CO_3^{2-} is titrated to CO_2). In other words, the effective molarity of the base has been increased, owing to further reaction of the HCO_3^-.

Sodium carbonate is essentially insoluble in nearly saturated sodium hydroxide. It is conveniently removed by dissolving the weighed NaOH in a volume (milliliters) of water equal to its weight in grams. The insoluble Na_2CO_3 can be allowed to settle for several days, and then the clear supernatant liquid can be carefully decanted,[1] or it can be filtered in a Gooch crucible with an asbestos mat (do not wash the filtered Na_2CO_3). The former procedure is preferred because of the carcinogenic nature of asbestos. This procedure does not work with KOH because K_2CO_3 remains soluble.

Water dissolves CO_2 from the air. In many routine determinations not requiring the highest degree of accuracy, carbonate or CO_2 impurities in the water will result in an error that is small enough to be considered negligible. For the highest accuracy, however, CO_2 should be removed from all water used to prepare solutions for acid–base titrations, particularly the alkaline solutions. This is conveniently done by boiling and then cooling under the cold-water tap.

Sodium hydroxide is usually standardized by titrating a weighed quantity of primary standard potassium acid phthalate (KHP), which is a moderately weak acid ($K_a = 4 \times 10^{-6}$), approximately like acetic acid; a phenolphthalein end point is used. The sodium hydroxide solution should be stored in a plastic bottle to prevent absorption of CO_2 from the air. If the bottle must be open (e.g., a siphon bottle), the opening is protected with an **Ascarite** (asbestos impregnated with NaOH) tube.

Remove Na_2CO_3 by preparing a saturated solution of NaOH.

See Experiment 6 for preparing and standardizing sodium hydroxide.

---

[1]The solution must be kept in a test tube stoppered with a material other than glass, or other appropriate vessel, to keep out atmospheric carbon dioxide, which would continue to react with the sodium hydroxide to produce sodium carbonate. Use a rubber stopper since concentrated alkali causes glass joints to "freeze."

2.6 Preparation of Standard Acid Solutions

Hydrochloric acid is the usual titrant for the titration of bases. Most chlorides are soluble, and few side reactions are possible with this acid. It is convenient to handle. It is not a primary standard (although constant-boiling HCl, which is a primary standard, can be prepared), and an approximate concentration is prepared simply by diluting the concentrated acid. For most accurate work, the water used to prepare the solution should be boiled, although use of boiled water is not so critical as with NaOH; CO_2 will have a low solubility in strongly acidic solutions and will tend to escape during shaking of the solution.

See Experiment 7 for preparing and standardizing hydrochloric acid.

Primary standard sodium carbonate is usually used to standardize HCl solutions. Its disadvantage is that the end point is not sharp unless methyl red, methyl purple, and so forth is used as the indicator and the solution is boiled at the end point. A modified methyl orange end point may be used without boiling, but this is not so sharp. Another disadvantage is the low formula weight of Na_2CO_3. Tris-(hydroxymethyl)aminomethane (THAM), $(HOCH_2)_3CNH_2$, is another primary standard that is more convenient to use. It is nonhygroscopic, but it is still a fairly weak base ($K_b = 1.3 \times 10^{-6}$) with a low molecular weight. The end point is not complicated by released CO_2, and it is recommended as the primary standard unless the HCl is being used to titrate carbonate samples.

A secondary standard is less accurate than a primary standard.

If a standardized NaOH solution is available, the HCl can be standardized by titrating an aliquot with the NaOH. The end point is sharp and the titration is more rapid. The NaOH solution is a **secondary standard**. Any error in standardizing this will be reflected in the accuracy of the HCl solution. The HCl is titrated with the base, rather than the other way around, to minimize absorption of CO_2 in the titration flask. Phenolphthalein or bromothymol blue can be used as indicator.

2.7 Other Apparatus—Handling and Treating Samples

Besides apparatus for measuring mass and volume, there are a number of other items of equipment commonly used in analytical procedures.

BLOOD SAMPLERS

Syringes are used to collect blood samples.[2] Stainless steel or aluminum needles are generally used with glass or plastic syringes. These usually present no problem of contamination, although special precautions may be required for certain trace element analyses. **Vacutainers** or similar devices are often used in place of syringes. These are evacuated test tubes with a rubber cap. The needle is pushed through the cap after the other end has been inserted into the vein, and the blood is drawn into the evacuated tube. The tube may contain an anticoagulating agent to prevent clotting of the blood if plasma or whole blood samples are to be analyzed.

A finger puncture, instead of a venipuncture, is used when small quantities of blood are to be collected for microprocedures. Up to 0.5 mL or more blood can be squeezed from the finger into a small collection tube by puncturing the finger with a sterilized sharp-pointed knifelike object.

---

[2]You should *not* attempt to collect a blood sample unless you have been specifically trained to do so. A trained technician will generally be assigned to this job.

Fig. 2.16. Desiccator and desiccator plate.

DESICCATORS

A **desiccator** is used to keep samples dry while they are cooling and before they are weighed and, in some cases, to dry the wet sample. Dried or ignited samples and vessels are cooled in the desiccator. A typical glass desiccator is shown in Figure 2.16. A desiccator is an airtight container that maintains an atmosphere of low humidity. A desiccant such as calcium chloride is placed in the bottom to absorb the moisture. This desiccant will have to be changed periodically as it becomes "spent." It will usually become wet in appearance or caked from the moisture when it is time to be changed. A porcelain plate is usually placed in the desiccator to support weighing bottles, crucibles, and other vessels. An airtight seal is made by application of stopcock grease to the ground-glass rim on the top of the desiccator. A **vacuum desiccator** has a side arm on the top for evacuation so that the contents can be kept in a vacuum rather than just an atmosphere of dry air.

> Oven-dried samples or reagents are cooled in a desiccator before weighing.

The top of a desiccator should not be removed any more than necessary since the removal of moisture from the air introduced is rather slow, and continued exposure will limit the lifetime of the desiccant. A red-hot crucible or other vessel should be allowed to cool in the air about 60 s before it is placed in the desiccator. Otherwise, the air in the desiccator will be heated appreciably before the desiccator is closed, and as the air cools, a partial vacuum will be created. This will result in a rapid inrush of air when the desiccator is opened and possible spilling or loss of sample as a consequence. A hot weighing bottle should not be stoppered when placed in a desiccator because on cooling, a partial vacuum is created and the stopper may seize. The stopper should be placed in the desiccator with the weighing bottle.

Table 2.5 lists some commonly used desiccants and their properties. Aluminum oxide, magnesium perchlorate, calcium oxide, calcium chloride, and silica gel can be regenerated by heating at 150, 240, 500, 275, and 150°C, respectively.

Table 2.5
Some Common Drying Agents

Agent	Capacity	Deliquescent[a]	Trade Name
$CaCl_2$ (anhydrous)	High	Yes	
$CaSO_4$	Moderate	No	Drierite (W. A. Hammond Drierite Co.)
CaO	Moderate	No	
$MgClO_4$ (anhydrous)	High	Yes	Anhydrone (J. T. Baker Chemical Co.); Dehydrite (Arthur H. Thomas Co.)
Silica gel	Low	No	
Al_2O_3	Low	No	
P_2O_5	Low	Yes	

[a]Becomes liquid by absorbing moisture.

Fig. 2.17.   Muffle furnace. (Courtesy of
Arthur H. Thomas Company.)

Fig. 2.18.   Drying oven. (Courtesy
of Arthur H. Thomas Company.)

Laminar-flow hoods provide clean
work areas.

FURNACES AND OVENS

A **muffle furnace** (Figure 2.17) is used to ignite samples to high temperatures,
either to convert precipitates to a weighable form or to burn organic materials prior
to inorganic analysis. There should be some means of regulating the temperature
since losses of some metals may occur at temperatures in excess of 500°C. Temperatures up to about 1200°C can be reached with muffle furnaces.

A **drying oven** is used to dry samples prior to weighing. A typical drying
oven is shown in Figure 2.18. These ovens are well ventilated for uniform heating.
The usual temperature employed is about 110°C, but temperatures of 200 to 300°C
may be obtained.

HOODS

A **fume hood** is used when chemicals or solutions are to be evaporated. When perchloric acid or acid solutions of perchlorates are to be evaporated, the fumes should
be collected, or the evaporation should be carried out in fume hoods specially designed for perchloric acid work (i.e., constructed from components resistant to attack by perchloric acid).

When performing trace analysis, as in trace metal analysis, care must be taken
to prevent contamination. The conventional fume hood is one of the "dirtiest" areas
of the laboratory since laboratory air is drawn into the hood and over the sample.
Laminar-flow hoods or workstations are available for providing very clean work
areas. Rather than drawing unfiltered laboratory air into the work area, the air is
prefiltered and then flows over the work area and out into the room to create a positive pressure and prevent unfiltered air from flowing in. A typical laminar-flow
workstation is shown in Figure 2.19. The high-efficiency particulate air (HEPA)
filter removes all particles larger than 0.3 μm from the air. Vertical laminar-flow
stations are preferred when fumes are generated that should not be blown over the
operator. Facilities are available to exhaust noxious fumes.

Fig. 2.19. Laminar-flow workstation. (Courtesy of Dexion, Inc., 344 Beltine Boulevard, Minneapolis, MN.)

WASH BOTTLES

A **wash bottle** of some sort should be handy in any analytical laboratory, to be used for quantitative transfer of precipitates and solutions and to wash precipitates. These are commercially available in a variety of shapes and sizes, as seen in Figure 2.20. Alternatively they may be constructed from a Florence flask and glass tubing, as in Figure 2.20*b*.

CENTRIFUGES AND FILTERS

A **centrifuge** has many useful applications, particularly in the clinical laboratory, where blood may have to be separated into fractions such as serum or plasma, and proteins may have to be separated by precipitation followed by centrifuging.

Filters for filtering precipitates (e.g., in gravimetric analysis) are of various types. The Gooch crucible, sintered-glass crucible, and porcelain filter crucible are illustrated in Figure 2.21. The **Gooch crucible** is porcelain and has holes in the bottom; a glass filter disk is supported on top of it. In the past, an asbestos mat

(a)   *(b)*

Fig. 2.20. Wash bottles: (*a*) polyethylene, squeeze type; (*b*) glass, blow type.

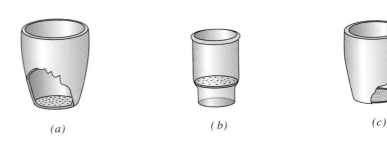

Fig. 2.21. Filtering crucibles:
(*a*) Gooch crucible; (*b*) sintered-glass crucible; (*c*) porcelain filter crucible.

was usually prepared, but this is inconvenient and potentially hazardous since asbestos fibers in the air are carcinogenic, and the filter disk will handle fine precipitates. The **sintered-glass crucible** contains a sintered-glass bottom, which is available in fine (F), medium (M), or coarse (C) porosity. The **porcelain filter crucible** contains a porous unglazed bottom. Glass filters are not recommended for concentrated alkali solutions because of the possibility of attack by these solutions. See Table 2.1 for maximum working temperatures for different types of crucible materials.

Gelatinous precipitates such as hydrous iron oxide should not be filtered in filter crucibles because they clog the pores. Even with filter paper, the filtration of the precipitates can be slow.

Filter crucibles are used with a **crucible holder** mounted on a filtering flask (Figure 2.22). A safety bottle is connected between the flask and the aspirator.

Ashless filter paper is generally used for quantitative work in which the paper is ignited away and leaves a precipitate suitable for weighing (see Chapter 10). There are various grades of filter papers for different types of precipitates. These are listed in Table 2.6 for Whatman (www.whatman.plc.uk) and for Schliecher and Schuell (www.s-und-s.de/english-index.html) papers.

TECHNIQUES OF FILTRATION

By proper fitting of the filter paper, the rate of filtration can be increased. A properly folded filter paper is illustrated in Figure 2.23. The filter paper is folded in the

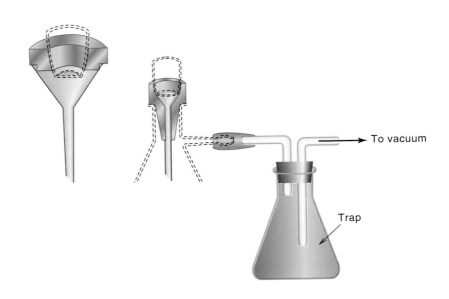

Fig. 2.22. Crucible holders.

Table 2.6
Types of Filter Papers

Precipitate	Whatman	Schliecher and Schuell
Very fine (e.g., $BaSO_4$)	No. 42 (2.5 μm)	No. 589/2 or 5, Blue or Red Band (2–4 μm)
Small or medium (e.g., AgCl)	No. 40 (8 μm)	No. 589/2, White Band (4–12 μm)
Gelatinous or large crystals (e.g., $Fe_2O_3 \cdot xH_2O$)	No. 41 (20–25 μm)	No. 589/1, Black Band (>12–25 μm)

shape of a cone, with the overlapped edges of the two quarters not quite meeting ($\frac{1}{8}$ inch apart). About $\frac{1}{4}$ inch is torn away from the corner of the inside edge. This will allow a good seal against the funnel to prevent air bubbles from being drawn in. After the folded paper is placed in the funnel, it is wetted with distilled water. The stem is filled with water and the top of the wet paper is pressed against the funnel to make a seal. With a proper fit, no air bubbles will be sucked into the funnel, and the suction supplied by the weight of the water in the stem will increase the rate of filtration. The filtration should be started immediately. The precipitate should occupy not more than one-third to one-half of the filter paper in the funnel because many precipitates tend to "creep." Do not allow the water level to go over the top of the paper.

Fig. 2.23. Properly folded filter paper.

The precipitate should be allowed to settle in the beaker before filtration is begun. The bulk of the clear liquid can then be decanted and filtered at a rapid rate before the precipitate fills the pores of the filter paper.

Let the precipitate settle before filtering.

Care must be taken in the decanting and the transferring of the precipitate to avoid losses. This is properly done by use of a stirring rod and a wash bottle, as illustrated in Figure 2.24. The solution is decanted by pouring it down the glass rod, which guides it into the filter without splashing. The precipitate is most readily washed while still in the beaker. After the mother liquor has been decanted off, wash the sides of the beaker down with several milliliters of the wash liquid, and then allow the precipitate to settle as before. Decant the wash liquid into the filter and repeat the washing operation two or three times. The precipitate is then transferred to the filter by holding the glass rod and beaker in one hand, as illustrated, and washing it out of the beaker with wash liquid from the wash bottle.

Wash the precipitate while it is in the beaker.

If the precipitate must be collected quantitatively, as in gravimetric analysis, the last portions of precipitate are removed by scrubbing the walls with a moistened **rubber policeman** (Figure 2.25). Wash the remainder of loosened precipitate from the beaker and from the policeman. If the precipitate is being collected in a filter paper, then instead of a rubber policeman, a small piece of the ashless filter paper can be rubbed on the beaker walls to remove the last bits of precipitate and added to the filter. This should be held with a pair of forceps.

After the precipitate is transferred to the filter, it is washed with five or six small portions of wash liquid. This is more effective than adding one large volume. Divert the liquid around the top edge of the filter to wash the precipitate down into the cone. Each portion should be allowed to drain before adding the next one. Check for completeness of washing by testing for the precipitating agent in the last few drops of the washings.

Test for completeness of washing.

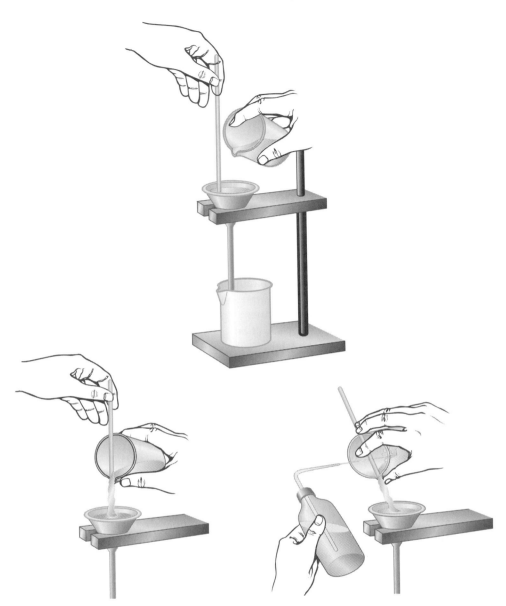

Fig. 2.24.   Proper technique for transfer of a precipitate.

Fig. 2.25.   Rubber policeman.

2.8 Igniting Precipitates—Gravimetric Analysis

If a precipitate is to be ignited in a porcelain filter crucible, the moisture should be driven off first at a low heat. The ignition may be done in a muffle furnace or by heating with a burner. If a burner is to be used, the filter crucible should be placed in a porcelain or platinum crucible to prevent reducing gases of the flame diffusing through the pores of the filter.

When precipitates are collected on filter paper, the cone-shaped filter containing the precipitate is removed from the funnel, the upper edge is flattened, and the corners are folded in. Then, the top is folded over and the paper and contents are placed in a crucible with the bulk of the precipitate on the bottom. The paper must now be dried and charred off. The crucible is placed at an angle on a triangle support with the crucible cover slightly ajar, as illustrated in Figure 2.26. The moisture is removed by low heat from the burner, with care taken to avoid splattering. The heating is gradually increased as moisture is evolved and the paper begins to char. Care should be taken to avoid directing the reducing portion of the flame into the crucible. A sudden increase in the volume of smoke evolved indicates that the paper is about to burst into flame, and the burner should be removed. If it does burst into flame, it should be smothered quickly by replacing the crucible cover. Carbon particles will undoubtedly appear on the cover, and these will ultimately have to be ignited. Finally, when no more smoke is detected, the charred paper is burned off by gradually increasing the flame temperature. The carbon residue should glow but should not flame. Continue heating until all the carbon and tars on the crucible and its cover are burned off. The crucible and precipitate are now ready for igniting. The ignition can be continued with the burner used at highest temperature or with the muffle furnace.

Do the initial ignition slowly.

Before a precipitate is collected in a filter crucible or transferred to a crucible, the crucible should be dried to constant weight (e.g., 1 h of heating, followed by $\frac{1}{2}$-h heatings) if the precipitate is to be dried, or it should be ignited to constant weight if the precipitate is to be ignited. Constant weight is considered to have been achieved with an analytical student balance when successive weighings agree within about 0.3 or 0.4 mg. The crucible plus the precipitate are heated to constant weight in a similar manner. After the first heating, the time of heating can be reduced by half. The crucible should be allowed to cool in a desiccator for at least $\frac{1}{2}$ h before weighing. Red-hot crucibles should be allowed to cool below redness before placing them in the desiccator (use crucible tongs—usually nickel plated or stainless steel to minimize contamination from rust). Before weighing a covered crucible, check for any radiating heat by placing your hand near it (don't touch).

Dry and weigh the crucible before adding the precipitate!

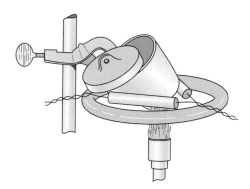

Fig. 2.26. Crucible and cover supported on a wire triangle for charring off paper.

2.9 Obtaining the Sample—Is It Solid, Liquid, or Gas?

See Chapter 3 for important statistical considerations in sampling.

Collecting a representative sample is an aspect of analytical chemistry that the beginning analytical student is often not concerned with because the samples handed to him or her are assumed to be homogeneous and representative. Yet this process can be the most critical aspect of an analysis. The significance and accuracy of measurements can be limited by the sampling process. Unless sampling is done properly, it becomes the weak link in the chain of the analysis. A life could sometimes depend on the proper handling of a blood sample during and after sampling. If the analyst is given a sample and does not actively participate in the sampling process, then the results obtained can only be attributed to the sample "as it was received." And the chain of custody as mentioned earlier must be documented.

Many professional societies have specified definite instructions for sampling given materials [e.g., the American Society for Testing and Materials (ASTM: www.astm.org), the Association of Official Analytical Chemists International (AOAC International: www.aoac.org), and the American Public Health Association (APHA: www.apha.org)]. By appropriate application of experience and statistics, these materials can be sampled as accurately as the analysis can be performed. Often, however, the matter is left up to the analyst. The ease or complexity of sampling will, of course, depend on the nature of the sample.

The problem involves obtaining a sample that is representative of the whole. This sample is called the **gross sample**. Its size may vary from a few grams or less to several pounds, depending on the type of bulk material. Once a representative gross sample is obtained, it may have to be reduced to a sufficiently small size to be handled. This is called the **sample**. Once the sample is obtained, an aliquot, or portion, of it will be analyzed. This aliquot is called the **analysis sample**. Several replicate analyses on the same sample may be performed by taking separate aliquots.

In the clinical laboratory, the gross sample is usually satisfactory for use as the sample because it is not large and it is homogeneous (e.g., blood and urine samples). The analysis sample will usually be from a few milliliters to a fraction of a drop (a few microliters) in quantity.

Some of the problems associated with obtaining gross samples of solids, liquids, and gases are considered below.

1. Solids. Inhomogeneity of the material, variation in particle size, and variation within the particle make sampling of solids more difficult than other materials. The easiest but usually most unreliable way to sample a material is the **grab sample**, which is one sample taken at random and assumed to be representative. The grab sample will be satisfactory only if the material from which it is taken is homogeneous. For most reliable results, it is best to take 1/50 to 1/100 of the total bulk for the gross sample, unless the sample is fairly homogeneous. The larger the particle size, the larger the gross sample should be.

The easiest and most reliable time to sample large bodies of solid materials is while they are being moved. In this way any portion of the bulk material can usually be exposed for sampling. Thus, a systematic sampling can be performed to obtain aliquots representing all portions of the bulk. Some samples follow.

In the loading or unloading of bags of cement, a representative sample can be obtained by taking every fiftieth or so bag or by taking a sample from each bag. In the moving of grain by wheelbarrow, representative wheelbarrow loads or a shovelful from each wheelbarrow can be taken. All of these aliquots are combined to form the gross sample.

Sampling (From the journals collection of the Chemical Heritage Foundations' Othmer Library.)

2. Liquids. These materials tend to be homogeneous and are much easier to sample. Liquids mix by diffusion only very slowly and must be shaken to obtain a homogeneous mixture. If the material is indeed homogeneous, a simple grab (single random) sample will suffice. For all practical purposes, this method is satisfactory for taking blood samples. Twenty-four-hour urine sample collections are generally more reliable than single specimens.

The timing of sampling of biological fluids is, however, very important. The composition of blood varies considerably before and after meals, and for many analyses a sample is collected after the patient has fasted for a number of hours. Preservatives such as sodium fluoride for glucose preservation and anticoagulants may be added to blood samples when they are collected.

Blood samples may be analyzed as *whole blood*, or they may be separated to yield *plasma* or *serum* according to the requirements of the particular analysis. Most commonly, the concentration of the substance external to the red cells (the extracellular concentration) will be a significant indication of physiological condition, and so serum or plasma is taken for analysis.

See Chapter 24 for more on sampling biological fluids.

If liquid samples are not homogeneous, and if they are small enough, they can be shaken and sampled immediately. For example, there may be particles in the liquid that have tended to settle. Large bodies of liquids are best sampled after a transfer or, if in a pipe, after passing through a pump when they have undergone the most thorough mixing. Large stationary liquids can be sampled with a *"thief"* sampler, which is a device for obtaining aliquots at different levels. It is best to take the sample at different depths at a diagonal, rather than straight down. The separate aliquots of liquids can be analyzed individually and the results combined, or the aliquots can be combined into one gross sample and replicate analyses performed. This latter procedure is probably preferred because the analyst will then have some idea of the precision of the analysis.

3. Gases. The usual method of sampling gases involves displacement of a liquid. The liquid must be one in which the sample has little solubility and with which it does not react. Mercury is the liquid employed most commonly. The mercury is allowed to trickle from the bottom of the container, whereupon the gas is pulled in at the top. Such a procedure allows collection of an average sample over a relatively long period of time. A grab-type sample is satisfactory in some cases. In the collecting of a breath sample, for example, the subject could blow into an evacuated bag. Auto exhaust could be collected in a large evacuated plastic bag.

See Chapter 26 for more on sampling environmental samples.

The volume of gross gas sample collected may or may not need to be known. Often, the *concentration* of a certain analyte in the gas sample is measured, rather than the *amount*. The temperature and pressure of the sample will, of course, be important in determining the volume and hence the concentration.

The gas sampling mentioned here does not concern gas constituents dissolved in liquids, such CO_2 or O_2 in blood. These are treated as liquid samples and are then handled accordingly to measure the gas in the liquid or to release it from the liquid for measuring.

2.10 Operations of Drying and Preparing a Solution of the Analyte

After a sample has been collected, a solution of the analyte must usually be prepared before the analysis can be continued. Drying of the sample may be required, and it must be weighed or the volume measured. If the sample is already a solution (e.g., serum, urine, or water), then extraction, precipitation, or concentration of the analyte may be in order, and this may also be true with other samples.

In this section we describe common means for preparing solutions of inorganic and organic materials. Included are the dissolution of metals and inorganic compounds in various acids or in basic fluxes (fusion), the destruction of organic and biological materials for determination of inorganic constituents (using wet digestion or dry ashing), and the removal of proteins from biological materials so they do not interfere in the analysis of organic or inorganic constituents.

DRYING THE SAMPLE

Solid samples will usually contain variable amounts of adsorbed water. With inorganic materials, the sample will generally be dried before weighing. This is accomplished by placing it in a drying oven at 105 to 110°C for 1 or 2 h. Other nonessential water, such as that entrapped within the crystals, may require higher temperatures for removal.

Decomposition or side reactions of the sample must be considered during drying. Material unstable under conditions of heat can be dried by setting it in a desiccator; using a vacuum desiccator will hasten the drying process. If the sample is weighed without drying, the results will be on an "as is" basis and should be reported as such.

Plant and tissue samples can usually be dried by heating. See Chapter 1 for a discussion of the various weight bases (wet, dry, ash) used in connection with reporting analytical results for these samples.

SAMPLE DISSOLUTION

Before the analyte can be measured, some sort of sample alteration is generally necessary to get the analyte into solution or, for biological samples, to rid it of interfering organic substances, such as proteins. There are two types of sample preparation: those that totally destroy the sample matrix and those that are nondestructive or only partially destructive. The former type can generally be used only when the analyte is inorganic or can be converted to an inorganic derivative for measurement (e.g., Kjeldahl analysis, in which organic nitrogen is converted to ammonium ion—see below). The latter type must be used if the analyte to be measured is an organic substance.

DISSOLVING INORGANIC SOLIDS

Strong mineral acids are good solvents for many inorganics. *Hydrochloric acid* is a good general solvent for dissolving metals that are above hydrogen in the electromotive series. *Nitric acid* is a strong oxidizing acid and will dissolve most of the common metals, nonferrous alloys, and the "acid-insoluble" sulfides.

Perchloric acid, when heated to drive off water, becomes a very strong and efficient oxidizing acid in the dehydrated state. It dissolves most of the common metals and destroys traces of organic matter. It must be used with extreme caution because it will react explosively with many easily oxidizable substances, especially organic matter.

Fusions are used when acids do not dissolve the sample.

Some inorganic materials will not dissolve in acids, and **fusion** with an acidic or basic **flux** in the molten state must be employed to render them soluble. The sample is mixed with the flux in a sample-to-flux ratio of about 1 to 10 or 20, and the combination is heated in an appropriate crucible until the flux becomes molten.

When the melt becomes clear, usually in about 30 min, the reaction is complete. The cooled solid is then dissolved in dilute acid or in water. During the fusion process, insoluble materials react with the flux to form a soluble product. Sodium carbonate is one of the most useful basic fluxes, and acid-soluble carbonates are produced.

DESTRUCTION OF ORGANIC MATERIALS FOR INORGANIC ANALYSIS—BURNING OR ACID OXIDATION

Animal and plant tissue, biological fluids, and organic compounds are usually decomposed by **wet digestion** with a boiling oxidizing acid or mixture of acids, or by **dry ashing** at a high temperature (400 to 700°C) in a muffle furnace. In wet digestion, the acids oxidize organic matter to carbon dioxide, water, and other volatile products, which are driven off, leaving behind salts or acids of the inorganic constituents. In dry ashing, atmospheric oxygen serves as the oxidant; that is, the organic matter is burned off, leaving an inorganic residue. Oxidizing aids may be employed in dry ashing.

1. Dry Ashing. Although various types of dry ashing and wet digestion combinations are used with about equal frequency by analysts for organic and biological materials, simple dry ashing with no chemical aids is probably the most commonly employed technique. Lead, zinc, cobalt, antimony, chromium, molybdenum, strontium, and iron traces can be recovered with little loss by retention or volatilization. Usually a porcelain crucible can be used. Lead is volatilized at temperatures in excess of about 500°C, especially if chloride is present, as in blood or urine. Platinum crucibles are preferred for lead for minimal retention losses.

In dry ashing, the organic matter is burned off.

If an oxidizing material is added to the sample, the ashing efficiency is enhanced. Magnesium nitrate is one of the most useful aids, and with this it is possible to recover arsenic, copper, and silver, in addition to the above-listed elements.

Liquids and wet tissues are dried on a steam bath or by gentle heat before they are placed in a muffle furnace. The heat from the furnace should be applied gradually up to full temperature to prevent rapid combustion and foaming.

After dry ashing is complete, the residue is usually leached from the vessel with 1 or 2 mL hot concentrated or 6 M hydrochloric acid and transferred to a flask or beaker for further treatment.

Another dry technique is that of **low-temperature ashing**. A radio-frequency discharge is used to produce activated oxygen radicals, which are very reactive and will attack organic matter at low temperatures. Temperatures of less than 100°C can be maintained, and volatility losses are minimized. Introduction of elements from the container and the atmosphere is reduced, and so are retention losses. Radiotracer studies have demonstrated that 17 representative elements are quantitatively recovered after complete oxidation of organic substrate.

Elemental analysis in the case of organic compounds (e.g., for carbon or hydrogen) is usually performed by **oxygen combustion** in a tube, followed by an absorption train. Oxygen is passed over the sample in a platinum boat, which is heated and quantitatively converts carbon to CO_2 and hydrogen to H_2O. These combustion gases pass into the absorption train, where they are absorbed in preweighed tubes containing a suitable absorbent. For example, **Ascarite** (sodium hydroxide on asbestos) is used to absorb the CO_2, and **Dehydrite** (magnesium perchlorate) is used to absorb the H_2O. The gain in weight of the absorption tubes is a measure of the CO_2 and H_2O liberated from the sample. Details of this technique are important, and, should you have occasion to use it, you are referred to more comprehensive texts on elemental analysis.

In wet ashing, the organic matter is oxidized with an oxidizing acid.

2. Wet Digestion. Wet digestion with a mixture of nitric and sulfuric acids is the second most often used oxidation procedure. Usually a small amount (e.g., 5 mL) of sulfuric acid is used with larger volumes of nitric acid (20 to 30 mL). Wet digestions are usually performed in a Kjeldahl flask (Figure 2.28). The nitric acid destroys the bulk of the organic matter, but it does not get hot enough to destroy the last traces. It is boiled off during the digestion process until only sulfuric acid remains and dense, white SO_3 fumes are evolved and begin to reflux in the flask. At this point, the solution gets very hot, and the sulfuric acid acts on the remaining organic material. Charring may occur at this point if there is considerable or very resistant organic matter left. If the organic matter persists, more nitric acid may be added. Digestion is continued until the solution clears. All digestion procedures must be performed in a fume hood.

A much more efficient digestion mixture employs a mixture of nitric, perchloric, and sulfuric acids in a volume ratio of about $3:1:1$. Ten milliliters of this mixture will usually suffice for 10 g fresh tissue or blood. The perchloric acid is an extremely efficient oxidizing agent when it is dehydrated and hot and will destroy the last traces of organic matter with relative ease. Samples are heated until nitric acid is boiled off and perchloric acid fumes, which are less dense than SO_3 but which fill the flask more readily, appear. The hot perchloric acid is boiled, usually until fumes of SO_3 appear, signaling the evaporation of all the perchloric acid. Sufficient nitric acid must be added at the beginning to dissolve and destroy the bulk of organic matter, and there must be sulfuric acid present to prevent the sample from going to dryness, or else there is danger of explosion from the perchloric acid. A hood specially designed for perchloric acid work should be used for all digestions incorporating perchloric acid.

This mixture is even more efficient if a small amount of molybdenum(VI) catalyst is added. As soon as water and nitric acid are evaporated. oxidation proceeds vigorously with foaming, and the digestion is complete in a few seconds. The digestion time is reduced considerably.

Perchloric acid must be used with caution!

A mixture of nitric and perchloric acids is also commonly used. The nitric acid boils off first, and care must be taken to prevent evaporation of the perchloric acid to near dryness, or a violent explosion may result; this procedure *is not recommended* unless you have considerable experience in digestion procedures. **Perchloric acid should never be added directly to organic or biological material.** Always add an excess of nitric acid first. Explosions with perchloric acid are generally associated with formation of peroxides, and the acid turns dark in color (e.g., yellowish brown) prior to explosion. Certain organic compounds such as ethanol, cellulose, and polyhydric alcohols can cause hot concentrated perchloric acid to explode violently; this is presumably due to formation of ethyl perchlorate.

A mixture of nitric, perchloric, and sulfuric acids allows zinc, selenium, arsenic, copper, cobalt, silver, cadmium, antimony, chromium, molybdenum, strontium, and iron to be quantitatively recovered. Lead is often lost if sulfuric acid is used. The mixture of nitric and perchloric acids can be used for lead and all the above elements. Perchloric acid must be present to prevent losses of selenium. It maintains strong oxidizing conditions and prevents charring that would result in formation of volatile compounds of lower oxidation states of selenium. Samples containing mercury cannot be dry ashed. Wet digestion with heat applied must be done using a reflux apparatus because of the volatile nature of mercury and its compounds. Cold or room temperature procedures are often preferred to obtain partial destruction of organic matter. For example, in urine samples, which contain a relatively small amount of organic matter compared with blood, mercury can be reduced to the element with copper(I) and hydroxylamine hydrochloride and the organic matter destroyed by potassium permanganate at room temperature. The mercury can then be dissolved and the analysis continued.

Many nitrogen-containing compounds can be determined by **Kjeldahl digestion** to convert the nitrogen to ammonium sulfate. The digestion mixture consists of sulfuric acid plus potassium sulfate to increase the boiling point of the acid and thus increase its efficiency. A catalyst is also added (such as copper or selenium). After destruction of the organic matter, sodium hydroxide is added to make the solution alkaline, and the ammonia is distilled into an excess of standard hydrochloric acid. The excess acid is back-titrated with standard alkali to determine the amount of ammonia collected. With a knowledge of the percent nitrogen composition in the compound of interest, the amount of the compound can be calculated from the amount of ammonia determined. This is the most accurate method for determining protein. Protein contains a definite percentage of nitrogen, which is converted to ammonium sulfate during the digestion. See Chapter 8 for further details.

> In Kjeldahl digestions, nitrogen is converted to ammonium ion, which is then distilled as ammonia and titrated.

The relative merits of the oxidation methods have been studied extensively. However, there is still no agreement as to which is to be preferred. Dry ashing is recommended for its simplicity and relative freedom from positive errors (contamination) since few or no reagents are added. The potential errors of dry oxidation are volatilization of elements and losses by retention on the walls of the vessel. Adsorbed metals on the vessel may in turn contaminate future samples. Wet digestion is considered superior in terms of rapidity (although it does require more operator attention), low level of temperature maintained, and freedom from loss by retention. The chief error attributed to wet digestion is the introduction of impurities from the reagents necessary for the reaction. This problem has been minimized as commercial reagent-grade acids have become available in greater purity and specially prepared high-purity acids can now be obtained commercially. The time required for ashing or digestion will vary with the sample and the technique employed. Two to 4 h are common for dry ashing and $\frac{1}{2}$ to 1 h is common for wet digestion.

> Dry and wet ashing each has advantages and limitations.

MICROWAVE PREPARATION OF SAMPLES

Microwave ovens are now widely used for rapid and efficient drying and acid decomposition of samples. Laboratory ovens are specially designed to overcome limitations of household ovens, and these are discussed below. Advantages of microwave digestions include reduction of dissolution times from hours to minutes and low blank levels due to reduced amounts of reagents required.

1. How Do Microwaves Heat? Microwaves occur between infrared radiation and radio waves in the electromagnetic spectrum, in the frequency range of 300 to 300,000 MHz (3×10^8 to 3×10^{11} Hz, or beginning at about 1000 μm wavelength—see Figure 16.2). Microwaves consist of an electric field and a magnetic field perpendicular to the electric field. The electric field is responsible for energy transfer between the microwave source and the irradiated sample. Microwave energy affects molecules in two ways: dipole rotation and ionic conduction. The first is the more important. When the microwave energy passes through the sample, the molecules having dipole moments will try to align with the electric field, and the more polar ones will have the stronger interaction with the field. This molecular motion (rotation) results in heating. The energy transfer, a function of the dipole moment and the dielectric constant, is most efficient when the molecules are able to relax quickly, that is, when the relaxation time matches the microwave frequency. Large molecules such as polymers relax slowly, but once the temperature increases and they relax more rapidly, they can absorb the energy more efficiently. Small molecules such as water, though, relax more quickly than the resonating microwave energy, and they move farther away from the resonance frequency and absorb less energy as they heat up.

The ionic conduction effect arises because ionic species in the presence of an electric field will migrate in one direction or the other. Energy is transferred from the electric field, causing ionic interactions that speed up the heating of a solution. Ionic absorbers become stronger absorbers of microwave energy as they are heated since ionic conductance increases with temperature. Deionized water heats slowly, but if salt is added, it heats rapidly. Acids, of course, are good conductors and heat rapidly.

Microwaves heat by causing molecules to rotate and ions to migrate.

So microwave energy heats by causing movement of molecules due to dipole rotation and movement of ions due to ionic conductance. The microwave energy interacts with different materials in different ways. Reflective materials such as metals are good heat conductors: They do not heat and instead will reflect the microwave energy. Transparent materials are insulators because they transmit the microwave energy and also do not heat. The absorptive materials, the molecules and ions discussed above, are the ones that receive microwaves and are heated. Microwave energy is too low to break chemical bonds (a feature that has generated interest in using microwave energy to speed up chemical reactions in syntheses). The properties of reflective and insulator materials are utilized in designing microwave digestion systems.

Household microwave ovens don't work for small sample heating.

2. Design of Laboratory Microwave Ovens. Home microwave ovens were initially used for laboratory purposes, but it soon became apparent that modifications were needed. Laboratory samples are usually much smaller than food samples that are cooked and do not absorb much of the energy produced by the magnetron of the oven. The energy not absorbed by the sample is bounced back to the magnetron, causing it to overheat and burn out. Also, arcing could occur. So laboratory ovens are designed to protect the magnetron from stray energy. The main components of these ovens (Figure 2.27) include the magnetron, an isolator, a waveguide, the cavity, and a mode stirrer. Microwave energy generated by the magnetron is propagated down the waveguide into the cavity. The stirrer distributes the energy in different directions. The isolator, made of a ferromagnetic material and placed between the magnetron and the waveguide, deflects the microwave energy returning from the cavity into a fan-cooled ceramic load, keeping it away from the magnetron.

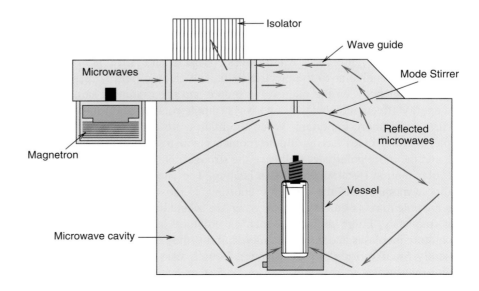

Fig. 2.27. Schematic of a microwave system. [From G. Le Blanc, *LC/GC Suppl.*, **17**(6S) (1999) S30.] (Courtesy of *LC/GC Magazine*.

The frequency used for cooking turns out to be good for chemistry as well, and the standard is 2450 MHz. Powers of 1200 W are typically used.

3. Acid Digestions. Digestions are normally done in closed plastic containers, either Teflon PFA (perfluoroalkoxy ethylene) or polycarbonate (insulators). This is to avoid acid fumes in the oven. It provides additional advantages. Pressure is increased and the boiling point of the acid is raised (the acid is superheated). So digestions occur more rapidly. Also, volatile metals are not lost. Modern ovens provide for control of pressure and temperature (fiber-optic temperature probe, transparent to microwave energy). Temperature control has enabled the use of the oven for microwave-assisted molecular extractions, by maintaining the temperature low enough to avoid molecular decomposition.

PARTIAL DESTRUCTION OR NONDESTRUCTION OF SAMPLE MATRIX

Obviously, when the substance to be determined is organic in nature, nondestructive means of preparing the sample must be employed. For the determination of metallic elements, it is also sometimes unnecessary to destroy the molecular structure of the sample, particularly with biological fluids. For example, several metals in serum or urine can be determined by atomic absorption spectroscopy by direct aspiration of the sample or a diluted sample into a flame. Constituents of solid materials such as soils can sometimes be extracted by an appropriate reagent. Thorough grinding, mixing, and refluxing are necessary to extract the analyte. Many trace metals can be extracted from soils with 1 M ammonium chloride or acetic acid solution. Some, such as selenium, can be distilled as the volatile chloride or bromide.

PROTEIN-FREE FILTRATES

Proteins in biological fluids interfere with many analyses and must be removed nondestructively. Several reagents will precipitate (coagulate) proteins. Trichloroacetic acid (TCA), tungstic acid (sodium tungstate plus sulfuric acid), and barium hydroxide plus zinc sulfate (a neutral mixture) are some of the common ones. A measured volume of sample (e.g., serum) is usually treated with a measured volume of reagent. Following precipitation of the protein (approximately 10 min), the sample is filtered through dry filter paper without washing, or else it is centrifuged. An aliquot of the **protein-free filtrate** (PFF) is then taken for analysis. Details for preparing specific types of protein-free filtrates are given in Chapter 22 (under Collection and Preservation of Samples) as well as in experiments requiring them.

See Chapter 24 for the preparation of protein-free filtrates.

LABORATORY TECHNIQUES—FOR DRYING AND DISSOLVING

When a solid sample is to be dried in a weighing bottle, the cap is removed from the bottle and, to avoid spilling, both are placed in a beaker and covered with a ribbed watch glass. Some form of identification should be placed on the beaker.

The weighed sample may be dissolved in a beaker or Erlenmeyer flask. If there is any fizzing action, cover the vessel with a watch glass. After dissolution is complete, wash the walls of the vessel down with distilled water. Also wash the watch glass so the washings fall into the vessel. You may have to evaporate the solution to decrease the volume. This is best done by covering the beaker with a

Take care in drying or dissolving samples.

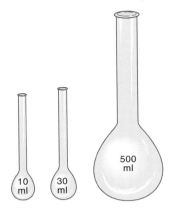

Fig. 2.28. Kjeldahl flasks.

ribbed watch glass to allow space for evaporation. Low heat should be applied to prevent bumping; a steam bath or variable-temperature hot plate is satisfactory.

Use of a **Kjeldahl flask** for dissolution will avoid some of the difficulties of splattering or bumping. Kjeldahl flasks are also useful for performing digestions. They derive their name from their original use in digesting samples for Kjeldahl nitrogen analysis. They are well suited to all types of wet digestions of organic samples and acid dissolution of metals. Kjeldahl flasks come in assorted sizes from 10 to 800 mL. Some of these are shown in Figure 2.28. The sample and appropriate acids are placed in the round bottom of the flask and the flask is tilted while it is heated. In this way the acid can be boiled or refluxed with little danger of loss by "bumping." The flask may be heated with a flame or in special electrically heated Kjeldahl digestion racks, which heat several samples simultaneously.

2.11 Laboratory Safety

You *must* familiarize yourself with laboratory safety rules and procedures before conducting experiments! Read Appendix D and the material provided by your instructor. Get a free copy of Ref. 32.

Before beginning any of the experiments, you must familiarize yourself with laboratory safety procedures. Appendix D discusses general safety rules. You should read this material before beginning experiments. Your instructor will provide you with specific guidelines and rules for operation in the laboratory and the disposal of chemicals. For a more complete discussion of safety in the laboratory, you are referred to *Safety in Academic Chemistry Laboratories*, published by The American Chemical Society (Ref. 32). This guide discusses personal protection and laboratory protocol, recommended laboratory techniques, chemical hazards, instructions on reading and understanding material safety data sheets (MSDSs), and safety equipment and emergency procedures. Rules are given for waste disposal, waste classification terminology, Occupational Safety and Health Administration (OSHA) laboratory standards for exposures to hazardous chemicals, and EPA requirements. The handling and treatment of inorganic and organic peroxides are discussed in detail, and an extensive list of incompatible chemicals is given,

(Courtesy of Merck KGaA. Reproduced by permission.)

Always wear eye protection in the laboratory!

and maximum allowable container capacities for flammable and combustible liquids are listed. This resourceful booklet is recommended reading for students and instructors. It is available for free (one copy) from The American Chemical Society, Washington, DC (1-800-227-5558).

The Waste Management Manual for Laboratory Personnel, also published by The American Chemical Society, provides an overview of government regulations (Ref. 33).

Learning Objectives

WHAT ARE SOME OF THE KEY THINGS WE LEARNED FROM THIS CHAPTER?

- Keep a good notebook, p. 20
- Use reagent-grade chemicals, p. 24
- How to use the analytical balance, p. 24
- Volumetric glassware and how to use it, pp. 32, 36
- How to calibrate glassware, p. 39
- How to prepare standard acid and base solutions, pp. 43, 44
- Common laboratory apparatus for handling and treating samples, p. 44
- How to filter and prepare precipitates for gravimetric analysis, p. 48
- How to sample solids, liquids, and gases, p. 52
- How to prepare a solution of the analyte, p. 53

Questions

1. Describe the basic pieces of apparatus used for volumetric measurements. List whether each is designed to contain or to deliver the specified volume.
2. Describe the principle and operation of the analytical balance.
3. Why is a microbalance more sensitive than an analytical balance?
4. What does TD on glassware mean? TC?
5. Explain weighing by difference.
6. Distinguish between the zero point and the rest point of a balance.
7. List the general rules for the use of the balance.
8. Describe the preparation of a standard HCl solution and a standard NaOH solution.
9. Describe the principles of dry ashing and wet digestion of organic and biological materials. List the advantages of each.
10. What are the two principal means of dissolving inorganic materials?
11. What is a PFF? How would you prepare it?
12. What set of conditions must be carefully avoided to use perchloric acid safely for digesting organic materials?
13. What is a gross sample? Sample? Analysis sample? Grab sample?
14. What happens when microwave energy heats samples?

Problems

GLASSWARE CALIBRATION/TEMPERATURE CORRECTIONS

15. You calibrate a 25-mL volumetric flask by filling to the mark with distilled water, equilibrated at 22°C. The dry stoppered flask weighs 27.278 g and the filled flask and stopper is 52.127 g. The balance uses stainless steel weights. What is the volume of the flask? What is it at the standard 20°C. Also insert the weight in air at 22°C into Table 2.4 in your CD, and compare the values obtained.

16. You calibrate a 25-mL pipet at 25°C using steel weights. The weight of the delivered volume of water is 24.971 g. What is the volume of the pipet at 25 and 20°C?

17. You calibrate a 50-mL buret in the winter time at 20°C, with the following corrections:

Buret Reading (mL)	Correction (mL)
10	+0.02
20	+0.03
30	0.00
40	−0.04
50	−0.02

You use the buret on a hot summer day at 30°C. What are the corrections then?

18. You prepare a standard solution at 21°C, and use it at 29°C. If the standardized concentration is 0.05129 M, what is it when you use the solution?

Recommended References

REAGENTS AND STANDARDS

1. *Reagent Chemicals, American Chemical Society Specifications*, 9th ed., Washington, DC: American Chemical Society, 2000.

2. *Dictionary of Analytical Reagents*, London: Chapman & Hall, 1993. Data on over 10,000 reagents.

3. J. R. Moody and E. S. Beary, "Purified Reagents for Trace Metal Analysis," *Talanta*, **29** (1982), 1003. Describes sub-boiling preparation of high-purity acids.

4. R. C. Richter, D. Link, and H. M. Kingston, "On-Demand Production of High-Purity Acids in the Analytical Laboratory," *Spectroscopy*, **15**(1) (2000) 38. Describes preparation using a commercial sub-boiling system.

5. www.thornsmithlabs.com. Thorn Smith Laboratories provides prepackaged materials for student analytical chemistry experiments (unknowns and standards); www.sigma-aldrich.com/analytical. A supplier of standard solutions.

ANALYTICAL BALANCES

6. D. F. Rohrbach and M. Pickering, "A Comparison of Mechanical and Electronic Balances," *J. Chem. Ed.*, **59** (1982) 418.

7. R. M. Schoonover, "A Look at the Electronic Analytical Balance," *Anal. Chem.*, **54** (1982) 973A.

8. K. M. Lang, "Time-Saving Applications of Electronic Analytical Balances," *Am. Lab.*, March (1983) 72.

CALIBRATION OF WEIGHTS

9. W. D. Abele, "Laboratory Note: Time-Saving Applications of Electronic Balances," *Am. Lab.*, **13** (1981) 154. Calibration of weights using mass standards calibrated by the National Institute of Standards and Technology is discussed.

10. D. F. Swinehart, "Calibration of Weights in a One-Pan Balance," *Anal. Lett.*, **10** (1977) 1123.

CALIBRATION OF VOLUMETRIC WARE

11. G. D. Christian, "Coulometric Calibration of Micropipets," *Microchem. J.*, **9** (1965) 16.

12. M. R. Masson, "Calibration of Plastic Laboratory Ware," *Talanta*, **28** (1981) 781. Tables for use in calibration of polypropylene vessels are presented.

13. W. Ryan, "Titrimetric and Gravimetric Calibration of Pipettors: A Survey," *Am. J. Med. Technol.*, **48** (1982) 763. Calibration of pipettors of 1 to 500 μL is described.

14. M. R. Winward, E. M. Woolley, and E. A. Butler, "Air Buoyancy Corrections for Single-Pan Balances," *Anal. Chem.*, **49** (1977) 2126.

15. R. M. Schoonover and F. E. Jones, "Air Buoyancy Corrections in High-Accuracy Weighing on Analytical Balances," *Anal. Chem.*, **53** (1981) 900.

CLEAN LABORATORIES

16. J. R. Moody, "The NBS Clean Laboratories for Trace Element Analysis," *Anal. Chem.*, **54** (1982) 1358A.

SAMPLING

17. J. A. Bishop, "An Experiment in Sampling," *J. Chem. Ed.*, **35** (1958) 31.

18. J. R. Moody, "The Sampling, Handling and Storage of Materials for Trace Analysis," *Phil. Trans. Roy. Soc. London, Ser. A*, **305** (1982) 669.

19. G. E. Baiulescu, P. Dumitrescu, and P. Gh. Zugravescu, *Sampling*, Chichester: Ellis Horwood, 1991.

SAMPLE PREPARATION AND DISSOLUTION

20. G. D. Christian, "Medicine, Trace Elements, and Atomic Absorption Spectroscopy," *Anal. Chem.*, **41** (1) (1969) 24A. Describes the preparation of solutions of biological fluids and tissues.

21. G. D. Christian, E. C. Knoblock, and W. C. Purdy, "Polarographic Determination of Selenium in Biological Materials," *J. Assoc. Offic. Agric. Chemists*, **48** (1965) 877; R. K. Simon, G. D. Christian, and W. C. Purdy, "Coulometric Determination of Arsenic in Urine," *Am. J. Clin. Pathol.*, **49** (1968) 207. Describes use of Mo(VI) catalyst in digestions.

22. T. T. Gorsuch, "Radiochemical Investigations on the Recovery for Analysis of Trace Elements in Organic and Biological Materials," *Analyst*, **84** (1959) 135.

23. S. Nobel and D. Nobel, "Determination of Mercury in Urine," *Clin. Chem.*, **4** (1958) 150. Describes room temperature digestion.

24. G. Knapp, "Mechanical Techniques for Sample Decomposition and Element Preconcentration," *Mikrochim. Acta*, **II** (1991) 445. Lists acid mixtures and uses for digestions and other decomposition methods.

MICROWAVE DIGESTION

25. H. M. Kingston and L. B. Jassie, eds., *Introduction to Microwave Sample Preparation: Theory and Practice.* Washington, D.C.: American Chemical Society, 1988.

26. H. M. Kingston and S. J. Haswell, eds., *Microwave-Enhanced Chemistry: Fundamentals, Sample Preparation, and Applications*, Washington, DC: American Chemical Society, 1997.

27. H. M. Kingston and L. B. Jassie, "Microwave Energy for Acid Decomposition at Elevated Temperatures and Pressures Using Biological and Botanical Samples," *Anal. Chem.*, **58** (1986) 2534.

28. R. A. Nadkarni, "Applications of Microwave Oven Sample Dissolution in Analysis," *Anal. Chem.*, **56** (1984) 2233.

29. B. D. Zehr, "Development of Inorganic Microwave Dissolutions," *Am. Lab.*, December (1992) 24. Also lists properties of useful acid mixtures.

30. J. K. Poudrier, "The Microwave Oven as a Specialized Instrument," *Today's Chemist at Work* (American Chemical Society), July/August (1995) 29.

31. www.cemx.com. CEM is a manufacturer of laboratory microwave ovens.

LABORATORY SAFETY

32. *Safety in Academic Chemistry Laboratories*, 6th ed., American Chemical Society, Committee on Chemical Safety. Washington, D.C.: American Chemical Society, 1995.

33. *The Waste Management Manual for Laboratory Personnel*, Task Force on RCRA, American Chemical Society, Department of Government Relations and Science Policy. Washington, DC: American Chemical Society, 1990.

34. A. K. Furr, ed., *CRC Handbook of Laboratory Safety*, 4th ed. Boca Raton, FL: CRC Press, 1993.

35. M.-A. Armour, *Hazardous Laboratory Chemicals Disposal Guide*, Boca Raton, FL: CRC Press, 1990.

36. R. J. Alaimo, ed., *Handbook of Chemical Health and Safety*, Washington, DC: American Chemical Society, 2001.

37. M. S. Lesny, "It's a Bad, Bad, Bad, Bad World," *Today's Chemist at Work* (American Chemical Society), December (2000) 26. Discusses and lists numerous government, nongovernment, and university health and safety websites.

MATERIAL SAFETY DATA SHEETS

38. http://siri.org/msds. Online searchable database, with links to other MSDS and hazardous chemical sites.

39. www.env-sol.com. Solutions Software Corporation. MSDS database available on DVD or CD-ROM.

Chapter Three

DATA HANDLING AND SPREADSHEETS IN ANALYTICAL CHEMISTRY

"Facts are stubborn, but statistics are much more pliable."
—Mark Twain

"43.8% of all statistics are worthless."
—Anonymous

Although data handling normally follows the collection of data in an analysis, it is treated early in the text because a knowledge of statistical analysis will be required as you perform experiments in the laboratory. Also, statistics are necessary to understand the significance of the data that are collected and therefore to set limitations on each step of the analysis. The design of experiments (including size of sample required, accuracy of measurements required, and number of analyses needed) is determined from a proper understanding of what the data will represent.

The availability of spreadsheets to process data has made statistical and other calculations very efficient. You will first be presented with the details of various calculations throughout the text, which are necessary for full understanding of the principles. But spreadsheet calculations will also be introduced throughout to illustrate how to take advantage of this software for routine calculations. We will introduce the principles of the use of spreadsheets in this chapter.

3.1 Accuracy and Precision: There Is a Difference

Accuracy is the degree of agreement between the measured value and the true value. An absolute true value is seldom known. A more realistic definition of accuracy, then, would assume it to be the agreement between a measured value and the *accepted* true value.

We can, by good analytical technique, such as making comparisons against a known standard sample of similar composition, arrive at a reasonable assumption about the accuracy of a method, within the limitations of the knowledge of

Accuracy is how close you get to the bullseye. *Precision* is how close the repetitive shots are to one another. It is nearly impossible to have accuracy without good precision.

Fig. 3.1. *Accuracy versus precision.*

the "known" sample (and of the measurements). The accuracy to which we know the value of the standard sample is ultimately dependent on some measurement that will have a given limit of certainty in it.

Good precision does not guarantee accuracy.

Precision is defined as the degree of agreement between replicate measurements of the same quantity. That is, it is the repeatability of a result. The precision may be expressed as the standard deviation, the coefficient of variation, the range of the data, or as a confidence interval (e.g., 95%) about the mean value. Good precision does not assure good accuracy. This would be the case, for example, if there were a systematic error in the analysis. A weight used to measure each of the samples may be in error. This error does not affect the precision, but it does affect the accuracy. On the other hand, the precision can be relatively poor and the accuracy, more or less by chance, may be good. Since all real analyses are unknown, the higher the degree of precision, the greater the chance of obtaining the true value. It is fruitless to hope that a value is accurate if the precision is poor, and the analytical chemist strives for repeatable results to assure the highest possible accuracy.

"To be sure of hitting the target, shoot first, and call whatever you hit the target."—Ashleigh Brilliant

These concepts can be illustrated with a target, as in Figure 3.1. Suppose you are at target practice and you shoot the series of bullets that all land in the bulls-eye (left target). You are both precise and accurate. In the middle target, you are precise (steady hand and eye), but inaccurate. Perhaps the sight on your gun is out of alignment. In the right target you are imprecise and therefore probably inaccurate. So we see that good precision is needed for good accuracy, but it does not guarantee it.

As we shall see later, the more measurements that are made, the more reliable will be the measure of precision. The number of measurements required will depend on the accuracy required and on the known reproducibility of the method.

3.2 Determinate Errors—They Are Systematic

Determinate or systematic errors are nonrandom and occur when something is wrong with the measurement.

Two main classes of errors can affect the accuracy or precision of a measured quantity. **Determinate errors** are those that, as the name implies, are determinable and that presumably can be either avoided or corrected. They may be constant, as in the case of an uncalibrated weight that is used in all weighings. Or, they may be variable but of such a nature that they can be accounted for and corrected, such as a buret whose volume readings are in error by different amounts at different volumes.

The error can be proportional to sample size or may change in a more complex manner. More often than not, the variation is unidirectional, as in the case of solubility loss of a precipitate (negative error). It can, however, be random in sign. Such an example is the change in solution volume and concentration occurring with changes in temperature. This can be corrected for by measuring the solution temperature. Such measurable determinate errors are classed as **systematic errors**.

Some common determinate errors are:

1. *Instrumental errors.* These include faulty equipment, uncalibrated weights, and uncalibrated glassware.

2. *Operative errors.* These include personal errors and can be reduced by experience and care of the analyst in the physical manipulations involved. Operations in which these errors may occur include transfer of solutions, effervescence and "bumping" during sample dissolution, incomplete drying of samples, and so on. These are difficult to correct for. Other personal errors include mathematical errors in calculations and prejudice in estimating measurements.

3. *Errors of the method.* These are the most serious errors of an analysis. Most of the above errors can be minimized or corrected for, but errors that are inherent in the method cannot be changed unless the conditions of the determination are altered. Some sources of methodic errors include coprecipitation of impurities, slight solubility of a precipitate, side reactions, incomplete reactions, and impurities in reagents. Sometimes correction can be relatively simple, for example, by running a **reagent blank**. A blank determination is an analysis on the added reagents only. It is standard practice to run such blanks and to subtract the results from those for the sample. When errors become intolerable, another approach to the analysis must be made. Sometimes, however, we are forced to accept a given method in the absence of a better one.

Determinate errors may be *additive* or *multiplicative*, depending on the nature of the error or how it enters into the calculation. In order to detect systematic errors in an analysis, it is common practice to add a known amount of standard to a sample and measure its recovery (see Validation of a Method in Chapter 1). The analysis of reference samples also helps guard against method errors or instrumental errors.

3.3 Indeterminate Errors—They Are Random

The second class of errors includes the **indeterminate errors**, often called accidental or random errors, which represent the experimental uncertainty that occurs in any measurement. These errors are revealed by small differences in successive measurements made by the same analyst under virtually identical conditions, and they cannot be predicted or estimated. These accidental errors will follow a random distribution; therefore, mathematical laws of probability can be applied to arrive at some conclusion regarding the most probable result of a series of measurements.

It is beyond the scope of this text to go into mathematical probability, but we can say that indeterminate errors should follow a **normal distribution**, or **Gaussian curve**. Such a curve is shown in Figure 3.2. The symbol σ represents the *standard deviation* of an infinite population of measurements, and this measure of precision defines the spread of the normal population distribution as shown in

Indeterminate errors are random and cannot be avoided.

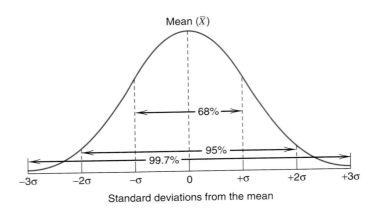

Fig. 3.2. Normal error curve.

"Undetectable errors are infinite in variety, in contrast to detectable errors, which by definition are limited."—Tom Gibb

Figure 3.2. It is apparent that there should be few very large errors and that there should be an equal number of positive and negative errors.

Indeterminate errors really originate in the limited ability of the analyst to control or make corrections for external conditions, or the inability to recognize the appearance of factors that will result in errors. Some random errors stem from the more statistical nature of things, for example, nuclear counting errors. Sometimes, by changing conditions, some unknown error will disappear. Of course, it will be impossible to eliminate all possible random errors in an experiment, and the analyst must be content to minimize them to a tolerable or insignificant level.

3.4 Significant Figures: How Many Numbers Do You Need?

The last digit of a measurement has some uncertainty. You can't include any more digits.

The weak link in the chain of any analysis is the measurement that can be made with the least accuracy. It is useless to extend an effort to make the other measurements of the analysis more accurately than this limiting measurement. The number of significant figures can be defined as **the number of digits necessary to express the results of a measurement consistent with the measured precision**. Since there is uncertainty (imprecision) in any measurement of at least ± 1 in the last significant figure, the number of significant figures includes all of the digits that are known, plus the first uncertain one. Each digit denotes the actual quantity it specifies. For example, in the number 237, we have 2 hundreds, 3 tens, and 7 units.

The digit 0 can be a significant part of a measurement, or it can be used merely to place the decimal point. The number of significant figures in a measurement is independent of the placement of the decimal point. Take the number 92,067. This number has five significant figures, regardless of where the decimal point is placed. For example, 92,067 μm, 9.2067 cm, 0.92067 dm, and 0.092067 m all have the same number of significant figures. They merely represent different ways (units) of expressing one measurement. The zero between the decimal point and the 9 in the last number is used only to place the decimal point. There is no doubt whether any zero that *follows* a decimal point is significant or is used to place the decimal point. In the number 727.0, the zero is not used to locate the decimal point but is a significant part of the figure. Ambiguity can arise if a zero *precedes* a decimal point. If it falls between two other nonzero integers, then it will be significant. Such was the case with 92,067. In the number 936,600, it is impossible to determine whether one or both or neither of the zeros is used merely to place the decimal point or whether they are a part of the measurement. It is best in cases like this to write only the significant figures you are sure about and then to locate the decimal point by scientific notation. Thus, 9.3660×10^5 has five significant figures, but 936,600 contains six digits, one to place the decimal.

Example 3.1

List the proper number of significant figures in the following numbers and indicate which zeros are significant.

0.216; 90.7; 800.0; 0.0670

Solution

0.216	three significant figures
90.7	three significant figures; zero is significant
800.0	four significant figures; all zeros are significant
0.0670	three significant figures; only the last zero is significant

If a number is written as 500, it could represent 500 ± 100. If it is written as 500., then it is 500 ± 1.

The significance of the last digit of a measurement can be illustrated as follows. Assume that each member of a class measures the length of a rod, using the same meter stick. Assume further that the meter stick is graduated in 1-mm increments. The measurements can be estimated to the nearest 0.1 division (0.1 mm) by interpolation, but the last digit is uncertain since it is only an estimation. A series of class readings, for example, might be

36.4 mm
36.8 mm
36.0 mm
37.1 mm

36.6 mm (average)

MULTIPLICATION AND DIVISION—THINK RELATIVE

In many measurements, one estimated digit that is uncertain will be included (e.g., tenth millimeter digit above). This is the last significant figure in the measurement; any digits beyond it are meaningless. In multiplication and division, the uncertainty of this digit is carried through the mathematical operations, thereby limiting the number of certain digits in the answer. There is at least the degree of relative uncertainty (the uncertainty as a function of the answer) in the answer of a multiplication or division as there is in the operator with the least degree of certainty, that is, the one with the least number of significant figures. We shall designate this limiting number as the **key number**. If there is more than one operator with the same lowest number of significant figures, then the one with the smallest absolute magnitude without regard to the decimal point (units) is the key number (since its uncertainty is the greatest). For example, the absolute uncertainty without regard to the decimal point of 0.0344 is 344, and of 5.39 is 539.

The answer of a multiplication or division can be no more accurate than the least accurately known operator.

Example 3.2

In the following pairs of numbers, pick the one that would represent the key number in a multiplication or division. (a) 42.67 or 0.0967; (b) 100.0 or 0.4570; (c) 0.0067 or 0.10.

Solution

(a) 0.0967 (has three significant figures)

(b) 100.0 (both have four significant figures, but the uncertainty here is 1 part per thousand versus about 1 part in 4600)

(c) 0.10 [both have two significant figures, but the uncertainty here is 10% (1 part in 10) versus about 1 part in 70]

Example 3.3

Give the answer of the following operation to the maximum number of significant figures and indicate the key number.

$$\frac{35.63 \times 0.5481 \times 0.05300}{1.1689} \times 100\% = 88.5470578\%$$

Solution

The key number is 35.63. The answer is therefore 88.55%, and it is meaningless to carry the operation out to more than five figures (the fifth figure is used to round off the fourth). The 100% in this calculation is an absolute number since it is used only to move the decimal point, and it has an infinite number of significant figures. Note that the key number has a relative uncertainty at best of 1 part in 3600, and so the answer has a relative uncertainty at best of 1 part in 3600; thus, the answer has a relative uncertainty at least of this magnitude (i.e., about 2.5 parts in 8900). The objective in a calculation is to express the answer to at least the precision of the least certain number, but to recognize the magnitude of its uncertainty. *The final number is determined by the measurement of significant figures.* (Similarly, in making a series of measurements, one should strive to make each to about the same degree of relative uncertainty.)

A subscript number is used to indicate an added degree of uncertainty. It is used when the number is smaller than the key number.

If the magnitude of the answer without regard to decimal or sign is *smaller* than that of the key number, *one additional figure may be carried in the answer in order to express the minimum degree of uncertainty*, but it is written as a subscript to indicate that it is more doubtful.

Example 3.4

Give the answer of the following operation to the maximum number of significant figures and indicate the key number.

$$\frac{42.68 \times 891}{132.6 \times 0.5247} = 546.57$$

Solution

The key number is 891. Since the absolute magnitude of the answer (its magnitude without respect to units) is less than the key number, it becomes $546_{.6}$. The last 6 is written as a subscript to indicate it is more doubtful. Again, the key number has a relative uncertainty of about 1 part in 900, so the answer has an uncertainty of at least 6 parts in 5500 (0.6 parts in 550).

• •

In multiplication and division, the answer from each step of a series of operations can statistically be rounded to the number of significant figures to be retained in the final answer. But for consistency in the final answer, it is preferable to carry one additional figure until the end and then round off.

ADDITION AND SUBTRACTION—THINK ABSOLUTE

Additions and subtractions are handled in a somewhat different manner. We deal with absolute numbers rather than relative numbers. Here, we do not have a key number, and the placing of the decimal point is important in determining how many figures will be significant. Suppose you wish to calculate the formula weight of Ag_2MoO_4 from the individual atomic weights:

The answer of an addition or subtraction is known to the same number of units as the number containing the least significant unit.

Ag	107.87	0
Ag	107.87	0
Mo	95.94	
O	15.99	94
O	15.99	94
O	15.99	94
O	15.99	94
	375.67	76

The atomic weight of molybdenum is known only to the nearest 0.01 atomic unit. Since this unit has an element of uncertainty in it, we cannot justifiably say that we know the formula weight of a compound containing molybdenum to any closer than 0.01 atomic unit. Therefore, the most accurately known value for the atomic weight of Ag_2MoO_4 is 375.68. All numbers being added or subtracted can be rounded to the least significant unit before adding or subtracting. But again, for consistency in the answer, one additional figure should be carried out and then the answer rounded to one less figure.

Summarizing the importance of significant figures, there are two questions to ask. First, how accurately do you have to *know* a particular result? If you only want to learn whether there is 12 or 13% of a substance in the sample, then you need only make all required measurements to two significant figures. If the sample weighs about 2 g, there is no need to weigh it to more than 0.1 g. The second question is, how accurately can you *make* each required measurement? Obviously, if you can read the absorbance of light by a colored solution to only three figures (e.g., $A = 0.447$), it would be useless to weigh the sample to more than three figures (e.g., 6.67 g).

When a number in a measurement is small (without regard to the decimal point) compared with those of the other measurements, there is some justification in making the measurement to one additional figure. This can be visualized as follows. Suppose you wish to weigh two objects of essentially the same mass, and you wish to weigh them with the same precision, for example, to the nearest 0.1 mg, or 1 part per thousand. The first object weighs 99.8 mg, but the second weighs

100.1 mg. You have weighed both objects with equal accuracy, but you have retained an additional significant figure in one of them. This analogy can also be related to the justification for adding an additional significant figure when the answer of a mathematical operation is less than the key number.

When the key number in a series of measurements is known, then, the overall accuracy can be improved if desired either by making the key number larger (e.g., by increasing the sample size) or by making the measurement to an additional figure if possible (e.g., by weighing to one additional figure). This would be desirable when the number is small in magnitude compared to those of the other measurements (without regard to the decimal) in order to bring its uncertainty closer to that of the others.

In carrying out analytical operations, then, you should measure quantities to the same *absolute* uncertainty when adding or subtracting and to the same *relative* uncertainty when multiplying or dividing.

> It is good practice to keep an extra figure during stepwise calculations and then drop it in the final number.

> "Check the answer you have worked out once more—before you tell it to anybody."—Edmund C. Berkely

If a computation involves both multiplication/division and addition/subtraction, then the individual steps must be treated separately. As good practice, retain one extra figure in the intermediate calculations until the final result (unless it drops out in a subsequent step). When a calculator is used, all digits can be kept in the calculator until the end. Do not assume that a number spit out by a calculator is correct. Always try to estimate the size of the answer you expect. If you expect 2% and the reading is 0.02, you probably forgot to multiply by 100. Or if you expect 20% and the answer is 4.3, you probably made a calculation error or perhaps a measurement error.

Example 3.5

Give the answer of the following computation to the maximum number of significant figures:

$$\frac{\left(\dfrac{97.7}{32.42} \times 100.0\right) + 36.04}{687}$$

Solution

$$\frac{301_{.36} + 36.04}{687} = \frac{337_{.4}}{687} = .0491_1$$

In the first operation, the key number is 97.7 and the result is $301_{.36}$. We carried an additional fifth figure until the addition step and then rounded to four figures since the division has only three significant figures. In the division step, the key number is 687; but since the absolute magnitude of the answer is less, we carry one more figure. Note that if in the first step we had rounded to $301_{.4}$, the numerator would have become $337_{.5}$ and the final answer would be 0.491_3 (still within the experimental uncertainty).

LOGARITHMS—THINK MANTISSA

> In logarithms, it is the mantissa that determines the number of significant figures.

In changing from logarithms to antilogarithms, and vice versa, the number being operated on and the logarithm mantissa have the same number of significant figures. (See Appendix B for a review of the use of logarithms.) All zeros in the mantissa are significant. Suppose, for example, we wish to calculate the pH of a 2.0×10^{-3} M solution of HCl from pH $= -\log[H^+]$. Then,

$$pH = -\log 2.0 \times 10^{-3} = -(-3 + 0.30) = 2.70$$

The -3 is the characteristic (from 10^{-3}), a pure number determined by the position of the decimal. The 0.30 is the mantissa from the logarithm of 2.0 and therefore has only two digits. So, even though we know the concentration to two figures, the pH (the logarithm) has three figures. If we wish to take the antilogarithm of a mantissa, the corresponding number will likewise have the same number of digits as the mantissa. The antilogarithm of 0.072 (contains three figures in mantissa .072) is 1.18, and the logarithm of 12.1 is 1.083 (1 is the characteristic, and the mantissa has three digits, .083).

3.5 Rounding Off

If the digit following the last significant figure is greater than 5, the number is rounded up to the next higher digit. If it is less than 5, the number is rounded to the present value of the last significant figure:

Always round to the even number, if the last digit is a 5.

$$9.47 = 9.5$$
$$9.43 = 9.4$$

If the last digit is a 5, the number is rounded off to the nearest even digit:

$$8.65 = 8.6$$
$$8.75 = 8.8$$
$$8.55 = 8.6$$

This is based on the statistical prediction that there is an equal chance that the last significant figure before the 5 will be even or odd. That is, in a suitably large sampling, there will be an equal number of even and odd digits preceding a 5. All nonsignificant digits should be rounded off all at once. The even-number rule applies only when the digit dropped is exactly 5 (not . . . 51, e.g.).

3.6 Ways of Expressing Accuracy

There are various ways and units in which the accuracy of a measurement can be expressed, an accepted true value for comparison being assumed.

ABSOLUTE ERRORS

The difference between the true value and the measured value, with regard to the sign, is the **absolute error**, and it is reported in the same units as the measurement. If a 2.62-g sample of material is analyzed to be 2.52 g, the absolute error is -0.10 g. If the measured value is the average of several measurements, the error is called the **mean error**. The mean error can also be calculated by taking the average difference, with regard to sign, of the *individual* test results from the true value.

RELATIVE ERROR

The absolute or mean error expressed as a percentage of the true value is the **relative error**. The above analysis has a relative error of $(-0.10/2.62) \times 100\% = -3.8\%$. The **relative accuracy** is the measured value or mean expressed as a percentage of the true value. The above analysis has a relative accuracy of $(2.52/2.62) \times 100\% = 96.2\%$. We should emphasize that neither number is known to be "true," and the relative error or accuracy is based on the mean of two sets of measurements.

The relative error can be expressed in units other than percentages. In very accurate work, we are usually dealing with relative errors of less than 1%, and it is convenient to use a smaller unit. A 1% error is equivalent to 1 part in 100. It is also equivalent to 10 parts in 1000. This latter unit is commonly used for expressing small uncertainties. That is, the uncertainty is expressed in **parts per thousand**, written as ppt. The number 23 expressed as parts per thousand of the number 6725 would be 23 parts per 6725, or 3.4 ppt. Parts per thousand is often used in expressing precision of measurement.

 ### Example 3.6

The results of an analysis are 36.97 g, compared with the accepted value of 37.06 g. What is the relative error in parts per thousand?

Solution

$$\text{Absolute error} = 36.97 \text{ g} - 37.06 \text{ g} = -0.09 \text{ g}$$

$$\text{Relative error} = \frac{-0.09}{37.06} \times 1000\text{‰} = -2.4 \text{ ppt}$$

‰ indicates parts per thousand, just as % indicates parts per hundred.

3.7 Standard Deviation—The Most Important Statistic

Each set of analytical results should be accompanied by an indication of the **precision** of the analysis. Various ways of indicating precision are acceptable.

The standard deviation σ of an infinite set of experimental data is theoretically given by

$$\sigma = \sqrt{\frac{\sum(x_i - \mu)^2}{N}} \tag{3.1}$$

where x_i represents the individual measurements and μ the mean of the infinite number of measurements (which should represent the "true" value). This equation holds strictly only as $N \to \infty$, where N is the number of measurements. In practice, we must calculate the individual deviations from the mean of a limited number of measurements, $\bar{x}$, in which it is anticipated that $\bar{x} \to \mu$, although we have no assurance this will be so; $\bar{x}$ is given by $\sum(x_i/N)$.

For a set of N measurements, it is possible to calculate N independently variable deviations from some reference number. But if the reference number chosen is the estimated mean, $\bar{x}$, the sum of the individual deviations (retaining signs) must necessarily add up to zero, and so values of $N - 1$ deviations are adequate to define the Nth value. That is, there are only $N - 1$ independent deviations from the mean; when $N - 1$ values have been selected, the last is predetermined. We have, in effect, used one degree of freedom of the data in calculating the mean, leaving $N - 1$ **degrees of freedom** for calculating the precision.

As a result, the **estimated standard deviation s of a finite set of experimental data** (generally $N < 30$) more nearly approximates σ if the number of degrees of freedom is substituted for N ($N - 1$ adjusts for the difference between $\bar{x}$ and μ).

See Section 3.15 and Equation 3.17 for another way of estimating s for four or less numbers.

$$s = \sqrt{\frac{\sum(x_i - \bar{x})^2}{N - 1}}$$ (3.2)

The value of s is only an estimate of σ, then, and will more nearly approach σ as the number of measurements increases. Since we deal with small numbers of measurements in an analysis, the precision is necessarily represented by s.

Example 3.7

Calculate the mean and the standard deviation of the following set of analytical results: 15.67, 15.69, and 16.03 g.

Solution

x_i	$x_i - \bar{x}$	$(x_i - \bar{x})^2$
15.67	0.13	0.0169
15.69	0.11	0.0121
16.03	0.23	0.0529
$\sum$ 47.39	$\sum$ 0.47	$\sum$ 0.0819

$$\bar{x} = \frac{\sum x_i}{N} = \frac{47.39}{3} = 15.80$$

$$s = \sqrt{\frac{0.0819}{3 - 1}} = 0.20 \text{ g}$$

The standard deviation may be calculated also using the following equivalent equation:

$$s = \sqrt{\frac{\sum x_i^2 - (\sum x_i)^2/N}{N - 1}}$$ (3.3)

This is useful for computations with a calculator. Many calculators, in fact, have a standard deviation program that automatically calculates the standard deviation from entered individual data.

Example 3.8

Calculate the standard deviation for the data in Example 3.7 using Equation 3.3.

Solution

x_i	x_i^2
15.67	245.55
15.69	246.18
16.03	256.96
$\sum$ 47.39	$\sum$ 748.69

$$s = \sqrt{\frac{748.69 - (47.39)^2/3}{3 - 1}} = 0.21 \text{ g}$$

The difference of 0.01 g from Example 3.7 is not statistically significant since the variation is at least ± 0.2 g. In applying this formula, it is important to keep an extra digit or even two in x_i^2 for the calculation.

The precision improves as the square root of the number of measurements.

The standard deviation calculation considered so far is an estimate of the probable error of a single measurement. The arithmetical mean of a series of N measurements taken from an infinite population will show less scatter from the "true value" than will an individual observation. The scatter will decrease as N is increased; as N gets very large the sample average will approach the population average μ, and the scatter approaches zero. The arithmetical mean derived from N measurements can be shown to be $\sqrt{N}$ times more reliable than a single measurement. Hence, the random error in the mean of a series of four observations is one-half that of a single observation. In other words, the **precision of the mean** of a series of N measurements is inversely proportional to the square root of N of the deviation of the individual values. Thus,

$$\boxed{\text{Standard deviation of the mean} = s_{\text{mean}} = \frac{s}{\sqrt{N}}} \qquad (3.4)$$

The standard deviation of the mean is sometimes referred to as the *standard error*.

The standard deviation is sometimes expressed as the **relative standard deviation** (rsd), which is just the standard deviation expressed as a fraction of the mean; usually it is given as the *percentage* of the mean (% rsd), which is often called the **coefficient of variation**.

Example 3.9

The following replicate weighings were obtained: 29.8, 30.2, 28.6, and 29.7 mg. Calculate the standard deviation of the individual values and the standard deviation of the mean. Express these as absolute (units of the measurement) and relative (% of the measurement) values.

Solution

x_i	$x_i - \bar{x}$	$(x_i - \bar{x})^2$
29.8	0.2	0.04
30.2	0.6	0.36
28.6	1.0	1.00
29.7	0.1	0.01
$\sum$ 118.3	$\sum$ 1.9	$\sum$ 1.41

$$\bar{x} = \frac{118.3}{4} = 29.6$$

$$s = \sqrt{\frac{1.41}{4 - 1}} = 0.69 \text{ mg (absolute)}; \quad \frac{0.69}{29.6} \times 100\% = 2.3\% \text{ (coefficient of variation)}$$

$$s_{\text{mean}} = \frac{0.69}{\sqrt{4}} = 0.34 \text{ mg (absolute)}; \quad \frac{0.34}{29.6} \times 100\% = 1.1\% \text{ (relative)}$$

• •

The precision of a measurement can be improved by increasing the number of observations. In other words, the spread $\pm s$ of the normal curve in Figure 3.2 becomes smaller as the number of observations is increased and would approach zero as the number of observations approached infinity. However, as seen above (Equation 3.4), the deviation of the mean does not decrease in direct proportion to the number of observations, but instead it decreases as the square root of the number of observations. A point will be reached where a slight increase in precision will require an unjustifiably large increase in the number of observations. For example, to decrease the standard deviation by a factor of 10 requires 100 times as many observations.

The practical limit of useful replication is reached when the standard deviation of the random errors is comparable to the magnitude of the determinate or systematic errors (unless, of course, these can be identified and corrected for). This is because the systematic errors in a determination cannot be removed by replication.

"Randomness is required to make statistical calculations come out right."—Anonymous

The significance of s in relation to the normal distribution curve is shown in Figure 3.2. The mathematical treatment from which the curve was derived reveals that 68% of the individual deviations fall within one standard deviation (for an infinite population) from the mean, 95% are less than twice the standard deviation, and 99% are less than 2.5 times the standard deviation. So, a good approximation is that 68% of the individual values will fall within the range $\bar{x} \pm s$, 95% will fall within $\bar{x} \pm 2s$, 99% will fall within $\bar{x} \pm 2.5s$, and so on.

Actually, these percentage ranges were derived assuming an infinite number of measurements. There are then two reasons why the analyst cannot be 95% certain that the true value falls within $\bar{x} \pm 2s$. First, one makes a limited number of measurements, and the fewer the measurements, the less certain one will be. Second, the normal distribution curve assumes no determinate errors, but only random errors. Determinate errors, in effect, shift the normal error curve from the true value. An estimate of the actual certainty a number falls within s can be obtained from a calculation of the *confidence limit* (see below).

The true value will fall within $\bar{x} \pm 2s$ 95% of the time for an infinite number of measurements. See the confidence limit and Example 3.15.

It is apparent that there are a variety of ways in which the precision of a number can be reported. Whenever a number is reported as $\bar{x} \pm x$, you should always qualify under what conditions this holds, that is, how you arrived at $\pm x$. It may, for example, represent s, $2s$, s (mean), or the coefficient of variation.

The variance equals s^2.

A term that is sometimes useful in statistics is the **variance**. This is the square of the standard deviation, s^2. We shall use this in determining the propagation of error and in the F test below (Section 3.13).

3.8 Use of Spreadsheets in Analytical Chemistry

A spreadsheet is a powerful software program that can be used for a variety of functions, such as data analysis and plotting. Spreadsheets are useful for organizing data, doing repetitive calculations, and displaying the calculations graphically or in chart form. They have built-in functions, for example, standard deviation and other statistical functions, for carrying out computations on data that are input by the user. Popular spreadsheet programs include Microsoft Excel, Lotus 1-2-3, and Quattro Pro. All operate basically the same but differ somewhat in specific commands and syntax. Because of its widespread availability and popularity, we will use Excel in our illustrations.

You probably have used a spreadsheet program before and are familiar with the basic functions. But we will summarize here the most useful aspects for analytical chemistry applications. You should refer to the spreadsheet manual for more detailed information. Also, the Exel Help on the tool bar provides specific information.

You are referred to the excellent tutorial on using the Excel spreadsheet prepared by faculty at California State University at Stanislaus: http://science.csustan. edu/tutorial/Excel/index.htm. The basic functions in the spreadsheet are described, including entering data and formulas, formatting cells, graphing, and regression analysis. You will find this very helpful and should definitely read it before continuing. The website www.wku.edu/~conteed/CHEM330 at Western Kentucky University gives summary instructions for graphing using either Microsoft Excel or Lotus 1-2-3. Go to the excelhandout.html and lotushandout.html links.

A spreadsheet consists of **cells** arranged in columns (labeled A, B, C, . . .) and **rows** (numbered 1, 2, 3, . . .). An individual cell is identified by its column letter and row number, for example, B3. Figure 3.3 has the identifiers typed into some of the cells to illustrate. When the mouse pointer (the cross) is clicked on an individual cell, it becomes the **active cell** (dark lines around it), and the active cell is indicated at the top left of the formula bar, and the contents of the cell are listed to the right of the equal sign on the bar.

FILLING THE CELL CONTENTS

You may enter *text*, *numbers*, or *formulas* in specific cells. Formulas are the key to the utility of spreadsheets, allowing the same calculation to be applied to many numbers. We will illustrate with calculations of the weights of water delivered by two different 20-mL pipets, from the difference in the weights of a flask plus water and the empty flask. Refer to Figure 3.4 as you go through the steps.

	A	B	C	D	E
1	A1	B1	C1	etc.	
2	A2	B2	C2	etc.	
3	A3	B3	C3	etc.	
4					
5					

Fig. 3.3. Spreadsheet cells.

	A	B	C	D
1	Net weights			
2				
3	Pipet	1	2	
4	Weight of flask + water, g	47.702	49.239	
5	Weight of flask, g	27.687	29.199	
6	Weight of water, g	20.015	20.040	
7				
8	cell B6=B4-B5			
9	cell C6=C4-C5			
10				

Fig. 3.4. Filling cell contents.

Open an Excel spreadsheet by clicking on the Excel icon (or the Microsoft Excel program under Start: Programs). You will enter text, numbers, and formulas. Double click on the specific cell to activate it. Enter as follows (information typed into a cell is entered by depressing the Enter key):

Cell A1: Net weights

Cell A3: Pipet

Cell A4: Weight of flask + water, g

Cell A5: Weight of flask, g

Cell A6: Weight of water, g

You may make corrections by double clicking on a cell; then edit the text. (You can also edit the text in the formula bar.) If you single click, new text replaces the old text. You will have to widen the A cells to accommodate the lengthy text. Do so by placing the mouse pointer on the line between A and B on the row at the top, and dragging it to the right till all the text shows. This moves the other cells to the right.

Cell B3: 1

Cell C3: 2

Cell B4: 47.702

Cell C4: 49.239

Cell B5: 27.687

Cell C5: 29.199

Cell B6: =B4-B5

You can also enter the formula by typing =, then click on B4, then type −, and click on B5. You need to format the cells B4 to C6 to three decimal places. Highlight that block of cells by clicking on one corner and dragging to the opposite corner of the block. In the Menu bar, click on Format:Cells:Number. For Decimal places: type 3, and click OK.

You need to add the formula to cell C6. You can retype it. But there is an easier way, by copying (filling) the formula in cell B6. Place the mouse pointer on the lower right corner of cell B6 and drag it to cell C6. This fills the formula into C6 (or additional cells to the right if there are more pipet columns). You may also fill formulas into highlighted cells by clicking on Edit:Fill:Down(or Right).

Double click on B6. This shows the formula in the cell and outlines the other cells contained in the formula. Do the same for C6. Note that when you activate the cell by either single or double clicking on it, the formula is shown in the formula bar.

SAVING THE SPREADSHEET

Save the spreadsheet you have just created by clicking on File:SaveAs. I like to save documents to the desktop first. Then they can be dragged to whatever file you wish, for example, My Documents. That way they don't get lost. So select Desktop at the top. Give the document a File Name at the bottom, for example, Pipet Calibration. Then click Save. If you wish to place the saved document on a disk, you can drag it from the desktop to the opened disk.

PRINTING THE SPREADSHEET

Click File:Page Setup. Normally, a sheet is printed in the Portrait format, that is, vertically on the 8½- × 11-inch paper. If there are many columns, you may wish to print in Landscape, that is, horizontally. If you want gridlines to print, click on Sheet:Gridlines. Now you are ready to print. Click on Print:OK. Just the working area of the spreadsheet will print, not the column and row identifiers.

RELATIVE VS. ABSOLUTE CELL REFERENCES

In the example above, we used *relative* cell references in copying the formula. The formula in cell B6 said subtract the cell above from the one above it. The copied formula in C6 said the same for the cells above it.

Sometimes we need to include a specific cell in each calculation, containing say, a constant. To do this, we need to identify it in the formula as an *absolute* reference. This is accomplished by placing a $ sign in front of the column and row cell identifiers, for example, $B$2. Placing the sign in front of both assures that whether we move across columns or rows, it will remain an absolute reference.

We can illustrate this by creating a spreadsheet to calculate the means of different series of numbers. Fill in the spreadsheet as follows (refer to Figure 3.5):

A1: Titration means

A3: Titn. No.

B3: Series A, mL

C3: Series B, mL

B4: 39.27

B5: 39.18

B6: 39.30

B7: 39.20

C4: 45.59

C5: 45.55

C6: 45.63

C7: 45.66

A4: 1

We can type in each of the titration numbers (1 through 4), but there are automatic ways of incrementing a string of numbers. Click on Edit:Fill:Series. Check Columns and Linear, and leave Step Value at 1. For Stop Value, enter 4 and click OK. The numbers 2 through 4 are inserted in the spreadsheet. You could also first highlight the cells you want filled (beginning with cell A4). Then you do not have to insert a Stop Value. Another way of incrementing a series is to do it by formula. In cell A5, type =A4+1. Then you can fill down by highlighting from A5 down, and clicking on Edit:Fill:Down. (This is a relative reference.) Or, you can highlight cell A5, click

	A	B	C	D
1	Titration means			
2				
3	Titn. No.	Series A, mL	Series B, mL	
4	1	39.27	45.59	
5	2	39.18	45.55	
6	3	39.30	45.63	
7	4	39.22	45.66	
8	Mean:	**39.24**	45.61	
9	Std.Dev.	**0.053150729**	0.047871355	
10				
11	**Cell B8=**	SUM(B4:B7)/$A$7 Copy right to Cell C8.		
12	**Cell B9=**	STDEV(B4:B7) Copy right to Cell C9.		
13	We have boldfaced the cells with formulas entered.			

Fig. 3.5. Relative and absolute cell references.

on its lower right corner, and drag it to cell A7. This automatically copies the formula in the other cells.

Now we wish to insert a formula in cell B8 to calculate the mean. This will be the sum divided by the number of titrations (cell A7).

$$B8: \ =sum(B4:B7)/\$A\$7$$

We place the $ signs in the devisor because it will be an *absolute reference* that we wish to copy to the right in cell C8. Placing a $ before both the column and row addresses assures that the cell will be treated as absolute whether it is copied horizontally or vertically. The sum(B4:B7) is a **syntax** in the program for summing a series of numbers, from cell B4 through cell B7. Instead of typing in the cell addresses, you can also type "=(", then click on cell B4 and drag cell B7, and type ")". We have now calculated the mean for series A. We wish to do the same for series B. Highlight cell B8, click on its lower right corner, and drag it to cell C8. *Voilà*, the next mean is calculated! Double click on cell C8, and you will see that the formula has the same divisor (*absolute reference*), but the sum is a *relative reference*. If we had not typed in the $ signs to make the divisor absolute, the formula would have assumed it was relative, and the divisor in cell C8 would be cell B7.

USE OF EXCEL STATISTICAL FUNCTIONS

Excel has a large number of mathematical and statistical functions that can be used for calculations in lieu of writing your own formulas. Let's try the statistical functions to automatically calculate the mean. Highlight an empty cell and click f_x on the tool bar. The Paste Function window appears. Select Statistical in the Function category. The following window appears:

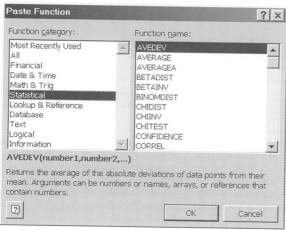

Select AVERAGE for the Function name. Click OK, and for Number, type B4:B7, and click OK. The same average is calculated as you obtained with your own formula. You can also type in the activated cell the syntax =average(B4:B7). Try it.

Let's calculate the standard deviation of the results. Highlight cell B9. Under the Statistical function, select STDEV for the Function name. Alternatively, you can type the syntax into cell B9, =stdev(B4:B7). Now copy the formula to cell C9. Perform the standard deviation calculation using Equation 3.2 and compare with the Exel values. The calculation for series A is ±0.05 mL. The value in the spreadsheet, of course, should be rounded to ±0.05 mL.

USEFUL SYNTAXES

Excel has numerous mathematical and statistical functions or syntaxes that can be used to simplify setting up calculations. Peruse the Function names for the Math & Trig and the Statistical function categories under f_x in the toolbar. Some you will find useful for this text are:

Math and trig functions

LOG10	Calculates the base-10 logarithm of a number
PRODUCT	Calculates the products of a series of numbers
POWER	Calculates the result of a number raised to a power
SQRT	Calculates the square root of a number

Statistical functions

AVERAGE	Calculates the mean of a series of numbers
MEDIAN	Calculates the median of a series of numbers
STDEV	Calculates the standard deviation of a series of numbers
TTEST	Calculates the probability associated with Student's *t* test
VAR	Calculates the variance of a series of numbers

The syntaxes may be typed, followed by the range of cells in parentheses, as we did above.

This tutorial should provide you the basics for other spreadsheet applications. You can write any formula that is in this book into an active cell, and insert appropriate data for calculations. And, obviously, we can perform a variety of data analyses. We can prepare plots and charts of the data, for example, a calibration curve of instrument response versus concentration, along with statistical information. We will illustrate this later in the chapter.

3.9 Propagation of Errors—Not Just Additive

When discussing significant figures earlier, we stated that the relative uncertainty in the answer to a multiplication or division operation could be no better than the relative uncertainty in the operator that had the poorest relative uncertainty. Also, the absolute uncertainty in the answer of an addition or subtraction could be no better than the absolute uncertainty in the number with the largest absolute uncertainty. Without specific knowledge of the uncertainties, we assumed an uncertainty of at least ±1 in the last digit of each number.

From a knowledge of the uncertainties in each number, it is possible to estimate the actual uncertainty in the answer. The errors in the individual numbers will propagate throughout a series of calculations, in either a relative or an absolute fashion, depending on whether the operation is a multiplication or division or whether it is an addition or a subtraction.

ADDITION AND SUBTRACTION—THINK ABSOLUTE VARIANCES

Consider the addition and subtraction of the following numbers:

$$(65.06 \pm 0.07) + (16.13 \pm 0.01) - (22.68 \pm 0.02) = 58.51 \ (\pm?)$$

The absolute variances of additions and subtractions are additive.

The uncertainties listed represent the random or indeterminate errors associated with each number, expressed as standard deviations of the numbers. The maximum error of the summation, expressed as a standard deviation, would be ± 0.10; that is, it could be either $+0.10$ or -0.10 if all uncertainties happened to have the same sign. The minimum uncertainty would be 0.00 if all combined by chance to cancel. Both of these extremes are not highly likely, and statistically the uncertainty will fall somewhere in between. For addition and subtraction, *absolute uncertainties* are additive. The most probable error is represented by the square root of the sum of the *absolute variances*. That is, the absolute variance of the answer is the sum of the individual variances. For $a = b + c - d$,

$$s_a^2 = s_b^2 + s_c^2 + s_d^2 \tag{3.5}$$

$$s_a = \sqrt{s_b^2 + s_c^2 + s_d^2} \tag{3.6}$$

In the above example,

$$s_a = \sqrt{(\pm 0.07)^2 + (\pm 0.01)^2 + (\pm 0.02)^2}$$
$$= \sqrt{(\pm 49 \times 10^{-4}) + (\pm 1 \times 10^{-4}) + (\pm 4 \times 10^{-4})}$$
$$= \sqrt{\pm 54 \times 10^{-4}} = \pm 7.3 \times 10^{-2}$$

So the answer is 58.51 ± 0.07. The number ± 0.07 represents the absolute uncertainty. If we wish to express it as a relative uncertainty, this would be

$$\frac{\pm 0.07}{58.51} \times 100\% = \pm 0.1_2\%$$

Example 3.10

You have received three shipments of uranium ore of equal weight. Analysis of the three ores indicated contents of $3.978 \pm 0.004\%$, $2.536 \pm 0.003\%$, and $3.680 \pm 0.003\%$, respectively. What is the average uranium content of the ores and what are the absolute and relative uncertainties?

Solution

$$\bar{x} = \frac{(3.978 \pm 0.004\%) + (2.536 \pm 0.003\%) + (3.680 \pm 0.003\%)}{3}$$

The uncertainty in the summation is

$$
\begin{aligned}
s_a &= \sqrt{(\pm 0.004)^2 + (\pm 0.003)^2 + (\pm 0.003)^2} \\
&= \sqrt{(\pm 16 \times 10^{-6}) + (\pm 9 \times 10^{-6}) + (\pm 9 \times 10^{-6})} \\
&= \sqrt{\pm 34 \times 10^{-6}} = \pm 5.8 \times 10^{-3}\% \text{ U}
\end{aligned}
$$

Hence, the absolute uncertainty is

$$
\bar{x} = \frac{10.194}{3} \pm 0.006\% = 3.398 \pm 0.006\% \text{ U}
$$

Note that since there is no uncertainty in the divisor 3, the *relative* uncertainty in the uranium content is

$$
\frac{5.8 \times 10^{-3}\% \text{ U}}{3.398\% \text{ U}} = 2 \times 10^{-3} \quad \text{or} \quad 0.2\%
$$

• •

MULTIPLICATION AND DIVISION—THINK RELATIVE VARIANCES

Consider the following operation:

The relative variances of multiplication and division are additive.

$$
\frac{(13.67 \pm 0.02)(120.4 \pm 0.2)}{4.623 \pm 0.006} = 356.0 \ (\pm ?)
$$

Here, the *relative uncertainties* are additive, and the most probable error is represented by the square root of the sum of the relative variances. That is, the relative variance of the answer is the sum of the individual relative variances.

For $a = bc/d$,

$$
(s_a^2)_{\text{rel}} = (s_b^2)_{\text{rel}} + (s_c^2)_{\text{rel}} + (s_d^2)_{\text{rel}} \tag{3.7}
$$

$$
(s_a)_{\text{rel}} = \sqrt{(s_b^2)_{\text{rel}} + (s_c^2)_{\text{rel}} + (s_d^2)_{\text{rel}}} \tag{3.8}
$$

In the above example,

$$
(s_b)_{\text{rel}} = \frac{\pm 0.02}{13.67} = \pm 0.0015
$$

$$
(s_c)_{\text{rel}} = \frac{\pm 0.2}{120.4} = \pm 0.0017
$$

$$
(s_d)_{\text{rel}} = \frac{\pm 0.006}{4.623} = \pm 0.0013
$$

$$
\begin{aligned}
(s_a)_{\text{rel}} &= \sqrt{(\pm 0.0015)^2 + (\pm 0.0017)^2 + (\pm 0.0013)^2} \\
&= \sqrt{\pm 2.2 \times 10^{-6}) + (\pm 2.9 \times 10^{-6}) + (\pm 1.7 \times 10^{-6})} \\
&= \sqrt{(\pm 6.8 \times 10^{-6})} = \pm 2.6 \times 10^{-3}
\end{aligned}
$$

The absolute uncertainty is given by

$$
\begin{aligned}
s_a &= a \times (s_a)_{\text{rel}} \\
&= 356.0 \times (\pm 2.6 \times 10^{-3}) = \pm 0.93
\end{aligned}
$$

So the answer is 356.0 ± 0.9.

Example 3.11

Calculate the uncertainty in the number of millimoles of chloride contained in 250.0 mL of a sample when three equal aliquots of 25.00 mL are titrated with silver nitrate with the following results: 36.78, 36.82, and 36.75 mL. The molarity of the $AgNO_3$ solution is 0.1167 ± 0.0002 M.

Solution

The mean volume is

$$\frac{36.78 + 36.82 + 36.75}{3} = 36.78 \text{ mL}$$

The standard deviation is

x_i	$x_i - \bar{x}$	$(x_i - \bar{x})^2$
36.78	0.00	0.0000
36.82	0.04	0.0016
36.75	0.03	0.0009
	Σ	0.0025

$$s = \sqrt{\frac{0.0025}{3-1}} = 0.035 \qquad \text{Mean volume} = 36.78 \pm 0.04 \text{ mL}$$

$$\text{mmol Cl}^- \text{ titrated} = (0.1167 \pm 0.0002 \text{ mmol/mL})(36.78 \pm 0.04 \text{ mL}) = 4.292\ (\pm?)$$

$$(s_b)_{\text{rel}} = \frac{\pm 0.0002}{0.1167} = \pm 0.0017$$

$$(s_c)_{\text{rel}} = \frac{\pm 0.035}{36.78} = \pm 0.00095$$

$$(s_a)_{\text{rel}} = \sqrt{(\pm 0.0017)^2 + (\pm 0.00095)^2}$$
$$= \sqrt{(\pm 2.9 \times 10^{-6}) + (\pm 0.90 \times 10^{-6})}$$
$$= \sqrt{+3.8 \times 10^{-6}}$$
$$= \pm 1.9 \times 10^{-3}$$

The absolute uncertainty in the millimoles of Cl^- is

$$4.292 \times (\pm 0.0019) = \pm 0.0082 \text{ mmol}$$
$$\text{mmol Cl}^- \text{ in 25 mL} = 4.292 \pm 0.0082 \text{ mmol}$$
$$\text{mmol Cl}^- \text{ in 250 mL} = 10(4.292 \pm 0.0082) = 42.92 \pm 0.08 \text{ mmol}$$

Note that we retained one extra figure in computations until the final answer. Here, the absolute uncertainty determined is proportional to the size of the sample; it would not remain constant for twice the sample size, for example.

If there is a combination of multiplication/division and addition/subtraction in a calculation, the uncertainties of these must be combined.

Example 3.12

You have received three shipments of iron ore of the following weights: 2852, 1578, and 1877 lb. There is an uncertainty in the weights of ± 5 lb. Analysis of the ores gives $36.28 \pm 0.04\%$, $22.68 \pm 0.03\%$, and $49.23 \pm 0.06\%$, respectively. You are to pay \$300 per ton of iron. What should you pay for these three shipments and what is the uncertainty in the payment?

Solution

We need to calculate the weight of iron in each shipment, with the uncertainties, and then add these together to obtain the total weight of iron and the uncertainty in this. The relative uncertainties in the weights are

$$\frac{\pm 5}{2852} = \pm 0.0017 \qquad \frac{\pm 5}{1578} = \pm 0.0032 \qquad \frac{\pm 5}{1877} = \pm 0.0027$$

The relative uncertainties in the analyses are

$$\frac{\pm 0.04}{36.28} = \pm 0.0011 \qquad \frac{\pm 0.03}{22.68} = \pm 0.0013 \qquad \frac{\pm 0.06}{49.23} = \pm 0.0012$$

The weights of iron in the shipments are

$$\frac{(2852 \pm 5 \text{ lb})(36.28 \pm 0.04\%)}{100} = 1034.7(\pm ?) \text{ lb Fe}$$

$$(s_a)_{\text{rel}} = \sqrt{(\pm 0.0017)^2 + (\pm 0.0011)^2} = \pm 0.0020$$

$$s_a = 1034.7 \times (\pm 0.0020) = \pm 2.1 \text{ lb}$$

$$\text{lb Fe} = 1034.7 \pm 2.1$$

(We will carry an additional figure throughout.)

$$\frac{(1578 \pm 5 \text{ lb})(22.68 \pm 0.03\%)}{100} = 357.89 \ (\pm ?) \text{ lb Fe}$$

$$(s_a)_{\text{rel}} = \sqrt{(\pm 0.0032)^2 + (\pm 0.0013)^2} = \pm 0.0034$$

$$s_a = 357.89 \times (\pm 0.0034) = \pm 1.2 \text{ lb}$$

$$\text{lb Fe} = 357.9 \pm 1.2 \text{ lb}$$

$$\frac{(1877 \pm 5 \text{ lb})(49.23 \pm 0.06\%)}{100} = 924.05 \ (\pm ?) \text{ lb Fe}$$

$$(s_a)_{\text{rel}} = \sqrt{(\pm 0.0027)^2 + (\pm 0.0012)^2} = \pm 0.0030$$

$$s_a = 924.05 \times (\pm 0.0030) = \pm 2.8 \text{ lb}$$

$$\text{lb Fe} = 924.0 \pm 2.8 \text{ lb}$$

$$\text{Total Fe} = (1034.7 \pm 2.1 \text{ lb}) + (357.9 \pm 1.2 \text{ lb}) + (924.0 \pm 2.8 \text{ lb})$$

$$= 2316.6 \ (\pm ?) \text{ lb}$$

$$s_a = \sqrt{(\pm 2.1)^2 + (\pm 1.2)^2 + (\pm 2.8)^2} = \pm 3.7 \text{ lb}$$

$$\text{Total Fe} = 2317 \pm 4 \text{ lb}$$

$$\text{Price} = (2316.6 \pm 3.7 \text{ lb})(\$0.15/\text{lb}) = \$347.49 \pm 0.56$$

Hence, you should pay \$347.50 $\pm$ 0.60.

Example 3.13

You determine the acetic acid content of vinegar by titrating with a standard (known concentration) solution of sodium hydroxide to a phenolphthalein end point. An approximately 5-mL sample of vinegar is weighed on an analytical balance in a weighing bottle (the increase in weight represents the weight of the sample) and is found to be 5.0268 g. The uncertainty in making a single weighing is ±0.2 mg. The sodium hydroxide must be accurately standardized (its concentration determined) by titrating known weights of high-purity potassium acid phthalate, and three such titrations give molarities of 0.1167, 0.1163, and 0.1164 M. A volume of 36.78 mL of sodium hydroxide is used to titrate the sample. The uncertainty in reading the buret is ±0.02 mL. What is the percent acetic acid in the vinegar, and what is its uncertainty?

Solution

Two weighings are required to obtain the weight of the sample: that of the empty weighing bottle and that of the bottle plus sample. Each has an uncertainty of ±0.2 mg, and so the uncertainty of the net sample weight (the difference of the two weights) is

$$s_{wt} = \sqrt{(\pm 0.2)^2 + (\pm 0.2)^2} = \pm 0.3 \text{ mg}$$

The mean of the molarity of the sodium hydroxide is 0.1165 M, and its standard deviation is ±0.0002 M. Similarly, two buret readings (initial and final) are required to obtain the volume of base delivered, and the total uncertainty is

	A	B
1	N1	0.1167
2	N2	0.1163
3	N3	0.1164
4	STDEV:	0.000208
5	Cell B4:	
6	STDEV(B1:B3)	

$$s_{vol} = \sqrt{(\pm 0.02)^2 + (\pm 0.02)^2} = \pm 0.03 \text{ mL}$$

The moles of acetic acid are equal to the moles of sodium hydroxide used to titrate it, so the percent of acetic acid is

%HOAc

$$= \frac{(0.1165 \pm 0.0002)\text{mmol mL}^{-1}(36.78 \pm 0.03)\text{mL} \times 60.05(\text{mg mmol}^{-1} \text{ acetic acid})}{(5026.8 \pm 0.3) \text{ mg}}$$

$\times 100\%$

$= 5.119 \pm ?\%$

The uncertainty in the formula weight of acetic acid is assumed to be negligible (we could actually calculate it to six figures to be exact).

$$(s_M)_{rel} = \frac{\pm 0.0002}{0.1165} = \pm 0.0017$$

$$(s_{vol})_{rel} = \frac{\pm 0.03}{36.78} = \pm 0.00082$$

$$(s_{wt})_{rel} = \frac{\pm 0.3}{5026.8} = \pm 0.000060$$

The uncertainty in the analysis is

$$(s_{total})_{rel} = \sqrt{(\pm 0.0017)^2 + (\pm 0.00082)^2 + (\pm 0.00060)^2} = \pm 0.0020$$
$$s_{total} = 5.119 \times 0.0020 = \pm 0.010\% \text{ acetic acid}$$

Hence, the acetic acid content is $5.119 \pm 0.010\%$. The relative uncertainty is 0.4 ppt.

The factor that limited the uncertainty the most was the variance in the molarity of the sodium hydroxide solution. This illustrates the importance of careful calibration, which is discussed in Chapter 2.

• •

3.10 Significant Figures and Propagation of Error

We noted earlier that the total uncertainty in a computation determines how accurately we can know the answer. In other words, the uncertainty sets the number of significant figures. Take the following example:

$$(73.1 \pm 0.2)(2.245 \pm 0.008) = 164.1 \pm 0.7$$

The number of significant figures in an answer is determined by the uncertainty due to propagation of error.

We are justified in keeping four figures, even though the key number has three. Here, we don't have to carry the additional figure as a subscript since we have indicated the actual uncertainty in it. Note that the greatest relative uncertainty in the multipliers is 0.0036, while that in the answer is 0.0043; so, due to the propagation of error, we know the answer somewhat less accurately than the key number. The key number (the one with the greatest uncertainty), when actual uncertainties are known, may not necessarily be the one with the smallest number of digits. For example, the relative uncertainty in 78.1 ± 0.2 is 0.003, while that in 11.21 ± 0.08 is 0.007.

Suppose we have the following calculation:

$$(73.1 \pm 0.9)(2.245 \pm 0.008) = 164.1 \pm 2.1 = 164 \pm 2$$

Now the uncertainty in the answer is the units place, and so figures beyond that are meaningless. In this instance, the uncertainty in the key number and the answer are similar (± 0.012) since the uncertainty in the other multiplier is significantly smaller.

 •

Example 3.14

Provide the answers to the following calculations to the proper number of significant figures:

(a) $$(38.68 \pm 0.07) - (6.16 \pm 0.09) = 32.52$$

(b) $$\frac{(12.18 \pm 0.08)(23.04 \pm 0.07)}{3.247 \pm 0.006} = 86.43$$

Solution

(a) The calculated absolute uncertainty in the answer is ± 0.11. Therefore, the answer is 32.5 ± 0.1.

(b) The calculated relative uncertainty in the answer is 0.0075, so the absolute uncertainty is $0.0075 \times 86.43 = 0.65$. Therefore, the answer is 86.4 ± 0.6, even though we know all the other numbers to four figures; there is substantial uncertainty in the fourth digit, which leads to the uncertainty in the answer. The relative uncertainty in that answer is 0.0075, and the largest relative uncertainty in the other numbers is 0.0066, very similar.

• •

3.11 Control Charts

A **quality control chart** is a time plot of a measured quantity that is assumed to be constant (with a Gaussian distribution) for the purpose of ascertaining that the measurement remains within a statistically acceptable range. It may be a day-to-day plot of the measured value of a standard that is run intermittently with samples. The control chart consists of a central line representing the known or assumed value of the control and either one or two pairs of limit lines, the **inner** and **outer control limits**. Usually the standard deviation of the procedure is known (a good estimate of σ), and this is used to establish the control limits.

An example of a control chart is illustrated in Figure 3.6, representing a plot of day-to-day results of the analysis of a pooled serum calcium or a control sample that is run randomly and blindly with samples each day. A useful inner control limit is two standard deviations since there is only 1 chance in 20 that an individual measurement will exceed this purely by chance. This might represent a warning limit. The outer limit might be 2.5 or 3σ, in which case there is only 1 chance in 100 or 1 chance in 500 a measurement will fall outside this range in the absence of systematic error. Usually, one control is run with each batch of samples (e.g., 20 samples), so several control points may be obtained each day. The mean of these may be plotted each day. The random scatter of this would be expected to be smaller by $\sqrt{N}$, compared to individual points.

Particular attention should be paid to trends in one direction; that is, the points lie largely on one side of the central line. This would suggest that either the control is in error or there is a systematic error in the measurement. A tendency for points to lie outside the control limits would indicate the presence of one or more determinate errors in the determination, and the analyst should check for deterioration of reagents, instrument malfunction, or environmental and other effects.

> A control chart is constructed by periodically running a "known" control sample.

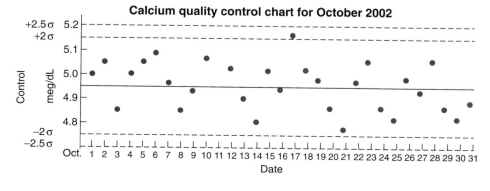

Fig. 3.6.

Typical quality control chart.

Trends should signal contamination of reagents, improper calibration or erroneous standards, or change in the control lot.

3.12 The Confidence Limit—How Sure Are You?

The true value falls within the confidence limit, estimated using t at the desired confidence level.

Calculation of the standard deviation for a set of data provides an indication of the precision inherent in a particular procedure or analysis. But unless there is a large number of data, it does not by itself give any information about how close the experimentally determined mean $\bar{x}$ might be to the true mean value μ. Statistical theory, though, allows us to estimate the range within which the true value might fall, within a given probability, defined by the experimental mean and the standard deviation. This range is called the **confidence interval**, and the limits of this range are called the **confidence limit**. The likelihood that the true value falls within the range is called the **probability**, or **confidence level**, usually expressed as a percent. The confidence limit is given by

$$\text{Confidence limit} = \bar{x} \pm \frac{ts}{\sqrt{N}} \tag{3.9}$$

where t is a statistical factor that depends on the number of degrees of freedom and the confidence level desired. The number of degrees of freedom is one less than the number of measurements. Values of t at different confidence levels and degrees of freedom v are given in Table 3.1. Note that the confidence limit is simply the product of t and the standard deviation of the mean ($s/\sqrt{N}$). (**The confidence limit for a single observation, x, when N is 1 is given by $x \pm ts$, being larger than that of the mean by a factor $\sqrt{N}$; t is for the number of measurements used to determine s.**)

Table 3.1
Values of t for v Degrees of Freedom for Various Confidence Levels[a]

v	Confidence Level			
	90%	95%	99%	99.5%
1	6.314	12.706	63.657	127.32
2	2.920	4.303	9.925	14.089
3	2.353	3.182	5.841	7.453
4	2.132	2.776	4.604	5.598
5	2.015	2.571	4.032	4.773
6	1.943	2.447	3.707	4.317
7	1.895	2.365	3.500	4.029
8	1.860	2.306	3.355	3.832
9	1.833	2.262	3.250	3.690
10	1.812	2.228	3.169	3.581
15	1.753	2.131	2.947	3.252
20	1.725	2.086	2.845	3.153
25	1.708	2.060	2.787	3.078
∞	1.645	1.960	2.576	2.807

[a]$v = N - 1 = $ degrees of freedom.

Example 3.15

A soda ash sample is analyzed in the analytical chemistry laboratory by titration with standard hydrochloric acid. The analysis is performed in triplicate with the following results: 93.50, 93.58, and 93.43% Na_2CO_3. Within what range are you 95% confident that the true value lies?

Solution

The mean is 93.50%. The standard deviation s is calculated to be 0.075% Na_2CO_3 (absolute—calculate it with a spreadsheet). At the 95% confidence level and two degrees of freedom, $t = 4.303$ and

$$\text{Confidence limit} = \bar{x} \pm \frac{ts}{\sqrt{N}}$$

$$= 93.50 \pm \frac{4.303 \times 0.075}{\sqrt{3}}$$

$$= 93.50 \pm 0.19\%$$

Too high a confidence level will give a wide range that may encompass nonrandom numbers. Too low a confidence level will give a narrow range and exclude valid random numbers. Confidence levels of 90 to 95% are generally accepted as reasonable.

So you are 95% confident that, in the absence of a determinate error, the true value falls within 93.31 to 93.69%. Note that for an infinite number of measurements, we would have predicted with 95% confidence that the true value falls within two standard deviations (Figure 3.2); we see that for $v = \infty$, t is actually 1.96 (Table 3.1), and so the confidence limit would indeed be about twice the standard deviation of the mean (which approaches σ for large N).

Compare with Figure 3.2 where 95% of the values fall within 2s.

Remember from Section 3.7 and Figure 3.2 that we are 68% confident that the true value falls within $\pm 1\sigma$, 95% confident it will fall within $\pm 2\sigma$, and 99% confident it will fall within $\pm 2.5\sigma$. Note that it is possible to estimate a standard deviation from a stated confidence interval, and vice versa a confidence interval from a standard deviation. If a mean value is 27.37 ± 0.06 g at the 95% confidence interval, then since this is two standard deviations for a suitably large number of measurements, the standard deviation is 0.03 g. If we know the standard deviation is 0.03 g, then this is the confidence interval at the 68% confidence level, or it is 0.06 g at the 95% confidence level. For small numbers of measurements, t will be larger, which proportionately changes these numbers.

As the number of measurements increases, both t and $s/\sqrt{N}$ decrease, with the result that the confidence interval is narrowed. So the more measurements you make, the more confident you will be that the true value lies within a given range or, conversely, that the range will be narrowed at a given confidence level. However, t decreases exponentially with an increase in N, just as the standard deviation of the mean does (see Table 3.1), so a point of diminishing returns is eventually reached in which the increase in confidence is not justified by the increase in the multiple of samples analyses required.

3.13 Tests of Significance—Is There a Difference?

In developing a new analytical method, it is often desirable to compare the results of that method with those of an accepted (perhaps standard) method. How, though, can one tell if there is a significant difference between the new method and the accepted one? Again, we resort to statistics for the answer.

Deciding whether one set of results is significantly different from another depends not only on the difference in the means but also on the amount of data available and the spread. There are statistical tables available that show how large a difference needs to be in order to be considered not to have occurred by chance. The F test evaluates differences between the spread of results, while the t test looks at differences between means.

THE F TEST

The F test is used to determine if two variances are statistically different.

This is a test designed to indicate whether there is a significant difference between two methods based on their standard deviations. F is defined in terms of the variances of the two methods, where the **variance** is the square of the standard deviation:

$$F = \frac{s_1^2}{s_2^2} \qquad (3.10)$$

where $s_1^2 > s_2^2$. There are two different degrees of freedom, ν_1 and ν_2, where degrees of freedom is defined as $N - 1$ for each case.

If the calculated F value from Equation 3.10 exceeds a tabulated F value at the selected confidence level, then there is a significant difference between the variances of the two methods. A list of F values at the 95% confidence level is given in Table 3.2.

Table 3.2
Values of F at the 95% Confidence Level

	$\nu_1 = 2$	3	4	5	6	7	8	9	10	15	20	30
$\nu_2 = 2$	19.0	19.2	19.2	19.3	19.3	19.4	19.4	19.4	19.4	19.4	19.4	19.5
3	9.55	9.28	9.12	9.01	8.94	8.89	8.85	8.81	8.79	8.70	8.66	8.62
4	6.94	6.59	6.39	6.26	6.16	6.09	6.04	6.00	5.96	5.86	5.80	5.75
5	5.79	5.41	5.19	5.05	4.95	4.88	4.82	4.77	4.74	4.62	4.56	4.50
6	5.14	4.76	4.53	4.39	4.28	4.21	4.15	4.10	4.06	3.94	3.87	3.81
7	4.74	4.35	4.12	3.97	3.87	3.79	3.73	3.68	3.64	3.51	3.44	3.38
8	4.46	4.07	3.84	3.69	3.58	3.50	3.44	3.39	3.35	3.22	3.15	3.08
9	4.26	3.86	3.63	3.48	3.37	3.29	3.23	3.18	3.14	3.01	2.94	2.86
10	4.10	3.71	3.48	3.33	3.22	3.14	3.07	3.02	2.98	2.85	2.77	2.70
15	3.68	3.29	3.06	2.90	2.79	2.71	2.64	2.59	2.54	2.40	2.33	2.25
20	3.49	3.10	2.87	2.71	2.60	2.51	2.45	2.39	2.35	2.20	2.12	2.04
30	3.32	2.92	2.69	2.53	2.42	2.33	2.27	2.21	2.16	2.01	1.93	1.84

Example 3.16

You are developing a new colorimetric procedure for determining the glucose content of blood serum. You have chosen the standard Folin-Wu procedure with which to compare your results. From the following two sets of replicate analyses on the same sample, determine whether the variance of your method differs significantly from that of the standard method.

Your Method (mg/dL)	Folin-Wu Method (mg/dL)
127	130
125	128
123	131
130	129
131	127
126	125
129	
mean ($\bar{x}_1$) 127	mean ($\bar{x}_2$) 128

Solution

$$s_1^2 = \frac{\sum(x_{i1} - \bar{x}_1)^2}{N_1 - 1} = \frac{50}{7 - 1} = 8.3$$

$$s_2^2 = \frac{\sum(x_{i2} - \bar{x}_2)^2}{N_2 - 1} = \frac{24}{6 - 1} = 4.8$$

$$F = \frac{8.3}{4.8} = 1.7_3$$

The variances are arranged so that the F value is >1. The tabulated F value for $v_1 = 6$ and $v_2 = 5$ is 4.95. Since the calculated value is less than this, we conclude that there is no significant difference in the precision of the two methods, that is, the standard deviations are from random error alone and don't depend on the sample.

THE STUDENT T TEST—ARE THERE DIFFERENCES IN THE METHODS?

Frequently, the analyst wishes to decide whether there is a statistical difference between the results obtained using two different procedures, that is, whether they both indeed measure the same thing. The t test is very useful for such comparisons.

    In this method, comparison is made between two sets of replicate measurements made by two different methods; one of them will be the *test method*, and the other will be an *accepted* method. A statistical t value is calculated and compared with a tabulated value for the given number of tests at the desired confidence level (Table 3.1). If the calculated t value *exceeds* the tabulated t value, then there is a *significant difference* between the results by the two methods at that confidence level. If it does not exceed the tabulated value, then we can predict that there is no

The t test is used to determine if two sets of measurements are statistically different.

significant difference between the methods. This in no way implies that the two results are identical.

Three ways in which a *t* test can be used will be described. If an accepted value of μ is available (from other measurements), then the test can be used to determine if a particular analysis method gives results statistically equal to μ at a given confidence level. If an accepted value is not available, then a series of replicate analyses on a single sample may be performed using two methods, or a series of analyses may be performed on a set of different samples by the two methods. One method should be an accepted method. We will describe these various uses of the *t* test.

1. *t* Test When an Accepted Value Is Known. Note that Equation 3.9 is a representation of the true value μ. We can write that

$$\mu = \bar{x} \pm \frac{ts}{\sqrt{N}} \tag{3.11}$$

It follows that

$$\pm t = (\bar{x} - \mu)\frac{\sqrt{N}}{s} \tag{3.12}$$

If a good estimate of the "true" value is available from other analyses, for example, from a National Institute of Standards and Technology (NIST) standard reference material (or the ultimate in chemical analysis, an atomic weight), then Equation 3.12 can be used to determine whether the value obtained from a test method is statistically equal to the accepted value.

· ·

 Example 3.17

You are developing a procedure for determining traces of copper in biological materials using a wet digestion followed by measurement by atomic absorption spectrophotometry. In order to test the validity of the method, you obtain an NIST orchard leaves standard reference material and analyze this material. Five replicas are sampled and analyzed, and the mean of the results is found to be 10.8 ppm with a standard deviation of ± 0.7 ppm. The listed value is 11.7 ppm. Does your method give a statistically correct value at the 95% confidence level?

Solution

$$\pm t = (\bar{x} - \mu)\frac{\sqrt{N}}{s}$$

$$= (10.8 - 11.7)\frac{\sqrt{5}}{0.7}$$

$$= 2_{.9}$$

There are five measurements, so there are four degrees of freedom $(N - 1)$. From Table 3.1, we see that the tabulated value of *t* at the 95% confidence level is 2.776.

This is *less* than the calculated value, so there is a determinate error in the new procedure. That is, there is a 95% probability that the difference between the reference value and the measured value is not due to chance.

• •

Note from Equation 3.12 that as the precision is improved, that is, as s becomes smaller, the calculated t becomes larger. Thus, there is a greater chance that the tabulated t value will be less than this. That is, as the precision improves, it is easier to distinguish nonrandom differences. Looking again at Equation 3.12, this means as s decreases, so must the difference between the two methods $(\bar{x} - \mu)$ in order for the difference to be ascribed only to random error. What this means is that comparing very large sets of samples, with a smaller s, will nearly always lead to a statistically significant difference, but a statistically significant result is not necessarily important because of the large number of samples that better describe the population.

2. Comparison of the Means of Two Samples. When the t test is applied to two sets of data, μ in Equation 3.12 is replaced by the mean of the second set. The reciprocal of the standard deviation of the mean $(\sqrt{N}/s)$ is replaced by that of the differences between the two, which is readily shown to be

$$\sqrt{\frac{N_1 N_2}{N_1 + N_2}} \Big/ s_p$$

where s_p is the pooled standard deviation of the individual measurements of two sets:

$$\pm t = \frac{\bar{x}_1 - \bar{x}_2}{s_p} \sqrt{\frac{N_1 N_2}{N_1 + N_2}} \qquad (3.13)$$

The pooled standard deviation, defined below, is sometimes used to obtain an improved estimate of the precision of a method, and it is used for calculating the precision of the two sets of data in a paired t test. That is, rather than relying on a single set of data to describe the precision of a method, it is sometimes preferable to perform several sets of analyses, for example, on different days, or on different samples with slightly different compositions. If the indeterminate (random) error is assumed to be the same for each set, then the data of the different sets can be pooled. This provides a more reliable estimate of the precision of a method than is obtained from a single set. The **pooled standard deviation** s_p is given by

$$s_p = \sqrt{\frac{\sum(x_{i1} - \bar{x}_1) + \sum(x_{i2} - \bar{x}_2)^2 + \cdots + \sum(x_{ik} - \bar{x}_k)^2}{N - k}} \qquad (3.14)$$

where $\bar{x}_1, \bar{x}_2, \ldots, \bar{x}_k$ are the means of each of k sets of analyses, and $x_{i1}, x_{i2}, \ldots, x_{ik}$ are the individual values in each set. N is the total number of measurements and is equal to $(N_1 + N_2 + \cdots + N_k)$. If five sets of 20 analyses each are performed, $k = 5$ and $N = 100$. (The number of samples in each set need not be equal.) $N - k$ is the degrees of freedom obtained from $(N_1 - 1) + (N_2 - 1) + \cdots + (N_k - 1)$; one degree of freedom is lost for each subset. This equation represents a combination of the equations for the standard deviations of each set of data.

The F test can be applied to the variances of the two methods rather than assuming they are statistically equal before applying the t test.

In applying the t test between two methods, it is assumed that both methods have essentially the same standard deviation, that is, each represents the precision of the population (the same σ). This can be verified using the F test above.

Example 3.18

A new gravimetric method is developed for iron(III) in which the iron is precipitated in crystalline form with an organoboron "cage" compound. The accuracy of the method is checked by analyzing the iron in an ore sample and comparing with the results using the standard precipitation with ammonia and weighing of Fe_2O_3. The results, reported as % Fe for each analysis, were as follows:

Test Method	Reference Method
20.10%	18.89%
20.50	19.20
18.65	19.00
19.25	19.70
19.40	19.40
19.99	$\bar{x}_2 = 19.24\%$
$\bar{x}_1 = 19.65\%$	

Is there a significant difference between the two methods?

Solution

x_{i1}	$x_{i1} - \bar{x}_1$	$(x_{i1} - \bar{x}_1)^2$	x_{i2}	$x_{i2} - \bar{x}_2$	$(x_{i2} - \bar{x}_2)^2$
20.10	0.45	0.202	18.89	0.35	0.122
20.50	0.85	0.722	19.20	0.04	0.002
18.65	1.00	1.000	19.00	0.24	0.058
19.25	0.40	0.160	19.70	0.46	0.212
19.40	0.25	0.062	19.40	0.16	0.026
19.99	0.34	0.116		$\sum(x_{i2} - \bar{x}_2)^2 = 0.420$	
	$\sum(x_{i1} - \bar{x}_1)^2 = 2.262$				

$$F = \frac{s_1^2}{s_2^2} = \frac{0.262/5}{0.420/4} = 4.31$$

This is less than the tabulated value (6.26), so the two methods have comparable standard deviations and the t test can be applied

$$s_p = \sqrt{\frac{\sum(x_{i1} - \bar{x}_1)^2 + \sum(x_{i2} - \bar{x}_2)^2}{N_1 + N_2 - 2}}$$

$$= \sqrt{\frac{2.262 + 0.420}{6 + 5 - 2}} = 0.546$$

$$\pm t = \frac{19.65 - 19.24}{0.546} \sqrt{\frac{(6)(5)}{6 + 5}} = 1.2_3$$

The tabulated t for nine degrees of freedom ($N_1 + N_2 - 2$) at the 95% confidence level is 2.262, so there is no statistical difference in the results by the two methods.

Rather than comparing two methods using one sample, two samples could be compared for comparability using a single analysis method in a manner identical to the above examples.

3. Paired t Test. In the clinical chemistry laboratory, a new method is frequently tested against an accepted method by analyzing several different samples of slightly varying composition (within physiological range). In this case, the t value is calculated in a slightly different form. The difference between each of the paired measurements on each sample is computed. An average difference $\overline{D}$ is calculated and the individual deviations of each from $\overline{D}$ are used to compute a standard deviation, s_d. The t value is calculated from

$$t = \frac{\overline{D}}{s_d}\sqrt{N} \qquad\qquad (3.15)$$

$$s_d = \sqrt{\frac{\Sigma(D_i - \overline{D})^2}{N - 1}} \qquad\qquad (3.16)$$

where D_i is the individual difference between the two methods for each sample, with regard to sign; and $\overline{D}$ is the mean of all the individual differences.

Example 3.19

You are developing a new analytical method for the determination of blood urea nitrogen (BUN). You want to determine whether your method differs significantly from a standard one for analyzing a range of sample concentrations expected to be found in the routine laboratory. It has been ascertained that the two methods have comparable precisions. Following are two sets of results for a number of individual samples.

Sample	Your Method (mg/dL)	Standard Method (mg/dL)	D_i	$D_i - \overline{D}$	$(D_i - \overline{D})^2$
A	10.2	10.5	−0.3	−0.6	0.36
B	12.7	11.9	0.8	0.5	0.25
C	8.6	8.7	−0.1	−0.4	0.16
D	17.5	16.9	0.6	0.3	0.09
E	11.2	10.9	0.3	0.0	0.00
F	11.5	11.1	0.4	0.1	0.01
			Σ 1.7		Σ 0.87
			$\overline{D} = 0.28$		

Solution

$$s_d = \sqrt{\frac{0.87}{6 - 1}} = 0.42$$

$$t = \frac{0.28}{0.42} \times \sqrt{6} = 1.6_3$$

The tabulated t value at the 95% confidence level for five degrees of freedom is 2.571. Therefore, $t_{calc} < t_{table}$, and there is no significant difference between the two methods at this confidence level.

Usually, a test at the 95% confidence level is considered significant, while one at the 99% level is highly significant. That is, the smaller the calculated t value, the more confident you are that there is no significant difference between the two methods. If you employ too low a confidence level (e.g., 80%), you are likely to conclude erroneously that there is a significant difference between two methods (type I error). On the other hand, too high a confidence level will require too large a difference to detect (type II error). If a calculated t value is near the tabular value at the 95% confidence level, more tests should be run to ascertain definitely whether the two methods are significantly different.

3.14 Rejection of a Result: The Q Test

Finagle's third law: In any collection of data, the figure most obviously correct, beyond all checking, is the mistake.

Frequently, when a series of replicate analyses is performed, one of the results will appear to differ markedly from the others. A decision will have to be made whether to reject the result or to retain it. Unfortunately, there are no uniform criteria that can be used to decide if a suspect result can be ascribed to accidental error rather than chance variation. It is tempting to delete extreme values from a data set because they will alter the calculated statistics in an unfavorable way, that is, increase the standard deviation and variance (measures of spread), and they may substantially alter the reported mean. The only reliable basis for rejection occurs when it can be decided that some specific error may have been made in obtaining the doubtful result. No result should be retained in cases where a known error has occurred in its collection.

Experience and common sense may serve as just as practical a basis for judging the validity of a particular observation as a statistical test would be. Frequently, the experienced analyst will gain a good idea of the precision to be expected in a particular method and will recognize when a particular result is suspect.

Additionally, an analyst who knows the standard deviation expected of a method may reject a data point that falls outside $2s$ or $2.5s$ of the mean because there is about 1 chance in 20 or 1 chance in 100 this will occur.

The Q test is used to determine if an "outlier" is due to a determinate error. If it is not, then it falls within the expected random error and should be retained.

$$Q = \frac{a}{w}$$

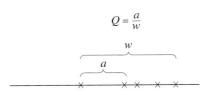

"And now the sequence of events in no particular order."—Dan Rather, television news anchor.

A wide variety of statistical tests have been suggested and used to determine whether an observation should be rejected. In all of these, a range is established within which statistically significant observations should fall. The difficulty with all of them is determining what the range should be. If it is too small, then perfectly good data will be rejected; and if it is too large, then erroneous measurements will be retained too high a proportion of the time. The **Q test** is, among the several suggested tests, one of the most statistically correct for a fairly small number of observations and is recommended when a test is necessary. The ratio Q is calculated by arranging the data in decreasing order of numbers. The difference between the suspect number and its nearest neighbor (a) is divided by the range (w), that is, the difference between the highest number and the lowest number. Referring to the figure in the margin, $Q = a/w$. This ratio is compared with tabulated values of Q. If it is equal to or greater than the tabulated value, the suspected observation can be rejected. The tabulated values of Q at the 90, 95, and 99% confidence levels are given in Table 3.3. If Q exceeds the tabulated value for a given number of observations and a given confidence level, the questionable measurement may be rejected with, for example, 95% confidence that some definite error is in this measurement.

Table 3.3

Rejection Quotient, Q, at Different Confidence Limits[a]

No. of Observations	Confidence Level		
	Q_{90}	Q_{95}	Q_{99}
3	0.941	0.970	0.994
4	0.765	0.829	0.926
5	0.642	0.710	0.821
6	0.560	0.625	0.740
7	0.507	0.568	0.680
8	0.468	0.526	0.634
9	0.437	0.493	0.598
10	0.412	0.466	0.568
15	0.338	0.384	0.475
20	0.300	0.342	0.425
25	0.277	0.317	0.393
30	0.260	0.298	0.372

[a]Adapted from D. B. Rorabacher, *Anal. Chem.*, **63** (1991) 139.

Example 3.20

The following set of chloride analyses on separate aliquots of a pooled serum were reported: 103, 106, 107, and 114 meq/L. One value appears suspect. Determine if it can be ascribed to accidental error, at the 95% confidence level.

Solution

The suspect result is 114 meq/L. It differs from its nearest neighbor, 107 meq/L, by 7 meq/L. The range is 114 to 103, or 11 meq/L. Q is therefore 7/11 = 0.64. The tabulated value for four observations is 0.829. Since the calculated Q is less than the tabulated Q, the suspected number may be ascribed to random error and should not be rejected.

For a small number of measurements (e.g., three to five), the discrepancy of the measurement must be quite large before it can be rejected by this criterion. and it is likely that erroneous results may be retained. This would cause a significant, change in the arithmetic mean because the mean is greatly influenced by a discordant value. For this reason, it has been suggested that the median rather than the mean be reported when a discordant number cannot be rejected from a small number of measurements. The **median** is the middle result of an odd number of results, or the average of the central pair for an even number, when they are arranged in order of magnitude. The median has the advantage of not being unduly influenced by an outlying value. In the above example, the median could be taken as the average of the two middle values [= (106 + 107)/2 = 106]. This compares with a mean of 108, which is influenced more by the suspected number.

The following procedure is suggested for interpretation of the data of three to five measurements if the precision is considerably poorer than expected and if one of the observations is considerably different from the others of the set.

Consider reporting the median when an outlier cannot quite be rejected.

1. Estimate the precision that can reasonably be expected for the method in deciding whether a particular number actually is questionable. Note that for three measurements with two of the points very close, the Q test is likely to fail. (See the paragraph below.)
2. Check the data leading to the suspected number to see if a definite error can be identified.
3. If new data cannot be collected, run a Q test.
4. If the Q test indicates retention of the outlying number, consider reporting the median rather than the mean for a small set of data.
5. As a last resort, run another analysis. Agreement of the new result with the apparently valid data previously collected will lend support to the opinion that the suspected result should be rejected. You should avoid, however, continually running experiments until the "right" answer is obtained.

The Q test should not be applied to three data points if two are identical. In that case, the test always indicates rejection of the third value, regardless of the magnitude of the deviation, because a is equal to w and Q_{calc} is always equal to 1. The same obviously applies for three identical data points in four measurements, and so forth.

3.15 Statistics for Small Data Sets

Large population statistics do not strictly apply for small populations.

We have discussed, in previous sections, ways of estimating, for a normally distributed population, the central value (mean, $\bar{x}$), the spread of results (standard deviation, s), and the confidence limits (t test). These statistical values hold strictly for a large population. In analytical chemistry, we typically deal with fewer than 10 results, and for a given analysis, perhaps 2 or 3. For such small sets of data, other estimates may be more appropriate.

The Q test in the previous section is designed for small data sets, and we mentioned there some rules for dealing with suspect results.

THE MEDIAN MAY BE BETTER THAN THE MEAN

The median may be a better representative of the true value than the mean, for small numbers of measurements.

The median M may be used as an estimate of the central value. It has the advantage that it is not markedly influenced by extraneous (outlier) values, as is the mean, $\bar{x}$. The efficiency of M, defined as the ratio of the variances of sampling distributions of these two estimates of the "true" mean value and denoted by E_M, is given in Table 3.4. It varies from 1 for only two observations (where the median is necessarily identical with the mean) to 0.64 for large numbers of observations. The numerical value of the efficiency implies that the median from, for example, 100 observations where the efficiency is essentially 0.64, conveys as much information about the central value of the population as does the mean calculated from 64 observations. The median of 10 observations is as efficient conveying the information as is the mean from $10 \times 0.71 = 7$ observations. It may be desirable to use the median in order to avoid deciding whether a gross error is present, that is, using the Q test. It has been shown that for three observations from a normal population, the median is better than the mean of the best two out of three (the two closest) values.

Table 3.4

Efficiencies and Conversion Factors for 2 to 10 Observations[a]

No. of Observations	Efficiency		Range Deviation Factor, K_R	Range Confidence Factor (t)	
	Of Median, E_M	Of Range, E_R		$t_{r\,0.95}$	$t_{r\,0.99}$
2	1.00	1.00	0.89	6.4	31.83
3	0.74	0.99	0.59	1.3	3.01
4	0.84	0.98	0.49	0.72	1.32
5	0.69	0.96	0.43	0.51	0.84
6	0.78	0.93	0.40	0.40	0.63
7	0.67	0.91	0.37	0.33	0.51
8	0.74	0.89	0.35	0.29	0.43
9	0.65	0.87	0.34	0.26	0.37
10	0.71	0.85	0.33	0.23	0.33
∞	0.64	0.00	0.00	0.00	0.00

[a]Adapted from R. B. Dean and W. J. Dixon, *Anal. Chem.*, **23** (1951) 636.

RANGE INSTEAD OF THE STANDARD DEVIATION

The range R for a small set of measurements, is highly efficient for describing the spread of results. The efficiency of the range, E_R, shown in Table 3.4, is virtually identical to that of the standard deviation for four or fewer measurements. This high relative efficiency arises from the fact that the standard deviation is a poor estimate of the spread for a small number of observations, although it is still the best known estimate for a given set of data. To convert the range to a measure of spread that is independent of the number of observations, we must multiply it by the **deviation factor, K**, given in Table 3.4. This factor adjusts the range R so that on average it reflects the standard deviation of the population, which we represent by s_r:

$$s_r = RK_R \tag{3.17}$$

The range is as good a measure of the spread of results as is the standard deviation for four or less measurements.

In Example 3.9 the standard deviation of the four weights is 0.69 mg. The range is 1.6 mg. Multiplying by K_R for four observations, $s_r = 1.6$ mg $\times$ 0.49 = 0.78 mg. As N increases, the efficiency of the range decreases relative to the standard deviation.

The median M may be used in computing the standard deviation, in order to minimize the influence of extraneous values. Taking Example 3.9 again, the standard deviation calculated using the median, 29.8, in place of the mean in Equation 3.2, is 0.73 mg, instead of 0.69 mg.

CONFIDENCE LIMITS USING THE RANGE

Confidence limits could be calculated using s_r obtained from the range, in place of s in Equation 3.9, and a corresponding but different t table. It is more convenient, though, to calculate the limits directly from the range as

$$\boxed{\text{Confidence limit} = \bar{x} \pm Rt_r} \tag{3.18}$$

The factor for converting R to s_r has been included in the quantity, t_r, which is tabulated in Table 3.4 for 99 and 95% confidence levels. The calculated confidence limit at the 95% confidence level in Example 3.15 using Equation 3.18 is 93.50 $\pm$ 0.19 (1.3) = 93.50 $\pm$ 0.25% Na_2CO_3.

3.16 Linear Least Squares—How to Plot the Right Straight Line

The analyst is frequently confronted with plotting data that fall on a straight line, as in an analytical calibration curve. Graphing, that is, curve fitting, is critically important in obtaining accurate analytical data. It is the calibration graph that is used to calculate the unknown concentration. Straight-line predictability and consistency will determine the accuracy of the unknown calculation. All measurements will have a degree of uncertainty, and so will the plotted straight line. Graphing is often done intuitively, that is, by simply "eyeballing" the best straight line by placing a ruler through the points, which invariably have some scatter. A better approach is to apply statistics to define the most probable straight-line fit of the data. The availability of statistical functions in spreadsheets today make it straightforward to prepare straight-line, or even nonlinear, fits. We will first learn the computations that are involved in curve fitting and statistical evaluation.

If a straight-line relationship is assumed, then the data fit the equation

$$y = mx + b \tag{3.19}$$

where y is the *dependent variable*, x is the *independent variable*, m is the *slope*, of the curve, and b is the *intercept* on the ordinate (y axis); y is usually the measured variable, plotted as a function of changing x (see Figure 3.7). In a spectrophotometric calibration curve, y would represent the measured absorbances and x would be the concentrations of the standards. Our problem, then, is to establish values for m and b.

LEAST-SQUARES PLOTS

It can be shown statistically that the best straight line through a series of experimental points is that line for which the *sum of the squares of the deviations (the residuals) of the points from the line is minimum*. This is known as the **method of least squares**. If x is the fixed variable (e.g., concentration) and y is the measured variable (absorbance in a spectrophotometric measurement, the peak area in a chromatographic measurement, etc.), then the deviation of y vertically from the line at a given value of x (x_i) is of interest. If y_1 is the value *on the line*, it is equal to $mx_i + b$. The square of the sum of the differences, S, is then

$$S = \sum(y_i - y_l)^2 = \sum[y_i - (mx_i + b)]^2 \tag{3.20}$$

This equation assumes no error in x, the independent variable.

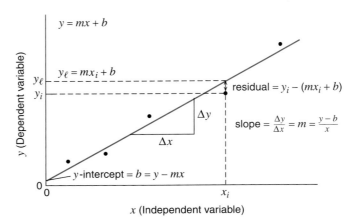

Fig. 3.7. Straight-line plot.

The best straight line occurs when S goes through a minimum. This is obtained by use of differential calculus by setting the derivatives of S with respect to m and b equal to zero and solving for m and b. The result is

The least-squares slope and intercept define the most probable straight line.

$$m = \frac{\sum(x_i - \bar{x})(y_i - \bar{y})}{\sum(x_i - \bar{x})^2} \qquad (3.21)$$

$$b = \bar{y} - m\bar{x} \qquad (3.22)$$

where $\bar{x}$ is the mean of all the values of x_i and $\bar{y}$ is the mean of all the values of y_i. The use of differences in calculations is cumbersome, and Equation 3.21 can be transformed into an easier to use form, especially if a calculator is available:

$$m = \frac{\sum x_i y_i - [(\sum x_i \sum y_i)/n]}{\sum x_i^2 - [(\sum x_i)^2/n]} \qquad (3.23)$$

where n is the number of data points.

Example 3.21

Riboflavin (vitamin B_2) is determined in a cereal sample by measuring its fluorescence intensity in 5% acetic acid solution. A calibration curve was prepared by measuring the fluorescence intensities of a series of standards of increasing concentrations. The following data were obtained. Use the method of least squares to obtain the best straight line for the calibration curve and to calculate the concentration of riboflavin in the sample solution. The sample fluorescence intensity was 15.4.

Riboflavin, $\mu g/mL$ (x_i)	Fluorescence Intensity, Arbitrary Units (y_i)	x_i^2	$x_i y_i$
0.000	0.0	0.0000	0.00
0.100	5.8	0.0100	0.58
0.200	12.2	0.0400	2.44
0.400	22.3	0.160_0	8.92
0.800	43.3	0.640_0	34.6_4
$\sum x_i = 1.500$	$\sum y_i = 83.6$	$\sum x_i^2 = 0.850_0$	$\sum x_i y_i = 46.5_8$

$$(\sum x_i)^2 = 2.250 \qquad n = 5$$

$$\bar{x} = \frac{\sum x_i}{n} = 0.300_0 \qquad \bar{y} = \frac{\sum y_i}{n} = 16.7_2$$

Solution

Using Equations 3.23 and 3.22.

$$m = \frac{46.5_8 - [(1.500 \times 83.6)/5]}{0.850_0 - 2.250/5} = 53.7_5 \text{ fluor. units/ppm}$$

$$b = 16.7_2 - (53.7_5 \times 0.300_0) = 0.6_0 \text{ fluor. units}$$

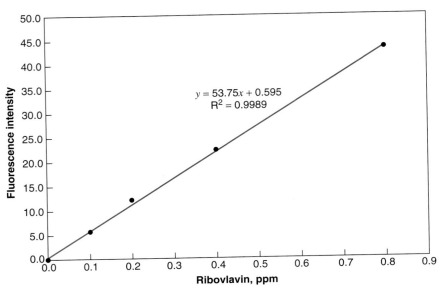

Fig. 3.8. Least-squares plot of data from Example 3.21.

We have retained the maximum number of significant figures in computation. Since the experimental values of y are obtained to only the first decimal place, we can round m and b to the first decimal. The equation of the straight line is (FU = fluorescence units; ppm = μg/mL)

$$y(FU) = 53.8(FU/ppm)x(ppm) + 0.6(Fu)$$

The sample concentration is

$$15.4 = 53.8x + 0.6$$
$$x = 0.275 \ \mu g/mL$$

To prepare an actual plot of the line, take two arbitrary values of x sufficiently far apart and calculate the corresponding y values (or vice versa) and use these as points to draw the line. The intercept $y = 0.6$ (at $x = 0$) could be used as one point. At 0.500 μg/mL, $y = 27.5$. A plot of the experimental data and the least-squares line drawn through them is shown in Figure 3.8. This was plotted using Excel, with the equation of the line and the square of the correlation coefficient (a measure of agreement between the two variables—ignore this for now, we will discuss it later). The program automatically gives additional figures, but note the agreement with our calculated values for the slope and intercept.

• •

STANDARD DEVIATIONS OF THE SLOPE AND INTERCEPT—THEY DETERMINE THE UNKNOWN UNCERTAINTY

The standard deviations of m and b give an equation from which the uncertainty in the unknown is calculated, using propagation of errors.

Each data point on the least-squares line exhibits a normal (Gaussian) distribution about the line on the y axis. The deviation of each y_i from the line is $y_i - y_l = y - (mx + b)$, as in Equation 3.20. The standard deviation of each of these y-axis deviations is given by an equation analogous to Equation 3.2, except that there are two less degrees of freedom since two are used in defining the slope and the intercept:

$$s_y = \sqrt{\frac{\sum[y_i - (mx_i + b)]^2}{N - 2}} = \sqrt{\frac{[\sum y_i^2 - (\sum y_i)^2/N] - m^2[\sum x_i^2 - (\sum x_i)^2/N]}{N - 2}} \quad (3.24)$$

This quantity is also called the *standard deviation of regression*, sr. The s_y value can be used to obtain uncertainties for the slope, m, and intercept, b, of the least-squares line since they are related to the uncertainty in each value of y. For the slope:

$$s_m = \sqrt{\frac{s_y^2}{\sum(\bar{x} - x_i)^2}} = \sqrt{\frac{s_y^2}{\sum x_i^2 - (\sum x_i)^2/N}} \quad (3.25)$$

where $\bar{x}$ is the mean of all x_i values. For the intercept:

$$s_b = s_y\sqrt{\frac{s_y\sum x_i^2}{N\sum x_i^2 - (\sum x_i)^2}} = s_y\sqrt{\frac{1}{N - (\sum x_i)^2/\sum x_i^2}} \quad (3.26)$$

In calculating an unknown concentration, x_i, from Equation 3.19, representing the least-squares line, the uncertainties in y, m, and b are all propagated in the usual manner, from which we can determine the uncertainty in the unknown concentration.

Example 3.22

Estimate the uncertainty in the slope, intercept, and y for the least-squares plot in Example 3.21, and the uncertainty in the determined riboflavin concentration.

Solution

In order to solve for all the uncertainties, we need values for $\sum y_i^2$, $(\sum y_i)^2$, $\sum x_i^2$, $(\sum x_i)^2$, and m^2. From Example 3.21, $(\sum y_i)^2 = (83.6)^2 = 6989.0$; $\sum x_i^2 = 0.850_0$; $(\sum x_i)^2 = 2.250$, and $m^2 = (53.7_5)^2 = 2.88_9$. The $(y_i)^2$ values are $(0.0)^2$, $(5.8)^2$, $(12.2)^2$, $(22.3)^2$, and $(43.3)^2 = 0.0$, 33.6, 148.8, 497.3, and 1874.9, and $\sum y_i^2 = 2554.6$ (carrying extra figures). From Equation 3.24,

$$s_y = \sqrt{\frac{(2554.6 - 6989.0/5) - (53.7_5)^2(0.850_0 - 2.250/5)}{5 - 2}} = \pm 0.6_3 \text{ FU}$$

From Equation 3.25,

$$s_m = \sqrt{\frac{(0.6_3)^2}{0.850_0 - 2.250/5}} = \pm 1.0 \text{ FU/ppm}$$

From Equation 3.26,

$$s_b = 0.6_3\sqrt{\frac{0.850_0}{5(0.850_0) - 2.250}} = \pm 0.4_1 \text{ FU}$$

Therefore, $m = 53._8 + 1._0$ and $b = 0.6 \pm 0.4$.

The unknown riboflavin concentration is calculated from

$$x = \frac{(y \pm s_y) - (b \pm s_b)}{m \pm s_m} = \frac{(15.4 \pm 0.6) - (0.6 \pm 0.4)}{53._8 \pm 1._0} = 0.27_5 \pm \, ?$$

Applying the principles of propagation of error (absolute variances in numerator additive, relative variances in the division step additive), we calculate that $x = 0.27_5 \pm 0.01_4$ ppm.

• •

See Chapter 16 for the spreadsheet calculation of the standard deviation of regression and the standard deviation of an unknown for this.

3.17 Correlation Coefficient and Coefficient of Determination

The **correlation coefficient** is used as a measure of the correlation between two variables. When variables x and y are correlated rather than being functionally related (i.e., are not directly dependent upon one another), we do not speak of the "best" y value corresponding to a given x value, but only of the most "probable" value. The closer the observed values are to the most probable values, the more definite is the relationship between x and y. This postulate is the basis for various numerical measures of the degree of correlation.

The **Pearson correlation coefficient** is one of the most convenient to calculate. This is given by

$$r = \sum \frac{(x_i - \bar{x})(y_i - \bar{y})}{n s_x s_y} \tag{3.27}$$

where r is the correlation coefficient, n is the number of observations, s_x is the standard deviation of x, s_y is the standard deviation of y, x_i and y_i are the individual values of the variables x and y, respectively, and $\bar{x}$ and $\bar{y}$ are their means. The use of differences in the calculation is frequently cumbersome, and the equation can be transformed to a more convenient form:

$$\begin{aligned} r &= \frac{\sum x_i y_i - n \bar{x} \bar{y}}{\sqrt{(\sum x_i^2 - n\bar{x}^2)(\sum y_i^2 - n\bar{y}^2)}} \\ &= \frac{n\sum x_i y_i - \sum x_i y_i}{\sqrt{[n\sum x_i^2 - (\sum x_i)^2][n\sum y_i^2 - (\sum y_i)^2]}} \end{aligned} \tag{3.28}$$

Despite its formidable appearance, Equation 3.28 is probably the most convenient for calculating r, particularly if a calculator is available.

The maximum value of r is 1. When this occurs, there is exact correlation between the two variables. When the value of r is zero (this occurs when xy is equal to zero), there is complete independence of the variables. The minimum value of r is -1. A negative correlation coefficient indicates that the assumed dependence is opposite to what exists and is therefore a positive coefficient for the reversed relation.

A correlation coefficient near 1 means there is a direct relationship between two variables, e.g., absorbance and concentration.

Example 3.23

Calculate the correlation coefficient for the data in Example 3.19, taking your method as x and the standard method as y.

Solution

We calculate that $\sum x_i^2 = 903.2$, $\sum y_i^2 = 855.2$, $\bar{x} = 12.0$, $\bar{y} = 11.7$, and $\sum x_i y_i = 878.5$. Therefore, from Equation 3.28,

$$r = \frac{878.5 - (6)(12.0)(11.7)}{\sqrt{[903.2 - (6)(12.0)^2][855.2 - (6)(11.7)^2]}} = 0.991$$

A correlation coefficient can be calculated for a calibration curve to ascertain the degree of correlation between the measured instrumental variable and the sample concentration. As a general rule, $0.90 < r < 0.95$ indicates a fair curve, $0.95 < r < 0.99$ a good curve, and $r > 0.99$ indicates excellent linearity. An $r > 0.999$ can sometimes be obtained with care.

The correlation coefficient gives the dependent and independent variables equal weight; which is usually not true in scientific measurements. The r value tends to give more confidence in the goodness of fit than warranted. The fit must be quite poor before r becomes smaller than about 0.98 and is really very poor when less than 0.9.

A more conservative measure of closeness of fit is the square of the correlation coefficient, r^2, and this is what most statistical programs calculate (including Excel—see Figure 3.8). An r value of 0.90 corresponds to an r^2 value of only 0.81, while an r of 0.95 is equivalent to an r^2 of 0.90. The goodness of fit is judged by the number of 9's. So three 9's (0.999) or better represents an excellent fit. We will use r^2 as a measure of fit. This is also called the **coefficient of determination**.

The coefficient of determination (r^2) is a better measure of fit.

It should be mentioned that it is possible to have a high degree of correlation between two methods (r^2 near unity) but to have a statistically significant difference between the results of each according to the t test. This would occur, for example, if there were a constant determinate error in one method. This would make the differences significant (not due to chance), but there would be a direct correlation between the results [r^2 would be near unity, but the slope (m) may not be near unity or the intercept (b) not near zero]. In principle, an empirical correction factor (a constant) could be applied to make the results by each method the same over the concentration range analyzed.

3.18 Using Spreadsheets for Plotting Calibration Curves

The availability of spreadsheets makes it unnecessary to plot data on graph paper and do hand calculations for the least-squares regression analysis and statistics. We will use the data in Example 3.21 to prepare the plot shown in Figure 3.8, using Excel.

Open a new spreadsheet and enter:

Cell A1: Riboflavin, ppm (adjust the column width to incorporate the text)
Cell B 1: Fluorescence intensity

Cell A3: 0.000

Cell A4: 0.100

Cell A5: 0.200

Cell A6: 0.400

Cell A7: 0.800

Cell B3: 0.0

Cell B4: 5.8

Cell B5: 12.2

Cell B6: 22.3

Cell B7: 43.3

Format the cell numbers to have three decimal places for column A and one for column B.

Click on the Chart Wizard icon on the toolbar (the one with the vertical bars). Step 1—Chart Type—of the Chart Wizard will appear.

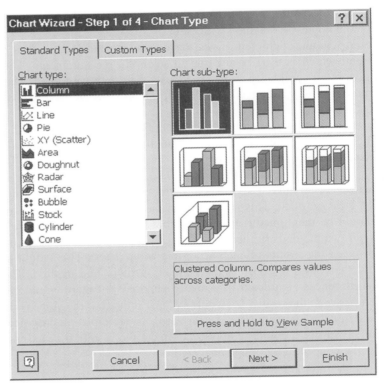

Follow the following sequences:

Select XY (scatter), and Scatter (no line) for Chart subtype

Next

Data Range: enter A3:B7 (click on Series, and note the X values and Y values addresses)

Check Columns (after going back to Data Range)

Next

Chart title: enter Calibration Curve

Value (X) axis: enter Riboflavin

Value (Y) axis: enter Fluorescence intensity

Gridlines: uncheck Major gridlines

Legend: Delete Show legend

Data labels: None (Try Show Value, and note the data entered on each point on the line)

Next

Click on As New sheet: Chart 1

Finish

The calibration graph is plotted on a new Excel sheet.

Now we wish to enter the least-squares equation line and the r^2 value. Click on the figure, and Chart will appear in the toolbar. Click on it and continue:

Add Trendline

Linear

Options

Display equation on chart

Display R-squared value on chart

OK

Now look at the chart. Click on it to remove the end markers. You can move the equation on the line toward the left and enlarge it. Click on the equation. It is highlighted with small squares. Click on a corner and drag it to the left, down the line. You can increase the font size. Click on Format: Select Data Lables:Font. Select size 14, then OK. Drag the equation closer to the line. You can also increase the font size of the axis labels by highlighting them and doing the same, as well as the title.

Let's get rid of the gray background. Click on the gray area, then Format: Select Plot Area. Click on the white color square, then OK. The chart you have now prepared should look similar to Figure 3.8.

When you prepare the graph, you can initially highlight the cells (A3:B7) that you want to graph, and the addresses will automatically be placed in the Data Range. Instead of placing the graph on a new sheet, you could have selected As object in: Sheet 1. This would have placed it in the spreadsheet in which you entered the data. You can adjust its position and size by clicking on it, and dragging the corners. Figure 3.9 shows the graph inserted into the spreadsheet. Try doing this. Once you have the graph inserted in the spreadsheet, this becomes a generic plot for new data, that is, if you change the data in columns A and B, a new line is automatically charted. Try this. (You should save your original spreadsheet/graph and rename the new one.) You may print only the graph by first clicking on it to highlight it.

3.19 Slope, Intercept, and Coefficient of Determination

We can use the Excel statistical functions to calculate the slope and intercept for a series of data, and the R^2 value, without a plot. Open a new spreadsheet and enter the calibration data from Example 3.21, as in Figure 3.9, in cells A3:B7. In cell A9 type Intercept, in cell A10, Slope, and in cell A11, R^2 Highlight cell B9, click on f_x:Statistical, and scroll down to INTERCEPT under Function name, and click OK. For Known_x's, enter the array A3:A7, and for Known_y's, enter B3:B7. Click

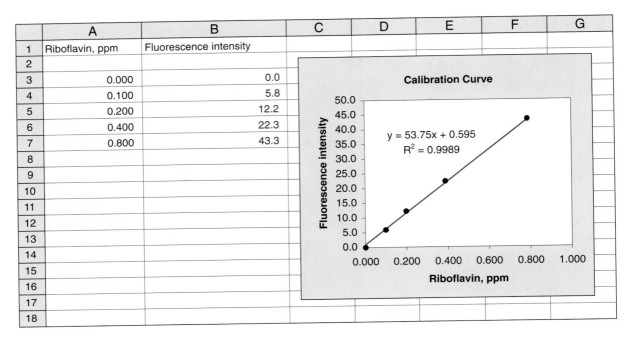

	A	B	C	D	E	F	G
1	Riboflavin, ppm	Fluorescence intensity					
2							
3	0.000	0.0					
4	0.100	5.8					
5	0.200	12.2					
6	0.400	22.3					
7	0.800	43.3					
8							
9							
10							
11							
12							
13							
14							
15							
16							
17							
18							

Fig. 3.9. Calibration graph inserted in spreadsheet (Sheet 1).

OK, and the intercept is displayed in cell B9. Now repeat, highlighting cell B10, scrolling to Slope, and entering the same arrays. The slope appears in cell B10. Repeat again, highlighting cell B11, and scrolling to RSQ. R^2 appears in cell B11. Compare with the values in Figure 3.9.

3.20 LINEST for Additional Statistics

The LINEST program of Excel allows us to quickly obtain several statistical functions for a set of data, in particular, the slope and its standard deviation, the intercept and its standard deviation, the coefficient of determination, and the standard error of the estimate, besides others we will not discuss now. Linest will automatically calculate a total of 10 functions in 2 columns of the spreadsheet.

Open a new spreadsheet, and enter the calibration data from Example 3.21 as you did above, in cells A3:B7. Refer to Figure 3.10. The statistical data will be placed in 10 cells, so let's label them now. We will place them in cells B9:C13. Type labels as follows:

Cell A9: slope

Cell A10: std. devn.

Cell A11: R^2

Cell A12: F

Cell A13: sum sq. regr.

Cell D9: intercept

Cell D10: std. devn.

Cell D11: std. error of estim.

Cell D12: d.f.

Cell D13: sum sq. resid.

	A	B	C	D	E
1	Riboflavin, ppm	Fluorescence intensity			
2					
3	0.000	0.0			
4	0.100	5.8			
5	0.200	12.2			
6	0.400	22.3			
7	0.800	43.3			
8					
9	slope	53.75	0.595	intercept	
10	std. devn.	1.017759	0.419633	std. devn.	
11	R^2	0.998926	0.643687	std. error of estim.	
12	F	2789.119	3	d.f.	
13	sum sq. regr.	1155.625	1.243	sum sq. resid.	

Fig. 3.10. Using LINEST for statistics.

Highlight cells B9:C13, and click on f_x. From the Statistical function, scroll down to LINEST and click OK. For Known_y's, enter the array B3:B7, and for Known_x's, enter A3:A7. Then in each of the boxes labeled Const and Stats, type "true". Now we have to use the keyboard to execute the calculations. Depress Shift, Control, and Enter, and release. The statistical data are entered into the highlighted cells. This keystroke combination must be used whenever performing a function on an array of cells, like here. The slope is in cell B9 and its standard deviation in cell B10. The intercept is in cell C9 and its standard deviation in cell C10. The coefficient of determination is in cell B11. Compare the standard deviations with those calculated in Example 3.22, and the slope, intercept, and R^2 with Example 3.21 or Figure 3.8.

Cell C11 contains the standard error of the estimate (or standard deviation of the regression) and is a measure of the error in estimating values of y. The smaller it is, the closer the numbers are to the line. The other cells contain data we will not consider here: Cell B12 is the F value, cell C12 the degrees of freedom (used for F), cell B13 the sum of squares of the regression, and cell C13 the sum of squares of the residuals.

How many significant figures should we keep for the least-squares line? The standard deviations give us the answer. The slope has a standard deviation of 1.0, and so we write the slope as $53.8 \pm 1._0$ at best. The intercept standard deviation is ± 0.42, so for the slope we write 0.6 ± 0.4. See also Example 3.22.

3.21 Statistics Software Packages

Excel offers a number of statistical functions, listed under the Tools menu. Go to Add-Ins, and check Analysis ToolPak. Click OK and return to the spreadsheet. Now when you go to the Tools menu, you will see Data Analysis. Go to that, and you will see 19 statistical programs listed. As you experiment with these, you will find some very useful. One Add-In that is very useful is Solver, for solving complicated formulas. Its use is described in Chapter 6. See also the text website www.wiley.com/college/christian for a list of some commercial software packages for performing basic as well as more advanced statistical calculations.

3.22 Detection Limits—There Is No Such Thing as Zero

The previous discussions have dealt with statistical methods to estimate the reliability of analyses at specific confidence levels, these being ultimately determined

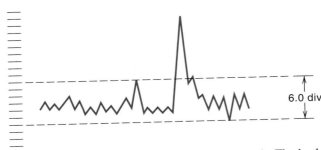

Fig. 3.11. Peak-to-peak noise level as a basis for detection limit. The background fluctuations represent continuously recorded background signals, with the analyte measurement represented by the peak signal. A "detectable" analyte signal would be 12 divisions above a line drawn through the average of the baseline fluctuations.

by the precision of the method. All instrumental methods have a degree of noise associated with the measurement that limits the amount of analyte that can be detected. The noise is reflected in the precision of the blank or background signal, and noise may be apparent even when there is no significant blank signal. This may be due to fluctuation in the dark current of a photomultiplier tube, flame flicker in an atomic absorption instrument, and other factors.

The limit of detection is the lowest concentration level that can be determined to be statistically different from an analyte blank. There are numerous ways that detection limits have been defined. For example, the concentration that gives twice the peak-to-peak noise of a series of background signal measurements (or of a continuously recorded background signal) may be taken as the detection limit (see Figure 3.11). A generally accepted detection limit is the concentration that gives a signal three times the standard deviation of the background signal.

> The concentration that gives a signal equal to three times the standard deviation of the background is generally taken as the detection limit.

Example 3.24

A series of sequential baseline absorbance measurements are made in a spectrophotometric method, for determining the purity of aspirins in tablets using a blank solution. The absorbance readings are 0.002, 0.000, 0.008, 0.006, and 0.003. A standard 1 ppm aspirin solution gives an absorbance reading of 0.051. What is the detection limit?

Solution

The standard deviation of the blank readings is ±0.0032 absorbance units, and the mean of the blank readings is 0.004 absorbance units. The detection limit is that concentration of analyte that gives a reading of 3 × 0.0032 = 0.0096 absorbance reading, above the blank signal. The net reading for the standard is 0.051 − 0.004 = 0.047. The detection limit would correspond to 1 ppm (0.0096/0.047) = 0.2 ppm and would give a total absorbance reading of 0.0096 + 0.004 = 0.014.

The precision at the detection limit is by definition about 33%. For quantitative measurements, concentrations should be at least 10 times the detection limit (2 ppm in the above example).

There have been various attempts to place the concept of detection limit on a more firm statistical ground. The International Conference on Harmonization (ICH; see Chapter 4) of Technical Requirements for Registration of Pharmaceuticals for Human Use has proposed guidelines for analytical method validation (Ref. 18). The ICH Q2B guideline on validation methodology suggests calculation based on the standard deviation, s, of the response and the slope or sensitivity, S, of the calibration curve at levels approaching the limit. For the limit of detection (LOD),

$$LOD = 3.3(s/S) \qquad (3.29)$$

And for limit of quantitation (LOQ)

$$LOQ = 10(s/S) \qquad (3.30)$$

The standard deviation of the response can be determined based on the standard deviation of either the blank, the residual standard deviation of the least-squares regression line, or the standard deviation of the y intercept of the regression line. The Excel statistical function can be used to obtain the last two.

The International Union of Pure and Applied Chemistry (IUPAC) uses a value of 3 in Equation 3.29 (for blank measurements), derived from a confidence level of 95% for a reasonable number of measurements. The confidence level, of course, varies with the number of measurements, and 7 to 10 measurements should be taken. The bottom line is that one should regard a detection limit as an approximate guide to performance and not make efforts to determine it too precisely.

3.23 Statistics of Sampling—How Many Samples, How Large?

The acquiring of a valid analytical sample is perhaps the most critical part of any analysis. The physical sampling of different types of materials (solids, liquids, gases) is discussed in Chapter 2. We describe here some of the statistical considerations in sampling.

THE PRECISION OF A RESULT—SAMPLING IS THE KEY

More often than not, the accuracy and precision of an analysis is limited by the sampling rather than the measurement step. The overall variance of an analysis is the sum of the sampling variance and the variance of the remaining analytical operations, that is,

$$s_o^2 = s_s^2 + s_a^2 \qquad (3.31)$$

If the variance due to sampling is known (e.g., by having performed multiple samplings of the material of interest and analyzing it using a precise measurement technique), then there is little to be gained by reduction of s_a to less than $\frac{1}{3}s_s$. For example, if the absolute standard deviation for sampling is 3.0% and that of the analysis is 1.0%, then $s_o^2 = (1.0)^2 + (3.0)^2 = 10.0$, or $s_o = 3.2\%$. Here, 94% of the imprecision is due to sampling and only 6% is due to measurement (s_o is increased from 3.0 to 3.2%, so 0.2% is due to the measurement). If the sampling imprecision is relatively large, it is better to use a rapid, lower precision method and analyze more samples.

Little is gained by improving the analytical variance to less than one-third the sampling variance. It is better to analyze more samples using a faster, less precise method.

We are really interested in the value and variance of the true value. The total variance is $s_{total}^2 = s_g^2 + s_s^2 + s_a^2$, where s_g^2 describes the "true" variability of the analyte *in the system*, the value of which is the goal of the analysis. For reliable interpretation of the chemical analysis, the combined sampling and analytical variance should not exceed 20% of the total variance. [See M. H. Ramsey, "Appropriate Precision: Matching Analytical Precision Specifications to the Particular Application," *Anal. Proc.*, **30** (1993) 110.]

THE "TRUE VALUE"

The range in which the true value falls for the analyte content in a bulk material can be estimated from a *t* test at a given confidence level (Equation 3.11). Here, $\bar{x}$ is the average of the analytical results for the particular material analyzed, and *s* is the standard deviation that is obtained previously from analysis of similar material samples or from the present analysis if there are sufficient samples.

MINIMUM SAMPLE SIZE

Statistical guidelines have been developed for the proper sampling of heterogeneous materials, based on the sampling variance. The minimum size of individual increments for a well-mixed population of different kinds of particles can be estimated from **Ingamell's sampling constant, K_s:**

$$\boxed{wR^2 = K_s} \qquad (3.32)$$

The greater the sample size, the smaller the variance.

where w is the weight of sample analyzed and R is the percent *relative* standard deviation of the sample composition. K_s represents the weight of sample for 1% sampling uncertainty at a 68% confidence level and is obtained by determining the standard deviation from the measurement of a series of samples of weight w. This equation, in effect, says that the sampling variance is inversely proportional to the sample weight.

Example 3.25

Ingamell's sampling constant for the analysis of the nitrogen content of wheat samples is 0.50 g. What weight sample should be taken to obtain a sampling precision of 0.2% rsd in the analysis?

Solution

$$w \, (0.2)^2 = 0.50 \text{ g}$$
$$w = 12.5 \text{ g}$$

Note that the entire sample is not likely to be analyzed. The 12.5-g gross sample will be finely ground, and a few hundred milligrams of the homogeneous material analyzed. If the sample were not made homogeneous, then the bulk of it would have to be analyzed.

MINIMUM NUMBER OF SAMPLES

The number of individual sample increments needed to achieve a given level of confidence in the analytical results is estimated by

$$n = \frac{t^2 s_s^2}{r^2 \bar{x}^2} \tag{3.33}$$

where t is the Student t value for the confidence level desired, s_s^2 is the sampling variance, r is the *acceptable* relative standard deviation of the average of the analytical results, $\bar{x}$; s_s is the *absolute* standard deviation, in the same units as $\bar{x}$, and so n is unitless. Values of s_s and $\bar{x}$ are obtained from preliminary measurements or prior knowledge. Since r is equal to $s_x/\bar{x}$, we can write that

$$n = \frac{t^2 s_s^2}{s_x^2} \tag{3.34}$$

s_s and s_x can then be expressed in *either* absolute or relative standard deviations, so long as they are both expressed the same. Since n is initially unknown, the t value for the given confidence level is initially estimated and an iterative procedure is used to calculate n.

Example 3.26

The iron content in a blended lot of bulk ore material is about 5% (wt/wt), and the relative standard deviation of sampling, s_s, is 0.021 (2.1% rsd). How many samples should be taken in order to obtain a relative standard deviation, r, of 0.016 (1.6% rsd) in the results at the 95% confidence level [i.e., the standard deviation, s_x, for the 5% iron content is 0.08% (wt/wt)]?

Solution

We can use either Equation (3.33) or (3.34). We will use the latter. Set $t = 1.96$ (for $n = \infty$, Table 3.1) at the 95% confidence level. Calculate a preliminary value of n. Then use this n to select a closer t value, and recalculate n; continue iteration to a constant n.

$$n = \frac{(1.96)^2(0.021)^2}{(0.016)^2} = 6.6$$

For $n = 7$, $t = 2.365$.

$$n = \frac{(2.365)^2(0.021)^2}{(0.016)^2} = 9.6$$

For $n = 10$, $t = 2.23$

$$n = \frac{(2.23)^2(0.021)^2}{(0.016)^2} = 8.6 \equiv 9$$

See if you get the same result using Equation (3.33).

Equation 3.33 holds for a **Gaussian distribution** of analyte concentration within the bulk material, that is, it will be centered around x with 68% of the values falling within one standard deviation, or 95% within two standard deviations. In this case, the variance of the population, σ^2, is small compared to the true value. If the concentration follows a **Poisson distribution**, that is, follows a random distribution in the bulk material such that the true or mean value $\bar{x}$ approximates the variance, s_s^2, of the population, then Equation 3.33 is somewhat simplified:

$$n = \frac{t^2}{r^2\bar{x}} \cdot \frac{s_s^2}{\bar{x}} = \frac{t^2}{r^2\bar{x}} \qquad (3.35)$$

Note that since s_s^2 is equal to $\bar{x}$, the right-hand part of the expression becomes equal to 1, but the units do not cancel. In this case, when the concentration distribution is broad rather than narrow, many more samples are required to get a representative result from the analysis.

If the analyte occurs in clumps or patches, the sampling strategy becomes more complicated. The patches can be considered as separate strata and sampled separately. If bulk materials are segregated or stratified, and the average composition is desired, then the number of samples from each stratum should be in proportion to the size of the stratum.

Learning Objectives

WHAT ARE SOME OF THE KEY THINGS WE LEARNED FROM THIS CHAPTER?

- Accuracy and precision, p. 65
- Types of errors in measurements, p. 68
- Significant figures in measurements and calculations, pp. 66, 67
- Standard deviation, p. 74
- How to use spreadsheets, p. 78
- Propagation of errors, p. 82
- Control charts, p. 89
- Statistics: confidence limits, t tests, F tests, p. 90
- Rejection of a result, p. 98
- Least-squares plots and coefficient of determination, pp. 102, 106
- Using spreadsheets for plotting calibration curves, p. 107
- Detection limits, p. 111
- Statistics of sampling, p. 113

Questions

1. Distinguish between accuracy and precision.
2. What is determinate error? An indeterminate error?
3. The following is a list of common errors encountered in research laboratories. Categorize each as a determinate or an indeterminate error, and further categorize determinate errors as instrumental, operative, or methodic: (a) An unknown being weighed is hygroscopic. (b) One component of a mixture being

analyzed quantitatively by gas chromatography reacts with the column packing. (c) A radioactive sample being counted repeatedly without any change in conditions yields a slightly different count at each trial. (d) The tip of the pipet used in the analysis is broken. (e) In measuring the same peak heights of a chromatogram, two technicians each report different heights.

Problems

For the statistical problems, do the calculations manually first and then use the Excel statistical functions and see if you get the same answers. See the CD, Problems 14–18, 20, 21, 25–30, and 37–40.

SIGNIFICANT FIGURES

4. How many significant figures does each of the following numbers have? (a) 200.06, (b) 6.030×10^{-4}, and (c) 7.80×10^{10}.

5. How many significant figures does each of the following numbers have? (a) 0.02670, (b) 328.0, (c) 7000.0, and (d) 0.00200.

6. Calculate the formula weight of $LiNO_3$ to the correct number of significant figures.

7. Calculate the formula weight of $PdCl_2$ to the correct number of significant figures.

8. Give the answer to the following problem to the maximum number of significant figures: $50.00 \times 27.8 \times 0.1167$.

9. Give the answer of the following to the maximum number of significant figures: $(2.776 \times 0.0050) - (6.7 \times 10^{-3}) + (0.036 \times 0.0271)$.

10. An analyst wishes to analyze spectrophotometrically the copper content in a bronze sample. If the sample weighs about 5 g and if the absorbance (A) is to be read to the nearest 0.001 absorbance unit, how accurately should the sample be weighed? Assume the volume of the measured solution will be adjusted to obtain minimum error in the absorbance, that is, so that $0.1 < A < 1$.

EXPRESSIONS OF RESULTS

11. A standard serum sample containing 102 meq/L chloride was analyzed by coulometric titration with silver ion. Duplicate results of 101 and 98 meq/L were obtained. Calculate (a) the mean value, (b) the absolute error of the mean, and (c) the relative error in percent.

12. A batch of nuclear fuel pellets was weighed to determine if they fell within control guidelines. The weights were 127.2, 128.4, 127.1, 129.0, and 128.1 g. Calculate (a) the mean, (b) the median, and (c) the range.

13. Calculate the absolute error and the relative error in percent and in parts per thousand in the following:

	Measured Value	Accepted Value
(a)	22.62 g	22.57 g
(b)	45.02 mL	45.31 mL
(c)	2.68%	2.71%
(d)	85.6 cm	85.0 cm

STANDARD DEVIATION

14. The tin and zinc contents of a brass sample are analyzed with the following results: (a) Zn: 33.27, 33.37, and 33.34% and (b) Sn: 0.022, 0.025, and 0.026%. Calculate the standard deviation and the coefficient of variation for each analysis.

15. Replicate water samples are analyzed for water hardness with the following results; 102.2, 102.8, 103.1, and 102.3 ppm $CaCO_3$. Calculate (a) the standard deviation, (b) the relative standard deviation, (c) the standard deviation of the mean, and (d) the relative standard deviation of the mean.

16. Replicate samples of a silver alloy are analyzed and determined to contain 95.67, 95.61, 95.71, and 95.60% Ag. Calculate (a) the standard deviation, (b) the standard deviation of the mean, and (c) the relative standard deviation of the mean (in percent) of the individual results.

PROPAGATION OF ERROR

17. Calculate the uncertainty in the answers of the following: (a) $(128 \pm 2) + (1025 \pm 8) - (636 \pm 4)$, (b) $(16.25 \pm 0.06) - (9.43 \pm 0.03)$, (c) $(46.1 \pm 0.4) + (935 \pm 1)$.

18. Calculate the absolute uncertainty in the answers of the following: (a) $(2.78 \pm 0.04)(0.00506 \pm 0.00006)$, (b) $(36.2 \pm 0.4)/(27.1 \pm 0.6)$, (c) $(50.23 \pm 0.07)(27.86 \pm 0.05)/(0.1167 \pm 0.0003)$.

19. Calculate the absolute uncertainty in the answer of the following: $[(25.0 \pm 0.1)(0.0215 \pm 0.0003) - (1.02 \pm 0.01)(0.112 \pm 0.001)](17.0 \pm 0.2)/(5.87 \pm 0.01)$.

CONFIDENCE LIMIT

20. The following molarities were calculated from replicate standardization of a solution: 0.5026, 0.5029, 0.5023, 0.5031, 0.5025, 0.5032, 0.5027, and 0.5026 M. Assuming no determinate errors, within what range are you 95% certain that the true mean value of the molarity falls?

21. Determination of the sodium level in separate portions of a blood sample by ion-selective electrode measurement gave the following results: 139.2, 139.8, 140.1, and 139.4 meq/L. What is the range within which the true value falls, assuming no determinate error (a) at the 90% confidence level, (b) at the 95% confidence level, and (c) at the 99% confidence level?

22. Lead on leaves by a roadside was measured spectrophotometrically by reaction with dithizone. The standard deviation for a triplicate analysis was 2.3 ppm. What is the 90% confidence limit?

23. The standard deviation established for the determination of blood chloride by coulometric titration is 0.5 meq/L. What is the 95% confidence limit for a triplicate determination?

24. Estimate the range of the true molarity of the solution at the 90% confidence level from the standardization in Problem 31.

TESTS OF SIGNIFICANCE

25. A study is being performed to see if there is a correlation between the concentration of chromium in the blood and a suspected disease. Blood samples from a series of volunteers with a history of the disease and other indicators of susceptibility are analyzed and compared with the results from the analysis of samples from healthy control subjects. From the following results, determine whether the differences between the two groups can be ascribed to chance or whether they are real. Control group (ppb Cr): 15, 23, 12, 18, 9, 28, 11, 10. Disease group: 25, 20, 35, 32, 15, 40, 16, 10, 22, 18.

26. An enzymatic method for determining alcohol in wine is evaluated by comparison with a gas-chromatographic (GC) method. The same sample is analyzed several times by both methods with the following results (% ethanol). Enzymatic method: 13.1, 12.7, 12.6, 13.3, 13.3. GC method: 13.5, 13.3, 13.0, 12.9. Does the enzymatic method give the same value as the GC method at the 95% confidence level?

27. Your laboratory is evaluating the precision of a colorimetric method for creatinine in serum in which the sample is reacted with alkaline picrate to produce a color. Rather than perform one set of analyses, several sets with different samples are performed over several days, in order to get a better estimate of the precision of the method. From the following absorbance data, calculate the pooled standard deviation.

Day 1 (Sample A)	Day 2 (Sample B)	Day 3 (Sample C)
0.826	0.682	0.751
0.810	0.655	0.702
0.880	0.661	0.699
0.865		0.724
$\bar{x}_A = 0.845$	$\bar{x}_B = 0.666$	$\bar{x}_C = 0.719$

28. The following replicate calcium determinations on a blood sample using atomic absorption spectrophotometry (AAS) and a new colorimetric method were reported. Is there a significant difference in the precision of the two methods?

AAS (mg/dL)	Colorimetric (mg/dL)
10.9	9.2
10.1	10.5
10.6	9.7
11.2	11.5
9.7	11.6
10.0	9.3
Mean 10.4	10.1
	11.2
	Mean 10.4

29. Potassium dichromate is an oxidizing agent that is used for the volumetric determination of iron by titrating iron(II). Although potassium dichromate is a high-purity material that can be used for the direct preparation of a standard solution of known concentration, the solution is frequently standardized by titrating a known amount of iron(II) prepared from high-purity iron wire or electrolytic iron, using the same procedure as for the sample. This is because the color of the iron(III) product of the titration tends to mask the indicator color (used to detect the end of the titration), causing a slight error. A solution prepared to be 0.1012 M was standardized with the following results:

0.1017, 0.1019, 0.1016, 0.1015 M. Is the supposition that the titration values are statistically different from the actual prepared concentration valid?

30. In the nuclear industry, detailed records are kept of the quantity of plutonium received, transported, or used. Each shipment of plutonium pellets received is carefully analyzed to check that the purity and hence the total quantity is as the supplier claims. A particular shipment is analyzed with the following results: 99.93, 99.87, 99.91, and 99.86%. The listed purity as received from the supplier is 99.95%. Is the shipment acceptable?

Q TEST

31. The following replicate molarities were obtained when standardizing a solution: 0.1067, 0.1071, 0.1066, and 0.1050. Can one of the results be discarded as due to accidental error at the 95% confidence level?

32. Can any of the data in Problem 14 be rejected at the 95% confidence level?

33. The precision of a method is being established, and the following data are obtained: 22.23. 22.18, 22.25, 22.09, and 22.17%. Is 22.09% a valid measurement at the 95% confidence level?

STATISTICS FOR SMALL SETS OF DATA

34. For Problem 15, estimate the standard deviation from the range. Compare with the standard deviation calculated in the problem.

35. For Problem 20, use the range to estimate the confidence limit at the 95% confidence level, and compare with the value calculated in the problem using the standard deviation.

36. For Problem 21, use the range to estimate the confidence limits at the 95 and 99% confidence levels, and compare with the values calculated in the problem using standard deviation.

LEAST SQUARES

37. Calculate the slope of the line in Example 3.21, using Equation 3.22. Compare with the value calculated using Equation 3.23.

38. A calibration curve for the colorimetric determination of phosphorous in urine is prepared by reacting standard solutions of phosphate with molybdenum(VI) and reducing the phosphomolybdic acid complex to produce the characteristic blue color. The measured absorbance A is plotted against the concentration of phosphorous. From the following data, determine the linear least-squares line and calculate the phosphorous concentration in the urine sample:

ppm P	A
1.00	0.205
2.00	0.410
3.00	0.615
4.00	0.820
Urine sample	0.625

39. Calculate the uncertainties in the slope and intercept of the least-squares line in Problem 38, and the uncertainty in the phosphorus concentration in the urine sample.

CORRELATION COEFFICIENT

40. From the data given below, determine the correlation coefficient between the amount of toxin produced by a fungus and the percent of yeast extract in the growth medium.

Sample	% Yeast Extract	Toxin (mg)
(a)	1.000	0.487
(b)	0.200	0.260
(c)	0.100	0.195
(d)	0.010	0.007
(e)	0.001	0.002

41. The cultures described in Problem 40 had the following fungal dry weights: sample (a) 116 mg, (b) 53 mg, (c) 37 mg, (d) 8 mg, and (e) 1 mg. Determine the correlation coefficient between the dry weight and the amount of toxin produced.

42. A new method for the determination of cholesterol in serum is being developed in which the rate of depletion of oxygen is measured with an oxygen electrode upon reaction of the cholesterol with oxygen, when catalyzed by the enzyme cholesterol oxidase. The results for several samples are compared with those of the standard Lieberman colorimetric method. From the following data, determine by the t test if there is a statistically significant difference between the two methods and calculate the correlation coefficient. Assume the two methods have similar precisions.

Sample	Enzyme Method (mg/dL)	Colorimetric Method (mg/dL)
1	305	300
2	385	392
3	193	185
4	162	152
5	478	480
6	455	461
7	238	232
8	298	290
9	408	401
10	323	315

DETECTION LIMIT

43. You are determining aluminum in plants by a fluorometric procedure. Seven prepared blanks give fluorescence readings of 0.12, 0.18, 0.25, 0.11, 0.16, 0.26, and 0.16 units. A 1.0 aluminum standard solution gave a reading of 1.25. What is the detection limit? What would be the total reading at this level?

SAMPLING STATISTICS

44. Four-tenth gram samples of paint from a bridge, analyzed for the lead content by a precise method (<1% rsd), gives a relative sampling precision, R, of 5%. What weight sample should be taken to improve this to 2.5%?

45. Copper in an ore sample is at a concentration of about 3% (wt/wt). How many samples should be analyzed to obtain a percent relative standard deviation of 5% in the analytical result at the 95% confidence level, if the sampling precision is 0.15% (wt/wt)?

Recommended References

STATISTICS

1. P. C. Meier and R. E. Zund, *Statistical Methods in Analytical Chemistry*, 2nd ed. New York: Wiley, 2000.

2. J. C. Miller and J. N. Miller, *Statistics and Chemometrics for Analytical Chemistry*, 4th ed. Englewood Cliffs, NJ: Prentice Hall, 2000.

3. J. C. Miller and J. N. Miller, "Basic Statistical Methods for Analytical Chemistry. A Review. Part 1. Statistics of Repeated Measurements." "Part 2. Calibration and Regression Methods," *Analyst*, **113**(1988) 1351; **116**(1991)3.

4. M. Otto, *Chemometrics: Statistics and Computer Application in Analytical Chemistry*. London: Wiley, 1999.

Q TEST

5. R. B. Dean and W. J. Dixon, "Simplified Statistics for Small Numbers of Observations," *Anal. Chem.*, **23** (1951) 636.

6. W. J. Blaedel, V. W. Meloche, and J. A. Ramsay, "A Comparison of Critiera for the Rejection of Measurements," *J. Chem. Educ.*, **28** (1951) 643.

7. D. B. Rorabacher, "Statistical Treatment for Rejection of Deviant Values; Critical Values of Dixon's 'Q' Parameter and Related Subrange Ratios at the 95% Confidence Level," *Anal. Chem.*, **63** (1991) 139.

8. C. E. Efstathiou, "A Test for the Simultaneous Detection of Two Outliers Among Extreme Values of Small Data Sets," *Anal. Lett.*, **26** (1993) 379.

QUALITY CONTROL

9. J. K. Taylor, "Quality Assurance of Chemical Measurements," *Anal. Chem.*, **53** (1981) 1588A.

10. J. K. Taylor, *Quality Assurance of Chemical Measurements*. Boca Raton, FL: CRC Press/Lewis, 1987.

11. J. K. Taylor, "Validation of Analytical Methods," *Anal. Chem.*, **55** (1983) 600A.

12. J. O. Westgard, P. L. Barry, and M. R. Hunt, "A Multi-Rule Shewhart Chart for Quality Control in Clinical Chemistry," *Clin. Chem.*, **27** (1981) 493.

LEAST SQUARES

13. P. Galadi and B. R. Kowalski, "Partial Least Squares Regression (PLS): A Tutorial," *Anal. Chim. Acta*, **185** (1986) 1.

DETECTION LIMITS

14. G. L. Long and J. D. Winefordner, "Limit of Detection. A Closer Look at the IUPAC Definition," *Anal. Chem.*, **55** (1983) 712A.

15. J. P. Foley and J. G. Dorsey, "Clarification of the Limit of Detection in Chromatography," *Chromatographia*, **18** (1984) 503.

16. J. E. Knoll, "Estimation of the Limit of Detection in Chromatography," *J. Chromatogr. Sci.*, **23** (1985) 422.

17. Analytical Methods Committee, "Recommendations for the Definition, Estimation and Use of the Detection Limit," *Analyst*, **112** (1987) 199.

18. *ICH-Q2B Validation of Analytical Procedures: Methodology* (International Conference on Harmonization of Technical Requirements for Registration of Pharmaceuticals for Human Use, Geneva, Switzerland, November 1996).

SAMPLING STATISTICS

19. B. Kratochvil and J. K. Taylor, "Sampling for Chemical Analysis," *Anal. Chem.*, **53** (1981) 924A.

20. M. H. Ramsey, "Sampling as a Source of Measurement Uncertainty: Techniques for Quantification and Comparison with Analytical Sources," *J. Anal. Atomic Spectrosc.*, **13** (1998) 97.

21. G. Brands, "Theory of Sampling. I. Uniform Inhomogeneous Material," *Fresenius' Z. Anal. Chem.*, **314** (1983) 6; II. "Sampling from Segregated Material," *Z. Anal. Chem.*, **314** (1983) 646.

22. N. T. Crosby and I. Patel, eds., *General Principles of Good Sampling Practice*. Cambridge, UK: Royal Society of Chemistry, 1995.

23. S. K. Thompson, *Sampling*, New York: Wiley, 1992.

SPREADSHEETS

24. D. Diamond and V. C. A. Hanratty, *Spreadsheet Applications in Chemistry Using Microsoft Excel*. New York: Wiley, 1997.

25. H. Freiser, *Concepts and Calculations in Analytical Chemistry: A Spreadsheet Approach*. Boca Rato, FL: CRC Press, 1992.

26. R. De Levie, *How to Use Excel in Analytical Chemistry and in General Scientific Data Analysis*. Cambridge, UK: Cambridge University Press, 2001.

Chapter Four

GOOD LABORATORY PRACTICE: QUALITY ASSURANCE OF ANALYTICAL MEASUREMENTS

"We can lick gravity, but sometimes the paperwork is overwhelming."
—Werner von Braun

We described in Chapter 1 the general principles of performing quantitative analyses, and in Chapters 2 and 3 we discussed aspects of sampling methodology and statistics and proper data handling and analysis. When you, as an analyst, adhere to these general guidelines, you will generally perform measurements properly, and if well-established methods are used, chances are good you will achieve acceptable (accurate) results. But, depending on what the results are to be used for, this may not be enough to satisfy the client. This is especially so if the measurements are for regulatory purposes or forensic analyses, all of which may have to be defended in court. As a result, the concepts of *good laboratory practice* (GLP), *method validation*, and *quality assurance* for testing laboratories have evolved as an approach to assure, to the extent possible, that reported analyzed results are correct within prescribed or documented limits. Various government agencies [the Environmental Protection Agency (EPA), the Food and Drug Administration (FDA)] and private agencies (e.g., AOAC International, ASTM) have promulgated their own specific guidelines for GLP or method validation and quality assurance. We will briefly describe some of these below, but they all have elements in common. We will first describe the basic elements for GLP.

The bottom line is that the lab management and analysts should use common sense in judging what quality assurance procedure should be implemented, based on the goal of the analysis, experience, available methods, time and cost constraints, and the like. But the closer you can adhere to accepted guidelines, the more confident you (and others) will be in your results. Remember, a proper analysis is more than simply receiving a sample and performing a one-shot analysis. If it is not properly documented, the analysis effort, time, and cost may be wasted.

Why Have Good Laboratory Practice?

The answer to this question is probably obvious. But it can be illustrated by the embarrassment faced by one of the premier analytical laboratories in the world, the Federal Bureau of Investigation (FBI) laboratory. In 1995, it was involved in a high-profile case, the bombing of the Alfred P. Murrah federal building in Oklahoma City, which partially leveled the building and killed 168 people and injured hundreds of others. The FBI lab had performed analyses for explosives at the scene and provided key evidence at trial. The jury found Timothy McVeigh guilty on all counts of conspiracy, bombing, and first-degree murder. But McVeigh's legal defense team, looking for weaknesses in the prosecution's case, introduced a 157-page Justice Department report on the FBI lab that had recently been released that listed a number of purported shoddy policies and practices (only 3 pages were admitted into evidence). The report was the result of an 18-month investigation that was triggered by a whistleblower in the lab who filed hundreds of complaints claiming contamination in the explosives unit lab, among hundreds of other accusations. The whistleblower even testified for the defense at the trial! While the Justice Department team found no evidence of contamination, and most of the whistleblower's allegations were not substantiated, the team did find evidence of insufficient documentation of test results, improper preparation of lab reports, and inadequate record management and record retention system. And the Justice Department concluded that management had failed to establish and enforce validated procedures and protocols. The outcome of the investigation was some 40 systemic suggestions for correcting or improving lab practices and procedures, including pursuing accreditation by the American Society of Crime Laboratory Directors/Laboratory Accreditation Board (ASCLD/LAB). Some practices the lab was to institute include:

- Each examiner who analyzes evidence should prepare and sign a separate report.
- All case files should contain notes, printouts, charts, and other data records used to reach conclusions.
- The lab must develop a record retention and retrieval system.
- Written procedures for handling evidence and for avoiding contamination should be refined.

Now, many of these concerns relevant to the FBI lab are not applicable to many other laboratories. But they illustrate the importance of instituting good laboratory practices. Had the FBI lab been more diligent in the care of its practices, it may have avoided the turmoil of this investigation.

4.1 What Is Good Laboratory Practice?

The exact definition of good laboratory practice depends on who is defining it and for what purpose. A broad definition encompasses such issues as organization of the laboratory, management, personnel, facilities, equipment, operations, method validation, quality assurance, and record keeping. The goal is to certify that every step of the analysis is valid. The aspects that need to be particularly addressed will vary by laboratory.

Good laboratory practices have been established by worldwide bodies such as the Organization for Economic Cooperation and Development (OECD) and the International Organization for Standardization (ISO). Government agencies have adopted them for their purposes as rules that must be followed for laboratories involved in analyzing substances that require regulation. Examples are pharmaceutical formulations, foods, and environmentally important samples.

GLPs can be defined as "a body of rules, operating procedures, and practices established by a given organization that are considered to be mandatory with a view to ensuring quality and correctness in the results produced by a laboratory" (M. Valcarcel, *Principles of Analytical Chemistry*, Berlin: Springer, 2000, p. 323). They all contain two common elements: standard operating procedures (SOPs) and a quality assurance unit (QAU). **Standard operating procedures** provide detailed descriptions of activities performed by the laboratory. Examples are sample custody chain, sample handling and preparation, the analytical method, instrument maintenance, archiving (record keeping), and the like. Detailed sample analysis procedures are provided for laboratory analysts or technicians to follow. These are generally more detailed than given in scientific publications of developed methods since the level of training and experience of different laboratory personnel will vary, even though highly trained analytical chemists may need less direction.

The **quality assurance unit** is generally independent from the laboratory and answers to the manager of the organization with which the laboratory is affiliated. The QAU is responsible for implementing quality procedures and assessing them on a continuing basis; this will include audits of the laboratory from time to time.

GLP ensures correct results are reported.

The laboratory should have SOPs for every method.

The QAU is responsible for assuring good laboratory practices are implemented. Everyone in the lab is responsible for following them.

4.2 Validation of Analytical Methods

Method validation is the process of documenting or proving that an analytical method provides analytical data acceptable for the intended use.

The basic concept of the validation process encompasses two aspects:

- The **problem** and the data requirements
- The **method** and its performance characteristics

First identify the problem and the requirements, then select the method to meet those requirements.

As mentioned in Chapter 1, the analytical process benefits when the analyst can be involved in defining the problem, that is, in making sure the proper questions are posed. When data requirements are poorly conceived or unrealistic, analytical measurements can be unnecessarily expensive if the method selected is more accurate than needed. Or, it may be inadequate if the method is less accurate than required, or of questionable value if the accuracy of the method is unknown. The first step in method development and validation is setting minimum requirements, which essentially are the specifications of the method for the intended purpose. How accurate and precise does it have to be? What is the target concentration?

HIERARCHY OF METHODOLOGY

We described in Chapter 1 the general procedure for establishing how an analysis will proceed. The hierarchy of methodology (Table 4.1) may be considered as follows:

Technique → method → procedure → protocol

Table 4.1

Hierarchy of Analytical Methodology[a]

	Definition	Example
Technique	Scientific principle useful for providing compositional information	Spectrophotometry
Method	Distinct adaptation of a technique for a selected measurement purpose	Pararosaniline method for measurement of sulfur dioxide
Procedure	Written directions necessary to use a method	ASTM D2914—Standard Test Method for the Sulfur Dioxide Content of the Atmosphere (West-Gaeke Method)
Protocol	Set of definitive directions that must be followed, without exception, if the analytical results are to be accepted for a given purpose	EPA Reference Method for the Determination of Sulfur Dioxide in the Atmosphere (Pararosaniline Method)

[a] Reprinted from J. K. Taylor, *Anal. Chem.*, **55** (1983) 600A. Published 1983 by the American Chemical Society.

These are critical steps in developing a method for a specific purpose that eventually leads to a validated method and address the list of validation characteristics above. The level of the hierarchy reached or used will depend on the need.

A **technique** is the scientific principle selected for providing compositional information. Spectrophotometry gives information about concentration, from the amount of light absorbed by the prepared sample solution. A **method** is the adaptation of the technique (using the appropriate chemistry), so it is selective for a given analyte. A **procedure** consists of the written directions necessary to utilize the method. (This is where we enter the broader area of GLP.) It does not necessarily reach the status of a standard method. Finally, a **protocol** is a set of specifically prescribed directions that must be followed, without exception, if the results are to be accepted for a given purpose, for example, for EPA regulations or action; the method has been validated to provide accurate results for the specified analyte in the specified matrix, and it is then a **reference method**.

VALIDATION PROCESS

The need to validate a method and the procedure to be followed are matters of professional judgment, although fairly well-prescribed procedures and guidelines are now available that aid in decision making.

Government and international agencies have issued guidelines for appropriate method validation, particularly for methods for regulatory submission. Generally, they include studies on:

- Selectivity
- Linearity
- Accuracy
- Precision
- Sensitivity
- Range
- Limit of detection
- Limit of quantitation
- Ruggedness or robustness

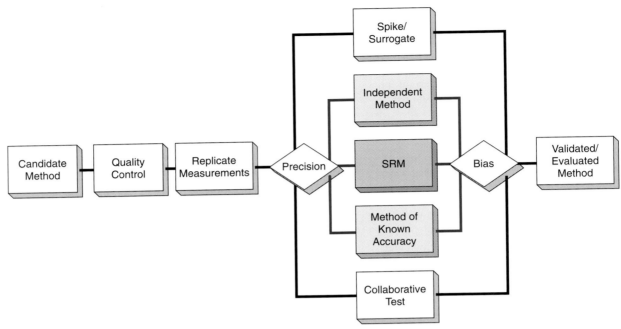

Fig. 4.1.   General process for evaluation/validation of methodology. [Reprinted from J. K. Taylor, *Anal. Chem.*, **55** (1983) 600A. Published 1983 by the American Chemical Society.]

These are best done during the development of a method. (If a method does not possess the required sensitivity, why proceed?) Figure 4.1 gives an overall view of the validation process. Different aspects are discussed in the following paragraphs.

SELECTIVITY

Selectivity is the extent that the method can measure the analyte of interest in the matrices of the samples being analyzed without interference from the matrix (including other analytes). Matrix effects may be either positive or negative. The analytical response of the analyte in the presence of potential sample components is compared with the response of a solution containing only the analyte. The selection of an appropriate measurement methodology is a key consideration. Methods, even previously validated in general terms, may not be assured to be valid for a particular sample matrix.

LINEARITY

A linearity study verifies that the response is linearly proportional to the analyte concentration in the concentration range of sample solutions. The study should be performed using standard solutions at five concentration levels, in the range of 50 to 150% of the target analyte concentration. Five concentration levels should allow detection of curvature on the calibration curve. Each standard should be measured at least three times.

   Linearity data are often judged from the coefficient of determination (r^2) and the y intercept of the linear regression line. An r^2 value of >0.998 is considered as evidence of acceptable fit of the data to the regression line. The y intercept should be a small percentage of the analyte target concentration, for example, $<2\%$. While these statistical evaluations are a practical way to assess linearity, they do not guarantee it. You should always do a visual inspection of the calibration curve. The linearity will often deviate somewhat at high and low values. (This is the reason *weighted* least-squares plots may be preferred, where the points on the curve with the least deviation are given more weight in the regression line.) One way of evaluating the range of linearity is to plot a *response factor* (RF) versus concentration.

$$\text{Response factor} = (\text{signal} - y \text{ intercept})/\text{concentration} \qquad (4.1)$$

The response per unit concentration should be nearly constant for good linearity.

If a plot with zero slope is obtained, this indicates that a linear response is obtained over this concentration range. A response factor change over the calibration concentration range within, for example, 2 to 3% of the target-level response factor or the average RF may be considered acceptable linearity. In Figure 3.8 the regression line is $y = mx + x$. The y intercept is 0.595. A plot of the response factor versus concentration is shown in Figure 4.2. The slope of the line is -1.48 RF/1 ppm. This corresponds to -1.0 over the concentration range 0.1 to 0.8 ppm, which is 1.8% of the average RF value of 54.4. This is acceptable linearity.

If a calibration curve deviates from linearity over the 50 to 150% target-level range, selection of a narrower range of, for example, 80 to 120% may provide the desired linearity.

ACCURACY

Accuracy of a method is the closeness of the obtained value to the true value for the sample. This is probably the most difficult parameter to validate. One should consider the sampling and sample treatment, in addition to the measurement method accuracy. Accuracy of the method can be determined in one of three ways. In increasing order of importance, these are:

- Recovery studies
- Comparison with results using another method known to be accurate
- Analysis of a reference material

Recovery studies are performed by spiking (adding) a known amount of the analyte either to a blank matrix (a sample that has an unmeasurable level of the test

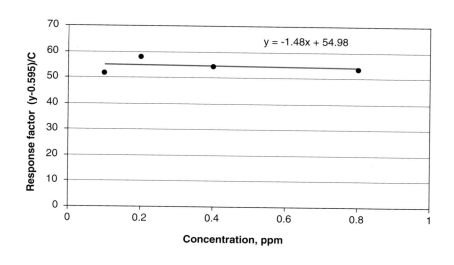

Fig. 4.2. Response factor plot for Figure 3.8. The *y*-intercept value is subtracted from the corresponding fluorescence intensity for each concentration (0.1, 0.2, 0.4, and 0.8 ppm) and divided by each concentration.

analyte) or by spiking a sample in which the background analyte is measured by the same procedure and subtracting from the total (sample + spike) value to obtain the recovery. The spiked samples should be prepared at three levels, the extremes and the midrange. They should be prepared at least in triplicate.

A better validation method is to perform the analysis by *two independent methods*, in which the second method is known to be accurate for the sample matrix of interest. Ideally, even the sample treatment should be different. You can often find in the scientific literature (journals, reference books, standard methods books) a method that is applicable to your sample (but that may not be appropriate to use because of expense, unavailability of equipment, etc.). If none can be found that has been applied to your sample matrix, but one is known to be generally applicable and accurate, then use this. If results by your method and the second method agree, that is good evidence they both work for your sample. If there is disagreement, then it is not possible to draw any conclusions since either may give erroneous results with your particular sample, although, it is probably more likely your new method is the culprit.

The ideal way to validate a method is to analyze a *reference material* identical in composition to your sample. The National Institute of Standards (NIST) has the goal of ensuring accurate and compatible measurements through the development, certification, and distribution of *standard reference materials* (SRMs). The SRM program has over 1000 SRMs available for use in (1) basic measurements in science and meteorology, (2) environmental analysis, (3) health measurements, and (4) industrial materials and production. The NIST has standards for chemical composition, physical properties, engineering materials, and the like (http://ts.nist.gov/ts/htdocs/230/232/info/index.htm). It serves as the main contact point for interfacing with similar efforts in the private sector, other federal agencies, and internationally. Other programs include the American Society for Testing and Materials (ASTM), the American Association for Clinical Chemistry (AACC), the International Union of Pure and Applied Materials (IUPAC), the International Organization for Standardization (ISO), and the European Union (EU). You can obtain information about each of these from their websites.

Chemical composition standards are certified for given concentrations, with a statistical (standard deviation) range given. If your method falls two standard deviations from the certified value, there is a 95% chance there is a significant (nonrandom) difference between the results. Depending on the concentration levels being measured, you may establish that the measurement should be within, for example, $\pm 2\%$ of the certified value, or perhaps $\pm 10\%$ if it is a trace analysis, and so forth.

There may not be reference materials available that are identical in composition to your sample, but similar. These will still provide a high level of confidence in the validation.

When performing measurements on reference materials or by comparison with, another method, statistical considerations suggest that at least six degrees of freedom (seven measurements) be made for proper validation.

> Accuracy is best determined by analysis of a standard reference material.

> Perform at least seven measurements for statistical validation.

> Repeatability is intralaboratory precision.

PRECISION

The precision of an analytical method is obtained from multiple analyses of a homogeneous sample. You can determine overall precision of the method, including sample preparation. Such precision data are obtained by one laboratory on one day, using aliquots of the homogeneous sample that have been independently prepared. Such interlaboratory precision is called *repeatability*. Interlaboratory precision, if appropriate, is also determined as part of a measurement of *reproducibility* or *robustness* of the method (see below).

You can also determine the precision of different steps of the analysis, for example, the precision of injecting a sample into a gas chromatograph determined from multiple injections of the same sample solution. Again, statistical considerations dictate that at least seven measurements should be made for each evaluation step.

SENSITIVITY

The sensitivity is the ability to distinguish two different concentrations and is determined by the slope of the calibration curve. You can measure the slope or measure samples of closely related concentrations at high, intermediate, and low concentrations. The sensitivity and precision will govern how many significant figures should be reported in a measurement. Do not report 11.25% when the method is lucky to distinguish 0.1% differences (1% relative differences).

RANGE

The working range of a method is the concentration range over which acceptable accuracy and precision are obtained. Usually it also includes linearity. The acceptable accuracy and precision are as defined in establishing criteria for the method. The precision will, of course, vary with the concentration, becoming poorer at low concentrations (Figure 4.3), as well as sometimes at high concentrations, as in spectrophotometric measurements.

LIMIT OF DETECTION

The limit of detection should be determined using a definition given in Chapter 3. Typically, replicate blanks of the sample matrix are analyzed to determine the mean blank value and its standard deviation. Then a matrix is spiked with analyte near the detection limit (e.g., to give a signal 10 times the standard deviation above the blank mean signal). The limit of detection is the concentration calculated to give a response equal to the blank signal plus three standard deviations.

LIMIT OF QUANTITATION

This is the lowest concentration of analyte that can be measured in the sample matrix at an acceptable level of precision and accuracy. An acceptable precision is typically 10 to 20% relative standard deviation, depending on the concentration levels measured. Absent a specified precision, then the concentration that gives a signal 10 standard deviations above the blank is used.

RUGGEDNESS/ROBUSTNESS

We have defined the precision of a method. *Repeatability* is the long-term precision over several weeks. *Ruggedness* refers to the precision of one lab over multiple days,

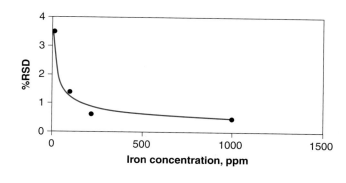

Fig. 4.3. Dependence of relative standard deviation on concentration.

which may include multiple analysts, multiple instruments, different sources of reagents, different chromatographic columns, and the like. A ruggedness study will identify those factors that will contribute to variability of the results and should not be changed. This is related to *robustness* or *reliability* of the method, which refers to how sensitive it is to deliberate or uncontrolled small changes in parameters, such as the size of the sample, the temperature, pH of the solution, reagent concentration, time of reaction, and so forth. It includes an evaluation of the stability of reagents, standards, and samples with time. Each parameter should be tested separately, unless statistically more sophisticated factorial analysis experiments for varying several parameters at once are designed, which we will not go into here.

Reproducibility (or *transferability*) is the analysis of the same sample between labs, in which a homogeneous sample is analyzed by multiple labs with one lab serving as the primary comparison lab. A reproducibility study generally focuses on *bias* between labs, besides precision. One strives for a bias that is within defined acceptable limits.

Interlaboratory variability is about double intralaboratory variability.

Reality Check on Interlaboratory Variability

Is interlaboratory variability significant? Is it different than intralaboratory variability? The answer on both accounts is yes. And it varies with concentration. William Horwitz and collaborators documented interlaboratory variability over two decades by examining over 10,000 interlaboratory data sets (see R. H. Albert, *Chemical & Engineering News*, September 13, 1999, p. 2). They developed an expression relating interlaboratory standard deviations of results, s_R, to concentration, C (expressed as decimal fraction, e.g., 1 mg/kg = 10^6). They found $s_R = 0.02C^{0.85}$, or as a relative standard deviation among laboratories, rsd(%) = $2C^{-0.15}$. These expressions indicate that starting with pure materials ($C = 1$) with an s_R of 2%, the interlaboratory precision increases by a factor of 2 for every two-decade decrease in concentration. This holds, irrespective of the analyte, method, matrix, or date. The precision of different types of analyses, whether agricultural, geological, or pharmaceutical, did not change for a half-century, even with the advent of modern instrumentation. Thus, the relative standard deviation for pesticide residue levels of 1 ppm (1 mg/kg; 10^{-6}) is 16%. (Do the calculation using either formula. Putting the equation in Excel makes it easy.)

The fact that this empirical function follows collaborative study statistics was substantiated by a separate study [M. Thompson and P. J. Lowthian, *J. AOAC Int.*, **80** (1997) 6786] in which it was shown that interlaboratory variability is about double that of intralaboratory variability. The variability for EPA-tested analytes with extensive quality assurance built into the procedure was somewhat better than predicted by the above [see *J. AOC Int.*, **79** (1996) 589], but the analyses cost approximately $1000 each, besides being slow. There is an obvious trade-off between quality assurance effort and cost and time.

Review Figure 4.1, which places in context most of the validation concepts and steps for a candidate method that we have discussed. We will discuss quality control in the context of quality assurance below.

4.3 Quality Assurance—Does the Method Still Work?

Once a method has been validated, an important aspect of applying it is to assure that it is working properly. **Quality assurance** (QA) is the implementation of procedures to ensure and document that the method continues to perform as required and is part of the responsibility of the quality assurance unit. It includes written documentation of validation of the method, procedures followed, and the sample custody chain. A number of **quality control** procedures are implemented, based on quantitative measurements. Typical quality control activities are listed below.

QA is an ongoing checking procedure to assure proper performance of a method.

CONTROL CHARTS

The laboratory should maintain a continuing quality control chart (Figure 3.6) for each method. A reference material of known analyte content is blindly and randomly run each day, or preferably with each batch of samples. If measured values fall outside prescribed standard deviation limits, then you should check for some systematic error such as reagent deterioration or instrument drift (needs recalibration).

DOCUMENTING AND ARCHIVING

This is a tedious and time-consuming, but critical, part of quality assurance. All activities performed by the laboratory dealing with quality assurance should be documented in written form. This includes recording the chain of custody of the sample, the calibration and performance of instruments, standard operating procedures, original measurement data, results, and reports. Documents should be traced to individuals, meaning they should be signed and dated by the individual creating or responsible for them.

PROFICIENCY TESTING

One way of documenting performance of the laboratory is to participate in collaborative interlaboratory studies. An official body provides aliquots of the same homogeneous material to laboratories for analysis. The goal is to compare results among laboratories and the uncertainties in the results. The mean of the results of the participating laboratories can be used as the reference, if the actual concentration is not known. Or better, a certified reference material whose concentration and certainty is known (not to the participating laboratories) is used. The latter is particularly more informative if the laboratories use different methods.

One way of expressing the results of a collaborative testing exercise is to report the laboratory's z score, a measure of its deviation from the standard deviation of the known concentration:

$$z = \frac{\overline{X}_i - \hat{X}}{s} \tag{4.2}$$

where $\overline{X}_i$ is the mean of i replicate measurements by the laboratory, $\hat{X}$ is the accepted concentration, and s is the standard deviation of the accepted concentration.

Example 4.1

You agree to participate in a collaborative study for the determination of calcium in serum. A sample containing 5.2 meq/dL, with a standard deviation of ± 0.2 meq/dL, is sent to 10 laboratories for analysis using atomic absorption spectroscopy. You obtain triplicate results of 5.0, 4.7, and 4.8 meq/dL. What is the z value for your laboratory? What do the results imply?

Solution

The mean is 4.8 meq/dL with a standard deviation of ± 0.15 meq/dL. The z value is

$$z = \frac{4.8 - 5.2}{0.2} = -2.0$$

This means your reported mean is low by two standard deviations away from the accepted value. There is a 95% chance this difference is due to a nonrandom systematic error. Also, the range for one standard deviation of your measurements is 4.6 to 5.0 meq/dL. The accepted range is 5.0 to 5.4 for one standard deviation. There is just a 68% chance that your value overlaps the accepted range.

Figure 4.4 shows the results of a representative proficiency collaborative test. It is not uncommon for several laboratories to be outside acceptable ranges.

4.4 Laboratory Accreditation

Another form of external evaluation is laboratory accreditation by a formal organization or government agency. This is generally voluntary but may be required for laboratories dealing with regulatory measurements. Accreditation is a procedure by which an authoritative body gives formal recognition that the laboratory is competent to carry out specific tasks. The accreditation procedure may take the form of qualitative inspection of the laboratory operations, to verify that good laboratory practice policies are followed, that is, proper documentation and record keeping, validation, proficiency testing, and the like. Or it may include measurement of

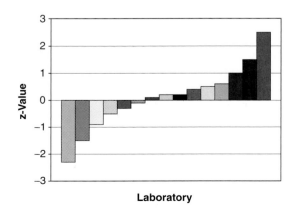

Fig. 4.4. Representative z-value distributions for proficiency tests with a series of laboratories.

submitted reference materials. In any event, certification will involve periodic laboratory audits, which may be unannounced.

4.5 Electronic Records and Electronic Signatures: 21 CFR, Part 11

Most laboratory tasks are dependent on computers, from sample log-in to reporting. The traditional way of maintaining records for audits, regulatory actions, and the like has been by printing hardcopy for signature, submission, and archiving. This process is time consuming, requires storage facilities, records can be lost or misplaced, and in part it defeats the purpose of computers. If acceptable records and signatures could be achieved electronically, this would improve efficiency. It would provide faster access to documents, the ability to search databases and view information from multiple perspectives, to determine trends or patterns. The Food and Drug Administration (FDA) worked for 6 years with the pharmaceutical industry to develop procedures to accommodate paperless record systems under the current good manufacturing practices (GMP) regulations. In 1997, the FDA issued the Final Rule on electronic records, signatures, and submissions, known as 21 *Code of Federal Regulations* (CFR), Part 11 ["Electronic Records; Electronic Signatures," *Fed. Reg.*, **62** (1997) 1000, 13,230; **64** (1999) 41442]. You can find it at www.fda.gov/ora under Compliance References. The main concern and challenge is that electronic records can be too easily changed or falsified, either accidentally or intentionally. The Final Rule provides criteria under which the FDA will consider electronic records to be equivalent to paper records, and electronic signatures equivalent to handwritten signatures, to ensure the integrity, accuracy, and authenticity of information stored in the systems.

ELECTRONIC RECORDS

Electronic validation will need to be done to document data integrity, backup and recovery, archiving and restoring, and how electronic signatures are used. A validated system must be for the life cycle of the software. If it is changed or updated, the data must be transferable.

One problem is that electronic records are comprised of databases, which are dynamic, that is, the content changes as new information is added. Worse, data can be changed or deleted, with no evidence and in a manner that destroys the original data. System access must be limited to authorized individuals. There must be regular system checks. There must be time- and date-stamped audit trails. If changes are made in the database, the audit trail must show who made the changes, when, what the old and new values are, and why the data were modified.

ELECTRONIC SIGNATURES

Access to the system must be limited to authorized persons. The type of security will depend on whether the system is open or closed. Electronic signature technologies include identification codes (user names, passwords) or more sophisticated biometric systems (based on measurement of physical features such as palm prints, finger prints, or iris or retinal pattern scanners). The latter is expensive and less likely to be implemented, especially for multiple users. User names and passwords must be unique and never reassigned. Passwords should be changed periodically.

21 CRF, Part 11, permits but does not require the use of electronic records and signatures. As more systems become validated and accepted, and as more

instrument manufacturers incorporate validated systems, this will become more commonplace. It is likely other agencies will adopt similar standards.

EPA: CROMERRR

The EPA's Office of Environmental Information (OEI) has defined a Cross Media Electronic Reporting and Record-Keeping Rule (COMERRR) to remove existing regulatory obstacles to electronic reporting and record keeping across a broad spectrum of EPA programs. The 80-page document can be found at the URL at www.epa.gov/cdx/cromerrr_rule.pdf (www.epa.gov/cdx is EPA's central data exchange site). CROMERRR will require criteria for electronic records that are consistent with 21 CFR, Part 11.

> QA costs are about a quarter of laboratory costs. Do it well!

What about Cost?

With quality assurance, while it is no guarantee of accurate results (see the box Reality Check on Interlaboratory Variability), it is necessary to have reasonable documentation of what the accuracy is and to identify areas where significant contributors to inaccuracies may occur and actions taken to minimize them. This, of course, does not come without cost. Implementing a quality assurance program will involve substantial initial investment, both in expense and time. It has been estimated that ongoing quality assurance costs amount to 20 to 30% of the laboratory budget. So it is important that the system be properly set up, as efficiently as possible (which will require an understanding by management of what is needed—could that be you?), and that it be taken seriously by all laboratory personnel (which certainly includes you).

4.6 Some Official Organizations

A number of government agencies and national and international organizations have established their own guidelines for method validation and good laboratory practice. Most are based on principles espoused by multinational organizations. Some of the major ones are listed below. Detailed information on each is given on the text website. Do take a look at these. You can obtain more (a lot more!) information on each by browsing their websites. It gives a flavor of the real world of standardization and regulation.

International Organization for Standardization (ISO): www.iso.ch

International Conference on Harmonization (ICH): www.ich.com

Organization for Economic Cooperation and Development (OECD): www.oecd.org

Food and Drug Administration (FDA): www.fda.gov/cder

Environmental Protection Agency (EPA): www.epa.gov

Office of Enforcement and Compliance: Laboratory Data Integrity Branch: http://es.epa.gov/labdata.html

Office of Solid Waste: www.epa.gov/osw

US-EPA, Region 4, Science and Ecosystem Support Division: www.epa.gov/region4/sesd

American Association for Clinical Chemistry (AACC): www.aacc.org

American Association of Cereal Chemistry (AACC): www.scisoc.org/aacc

American Oil Chemists Society (AOCS): www.aocs.org

The Society of Quality Assurance (SQA): www.sqa.org

American Society for Testing and Materials (ASTM): www.astm.org

Association of Official Analytical Chemists International (AOAC International): www.aoac.org

National Institute of Standards and Technology (NIST): www.nist.gov

PRACTICING GLP PROCEDURES

Experiment 39 provides practice in method validation and quality control, and Experiment 40 is an exercise in proficiency testing. These are class team experiments. Read these, even if they are not part of your assigned laboratory exercises.

Learning Objectives

WHAT ARE SOME OF THE KEY THINGS WE LEARNED FROM THIS CHAPTER?

- Good laboratory—what it is, how to apply it, p. 125
- How to validate a method: selectivity, linearity, accuracy, precision, sensitivity, range, LOD, LOQ, ruggedness, p. 126
- Quality assurance: control charts, documenting, proficiency testing, p. 133
- Electronic records, p. 135
- Official organizations that provide GLP information, p. 136

Questions

GOOD LABORATORY PRACTICE

1. What is good laboratory practice?
2. What are the common elements of GLP implementation?
3. What are SOPs?
4. What are the characteristics of a quality assurance unit?

METHOD VALIDATION

5. What are the two aspects of the validation process?
6. What is the first step in a method development?
7. Distinguish a technique, a method, a procedure, and a protocol.
8. What are the essential features of most method validation processes?

9. What is the response factor?

10. What are ways of assessing calibration linearity?

11. What are the main ways of assessing accuracy of a method?

12. How many measurements should you make to obtain reasonable statistical validation?

13. Distinguish among repeatability, ruggedness, robustness, and reproducibility of a method.

14. What are the main requirements for validation of electronic records and signatures?

QUALITY ASSURANCE

15. What is quality assurance? Quality control?

16. What are some typical quality control procedures?

17. What is a z score?

18. What is laboratory accreditation?

Problems

VALIDATION

19. You prepare a calibration curve for the measurements of blood ethanol by gas chromatography. The recorded peak area as a function of concentration is:

Concentration, % (wt/vol)	Peak Area (arbitrary units)
0	0.0
0.020	43
0.040	80
0.080	155
0.120	253
0.160	302
0.200	425

Plot the calibration curve using Excel and determine the least-squares line, indicating the y intercept and slope. Calculate the response factor and the slope of its plot versus concentration. What percent of the average response factor is the RF change over the calibration range?

20. Calculate the 16% relative standard deviation interlaboratory variation listed for 1 ppm pesticide residue levels in the box Reality Check on Interlaboratory Variability. Perform the calculation using both formulas given. Place each formula in an Excel spreadsheet cell to perform the calculations.

QUALITY ASSURANCE

21. You participate in a collaborative study for measuring lead in leaves. A homogeneous standard reference material of ground leaves, certified to contain 10.3 ± 0.5 ppm lead, is given to the participating labs. You analyze the sample, using acid digestion and atomic absorption spectrometry. You report 9.8 ± 0.3 ppm for seven analyzed aliquots. What is the z value for your laboratory?

WEB EXERCISE

22. Look up the websites of at least three of the government agencies and professional societies listed in Section 4.6, and link to their pages dealing with method validation. Document the similarities and differences between them.

Recommended References

WEBSITES

1. www.labcompliance.com/index.htm. Contains a wide variety of information related to compliance issues. Check the Regulations link. Defines terminology used. Excellent tutorial on GLP.

2. http://21cfrpart11.com. This commercial site dealing with compliance-related issues includes useful links.

3. www.PDA.org. Features an online conference on computer validation issues and 21CFR, Part 11.

4. www.isric.nl/GLP.htm. International soil reference and information center. Provides FAO Soils Bulletin 74, "Guidelines for Quality Management in Soil and Plant Laboratories."

5. www.waters.com. This company website has a link to Introduction to Validation (under Applications), which gives a nice summary of the validation process, for USP method validation and for ICH method validation guidelines.

GOOD LABORATORY PRACTICES

6. J. M. Miller and J. B. Crowther, eds., *Analytical Chemistry in a GMP Environment: A Practical Guide.* New York: Wiley, 2000.

7. J. Kenkel, *A Primer on Quality in the Analytical Laboratory.* Boca Raton, FL: CRC Press, 2000.

8. W. Garner, M. S. Barge, and P. Ussary, eds., *Good Laboratory Practice Standards.* Washington, DC: American Chemical Society (Oxford), 1992.

QUALITY ASSURANCE/QUALITY CONTROL

9. H. Marchandise, "Quality and Accuracy in Analytical Chemistry," *Fresenius' J. Anal. Chem.*, **345** (1993) 82.

10. H. Y. Aboul-Enein, R-I. Stefan, and G-E. Baiulescu, *Quality and Reliability in Analytical Chemistry.* Boca Raton, FL: CRC Press, 2000.

11. F. E. Prichard, *Quality in the Analytical Chemistry Laboratory.* New York: Wiley, 1999.

12. M. Sargent and G. MacKay, eds., *Guidelines for Achieving Quality in Trace Analysis.* Cambridge, UK: Royal Society of Chemists, 1995. (Available from the American Chemical Society)

13. J. M. Green, "A Practical Guide to Analytical Method Validation," *Anal. Chem.*, **68** (1996) 305A.

14. M. Swartz and I. S. Krull, *Analytical Method Development and Validation.* New York: Marcel Dekker, 1997.

15. L. Huber, *Validation and Qualification in Analytical Laboratories.* Buffalo Grove, IL: Interpharm, 1999.

16. M. Stoeppler, W. R. Wolf, and P. S. Jenks, eds., *Reference Materials for Chemical Analysis: Certification, Availability and Proper Usage.* New York: Wiley, 2001.

17. D. G. Rhoads, *Lab Statistics Fun and Easy: A Practical Approach to Method Validation.* Washington, DC: AACC (American Association for Clinical Chemistry), 1999.

18. D. A. Sanders, *Passing Your ISO 9000/QS-9000 Audit: A Step-by-Step Guide.* Washington, DC: AACC, 1996.

19. R. D. McDowall, "Validation of Spectrometry Software. Part II: Roles of Validation Plan and User Requirement Specifications," *Spectroscopy,* **16**(7) (2001) 30, www.spectroscopyonline.com. Discusses the technical aspects of the validation plan and what it must include.

20. J. Kenkel, *A Primer on Quality in the Analytical Laboratory.* Boca Raton, FL: Lewis, 1999.

Chapter Five

STOICHIOMETRIC CALCULATIONS: THE WORKHORSE OF THE ANALYST

Analytical chemistry deals with measurements of solids and solution concentrations of them, from which we calculate masses. Thus, we prepare solutions of known concentrations that can be used to calibrate instruments or to titrate sample solutions. We calculate the mass of analyte in a solution from its concentration and the volume. We calculate the mass of product expected from the mass of reactants. All of these calculations require a knowledge of **stoichiometry**, that is, the ratios in which chemicals react, from which we apply appropriate conversion factors to arrive at the desired calculated results.

Stoichiometry deals with the ratios in which chemicals react.

In this chapter we review the fundamental concepts of mass, moles, and equivalents; the ways in which analytical results may be expressed for solids and liquids; and the principles of volumetric analysis and how stoichiometric relationships are used in titrations to calculate the mass of analyte.

5.1 Review of the Fundamentals

Quantitative analysis is based on a few fundamental atomic and molecular concepts, which we review below. You have undoubtedly been introduced to these in your general chemistry course, but we briefly review them here since they are so fundamental to quantitative calculations.

THE BASICS: ATOMIC, MOLECULAR, AND FORMULA WEIGHTS

The gram-atomic weight for any element is the weight of a specified number of atoms of that element, and that number is the same from one element to another. A gram-atomic weight of any element contains exactly the same number of atoms of that element as there are carbon atoms in exactly 12 g of carbon 12. This number is Avogadro's number, 6.022×10^{23} atoms/g-at wt.

Since naturally occurring elements consist of mixtures of isotopes, the chemical atomic weights will be an average of the isotope weights of each element, taking into account their relative naturally occurring abundances. Thus, none of the elements has an integral weight. For example, bromine has two isotopes: ^{79}Br with atomic weight 78.981338 at a relative abundance of 100.0, and ^{81}Br with atomic weight 80.9162921 with relative abundance of 97.28 (i.e., 50.69 and 49.31%, respectively). These average to 79.904, the natural atomic weight we use in chemical calculations. Another measurement used by chemists is **gram-molecular weight** (gmw), defined as the sum of the atomic weights of the atoms that make up a compound. The term **gram-formula weight** (gfw) is a more accurate description for substances that don't exist as molecules but exist as ionic compounds (strong electrolytes—acids, bases, salts). We sometimes use the term **molar mass** in place of gram-formula weight. We will simply use the term **formula weight** throughout our calculations.

We will use formula weight (f wt) to express grams per mole.

WHAT IS A DALTON?

Biologists and biochemists sometimes use the unit **dalton** to report masses of large proteinaceous substances such as chromosomes, ribosomes, viruses, and mitochondria, where the term molecular weight would be inappropriate. The mass of a single carbon-12 atom is equivalent to 12 daltons, and 1 dalton is therefore 1.661×10^{-24} g, the reciprocal of Avogadro's number. The number of daltons in a single molecule is numerically equivalent to the molecular weight (g/mol). Strictly speaking, it is not correct to use the dalton as a unit of molecular weight, and it should be reserved for the types of substances mentioned above. For example, the mass of an *Escherichia coli* bacterium cell is about 1×10^{-12} g, or 6×10^{11} daltons.

MOLES: THE BASIC UNIT FOR EQUATING THINGS

The chemist knows that atoms and molecules react in definite proportions. Unfortunately, he or she cannot conveniently count the number of atoms or molecules that participate in a reaction. But since the chemist has determined their relative masses, he or she can describe their reactions on the basis of the relative masses of atoms and molecules reacting, instead of the number of atoms and molecules reacting. For example, in the reaction

$$Ag^+ + Cl^- \rightarrow \underline{AgCl}$$

we know that one silver ion will combine with one chloride ion. We know further, since the atomic weight of silver is 107.870 and the atomic weight of chlorine is 35.453, that 107.870 mass units of the silver will combine with 35.453 mass units of chlorine. To simplify calculations, chemists have developed the concept of the **mole**, which is Avogadro's number (6.022×10^{23}) of atoms, molecules, ions, or other species. Numerically, it is the atomic, molecular, or formula weight of a substance expressed in **grams**.[1]

There are 6.022×10^{23} atoms in a mole of atoms.

Now, since a mole of any substance contains the same number of atoms or molecules as a mole of any other substance, atoms will react in the same mole ratio as their atom ratio in the reaction. In the above example, one silver ion reacts with one chloride ion, and so each mole of silver ion will react with one mole of chloride ion. (Each 107.87 g will react with 35.453 g.)

---

[1] Actually, the term *gram-atomic weight* is more correct for atoms, *gram-formula weight* for ionic substances, and *gram-molecular weight* for molecules, but we will use *moles* in a broad sense to include all substances. In place of gram-formula weight we will simply use *formula weight* (f wt)

Example 5.1

Calculate the number of grams in one mole of $CaSO_4 \cdot 7H_2O$.

Solution

One mole is the formula weight expressed in grams. The formula weight is

Ca	40.08	
S	32.06	
11 O	176.00	
14 H	14.11	
	262.25 g/mol	

The number of moles of a substance is calculated from

$$\text{Moles} = \frac{\text{grams}}{\text{formula weight (g/mol)}} \qquad (5.1)$$

where formula weight represents the atomic or molecular weight of the substance. Thus,

$$\text{Moles Na}_2\text{SO}_4 = \frac{g}{f\ wt} = \frac{g}{142.04\ \text{g/mol}}$$

$$\text{Moles Ag}^+ = \frac{g}{f\ wt} = \frac{g}{107.870\ \text{g/mol}}$$

Since many experiments deal with very small quantities, a more convenient form of measurement is the **millimole**. The formula for calculating millimoles is

$$\text{Millimoles} = \frac{\text{milligrams}}{\text{formula weight (mg/mmol)}} \qquad (5.2)$$

g/mol = mg/mmol = formula weight; g/L = mg/mL; mol/L = mmol/mL = molarity.

Just as we can calculate the number of moles from the gram of material, we can likewise calculate the grams of material from the number of moles:

$$g\ \text{Na}_2\text{SO}_4 = \text{moles} \times f\ wt = \text{moles} \times 142.04\ \text{g/mol}$$

$$g\ \text{Ag} = \text{moles} \times f\ wt = \text{moles} \times 107.870\ \text{g/mol}$$

Again, we usually work with millimole quantities, so

$$\text{Milligrams} = \text{millimoles} \times \text{formula weight (mg/mmol)} \qquad (5.3)$$

Note that g/mol is the same as mg/mmol, g/L the same as mg/mL, and mol/L the same as mmol/mL.

Example 5.2

Calculate the number of moles in 500 mg Na_2WO_4 (sodium tungstate).

Solution

$$\frac{500 \text{ mg}}{293.8 \text{ mg/mmol}} \times 0.001 \text{ mol/mmol} = 0.00170 \text{ mol}$$

Example 5.3

How many milligrams are in 0.250 mmol Fe_2O_3 (ferric oxide)?

Solution

$$0.250 \text{ mmol} \times 159.7 \text{ mg/mmol} = 39.9 \text{ mg}$$

5.2 How Do We Express Concentrations of Solutions?

Chemists express solution concentrations in a number of ways. Some are more useful than others in quantitative calculations. We will review here the common concentration units that chemists use. Their use in quantitative volumetric calculations is treated in more detail below.

MOLARITY—THE MOST WIDELY USED

The mole concept is useful in expressing concentrations of solutions, especially in analytical chemistry, where we need to know the volume ratios in which solutions of different materials will react. A one-**molar** solution is defined as one that contains one mole of substance in each liter of a solution. It is prepared by dissolving one mole of the substance in the solvent and diluting to a final volume of one liter in a volumetric flask; or a faction or multiple of the mole may be dissolved and diluted to the corresponding fraction or multiple of a liter (e.g., 0.01 mol in 10 mL). More generally, the **molarity** of a solution is expressed as moles per liter or as millimoles per milliliter. Molar is abbreviated as M, and we talk of the *molarity* of a solution when we speak of its concentration. A one-molar solution of silver nitrate and a one-molar solution of sodium chloride will react on an equal-volume basis, since they react in a $1:1$ ratio: $Ag^+ + Cl^- \rightarrow AgCl$. We can be more general and calculate the moles of substance in any volume of the solution.

$$\text{Moles} = (\text{moles/liter}) \times \text{liters}$$
$$= \text{molarity} \times \text{liters}$$

(5.4)

The liter is an impractical unit for the relatively small quantities encountered in titrations, and we normally work with milliliters. This is what your buret reads. So,

$$\boxed{\begin{array}{c} \text{Millimoles} = \text{molarity} \times \text{milliliters} \\ \text{(or mmol} = M \times \text{mL)} \end{array}} \qquad (5.5)$$

We often work with millimoles in analytical chemistry. Remember this formula!

Example 5.4

A solution is prepared by dissolving 1.26 g $AgNO_3$ in a 250-mL volumetric flask and diluting to volume. Calculate the molarity of the silver nitrate solution. How many millimoles $AgNO_3$ were dissolved?

Solution

$$M = \frac{1.26\ \text{g}/169.9\ \text{g/mol}}{0.250\ \text{L}} = 0.0297\ \text{mol/L (or 0.0297 mmol/mL)}$$

Then,

$$\text{Millimoles} = (0.0297\ \text{mmol/mL})(250\ \text{mL}) = 7.42\ \text{mmol}$$

Always remember that *the units in a calculation must combine to give the proper units in the answer*. Thus, in this example, grams cancel to leave the proper unit, moles/liter, or molarity. Using units in the calculation to check if the final units are proper is called **dimensional analysis**. Accurate use of dimensional analysis is essential to properly setting up computations.

Always use dimensional analysis to set up a calculation properly. Don't just memorize a formula.

Example 5.5

How many grams per milliliter of NaCl are contained in a 0.250 M solution?

Solution

0.250 mol/L = 0.250 mmol/mL

0.250 mmol/mL $\times$ 58.4 mg/mmol $\times$ 0.001 g/mg = 0.0146 g/mL

Example 5.6

How many grams Na_2SO_4 should be weighed out to prepare 500 mL of a 0.100 M solution?

Solution

500 mL $\times$ 0.100 mmol/mL = 50.0 mmol

50.0 mmol $\times$ 142 mg/mmol $\times$ 0.001 g/mg = 7.10 g

Example 5.7

Calculate the concentration of potassium ion in grams per liter after mixing 100 mL of 0.250 M KCl and 200 mL of 0.100 M K_2SO_4.

Solution

$$\text{mmol } K^+ = \text{mmol KCl} + 2 \times \text{mmol } K_2SO_4$$
$$= 100 \text{ mL} \times 0.250 \text{ mmol/mL}$$
$$+ 2 \times 200 \text{ mL} \times 0.100 \text{ mmol/mL}$$
$$= 65.0 \text{ mmol in } 300 \text{ mL}$$
$$\frac{65.0 \text{ mmol} \times 39.1 \text{ mg/mmol} \times 0.001 \text{ g/mg} \times 1000 \text{ mL/L}}{300 \text{ mL}} = 8.47 \text{ g/L}$$

NORMALITY

The equivalent weight (or the number of reacting units) depends on the chemical reaction. It may vary most often in redox reactions, when different products are obtained.

Although molarity is widely used in chemistry, some chemists use a unit of concentration in quantitative analysis called **normality** (N). A one-**normal** solution contains one equivalent per liter. An **equivalent** represents the mass of material providing Avogadro's *number of reacting units*. A reacting unit is a *proton* or an *electron* (see Section 5.6, Equations 5.40 and 5.41). The number of equivalents is given by the number of moles multiplied by the number of reacting units per molecule or atom; the **equivalent weight** is the formula weight divided by the number of reacting units. Table 5.1 lists the reacting units used for different types of reactions. For acids and bases, the number of reacting units is based on the number of protons (i.e., hydrogen ions) an acid will furnish or a base will react with. For oxidation–reduction reactions it is based on the number of electrons an oxidizing or reducing agent will take on or supply. Thus, for example, sulfuric acid, H_2SO_4, has two reacting units of protons; that is, there are two equivalents of protons in each mole. Therefore,

$$\text{Equivalent weight} = \frac{98.08 \text{ g/mol}}{2 \text{ eq/mol}} = 49.04 \text{ g/eq}$$

g/eq = mg/meq = equivalent weight; eq/L = meq/mL = normality.

So, the normality of a sulfuric acid solution is twice its molarity, that is, $N = $ (g/eq wt)/L. The number of equivalents is given by

$$\boxed{\text{Number of equivalents (eq)} = \frac{\text{wt (g)}}{\text{eq wt (g/eq)}} = \text{normality (eq/L)} \times \text{volume (L)}} \quad (5.6)$$

Table 5.1
Reacting Units in Different Reactions

Reaction Type	Reacting Unit
Acid–base	H^+
Oxidation–reduction	Electron

As with molarity, we usually work with millequivalent quantities, and

$$meq = \frac{mg}{eq\ wt\ (mg/meq)} = normality\ (meq/mL) \times mL \qquad (5.7)$$

In clinical chemistry, equivalents are frequently defined in terms of the number of charges on an ion rather than on the number of reacting units. Thus, for example, the equivalent weight of Ca^{2+} is one-half its atomic weight, and the number of equivalents is twice the number of moles. This use is convenient for electroneutrality calculations. We discuss equivalents in more detail in Section 5.3.

There is no ambiguity in a molar concentration.

While normality has been used extensively in the past and is found in the scientific literature, it is not as widely used today as molarity. We discuss normality in Section 5.6 for those who do make use of it. We will use moles and molarity throughout most of this text so there will be no ambiguity about what the concentration represents. Molarity calculations require a knowledge of the stoichiometry of reactions, that is, the ratio in which substances react. The journal *Analytical Chemistry* does not allow normality in articles it publishes, but other publications do.

FORMALITY—INSTEAD OF MOLARITY

Chemists sometimes use the term **formality** for solutions of ionic salts that do not exist as molecules in the solid or in solution. The concentration is given as **formal** (*F*). Operationally, formality is identical to molarity: The former is sometimes reserved for describing makeup concentrations of solutions (i.e., total analytical concentration), and the latter for equilibrium concentrations. For convenience, we shall use molarity exclusively, a common practice.

Formality is numerically the same as molarity.

MOLALITY—THE TEMPERATURE-INDEPENDENT CONCENTRATION

In addition to molarity and normality, another useful concentration unit is **molality**, *m*. A one-**molal** solution contains one mole per 1000 g of **solvent**. The molal concentration is convenient in physicochemical measurements of the colligative properties of substances, such as freezing point depression, vapor pressure lowering, and osmotic pressure because colligative properties depend solely on the number of solute particles present in solution per mole of solvent. Molal concentrations are not temperature dependent as molar and normal concentrations are (since the solvent volume in molar and normal concentrations is temperature dependent).

Molality does not change with temperature.

DENSITY CALCULATIONS—HOW DO WE CONVERT TO MOLARITY?

The concentrations of many fairly concentrated commercial acids and bases are usually given in terms of percent by weight. It is frequently necessary to prepare solutions of a given approximate molarity from these substances. In order to do so, we must know the density in order to calculate the molarity. **Density** is the weight per unit volume at the specified temperature, usually g/mL or g/cm^3 at 20°C. (One milliliter is the volume occupied by 1 cm^3.)

Sometimes substances list **specific gravity** rather than density. Specific gravity is defined as the ratio of the mass of a body (e.g., a solution), usually at 20°C, to the mass of an equal volume of water at 4°C (or sometimes 20°C). That is, specific gravity is the *ratio of the densities of the two substances;* it is a dimensionless quantity. Since the density of water is 4°C is 1.00000 g/mL, density and specific gravity are equal when referred to water at 4°C. When specific gravity is referred to water at 20°C, density is equal to specific gravity $\times$ 0.99821 (the density of water is 0.99821 g/mL at 20°C).

For 4°C reference:

$$\text{Specific gravity} = \frac{\text{g/mL of solution } (T°C)}{\text{g/mL of H}_2\text{O } (4°C)(\text{i.e., } 1.0000 \text{ g/mL})}$$

$$= 1.0000 \text{ at } 4°C \text{ (for dilute solutions with density close to that of water: } D_{\text{H}_2\text{O}} = 1.0000 \text{ g/mL)}$$

$$= 0.99820 \text{ at } 20°C \ (D_{\text{H}_2\text{O}} = 0.99820 \text{ g/mL})$$

(For $T°C$, density of solution = specific gravity $\times$ 1.0000)

For 20°C reference:

$$\text{Specific gravity} = \frac{\text{g/mL of solution } (T°C)}{\text{g/mL of H}_2\text{O } (20°C)(\text{i.e., } 0.99820 \text{ g/mL})}$$

$$= 1.0000 \text{ at } 20°C \text{ (dilute solutions: } D_{\text{H}_2\text{O}} = 0.99820 \text{ g/mL)}$$

$$= 1.0018 \text{ at } 4°C \ (D_{\text{H}_2\text{O}} = 1.0000 \text{ g/mL})$$

(For $T°C$, density of solution = specific gravity $\times$ 0.99820)

Example 5.8

How many milliliters of concentrated sulfuric acid, 94.0% (g/100 g solution), density 1.831 g/cm^3, are required to prepare 1 liter of a 0.100 M solution?

Solution

Consider 1 cm^3 = 1 mL. The concentrated acid contains 0.940 g H$_2$SO$_4$ per gram of solution, and the solution weighs 1.831 g/mL. The product of these two numbers, then, gives the gram H$_2$SO$_4$ per milliliter of solution:

$$M = \frac{(0.940 \text{ g H}_2\text{SO}_4/\text{g solution})(1.831 \text{ g/mL})}{98.1 \text{ g/mol}} \times 1000 \text{ mL/L}$$

$$= 17.5 \text{ mol H}_2\text{SO}_4/\text{L solution}$$

We must dilute this solution to prepare 1 liter of a 0.100 M solution. The same number of millimoles of H$_2$SO$_4$ must be taken as will be contained in the final solution. Since mmol = $M \times$ mL and mmol dilute acid = mmol concentrated acid.

$$0.100 \ M \times 1000 \text{ mL} = 17.5 \ M \times \text{mL}$$

$$x = 5.71 \text{ mL concentrated acid to be diluted to } 1000 \text{ mL}$$

Molarity and normality are the most useful concentrations in quantitative analysis. Calculations using these for volumetric analysis are discussed in more detail below.

See Sections 5.5 and 5.6 for volumetric calculations using molarity or normality.

ANALYTICAL AND EQUILIBRIUM CONCENTRATIONS—THEY ARE NOT THE SAME

Analytical chemists prepare solutions of known analytical concentrations, but the dissolved substances may partially or totally dissociate to give equilibrium concentrations of different species. Acetic acid, for example, is a weak acid that dissociates a few percent depending on the concentration.

$$HOAc \rightleftharpoons H^+ + OAc^-$$

to give equilibrium amounts of the proton and the acetate ion. The more dilute the solutions, the greater the dissociation. We often use these equilibrium concentrations in calculations involving equilibrium constants (Chapter 6), usually using molarity concentrations. The **analytical molarity** is given by the notation C_X, while **equilibrium molarity** is given by [X]. A solution of 1 M $CaCl_2$ (analytical molarity) gives at equilibrium, 0 M $CaCl_2$, 1 M Ca^{2+}, and 2 M Cl^- (equilibrium molarities). Hence, we say the solution is 1 M in Ca^{2+}.

The analytical concentration represents the concentration of total dissolved substance, i.e., the sum of all species of the substance in solution = C_X.

An equilibrium concentration is that of a given dissolved form of the substance = [X].

DILUTIONS—PREPARING THE RIGHT CONCENTRATION

We often must prepare dilute solutions from more concentrated stock solutions. For example, we may prepare a dilute HCl solution from concentrated HCl to be used for titrations (following standardization). Or, we may have a stock standard solution from which we wish to prepare a series of more dilute standards. The millimoles of stock solution taken for dilution will be identical to the millimoles in the final diluted solution.

The millimoles taken for dilution will be the same as the millimoles in the diluted solution, i.e.,
$M_{stock} \times mL_{stock} = M_{diluted} \times mL_{diluted}$

 Example 5.9

You wish to prepare a calibration curve for the spectrophotometric determination of permanganate. You have a stock 0.100 M solution of $KMnO_4$ and a series of 100-mL volumetric flasks. What volumes of the stock solution will you have to pipet into the flasks to prepare standards of 1.00, 2.00, 5.00, and 10.0 $\times$ 10^{-3} M $KMnO_4$ solutions?

Solution

A 100-mL solution of 1.00 $\times$ 10^{-3} M $KMnO_4$ will contain

$$100 \text{ mL} \times 1.00 \times 10^{-3} \text{ mmol/mL} = 0.100 \text{ mmol } KMnO_4$$

We must pipet this amount from the stock solution.

$$0.100 \text{ mmol/mL} \times x \text{ mL} = 0.100 \text{ mmol}$$
$$x = 1.00 \text{ mL stock solution}$$

Similarly, for the other solutions we will need 2.00, 5.00, and 10.0 mL of the stock solution, which will be diluted to 100 mL.

Example 5.10

You are analyzing for the manganese content in an ore sample by dissolving it and oxidizing the manganese to permanganate for spectrophotometric measurement. The ore contains about 5% Mn. A 5-g sample is dissolved and diluted to 100 mL, following the oxidation step. By how much must the solution be diluted to be in the range of the calibration curve prepared in Example 5.9, that is, about 3×10^{-3} M permanganate?

Solution

The solution contains 0.05×5-g sample $= 0.25$ g Mn. This corresponds to 0.25 g/$(55$ g Mn/mol$) = 4.5 \times 10^{-3}$ mol MnO_4^-/100 mL $= 4.5 \times 10^{-2}$ M. For 3×10^{-3} M, we must dilute it by $4.5 \times 10^{-2}/3 \times 10^{-3} = 15$-fold. If we have a 100-mL volumetric flask,

$$4.5 \times 10^{-2} \ M \times x \ \text{mL} = 3 \times 10^{-3} \ M \times 100 \ \text{mL}$$
$$x = 6.7 \ \text{mL needed for dilution to 100 mL}$$

Since we need to pipet accurately, we could probably take an accurate 10-mL aliquot, which would give about 4.5×10^{-3} M permanganate for measurement.

MORE DILUTION CALCULATIONS

We can use the relationship $M \times \text{mL} = \text{mmol}$ to calculate the dilution required to prepare a certain concentration of a solution from a more concentrated solution. For example, if we wish to prepare 500 mL of a 0.100 M solution by diluting a more concentrated solution, we can calculate the millimoles of the solution that must be taken. From this we can calculate the volume of the more concentrated solution to be diluted to 500 mL.

Remember, the millimoles before and after diluting are the same.

 **Example 5.11**

You wish to prepare 500 mL of a 0.100 M $K_2Cr_2O_7$ solution from a 0.250 M solution. What volume of the 0.250 M solution must be diluted to 500 mL?

Solution

$$M_{\text{final}} \times \text{mL}_{\text{final}} = M_{\text{original}} \times \text{mL}_{\text{original}}$$
$$0.100 \ \text{mmol/mL} \times 500 \ \text{mL} - 0.250 \ \text{mmol/mL} \times \text{mL}_{\text{original}}$$
$$\text{mL}_{\text{original}} = 200 \ \text{mL}$$

 **Example 5.12**

What volume of 0.40 M $Ba(OH)_2$ must be added to 50 mL of 0.30 M NaOH to give a solution 0.50 M in OH^-?

Solution

Let x = mL $Ba(OH)_2$. The final volume is $(50 + x)$ mL

$$mmol\ OH^- = mmol\ NaOH + 2 \times mmol\ Ba(OH)_2$$
$$0.50\ M \times (50 + x)mL = 0.30\ M\ NaOH \times 50\ mL + 2 \times 0.40\ M\ Ba(OH)_2 \times x\ mL$$
$$x = 33\ mL\ Ba(OH)_2$$

Often, the analyst is confronted with serial dilutions of a sample or standard solution. Again, obtaining the final concentration simply requires keeping track of the number of millimoles and the volumes.

Example 5.13

You are to determine iron in a sample by spectrophotometry by reacting Fe^{2+} with 1,10-phenanthroline to form an orange color. This requires preparation of a series of standards against which to compare absorbances or color intensities (i.e., to prepare a calibration curve). A stock standard solution of $1.000 \times 10^{-3}\ M$ iron is prepared from ferrous ammonium sulfate. Working standards A and B are prepared by adding with pipets 2.000 and 1.000 mL, respectively, of this solution to 100-mL volumetric flasks and diluting to volume. Working standards C, D, and E are prepared by adding 20.00, 10.000, and 5.000 mL of working standard A to 100-mL volumetric flasks and diluting to volume. What are the concentrations of the prepared working solutions?

Solution

Solution A: $M_{stock} \times mL_{stock} = M_A \times mL_A$

                        $(1.000 \times 10^{-3}\ M)(2.000\ mL) = M_A \times 100.0\ mL$

                        $M_A = 2.000 \times 10^{-5}\ M$

Solution B: $(1.000 \times 10^{-3}\ M)(1.000\ mL) = M_B \times 100.0\ mL$

                        $M_B = 1.000 \times 10^{-5}\ M$

Solution C: $M_A \times mL_A = M_C \times mL_C$

                        $(2.000 \times 10^{-5}\ M)(20.00\ mL) = M_C \times 100.0\ mL$

                        $M_C = 4.000 \times 10^{-6}\ M$

Solution D: $(2.000 \times 10^{-5}\ M)(10.00\ mL) = M_D \times 100.0\ mL$

                        $M_D = 2.000 \times 10^{-6}\ M$

Solution E: $(2.000 \times 10^{-5}\ M)(5.000\ mL) = M_E \times 100.0\ mL$

                        $M_E = 1.000 \times 10^{-6}\ M$

The above calculations apply to all types of reactions, including acid–base, redox, precipitation, and complexometric reactions. The primary requirement before making calculations is to know the ratio in which the substances react, that is, start with a balanced reaction.

Solution preparation procedures in the chemical literature often call for the dilution of concentrated stock solutions, and authors may use different terms. For example, a procedure may call for 1 + 9 dilution (solute + solvent) of sulfuric

The solute + solvent method of dilution should not be used for quantitative dilutions.

acid. Or a 1 : 10 dilution (original volume:final volume) may be indicated. The first procedure calls for diluting a concentrated solution to 1/10th of its original concentration by adding 1 part to 9 parts of solvent; the second procedure by diluting to 10 times the original volume. The first procedure does not give an exact 10-fold dilution because volumes are not completely additive, whereas the second procedure does (e.g., adding 10 mL with a pipet to a 100-mL volumetric flask and diluting to volume—fill the flask partially with water before adding sulfuric acid!). The solute + solvent approach is fine for reagents whose concentrations need not be known accurately.

5.3 Expressions of Analytical Results—So Many Ways

We can report the results of analysis in many ways, and the beginning analytical chemist should be familiar with some of the common expressions and units of measure employed. Results will nearly always be reported as *concentration*, on either a weight or a volume basis: the quantity of analyte per unit weight or per volume of sample. The units used for the analyte will vary.

We shall first review the common units of weight and volume in the metric system and then describe methods of expressing results. The gram (g) is the basic unit of mass and is the unit employed most often in macro analyses. For small samples or trace constituents, chemists use smaller units. The milligram (mg) is 10^{-3} g, the microgram (μg) is 10^{-6} g, and the nanogram (ng) is 10^{-9} g. The basic unit of volume is the liter (L). The milliliter (mL) is 10^{-3} L and is used commonly in volumetric analysis. The microliter (μL) is 10^{-6} L (10^{-3} mL), and the nanoliter (nL) is 10^{-9} L (10^{-6} mL). (Prefixes for even smaller quantities include pico for 10^{-12} and femto for 10^{-15}.)

Y = yotta = 10^{24}
Z = zetta = 10^{21}
E = exa = 10^{18}
P = peta = 10^{15}
T = tera = 10^{12}
G = giga = 10^{9}
M = mega = 10^{6}
k = kilo = 10^{3}
d = deci = 10^{-1}
c = centi = 10^{-2}
m = milli = 10^{-3}
μ = micro = 10^{-6}
n = nano = 10^{-9}
p = pico = 10^{-12}
f = femto = 10^{-15}
a = atto = 10^{-18}
z = zepto = 10^{-21}
y = yocto = 10^{-24}

Mass and weight are really different. See Chapter 2. We deal with masses but will use mass and weight interchangeably.

SOLID SAMPLES

Calculations for solid samples are based on weight.[2] The most common way of expressing the results of macro determinations is to give the weight of analyte as a **percent** of the weight of sample (weight/weight basis). The weight units of analyte and sample are the same. For example, a limestone sample weighing 1.267 g and containing 0.3684 g iron would contain

$$\frac{0.3684 \text{ g}}{1.267 \text{ g}} \times 100\% = 29.08\% \text{ Fe}$$

The general formula for calculating percent on a weight/weight basis, which is the same as parts per hundred, then is

$$\% \text{ (wt/wt)} = \left[\frac{\text{wt solute (g)}}{\text{wt sample (g)}} \right] \times 10^2 \text{ (\%/g solute/g sample)} \qquad (5.8)$$

It is important to note that in such calculations, grams of solute do *not* cancel with grams of solution; the fraction represents grams of solute per gram of sample. Multiplication by 10^2 (grams of sample per gram of solute) converts to grams of solute per 100 g of sample. Since the conversion factors for converting weight of solute

---

[2] They are really based on mass, but the term *weight* is commonly used. See Chapter 2 for a description and determination of mass and weight.

and weight of sample (weights expressed in any units) to grams of solute and grams of sample are always the same, the conversion factors will always cancel. Thus, we can use any weight in the definition.

Trace concentrations are usually given in smaller units, such as **parts per thousand** (ppt, ‰), **parts per million** (ppm), or **parts per billion** (ppb). These are calculated in a manner similar to parts per hundred (%):

$$\text{pt (wt/wt)} = \left[\frac{\text{wt solute (g)}}{\text{wt sample (g)}} \right] \times 10^3 \text{ (ppt/g solute/g sample)} \qquad (5.9)$$

$$\text{ppm (wt/wt)} = \left[\frac{\text{wt solute (g)}}{\text{wt sample (g)}} \right] \times 10^6 \text{ (ppm/g solute/g sample)} \qquad (5.10)$$

$$\text{ppb (wt/wt)} = \left[\frac{\text{wt solute (g)}}{\text{wt sample (g)}} \right] \times 10^9 \text{ (ppb/g solute/g sample)} \qquad (5.11)$$

1 ppt (thousand) = 1000 ppm = 1,000,000 ppb; 1 ppm = 1000 ppb = 1,000,000 ppt (trillion).

You can use any weight units in your calculations so long as both analyte and sample weights are in the same units. **Parts per trillion** (parts per 10^{12} parts) is also abbreviated ppt, so be careful to define which one you mean. In the above example, we have 29.08 parts per hundred of iron in the sample, or 290.8 parts per thousand; 2908 parts per ten thousand; 29,080 parts per hundred thousand; and 290,800 parts per million (290,800 g or iron per 1 million grams of sample, 290,800 lb of iron per 1 million pounds of sample, etc.) Working backward, 1 ppm corresponds to 0.0001 part per hundred, or 10^{-4}%. Table 5.2 summarizes the concentration relationships for ppm and ppb. Note that ppm is simply mg/kg or μg/g and that ppb is μg/kg, or ng/g.

ppt = mg/g = g/kg
ppm = μg/g = mg/kg
ppb = ng/g = μg/kg

Example 5.14

A 2.6-g sample of plant tissue was analyzed and found to contain 3.6 μg zinc. What is the concentration of zinc in the plant in ppm? In ppb?

Solution

$$\frac{3.6 \ \mu g}{2.6 \ g} = 1.4 \ \mu g/g \equiv 1.4 \text{ ppm}$$

$$\frac{3.6 \times 10^3 \text{ ng}}{2.6 \ g} = 1.4 \times 10^3 \text{ ng/g} \equiv 1400 \text{ ppb}$$

One ppm is equal to 1000 ppb. One ppb is equal to 10^{-7}%.

Table 5.2
Common Units for Expressing Trace Concentrations

Unit	Abbreviation	wt/wt	wt/vol	vol/vol
Parts per million ($1 \text{ ppm} = 10^{-4}$%)	ppm	mg/kg μg/g	mg/L μg/mL	μL/L nL/mL
Parts per billion ($1 \text{ ppb} = 10^{-7}$% = 10^{-3} ppm)	ppb	μg/kg ng/g	μg/L ng/mL	nL/L pL/mL[a]
Milligram percent	mg%	mg/100 g	mg/100 mL	

[a] pL = picoliter = 10^{-12} L.

Clinical chemists sometimes prefer to use the unit **milligram percent** (mg%) rather than ppm for small concentrations. This is defined as milligrams of analyte per 100 g of sample. The sample in Example 5.14 would then contain $(3.6 \times 10^{-3}$ mg/2.6 g$) \times 100$ mg% = 0.14 mg% zinc.

LIQUID SAMPLES

A deciliter is 0.1 L or 100 mL.

You can report results for liquid samples on a weight/weight basis, as above, or they may be reported on a **weight/volume basis**. The latter is probably more common, at least in the clinical laboratory. The calculations are similar to those above. Percent on a weight/volume basis is equal to grams of analyte per 100 mL of sample, while mg% is equal to milligrams of analyte per 100 mL of sample. This latter unit is often used by clinical chemists for biological fluids, and their accepted terminology is *milligrams per deciliter (mg/dL)* to distinguish from mg% on a weight/weight basis. Whenever a concentration is expressed as a percentage, it should, if not clear, be specified whether this is wt/vol or wt/wt. Parts per million, parts per billion, and parts per trillion can also be expressed on a weight/volume basis; ppm is calculated from mg/L or μg/mL; ppb is calculated from μg/L or ng/mL; and ppt is calculated from pg/mL or mg/L. Alternatively, the following fundamental calculations may be used:

ppm = μg/mL = mg/L
ppb = ng/mL = μg/L
ppt = pg/mL = ng/L

$$\% \text{ (wt/vol)} = \left[\frac{\text{wt solute (g)}}{\text{vol sample (mL)}} \right] \times 10^2 \text{ (\%/g solute/mL sample)} \qquad (5.12)$$

$$\text{ppm (wt/vol)} = \left[\frac{\text{wt solute (g)}}{\text{vol sample (mL)}} \right] \times 10^6 \text{ (ppm/g solute/mL sample)} \qquad (5.13)$$

$$\text{ppb (wt/vol)} = \left[\frac{\text{wt solute (g)}}{\text{vol sample (mL)}} \right] \times 10^9 \text{ (ppb/g solute/mL sample)} \qquad (5.14)$$

$$\text{ppt (wt/vol)} = \left[\frac{\text{wt solute (g)}}{\text{vol sample (mL)}} \right] \times 10^{12} \text{ ppt/g solute/mL sample} \qquad (5.15)$$

Note that % (wt/vol) is not pounds/100 gal of solution; the units must be expressed in grams of solute and milliliters of solution.

 Example 5.15

A 25.0-μL serum sample was analyzed for glucose content and found to contain 26.7 μg. Calculate the concentration of glucose in ppm and in mg/dL.

Solution

$$25.0 \ \mu\text{L} \times \frac{1 \text{ mL}}{1000 \ \mu\text{L}} = 2.50 \times 10^{-2} \text{ mL}$$

$$26.7 \ \mu\text{g} \times \frac{1 \text{ g}}{10^5 \ \mu\text{g}} = 2.67 \times 10^{-5} \text{ g}$$

$$\text{ppm} = \frac{2.67 \times 10^{-5} \text{ g glucose}}{2.50 \times 10^{-2} \text{ mL serum}} \times 10^6 \ \mu\text{g/g} = 1.07 \times 10^3 \ \mu\text{g/mL}$$

or

$$\frac{26.7 \ \mu g \ glucose}{0.0250 \ mL \ serum} = 1.07 \times 10^3 \ \mu g/mL = ppm$$

$$mg/dL = \frac{26.7 \ \mu g \ glucose \times 10^{-3} \ mg/\mu g}{0.025 \ mL \ serum} \times 100 \ ml/dL = 107 \ mg/dL$$

[Note the relationship between ppm (wt/vol) and mg/dL.]

• •

What does 1 ppm represent in terms of moles per liter? It depends on the formula weight, but the *approximate* relationship between concentrations in parts per million (or parts per billion) and in moles per liter can be seen by assuming a formula weight of 100 for an analyte. Then, since 1 ppm $= 10^{-3}$ g/L, it is equal to $(10^{-3}$ g/L$)(10^2$ g/mol$) = 10^{-5}$ mol/L. Similarly, 1 ppb $= 10^{-8}$ mol/L. Note that this latter concentration is smaller than the hydrogen ion concentration in pure water $(10^{-7}$ mol/L)! Of course, this relationship is approximate and will vary with the formula weight. One part per million solutions of zinc and copper, for example, will not be the same molarity. Conversely, equal molar solutions of different species will not be equal in terms of ppm unless the formula weights are equal. The former concentration is based on the number of molecules per unit volume, while the latter is based on the weight of the species per unit volume.

For a formula weight of 100:
1 ppm $= 10^{-5}$ M
1 ppb $= 10^{-8}$ M

Let's do some actual conversions using real formula weights. We begin with a solution that contains 2.5 ppm benzene. The formula weight (C_6H_6) is 78.1. The concentration in moles per liter is $(2.5 \times 10^{-3}$ g/L$)/(78$ g/mol$) = 3.8 \times 10^{-5}$ g/mol. Another solution contains 5.8×10^{-8} M lead. The concentration in parts per billion is $(5.8 \times 10^{-8}$ mol/L$)(207$ g/mol$) = 1.2_0 \times 10^{-5}$ g/L. For parts per billion (μg/L), then $(1.2_0 \times 10^{-5}$ g/L$)/(10^6 \ \mu g/g) = 1.2_0 \times 10^1$ mg/L = 12 ppb. A drinking water sample that contains 350 pg/L of carbon tetrachloride has a parts per trillion (ng/L) concentration of $(350 \times 10^{-12}$ g/L$)/(10^9$ ng/g$) = 350 \times 10^{-3}$ ng/L = 0.35 ppt. The molar concentration is $(350 \times 10^{-12}$ g/L$)/(154$ g/mol$) = 2.3 \times 10^{-12}$ mol/L. (Chlorine-treated water may contain traces of chlorinated hydrocarbons—this is very low.)

A key point to remember is that solutions of equal concentrations on a weight/weight or weight/volume basis do not have the same number of molecules or reaction species, but solutions of the same molarity do.

 •

Example 5.16

(a) Calculate the molar concentrations of 1.00 ppm solutions each of Li^+ and Pb^{2+}.
(b) What weight of $Pb(NO_3)_2$ will have to be dissolved in 1 liter of water to prepare a 100 ppm Pb^{2+} solution?

Solution

(a) Li concentration = 1.00 ppm = 1.00 mg/L   Pb concentration = 1.00 ppm = 1.00 mg/L

$$M_{Li} = \frac{1.00 \ mg \ Li/L \times 10^{-3} \ g/mg}{6.94 \ g \ Li/mol} = 1.44 \times 10^{-4} \ mol/L \ Li$$

$$M_{Pb} = \frac{1.00 \ mg \ Pb/L \times 10^{-3} \ g/mg}{207 \ g \ Pb/mol} = 4.83 \times 10^{-6} \ mol/L \ Pb$$

Because lead is much heavier than lithium, a given weight contains a smaller number of moles and its molar concentration is less.

(b)
$$100 \text{ ppm Pb}^{2+} = 100 \text{ mg/L} = 0.100 \text{ g/L}$$

$$\frac{0.100 \text{ g Pb}}{207 \text{ g/mol}} = 4.83 \times 10^{-4} \text{ mol Pb}$$

Therefore, we need 4.83×10^{-4} mol $Pb(NO_3)_2$.

$$4.83 \times 10^{-4} \text{ mol} \times 283.2 \text{ g Pb(NO}_3)_2/\text{mol} = 0.137 \text{ g Pb(NO}_3)_2$$

• •

For dilute solutions, wt/wt ≈ wt/vol.

The concentration units wt/wt and wt/vol are related through the density. They are numerically the same for dilute aqueous solutions with a density of 1 g/mL.

If the analyte is a liquid dissolved in another liquid, the results may be expressed on a **volume/volume** basis, but you will likely encounter this formula only in rare situations. You would handle the calculations in the same manner as those above, using the same volume units for solute and sample. Gas analyses may be reported on a weight/weight, weight/volume, or volume/volume basis.

Alcohol in wine and liquor is expressed as vol/vol (200 proof = 100% vol/vol). Since the specific gravity of alcohol is 0.8, wt/vol concentration = 0.8 × (vol/vol) = 0.4 × proof

It is always best to specify which form you mean. In the absence of clear labels, it is best to assume that solids are usually reported wt/wt, gases are usually vol/vol, and liquids may be wt/wt (concentrated acid and base reagents), wt/vol (most dilute aqueous solutions), or vol/vol (the U.S. alcoholic beverage industry).

Clinical chemists frequently prefer to use a unit other than weight for expressing the amount of major electrolytes in biological fluids (Na^+, K^+, Ca^{2+}, Mg^{2+}, Cl^-, $H_2PO_4^-$, etc.). This is the unit **milliequivalent** (meq). In this context, milliequivalent is defined as the number of millimoles of analyte multiplied by the charge on the analyte ion. Results are generally reported as meq/L. This concept is employed to give an overall view of the electrolyte balance. The physician can tell at a glance if total electrolyte concentration has increased or decreased markedly. Obviously, the milliequivalents of cations will be equal to the milliequivalents of anions. One mole of a monovalent (+1) cation (1 eq) and half a mole of a divalent (−2) anion (1 eq) have the same number of positive and negative charges (one mole each). As an example of electrolyte or charge balance, Table 5.3 summarizes the averages of major electrolyte compositions normally present in human blood plasma and urine. Chapter 24 discusses the ranges and physiological significant of some chemical constituents of the human body.

Table 5.3
Major Electrolyte Composition of Normal Human Plasma[a]

The equivalents of cations and anions must be equal.

Cations	meq/L	Anions	meq/L
Na^+	143	Cl^-	104
K^+	4.5	HCO^{3-}	29
Ca^{2+}	5	Protein	16
Mg^{2+}	2.5	HPO_4^-	2
		SO_4^{2-}	1
		Organic acids	3
Total	155	Total	155

[a] Reproduced from Joseph S. Annino, *Clinical Chemistry*, 3rd ed., by Boston: LIttle, Brown, 1964.

We can calculate the milliequivalents of a substance from its weight in milligrams simply as follows (similar to how we calculate millimoles):

$$\text{meq} = \frac{\text{mg}}{\text{eq wt (mg/meq)}} = \frac{\text{mg}}{\text{f wt (mg/mmol)}/n \text{ (meq/mmol)}} \qquad (5.16)$$

$$n = \text{charge on ion}$$

The equivalent weight of Na^+ is 23.0 (mg/mmol)/1 (meq/mmol) = 23.0 mg/meq. The equivalent weight of Ca^{2+} is 40.1 (mg/mmol)/2 (meq/mmol) = 20.0 mg/meq.

Example 5.17

The concentration of zinc ion in blood serum is about 1 ppm. Express this as meq/L.

Solution

$$1 \text{ ppm} = 1 \ \mu g/mL = 1 \text{ mg/L}$$

The equivalent weight of Zn^{2+} is 65.4 (mg/mmol)/2 (meq/mmol) = 32.7 mg/meq. Therefore,

$$\frac{1 \text{ mg Zn/L}}{32.7 \text{ mg/meq}} = 3.06 \times 10^{-2} \text{ meq/L Zn}$$

This unit is actually used for the major electrolyte constituents as in Table 5.3 rather than the trace constituents, as in the example here.

REPORTING CONCENTRATIONS
AS DIFFERENT CHEMICAL SPECIES

Thus far, we have more or less implied that the analyte is determined in the form it exists or for which we want to report the results. However, this is often not true. In the analysis of the iron content of an ore, for example, we may determine the iron in the form of Fe_2O_3 and then report it as % Fe. Or we may determine the iron in the form of Fe^{2+} (e.g., by titration) and report it as % Fe_2O_3. This is perfectly proper so long as we know the relationship of what is measured to what we wish. Hence, if we analyze for the calcium content of water, we may wish to report it as parts per million (mg/L) of $CaCO_3$ (this is the typical way of expressing *water hardness*). We know that each gram of Ca^{2+} is equivalent to (or could be converted to) f wt $CaCO_3$/f wt Ca^{2+} grams of $CaCO_3$. That is, multiplying the milligrams of Ca^{2+} determined by 100.09/40.08 will give us the equivalent number of milligrams of $CaCO_3$. The calcium does not have to exist in this form (we may not even know in what form it actually exists); we simply have calculated the weight that could exist and will report the result as if it did. Specific operations necessary for calculating the weight of the desired constituent will be described below.

At this point we should mention some of the different weight criteria used for expressing results with biological tissues and solids. The sample may be weighed in one of three physical forms; wet, dry, or ashed. This can apply also to fluids, although fluid volume is usually employed. The wet weight is taken on the fresh,

We may express results in any form of the analyte.

Water hardness due to calcium ion is expressed as ppm $CaCO_3$.

untreated sample. The dry weight is taken after the sample has been dried by heating, desiccation, or freeze-drying. If the test substance is unstable to heat, the sample should not be dried by heating. The weight of the ash residue after the organic matter has been burned off is the third basis of weight. This can obviously be used only for mineral (inorganic) analysis.

5.4 Volumetric Analysis: How Do We Make Stoichiometric Calculations?

Volumetric or titrimetric analyses are among the most useful and accurate analytical techniques, especially for millimole amounts of analyte. They are rapid and can be automated, and they can be applied to smaller amounts of analyte when combined with a sensitive instrumental technique for detecting the completion of the titration reaction, for example, pH measurement. Manual titrations nowadays are used in situations that require high accuracy for relatively small numbers of samples. They are used, for example, to calibrate or validate more routine instrumental methods. Automated titrations are useful when large numbers of samples must be processed. (A titration may be automated, for instance, by means of a color change or a pH change that activates a motor-driven buret to stop delivery: The volume delivered may then be registered on a digital counter—an example is given at the end of Chapter 14.) In this section, the types of titrations that can be performed are described and the principles applicable to all types are given, including the principles and requirements of a titration and of standard solutions. The volumetric relationship described earlier in this chapter may be used for calculating quantitative information about the titrated analyte. Calculations in volumetric analysis are given in Sections 5.5–5.7, including molarity and normality titer, and back-titrations.

TITRATION—WHAT ARE THE REQUIREMENTS?

We calculate the moles of analyte titrated from the moles of titrant added and the ratio in which they react.

In a **titration**, the test substance (analyte) in a flash reacts with a reagent added from a buret as a solution of known concentration. This is referred to as a **standard solution**, and is called the **titrant**. The volume of titrant required to just completely react with the analyte is measured. Since we know the concentration as well as the reaction between the analyte and the reagent, we can calculate the amount of analyte. The requirements of a titration are as follows:

1. The reaction must be **stoichiometric**. That is, there must be a well-defined and known reaction between the analyte and the titrant. In the titration of acetic acid in vinegar with sodium hydroxide, for example, a well-defined reaction takes place:

$$HC_2H_3O_2 + NaOH \rightarrow NaC_2H_3O_2 + H_2O$$

2. The reaction should be *rapid*. Most ionic reactions, as above, are very rapid.

3. There should be *no side reactions*, and the reaction should be specific. If there are interfering substances, these must be removed. In the above example, there should be no other acids present.

4. There should be a *marked change in some property of the solution when the reaction is complete*. This may be a change in color of the solution or in some electrical or other physical property of the solution. In the titration of acetic acid with sodium hydroxide, there is a marked increase in the pH of the solution when the reaction is complete. A color change is usually brought about by addition of an **indicator**, whose color is dependent on the properties of the solution, for example, the pH.

5. The point at which an equivalent or stoichiometric amount of titrant is added is called the **equivalence point**. The point at which the reaction is *observed* to be complete is called the **end point**, that is, when a change in some property of the solution is detected. The end point should coincide with the equivalence point or be at a reproducible interval from it.

6. The reaction should be **quantitative**. That is, the equilibrium of the reaction should be far to the right so that a sufficiently *sharp* change will occur at the end point to obtain the desired accuracy. If the equilibrium does not lie far to the right, then there will be gradual change in the property marking the end point (e.g., pH) and this will be difficult to detect precisely.

> The *equivalence point* is the theoretical end of the titration where the number of moles = number of moles. The *end point* is the observed end of the titration.

STANDARD SOLUTIONS—THERE ARE DIFFERENT KINDS

A standard solution is prepared by dissolving an accurately weighed quantity of a highly pure material called a **primary standard** and diluting to an accurately known volume in a volumetric flask. Alternatively, if the material is not sufficiently pure, a solution is prepared to give approximately the desired concentration, and this is **standardized** by titrating a weighed quantity of a primary standard. For example, sodium hydroxide is not sufficiently pure to prepare a standard solution directly. It is therefore standardized by titrating a primary standard acid, such as potassium acid phthalate (KHP). Potassium acid phthalate is a solid that can be weighed accurately. Standardization calculations are treated below.

> A solution standardized by titrating a primary standard is itself a secondary standard. It will be less accurate than a primary standard solution due to the errors of titrations.

A **primary standard** should fulfill these requirements:

1. It should be *100.00% pure*, although 0.01 to 0.02% impurity is tolerable if it is accurately known.

2. It should be *stable to drying* temperatures, and it should be stable indefinitely at room temperature. The primary standard is always dried before weighing.[3]

3. It should be *readily available* and fairly inexpensive.

4. Although not necessary, it should have a *high formula weight*. This is so that a relatively large amount of it will have to be weighed to get enough to titrate. The relative error in weighing a greater amount of material will be smaller than that for a small amount.

5. If it is to be used in titration, it should possess the *properties required for a titration* listed above. In particular, the equilibrium of the reaction should be far to the right so that a very sharp end point will be obtained.

> A high formula weight means a larger weight must be taken for a given number of moles. This reduces the error in weighing.

---

[3] There are a few exceptions when the primary standard is a hydrate.

CLASSIFICATION OF VOLUMETRIC METHODS— WHAT KINDS ARE THERE?

There are four general classes of volumetric or titrimetric methods.

1. *Acid–Base.* Many compounds, both inorganic and organic, are either acids or bases and can be titrated with a standard solution of a strong base or a strong acid. The end points of these titrations are easy to detect, either by means of an indicator or by following the change in pH with a pH meter. The acidity and basicity of many organic acids and bases can be enhanced by titrating in a *nonaqueous solvent*. The result is a sharper end point, and weaker acids and bases can be titrated in this manner.

2. *Precipitation.* In the case of precipitation, the titrant forms an insoluble product with the analyte. An example is the titration of chloride ion with silver nitrate solution to form silver chloride precipitate. Again, indicators can be used to detect the end point, or the potential of the solution can be monitored electrically.

3. *Complexometric.* In complexometric titrations, the titrant is a reagent that forms a water-soluble complex with the analyte, a metal ion. The titrant is often a **chelating agent**.[4] The reverse titration may be carried out also. Ethylenediaminetetraacetic acid (EDTA) is one of the most useful chelating agents used for titration. It will react with a large number of elements, and the reactions can be controlled by adjustment of the pH. Indicators can be used to form a highly colored complex with the metal ion.

4. *Reduction–Oxidation.* These "redox" titrations involve the titration of an oxidizing agent with a reducing agent, or vice versa. An oxidizing agent gains electrons and a reducing agent loses electrons in a reaction between them. There must be a sufficiently large difference between the oxidizing and reducing capabilities of these agents for the reaction to go to completion and give a sharp end point; that is, one should be a fairly strong oxidizing agent (strong tendency to gain electrons) and the other a fairly strong reducing agent (strong tendency to lose electrons). You can use appropriate indicators for these titrations, or you may employ various electrometric means to detect the end point.

These different types of titrations and the means of detecting their end points will be treated separately in succeeding chapters.

5.5 Volumetric Calculations—Let's Use Molarity

We shall use molarity throughout the majority of the text for volumetric calculations. However, another useful concentration unit for volumetric calculations is **normality**, which uses the concepts of **equivalents** and **equivalent weights** in place of moles and formula weights. Normal concentration depends on the particular reaction, and the reaction should be specified. Some instructors prefer to introduce the concept of normality, and students are likely to encounter it in reference books. Therefore, a review of equivalents and normality is given following the discussion of calculations using molarity.

---

[4]A chelating agent (the term is derived from the Greek word for *clawlike*) is a type of complexing agent that contains two or more groups capable of complexing with a metal ion. EDTA has six such groups.

We reviewed some of the ways of expressing concentrations earlier in the chapter. Basic concepts are summarized below.

$$\text{Moles} = \frac{g}{\text{f wt (g/mol)}} \qquad \text{Millimoles} = \frac{mg}{\text{f wt (mg/mmol)}} \qquad (5.17)$$

$$\text{Molar concentration} = M = \frac{\text{mol}}{\text{L}} \qquad M = \frac{\text{mmol}}{\text{mL}} \qquad (5.18)$$

Learn these relationships well. They are the basis of all volumetric calculations, solution preparation, and dilutions. Think units!

By rearranging these equations, we obtain the expressions for calculating other quantities.

$$M \text{ (mol/L)} \times \text{L} = \text{mol} \qquad M \text{ (mmol/mL)} \times \text{mL} = \text{mmol} \qquad (5.19)$$

$$g = \text{mol} \times \text{f wt (g/mol)} \qquad \text{mg} = \text{mmol} \times \text{f wt (mg/mmol)} \qquad (5.20)$$

$$g = M \text{ (mol/L)} \times \text{L} \times \text{f wt (g/mol)} \qquad (5.21)$$

$$\text{mg} = M \text{ (mmol/mL)} \times \text{mL} \times \text{f wt (mg/mmol)}$$

We usually work with millimole (mmol) and milliliter (mL) quantities in titrations; therefore, the right-hand equations are more useful. Note that the expression for formula weight contains the same numerical value whether it be in g/mol or mg/mmol. Note also that care must be taken in utilizing "milli" quantities (millimoles, milligrams, milliliters). Incorrect use could result in calculations errors of 1000-fold.

Assume 25.0 mL of 0.100 M $AgNO_3$ is required to titrate a sample containing sodium chloride. The reaction is

$$Cl^- + Ag^+ \rightarrow AgCl$$

Since Ag^+ and Cl^- react on a 1:1 molar basis, the number of millimoles of Cl^- is equal to the number of millimoles of Ag^+ needed for titration. We can calculate the milligrams of NaCl as follows:

For 1:1 reactions, $mmol_{analyte} = mmol_{titrant}$.

$$mmol_{NaCl} = mL_{AgNO_3} \times M_{AgNO_3}$$
$$= 25.0 \text{ mL} \times 0.100 \text{ (mmol/mL)} = 2.50 \text{ mmol}$$
$$mg_{NaCl} = mmol \times \text{f wt}_{NaCl}$$
$$= 2.50 \text{ mmol} \times 58.44 \text{ mg/mmol} = 146 \text{ mg}$$

We can calculate the percentage of analyte A that reacts on a *1:1 mole basis* with the titrant using the following general formula:

$$\% \text{ A} = \text{fraction}_{analyte} \times 100\% = \frac{mg_{analyte}}{mg_{sample}} \times 100\%$$

$$= \frac{mmol \times \text{f wt}_{analyte}(\text{mg/mmol})}{mg_{sample}} \times 100\% \qquad (5.22)$$

$$= \frac{M_{titrant} \text{ (mmol/mL)} \times mL_{titrant} \times \text{f wt}_{analyte} \text{ (mg/mmol)}}{mg_{sample}} \times 100\%$$

Think units!

Note that this computation is a summary of the individual calculation steps taken to arrive at the fraction of analyte in the sample using proper dimensional analysis. You should use it in that sense rather than simply memorizing a formula.

Example 5.18

A 0.4671-g sample containing sodium bicarbonate was dissolved and titrated with standard 0.1067 M hydrochloric acid solution, requiring 40.72 mL. The reaction is

$$HCO_3^- + H^+ \rightarrow H_2O + CO_2$$

Calculate the percent sodium bicarbonate in the sample.

Solution

The millimoles of sodium bicarbonate are equal to the millimoles of acid used to titrate it, since they react in a 1:1 ratio.

$$mmol_{HCl} = 0.1067 \text{ mmol/mL} \times 40.72 \text{ mL} = 4.344_8 \text{ mmol}_{HCl} \equiv mmol \text{ NaHCO}_3$$

(Extra figures are carried so an identical answer is obtained when all steps are done together below.)

$$mg_{NaHCO_3} = 4.3448 \text{ mmol} \times 84.01 \text{ mg/mmol} = 365.0_1 \text{ mg NaHCO}_3$$

$$\% \text{ NaHCO}_3 = \frac{365.0_1 \text{ mg NaHCO}_3}{467.1 \text{ mg}_{sample}} \times 100\% = 78.14\% \text{ NaHCO}_3$$

Or, combining all the steps,

$$\% \text{ NaHCO}_3 = \frac{M_{HCl} \times mL_{HCl} \times f \text{ wt}_{NaHCO_3}}{mg_{sample}} \times 100\%$$

$$= \frac{0.1067 \text{ mmol HCl/mL} \times 40.72 \text{ mL HCl} \times 84.01 \text{ mg NaHCO}_3/\text{mmol}}{467.1 \text{ mg}} \times 100\%$$

$$= 78.14\% \text{ NaHCO}_3$$

SOME USEFUL THINGS TO KNOW FOR MOLARITY CALCULATIONS

When the reaction is not 1:1, a conversion factor is used to equate the moles of analyte and titrant.

Many substances do not react on a 1:1 mole basis, and so the simple calculation in the above example cannot be applied to all reactions. It is possible, however, to write a generalized formula for calculations applicable to *all* reactions based on the **balanced equation** for reactions.

Consider the general reaction

$$aA \times tT \rightarrow P \tag{5.23}$$

where A is the analyte, T is the titrant, and they react in the ratio a/t to give products P. Then, noting the units and using dimensional analysis,

Still think units! We have added $mmol_{analyte}/mmol_{titrant}$.

$$mmol_A = mmol_T \times \frac{a}{t} \text{ (mmol A/mmol T)} \tag{5.24}$$

$$mmol_A = M_T \text{ (mmol/mL)} \times mL_T \times \frac{a}{t} \text{ (mmol A/mmol T)} \tag{5.25}$$

$$mg_A = mmol_A \times f\ wt_A\ (mg/mmol) \qquad (5.26)$$

$$mg_A = M_T\ (mmol/mL) \times mL_T \times \frac{a}{t}\ (mmol\ A/mmol\ T)$$
$$\times\ f\ wt_A\ (mg/mmol) \qquad (5.27)$$

Note that the a/t factor serves to equate the analyte and titrant. To avoid a mistake in setting up the factor, it is helpful to remember that when you calculate the amount of analyte, you must multiply the amount of titrant by the a/t ratio (a comes first). Conversely, if you are calculating the amount of titrant (e.g., molarity) from a known amount of analyte titrated, you must multiply the amount of analyte by the t/a ratio (t comes first). The best way, of course, to ascertain the correct ratio is to always do a dimensional analysis to obtain the correct units.

In a manner similar to that used to derive Equation 5.22, we can list the steps in arriving at a general expression for calculating the percent analyte A in a sample determined by titrating a known weight of sample with a standard solution of titrant T:

$$\% A = fraction_{analyte} \times 100\% = \frac{mg_{analyte}}{mg_{sample}} \times 100\%$$

$$= \frac{mmol_{titrant} \times (a/t)(mmol_{analyte}/mmol_{titrant}) \times f\ wt_{analyte}\ (mg/mmol)}{mg_{sample}} \times 100\%$$

$$= \frac{M_{titrant}\ (mmol/mL) \times mL_{titrant} \times (a/t)(mmol_{analyte}/mmol_{titrant}) \times f\ wt_{analyte}\ (mg/mmol)}{mg_{sample}}$$
$$\times\ 100\% \qquad (5.28)$$

Again, note that we simply use dimensional analysis, that is, we perform stepwise calculations in which units cancel to give the desired units. In this general procedure, the dimensional analysis includes the stoichiometric factor a/t that converts millimoles of titrant to an equivalent number of millimoles of titrated analyte.

Example 5.19

A 0.2638-g soda ash sample is analyzed by titrating the sodium carbonate with the standard 0.1288 M hydrochloride solution, requiring 38.27 mL. The reaction is

$$CO_3^{2-} + 2H^+ \rightarrow H_2O + CO_2$$

Calculate the percent sodium carbonate in the sample.

Solution

The millimoles of sodium carbonate is equal to one-half the millimoles of acid used to titrate it, since they react in a 1:2 ratio ($a/t = \frac{1}{2}$).

$$mmol_{HCl} = 0.1288\ mmol/mL \times 38.27\ mL = 4.929\ mmol\ HCl$$

$$mmol_{NaCO_3} = 4.929\ mmol\ HCl \times \tfrac{1}{2}(mmol\ Na_2CO_3/mmol\ HCl) = 2.464_5\ mmol\ Na_2CO_3$$

$$mg_{Na_2CO_3} = 2.464_5 \text{ mmol} \times 105.99 \text{ mg } Na_2CO_3/\text{mmol} = 261.2_1 \text{ mg } Na_2CO_3$$

$$\% \ Na_2CO_3 = \frac{261.2_1 \text{ mg } Na_2CO_3}{263.8 \text{ mg}_{sample}} \times 100\% = 99.02\% \ Na_2CO_3$$

Or, combining all the steps,

$$\% \ Na_2CO_3 = \frac{M_{HCl} \times mL_{HCl} \times \frac{1}{2} (\text{mmol } Na_2CO_3/\text{mmol } HCl) \times f \ wt_{Na_2CO_3}}{mg_{sample}} \times 100\%$$

$$= \frac{0.1288 \text{ mmol } HCl \times 38.27 \text{ mL } HCl \times \frac{1}{2} (\text{mmol } Na_2CO_3/\text{mmol } HCl) \times 105.99 \ (\text{mg } Na_2CO_3/\text{mmol})}{263.8 \text{ mg}_{sample}} \times 100\%$$

$$= 99.02\% \ Na_2CO_3$$

Example 5.20

How many milliliters of 0.25 M solution of H_2SO_4 will react with 10 mL of a 0.25 M solution of NaOH?

Solution

The reaction is

$$H_2SO_4 + 2NaOH \rightarrow Na_2SO_4 + 2H_2O$$

One-half as many millimoles of H_2SO_4 as of NaOH will react, or

$$M_{H_2SO_4} \times mL_{H_2SO_4} = M_{NaOH} \times mL_{NaOH} \times \tfrac{1}{2} (\text{mmol } H_2SO_4/\text{mmol NaOH})$$

Therefore,

$$mL_{H_2SO_4} = \frac{0.25 \text{ mmol NaOH/mL} \times 10 \text{ mL NaOH} \times \frac{1}{2} (\text{mmol } H_2SO_4/\text{mmol NaOH})}{0.25 \text{ mmol } H_2SO_4/\text{mL}}$$

$$= 5.0 \text{ mL } H_2SO_4$$

Note that, in this case, we multiplied the amount of titrant by the *a/t* ratio (mmol analyte/mmol titrant).

Example 5.21

A sample of impure salicylic acid, $C_6H_4(OH)COOH$ (one titratable proton), is analyzed by titration. What size sample should be taken so that the percent purity is equal to five times the milliliters of 0.0500 M NaOH used to titrate it?

Solution

Let x = mL NaOH; % salicylic acid (HA) = $5x$:

$$\% \text{ HA} = \frac{M_{\text{NaOH}} \times \text{mL}_{\text{NaOH}} \times 1 \text{ (mmol HA/mmol NaOH)} \times \text{f wt}_{\text{HA}} \text{ (mg/mmol)}}{\text{mg}_{\text{sample}}}$$

$$\times 100\%$$

$$5x\% = \frac{0.0500 \text{ } M \times x \text{ mL NaOH} \times 1 \times 138 \text{ mg HA/mmol}}{\text{mg}_{\text{sample}}} \times 100\%$$

$$\text{mg}_{\text{sample}} = 138 \text{ mg}$$

• •

You can apply the above examples of acid–base calculations to the titrations described in Chapter 8.

STANDARDIZATION AND TITRATION CALCULATIONS— THEY ARE THE REVERSE OF ONE ANOTHER

When a titrant material of high or known purity is not available, the concentration of the approximately prepared titrant solution must be accurately determined by **standardization**; that is, by titrating an accurately weighed quantity (a known number of millimoles) of a primary standard. From the volume of titrant used to titrate the primary standard, we can calculate the molar concentration of the titrant.

In standardization, the concentration of the titrant is unknown and the moles of analyte (primary standard) are known.

Taking the analyte A in Equation 5.23 to be the primary standard,

$$\text{mmol}_{\text{standard}} = \frac{\text{mg}_{\text{standard}}}{\text{f wt}_{\text{standard}} \text{ (mg/mmol)}}$$

$$\text{mmol}_{\text{titrant}} = M_{\text{titrant}} \text{ (mmol/mL)} \times \text{mL}_{\text{titrant}}$$

$$= \text{mmol}_{\text{standard}} \times t/a \text{ (mmol}_{\text{titrant}}/\text{mmol}_{\text{standard}})$$

$$M_{\text{titrant}} \text{ (mmol/mL)} = \frac{\text{mmol}_{\text{standard}} \times t/a \text{ (mmol}_{\text{titrant}}/\text{mmol}_{\text{standard}})}{\text{mL}_{\text{titrant}}}$$

Or, combining all steps

$$\boxed{M_{\text{titrant}} \text{ (mmol/mL)} = \frac{\text{mg}_{\text{standard}}/\text{f wt}_{\text{standard}} \text{ (mg/mmol)} \times t/a \text{ (mmol}_{\text{titrant}}/\text{mmol}_{\text{standard}})}{\text{mL}_{\text{titrant}}}}$$

Units!

(5.29)

Note once again that dimensional analysis (cancellation of units) results in the desired units of mmol/mL.

Example 5.22

An approximate 0.1 M hydrochloric acid solution is prepared by 120-fold dilution of concentrated hydrochloric acid. It is standardized by titrating 0.1876 g of dried primary standard sodium carbonate:

$$CO_3^{2-} + 2H^+ \rightarrow H_2O + CO_2$$

The titration required 35.86 mL acid. Calculate the molar concentration of the hydrochloric acid.

Solution

The millimoles hydrochloric acid are equal to twice the millimoles of sodium carbonate titrated.

$$\text{mmol}_{Na_2CO_3} = 187.6 \text{ mg } Na_2CO_3/105.99 \text{ (mg } Na_2CO_3/\text{mmol)} = 1.770_0 \text{ mmol } Na_2CO_3$$
$$\text{mmol}_{HCl} = M_{HCl} \text{ (mmol/mL)} \times 35.86 \text{ mL HCl} = 1.770_0 \text{ mmol } Na_2CO_3$$
$$\times 2 \text{ (mmol HCl/mmol } Na_2CO_3)$$

$$M_{HCl} = \frac{1.770_0 \text{ mmol } Na_2CO_3 \times 2 \text{ (mmol HCl/mmol } Na_2CO_3)}{35.86 \text{ mL HCl}} = 0.09872 \ M$$

Or, combining all steps,

$$M_{HCl} = \frac{(mg_{Na_2CO_3}/f \ wt_{Na_2CO_3} \times (2/1)(\text{mmol HCl/mmol } Na_2CO_3)}{mL_{HCl}}$$

$$= \frac{[187.6 \text{ mg}/105.99 \text{ (mg/mmol)}] \times 2 \text{ (mmol HCl/mmol } Na_2CO_3)}{35.86 \text{ mL}}$$

$$= 0.09872 \text{ mmol/mL}$$

Note that we multiplied the amount of analyte, Na_2CO_3, by the t/a ratio (mmol titrant/mmol analyte). Note also that although all measurements were to four significant figures, we computed the formula weight of Na_2CO_3 to five figures. This is because with four figures, it would have become the key number with an uncertainty of about one part per thousand compared to 187.6 with an uncertainty of about half that. It is not bad practice, as a matter of routine, to carry the formula weight to one additional figure, particularly if you have access to a calculator.

• •

The following examples illustrate titration calculations for different types of reactions and stoichiometry.

Example 5.23

The iron(II) in an acidified solution is titrated with a 0.0206 M solution of potassium permanganate:

$$5Fe^{2+} + MnO_4^- + 8H^+ \rightarrow 5Fe^{3+} + Mn^{2+} + 4H_2O$$

If the titration required 40.2 mL, how many milligrams iron are in the solution?

Solution

There are five times as many millimoles of iron as there are of permanganate that react with it, so

$$\text{mmol}_{Fe} = \frac{mg_{Fe}}{f \ wt_{Fe}} = M_{KMnO_4} \times mL_{KMnO_4} \times \tfrac{5}{1} \text{ (mmol Fe/mmol } KMnO_4)$$

$$mg_{Fe} = 0.0206 \text{ mmol } KMnO_4/mL \times 40.2 \text{ mL } KMnO_4 \times 5 \text{ (mmol Fe/mmol } MnO_4^-)$$
$$\times 55.8 \text{ mg Fe/mmol}$$
$$= 231 \text{ mg Fe}$$

Calculations of this type are used for the redox titrations described in Chapter 14.

Following is a list of typical precipitation and complexometric titration reactions and the factors for calculating the milligrams of analyte from millimoles of titrant.[5]

$Cl^- + Ag^+ \rightarrow \underline{AgCl}$ $mg_{Cl^-} = M_{Ag^+} \times mL_{Ag^+} \times 1 \text{ (mmol } Cl^-/\text{mmol } Ag^+) \times f \text{ wt}_{Cl^-}$

$2Cl^- + Pb^{2+} \rightarrow \underline{PbCl_2}$ $mg_{Cl^-} = M_{Pb^{2+}} \times mL_{Pb^{2+}} \times 2 \text{ (mmol } Cl^-/\text{mmol } Pb^{2+}) \times f \text{ wt}_{Cl^-}$

$PO_4^{3-} + 3Ag^+ \rightarrow \underline{Ag_3PO_4}$ $mg_{PO_4^{3-}} = M_{Ag^+} \times mL_{Ag^+} \times \frac{1}{3} \text{ (mmol } PO_4^{3-}/\text{mmol } Ag^+) \times f \text{ wt}_{PO_4^{3-}}$

$2CN^- + Ag^+ \rightarrow Ag(CN)_2^-$ $mg_{CN^-} = M_{Ag^+} \times mL_{Ag^+} \times 2 \text{ (mmol } CN^-/\text{mmol } Ag^+) \times f \text{ wt}_{CN^-}$

$2CN^- + 2Ag^+ \rightarrow \underline{Ag[Ag(CN)_2]}$ $mg_{CN^-} = M_{Ag^+} \times mL_{Ag^+} \times 1 \text{ (mmol } CN^-/\text{mmol } Ag^+) \times f \text{ wt}_{CN^-}$

$Ba^{2+} + SO_4^{2-} \rightarrow \underline{BaSO_4}$ $mg_{Ba^{2+}} = M_{SO_4^{2-}} \times 1 \text{ (mmol } Ba^{2+}/\text{mmol } SO_4^{2-}) \times f \text{ wt}_{Ba^{2+}}$

$Ca^{2+} + H_2Y^{2-} \rightarrow CaY^{2-} + 2H^+$ $mg_{Ca^{2+}} = M_{EDTA} \times 1 \text{ (mmol } Ca^{2+}/\text{mmol EDTA)} \times f \text{ wt}_{Ca^{2+}}$

These formulas are useful calculations involving the precipitation and complexometric titrations described in Chapters 8 and 11.

Example 5.24

Aluminum is determined by titrating with EDTA:

$$Al^{3+} + H_2Y^{2-} \rightarrow AlY^- + 2H^+$$

A 1.00-g sample requires 20.5 mL EDTA for titration. The EDTA was standardized by titrating 25.0 mL of a 0.100 M $CaCl_2$ solution, requiring 30.0 mL EDTA. Calculate the percent Al_2O_3 in the sample.

Solution

Since Ca^{2+} and EDTA react on a 1:1 mole ratio,

$$M_{EDTA} = \frac{0.100 \text{ mmol } CaCl_2/mL \times 25.0 \text{ mL } CaCl_2}{30.0 \text{ mL EDTA}} = 0.0833 \text{ mmol/mL}$$

The millimoles Al^{3+} are equal to the millimoles EDTA used in the sample titration, but there are one-half this number of millimoles of Al_2O_3 (since $1Al^{3+} \rightarrow \frac{1}{2}Al_2O_3$). Therefore,

$$\% \ Al_2O_3 = \frac{M_{EDTA} \times mL_{EDTA} \times \frac{1}{2} \text{ (mmol } Al_2O_3/\text{mmol EDTA)} \times f \text{ wt}_{Al_2O_3}}{mg_{sample}} \times 100\%$$

$$\% \ Al_2O_3 = \frac{0.0833 \text{ mmol EDTA/mL} \times 20.5 \text{ mL EDTA} \times \frac{1}{2} \times 101.96 \text{ mg } Al_2O_3/\text{mmol}}{1000\text{-mg sample}}$$

$$\times 100\% = 8.71\% \ Al_2O_3$$

---

[5] H_4Y = EDTA in the last equation.

WHAT IF THE ANALYTE AND TITRANT CAN REACT IN DIFFERENT RATIOS?

As you might be aware from your introductory chemistry course, some substances can undergo reaction to different products. The factor used in calculating millimoles of such a substance from the millimoles of titrant reacted with it will depend on the specific reaction. Sodium carbonate, for example, can react as a diprotic or a monoprotic base:

$$CO_3^{2-} + 2H^+ \rightarrow H_2O + CO_2$$

or

$$CO_3^{2-} + H^+ \rightarrow HCO_3^-$$

In the first case, mmol Na_2CO_3 = mmol acid $\times \frac{1}{2}$ (mmol CO_3^{2-}/mmol H^+). In the second case, mmol Na_2CO_3 = mmol acid. Similarly, phosphoric acid can be titrated as a monoprotic or a diprotic acid:

$$H_3PO_2 + OH^- \rightarrow H_2PO_4^- + H_2O$$

or

$$H_3PO_4 + 2OH^- \rightarrow HPO_4^{2-} + 2H_2O$$

 Example 5.25

In acid solution, potassium permanganate reacts with H_2O_2 to form Mn^{2+}:

$$5H_2O_2 + 2MnO_4^- + 6H^+ \rightarrow 5O_2 + 2Mn^{2+} + 8H_2O$$

In neutral solution, it reacts with $MnSO_4$ to form MnO_2:

$$3Mn^{2+} + 2MnO_4^- + 4OH^- \rightarrow \underline{5MnO_2} + 2H_2O$$

Calculate the number of milliliters of 0.100 M $KMnO_4$ that will react with 50.0 mL of 0.200 M H_2O_2 and with 50.0 mL of 0.200 M $MnSO_4$.

Solution

Keep track of millimoles!

The number of millimoles of MnO_4^- will be equal to two-fifths of the number of millimoles of H_2O_2 reacted:

$$M_{MnO_4^-} \times mL_{MnO_4^-} = M_{H_2O_2} \times mL_{H_2O_2} \times \tfrac{2}{5} \text{ (mmol } MnO_4^-/\text{mmol } H_2O_2)$$

$$mL_{MnO_4^-} = \frac{0.200 \text{ mmol } H_2O_2/mL \times 50.0 \text{ mL } H_2O_2 \times \tfrac{2}{5}}{0.100 \text{ mmol } MnO_4^-/mL} = 40.0 \text{ mL } KMnO_4$$

The number of millimoles of MnO_4^- reacting with Mn^{2+} will be equal to two-thirds of the number of millimoles of Mn^{2+}:

$$M_{MnO_4^-} \times mL_{MnO_4^-} = M_{Mn^{2+}} \times \tfrac{2}{3} \text{ (mmol } MnO_4^-/\text{mmol } Mn^{2+})$$

$$mL_{MnO_4^-} = \frac{0.200 \text{ mmol } Mn^{2+}/mL \times 50.0 \text{ mL } Mn^{2+} \times \tfrac{2}{3}}{0.100 \text{ mmol } MnO_4^-/mL} = 66.7 \text{ mL } KMnO_4$$

Example 5.26

Oxalic acid, $H_2C_2O_4$, is a reducing agent that reacts with $KMnO_4$ as follows:

$$5H_2C_2O_4 + 2MnO_4^- + 6H^+ \rightarrow 10CO_2 + 2Mn^{2+} + 8H_2O$$

Its two protons are also titratable with a base. How many milliliters of 0.100 M NaOH and 0.100 M $KMnO_4$ will react with 500 mg $H_2C_2O_4$?

Solution

$$\text{mmol NaOH} = 2 \times \text{mmol } H_2C_2O_4$$

$$0.100 \text{ mmol/mL} \times x \text{ mL NaOH} = \frac{500 \text{ mg}H_2C_2O_4}{90.0 \text{ mg/mmol}} \times 2 \text{ (mmol OH}^-\text{/mmol } H_2C_2O_4)$$

$$x = 111 \text{ mL NaOH}$$

$$\text{mmol } KMnO_4 = \tfrac{2}{5} \times \text{mmol } H_2C_2O_4$$

$$0.100 \text{ mmol/mL} \times x \text{ mL } KMnO_4 = \frac{500 \text{ mg } H_2C_2O_4}{90.0 \text{ mg/mmol}} \times \tfrac{2}{5} \text{ (mmol } KMnO_4\text{/mmol } H_2C_2O_4)$$

$$x = 22.2 \text{ mL } KMnO_4$$

Example 5.27

Pure $Na_2C_2O_4$ plus $KHC_2O_4 \cdot H_2C_2O_4$ (three replaceable protons, KH_3A_2) are mixed in such a proportion that each gram of the mixture will react with equal volumes of 0.100 M $KMnO_4$ and 0.100 M NaOH. What is the proportion?

Solution

Assume 10.0-mL titrant, so there is 1.00 mmol NaOH or $KMnO_4$. The acidity is due to $KHC_2O_4 \cdot H_2C_2O_4(KH_3A_2)$:

$$\text{mmol } KH_3A_2 = \text{mmol NaOH} \times \tfrac{1}{3}(\text{mmol } KH_3A_2\text{/mmol OH}^-)$$
$$1.00 \text{ mmol NaOH} \times \tfrac{1}{3} = 0.333 \text{ mmol } KH_3A$$

From Example 5.26, each mmol $Na_2C_2O_4$ (Na_2A) reacts with $\tfrac{2}{5}$ mmol $KMnO_4$.

$$\text{mmol } KMnO_4 = \text{mmol } Na_2A \times \tfrac{2}{5} \text{ (mmol } MnO_4^-\text{/mmol } Na_2A) + \text{mmol } KH_3A_2$$
$$\times \tfrac{4}{5} \text{ (mmol } MnO_4^-\text{/mmol } KH_3A_2)$$
$$1.00 \text{ mmol } KMnO_4 = \text{mmol } Na_2A \times \tfrac{2}{5} + 0.333 \text{ mmol } KH_3A_2 \times \tfrac{4}{5}$$
$$\text{mmol } Na_2A = 1.8_3 \text{ mmol}$$

The ratio is 1.8_3 mmol Na_2A/0.333 mmol $KH_3A_2 = 5.5_0$ mmol Na_2A/mmol KH_3A_2. The weight ratio is

$$\frac{5.5_0 \text{ mmol } Na_2A \times 134 \text{ mg/mmol}}{218 \text{ kg } KH_3A_2\text{/mmol}} = 3.38 \text{ g } Na_2A\text{/g } KH_3A_2$$

IF THE REACTION IS SLOW, DO A BACK-TITRATION

In back-titrations, a known number of millimoles of reactant is taken, in excess of the analyte. The unreacted portion is titrated.

Sometimes a reaction is slow to go to completion, and a sharp end point cannot be obtained. One example is the titration of antacid tablets with a strong acid such as HCl. In these cases, a **back-titration** will often yield useful results. In this technique, a measured amount of the reagent, which would normally be the titrant, is added to the sample so that there is a slight excess. After the reaction with the analyte is allowed to go to completion, the amount of excess (unreacted) reagent is determined by titration with another standard solution; the analyte reaction also may be speeded up in the presence of excess reagent. So by knowing the number of millimoles of reagent taken and by measuring the number of millimoles remaining unreacted, we can calculate the number of millimoles of sample that reacted with the reagent:

> mmol reagent reacted = mmol taken − mmol back-titrated
>
> mg analyte = mmol reagent reacted × factor (mmol analyte/mmol reagent)
>
> × f wt analyte (mg/mmol)

 Example 5.28

Chromium(III) is slow to react with EDTA (H_4Y) and is therefore determined by back-titration. A pharmaceutical preparation containing chromium(III) is analyzed by treating a 2.63-g sample with 5.00 mL of 0.0103 M EDTA. Following reaction, the unreacted EDTA is back-titrated with 1.32 mL of 0.0122 M zinc solution. What is the percent chromium chloride in the pharmaceutical preparation?

Solution

Both Cr^{3+} and Zn^{2+} react in a $1:1$ ratio with EDTA:

$$Cr^{3+} + H_4Y \rightarrow CrY^- + 4H^+$$
$$Zn^{2+} + H_4Y \rightarrow ZnY^2 + 4H^+$$

The millimoles of EDTA taken is

$$0.0103 \text{ mmol EDTA/mL} \times 5.00 \text{ mL EDTA} = 0.0515 \text{ mmol EDTA}$$

The millimoles of unreacted EDTA is

$$0.0112 \text{ mmol } Zn^{2+}/\text{mL} \times 1.32 \text{ mL } Zn^{2+} = 0.0148 \text{ mmol unreacted EDTA}$$

The millimoles of reacted EDTA is

$$0.0515 \text{ mmol taken} - 0.0148 \text{ mmol left} = 0.0367 \text{ mmol EDTA} \equiv \text{mmol } Cr^{3+}$$

The milligrams of $CrCl_3$ titrated is

$$0.0367 \text{ mmol } CrCl_3 \times 158.4 \text{ mg/mmol} = 5.81 \text{ mg } CrCl_3$$

$$\% \, CrCl_3 = \frac{5.81 \text{ mg } CrCl_3}{2630\text{-mg sample}} \times 100\% = 0.221\% \, CrCl_3$$

Or, combining all steps,

% CrCl₃

$$= \frac{(M_{\text{EDTA}} \times \text{mL}_{\text{EDTA}} - M_{\text{Zn}} \times \text{mL}_{\text{Zn}^{2+}}) \times 1(\text{mmol CrCl}_3/\text{mmol EDTA}) \times \text{f wt}_{\text{CrCl}_3}}{\text{mg}_{\text{sample}}} \times 100\%$$

$$= \frac{(0.0103 \text{ mmol EDTA/mL} \times 5.00 \text{ mL EDTA} - 0.0112 \text{ mmol Zn}^{2+}/\text{mL} \times 1.32 \text{ mL Zn}^{2+}) \times 1 \times 158.4 \text{ mg CrCl}_3/\text{mmol}}{2630\text{-mg sample}}$$

$$\times 100\%$$

$$= 0.221\% \text{ CrCl}_3$$

● ●

 Example 5.29

A 0.200-g sample of pyrolusite is analyzed for manganese content as follows. Add 50.0 mL of a 0.100 M solution of ferrous ammonium sulfate to reduce the MnO_2 to Mn^{2+}. After reduction is complete, the excess ferrous ion is titrated in acid solution with 0.0200 M KMnO₄, requiring 15.0 mL. Calculate the percent manganese in the sample as Mn_3O_4 (only part or none of the manganese may exist in this form, but we can make the calculations on the assumption that it does).

Solution

The reaction between Fe^{2+} and MnO_4^- is

$$5Fe^{2+} + MnO_4^- + 8H^+ \rightarrow 5Fe^{3+} + Mn^{2+} + 4H_2O$$

The reactant may react in different ratios with the analyte and titrant.

and so there are five times as many millimoles of excess Fe^{2+} as of MnO_4^- that reacted with it.

The reaction between Fe^{2+} and MnO_2 is

$$MnO_2 + 2Fe^{2+} + 4H^+ \rightarrow Mn^{2+} + 2Fe^{3+} + 2H_2O$$

and there are one-half as many millimoles of MnO_2 as millimoles of Fe^{2+} that react with it. There are one-third as many millimoles of Mn_3O_4 as of MnO_2 ($1MnO_2 \rightarrow \frac{1}{3} Mn_3O_4$). Therefore,

mmol Fe²⁺ reacted = 0.100 mmol Fe²⁺/mL × 50.0 mL Fe²⁺ − 0.0200 mmol MnO₄⁻/mL

$$\times 15.0 \text{ mL MnO}_4^- \times 5 \text{ mmol Fe}^{2+}/\text{mmol MnO}_4^-$$

$$= 3.5 \text{ mmol Fe}^{2+} \text{ reacted}$$

mmol MnO₂ = 3.5 mmol Fe²⁺ × $\frac{1}{2}$ (mmol MnO₂/mmol Fe²⁺) = 1.75 mmol MnO₂

mmol Mn₃O₄ = 1.7₅ mmol MnO₂ × $\frac{1}{3}$ (mmol Mn₃O₄)/mmol MnO₂)

$$= 0.58_3 \text{ mmol Mn}_3\text{O}_4$$

$$\% \text{ Mn}_3\text{O}_4 = \frac{0.58_3 \text{ mmol Mn}_3\text{O}_4 \times 228.8 \text{ (mg Mn}_3\text{O}_4/\text{mmol)}}{200\text{-mg sample}} \times 100\%$$

$$= 66.7\% \text{ Mn}_3\text{O}_4$$

Or, combining all steps

$$\% \; Mn_3O_4 = \{[M_{Fe^{2+}} \times mL_{Fe^{2-}} - M_{MnO_4^-} \times mL_{MnO_4^-} \times 5(mmol \; Fe^{2+}/mmol \; MnO_4^-)$$
$$\times \tfrac{1}{2} (mmol \; MnO_2/mmol \; Fe^{2+}) \times \tfrac{1}{3} (mmol \; Mn_3O_4/mmol \; MnO_2)$$
$$\times f \; wt_{Mn_3O_4}]/mg_{sample}\} \times 100\%$$

$$= \frac{(0.100 \times 50.0 - 0.0200 \times 15.0 \times 5) \times \tfrac{1}{2} \times \tfrac{1}{3} \times 228.8 \; mg/mmol}{200}$$

$$\times 100\%$$

$$= 66.7\% \; Mn_3O_4$$

• •

5.6 Normality—A Different Way to Do Volumetric Calculations[6]

Many substances do not react on a 1:1 mole basis, and so solutions of equal molar concentration do not react on a 1:1 volume basis. By introducing the concepts of equivalents and normality, we can make calculations in these cases that are similar to molar calculations for 1:1 mole reactions. To do so, we define a new unit of concentration called **normality**. The symbol N stands for **normal**, just as M stands for molar. The normality of a solution is equal to the number of **equivalents** of material per liter of solution:

$$N = \frac{eq}{L} = \frac{meq}{mL} \tag{5.30}$$

where meq represents **milliequivalents**.

The number of reacting units will depend on the reaction. These often vary in different redox reactions.

Equivalents are based on the same concept as moles, but the number of equivalents will depend on the number of **reacting units** supplied by each molecule or the number with which it will react. For example, if we have one mole of HCl, we have one mole of H^+ to react as an acid. Therefore, we have one equivalent of H^+. If, on the other hand, we have one mole of H_2SO_4, we have two moles of the reacting unit H^+ and two equivalents of H^+. The number of equivalents can be calculated from the number of moles by

$$eq = mol \times no. \; of \; reacting \; units \; per \; molecule$$
$$meq = mmol \times no. \; or \; reacting \; units \; per \; molecule$$

WHAT IS EQUIVALENT WEIGHT?

The **equivalent weight** is that weight of a substance in grams that will furnish one mole of the reacting unit. Thus, for HCl, the equivalent weight is equal to the

---

[6] The concept of equivalents and normality, while historically useful, is not as widely used today for calculations as is molarity. Learn this section if your instructor asks that you do, otherwise you may choose to skip it.

formula weight. For H_2SO_4, it takes only one-half the number of molecules to furnish one mole of H^+, and so the equivalent weight is one-half the formula weight:

$$\text{eq wt HCl} = \frac{\text{f wt}_{HCl} \text{ (g/mol)}}{1 \text{ (eq/mol)}}$$

$$\text{eq wt } H_2SO_4 = \frac{\text{f wt}_{H_2SO_4} \text{ (g/mol)}}{2 \text{ (eq/mol)}}$$

 **Example 5.30**

Calculate the equivalent weights of the following substances: (a) NH_3, (b) $H_2C_2O_4$ (in reaction with NaOH), and (c) $KMnO_4$ [Mn(VII) is reduced to Mn^{2+}].

Solution

(a)
$$\text{eq wt} = \frac{NH_3 \text{ g/mol}}{1 \text{ eq/mol}} = \frac{17.03}{1} = 17.03 \text{ g/eq}$$

(b)
$$\text{eq wt} = \frac{H_2C_2O_4 \text{ g/mol}}{2 \text{ eq/mol}} = \frac{90.04}{2} = 45.02 \text{ g/eq}$$

(c) The Mn undergoes a five-electron change (the electron, e^-, is denoted), from valence +7 to +2:

$$MnO_4^- + 8H^+ + 5e^- = Mn^{2+} + 4H_2O$$

$$\text{eq wt} = \frac{KMnO_4 \text{ g/mol}}{5 \text{ eq/mol}} = \frac{158.04}{5} = 31.608 \text{ g/eq}$$

Just as we can calculate the number of moles from the number of grams by dividing by the formula weight, we can also calculate the number of equivalents of a substance by dividing by the equivalent weight:

$$\boxed{\text{eq} = \frac{g}{\text{eq wt (g/eq)}} \qquad \text{meq} = \frac{mg}{\text{eq wt (mg/meq)}}} \qquad (5.31)$$

The normality of a solution, then, is calculated from the number of equivalents and the volume:

$$\boxed{N = \frac{\text{eq}}{L} = \frac{\text{g/eq wt (g/eq)}}{L} \qquad N = \frac{\text{meq}}{mL} = \frac{\text{mg/eq wt (mg/meq)}}{mL}} \qquad (5.32)$$

Example 5.31

Calculate the normality of the solutions containing the following: (a) 5.300 g/L Na_2CO_3 (when the CO_3^{2-} reacts with two protons), (b) 5.267 g/L $K_2Cr_2O_7$ (the Cr is reduced to Cr^{3+}).

Solution

(a) CO_3^{2-} reacts with $2H^+$ to H_2CO_3:

$$N = \frac{5.300 \text{ g/L}}{(105.99/2) \text{ g/eq}} = 0.1000 \text{ eq/L}$$

(b) Each Cr(VI) is reduced to Cr^{3+}, a total change of $6e^-$/molecule $K_2Cr_2O_7$:

$$Cr_2O_7^{2-} + 14H^+ + 6e^- = 2Cr^3 + 7H_2O$$

$$N = \frac{5.267 \text{ g/L}}{(294.19/6) \text{ g/eq}} = 0.1074 \text{ eq/L}$$

It is not always necessary to use equivalent weights to calculate equivalents. We can also use a *stoichiometry factor*, n (units of eq/mol), to convert between moles and equivalents. Thus,

$$\text{Equivalents} = \text{moles} \times n \text{ (eq/mol)}$$
$$N \text{ (eq/L)} = M \text{ (mol/L)} \times n \text{ (eq/mol)}$$
$$\text{eq wt (g/eq)} = \frac{\text{f wt (g/mol)}}{n \text{ (eq/mol)}}$$

> You may use a balanced redox half-reaction, rather than the change in oxidation number, to determine the equivalent weight.

In oxidation–reduction reactions, we do not have to rely on oxidation numbers but can use the balanced half-reaction. For example, when dichromate is reduced to Cr^{3+}, the half-reaction is

$$Cr_2O_7^{2-} + 14H^+ + 6e^- \rightleftharpoons 2Cr^{3+} + 7H_2O$$

So $n = 6$ (6 electrons/mol $Cr_2O_7^{2-}$). Hence, a 0.1 M solutions is 0.1 $M \times 6$ (eq/mol) = 0.6 N.

> Normality calculations are treated like 1:1 reactions in molarity calculations.

The advantage of expressing concentrations in normality and quantities as equivalents is that one equivalent of substance A will **ALWAYS** *react with one equivalent of substance B.* Thus, one equivalent of NaOH ($\equiv 1$ mol) will react with one equivalent of HCl ($\equiv 1$ mol) or with one equivalent of H_2SO_4 ($\equiv \frac{1}{2}$ mol). We can, therefore, calculate the weight of analyte from the number of equivalents of titrant because the latter is equal to the equivalents of analyte.

It must be emphasized that the number of equivalents, or the normality, depends upon the specific reaction. Na_2CO_3, for example, may react with either one H^+ ($CO_3^{2-} + H^+ \rightarrow HCO_3^-$) or two H^+ ($CO_3^{2-} + 2H^+ \rightarrow H_2CO_3$). In the first case there is one reacting unit, compared to two in the second case. Thus, there is a danger in using the units normality, equivalents, or equivalent weight; that is, *they are entirely dependent on the particular reaction*, and this must be specified. If you encounter a 0.1000 N solution of $KMnO_4$, you may not know if this refers to re-duction of permanganate to Mn^{2+} (the usual reaction in acid solution) or to MnO_2 (occurs in neutral solution). The first reaction corresponds to a five-electron change (number of reacting units), and the second to a three-electron change. If the mo-larity is given, there is no question what the concentration is. So normality depends on the knowledge and availability of a balanced reaction, whereas molarity does not. A 1 M $KMnO_4$ solution is always 1 M, but it can be 5 N or 3 N, depending on how it is used.

So remember that the number of reacting units of a compound, and hence the number of equivalents, will depend on the reaction it is undergoing. If A is the sample and T is the titrant,

$$meq_A = meq_T \tag{5.33}$$

$$meq_A = \frac{mg_A}{eq\ wt_A\ (mg/meq)} = N_T\ (meq/mL) \times mL \tag{5.34}$$

$$mg_A = meq_T = eq\ wt_A\ (mg/meq) \tag{5.35}$$

$$mg_A = N_T\ (meq/mL) \times mL_T \times eq\ wt_A\ (mg/meq) \tag{5.36}$$

The equivalent weight of A is determined in the same way as it is for T; that is, how many reacting units does A liberate or react with per molecule?

A general equation for calculating the percent of a constituent in the sample can now be written (analogous to Equation 5.22):

$$\%\ A = \frac{N_T\ (meq/mL) \times mL_T \times 1\ (meq_A/meq_T) \times eq\ wt_A\ (mg/meq)}{mg_{sample}} \times 100\% \tag{5.37}$$

The factor meq_A/meq_T is always unity. Hence, it is understood in all dimensional analysis calculations.

Example 5.32

A 0.4671-g sample containing sodium bicarbonate (a monoacidic base) is dissolved and titrated with a standard solution of hydrochloric acid, requiring 40.72 mL. The hydrochloric acid was standardized by titrating 0.1876 g sodium carbonate, which required 37.86 mL acid (See Example 5.18 for reaction.) Calculate the percent so-dium bicarbonate in the sample.

Solution

Na_2CO_3 is a base that reacts with two H^+ per molecule:

$$N_{HCl} = \frac{meq_{Na_2CO_2}}{mL_{HCl}} = \frac{mg_{Na_2CO_3}/f\ wt_{Na_2CO_3}/2)}{mL_{HCl}}$$

$$= \frac{187.6\ mg\ Na_2CO_3/(105.99/2\ mg/meq)}{37.86\ mL\ HCl} = 0.09350\ meq/mL\ HCl$$

% NaHCO$_3$

$$= \frac{N_{HCl} \times mL_{HCl} \times (f\ wt_{NaHCO_3}/1)}{mg_{sample}} \times 100\%$$

$$= \frac{0.09350\ meq\ HCl/mL \times 40.72\ mL\ HCl \times (84.01/1\ mg\ NaHCO_3/meq)}{476.1\ mg_{sample}}$$

$$\times 100\%$$

$$= 67.18\%\ NaHCO_3$$

· ·

It is important to remember that one equivalent of a substance will always react with one equivalent of its counterpart. It is useful to recognize that, since

$$meq_A = meq_B \qquad (5.38)$$

we can calculate the volumes of two solutions that will react by

$$\boxed{N_A\,(meq/mL) \times mL_A = N_T\,(meq/mL) \times mL_T} \qquad (5.39)$$

 ## Example 5.33

How many milliliters of a 0.25 M solution of H$_2$SO$_4$ will react with 10 mL of a 0.25 M solution of NaOH?

Solution

Since there are two reacting units per molecule of H$_2$SO$_4$, the normality of this solution will be twice its molarity:

$$2NaOH + H_2SO_4 \rightarrow Na_2SO_4 + 2H_2O$$
$$N_{H_2SO_4} = 2\ (eq/mol) \times 0.25\ (mol/L) = 0.50\ eq/L$$

The normality of the NaOH will be the same as its molarity, since it will consume one reacting unit per molecule:

$$N_{NaOH} = 0.25\ eq/L$$
$$meq_{H_2SO_4} = meq_{NaOH}$$
$$N_{H_2SO_4} \times mL_{H_2SO_4} = N_{NaOH} \times mL_{NaOH}$$
$$\therefore 0.50\ meq/mL \times mL_{H_2SO_4} = 0.25\ meq/mL \times 10\ mL$$
$$mL_{H_2SO_4} = 5.0\ mL$$

· ·

An approach similar to Equation 5.29 can also be used to calculate the dilution required to prepare a certain normality of a solution from a more concentrated solution in a manner similar to molarity dilutions (see Example 5.22).

Table 5.4 summarizes the relationship between mole-based units and equivalent-based units.

Table 5.4

Comparison of Mole-Based and Equivalent-Based Units

$mol \times n(eq/mol) = eq$	$mmol \times n(meq/mmol) = meq$
$M(mol/L) \times n(eq/mol) = N(eq/L)$	$M(mmol/L) \times n(meq/mmol) = N(meq/mL)$
$f\ wt\ (g/mol) \div n(eq/mol) = eq\ wt\ (g/eq)$	$f\ wt\ (mg/mmol) \div n(meq/mmol) = eq\ wt\ (mg/meq)$

REACTING UNITS IN NORMALITY CALCULATIONS—PROTONS AND ELECTRONS

1. Acid–Base. As we have mentioned, the reacting unit for acids and bases is the proton H^+. If the substance reacts as an acid, we must determine the number of reactive protons it possesses per molecule. If it reacts as a base, we must determine the number of protons it will react with per molecule. Then,

$$eq\ wt = \frac{f\ wt}{no.\ of\ H^+} \qquad (5.40)$$

This may depend on the particular reaction we choose. With H_2SO_4, both protons are strongly ionized; thus there are two equivalents of reacting units per mole of H_2SO_4. These react together to give a single end point, which corresponds to the titration of both protons. In H_3PO_4, the first proton is fairly strongly ionized, the second weakly so, and the third is too weakly ionized to be titrated. So, we can titrate the first and second protons *stepwise* to obtain two separate end points, while the third is too weak to yield a detectable end point. If we choose to titrate only the first proton, then there will be only one reacting unit per molecule of H_3PO_4, *in this particular reaction*. The number of equivalents will, therefore, be equal to the number of moles of H_3PO_4. If we choose, however, to titrate two protons (to double the titration volume and improve the relative precision), then the number of equivalents will be twice the number of moles.

Na$_2$CO$_3$ is a strong base that will react with an acid to produce NaHCO$_3$. The NaHCO$_3$ is a weak base and can be titrated one step further to carbonic acid:

$$CO_3^{2-} + H^+ \rightleftharpoons HCO_3^-$$
$$HCO_3^- + H^+ \rightleftharpoons H_2O + CO_2$$

The number of reacting units (protons) depends on which reaction we choose.

Again, we can have two separate end points, and the number of equivalents (and the equivalent weight) will depend on which reaction we choose. Thus, it is essential that we know the reaction taking place before we begin a calculation.

 Example 5.34

A solution of sodium carbonate is prepared by dissolving 0.212 g Na$_2$CO$_3$ and diluting to 100 mL. Calculate the normality of the solution (a) if it is used as a monoacidic base, and (b) if it is used as a diacidic base.

Solution

(a) $N = \dfrac{mg_{Na_2CO_3}/(Na_2CO_3/1)}{mL} = \dfrac{212\ mg/(106.0/1\ mg/meq)}{100\ mL} = 0.0200\ meq/mL$

(b) $N = \dfrac{mg_{Na_2CO_3}/(Na_2CO_3/2)}{mL} = \dfrac{212\ mg/(106.0/2\ mg/meq)}{100\ mL} = 0.0400\ meq/mL$

2. Reduction–Oxidation. The reacting unit in a reduction–oxidation reaction is the **electron**. A reducing agent liberates electrons and is thereby oxidized, and an oxidizing agent takes on electrons and is thereby reduced. For example, in the reaction

$$5Fe^{2+} + MnO_4^- + 8H^+ \rightleftharpoons 5Fe^{3+} + Mn^{2+} + 4H_2O$$

each Fe^{2+} (reducing agent) loses one electron, and each MnO_4^- (oxidizing agent) gains five electrons in being reduced from Mn^{7+} to Mn^{2+}. We treat the number of reacting units just as we did with acids and bases. Thus,

$$\text{eq wt} = \frac{\text{f wt}}{\text{no. of moles of electrons gained or lost}} \qquad (5.41)$$

In this example, the number of equivalents of iron is equal to its number of moles, and the equivalent weight is equal to the atomic weight of iron. There are five times as many equivalents of the permanganate as moles, and its equivalent weight is one-fifth its formula weight. But just as with acids and bases, one equivalent of reducing agent will react with one equivalent of oxidizing agent.

Example 5.35

Iodine (I_2) is an oxidizing agent that in reactions with reducing agents is reduced to iodide ion (I^-). How many grams I_2 would you weigh out to prepare 100 mL of a 0.100 N I_2 solution?

Solution

Since each molecule of I_2 consumes two electrons,

$$I_2 + 2e^- \rightarrow 2I^-$$

The equivalent weight is one-half the formula weight:

$$\text{eq wt} = \frac{\text{f wt}_{I_2}\ (\text{g/mol})}{2\ (\text{eq/mol})}$$

$$N \times \text{mL} = \text{meq} = \frac{\text{mg}_{I_2}}{\text{eq wt}} = \frac{\text{mg}_{I_2}}{\text{f wt}_{I_2}/2}$$

$$0.100\ \text{meq/mL} \times 100\ \text{mL} = \frac{\text{mg}_{I_2}}{254/2\ \text{mg/meq}}$$

$$\text{mg}_{I_2} = 0.100\ \text{meq/mL} \times 100\ \text{mL} \times (254/2\ \text{mg/meq}) = 1270\ \text{mg}$$

We would, therefore, weigh out 1.27 g.

Example 5.36

Calculate the normality of a solution of 0.25 g/L $H_2C_2O_4$, both as an acid and as a reducing agent.

The normality depends on whether $H_2C_2O_4$ reacts as an acid or as a redox agent.

Solution

The equivalent weight as an acid is half the formula weight;

$$\therefore N_{acid} = \frac{250 \text{ mg}/(90.04/2 \text{ mg/meq})}{1000 \text{ mL}} = 0.00555 \text{ meq/mL}$$

Each oxalate ion gains two electrons in being oxidized to CO_2 (each carbon is oxidized from a valence of $+3$ to a valence of $+4$). Therefore, the equivalent weight is half the formula weight, and the normality as a reducing agent is the same as it is as an acid.

SUMMARY OF NORMALITY CALCULATIONS

Summarized here are the equations most frequently used in applying the concept of normality to volumetric titrations:

$$\text{meq} = \frac{\text{mg}}{\text{eq wt (mg/meq)}} \qquad (5.42)$$

$$N = \frac{\text{meq}}{\text{mL}} \qquad (5.43)$$

One of the following equations will usually be used to calculate the results of a titration in which substance A is titrated with substance T:

Think units!

$$N_T \text{ (meq/mL)} \times \text{mL}_T \times \text{eq wt}_A \text{ (mg/meq)} = \text{mg}_A \qquad (5.44)$$

$$\frac{N_T \text{ (meq/mL)} \times \text{mL}_T \times \text{eq wt}_A \text{ (mg/meq)}}{\text{mg}_{sample}} \times 100\% = \% \text{ A} \qquad (5.45)$$

5.7 Titer—How to Make Rapid Routine Calculations

For routine titrations, it is often convenient to calculate the **titer** of the titrant. The titer is the weight of analyte that is chemically equivalent to 1 mL of the titrant, usually expressed in milligrams. For example, if a potassium dichromate solution has a titer of 1.267 mg Fe, each milliliter potassium dichromate will react with 1.267 mg iron, and the weight of iron titrated is obtained by simply multiplying the volume of titrant used by the titer. The titer can be expressed in terms of any form of the analyte desired, for example, milligrams FeO or Fe_2O_3.

Titer = milligrams analyte that react with 1 mL of titrant.

Example 5.37

A standard solution of potassium dichromate contains 5.442 g/L. What is its titer in terms of milligrams Fe_3O_4?

Solution

The iron is titrated as Fe^{2+} and each $Cr_2O_7^{2-}$ will react with $6Fe^{2+}$ or the iron from $2Fe_3O_4$:

$$6Fe^{2+} + Cr_2O_7^{2-} + 14H^+ \rightarrow 6Fe^{3+} + 2Cr^{3+} + 7H_2O$$

The molarity of the $K_2Cr_2O_7$, solution is

$$M_{Cr_2O_7^{2-}} = \frac{g/L}{f\ wt_{K_2Cr_2O_7}} = \frac{5.442\ g/L}{294.19\ g/mol} = 0.01850\ mol/L$$

Therefore the titer is

$$0.01850\left(\frac{mmol\ K_2Cr_2O_7}{mL}\right) \times \frac{2}{1}\left(\frac{mmol\ Fe_3O_4}{mmol\ K_2Cr_2O_7}\right) \times 231.54\left(\frac{mg\ Fe_3O_4}{mmol\ Fe_3O_4}\right)$$

$$= 8.567\ mg\ Fe_3O_4/mL\ K_2Cr_2O_7$$

5.8 Weight Relationships—You Need These for Gravimetric Calculations

In the technique of gravimetric analysis (Chapter 7), the analyte is converted to an insoluble form, which is weighed. From the weight of the precipitate formed and the weight relationship between the analyte and the precipitate, we can calculate the weight of analyte. We review here some of the calculation concepts.

The analyte is almost always weighed in a form different from what we wish to report. We must, therefore, calculate the weight of the desired substance from the weight of the gravimetric precipitate. We can do this by using a direct proportion. For example, if we are analyzing for the percentage of chloride in a sample by weighing it as AgCl, we can write

$$Cl^- \xrightarrow{precipitating\ reagent} AgCl$$

We derive one mole AgCl from one mole Cl^-, so

$$\frac{g\ Cl^-}{g\ AgCl} = \frac{at\ wt\ Cl\ (g\ Cl/mol\ Cl)}{f\ wt\ AgCl\ (g\ AgCl/mol\ AgCl)} \times \frac{1\ mol\ Cl}{1\ mol\ AgCl}$$

or

$$g\ Cl^- = g\ AgCl \times \frac{at\ wt\ Cl}{f\ wt\ AgCl}\ (g\ Cl/g\ AgCl)$$

In other words, the weight of Cl contained in or used to create AgCl is equal to the weight of AgCl times the **fraction** of Cl in it.

Calculation of the corresponding weight of Cl_2 that would be contained in the sample would proceed thus:

$$Cl_2 \xrightarrow{\text{precipitating reagent}} 2AgCl$$

We derive two moles of AgCl from each mole of Cl_2, so

$$\frac{g\ Cl_2}{g\ AgCl} = \frac{f\ wt\ Cl_2\ (g\ Cl_2/mol\ Cl_2)}{f\ wt\ AgCl\ (g\ AgCl/mol\ AgCl)} \times \frac{1\ mol\ Cl_2}{2\ mol\ AgCl}$$

and

$$g\ Cl_2 = g\ AgCl \times \frac{f\ wt\ Cl_2}{2\ (f\ wt\ AgCl)}\ (g\ Cl_2/g\ AgCl)$$

or

$$g\ AgCl \times \frac{70.906\ g\ Cl_2/mol\ Cl_2}{(2\ mol\ AgCl/mol\ Cl_2)(143.32\ g\ AgCl/mol\ AgCl)} = g\ Cl_2$$

We may also write

$$g\ AgCl \times \frac{1\ mol\ AgCl}{143.32\ g\ AgCl} \times \frac{1\ mol\ Cl_2}{2\ mol\ AgCl} \times \frac{70.906\ g\ Cl_2}{1\ mol\ Cl_2} = g\ Cl_2$$

The **gravimetric factor** (GF) is the appropriate ratio of the formula weight of the substance *sought* to that of the substance *weighed*:

$$\boxed{GF = \text{gravimetric factor} = \frac{f\ wt\ of\ substance\ sought}{f\ wt\ of\ substance\ weighed} \times \frac{a}{b}\ (mol\ sought/mol\ weighed)}$$

$$(5.46)$$

where a and b are integers that make the formula weights in the numerator and denominator chemically equivalent. In the above examples, the gravimetric factors were (Cl/AgCl) $\times$ 1/1, and $(Cl_2/AgCl) \times \frac{1}{2}$. Note that one or both of the formula weights may be multiplied by an integer in order to keep the same number of atoms of the key element in the numerator and denominator.

The weight of the substance sought is obtained by multiplying the weight of the precipitate by the gravimetric factor:

$$\boxed{\text{weight (g)} \times \frac{f\ wt\ of\ substance\ sought}{f\ wt\ of\ substance\ weighed} \times \frac{a}{b} = \text{sought (g)}} \quad (5.47)$$

Note that the *species* and the *units* of the equation can be checked by dimensional analysis (canceling of like species and units). For example,

$$\cancel{g\ AgCl} \times \tfrac{1}{2}\ (\cancel{mol\ Cl_2}\ /\cancel{mol\ AgCl}) \times \frac{Cl_2\ (g\ Cl_2/\cancel{mol\ Cl_2})}{AgCl\ (\cancel{g\ AgCl}/\cancel{mol\ AgCl})} = g\ Cl_2$$

In gravimetric analysis, the moles of analyte is a multiple of the moles of precipitate formed (the moles of analyte contained in each mole of precipitate).

Remember to keep track of the units!

The gravimetric factor is the weight of analyte per unit weight of precipitate.

Note also that we have calculated the amount of Cl_2 gas *derivable* from the sample instead of the amount of Cl^- ion, the form in which it probably exists in the sample and the form in which it is weighed. If we precipitate the chloride as $PbCl_2$,

$$2Cl^- \xrightarrow{\text{precipitating agent}} PbCl_2$$

and $$Cl_2 \rightarrow PbCl_2$$

then,

$$g\ Cl^- = g\ PbCl_2 \times \frac{2(f\ wt\ Cl)}{f\ wt\ PbCl_2}\ (g\ Cl/g\ PbCl_2) = g\ PbCl_2 \times GF$$

or

$$g\ Cl_2 = g\ PbCl_2 \times \frac{f\ wt\ Cl_2}{f\ wt\ PbCl_2}\ (g\ Cl_2/g\ PbCl_2) = g\ PbCl_2 \times GF$$

Conversion from weight of one substance to the equivalent weight of another is done using dimensional analysis of the units to arrive at the desired weight. The gravimetric factor is one step of that calculation and is useful for routine calculations. That is, if we know the gravimetric factor, we simply multiply the weight of the precipitate by the gravimetric factor to arrive at the weight of the analyte.

> The grams of analyte = grams precipitate × GF.

Example 5.38

Calculate the weight of barium and the weight of Cl present in 25.0 g $BaCl_2$

Solution

$$5.0\ g\ BaCl_2 \times 1\ (mol\ Ba/mol\ BaCl_2) \times \frac{137.3\ (g\ Ba/mol\ Ba)}{208.2\ (g\ BaCl_2/mol\ BaCl_2)} = 16.5\ g\ Ba$$

$$25.0\ g\ BaCl_2 \times \tfrac{2}{1}\ (mol\ Cl/mol\ BaCl_2) \times \frac{35.45\ (g\ Cl/mol\ Cl)}{208.2\ (g\ BaCl_2/mol\ BaCl_2)} = 8.51\ g\ Cl$$

Example 5.39

Aluminum in an ore sample is determined by dissolving it and then precipitating with base as $Al(OH)_3$ and igniting to Al_2O_3, which is weighed. What weight of aluminum was in the sample if the ignited precipitate weighed 0.2385 g?

Solution

$$g\ Al_2O_3 \times \frac{2}{1}\ (mol\ Al/mol\ Al_2O_3) \times \frac{Al(g\ Al/mol\ Al)}{Al_2O_3\ (g\ Al_2O_3/mol\ Al_2O_3)} = g\ Al_2O_3$$

$$0.2385\ g\ Al_2O_3 = \frac{2}{1} \times \frac{26.982\ (g\ Al/mol\ Al)}{101.96\ (g\ Al_2O_3/mol\ Al_2O_3)} = 0.1262_3\ g\ Al$$

The gravimetric factor is

$$\frac{2 \text{ Al}}{\text{Al}_2\text{O}_3} \text{ (g Al/g Al}_2\text{O}_3) = \frac{2 \text{ (26.982 g Al/mol Al)}}{101.96 \text{ (g Al}_2\text{O}_3\text{/mol Al}_2\text{O}_3)} = 0.52927 \text{(g Al/g Al}_2\text{O}_3)$$

or

$$0.2385 \text{ g Al}_2\text{O}_3 \times 0.52927 \text{ (g Al/g Al}_2\text{O}_3) = 0.1262_3 \text{ g Al}$$

Following are some other examples of gravimetric factors:

Sought	Weighed	Gravimetric Factor
SO_3	$BaSO_4$	$\dfrac{SO_3 \text{ f wt}}{BaSO_4 \text{ f wt}}$
Fe_3O_4	Fe_2O_3	$\dfrac{2Fe_3O_4 \text{ f wt}}{3Fe_2O_3 \text{ f wt}}$
Fe	Fe_2O_3	$\dfrac{2Fe \text{ f wt}}{Fe_2O_3 \text{ f wt}}$
MgO	$Mg_2P_2O_7$	$\dfrac{2MgO \text{ f wt}}{Mg_2P_2O_7 \text{ f wt}}$
P_2O_5	$Mg_2P_2O_7$	$\dfrac{P_2O_5 \text{ f wt}}{Mg_2P_2O_7 \text{ f wt}}$

More examples of gravimetric calculations are given in Chapter 7.

The operations of gravimetric analyses are described in detail in Chapter 10.

Learning Objectives

WHAT ARE SOME OF THE KEY THINGS WE LEARNED FROM THIS CHAPTER?

- How to calculate molarities and moles (key equations: 5.4, 5.5), p. 144
- How to express analytical results, p. 152
- How to calculate weight and percent analyted from molarities, volumes, and reaction ratios (Key equations: 5.5, 5.17–5.22, 5.27), p. 160
- Normality as an alternative concentration unit (key equations 5.30–5.32; 5.36), p. 170
- Weight relationships for gravimetric analysis (key equation: 5.47), p. 180

Questions

1. Distinguish between the expression of concentration on weight/weight, weight/volume, and volume/volume bases.
2. Express ppm and ppb on weight/weight, weight/volume, and volume/volume bases.
3. Define the term "equivalent weight," used for electrolytes in clinical chemistry. Why is this used?
4. List the requirements for a titration. What are the four classes of titrations?
5. What is the equivalence point of a titration? The end point?

6. What is a standard solution? How is it prepared?

7. What are the requirements of a primary standard?

8. Why should a primary standard have a high formula weight?

Problems

WEIGHT/MOLE CALCULATIONS

9. Calculate the grams of substance required to prepare the following solutions: (a) 250 mL of 5.00% (wt/vol) $NaNO_3$; (b) 500 mL of 1.00% (wt/vol) NH_4NO_3, (c) 1000 mL of 10.0% (wt/vol) $AgNO_3$.

10. What is the wt/vol % of the solute in each of the following solutions? (a) 52.3 g Na_2SO_4/L, (b) 275 g KBr in 500 mL, (c) 3.65 g SO_2 in 200 mL.

11. Calculate the formula weights of the following substances: (a) $BaCl_2 \cdot 2H_2O$, (b) $KHC_2O_4 \cdot H_2C_2O_4$, (c) $Ag_2Cr_2O_7$, (d) $Ca_3(PO_4)_2$.

12. Calculate the number of millimoles contained in 500 mg of each of the following substances: (a) $BaCrO_4$, (b) $CHCl_3$, (c) $KIO_3 \cdot HIO_3$, (d) $MgNH_4PO_4$, (e) $Mg_2P_2O_7$, (f) $FeSO_4 \cdot C_2H_4(NH_3)_2SO_4 \cdot 4H_2O$.

13. Calculate the number of grams of each of the substances in Problem 12 that would have to be dissolved and diluted to 100 mL to prepare a 0.200 M solution.

14. Calculate the number of milligrams of each of the following substances you would have to weigh out in order to prepare the listed solutions: (a) 1.00 L of 1.00 M NaCl, (b) 0.500 L of 0.200 M sucrose $(C_{12}H_{22}O_{11})$, (c) 10.0 mL of 0.500 M sucrose, (d) 0.0100 L of 0.200 M Na_2SO_4, (e) 250 mL of 0.500 M KOH, (f) 250 mL of 0.900% NaCl (g/100 mL solution).

15. The chemical stockroom is supplied with the following stock solutions: 0.100 M HCl, 0.0200 M NaOH, 0.0500 M KOH, 10.0% HBr (wt/vol), and 5.00% Na_2CO_3 (wt/vol). What volume of stock solution would be needed to obtain the following amounts of solutes? (a) 0.0500 mol HCl, (b) 0.0100 mol NaOH, (c) 0.100 mol KOH, (d) 5.00 g HBr, (e) 4.00 g Na_2CO_3, (f) 1.00 mol HBr, (g) 0.500 mol Na_2CO_3.

MOLARITY CALCULATIONS

16. Calculate the molar concentrations of all the cations and anions in a solution prepared by mixing 10.0 mL each of the following solutions: 0.100 M $Mn(NO_3)_2$, 0.100 M KNO_3, and 0.100 M K_2SO_4.

17. A solution containing 10.0 mmol $CaCl_2$ is diluted to 1 L. Calculate the number of grams of $CaCl_2 \cdot 2H_2O$ per milliliter of the final solution.

18. Calculate the molarity of each of the following solations: (a) 10.0 g H_2SO_4 in 250 mL of solution, (b) 6.00 g NaOH in 500 mL of solution, (c) 25.0 g $AgNO_3$ in 1.00 L of solution.

19. Calculate the number of grams in 500 mL of each of the following solutions: (a) 0.100 M Na_2SO_4, (b) 0.250 M $Fe(NH_4)_2(SO_4)_2 \cdot 6H_2O$, (c) 0.667 M $Ca(C_9H_6ON)_2$.

20. Calculate the grams of each substance required to prepare the following solutions: (a) 250 mL of 0.100 M KOH, (b) 1.00 L of 0.0275 M $K_2Cr_2O_7$, (c) 500 mL of 0.0500 M $CuSO_4$.

21. How many milliliters of concentration hydrochloric acid, 38.0% (wt/wt), specific gravity 1.19, are required to prepare 1 L of a 0.100 M solution? (Assume density and specific gravity are equal within three significant figures.)

22. Calculate the molarity of each of the following commercial acid or base solutions: (a) 70.0% $HClO_4$, specific gravity 1.668, (b) 69.0% HNO_3, specific gravity 1.409, (c) 85.0% H_3PO_4, specific gravity 1.689, (d) 99.5% $HC_2H_3O_2$ (acetic acid), specific gravity 1.051, (e) 28.0% NH_3, specific gravity 0.898. (Assume density and specific gravity are equal within three significant figures.)

PPM CALCULATIONS

23. A solution contains 6.0 μmol Na_2SO_4 in 250 mL. How many ppm sodium does it contain? Of sulfate?

24. A solution (100 mL) containing 325 ppm K^+ is analyzed by precipitating it as the tetraphenyl borate, $K(C_6H_5)_4B$, dissolving the precipitate in acetone solution, and measuring the concentration of tetraphenyl borate ion, $(C_6H_5)_4B^-$, in the solution. If the acetone solution volume is 250 mL, what is the concentration of the tetraphenyl borate in ppm?

25. Calculate the molar concentrations of 1.00-ppm solutions of each of the following. (a) $AgNO_3$, (b) $Al_2(SO_4)_3$, (c) CO_2, (d) $(NH_4)_4Ce(SO_4)_4 \cdot 2H_2O$, (e) HCl, (f) $HClO_4$.

26. Calculate the ppm concentrations of 2.50×10^{-4} M solutions of each of the following. (a) Ca^{2+}, (b) $CaCl_2$, (c) HNO_3, (d) KCN, (e) Mn^{2+}, (f) MnO_4^-.

27. You want to prepare 1 L of a solution containing 1.00 ppm Fe^{2+}. How many grams ferrous ammonium sulfate, $FeSO_4 \cdot (NH_4)_2SO_4 \cdot 6H_2O$, must be dissolved and diluted in 1 L? What would be the molarity of this solution?

28. A 0.456-g sample of an ore is analyzed for chromium and found to contain 0.560 mg Cr_2O_3. Express the concentration of Cr_2O_3 in the sample as (a) percent, (b) parts per thousand, and (c) parts per million.

29. How many grams $NaCl$ should be weighed out to prepare 1 L of a 100-ppm solution of (a) Na^+ and (b) Cl^-?

30. You have a 250-ppm solution of K^+ as KCl. You wish to prepare from this a 0.00100 M solution of Cl^-. How many milliliters must be diluted to 1 L?

31. One liter of a 500-ppm solution of $KClO_3$ contains how many grams K^+?

DILUTION CALCULATIONS

32. A 12.5-mL portion of a solution is diluted to 500 mL, and its molarity is determined to be 0.125. What is the molarity of the original solution?

33. What volume of 0.50 M H_2SO_4 must be added to 65 mL of 0.20 M H_2SO_4 to give a final solution of 0.35 M? Assume volumes are additive.

34. How many milliliters of 0.10 M H_2SO_4 must be added to 50 mL of 0.10 M $NaOH$ to give a solution that is 0.050 M in H_2SO_4? Assume volumes are additive.

35. You are required to prepare working standard solutions of 1.00×10^{-5}, 2.00×10^{-5}, 5.00×10^{-5}, and 1.00×10^{-4} M glucose from a 0.100 M stock solution. You have available 100-mL volumetric flasks and pipets of 1.00-, 2.00-, 5.00-, and 10.00-mL volume. Outline a procedure for preparing the working standards.

36. A 0.500-g sample is analyzed spectrophotometrically for manganese by dissolving it in acid and transferring to a 250-mL flask and diluting to volume. Three aliquots are analyzed by transferring 50-mL portions with a pipet to 500-mL Erlenmeyer flasks and reacting with an oxidizing agent, potassium peroxydisulfate, to convert the manganese to permanganate. After reaction, these are quantitatively transferred to 250-mL volumetric flasks, diluted to volume, and measured spectrophotometrically. By comparison with standards, the average concentration in the final solution is determined to be 1.25×10^{-5} M. What is the percent manganese in the sample?

STANDARDIZATION CALCULATIONS

37. A preparation of soda ash is known to contain 98.6% Na_2CO_3. If a 0.678-g sample requires 36.8 mL of a sulfuric acid solution for complete neutralization, what is the molarity of the sulfuric acid solution?

38. A 0.1 M sodium hydroxide solution is to be standardized by titrating primary standard sulfamic acid (NH_2SO_3H). What weight of sulfamic acid should be taken so that the volume of NaOH delivered from the buret is about 40 mL?

ANALYSIS CALCULATIONS

39. A sample of USP-grade citric acid ($H_3C_6H_5O_7$, three titratable protons) is analyzed by titrating with 0.1087 M NaOH. If a 0.2678-g sample requires 38.31 mL for titration, what is the purity of the preparation? (USP requires 99.5%.)

40. Calcium in a 200-μL serum sample is titrated with 1.87×10^{-4} M EDTA solution, requiring 2.47 mL. What is the calcium concentration in the blood in mg/dL?

41. A 0.372-g sample of impure $BaCl_2 \cdot 2H_2O$ is titrated with 0.100 M $AgNO_3$, requiring 27.2 mL. Calculate (a) the percent Cl in the sample and (b) the percent purity of the compound.

42. An iron ore is analyzed for iron content by dissolving in acid, converting the iron to Fe^{2+}, and then titrating with standard 0.0150 M $K_2Cr_2O_7$ solution. If 35.6 mL is required to titrate the iron in a 1.68-g ore sample, how much iron is in the sample, expressed as percent Fe_2O_3? (See Example 5.37 for the titration reaction.)

43. Calcium in a 2.00-g sample is determined by precipitating CaC_2O_4, dissolving this in acid, and titrating the oxalate with 0.0200 M $KMnO_4$. What percent of CaO is in the sample if 35.6 mL $KMnO_4$ is required for titration? (The reaction is $5H_2C_2O_4 + 2MnO_4^- + 6H^+ \rightarrow 10CO_2 + 2Mn^{2+} + 8H_2O$.)

44. A potassium permanganate solution is prepared by dissolving 4.68 g $KMnO_4$ in water and diluting to 500 mL. How many milliliters of this will react with the iron in 0.500 g of an ore containing 35.6% Fe_2O_3? (See Example 5.29 for the titration reaction.)

45. A sample contains $BaCl_2$ plus inert matter. What weight must be taken so that when the solution is titrated with 0.100 $AgNO_3$, the milliliters of titrant will be equal to the percent $BaCl_2$ in the sample?

46. A 0.250-g sample of impure $AlCl_3$ is titrated with 0.100 M $AgNO_3$, requiring 48.6 mL. What volume of 0.100 M EDTA would react with a 0.350-g sample? (EDTA reacts with Al^{3+} in a 1:1 ratio.)

47. A 425.2-mg sample of a purified monoprotic organic acid is titrated with 0.1027 M NaOH, requiring 28.78 mL. What is the formula weight of the acid?

48. The purity of a 0.287-g sample of $Zn(OH)_2$ is determined by titrating with a standard HCl solution, requiring 37.8 mL. The HCl solution was standardized by precipitating AgCl in a 25.0-mL aliquot and weighing (0.462 g AgCl obtained). What is the purity of the $Zn(OH)_2$?

49. A sample of pure $KHC_2O_4 \cdot H_2C_2O_4 \cdot 2H_2O$ (three replaceable hydrogens) requires 46.2 mL of 0.100 M NaOH for titration. How many milliliters of 0.100 M $KMnO_4$ will the same-size sample react with? (See Problem 43 for reaction with $KMnO_4$.)

BACK-TITRATIONS

50. A 0.500-g sample containing Na_2CO_3 plus inert matter is analyzed by adding 50.0 mL of 0.100 M HCl, a slight excess, boiling to remove CO_2, and then back-titrating the excess acid with 0.100 M NaOH. If 5.6 mL NaOH is required for the back-titration, what is the percent Na_2CO_3 in the sample?

51. A hydrogen peroxide solution is analyzed by adding a slight excess of standard $KMnO_4$ solution and back-titrating the unreacted $KMnO_4$ with standard Fe^{2+} solution. A 0.587-g sample of the H_2O_2 solution is taken, 25.0 mL of 0.0215 M $KMnO_4$ is added, and the back-titration requires 5.10 mL of 0.112 M Fe^{2+} solution. What is the percent H_2O_2 in the sample? (See Examples 5.25 and 5.29 for the reactions.)

52. The sulfur content of a steel sample is determined by converting it to H_2S gas, absorbing the H_2S in 10.0 mL of 0.00500 M I_2, and then back-titrating the excess I_2 with 0.00200 M $Na_2S_2O_3$. If 2.6 mL $Na_2S_2O_3$ is required for the titration, how many milligrams of sulfur are contained in the sample? Reactions:

$$H_2S + I_2 \rightarrow S + 2I^- + 2H^+$$
$$I_2 + 2S_2O_3^{2-} \rightarrow 2I^- + S_4O_6^{2-}$$

TITER

53. Express the titer of a 0.100 M EDTA solution in mg BaO/mL.

54. Express the titer of a 0.0500 M $KMnO_4$ solution in mg Fe_2O_3/mL.

55. The titer of a silver nitrate solution is 22.7 mg Cl/mL. What is its titer in mg Br/mL?

EQUIVALENT WEIGHT CALCULATIONS

56. Calculate the equivalent weights of the following substances as acids or bases: (a) HCl, (b) $Ba(OH)_2$, (c) $KH(IO_3)_2$, (d) H_2SO_3, (e) $HC_2H_3O_2$ (acetic acid).

57. Calculate the molarity of a 0.250 N solution of each of the acids or bases in Problem 56.

EQUIVALENT WEIGHT

58. Calculate the equivalent weight of KHC_2O_4 (a) as an acid and (b) as a reducing agent in reaction with MnO_4^- ($5HC_2O_4^- + 2MnO_4^- + 11H^+ \rightarrow 10CO_2 + 2Mn^{2+} + 8H_2O$).

59. Mercuric oxide, HgO, can be analyzed by reaction with iodide and then titration with an acid: $HgO + 4I^- \rightarrow HgI_4^{2-} + 2OH^-$. What is its equivalent weight?

60. Calculate the grams of one equivalent each of the following for the indicated reaction: (a) $FeSO_4$ ($Fe^{2+} \rightarrow Fe^{3+}$), (b) H_2S ($\rightarrow S^0$), (c) H_2O_2 ($\rightarrow O_2$), (d) H_2O_2 ($\rightarrow H_2O$).

61. $BaCl_2 \cdot 2H_2O$ is to be used to titrate Ag^+ to yield AgCl. How many milliequivalents are contained in 0.5000 g $BaCl_2 \cdot 2H_2O$?

NORMALITY

62. A solution is prepared by dissolving 7.82 g NaOH and 9.26 g $Ba(OH)_2$ in water and diluting to 500 mL. What is the normality of the solution as a base?

63. What weight of arsenic trioxide, As_2O_3, is required to prepare 1 L of 0.1000 N arsenic(III) solution (arsenic 3+ is oxidized to 5+ in redox reactions)?

64. If 2.73 g $KHC_2O_4 \cdot H_2C_2O_4$ (three ionizable protons) having 2.0% inert impurities and 1.68 g $KHC_8H_4O_4$ (one ionizable proton) are dissolved in water and diluted to 250 mL, what is the normality of the solution as an acid, assuming complete ionization?

65. A solution of $KHC_2O_4 \cdot H_2C_2O_4 \cdot 2H_2O$ (three replaceable hydrogens) is 0.200 N as an acid. What is its normality as reducing agent? (See Problem 43 for its reaction as a reducing agent.)

66. $Na_2C_2O_4$ and $KHC_2O_4 \cdot H_2C_2O_4$ are mixed in such a proportion by weight that the normality of the resulting solution as a reducing agent is 3.62 times the normality as an acid. What is the proportion? (See Problem 43 for its reaction as a reducing agent.)

67. What weight of $K_2Cr_2O_7$ is required to prepare 1.000 L of 0.1000 N solution? (In reaction, $Cr_2O_7^{2-} + 14H^+ + 6e^- \rightleftharpoons 2Cr^3 + 7H_2O$.)

CHARGE EQUIVALENT CALCULATIONS

68. A chloride concentration is reported as 300 mg/dL. What is the concentration in meq/L?

69. A calcium concentration is reported as 5.00 meq/L. What is the concentration in mg/dL?

70. A urine specimen has a chloride concentration of 150 meq/L. If we assume that the chloride is present in urine as sodium chloride, what is the concentration of NaCl in g/L?

GRAVIMETRIC CALCULATIONS

71. What weight of manganese is present in 2.58 g of Mn_3O_4?

72. Zinc is determined by precipitating and weighing as $Zn_2Fe(CN)_6$.

    (a) What weight of zinc is contained in a sample that gives 0.348 g precipitate?

    (b) What weight of precipitate would be formed from 0.500 g of zinc?

73. Calculate the gravimetric factors for:

Substance Sought	Substance Weighed
Mn	Mn_3O_4
Mn_2O_3	Mn_3O_4
Ag_2S	$BaSO_4$
$CuCl_2$	$AgCl$
MgI_2	PbI_2

Recommended References

1. T. P. Hadjiioannou, G. D. Christian, C. E. Efstathiou, and D. Nikolelis, *Problem Solving in Analytical Chemistry*. Oxford: Pergamon, 1988.

2. Q. Fernando and M. D. Ryan, *Calculations in Analytical Chemistry*. New York: Harcourt Brace Jovanovich, 1982.

3. M. R. F. Ashworth, *Titrimetric Organic Analysis*. New York: Interscience, 1964.

Chapter Six

GENERAL CONCEPTS OF CHEMICAL EQUILIBRIUM

"The worst form of inequality is to try to make unequal things equal."
—Aristotle

Even though chemical reactions may go far to completion, the reactions *never* go in only one direction. In fact, reactions reach an equilibrium in which the rates of reactions in both directions are equal. In this chapter we review the equilibrium concept and the equilibrium constant and describe general approaches for calculations using equilibrium constants. We discuss the activity of ionic species along with the calculation of activity coefficients. These values are required for calculations using thermodynamic equilibrium constants, that is, for the diverse ion effect, described at the end of the chapter. They are also used in potentiometric calculations (Chapter 13).

6.1 Chemical Reactions: The Rate Concept

In 1863 Guldberg and Waage described what we now call the law of mass action, which states that the rate of a chemical reaction is proportional to the "active masses" of the reacting substances present at any time. The active masses may be concentrations or pressures. Guldberg and Waage derived an equilibrium constant by defining equilibrium as the condition when the rates of the forward and reverse reactions are equal. Consider the chemical reaction

$$a\mathrm{A} + b\mathrm{B} \rightleftharpoons c\mathrm{C} + d\mathrm{D} \tag{6.1}$$

According to Guldberg and Waage, the rate of the forward reaction is equal to a constant times the concentration of each species raised to the power of the number of molecules participating in the reaction: that is,[1]

$$\mathrm{Rate}_f = k_f[\mathrm{A}]^a[\mathrm{B}]^b \tag{6.2}$$

---

[1] [   ] represents moles/liter and here represents the effective concentration. The effective concentration will be discussed under the diverse ion effect, when we talk about activities.

when rate_f is the rate of the forward reaction and k_f is the **rate constant**, which is dependent on such factors as the temperature and the presence of catalysts. [A] and [B] represent the molar concentrations of A and B. Similarly, for the backward reaction, Guldberg and Waage wrote

$$\text{Rate}_b = k_b[\text{C}]^c[\text{D}]^d \tag{6.3}$$

and for a system at equilibrium, the forward and reverse rates are equal:

$$k_f[\text{A}]^a[\text{B}]^b = k_b[\text{C}]^c[\text{D}]^d \tag{6.4}$$

At equilibrium, the rate of the reverse reaction equals the rate of the forward reaction.

Rearranging these equations gives the **molar equilibrium constant** (which holds for dilute solutions) for the reaction, K:

$$\frac{[\text{C}]^c[\text{D}]^d}{[\text{A}]^a[\text{B}]^b} = \frac{k_f}{k_b} = K \tag{6.5}$$

The expression obtained here is the correct expression for the equilibrium constant, *but the method of derivation has no general validity.* This is because reaction rates actually depend on the *mechanism* of the reaction, determined by the number of colliding species, whereas the equilibrium constant expression depends only on the *stoichiometry* of the chemical reaction. The sum of the exponents in the rate constant gives the *order* of the reaction, and this may be entirely different from the stoichiometry of the reaction (see Chapter 22). An example is the rate of reduction of $S_2O_8^{2-}$ with I^-:

$$S_2O_8^{2-} + 3I^- \rightarrow 2SO_4^{2-} + I_3^-$$

The rate is actually given by $k_f[S_2O_8^{2-}][I^-]$ (a second-order reaction) and not $k_f[S_2O_8^{2-}][I^-]^3$, as might be expected from the balanced chemical reaction (a fourth-order reaction would be predicted). The only sound theoretical basis for the equilibrium constant comes from thermodynamic arguments. See Gibbs free energy in Section 6.3 for the thermodynamic computation of equilibrium constant values.

The value of K can be calculated empirically by measuring the concentrations of A, B, C, and D at equilibrium. Note that the more favorable the rate constant of the forward reaction relative to the backward reaction, the larger will be the equilibrium constant and the farther to the right the reaction will be at equilibrium.

The larger the equilibrium constant, the farther to the right is the reaction at equilibrium.

When the reaction between A and B is initiated, the rate of the forward reaction is large because the concentrations of A and B are large, whereas the backward reaction is slow because the concentrations of C and D are small (that rate is initially zero). As the reaction progresses, A and B decrease and C and D increase, so that the rate of the forward reaction diminishes while that for the backward reaction increases (Figure 6.1). Eventually, the two rates become equal, and

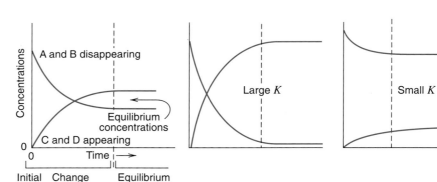

Fig. 6.1. Progress of a chemical reaction.

the system is in a state of equilibrium. At this point, the individual concentrations of A, B, C, and D remain constant (the relative values will depend on the reaction stoichiometry and how far the equilibrium lies to the right). However, the system remains in dynamic equilibrium, with the forward and backward reactions continuing at equal rates.

You will notice that the equilibrium constant expression is the ratio in which the concentrations of the products appear in the numerator and the concentrations of the reactants appear in the denominator. This is quite arbitrary, but it is the accepted convention. Hence, a large equilibrium constant indicates the equilibrium lies far to the right.

We should point out that although a particular reaction may have a rather large equilibrium constant, the reaction may proceed from *right* to *left* if sufficiently large concentrations of the *products* are initially present. Also, the equilibrium constant tells us nothing about how *fast* a reaction will proceed toward equilibrium. Some reactions, in fact, may be so slow as to be unmeasurable. The equilibrium constant merely tells us the tendency of a reaction to occur and in what direction, not whether it is fast enough to be feasible in practice. (See Chapter 22 on kinetic methods of analysis for the measurement of reaction rates and their application to analyses.)

> A large equilibrium constant does not assure the reaction will proceed at an appreciable rate.

For the reaction depicted in Equation 6.1, the rate at which equilibrium approached will likely be different for either the forward or the reverse reaction. That is, if we start with a mixture of C and D, the rate at which equilibrium is approached may be much slower or faster than for the converse reaction.

6.2 Types of Equilibria

We can write equilibrium constants for many types of chemical processes. Some of these equilibria are listed in Table 6.1. The equilibria may represent dissociation (acid/base, solubility), formation of products (complexes), reactions (redox), a distribution between two phases (water and nonaqueous solvent—solvent extraction; adsorption from water onto a surface, as in chromatography, etc.). We will describe some of these equilibria below and in later chapters.

> Equilibrium constants may be written for dissociations, associations, reactions, or distributions.

6.3 Gibbs Free Energy and the Equilibrium Constant

The tendency for a reaction to occur is defined thermodynamically from its change in **enthalpy** (ΔH) and **entropy** (ΔS). Enthalpy is the heat absorbed when an endothermic reaction occurs under constant pressure. When heat is given off

Table 6.1
Types of Equilibria

Equilibrium	Reaction	Equilibrium Constant
Acid–base dissociation	$HA + H_2O \rightleftharpoons H_3O^+ + A^-$	K_a, acidity constant
Solubility	$MA \rightleftharpoons M^{n+} + A^{n-}$	K_{sp}, solubility product
Complex formation	$M^{n+} + aL^{b-} \rightleftharpoons ML_a^{(n-ab)+}$	K_f, formation constant
Reduction–oxidation	$A_{red} + B_{ox} \rightleftharpoons A_{ox} + B_{red}$	K_{eq}, reaction equilibrium constant
Phase distribution	$A_{H_2O} \rightleftharpoons A_{organic}$	K_D, distribution coefficient

(exothermic reaction), ΔH is negative. Entropy is a measure of the disorder, or randomness, of a substance or system.

A system will always tend toward lower energy and increased randomness, that is, lower enthalpy and higher entropy. For example, a stone on a hill will tend to roll spontaneously down the hill (lower energy state), and a box of marbles ordered by color will tend to become randomly ordered when shaken. The combined effect of enthalpy and entropy is given by the **Gibbs free energy**, G:

$$G = H - TS \tag{6.6}$$

where T is the absolute temperature in kelvins; G is a measure of the energy of the system, and a system spontaneously tends toward lower energy states. The change in energy of a system at a constant temperature is

$$\Delta G = \Delta H - T\,\Delta S \tag{6.7}$$

So a process will be *spontaneous when ΔG is negative,* will be spontaneous in the reverse direction when ΔG is positive, and will be at equilibrium when ΔG is zero. Hence, a reaction is favored by heat given off (negative ΔH), as in exothermic reactions, and by increased entropy (positive ΔS). Both ΔH and ΔS can be either positive or negative, and the relative magnitudes of each and the temperature will determine whether ΔG will be negative so that the reaction will be spontaneous.

Standard enthalpy $H°$, standard entropy $S°$, and standard free energy $G°$ represent the thermodynamic quantities for one mole of a substance at standard state ($P = 1$ atm, $T = 298$ K, unit concentration). Then,

$$\Delta G° = \Delta H° - T\,\Delta S° \tag{6.8}$$

$\Delta G°$ is related to the equilibrium constant of a reaction by

$$\boxed{K = e^{-\Delta G°/RT}} \tag{6.9}$$

or

$$\boxed{\Delta G° = -RT \ln K = -2.303RT \log K} \tag{6.10}$$

where R is the gas constant (8.314 J K^{-1} mol^{-1}). Hence, if we know the standard free energy of a reaction, we can calculate the equilibrium constant. Obviously, the larger $\Delta G°$ (when negative), the larger will be K. Note that while $\Delta G°$ and ΔG give information about the spontaneity of a reaction, they say nothing about the *rate* at which it will occur.

6.4 Le Châtelier's Principle

The equilibrium concentrations of reactants and products can be altered by applying stress to the system, for example, by changing the temperature, the pressure, or the concentration of one of the reactants. The effects of such changes can be predicted from **Le Châtelier's principle**, which states that when stress is applied to a system at chemical equilibrium, the equilibrium will shift in a direction that tends to relieve or counteract that stress. The effects of temperature, pressure, and concentrations on chemical equilibria are considered below.

Margin notes:

Everything in the universe tends toward increased disorder (increased entropy) and lower energy (lower enthalpy).

A spontaneous reaction results in energy given off and a lower free energy. At equilibrium, the free energy does not change.

A large equilibrium constant results from a large negative free energy.

We can shift an unfavorable equilibrium by increasing the reactant concentration.

6.5 Temperature Effects on Equilibrium Constants

As we have mentioned, temperature influences the individual rate constants for the forward and backward reactions and therefore the equilibrium constant (more correctly, temperature affects the free energy—see Equation 6.10). An increase in temperature will displace the equilibrium in the direction that results in absorbing heat, since this removes the source of the stress. So an endothermic forward reaction (which absorbs heat) will be displaced to the right, with an increase in the equilibrium constant. The reverse will be true for an exothermic forward reaction, which releases heat. The extent of the displacement will depend on the magnitude of the heat of reaction for the system.

In addition to influencing the position of equilibrium, temperature has a pronounced effect on the rates of the forward and backward reactions involved in the equilibrium, and so it influences the *rate* at which equilibrium is approached. This is because the number and the energy of collisions among the reacting species increase with increasing temperature. The rates of many endothermic reactions increase about two- to threefold for every 10°C rise in temperature.

All equilibrium constants are temperature dependent, as are the rates of reactions.

6.6 Pressure Effects on Equilibria

Pressure can have a large influence on the position of chemical equilibrium for reactions occurring in the gaseous phase. An increase in pressure favors a shift in the direction that results in a reduction in the volume of the system. But for reactions occurring in solutions, normal pressure changes have a negligible effect on the equilibrium because liquids cannot be compressed the way gases can.

For solutions, pressure effects are usually negligible.

6.7 Effect of Concentrations on Equilibria

The value of an equilibrium constant is independent of the concentrations of the reactants and products. However, the *position* of equilibrium is very definitely influenced by the concentrations. The direction of change is readily predictable from Le Châtelier's principle. Consider the reaction of iron(III) with iodide:

$$3I^- + 2Fe^{3+} \rightleftharpoons I_3^- + 2Fe^{2+}$$

If the four components are in a state of equilibrium, as determined by the equilibrium constant, addition or removal of one of the components would cause the equilibrium to reestablish itself. For example, suppose we add more iron(II) to the solution. According to Le Châtelier's principle, the reaction will shift to the left to relieve the stress. Equilibrium will eventually be reestablished, and its position will still be defined by the same equilibrium constant.

Changes in concentration do not affect the equilibrium constant. They *do* affect the position of the equilibrium.

6.8 Catalysts

Catalysts either speed up or retard the rate at which an equilibrium is attained by affecting the rates of both the forward and the backward reactions. But catalysts affect both rates to the same extent and thus have no effect on the value of an equilibrium constant.

Catalysts do not affect the equilibrium constant or the position at equilibrium.

See Chapter 22 for analytical uses of enzyme catalysts.

Catalysts are very important to the analytical chemist in a number of reactions that are normally too slow to be analytically useful. An example is the use of an osmium tetroxide catalyst to speed up the titration reaction between arsenic(III) and cerium(IV), whose equilibrium is very favorable but whose rate is normally too slow to be useful for titrations. The measurement of the change in the rate of a kinetically slow reaction in the presence of a catalyst can actually be used for determining the catalyst concentration.

6.9 Completeness of Reactions

For quantitative analysis, equilibria should be at least 99.9% to the right for precise measurements. A reaction that is 75% to the right is still a "complete" reaction.

If the equilibrium of a reaction lies sufficiently to the right that the remaining amount of the substance being determined (reacted) is too small to be measured by the measurement technique, we say the reaction has gone to completion. If the equilibrium is not so favorable, then Le Châtelier's principle may be applied to make it so. We may either increase the concentration of a reactant or decrease the concentration of a product. Production of more product may be achieved by (1) allowing a gaseous product to escape, (2) precipitating the product, (3) forming a stable ionic complex of the product in solution, or (4) preferential extraction.

It is apparent from the above discussion that Le Châtelier's principle is the dominant concept behind most chemical reactions in the real world. It is particularly important in biochemical reactions, and external factors such as temperature can have a significant effect on biological equilibria. Catalysts (enzymes) are also key players in many biological and physiological reactions, as we shall see in Chapter 22.

6.10 Equilibrium Constants for Dissociating or Combining Species—Weak Electrolytes and Precipitates

Equilibrium constants are finite when dissociations are less than 100%.

When a substance dissolves in water, it will often partially or completely dissociate or ionize. Partially dissociated electrolytes are called *weak electrolytes*, and completely dissociated ones are *strong electrolytes*. For example, acetic acid only partially ionizes in water and is therefore a weak electrolyte. But hydrochloric acid is completely ionized and thus is a strong electrolyte. (Acid dissociations in water are really proton transfer reactions: $HOAc + H_2O \rightleftharpoons H_3O^+ + OAc^-$). Some substances completely ionize in water but have limited solubility; we call these *slightly soluble substances*. Substances may combine in solution to form a dissociable product, for example, a complex. An example is the reaction of copper(II) with ammonia to form the $Cu(NH_3)_4^{2+}$ species.

A weak electrolyte is only partially dissociated. A slightly soluble substance is a strong electrolyte because the portion that dissolves is totally ionized.

The dissociation of weak electrolytes or the solubility of slightly soluble substances can be quantitatively described by equilibrium constants. Equilibrium constants for completely dissolved and dissociated electrolytes are effectively infinite. Consider the dissociating species AB:

$$AB \rightleftharpoons A + B \tag{6.11}$$

The equilibrium constant for such a dissociation can be written generally as

$$\boxed{\frac{[A][B]}{[AB]} = K_{eq}} \tag{6.12}$$

The larger K_{eq}, the greater will be the dissociation. For example, the larger the equilibrium constant of an acid, the stronger will be the acid.

Some species dissociate stepwise, and an equilibrium constant can be written for each dissociation step. A compound A_2B, for example, may dissociate as follows:

$$A_2B \rightleftharpoons A + AB \qquad K_1 = \frac{[A][AB]}{[A_2B]} \qquad (6.13)$$

$$AB \rightleftharpoons A + B \qquad K_2 = \frac{[A][B]}{[AB]} \qquad (6.14)$$

The overall dissociation of the compound is the sum of these two equilibria:

Successive stepwise dissociation constants become smaller.

$$A_2B \rightleftharpoons 2A + B \qquad K_{eq} = \frac{[A]^2[B]}{[A_2B]} \qquad (6.15)$$

If we multiply Equations 6.13 and 6.14 together, we arrive at the overall equilibrium constant:

$$K_{eq} = K_1 K_2 = \frac{[A][AB]}{[A_2B]} \cdot \frac{[A][B]}{[AB]} \qquad (6.16)$$
$$= \frac{[A]^2[B]}{[A_2B]}$$

When chemical species dissociate in a stepwise manner like this, the successive equilibrium constants generally become progressively smaller. Note that in equilibrium calculations we always use mol/L for solution concentrations.

If a reaction is written in the reverse, the same equilibria apply, but the equilibrium constant is inverted. Thus, in the above example, for $A + B \rightleftharpoons AB$, $K_{eq(reverse)} = [AB]/([A][B]) = 1/K_{eq(forward)}$. If K_{eq} for the forward reaction is 10^5, then K_{eq} for the reverse reaction is 10^{-5}.

$K_{forward} = 1/K_{backward}$

Similar concepts apply for combining species, except, generally, the equilibrium constant will be greater than unity rather than smaller, since the reaction is favorable for forming the product (e.g., complex). We will discuss equilibrium constants for acids, complexes, and precipitates in later chapters.

6.11 Calculations Using Equilibrium Constants—How Much Is in Equilibrium?

Equilibrium constants are useful for calculating the concentrations of the various species in equilibrium, for example, the hydrogen ion concentration from the dissociation of a weak acid. In this section we present the general approach for calculations using equilibrium constants. The applications to specific equilibria are treated in later chapters dealing with these equilibria.

CHEMICAL REACTIONS

It is sometimes useful to know the concentrations of reactants and products in equilibrium in a chemical reaction. For example, we may need to know the amount of reactant for the construction of a titration curve or for calculating the potential of an electrode in the solution. These are, in fact, applications we consider in later chapters. Some example calculations will illustrate the general approach to solving equilibrium problems.

Example 6.1

The chemicals A and B react as follows to produce C and D:

$$A + B \rightleftharpoons C + D \qquad K = \frac{[C][D]}{[A][B]}$$

The equilibrium constant K has a value of 0.30. Assume 0.20 mol of A and 0.50 mol of B are dissolved in 1.00 L, and the reaction proceeds. Calculate the concentrations of reactants and products at equilibrium.

Solution

The initial concentration of A is 0.20 M and that of B is 0.50 M, while C and D are initially 0 M. After the reaction has reached equilibrium, the concentrations of A and B will be decreased and those of C and D will be increased. Let x represent the equilibrium concentration of C or the moles/liter of A and B reacting. Since we get one mole of D with each mole of C, the concentration of D will also be x. We may represent the *initial* concentration of A and B as the **analytical concentrations**, C_A and C_B. The **equilibrium concentrations** are [A] and [B]. The concentrations of A and B will each be diminished by x, that is, $[A] = C_A - x$ and $[B] = C_B - x$. So the equilibrium concentrations are

The equilibrium concentration is the initial (analytical) concentration minus the amount reacted.

	$[A]$	$[B]$	$[C]$	$[D]$
Initial	0.20	0.50	0	0
Change (x = mmol/mL reacting)	$-x$	$-x$	$+x$	$+x$
Equilibrium	$0.20 - x$	$0.50 - x$	x	x

We can substitute these values in the equilibrium constant expression and solve for x:

$$\frac{(x)(x)}{(0.20 - x)(0.50 - x)} = 0.30$$

$$x^2 = (0.10 - 0.70x + x^2)0.30$$

$$0.70x^2 + 0.21x - 0.030 = 0$$

This is a quadratic equation and can be solved algebraically for x using the quadratic formula given in Appendix B:

$$x = \frac{-b \pm \sqrt{b^2 - 4ac}}{2a}$$

$$= \frac{-0.21 \pm \sqrt{(0.21)^2 - 4(0.70)(-0.030)}}{2(0.70)}$$

$$= \frac{-0.21 \pm \sqrt{0.044 + 0.084}}{1.40} = 0.11 \, M$$

$$[A] = 0.20 - x = 0.09 \ M$$
$$[B] = 0.50 - x = 0.39 \ M$$
$$[C] = [D] = x = 0.11 \ M$$

Instead of using the quadratic equation, we may use the **method of successive approximations**. In this procedure, we will first neglect x compared to the initial concentrations to simplify calculations, and calculate an initial value of x. Then we can use this first estimate of x to subtract from C_A and C_B to give an initial estimate of the equilibrium concentration of A and B, and calculate a new x. The process is repeated until x is essentially constant.

In successive approximations, we begin by taking the analytical concentration as the equilibrium concentration, to calculate the amount reacted. Then we repeat the calculation after subtracting the calculated reacted amount, until it is constant.

First calculation
$$\frac{(x)(x)}{(0.20)(0.50)} = 0.30$$
$$x = 0.173$$

The calculations converge more quickly if we keep an extra digit throughout.

Second calculation
$$\frac{(x)(x)}{(0.20 - 0.173)(0.50 - 0.173)} = 0.30$$
$$x = 0.051$$

Third calculation
$$\frac{(x)(x)}{(0.20 - 0.051)(0.50 - 0.051)} = 0.30$$
$$x = 0.14_2$$

Fourth calculation
$$\frac{(x)(x)}{(0.20 - 0.142)(0.50 - 0.142)} = 0.30$$
$$x = 0.079$$

Fifth calculation
$$\frac{(x)(x)}{(0.20 - 0.79)(0.50 - 0.079)} = 0.30$$
$$x = 0.12_4$$

Sixth calculation
$$\frac{(x)(x)}{(0.20 - 0.124)(0.50 - 0.124)} = 0.30$$
$$x = 0.093$$

Seventh calculation
$$\frac{(x)(x)}{(0.20 - 0.093)(0.50 - 0.093)} = 0.30$$
$$x = 0.11_4$$

Eighth calculation
$$\frac{(x)(x)}{(0.20 - 0.114)(0.50 - 0.114)} = 0.30$$
$$x = 0.10_4$$

Ninth calculation
$$\frac{(x)(x)}{(0.20 - 0.100)(0.50 - 0.100)} = 0.30$$
$$x = 0.11_0$$

We will take 0.11 as the equilibrium value of x since it essentially repeated the value of the seventh calculation. Note that in these iterations, x oscillates above and below the equilibrium value. The larger x is compared to C, the larger will be the

Shorten the number of iterations by taking the average of the first two for the next.

oscillations and the more iterations that will be required to reach an equilibrium value (as in this example—not the best for this approach). There is a more efficient way of completing the iteration. Take the average of the first and second for the third iteration, which should be close to the final value (in this case, 0.11_2). One or two more iterations will tell us we have reached the equilibrium value. Try it!

● ●

In Example 6.1, appreciable amounts of A remained, even though B was in excess, because the equilibrium constant was not very large. In fact, the equilibrium was only about halfway to the right since C and D were about the same concentration as A. In most reactions of analytical interest, the equilibrium constants are large, and the equilibrium lies far to the right. In these cases, the equilibrium concentrations of reactants that are not in excess are generally very small compared to the other concentrations. This simplifies our calculations.

EXCEL SOLVER FOR PROBLEM SOLVING—THE QUADRATIC EQUATION

Microsoft Excel has a very useful program called *Solver*. To use this program, we must first arrange an equation to give a formula equal to a known value. Solver effectively calculates the formula in reverse, by starting with the answer of the formula (the known value) and, by iteration, solving for the value or values of one or more variables in the formula that results in the correct answer. (Other spreadsheet programs have similar problem solvers.)

As previously stated, in order to apply the Solver program, we must arrange the equation to give a formula equal to a known value. This can be a value we want to minimize (as in least squares), maximize, set to zero, or any other fixed number. When we do this, the variable or variables (whose values we wish to calculate) are on one side of the formula. In Example 6.1, we arrange the equilibrium constant equation to give a formula equal to zero: $0.70x^2 + 0.21x - 0.30 = 0$. The variable is x.

The Solver program is generally found under Tools in the Excel spreadsheet. If you do not see it there, it is probably an Add-In, which you have to activate by checking the Solver Add-In box. Once you locate Solver, click on it and you will open the Solver Parameters dialogue box. The filled-in Solver dialogue box for Example 6.1 is shown here:

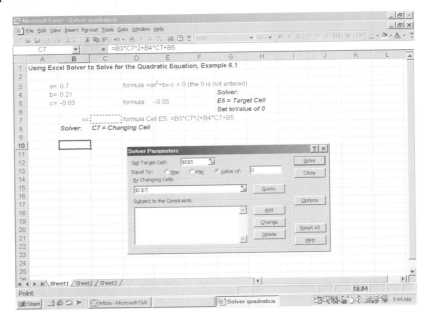

Three parameters need to be entered. The Set Target Cell: is where you enter the cell containing the entered formula. Entering $ signs for the cell is optional. If you click on the cell to enter (either before or after opening Solver), they are automatically entered. The Equal To: is where you enter the value that the equation is set to (0, in this case). And By Changing Cells: is where you enter the cell or cells containing the variable or variables (x, in this case). If you have more than one variable cell, insert a comma between each cell entry. Since we don't know the value of the x cell, we may leave it empty, or we may put an estimated initial value in the cell.

For the equation we have chosen, create a spreadsheet with cells that contain the constants a, b, and c, that is, 0.70, 0.21, and -0.030, that will be used in the formula; the variable x; the formula $(0.70x^2 + 0.21x - 0.30)$: =[0.70 cell]* [x cell]^2 + [0.21 cell]*[x cell] + [-0.030 cell]. Label each cell. In the spreadsheet above, the constants are in cells B3, B4, and B5. The variable x is in cell C7, and the formula is B3*C7^2 + B4*C7 + B5. The formula is entered in cell E5. Note that E5 shows a value of -0.03 (i.e., the value of c) since x does not yet have a value.

Click on Solver. When its parameters dialogue box appears in the spreadsheet, click on the Set Target Cell: (it should be empty, if not, delete any cell that is there), followed by clicking on cell E5 (or type E5). Do the same for By Changing Cells:, clicking on cell C7 to enter. In Equal To:, click Value of: and enter zero. You are now ready to solve the formula. But before you do, click on Options and note the Precision box, where the precision is entered as 0.000001 (10^{-6}). This number should be at least 100 times smaller than the smallest number being operated on (a, b, and c in this case) and the solution, x. This is the case for this problem, but if you should encounter problems where it is not, you should insert more zeros in the Precision number, for example, when the magnitude of the calculated answer is on the order of the entered precision (in which case, repeat the Solver calculation).

Close the Solver Options window and click on Solve and you receive a message that "Solver found a solution," and you see the answer for x is 0.10565:

	A	B	C	D	E	F	G	H
1	Using Excel Solver to Solve for the Quadratic Equation, Example 6.1							
2								
3	a=	0.7		formula =ax²+bx-c = 0 (the 0 is not entered)				
4	b=	0.21					Solver:	
5	c=	-0.03		formula:	-8E-08		E5 = Target Cell	
6							Set to Value of 0	
7		x=	0.10565	formula Cell E5: =B3*C7^2+B4*C7+B5				
8		Solver:	C7 = Changing Cell					

The formula is equal to $-8E-08$, a very small number resulting from the iterations (essentially equal to zero).

Mathematically, the quadratic equation also always has a negative answer for x (which we are not interested in since it makes no chemical sense). That value of x can be obtained by constraining the answer to be less than zero in the Subject to the Constraints box. Click on Add. For Cell Reference, click on C7. Adjust the arrow to move to <=. In the Constraint dialogue box, type 0. Then click OK. Click on Solve, and you see the answer $x = -0.40565$ (and formula = $-9E-07$).

If you click on Options in the Solver dialogue box, you can set limits that may slightly alter the final result. We will use Solver on other occasions to aid in solving tedious problems.

The spreadsheet setup for Example 6.1 is given in your CD, Chapter 6. You can use it for solving other quadratic equations by inserting the appropriate a, b,

c constants in the cells and then activating Solver to calculate x; insert the target cell (formula) and the changing cell (x).

Example 6.2

Assume that in Example 6.1 the equilibrium constant was 2.0×10^{16}. Calculate the equilibrium concentrations of A, B, C, and D.

Solution

Since K is very large, the reaction of A with B will be virtually complete to the right, leaving only traces of A at equilibrium. Let x represent the equilibrium concentration of A. An amount of B equal to A will have reacted to form an equivalent amount of C and D (about 0.20 M for each). We can summarize the equilibrium concentrations as follows:

$$[A] = x$$
$$[B] = (0.50 - 0.20) + x = 0.30 + x$$
$$[C] = 0.20 - x$$
$$[D] = 0.20 - x$$

Or, looking at the equilibrium,

$$
\begin{array}{ccccccc}
A & + & B & \rightleftharpoons & C & + & D \\
x & & 0.30 + x & & 0.20 - x & & 0.20 - x
\end{array}
$$

Basically, we have said that all of A is converted to a like amount of C and D, except for a small amount x. Now x will be very small compared to 0.20 and 0.30 and can be neglected. So we can say

$$[A] = x$$
$$[B] \approx 0.30$$
$$[C] \approx 0.20$$
$$[D] \approx 0.20$$

The only unknown concentration is [A]. Substituting these numbers in the equilibrium constant expression, we have

$$\frac{(0.20)(0.20)}{(x)(0.30)} = 2.0 \times 10^{16}$$

$$x = [A] = 6.7 \times 10^{-18}\ M \text{ (analytically undetectable)}$$

In this case the calculation was considerably simplified by neglecting x in comparison to other concentrations. If x should turn out to be significant compared to these concentrations, then the solution should be reworked using the quadratic formula, or the method of successive approximations starting with the first estimate of x. **Generally, if the value of x is less than about 5% of the assumed concentration, it can be neglected.** In this case, the error in x itself is usually 5%

If the equilibrium constant for a reaction is very large, x is very small compared to the analytical concentration, which simplifies calculations.

Neglect x compared to C (product) if $C \leq 0.01 K_{eq}$ in a reaction.

or less. **This simplification will generally hold if the product concentration is less than 1% at K_{eq}, that is $\leq 0.01\ K_{eq}$.**

Example 6.3

A and B react as follows:

$$A + 2B \rightleftharpoons 2C \qquad K = \frac{[C]^2}{[A][B]^2}$$

Assume 0.10 mol of A is reacted with 0.20 mol of B in a volume of 1000 mL; $K = 1.0 \times 10^{10}$. What are the equilibrium concentrations of A, B, and C?

Solution

We have stoichiometrically equal amounts of A and B, so both are virtually all reacted, with trace amounts remaining. Let x represent the equilibrium concentration of A. At equilibrium, we have

$$
\begin{array}{ccccc}
A & + & 2B & \rightleftharpoons & 2C \\
x & & 2x & & 0.20 - 2x \approx 0.20
\end{array}
$$

For each mole of A that either reacts or is produced, we produce or remove two moles of C, and consume or produce two moles of B. Substituting into the equilibrium constant expression,

$$\frac{(0.20)^2}{(x)(2x)^2} = 1.0 \times 10^{10}$$

$$\frac{0.040}{4x^3} = 1.0 \times 10^{10}$$

$$x = [A] = \sqrt[3]{\frac{4.0 \times 10^{-2}}{4.0 \times 10^{10}}} = \sqrt[3]{1.0 \times 10^{-12}} = 1.0 \times 10^{-4}\ M$$

$$B = 2x = 2.0 \times 10^{-4}\ M$$

(analytically detectable, but not appreciable compared to the starting concentration)

DISSOCIATION EQUILIBRIA

Calculations involving dissociating species are not much different from the example just given for chemical reactions.

Example 6.4

Calculate the equilibrium concentrations of A and B in a 0.10 M solution of a weak electrolyte AB with an equilibrium constant of 3.0×10^{-6}.

Solution

$$AB \rightleftharpoons A + B \qquad K_{eq} = \frac{[A][B]}{[AB]}$$

Both [A] and [B] are unknown and equal. Let x represent their equilibrium concentrations. The concentration of AB at equilibrium is equal to its initial analytical concentration minus x.

$$\begin{array}{ccc} AB & \rightleftharpoons & A + B \\ 0.10 - x & & x \quad x \end{array}$$

Neglect x compared to C (analytical concentration) if $C \geq 100 K_{eq}$ in a dissociation.

The value of K_{eq} is quite small, so we are probably justified in neglecting x compared to 0.10. Otherwise, we will have to use a quadratic equation. Substituting into the K_{eq} expression,

$$\frac{(x)(x)}{0.10} = 3.0 \times 10^{-6}$$

$$x = [A] = [B] = \sqrt{3.0 \times 10^{-7}} = 5.5 \times 10^{-4} \; M$$

6.12 The Common Ion Effect—Shifting the Equilibrium

Equilibria can be markedly affected by adding one or more of the species present, as is predicted from Le Châtelier's principle. Example 6.5 illustrates this principle.

Example 6.5

Recalculate the concentration of A in Example 6.4, assuming that the solution also contains 0.20 M B.

Solution

We can represent the equilibrium concentration as follows:

	$[AB]$	$[A]$	$[B]$
Initial	0.10	0	0.20
Change (x = mmol/mL of AB dissociated)	$-x$	$+x$	$+x$
Equilibrium	$0.10 - x$	x	$0.20 + x$
	≈ 0.10		≈ 0.20

The value of x will be smaller now than before because of the common ion effect of B, so we can certainly neglect it compared to the initial concentrations. Substituting in the equilibrium constant expression,

$$\frac{(x)(0.20)}{(0.10)} = 3.0 \times 10^{-6}$$

$$x = 1.5 \times 10^{-6} \; M$$

The concentration of A was decreased nearly 400-fold.

The common ion effect can be used to make analytical reactions more favorable or quantitative. The adjustment of acidity, for example, is frequently used to shift equilibria. Titrations with potassium dichromate, for example, are favored in acid solution, since protons are consumed in its reactions. Titrations with iodine, a weak oxidizing agent, are usually done in slightly alkaline solution to shift the equilibrium toward completion of the reaction, for example, in titrating arsenic(III):

$$H_3AsO_3 + I_2 + H_2O \rightleftharpoons H_3AsO_4 + 2I^- + 2H^+$$

Adjusting the pH is a common way of shifting the equilibrium.

6.13 Systematic Approach to Equilibrium Calculations—How to Solve Any Equilibrium Problem

Now that some familiarity has been gained with equilibrium problems, we will consider a systematic approach for calculating equilibrium concentrations that will work with all equilibria, no matter how complex. It consists of identifying the unknown concentrations involved and *writing a set of simultaneous equations equal to the number of unknowns.* Simplifying assumptions are made with respect to relative concentrations of species (not unlike the approach we have already taken) to shorten the solving of the equations. This approach involves writing expressions for **mass balance** of species and one for **charge balance** of species as part of our equations. We will first describe how to arrive at these expressions.

MASS BALANCE EQUATIONS

The principle of mass balance is based on the law of mass conservation, and it states that the number of atoms of an element remains constant in chemical reactions because no atoms are produced or destroyed. The principle is expressed mathematically by equating the concentrations, usually in molarities. The equations for all the pertinent chemical equilibria are written, from which appropriate relations between species concentrations are written.

 Example 6.6

Write the equation of mass balance for a 0.100 *M* solution of acetic acid.

Solution

The equilibria are

$$HOAc \rightleftharpoons H^+ + OAc^-$$
$$H_2O \rightleftharpoons H^+ + OH^-$$

We know that the analytical concentration of acetic acid is equal to the sum of the equilibrium concentrations of all its species:

$$C_{HOAc} = [HOAc] + [OAc^-] = 0.100 \ M$$

A second mass balance expression may be written for the equilibrium concentration of H^+, which is derived from both HOAc and H_2O. We obtain one H^+ for each OAc^- and one for each OH^-:

$$[H^+] = [OAc^-] + [OH^-]$$

In a mass balance, the analytical concentration is equal to the sum of the concentrations of the equilibrium species derived from the parent compound (or an appropriate multiple).

 **Example 6.7**

Write the equations of mass balance for a 1.00×10^{-5} M [Ag(NH$_3$)$_2$]Cl solution.

Solution

The equilibria are

$$[Ag(NH_3)_2]Cl \rightarrow Ag(NH_3)_2^+ + Cl^-$$
$$Ag(NH_3)_2^+ \rightleftharpoons Ag(NH_3)^+ + NH_3$$
$$Ag(NH_3)^+ \rightleftharpoons Ag^+ + NH_3$$
$$NH_3 + H_2O \rightleftharpoons NH_4^+ + OH^-$$
$$H_2O \rightleftharpoons H^+ + OH^-$$

The Cl$^-$ concentration is equal to the concentration of the salt that dissociated, that is, 1.00×10^{-5} M. Likewise, the sum of the concentrations of all silver species is equal to the concentration of Ag in the original salt that dissociated:

$$C_{Ag} = [Ag^+] + [Ag(NH_3)^+] + [Ag(NH_3)_2^+] = [Cl^-] = 1.00 \times 10^{-5} \ M$$

We have the following nitrogen-containing species:

$$NH_4^+ \qquad NH_3 \qquad Ag(NH_3)^+ \qquad Ag(NH_3)_2^+$$

The concentration of N from the last species is twice the concentration of Ag(NH$_3$)$_2^+$. The total concentration of the nitrogen is twice the concentration of the original salt, since there are two NH$_3$ per molecule. Hence, we can write

$$C_{NH_3} = [NH_4^+] + [NH_3] + [Ag(NH_3)^+] + 2[Ag(NH_3)_2^+] = 2.00 \times 10^{-5} \ M$$

Finally, we can write

$$[OH^-] = [NH_4^+] + [H^+]$$

Some of the equilibria and the concentrations derived from them may be insignificant compared to others and may not be needed in subsequent calculations, for example, the last mass balance.

We have seen that several mass balance expressions may be written. Some may not be needed for calculations (we may have more equations that unknowns), or some may be simplified or ignored due to the small concentrations involved compared to others. This will become apparent in the equilibrium calculations below.

CHARGE BALANCE EQUATIONS

According to the **principle of electroneutrality**, all solutions are electrically neutral; that is, there is no solution containing a detectable excess of positive or negative charge because the sum of the positive charges equals the sum of negative charges. We may write a *single* charge balance equation for a given set of equilibria.

Example 6.8

Write a charge balance equation for a solution of H_2S.

Solution

The equilibria are

> In a charge balance, the sum of the charge concentrations of cationic species equals the sum of charge concentrations of the anionic species in equilibrium.

$$H_2S \rightleftharpoons H^+ + HS^-$$
$$HS^- \rightleftharpoons H^+ + S^{2-}$$
$$H_2O \rightleftharpoons H^+ + OH^-$$

Dissociation of H_2S gives H^+ and two anionic species, HS^- and S^{2-}, and that of water gives H^+ and OH^-. The amount of H^+ from that portion of *completely* dissociated H_2S is equal to twice the amount of S^{2-} formed, and from *partial* (first step) dissociation is equal to the amount of HS^- formed. That is, for each S^{2-} formed, there are 2 H^+; for each HS^- formed, there is 1 H^+; and for each OH^- formed, there is 1 H^+. Now, for the singly charged species, the *charge* concentration is identical to the concentration of the *species*. But for S^{2-}, the charge concentration is twice the concentration of the species, so we must multiply the S^{2-} concentration by 2 to arrive at the charge concentration from it. According to the principle of electroneutrality, positive charge concentration must equal the negative charge concentration. Hence,

> The charge concentration is equal to the molar concentration times the charge of a species.

$$[H^+] = 2[S^{2-}] + [HS^-] + [OH^-]$$

Note that while there may be more than one source for a given species (H^+ in this case), the total charge concentrations from all sources is always equal to the net equilibrium concentration of the species multiplied by its charge.

Example 6.9

Write a charge balance expression for a solution containing KCl, $Al_2(SO_4)_3$, and KNO_3. Neglect the dissociation of water.

Solution

$$[K^+] + 3[Al^{3+}] = [Cl^-] + 2[SO_4^{2-}] + [NO_3^-]$$

(handwritten notes in margin:)
$K^+ + Cl^-$
$Ag^{3+} + SO_4^{2-}$
$K^+ \quad NO_3^-$

Example 6.10

Write a charge balance equation for a saturated solution of CdS.

Solution

The equilibria are

$$CdS \rightleftharpoons Cd^{2+} + S^{2-}$$
$$S^{2-} + H_2O \rightleftharpoons HS^- + OH^-$$

$$HS^- + H_2O \rightleftharpoons H_2S + OH^-$$
$$H_2O \rightleftharpoons H^+ + OH^-$$

Again, the charge concentration for the singly charged species (H^+, OH^-, HS^-) will be equal to the concentrations of the species. But for Cd^{2+} and S^{2-}, the charge concentration will be twice their concentrations. We must again equate the positive and negative charge concentrations.

$$2[Cd^{2+}] + [H^+] = 2[S^{2-}] + [HS^-] + [OH^-]$$

Example 6.11

Write a charge balance equation for Example 6.7.

Solution

$$[Ag^+] + [Ag(NH_3)^+] + [Ag(NH_3)_2^+] + [NH_4^+] + [H^+] = [Cl^-] + [OH^-]$$

Since all are singly charged species, the charge concentrations are equal to the molar concentrations.

EQUILIBRIUM CALCULATIONS USING THE SYSTEMATIC APPROACH—THE STEPS

In the systematic approach, a series of equations equal in number to the number of unknown species is written. These are simultaneously solved, using approximations to simplify.

We may now describe the systematic approach for calculating equilibrium concentrations in problems involving several equilibria. The basic steps can be summarized as follows:

1. Write the chemical reactions appropriate for the system.
2. Write the equilibrium constant expressions for these reactions.
3. Write all the mass balance expressions.
4. Write the charge balance expression.
5. Count the number of chemical species involved and the number of *independent* equations (from steps 2, 3, and 4). If the number of equations is greater than or equal to the number of chemical species, then a solution is possible. At this point, it is possible to proceed to an answer simply by brute (mathematical) force.
6. Make simplifying assumptions concerning the relative concentrations of chemical species. At this point you need to think like a chemist so that the *math* will be simplified.
7. Calculate the answer.
8. Check the validity of your assumptions!

Let us examine one of the examples worked before, but using this approach.

Example 6.12

Repeat the problem stated in Example 6.4 using the systematic approach outlined above.

Chemical reactions

$$AB = A + B$$

Equilibrium constant expressions

$$K_{eq} = \frac{[A][B]}{[AB]} = 3.0 \times 10^{-6} \tag{1}$$

Use equilibrium constant expressions plus mass and charge balance expressions to write the equations.

Mass balance expressions

$$C_{AB} = [AB] + [A] = 0.10 \ M \tag{2}$$

$$[A] = [B] \tag{3}$$

Remember that C represents the total analytical concentration of AB.

Charge balance expression

There is none because none of the species is charged.

Number of expressions versus number of unknowns

There are three unknowns ([AB], [A], and [B]) and three expressions (one equilibrium and two mass balance).

Simplifying assumptions: We want the equilibrium concentrations of A, B, and AB. Because K is small, very little AB will dissociate, so from (2):

$$[AB] = C_{AB} - [A] = 0.10 - [A] \approx 0.10 \ M$$

Use the same rules as before for simplifying assumptions ($C_A \geq 100 K_{eq}$ for dissociations, $C \leq 0.01 K_{eq}$ for reactions).

Calculate

[AB] was found above.

[A] can be found from (1) and (3).

$$\frac{[A][B]}{0.10} = 3.0 \times 10^{-6}$$

$$[A] = \sqrt{3.0 \times 10^{-7}} = 5.5 \times 10^{-4} \ M$$

[B] can be found from (3):

$$[B] = [A] = 5.5 \times 10^{-4} \ M$$

Check

$$[AB] = 0.10 - 5.5 \times 10^{-4} = 0.10 \ M \text{ (within significant figures)}$$

You see that the same answer was obtained as when the problem was worked intuitively as in Example 6.4. You may think that the systematic approach is excessively complicated and formal. For this extremely simple problem that may be a justified opinion. However, you should realize that the systematic approach will be applicable to *all* equilibrium calculations, regardless of the difficulty of the

The systematic approach is applicable to multiple equilibria.

problem. You may find problems involving multiple equilibria and/or many species to be hopelessly complicated if you use only an intuitive approach. Nevertheless, you should also realize that a good intuitive "feel" for equilibrium problems is an extremely valuable asset. You should attempt to improve your intuition concerning equilibrium problems. Such intuition comes from experience gained by working a number of problems of different varieties. As you gain experience you will be able to shorten some of the formalism of the systematic approach, and you will find it easier to make appropriate simplifying assumptions.

Example 6.13

Repeat the problem outlined in Example 6.5 using the systematic approach. Assume the charge on A is $+1$, the charge on B is -1, and that the extra B ($0.20\ M$) comes from MB; MB is completely dissociated.

Solution
Chemical reactions

$$AB = A^+ + B^-$$
$$MB \rightarrow M^+ + B^-$$

Equilibrium expressions

$$K_{eq} = \frac{[A^+][B^-]}{[AB]} = 3.0 \times 10^{-6} \qquad (1)$$

Mass balance expressions

$$C_{AB} = [AB] + [A^+] = 0.10\ M \qquad (2)$$
$$[B^-] = [A^+] + [M^+] = [A^+] + 0.20\ M \qquad (3)$$

Charge balance expression

$$[A^+] + [M^+] = [B^-] \qquad (4)$$

Number of expressions versus number of unknowns
There are three unknowns ([AB], [A$^+$], and [B$^-$]; the concentration of M$^+$ is known to be $0.20\ M$) and three independent expressions (one equilibrium and two mass balance; the charge balance is the same as the second mass balance).

Simplifying assumptions
(i) Because K_{eq} is small, very little AB will dissociate, so from (2).

$$[AB] = 0.10 - [A^+] \approx 0.10\ M$$

(ii) $[A] \ll [M]$ so from (3) or (4):

$$[B^-] = 0.20 + [A^+] \approx 0.20\ M$$

Calculate
[A] is now found from (1):

$$\frac{[A^+](0.20)}{0.10} = 3.0 \times 10^{-6}$$
$$[A^+] = 1.5 \times 10^{-6}\ M$$

Note that the charge balance is usually not needed.

Check

(i) $[AB] = 0.10 - 1.5 \times 10^{-6} = 0.10 \ M$

(ii) $[B] = 0.20 + 1.5 \times 10^{-6} = 0.20 \ M$

•••

We will in general use the approximation approaches given in Sections 6.10 and 6.11, which actually incorporate many of the equilibria and assumptions used in the systematic approach. The use of the systematic approach for problems involving multiple equilibria is discussed in Chapter 8.

We can now write some general rules for solving chemical equilibrium problems, using the approximation approach. These rules should be applicable to acid–base dissociation, complex formation, oxidation–reduction reactions, and others. That is, all equilibria can be treated similarly.

1. Write down the equilibria involved.

2. Write the equilibrium constant expressions and the numerical values.

3. From a knowledge of the chemistry involved, let x represent the equilibrium concentration of the species that will be unknown and small compared to other equilibrium concentrations; other species of unknown and small concentrations will be multiples of this.

4. List the equilibrium concentrations of all species, adding or subtracting the appropriate multiple of x from the analytical concentration where needed.

5. Make suitable approximations by neglecting x compared to finite equilibrium concentrations. This is generally valid if the finite concentration is about $100 \times K_{eq}$ or more. Also, if the calculated x is less than approximately 5% of the finite concentration, the assumption is valid.

6. Substitute the approximate representation of individual concentration into the equilibrium constant expression and solve for x.

7. If the approximations in step 5 are invalid, use the quadratic formula to solve for x, or use the method of successive approximations.

The application of these rules will become more apparent in subsequent chapters when we deal with specific equilibria in detail.

6.14 Heterogeneous Equilibria—Solids Don't Count

Equilibria in which all the components are in solution generally occur quite rapidly. If an equilibrium involves two phases, the rate of achieving equilibrium will generally be substantially slower than in the case of solutions. An example is the distribution equilibrium of an analyte between a chromatographic column (e.g., solid) and an eluent solvent. Because the equilibrium time is infinite, the rate of elution of the analyte down the chromatographic column must be slow enough for equilibrium to be achieved. The dissolution of a solid or formation of a precipitate will not be instantaneous.

Heterogeneous equilibria are slower than solution equilibria.

The "concentration" of a solid or pure liquid is unity.

Another way in which heterogeneous equilibria differ from homogeneous equilibria is the manner in which the different constituents offset the equilibrium. Guldberg and Waage showed that when a solid is a component of a reversible chemical process, its active mass can be considered constant, regardless of how much of the solid is present. That is, adding more solid does not bring about a shift in the equilibrium. So the expression for the equilibrium constant need not contain any concentration terms for substances present as solids. That is, the standard state of a solid is taken as that of the solid itself, or unity. Thus, for the equilibrium

$$CaF_2 \rightleftharpoons Ca^{2+} + 2F^-$$

we write that

$$K_{eq} = [Ca^{2+}][F^-]^2$$

The same is true for pure liquids (undissolved) in equilibrium, such as mercury. The standard state of water is taken as unity in dilute *aqueous solutions*, and water does not appear in equilibrium constant expressions.

6.15 Activity and Activity Coefficients— Concentration Is Not the Whole Story

The "effective concentration" of an ion is decreased by shielding it with other "inert" ions, and it represents the activity of the ion.

Generally, the presence of diverse salts (not containing ions common to the equilibrium involved) will cause an increase in dissociation of a weak electrolyte or in the solubility of a precipitate. Cations attract anions, and vice versa, and so the cations of the analyte attract anions of the diverse electrolyte and the anions of the analyte attract the cations. The attraction of the ions of the equilibrium reaction by the dissolved electrolyte effectively shields them, *deceasing their effective concentration* and shifting the equilibrium. As the charge on either the diverse salt or the ions of the equilibrium reaction is increased, the diverse salt effect generally increases. This effect on the equilibrium is not predicted by Le Châtelier's principle; but if you think in terms of the effective concentrations being changed, it is analogous in some ways to the common ion effect.

Activities are important in potentiometric measurements. See Chapter 13.

This "effective concentration" of an ion in the presence of an electrolyte is called the **activity** of the ion. Activity can be used to describe quantitatively the effects of salts on equilibrium constants (see the diverse salt effect below). Activity is also important in potentiometric measurements (see Chapter 13). In this section we describe how to estimate activity.

THE ACTIVITY COEFFICIENT

The **activity** of an ion a_i is defined by

$$a_i = C_i f_i \tag{6.17}$$

where C_i is the concentration of the ion i and f_i is its **activity coefficient**. The concentration is usually expressed as molarity, and the activity has the same units as the concentration. The activity coefficient is dimensionless, but numerical values for activity coefficients do depend on the choice of standard state. The activity coefficient varies with the total number of ions in the solution and with their charge, and it is a correction for interionic attraction. *In dilute solution, less than 10^{-4} M, the activity coefficient of a simple electrolyte is near unity*, and activity is approximately equal to the concentration. As the concentration of an electrolyte increases, or as an extraneous salt is added, the activity coefficient generally decreases, and the activity becomes less than the concentration.

IONIC STRENGTH

From the above discussion, we can see that the activity coefficient is a function of the total electrolyte concentration of the solution. The **ionic strength** is a measure of total electrolyte concentration and is defined by

For ionic strengths less than 10^{-4}, activity coefficients are near unity.

$$\mu = \tfrac{1}{2}\sum C_i Z_i^2 \qquad (6.18)$$

where μ is the ionic strength and Z_i is the charge on each individual ion. All cations and anions present in solution are included in the calculation. Obviously, for each positive charge there will be a negative charge.

 Example 6.14

Calculate the ionic strength of a 0.2 M solution of KNO_3 and a 0.2 M solution of K_2SO_4.

Solution
For KNO_3,

$$\mu = \frac{C_{K^+}Z_{K^+}^2 + C_{NO_3^-}Z_{NO_3^-}^2}{2}$$

$$[K^+] = 0.2\ M \qquad [NO_3^-] = 0.2\ M$$

$$\mu = \frac{0.2 \times (1)^2 + 0.2 \times (1)^2}{2} = 0.2$$

For K_2SO_4,

$$\mu = \frac{C_{K^+}Z_{K^+}^2 + C_{SO_4^{2-}}Z_{SO_4^{2-}}^2}{2}$$

$$[K^+] = 0.4\ M \qquad [SO_4^{2-}] = 0.2\ M$$

So,
$$\mu = \frac{0.4 \times (1)^2 + 0.2 \times (2)^2}{2} = 0.6$$

Note that due to the doubly charged SO_4^{2-}, the ionic strength of the K_2SO_4 is three times that of the KNO_3.

Higher charged ions contribute more to the ionic strength.

    If more than one salt is present, then the ionic strength is calculated from the total concentration and charges of all the different ions. For any given electrolyte, the ionic strength will be proportional to the concentration. Strong acids that are completely ionized are treated in the same manner as salts. If the acids are partially ionized, then the concentration of the ionized species must be estimated from the ionization constant before the ionic strength is computed. Very weak acids can usually be considered to be nonionized and do not contribute to the ionic strength.

Example 6.15

Calculate the ionic strength of a solution consisting of 0.30 M NaCl and 0.20 M Na$_2$SO$_4$.

$$\mu = \frac{C_{N^+}Z^2_{Na^+} + C_{Cl^-}Z^2_{Cl^-} + C_{SO_4^{2-}}Z^2_{SO_4^{2-}}}{2}$$

$$= \frac{0.70 \times (1)^2 + 0.30 \times (1)^2 + 0.20 \times (2)^2}{2}$$

$$= 0.90$$

CALCULATION OF ACTIVITY COEFFICIENTS

In 1923, Debye and Hückel derived a theoretical expression for calculating activity coefficients. The equation, known as the **Debye–Hückel equation**, is

This equation applies for ionic strengths up to 0.2.

$$-\log f_i = \frac{0.51 Z_i^2 \sqrt{\mu}}{1 + 0.33 \alpha_i \sqrt{\mu}} \tag{6.19}$$

The estimation of the ion size parameter places a limit on the accuracy of the calculated activity coefficient.

The numbers 0.51 and 0.33 are constants for water at 25°C, and the former includes the $-\frac{3}{2}$ power of both the dielectric constant of the solvent and the absolute temperature; α_i is the **ion size parameter**, which is the effective diameter of the hydrated ion in angstrom units, Å. An angstrom is 100 picometers (pm, 10^{-10} meter). A limitation of the Debye–Hückel equation is the accuracy to which α_i can be evaluated. For many singly charged ions, α_i is generally about 3 Å, and for practical purposes Equation 6.19 simplifies to

This equation may be used for ionic strengths less than 0.01.

$$-\log f_i = \frac{0.51 Z_i^2 \sqrt{\mu}}{1 + \sqrt{\mu}} \tag{6.20}$$

See Ref. 9 for a tabulation of α_i values.

For common multiply charged ions, α_i may become as large as 11 Å. But at ionic strengths less than 0.01, the second term of the denominator becomes small with respect to 1, so uncertainties in α_i become relatively unimportant, and Equation 6.20 can be applied at ionic strengths of 0.01 or less. Equation 6.19 can be applied up to ionic strengths of about 0.2. Ref. 9 at the end of the chapter lists values for α_i for different ions and also includes a table of calculated activity coefficients, using Equation 6.19, at ionic strengths ranging from 0.0005 to 0.1. The CD for Chapter 6 contains a list of ion size parameters for some common ions taken from this reference.

Example 6.16

Calculate the activity coefficients for K$^+$ and SO$_4^{2-}$ in a 0.0020 M solution of potassium sulfate.

Solution

The ionic strength is 0.0060, so we can apply Equation 6.20:

$$-\log f_{K^+} = \frac{0.51(1)^2\sqrt{0.0060}}{1 + \sqrt{0.0060}} = 0.037$$

$$f_{K^+} = 10^{-0.037} = 10^{-1} \times 10^{0.963} = 0.918$$

$$-\log f_{SO_4^{2-}} = \frac{0.51(2)^2\sqrt{0.0060}}{1 + \sqrt{0.0060}} = 0.14_7$$

$$f_{SO_4^{2-}} = 10^{-0.14_7} = 10^{-1} \times 10^{0.85_3} = 0.71_3$$

 Example 6.17

Calculate the activity coefficients for K^+ and SO_4^{2-} in a 0.020 M solution of potassium sulfate.

Solution

The ionic strength is 0.060, so we would use Equation 6.19. From Ref. 9, we find that $\alpha_{K^+} = 3$ Å and $\alpha_{SO_4^{2-}} = 4.0$ Å. For K^+, we can use Equation 6.20:

$$-\log f_{K^+} = \frac{0.51(1)^2\sqrt{0.060}}{1 + \sqrt{0.060}} = 0.10_1$$

$$f_{K^+} = 10^{-0.101} = 10^{-1} \times 10^{0.899} = 0.79_4$$

For SO_4^{2-}, use Equation 6.19:

$$-\log f_{SO_4^{2-}} = \frac{0.51(2)^2\sqrt{0.060}}{1 + 0.33 \times 4.0\sqrt{0.060}} = 0.37_8$$

$$f_{SO_4^{2-}} = 10^{-1} \times 10^{0.62_2} = 0.41_9$$

This latter compares with a calculated value of 0.39_6 using Equation 6.20. Note the decrease in the activity coefficients compared to 0.002 M K_2SO_4, especially for the SO_4^{2-} ion.

Spreadsheets for calculating activity coefficients using Equations 6.19 and 6.20 are given in the CD for Chapter 6.

For higher ionic strengths, a number of empirical equations have been developed. Perhaps one of the more useful is the **Davies modification** (see Ref. 8):

$$-\log f_i = 0.51Z_i^2\left(\frac{\sqrt{\mu}}{1 + \sqrt{\mu}} - 0.3\,\mu\right) \qquad (6.21)$$

Use this equation for ionic strengths of 0.2–0.5. It gives increasing activity coefficients compared to the Debye–Hückel equation.

It is valid up to ionic strengths of about 0.5.

A 0.01 *M* solution of HCl prepared in 8 *M* NaCl has an activity about 100 times that in water! Its pH is actually 0.0. See F. E. Critchfield and J. B. Johnson, *Anal. Chem.*, **30**, (1958) 1247 and G. D. Christian, *CRC Crit. Rev. in Anal. Chem.*, **5**(2) (1975) 119–153.

At very high electrolyte concentrations, activity coefficients may actually increase and become greater than unity. This is because the activity of the solvent, water, is decreased and solvated ionic species become partially desolvated. This increases their reactivity and hence their activity. Note that Equation 6.21 actually "corrects" the value of f_i to a larger value as M increases.

We can draw some general conclusions about the estimation of activity coefficients.

The greater the charge on diverse ions, the greater their effect on the activity.

---

1. The activity coefficients of ions of a given charge type are approximately the same in solutions of a given ionic strength, and this activity coefficient is the same regardless of their individual concentrations.
2. The behavior of ions become less ideal as the charge type increases, resulting in less confidence in calculated activity coefficients.
3. The calculated activity coefficient of an ion in a mixed electrolyte solution will be less accurate than in a single-electrolyte solution.
4. The activity coefficients of nonelectrolytes (uncharged molecules) can generally be considered equal to unity in ionic strengths up to 0.1, and deviations from this approximation are only moderate in ionic strengths as high as 1. Undissociated acids, HA, are nonelectrolytes whose activity coefficients can be taken as unity.

The activity of nonelectrolytes is the same as the concentration, up to ionic strengths of 1.

6.16 The Diverse Ion Effect: The Thermodynamic Equilibrium Constant and Activity Coefficients

We mentioned at the beginning of the last section on activity that the presence of diverse salts will generally increase the dissociation of weak electrolytes due to a shielding (or decrease in the activity) of the ionic species produced upon dissociation. We can quantitatively predict the extent of the effect on the equilibrium by taking into account the activities of the species in the equilibrium.

In our consideration of equilibrium constants thus far, we have assumed no diverse ion effect, that is, an ionic strength of zero and an activity coefficient of 1. Equilibrium constants should more exactly be expressed in terms of activities rather than concentrations. Consider the dissociation of AB. The **thermodynamic equilibrium constant** (i.e., the equilibrium constant extrapolated to the case of infinite dilution) K_{eq}° is

Thermodynamic equilibrium constants hold at all ionic strengths.

$$K_{eq}^{\circ} = \frac{a_A \cdot a_B}{a_{AB}} = \frac{[A]f_A \cdot [B]f_B}{[AB]f_{AB}} \tag{6.22}$$

Since the **concentration equilibrium constant** $K_{eq} = [A][B]/[AB]$, then

$$K_{eq}^{\circ} = K_{eq} \frac{f_A \cdot f_B}{f_{AB}} \tag{6.23}$$

or
$$K_{eq} = K_{eq}^{\circ} \frac{f_{AB}}{f_A \cdot f_B} \qquad (6.24)$$

The numerical value of K_{eq}° holds for all activities. $K_{eq} = K_{eq}^{\circ}$ at zero ionic strength, but at appreciable ionic strengths, a value for K_{eq} must be calculated for each ionic strength using Equation 6.24. The equilibrium constants listed in Appendix C are for zero ionic strength; that is, they are really thermodynamic equilibrium constants. (Experimental K_{eq} values are available at different ionic strengths and can be used for equilibrium calculations at the listed ionic strength, using molar concentrations without having to calculate activity coefficients.)

Concentration equilibrium constants must be corrected for ionic strength.

 Example 6.18

The weak electrolyte AB dissociates to A^+ and B^-, with a thermodynamic equilibrium constant K_{eq}° of 2×10^{-8} (a) Calculate the molar equilibrium constant K_{eq}. (b) Calculate the percent dissociation of a 1.0×10^{-4} M solution of AB in water and in the presence of a diverse salt of ionic strength 0.1, if the activity coefficients of A^+ and B^- are 0.6 and 0.7, respectively, at $\mu = 0.1$.

Solution

(a)
$$AB \rightleftharpoons A^+ + B^-$$

$$K_{eq} = \frac{[A^+][B^-]}{[AB]}$$

$$K_{eq}^{\circ} = \frac{a_{A^+} \cdot a_{B^-}}{a_{AB}} = \frac{[A^+]f_{A^+} \cdot [B^-]f_{B^-}}{[AB]f_{AB}}$$

The activity coefficient of a neutral species is unity, so

$$K_{eq}^{\circ} = \frac{[A^+][B^-]}{[AB]} \cdot f_{A^+} \cdot f_{B^-} = K_{eq}f_{A^+} \cdot f_{B^-}$$

$$K_{eq} = \frac{K_{eq}^{\circ}}{f_{A^+} \cdot f_{B^-}} = \frac{2 \times 10^{-8}}{(0.6)(0.7)} = 5 \times 10^{-8}$$

(b)
$$\begin{array}{cccc} AB & \rightleftharpoons & A^+ & + & B^- \\ 1 \times 10^{-4} - x & & x & & x \end{array}$$

In water, $f_{A^+} = f_{B^-} \approx 1$ (since $\mu < 10^{-4}$), $x \ll 10^{-4}$

$$\frac{[A^+][B^-]}{[AB]} = 2 \times 10^{-8}$$

$$\frac{(x)(x)}{1.0 \times 10^{-4}} = 2 \times 10^{-8}$$

$$x = 1.4 \times 10^{-6} \ M$$

$$\% \text{ dissociated} = \frac{1.4 \times 10^{-6} \ M}{1.0 \times 10^{-4} \ M} \times 100\% = 1.4\%$$

For 0.1 M salt,

$$\frac{[A^+][B^-]}{[AB]} = 5 \times 10^{-8}$$

$$\frac{(x)(x)}{1.0 \times 10^{-4}} = 5 \times 10^{-8}$$

$$x = 2.2 \times 10^{-6}$$

$$\% \text{ dissociated} = \frac{2.2 \times 10^{-6}}{1.0 \times 10^{-4}} \times 100\% = 2.2\%$$

which represents a 57% increase in dissociation.

· ·

We will generally ignore diverse salt effects.

Calculations using the diverse ion effect are illustrated in Chapter 7 for acid dissociation and in Chapter 10 for precipitate solubilities. For illustrative purposes throughout this book, we will in general neglect the diverse ion effects on equilibria. In most cases, we are interested in *relative* changes in equilibrium concentrations, and the neglect of activities will not change our arguments.

Learning Objectives

WHAT ARE SOME OF THE KEY THINGS WE LEARNED FROM THIS CHAPTER?

- The equilibrium constant (key equations: 6.12, 6.15), p. 194
- Calculation of equilibrium concentrations, p. 195
- Using Excel Solver to solve the quadratic equation, p. 198
- The systematic approach to equilibrium calculations: mass balance and charge balance equations, p. 203
- Activity and activity coefficients (key equation: 6.19), p. 210
- Thermodynamic equilibrium constants (key equation: 6.22), p. 214

Problems

EQUILIBRIUM CALCULATIONS

1. A and B react as follows: $A + B \rightleftharpoons C + D$. The equilibrium constant is 2.0×10^3. If 0.30 mol of A and 0.80 mol of B are mixed in 1 L, what are the concentrations of A, B, C, and D after reaction?
2. A and B react as follows: $A + B \rightleftharpoons 2C$. The equilibrium constant is 5.0×10^6. If 0.40 mol of A and 0.70 mol of B are mixed in 1 L, what are the concentrations of A, B, and C after reaction?
3. The dissociation constant for salicylic acid, $C_6H_4(OH)COOH$, is 1.0×10^{-3}. Calculate the percent dissociation of a 1.0×10^{-3} M solution. There is one dissociable proton. (See also Excel Problem 25 below.)
4. The dissociation constant for hydrocyanic acid, HCN, is 7.2×10^{-10}. Calculate the percent dissociation of a 1.0×10^{-3} M solution.
5. Calculate the percent dissociation of the salicylic acid in Problem 3 if the solution also contained 1.0×10^{-2} M sodium salicylate (the salt of salicylic acid).

6. Hydrogen sulfide, H_2S, dissociates stepwise, with dissociation constants of 9.1×10^{-8} and 1.2×10^{-15}, respectively. Write the overall dissociation reaction and the overall equilibrium constant.

7. Fe^{2+} and $Cr_2O_7^{2-}$ react as follows: $6Fe^{2+} + Cr_2O_7^{2-} + 14H^+ \rightleftharpoons 6\,Fe^{3+} + 2Cr^{3+} + 7H_2O$. The equilibrium constant for the reaction is 1×10^{57}. Calculate the equilibrium concentrations of the iron and chromium species if 10 mL each of 0.02 M $K_2Cr_2O_7$ in 1.14 M HCl and 0.12 M $FeSO_4$ in 1.14 M HCl are reacted.

SYSTEMATIC APPROACH TO EQUILIBRIUM CALCULATIONS

8. Write charge balance expressions for (a) a saturated solution of Bi_2S_3; (b) a solution of Na_2S.

9. Write the equations of mass balance and electroneutrality for a 0.100 M $[Cd(NH_3)_4]Cl_2$ solution.

10. Prove the following relations using the principles of electroneutrality and mass balance:

   (a) $[NO_2^-] = [H^+] - [OH^-]$ for 0.2 M HNO_2 solution

   (b) $[CH_3COOH] = 0.2 - [H^+] + [OH^-]$ for 0.2 M CH_3COOH solution

   (c) $[H_2C_2O_4] = 0.1 - [H^+] + [OH^-] - [C_2O_4^{2-}]$ for 0.1 M $H_2C_2O_4$ solution

   (d) $[HCN] = [OH^-] - [H^+]$ for 0.1 M KCN solution

   (e) $[H_2PO_4^-] = \dfrac{[OH^-] - [H^+] - [HPO_4^{2-}] - 3[H_3PO_4]}{2}$ for 0.1 M Na_3PO_4

   solution.

   (f) $[HSO_4^-] = 0.2 - [H^+] - [OH^-]$ for 0.1 M H_2SO_4 solution (assume that the dissociation of H_2SO_4 to H^+ and HSO_4^- is quantitative).

11. Write equations of mass balance for an aqueous saturated solution of BaF_2 containing the species F^-, HF, HF_2^-, and Ba^{2+}.

12. Write an equation of mass balance for an aqueous solution of $Ba_3(PO_4)_2$.

13. Calculate the pH of a 0.100 M solution of acetic acid using the charge/mass balance approach.

IONIC STRENGTH

14. Calculate the ionic strengths of the following solutions: (a) 0.30 M NaCl; (b) 0.30 M Na_2SO_4; (c) 0.30 M NaCl and 0.20 M K_2SO_4; (d) 0.20 M $Al_2(SO_4)_3$ and 0.10 M Na_2SO_4.

15. Calculate the ionic strengths of the following solutions: (a) 0.20 M $ZnSO_4$; (b) 0.40 M $MgCl_2$; (c) 0.50 M $LaCl_3$; (d) 1.0 M $K_2Cr_2O_7$; (e) 1.0 M $Tl(NO_3)_3$ + 1.0 M $Pb(NO_3)_2$.

ACTIVITY

16. Calculate the activity coefficients of the sodium and chloride ions for a 0.00100 M solution of NaCl.

17. Calculate the activity coefficients of each ion in a solution containing 0.0020 M Na_2SO_4 and 0.0010 M $Al_2(SO_4)_3$.

18. Calculate the activity of the NO_3^- ion in a solution of 0.0020 M KNO_3.

19. Calculate the activity of the CrO_4^{2-} ion in a 0.020 M solution of Na_2CrO_4.

THERMODYNAMIC EQUILIBRIUM CONSTANTS

20. Write thermodynamic equilibrium constant expressions for the following:

   (a) $HCN \rightleftharpoons H^+ + CN^-$

   (b) $NH_3 + H_2O \rightleftharpoons NH_4^+ + OH^-$

21. Calculate the pH of a solution of 5.0×10^{-3} M benzoic acid (a) in water and (b) in the presence of 0.05 M K_2SO_4.

EXCEL EXERCISES

22. Without reference to the CD, write a spreadsheet program for calculating activity coefficients using Equation 6.20. Then compare it with the one on the CD. Do a calculation with both to check the accuracy.

23. Calculate the activity coefficients for K^+ and SO_4^{2-} in Example 6.16 using the CD spreadsheet for Equation 6.20. Compare your results with the manually calculated values in the example.

24. Calculate the activity coefficients in Example 6.17 using the CD spreadsheets for Equations 6.19 and 6.20. Compare your results with the manually calculated values in the example.

25. Use Excel Solver to calculate the concentration, x, in Problem 3 above. (The problem requires use of the quadratic equation.)

26. Solve Problems 16 to 19 above using Excel.

Recommended References

EQUILIBRIA

1. A. J. Bard, *Chemical Equilibrium*. New York: Harper & Row, 1966.

2. T. R. Blackburn, *Equilibrium: A Chemistry of Solutions*. New York: Holt, Rinehart and Winston, 1969.

3. J. N. Butler. *Ionic Equilibrium. A Mathematical Approach*. Reading, MA: Addison-Wesley, 1964.

4. G. M. Fleck, *Ionic Equilibria in Solution*. New York: Holt, Rinehart and Winston, 1966.

5. H. Freiser and Q. Fernando, *Ionic Equilibria in Analytical Chemistry*. New York: Wiley, 1963.

6. A. E. Martell and R. J. Motekaitis, *The Determination and Use of Stability Constants*. New York: VCH, 1989.

METHOD OF SUCCESSIVE APPROXIMATIONS

7. S. Brewer, *Solving Problems in Analytical Chemistry*. New York: Wiley, 1980.

ACTIVITY

8. C. W. Davies, *Ion Association*. London: Butterworth, 1962.

9. J. Kielland, "Individual Activity Coefficients of Ions in Aqueous Solutions," *J. Am. Chem. Soc.*, **59** (1937) 1675.

10. K. S. Pitzer. *Activity Coefficients in Electrolyte Solutions*, 2nd ed. Boca Raton, FL: CRC Press, 1991.

11. P. C. Meier, "Two-Parameter Debye-Huckel Approximation for the Evaluation of Mean Activity Coefficients of 109 Electrolytes," *Anal. Chim. Acta*, **136** (1982) 363.

Chapter Seven

ACID–BASE EQUILIBRIA

"There are three side effects of acid. Enhanced long term memory, decreased short term memory, and I forget the third."
—Timothy Leary

The acidity or basicity of a solution is frequently an important factor in chemical reactions. The use of buffers of a given pH to maintain the solution pH at a desired level is very important. In addition, fundamental acid–base equilibria are important in understanding acid–base titrations and the effects of acids on chemical species and reactions, for example, the effects of complexation or precipitation. In Chapter 6, we described the fundamental concept of equilibrium constants. In this chapter, we consider in more detail various acid–base equilibrium calculations, including *weak acids* and *bases*, hydrolysis of *salts of weak acids and bases*, buffers, polyprotic acids and their salts, and *physiological buffers*. Acid–base theories and the basic pH concept are reviewed first.

7.1 Acid–Base Theories—Not All Are Created Equal

Several acid–base theories have been proposed to explain or classify acidic and basic properties of substances. You are probably most familiar with the **Arrhenius theory**, which is applicable only to water. Other theories are more general and are applicable to other solvents. We describe the common acid–base theories here.

The Arrhenius theory is restricted to aqueous solutions. See *J. Am. Chem. Soc.*, **36** (1912) 353 for his personal observations of the difficulty Arrhenius had in the acceptance of his theory.

ARRHENIUS THEORY—H⁺ AND OH⁻

Arrhenius, as a graduate student, introduced a radical theory in 1894 (for which he received the Nobel Prize) that an **acid** is any substance that ionizes (partially or completely) in water to give *hydrogen ions* (which associate with the solvent to give hydronium ions, H_3O^+):

$$HA + H_2O \rightleftharpoons H_3O^+ + A^-$$

A **base** ionizes in water to give *hydroxyl ions*. Weak (partially ionized) bases generally ionize as follows:

$$B + H_2O \rightleftharpoons BH^+ + OH^-$$

while strong bases such as metal hydroxides (e.g., NaOH) dissociate as

$$M(OH)_n \rightarrow M^{n+} + nOH^-$$

This theory is obviously restricted to water as the solvent.

THEORY OF SOLVENT SYSTEMS—SOLVENT CATIONS AND ANIONS

Franklin's theory is similar to the Arrhenius theory but is applicable also to other ionizable solvents.

In 1905, Franklin introduced the solvent system concept of acids and bases. This theory recognizes the ionization of a solvent to give a cation and an anion; for example, $2H_2O \rightleftharpoons H_3O^+ + OH^-$ or $2NH_3 \rightleftharpoons NH_4^+ + NH_2^-$. An **acid** is defined as a solute that yields the *cation of the solvent* while a **base** is a solute that yields the *anion of the solvent*. Thus, NH_4Cl is a strong acid in liquid ammonia (similar to HCl in water: $HCl + H_2O \rightarrow H_3O^+ + Cl^-$) while $NaNH_2$ is a strong base in ammonia (similar to NaOH in water); both of these compounds ionize to give the solvent cation and anion, respectively. Ethanol ionizes as follows: $2C_2H_5OH \rightleftharpoons C_2H_5OH_2^+ + C_2H_5O^-$. Hence, sodium ethoxide, $NaOC_2H_5$, is a strong base in this solvent.

BRØNSTED–LOWRY THEORY—TAKING AND GIVING PROTONS

The Brønsted–Lowry theory assumes a transfer of protons from an acid to a base, i.e., conjugate pairs.

The theory of solvent systems is suitable for ionizable solvents, but it is not applicable to acid–base reactions in nonionizable solvents such as benzene or dioxane. In 1923, Brønsted and Lowry separately described what is now known as the **Brønsted–Lowry** theory. This theory states that an **acid** is any substance that can *donate a proton*, and a **base** is any substance that can *accept a proton*. Thus, we can write a "half-reaction"

$$\boxed{acid = H^+ + base} \tag{7.1}$$

The acid and base of a half-reaction are called **conjugate pairs**. Free protons do not exist in solution, and there must be a proton acceptor (base) before a proton donor (acid) will release its proton. That is, there must be a combination of two half-reactions. Some acid–base reactions in different solvents are illustrated in Table 7.1. In the first example, acetate ion is the conjugate base of acetic acid and ammonium ion is the conjugate acid of ammonia. The first four examples represent ionization of an acid or a base in a solvent, while the others represent a neutralization reaction between an acid and a base in the solvent.

Table 7.1
Brønsted Acid–Base Reactions

Solvent	Acid$_1$	+	Base$_2$	$\rightarrow$	Acid$_2$	+	Base$_1$
NH_3 (liq.)	HOAc		NH_3		NH_4^+		OAc^-
H_2O	HCl		H_2O		H_3O^+		Cl^-
H_2O	NH_4^+		H_2O		H_3O^+		NH_3
H_2O	H_2O		OAc^-		HOAc		OH^-
H_2O	HCO_3^-		OH^-		H_2O		CO_3^{2-}
C_2H_5OH	NH_4^+		$C_2H_5O^-$		C_2H_5OH		NH_3
C_6H_6	H picrate		$C_6H_5NH_2$		$C_6H_5NH_3^+$		picrate$^-$

It is apparent from the above definition that a substance cannot act as an acid unless a base is present to accept the protons. Thus, acids will undergo complete or partial ionization in basic solvents such as water, liquid ammonia, or ethanol, depending on the basicity of the solvent and the strength of the acid. But in neutral or "inert" solvents, ionization is insignificant. However, ionization in the solvent is not a prerequisite for an acid–base reaction, as in the last example in the table, where picric acid reacts with aniline.

LEWIS THEORY—TAKING AND GIVING ELECTRONS

Also in 1923, G. N. Lewis introduced the electronic theory of acids and bases. In the **Lewis** theory, an acid is a substance that can accept an electron pair and a base is a substance that can donate an electron pair. The latter frequently contains an oxygen or a nitrogen as the electron donor. Thus, nonhydrogen-containing substances are included as acids. Examples of acid–base reactions in the Lewis theory are as follows:

> The Lewis theory assumes a donation (sharing) of electrons from a base to an acid.

$$H^+ \text{ (solvated)} + :NH_3 \rightarrow H:NH_3^+$$

$$AlCl_3 + :O\overset{R}{\underset{R}{\diagup\diagdown}} \rightarrow Cl_3Al:OR_2$$

$$\overset{H}{\underset{H}{\diagdown\diagup}}O: + H^+ \rightarrow H_2O:H^+$$

$$H^+ + :OH^- \rightarrow H:OH$$

In the second example, aluminum chloride is an acid and ether is a base.

7.2 Acid–Base Equilibria in Water

We see from the above that when an acid or base is dissolved in water, it will dissociate, or **ionize**, the amount of ionization being dependent on the strength of the acid. A "strong" electrolyte is completely dissociated, while a "weak" electrolyte is partially dissociated. Table 7.2 lists some common electrolytes, some strong and some weak. Other weak acids and bases are listed in Appendix C.

Table 7.2
Some Strong Electrolytes and Some Weak Electrolytes

Strong	Weak
HCl	$HC_2H_3O_2$ (acetic acid)
$HClO_4$	NH_3
$H_2SO_4{}^a$	C_6H_5OH (phenol)
HNO_3	$HCHO_2$ (formic acid)
NaOH	$C_6H_5NH_2$ (aniline)
$NaC_2H_3O_2$	

^{a} The first proton is completely ionized in dilute solution, but the second proton is partially ionized ($K_i = 10^{-2}$).

Hydrochloric acid is a strong acid, and its ionization is complete:

$$HCl + H_2O \rightarrow H_3O^+ + Cl^- \tag{7.2}$$

An equilibrium constant for Equation 7.2 would have a value of infinity. The proton H^+ exists in water as a hydrated ion, the **hydronium ion**, H_3O^+. Higher hydrates probably exist, particularly $H_9O_4^+$. The hydronium ion is written as H_3O^+ for convenience and to emphasize Brønsted behavior.

Acetic acid[1] is a weak acid, which ionizes only partially (a few percent):

$$HOAc + H_2O \rightleftharpoons H_3O^+ + OAc^- \tag{7.3}$$

We can write an **equilibrium constant** for this reaction:

$$K_a^\circ = \frac{a_{H_3O^+} \cdot a_{OAc^-}}{a_{HOAc} \cdot a_{H_2O}} \tag{7.4}$$

where K_a° is the **thermodynamic acidity constant** (see Section 6.16) and a is the **activity** of the indicated species. Salt cations or anions may also partially react with water after they are dissociated, for example, acetate ion from dissociated acetate salt, to give HOAc.

The activity can be thought of as representing the effective concentration of an ion (described in Chapter 6). The effects of protons in reactions are often governed by their activities, and it is the activity that is measured by the widely used pH meter (Chapter 13). Methods for predicting numerical values of activity coefficients were described in Chapter 6.

In dilute solutions, the activity of water remains essentially constant, and is taken as unity at standard state. Therefore, Equation 7.4 can be written as

$$K_a^\circ = \frac{a_{H_3O^+} \cdot a_{OAc^-}}{a_{HOAc}} \tag{7.5}$$

Autoprotolysis is the self-ionization of a solvent to give a cation and anion, e.g., $2CH_3OH \rightleftharpoons CH_3OH_2^+ + CH_3O^-$.

Pure water ionizes slightly, or undergoes **autoprotolysis**:

$$2H_2O \rightleftharpoons H_3O^+ + OH^- \tag{7.6}$$

The equilibrium constant for this is

$$K_w^\circ = \frac{a_{H_3O^+} \cdot a_{OH^-}}{a_{H_2O}^2} \tag{7.7}$$

Again, the activity of water is constant in dilute solutions (its concentration is essentially constant at ~55.3 M), so

$$K_w^\circ = a_{H_3O^+} \cdot a_{OH^-} \tag{7.8}$$

where K_w° is the **thermodynamic autoprotolysis**, or **self-ionization, constant**.

We will use H^+ in place of H_3O^+, for simplification. Also, molar concentrations will generally be used instead of activities.

Calculations are simplified if we neglect activity coefficients. This simplification results in only slight errors for dilute solutions, and we shall use molar concentrations in all our calculations. This will satisfactorily illustrate the equilibria involved. Most of the solutions we will be concerned with are rather dilute, and we will frequently be interested in relative changes in pH (and large ones) in which

---

[1] We shall use the symbol OAc^- to represent the acetate ion $CH_3-\overset{\overset{O}{\|}}{C}-O^-$.

case small errors are insignificant. We will simplify our expressions by using H^+ in place of H_3O^+. This is not inconsistent since the waters of solvation associated with other ions or molecules (e.g., metal ions) are not generally written and H_3O^+ is not an entirely accurate representation of the species present anyway.

Molar concentration will be represented by square brackets [ ] around the species. Simplified equations for the above reactions are

$$HCl \rightarrow H^+ + Cl^- \tag{7.9}$$

$$HOAc \rightleftharpoons H^+ + OAc^- \tag{7.10}$$

$$\boxed{K_a = \frac{[H^+][OAc^-]}{[HOAc]}} \tag{7.11}$$

$$H_2O \rightleftharpoons H^+ + OH^- \tag{7.12}$$

$$\boxed{K_w = [H^+][OH^-]} \tag{7.13}$$

K_a and K_w are the **molar equilibrium constants**.

At 25°C, $K_w = 1.0 \times 10^{-14}$. The product of the hydrogen ion concentration and the hydroxyl ion concentration in aqueous solution is *always* equal to 1.0×10^{-14} at room temperature:

$$\boxed{[H^+][OH^-] = 1.0 \times 10^{-14}} \tag{7.14}$$

Chemists (and especially students!) are lucky that nature made K_w an even unit number at room temperature. Imagine doing pH calculations with a K_w like 2.39×10^{-13}. However, see Section 7.4 where you indeed must (for other temperatures).

In pure water, then, the concentrations of these two species are equal since there are no other sources of H^+ or OH^- except H_2O dissociation:

$$[H^+] = [OH^-]$$

Therefore,

$$[H^+][H^+] = 1.0 \times 10^{-14}$$

$$[H^+] = 1.0 \times 10^{-7}\ M \equiv [OH^-]$$

If an acid is added to water, we can calculate the hydroxyl ion concentration if we know the hydrogen ion concentration from the acid. *Except when the hydrogen ion concentration from the acid is very small, 10^{-6} M or less, any contribution to [H$^+$] from the ionization of water can be neglected.*

Example 7.1

A $1.0 \times 10^{-3}\ M$ solution of hydrochloric acid is prepared. What is the hydroxyl ion concentration?

Solution

Since hydrochloric acid is a strong electrolyte and is completely ionized, the H^+ concentration is $1.0 \times 10^{-3}\ M$. Thus,

$$(1.0 \times 10^{-3})[OH^-] = 1.0 \times 10^{-14}$$

$$[OH^-] = 1.0 \times 10^{-11}\ M$$

7.3 The pH Scale

pScales are used to compress a range of numbers over several decades in magnitude.

The concentration of H^+ or OH^- in aqueous solution can vary over extremely wide ranges, from 1 M or greater to 10^{-14} M or less. To construct a plot of H^+ concentration against some variable would be very difficult if the concentration changed from, say, 10^{-1} M to 10^{-13} M. This range is common in a titration. It is more convenient to compress the acidity scale by placing it on a logarithm basis. The **pH** of a solution was defined by Sørenson as

$$\boxed{pH = -\log[H^+]} \tag{7.15}$$

pH is really $-\log a_{H^+}$. This is what a pH meter (glass electrode) measures—see Chapter 13.

The minus sign is used because most of the concentrations encountered are less than 1 M, and so this designation gives a positive number. (More strictly, pH is now defined as $-\log a_{H^+}$, but we will use the simpler definition of Equation 7.15.) In general, **pAnything = −log Anything**, and this method of notation will be used later for other numbers that can vary by large amounts, or are very large or small (e.g., equilibrium constants).

 Example 7.2

Calculate the pH of a 2.0×10^{-3} M solution of HCl.

Solution

HCl is completely ionized, so

$$[H^+] = 2.0 \times 10^{-3} \ M$$
$$pH = -\log(2.0 \times 10^{-3}) = 3 - \log 2.0 = 3 - 0.30 = 2.70$$

A similar definition is made for the hydroxyl ion concentration:

$$\boxed{pOH = -\log[OH^-]} \tag{7.16}$$

A 1 M HCl solution has a pH of 0 and pOH of 14. A 1 M NaOH solution has a pH of 14 and a pOH of 0.

Equation 7.13 can be used to calculate the hydroxyl ion concentration if the hydrogen ion concentration is known, and vice versa. The equation in logarithm form for a more direct calculation of pH or pOH is

$$-\log K_w = -\log[H^+][OH^-] = -\log[H^+] - \log [OH^-] \tag{7.17}$$

$$\boxed{pK_w = pH + pOH} \tag{7.18}$$

At 25°C,

$$\boxed{14.00 = pH + pOH} \tag{7.19}$$

Example 7.3

Calculate the pOH and the pH of a $5.0 \times 10^{-2}\ M$ solution of NaOH.

Solution

$$[OH^-] = 5.0 \times 10^{-2}\ M$$
$$pOH = -\log(5.0 \times 10^{-2}) = 2 - \log 5.0 = 2 - 0.70 = 1.30$$
$$pH + 1.30 = 14.00$$
$$pH = 12.70$$

or

$$[H^+] = \frac{1.0 \times 10^{-14}}{5.0 \times 10^{-2}} = 2.0 \times 10^{-13}\ M$$

$$pH = -\log(2.0 \times 10^{-13}) = 13 - \log 2.0 = 13 - 0.30 = 12.70$$

Example 7.4

Calculate the pH of a solution prepared by mixing 2.0 mL of a strong acid solution of pH 3.00 and 3.0 mL of a strong base of pH 10.00.

Keep track of millimoles!

Solution

$$[H^+]\ \text{of acid solution} = 1.0 \times 10^{-3}\ M$$
$$\text{mmol } H^+ = 1.0 \times 10^{-3}\ M \times 2.0\ \text{mL} = 2.0 \times 10^{-3}\ \text{mmol}$$
$$pOH\ \text{of base solution} = 14.00 - 10.00 = 4.00$$
$$[OH^-] = 1.0 \times 10^{-4}\ M$$
$$\text{mmol } OH^- = 1.0 \times 10^{-4}\ M \times 3.0\ \text{mL} = 3.0 \times 10^{-4}\ \text{mmol}$$

There is an excess of acid.

$$\text{mmol } H^+ = 0.0020 - 0.0003 = 0.0017\ \text{mmol}$$
$$[H^+] = 0.0017\ \text{mmol}/5.0\ \text{mL} = 3.4 \times 10^{-4}\ M$$
$$pH = -\log 3.4 \times 10^{-4} = 4 - 0.53 = 3.47$$

Example 7.5

The pH of a solution is 9.67. Calculate the hydrogen ion concentration in the solution.

Solution

$$-\log[H^+] = 9.67$$
$$[H^+] = 10^{-9.67} = 10^{-10} \times 10^{0.33}$$
$$[H^+] = 2.1 \times 10^{-10}\ M$$

$[H^+] = 10^{-pH}$.

When $[H^+] = [OH^-]$, then a solution is said to be **neutral**. If $[H^+] > [OH^-]$, then the solution is **acidic**. And if $[H^+] < [OH^-]$, the solution is **alkaline**. The hydrogen ion and hydroxyl ion concentrations in pure water at 25°C are each 10^{-7} M, and the pH of water is 7. A pH of 7 is therefore neutral. Values of pH that are greater than this are alkaline, and pH values less than this are acidic. The reverse is true of pOH values. A pOH of 7 is also neutral. Note that the product of $[H^+]$ and $[OH^-]$ is always 10^{-14} at 25°C, and the sum of pH and pOH is always 14. If the temperature is other than 25°C, then K_w is different from 1.00×10^{-14}, and a neutral solution will have other than 10^{-7} M H^+ and OH^- (see below).

A 10 M HCl solution should have a pH of -1 and pOH of 15.

Students are often under the illusion that it is impossible to have a **negative pH**. There is no theoretical basis for this. A negative pH only means that the hydrogen ion concentration is greater than 1 M. In actual practice, a negative pH is uncommon because of two reasons. First, even strong acids may become partially undissociated at high concentrations. For example, 100% H_2SO_4 is so weakly dissociated that it can be stored in iron containers; more dilute H_2SO_4 solutions would contain sufficient protons from dissociation to attack and dissolve the iron. The second reason has to do with the *activity*, which we have chosen to neglect for dilute solutions. Since pH is really $-\log a_{H^+}$ (this is what a pH meter reading is a measure of), a solution that is 1.1 M in H^+ may actually have a positive pH because the activity of the H^+ is less than 1.0 M.[2] This is because at these high concentrations, the activity coefficient is less than unity (although at still higher concentrations the activity coefficient may become greater than unity—see Chapter 6). Nevertheless, there is mathematically no basis for not having a negative pH (or a negative pOH), although it may be rare in analytical solutions encountered.

The pH of 10^{-9} M HCl is *not* 9!

If the concentration of an acid or base is much less than 10^{-7} M, then its contribution to the acidity or basicity will be negligible compared with the contribution from water. The pH of a 10^{-8} M sodium hydroxide solution would therefore not differ significantly from 7. If the concentration of the acid or base is around 10^{-7} M, then its contribution is not negligible and neither is that from water; hence the sum of the two contributions must be taken.

Example 7.6

Calculate the pH and pOH of a 1.0×10^{-7} M solution of HCl.

Solution

Equilibria:

$$HCl \rightarrow H^+ + Cl^-$$
$$H_2O \rightleftharpoons H^+ + OH^-$$
$$[H^+][OH^-] = 1.0 \times 10^{-14}$$
$$[H^+]_{H_2O\ diss.} = [OH^-]_{H_2O\ diss.} = x$$

Since the hydrogen ions contributed from the ionization of water are not negligible compared to the HCl added,

$$[H^+] = C_{HCl} + [H^+]_{H_2O\ diss.}$$

---

[2]As will be seen in Chapter 13, it is also difficult to *measure* the pH of a solution having a negative pH or pOH because high concentrations of acids or bases tend to introduce an error in the measurement by adding a significant and unknown liquid-junction potential in the measurements.

Then,

$$([H^+]_{HCl} + x)(x) = 1.0 \times 10^{-14}$$
$$(1.00 \times 10^{-7} + x)(x) = 1.0 \times 10^{-14}$$
$$x^2 + 1.00 \times 10^{-7}x - 1.0 \times 10^{-14} = 0$$

Using the quadratic equation to solve [see Appendix B and the use of Excel Solver (Section 6.11)],

$$x = \frac{-1.00 \times 10^{-7} \pm \sqrt{1.0 \times 10^{-14} + 4(1.0 \times 10^{-14})}}{2} = 6.2 \times 10^{-8} \ M$$

Therefore, the *total* H^+ concentration $= (1.00 \times 10^{-7} + 6.2 \times 10^{-8}) = 1.62 \times 10^{-7} \ M$:

$$pH = -\log 1.62 \times 10^{-7} = 7 - 0.21 = 6.79$$
$$pOH = 14.00 - 6.79 = 7.21$$

or, since $[OH^-] = x$,

$$pOH = -\log(6.2 \times 10^{-8}) = 8 - 0.79 = 7.21$$

Note that, owing to the presence of the added H^+, the ionization of water is suppressed by 38% by the common ion effect (Le Châtelier's principle). At higher acid (or base) concentrations, the suppression is even greater and the contribution from the water becomes negligible. This contribution can be considered negligible if the concentration of protons or hydroxyl ions from an acid or base is 10^{-6} M *or greater*.

The calculation in this example is more academic than practical because *carbon dioxide from the air dissolved in water exceeds these concentrations*. Since carbon dioxide in water forms an acid, extreme care would have to be taken to remove and keep this from the water, to have a solution of 10^{-7} M acid.

We usually neglect the contribution of water to the acidity in the presence of an acid since its ionization is suppressed in the presence of the acid.

● ●

7.4 pH at Elevated Temperatures: Blood pH

It is a convenient fact of nature for students and chemists who deal with acidity calculations and pH scales in aqueous solutions at room temperature that K_w is an integral number. At 100°C, for example, $K_w = 5.5 \times 10^{-13}$, and a *neutral solution* has

$$[H^+] = [OH^-] = \sqrt{5.5 \times 10^{-13}} = 7.4 \times 10^{-7} \ M$$
$$pH = pOH = 6.13$$
$$pK_w = 12.26 = pH + pOH$$

Not all measurements or interpretations are done at room temperature, however, and the temperature dependence of K_w must be taken into account (recall from Chapter 6 that equilibrium constants are temperature dependent). An important example is the pH of the body. The pH of blood at body temperature (37°C) is 7.35 to 7.45. This value represents a slightly more alkaline solution relative to neutral water than the same value would be at room temperature. At 37°C, $K_w = 2.5 \times 10^{-14}$ and $pK_w = 13.60$. The pH (and pOH) of a neutral solution is $13.60/2 = 6.80$. The hydrogen ion (and hydroxide ion) concentration is $\sqrt{2.5 \times 10^{-14}} =$

A neutral solution has pH < 7 above room temperature.

1.6×10^{-7} M. Since a neutral blood solution at 37°C would have pH 6.8, a blood pH of 7.4 is more alkaline at 37°C by 0.2 pH units than it would be at 25°C. This is important when one considers that a change of 0.3 pH units in the body is extreme.

The hydrochloric acid concentration in the stomach is about 0.1 to 0.02 M. Since pH = −log [H$^+$], the pH at 0.02 M would be 1.7. It will be the same *regardless of the temperature* since the hydrogen ion concentration is the same (neglecting solvent volume changes), and the same pH would be measured at either temperature. But, while the pOH would be 14.0 − 1.7 = 12.3 at 25°C, it is 13.6 − 1.7 = 11.9 at 37°C.

The pH of blood must be measured at body temperature to accurately reflect the status of blood buffers.

Not only does the temperature affect the ionization of water in the body and therefore change the pH of a neutral solution, it also affects the ionization constants of the acids and bases from which the buffer systems in the body are derived. As we shall see later in the chapter, this influences the pH of the buffers, and so a blood pH of 7.4 measured at 37°C will not be the same when measured at room temperature, in contrast to the stomach pH, whose value was determined by the concentration of a strong acid. For this reason, measurement of blood pH for diagnostic purposes is generally done at 37°C (see Chapter 13). (Neglecting changes in equilibrium constants of the blood buffer systems, the measured pH would be the same at 25 or 37°C—remembering to readjust the acidity scale at 37°C—but this is purely academic since the equilibrium constants do indeed change.)

7.5 Weak Acids and Bases—What Is the pH?

We have limited our calculations so far to strong acids and bases in which ionization is assumed to be complete. Since the concentration of H$^+$ or OH$^-$ is determined readily from the concentration of the acid or base, the calculations are straightforward. As seen in Equation 7.3, weak acids (or bases) are only partially ionized. While mineral (inorganic) acids and bases such as HCl, HClO$_4$, HNO$_3$, and NaOH are strong electrolytes that are totally ionized in water, most organic acids and bases, as found in clinical applications, are weak.

The ionization constant can be used to calculate the amount ionized and, from this, the pH. The acidity constant for acetic acid at 25°C is 1.75×10^{-5}:

$$\frac{[\text{H}^+][\text{OAc}^-]}{[\text{HOAc}]} = 1.75 \times 10^{-5} \tag{7.20}$$

When acetic acid ionizes, it dissociates to equal portions of H$^+$ and OAc$^-$ by such an amount that the computation on the left side of Equation 7.20 will always be equal to 1.75×10^{-5}:

$$\text{HOAc} \rightleftharpoons \text{H}^+ + \text{OAc}^- \tag{7.21}$$

If the original concentration of acetic acid is C and the concentration of ionized acetic acid species (H$^+$ and OAc$^-$) is x, then the final concentration for each species at equilibrium is given by

$$\text{HOAc} \rightleftharpoons \text{H}^+ + \text{OAc}^- \tag{7.22}$$
$$(C - x) \qquad x \qquad x$$

Example 7.7

Calculate the pH and pOH of a 1.00×10^{-3} M solution of acetic acid.

Solution

$$HOAc \rightleftharpoons H^+ + OAc^-$$

The concentrations of the various species are as follows:

	[HOAc]	[H⁺]	[OAc⁻]
Initial	1.00×10^{-3}	0	0
Change (x = mmol/mL HOAc ionized)	$-x$	$+x$	$+x$
Equilibrium	$1.00 \times 10^{-3} - x$	x	x

From Equation 7.20

$$\frac{(x)(x)}{1.00 \times 10^{-3} - x} = 1.75 \times 10^{-5}$$

The solution is that of a quadratic equation. If less than about 10 or 15% of the acid is ionized, the expression may be simplified by neglecting x compared with C (10^{-3} M in this case). This is an arbitrary (and not very demanding) criterion. The simplification applies if K_a *is smaller than about 0.01C*, that is, smaller than 10^{-4} at $C = 0.01$ M, 10^{-3} at $C = 0.1$ M, and so forth. Under these conditions, the error in calculation is 5% or less (results come out too high), and within the probable accuracy of the equilibrium constant. Our calculation simplifies to

If $C_{HA} > 100K_a$, x can be neglected compared to C_{HA}.

$$\frac{x^2}{1.00 \times 10^{-3}} = 1.75 \times 10^{-15}$$

$$x = 1.32 \times 10^{-4} \ M \equiv [H^+]$$

Therefore,

$$pH = -\log 1.32 \times 10^{-4} = 4 - \log 1.32 = 4 - 0.12 = 3.88$$
$$pOH = 14.00 - 3.88 = 10.12$$

The simplification in the calculation is not serious, particularly since equilibrium constants are often not known to a high degree of accuracy (frequently no better than $\pm 10\%$). In the above example, solution of the quadratic equation results in $[H^+] = 1.26 \times 10^{-4}$ M (5% less) and pH = 3.90. This pH is within 0.02 unit of that calculated using the simplification, which is near the limit of accuracy to which pH measurements can be made, and almost certainly as close a calculation as is justified in view of the experimental errors in K_a or K_b values and the fact that we are using concentrations rather than activities in the calculations. In our calculations, we also neglected the contribution of hydrogen ions from the ionization of water (which was obviously justified); this is generally permissible except for very dilute ($<10^{-6}$ M) or very weak ($K_a < 10^{-12}$) acids.

    Similar equations and calculations hold for weak bases.

The absolute accuracy of pH measurements is no better than 0.02 pH units. See Chapter 13.

Example 7.8

The basicity constant K_b for ammonia is 1.75×10^{-5} at 25°C. (It is only coincidental that this is equal to K_a for acetic acid.) Calculate the pH and pOH for a 1.00×10^{-3} M solution of ammonia.

Solution

$$NH_3 \quad\quad + H_2O \rightleftharpoons NH_4^+ + OH^-$$
$$(1.00 \times 10^{-3} - x) \quad\quad\quad x \quad\quad x$$

$$\frac{[NH_4^+][OH^-]}{[NH_3]} = 1.75 \times 10^{-5}$$

The same rule applies for the approximation applied for a weak acid. Thus,

$$\frac{(x)(x)}{1.00 \times 10^{-3}} = 1.75 \times 10^{-5}$$

$$x = 1.32 \times 10^{-4} \, M = [OH^-]$$

$$pOH = -\log 1.32 \times 10^{-4} = 3.88$$

$$pH = 14.00 - 3.88 = 10.12$$

7.6 Salts of Weak Acids and Bases—They Aren't Neutral

The hydrolysis of OAc$^-$ is no different than the "ionization" of NH$_3$ in Example 7.8.

The salt of a weak acid, for example, NaOAc, is a strong electrolyte, like (almost) all salts, and completely ionizes. In addition, the anion of the salt of a weak acid is a **Brønsted base**, which will accept protons. It partially hydrolyzes in water (a Brønsted acid) to form hydroxide ion and the corresponding undissociated acid. For example,

$$OAc^- + H_2O \rightleftharpoons HOAc + OH^- \quad\quad\quad (7.23)$$

The HOAc here is undissociated and therefore does not contribute to the pH. This ionization is also known as **hydrolysis** of the salt ion. Because it hydrolyzes, sodium acetate is a weak base (the conjugate base of acetic acid). The ionization constant for Equation 7.23 is equal to the basicity constant of the salt. The weaker the conjugate acid, the stronger the conjugate base, that is, the more strongly the salt will combine with a proton, as from the water, to shift the ionization in Equation 7.23 to the right. *Equilibria for these Brønsted bases are treated identically to the weak bases we have just considered.* We can write an equilibrium constant:

$$K_H = K_b = \frac{[HOAc][OH^-]}{[OAc^-]} \qu\quad\quad (7.24)$$

K_H is called the **hydrolysis constant** of the salt and is the same as the basicity constant. We will use K_b to emphasize that these salts are treated the same as for any other weak base.

The value of K_b can be calculated from K_a of acetic acid and K_w if we multiply both the numerator and denominator by $[H^+]$:

$$K_b = \frac{[HOAc]\,\overline{[OH^-]}}{[OAc^-]} \cdot \frac{[H^+]}{[H^+]}$$ (7.25)

The quantity inside the dashed line is K_w and the remainder is $1/K_a$. Hence,

$$K_b = \frac{K_w}{K_a} = \frac{1.0 \times 10^{-14}}{1.75 \times 10^{-5}} = 5.7 \times 10^{-10}$$ (7.26)

We see from the small K_b that the acetate ion is quite a weak base with only a small fraction of ionization. *The product of K_a of any weak acid and K_b of its conjugate base is always equal to K_w:*

$$\boxed{K_a K_b = K_w}$$ (7.27)

For any salt of a weak acid HA that hydrolyzes in water,

$$A^- + H_2O \rightleftharpoons HA + OH^-$$ (7.28)

$$\boxed{\frac{[HA][OH^-]}{[A^-]} = \frac{K_w}{K_a} = K_b}$$ (7.29)

The pH of such a salt (a Brønsted base) is calculated in the same manner as for any other weak base. When the salt hydrolyzes, it forms an equal amount of HA and OH^-. If the original concentration of A^- is C_{A^-}, then

$$\underset{(C_{A^-} - x)}{A^-} + H_2O \rightleftharpoons \underset{x}{HA} + \underset{x}{OH^-}$$ (7.30)

The quantity x can be neglected compared to C_{A^-} if $C_{A^-} > 100K_b$, which will generally be the case for such weakly ionized bases.

We can solve for the OH^- concentration using Equation 7.29:

$$\frac{[OH^-][OH^-]}{C_{A^-}} = \frac{K_w}{K_a} = K_b$$ (7.31)

Compare this with the algebraic setup in Example 7.8. They are identical:

$$\boxed{[OH^-] = \sqrt{\frac{K_w}{K_a} \cdot C_{A^-}} = \sqrt{K_b \cdot C_{A^-}}}$$ (7.32)

This equation holds only if $C_{A^-} > 100K_b$, and x can be neglected compared to C_{A^-}. If this is not the case, then the quadratic formula must be solved as for other bases in this situation.

Example 7.9

Calculate the pH of a 0.10 M solution of sodium acetate.

Solution

Compare this base "ionization" with that of NH_3, Example 7.8.

Write the equilibria

$$NaOAc \rightarrow Na^+ + OAc^- \text{ (ionization)}$$
$$OAc^- + H_2O \rightleftharpoons HOAc + OH^- \text{ (hydrolysis)}$$

Write the equilibrium constant

$$\frac{[HOAc][OH^-]}{[OAc^-]} = K_b = \frac{K_w}{K_a} = \frac{1.0 \times 10^{-14}}{1.75 \times 10^{-5}} = 5.7 \times 10^{-10}$$

Let x represent the concentration of HOAc and OH^- at equilibrium. Then, at equilibrium,

$$[HOAc] = [OH^-] = x$$
$$[OAc^-] = C_{OAc^-} - x = 0.10 - x$$

Since $C_{OAc^-} \gg K_b$, neglect x compared to C_{OAc^-}. Then,

$$\frac{(x)(x)}{0.10} = 5.7 \times 10^{-10}$$
$$x = \sqrt{5.7 \times 10^{-10} \times 0.10} = 7.6 \times 10^{-6} \ M$$

Compare this last step with Equation 7.32. Also, compare the entire setup and solution with those in Example 7.8. The HOAc formed is undissociated and does not contribute to the pH:

$$[OH^-] = 7.6 \times 10^{-6} \ M$$
$$[H^+] = \frac{1.0 \times 10^{-14}}{7.6 \times 10^{-6}} = 1.3 \times 10^{-9} \ M$$
$$pH = -\log 1.3 \times 10^{-9} = 9 - 0.11 = 8.89$$

Similar equations can be derived for the cations of salts of weak bases (the salts are completely dissociated). These are **Brønsted acids** and ionize (hydrolyze) in water:

$$BH^+ + H_2O \rightleftharpoons B + H_3O^+ \tag{7.33}$$

The B is undissociated and does not contribute to the pH. The acidity constant is

$$K_H = K_a = \frac{[B][H_3O^+]}{[BH^+]} \tag{7.34}$$

The acidity constant (hydrolysis constant) can be derived by multiplying the numerator and denominator by $[OH^-]$:

$$K_a = \frac{[B]}{[BH^+]} \boxed{\frac{[H_3O^+]}{} \cdot \frac{[OH^-]}{[OH^-]}} \tag{7.35}$$

Again, the quantity inside the dashed lines is K_w, while the remainder is $1/K_b$. Therefore,

$$\boxed{\frac{[B][H_3O^+]}{[BH^+]} = \frac{K_w}{K_b} = K_a} \tag{7.36}$$

and for NH_4^+,

$$K_a = \frac{K_w}{K_b} = \frac{1.0 \times 10^{-14}}{1.75 \times 10^{-5}} = 5.7 \times 10^{-10} \tag{7.37}$$

We could, of course, have derived K_a from Equation 7.27. It is again coincidence that the numerical value of K_a for NH_4^+ equals K_b for OAc^-.

The salt of a weak base ionizes to form equal amounts of B and H_3O^+ (H^+ if we disregard hydronium ion formation as was done previously). We can therefore solve for the hydrogen ion concentration (by assuming $C_{BH^+} > 100\,K_a$):

$$\frac{[H^+][H^+]}{C_{BH^+}} = \frac{K_w}{K_b} = K_a \tag{7.38}$$

$$\boxed{[H^+] = \sqrt{\frac{K_w}{K_b} \cdot C_{BH^+}} = \sqrt{K_a \cdot C_{BH^+}}} \tag{7.39}$$

Again, this equation only holds if $C_{BH^+} > 100 K_a$. Otherwise, the quadratic formula must be solved.

Example 7.10

Calculate the pH of a 0.25 M solution of ammonium chloride.

Solution

Write the equilibria

$$NH_4Cl \rightarrow NH_4^+ + Cl^- \quad \text{(ionization)}$$
$$NH_4^+ + H_2O \rightleftharpoons NH_4OH + H^+ \quad \text{(hydrolysis)}$$
$$(NH_4^+ + H_2O \rightleftharpoons NH_3 + H_3O^+)$$

Write the equilibrium constant

$$\frac{[NH_4OH][H^+]}{[NH_4^+]} = K_a = \frac{K_w}{K_b} = \frac{1.0 \times 10^{-14}}{1.75 \times 10^{-5}} = 5.7 \times 10^{-10}$$

Let x represent the concentration of [NH$_4$OH] and [H$^+$] at equilibrium. Then, at equilibrium,

$$[NH_4OH] = [H^+] = x$$
$$[NH_4^+] = C_{NH_4^+} - x = 0.25 - x$$

Since $C_{NH_4^+} \gg K_a$, neglect x compared to $C_{NH_4^+}$. Then,

$$\frac{(x)(x)}{0.25} = 5.7 \times 10^{-10}$$
$$x = \sqrt{5.7 \times 10^{-10} \times 0.25} = 1.2 \times 10^{-5} \, M$$

Compare this last step with Equation 7.39. Also, compare the entire setup and solution with those in Example 7.7. The NH$_4$OH formed is undissociated and does not contribute to the pH:

$$[H^+] = 1.2 \times 10^{-5} \, M$$
$$pH = -\log(1.2 \times 10^{-5}) = 5 - 0.08 = 4.92$$

• •

7.7 Buffers—Keeping the pH Constant (or Nearly So)

A **buffer** is defined as a solution that resists change in pH when a small amount of an acid or base is added or when the solution is diluted. This is very useful for maintaining the pH for a reaction at an optimum value. A buffer solution consists of a mixture of a weak acid and its conjugate base or a weak base and its conjugate acid at predetermined concentrations or ratios. That is, we have a mixture of a weak acid and its salt or a weak base and its salt. Consider an acetic acid–acetate buffer. The acid equilibrium that governs this system is

$$HOAc \rightleftharpoons H^+ + OAc^-$$

But now, since we have added a supply of acetate ions to the system (e.g., from sodium acetate), the hydrogen ion concentration is no longer equal to the acetate ion concentration. The hydrogen ion concentration is

$$[H^+] = K_a \frac{[HOAc]}{[OAc^-]} \tag{7.40}$$

Taking the negative logarithm of each side of this equation, we have

$$-\log[H^+] = -\log K_a - \log \frac{[HOAc]}{[OAc^-]} \tag{7.41}$$

$$pH = pK_a - \log \frac{[HOAc]}{[OAc^-]} \tag{7.42}$$

Upon inverting the last log term, it becomes positive:

$$pH = pK_a + \log \frac{[OAc^-]}{[HOAc]} \tag{7.43}$$

This form of the ionization constant equation is called the **Henderson–Hasselbalch equation**. It is useful for calculating the pH of a weak acid solution containing its salt. A general form can be written for a weak acid HA that ionizes to its salt, A^-, and H^+:

The pH of a buffer is determined by the ratio of the conjugate acid–base pair concentrations.

$$HA + \rightleftharpoons H^+ + A^- \qquad (7.44)$$

$$pH = pK_a + \log \frac{[A^-]}{[HA]} \qquad (7.45)$$

$$pH = pK_a + \log \frac{[\text{conjugate base}]}{[\text{acid}]} \qquad (7.46)$$

$$pH = pK_a + \log \frac{[\text{proton acceptor}]}{[\text{proton donor}]} \qquad (7.47)$$

 Example 7.11

Calculate the pH of a buffer prepared by adding 10 mL of 0.10 M acetic acid to 20 mL of 0.10 M sodium acetate.

Solution

We need to calculate the concentration of the acid and salt in the solution. The final volume is 30 mL:

$$M_1 \times \text{mL}_1 = M_2 \times \text{mL}_2$$

For HOAc,

$$0.10 \text{ mmol/mL} \times 10 \text{ mL} = M_{\text{HOAc}} \times 30 \text{ mL}$$
$$M_{\text{HOAc}} = 0.033 \text{ mmol/mL}$$

For OAc^-,

$$0.10 \text{ mmol/mL} \times 20 \text{ mL} = M_{\text{OAc}^-} \times 30 \text{ mL}$$
$$M_{\text{OAc}^-} = 0.067 \text{ mmol/mL}$$

Some of the HOAc dissociates to $H^+ + OAc^-$, and the equilibrium concentration of HOAc would be the amount added (0.033 M) minus the amount dissociated, while that of OAc^- would be the amount added (0.067 M) plus the amount of HOAc dissociated. However, *the amount of acid dissociated is very small*, particularly in the presence of the added salt (ionization suppressed by the common ion effect), and can be neglected. Hence, we can assume the added concentrations to be the equilibrium concentrations:

The ionization of the acid is suppressed by the salt and can be neglected.

$$pH = -\log K_a + \log \frac{[\text{proton acceptor}]}{[\text{proton donor}]}$$

$$pH = -\log(1.75 \times 10^{-5}) + \log \frac{0.067 \text{ mmol/mL}}{0.033 \text{ mmol/mL}}$$

$$= 4.76 + \log 2.0$$

$$= 5.06$$

We can use millimoles of acid and salt in place of molarity.

We could have shortened the calculation by recognizing that in the log term the volumes cancel. So we can take the ratio of millimoles only:

$$mmol_{HOAc} = 0.10 \text{ mmol/mL} \times 10 \text{ mL} = 1.0 \text{ mmol}$$

$$mmol_{OAc^-} = 0.10 \text{ mmol/mL} \times 20 \text{ mL} = 2.0 \text{ mmol}$$

$$H = 4.76 + \log \frac{2.0 \text{ mmol}}{1.0 \text{ mmol}} = 5.06$$

• •

The mixture of a weak acid and its salt may also be obtained by mixing an excess of weak acid with some strong base to produce the salt by neutralization, or by mixing an excess of salt with strong acid to produce the weak acid component of the buffer.

 **Example 7.12**

Calculate the pH of a solution prepared by adding 25 mL of 0.10 M sodium hydroxide to 30 mL of 0.20 M acetic acid (this would actually be a step in a typical titration).

Keep track of millimoles of reactants!   **Solution**

$$mmol \text{ HOAc} = 0.20 \text{ } M \times 30 \text{ mL} = 6.0 \text{ mmol}$$

$$mmol \text{ NaOH} = 0.10 \text{ } M \times 25 \text{ mL} = 2.5 \text{ mmol}$$

These react as follows:

$$HOAc + NaOH \rightleftharpoons NaOAc + H_2O$$

After reaction,

$$mmol \text{ NaOAc} = 2.5 \text{ mmol}$$

$$mmol \text{ HOAc} = 6.0 - 2.5 = 3.5 \text{ mmol}$$

$$pH = 4.76 + \log \frac{2.5}{3.5} = 4.61$$

• •

The **buffering mechanism** for a mixture of a weak acid and its salt can be explained as follows. The pH is governed by the logarithm of the ratio of the salt and acid:

$$pH = constant + \log \frac{[A^-]}{[HA]} \qquad (7.48)$$

Dilution does not change the ratio of the buffering species.

If the solution is diluted, the ratio remains constant, and so the pH of the solution does not change.[3] If a small amount of a strong acid is added, it will combine

_____

[3]In actuality, the pH will *increase* slightly because the activity coefficient of the salt has been increased by decreasing the ionic strength. The activity of an uncharged molecule (i.e., undissociated acid) is equal to its molarity (see Chapter 6), and so the ratio increases, causing a slight increase in pH. See the end of the chapter.

with an equal amount of the A^- to convert it to HA. That is, in the equilibrium $HA \rightleftharpoons H^+ + A^-$, Le Châtelier's principle dictates added H^+ will combine with A^- to form HA, with the equilibrium lying far to the left if there is an excess of A^-. The change in the ratio $[A^-]/[HA]$ is small and hence the change in pH is small. If the acid had been added to an unbuffered solution (e.g., a solution of NaCl), the pH would have decreased markedly. If a small amount of a strong base is added, it will combine with part of the HA to form an equivalent amount of A^-. Again, the change in the ratio is small.

The amount of acid or base that can be added without causing a large change in pH is governed by the **buffering capacity** of the solution. This is determined by the concentrations of HA and A^-. The higher their concentrations, the more acid or base the solution can tolerate. The buffer capacity (buffer intensity, buffer index) of a solution is defined as

The buffering capacity increases with the concentrations of the buffering species.

$$\beta = dC_{BOH}/d\text{pH} = -dC_{HA}/d\text{pH} \qquad (7.49)$$

where dC_{BOH} and dC_{HA} represent the number of moles per liter of strong base or acid, respectively, needed to bring about a pH change of dpH. The buffer capacity is a positive number. The larger it is, the more resistant the solution is to pH change. For weak acid/conjugate base buffer solutions of greater than 0.001 M, the buffer capacity is approximated by:

$$\beta = 2.303 \frac{C_{HA}C_{A^-}}{C_{HA} + C_{A^-}} \qquad (7.50)$$

where C_{HA} and C_{A^-} represent the analytical concentrations of the acid and its salt, respectively. Thus, if we have a mixture of 0.10 mol/L acetic acid and 0.10 mol/L sodium acetate, the buffer capacity is

$$\beta = 2.303 \frac{0.10 \times 0.10}{0.10 + 0.10} = 0.050 \text{ mol/L per pH}$$

If we add solid sodium hydroxide until it becomes 0.0050 mol/L, the change in pH is

$$d\text{pH} = dC_{BOH}/\beta = 0.0050/0.050 = 0.10 = \Delta\text{pH}$$

In addition to concentration, the buffering capacity is governed by the *ratio* of HA to A^-. It is *maximum* when the ratio is unity, that is, when the pH = pK_a:

The buffering capacity is maximum at pH = pK_a.

$$\text{pH} = \text{p}K_a + \log \frac{1}{1} = \text{p}K_a \qquad (7.51)$$

This corresponds to the midpoint of a titration of a weak acid. In general, the buffering capacity is satisfactory over a *pH range of p$K_a \pm 1$*. We will discuss the buffering capacity on a pictorial basis in Chapter 8, when the titration curves of weak acids are discussed.

Example 7.13

A buffer solution is 0.20 M in acetic acid and in sodium acetate. Calculate the change in pH upon adding 1.0 mL of 0.10 M hydrochloric acid to 10 mL of this solution.

Solution

Initially, the pH is equal to pK_a, because the ratio [OAc$^-$][HOAc] is unity. The pH is 4.76. To calculate the new pH, we need to determine the new concentrations of HOAc and OAc$^-$. We started with $10 \times 0.20 = 2.0$ mmol OAc$^-$ per 10 mL. We added $1.0 \times 0.10 = 0.10$ mmol of H$^+$ and therefore converted 0.10 mmol of OAc$^-$ to HOAc:

$$\text{mmol HOAc} = 2.0 + 0.1 = 2.1 \text{ mmol}$$
$$\text{mmol OAc}^- = 2.0 - 0.1 = 1.9 \text{ mmol}$$

These new amounts of acid and salt are contained in 11 mL but, again, the volumes cancel in our calculations:

$$\text{pH} = 4.76 + \log \frac{1.9 \text{ mmol/11 mL}}{2.1 \text{ mmol/11 mL}}$$

$$= 4.76 + \log \frac{1.9 \text{ mmol}}{2.1 \text{ mmol}} = 4.76 + \log 0.90$$

$$= 4.76 + (0.95 - 1) = 4.71$$

The change in pH is -0.05. This is rather small especially if we consider that had the HCl been added to an unbuffered neutral solution, the final concentration would have been approximately 10^{-2} M, and the pH would be 2.0

• •

Note that a buffer can resist a pH change, even when there is added an amount of strong acid or base greater (in moles) than the equilibrium amount of H$^+$ or OH$^-$ (in moles) in the buffer. For example, in Example 7.13, the pH of the buffer is 4.76 and [H$^+$] = 1.7×10^{-5} M, and millimoles H$^+$ = (1.7×10^{-5} mmol/mL) (10 mL) = 1.7×10^{-4} mmol (in equilibrium with the buffer components). We added 0.10 mmol H$^+$, well in excess of this. However, due to the reserve of buffer components (OAc$^-$ to react with H$^+$ in this case), the added H$^+$ is consumed so that the pH remains relatively constant, *so long as we do not exceed the amount of buffer reserve.*

Similar calculations apply for mixtures of a weak base and its salt. We can consider the equilibrium between the base B and its conjugate acid BH$^+$ and write a K_a for the conjugate (Brønsted) acid:

$$\text{BH}^+ = \text{B} + \text{H}^+ \tag{7.52}$$

$$K_a = \frac{[\text{B}][\text{H}^+]}{[\text{BH}^+]} = \frac{K_w}{K_b} \tag{7.53}$$

The logarithmic Henderson–Hasselbalch form is derived exactly as above:

$$[\text{H}^+] = K_a \cdot \frac{[\text{BH}^+]}{[\text{B}]} = \frac{K_w}{K_b} \cdot \frac{[\text{BH}^+]}{[\text{B}]} \tag{7.54}$$

$$-\log[\text{H}^+] = -\log K_a - \log \frac{[\text{BH}^+]}{[\text{B}]} = -\log \frac{K_w}{K_b} - \log \frac{[\text{BH}^+]}{[\text{B}]} \tag{7.55}$$

$$pH = pK_a + \log \frac{[B]}{[BH^+]} = (pK_w - pK_b) + \log \frac{[B]}{[BH^+]} \qquad (7.56)$$

$$pH = pK_a + \log \frac{[\text{proton acceptor}]}{[\text{proton donor}]} = (pK_w - pK_b) + \log \frac{[\text{proton acceptor}]}{[\text{proton donor}]} \qquad (7.57)$$

Since $pOH = pK_w - pH$, we can also write, by subtracting either Equation 7.56 or Equation 7.57 from pK_w,

$$pOH = pK_b + \log \frac{[BH^+]}{[B]} = pK_b + \log \frac{[\text{proton donor}]}{[\text{proton acceptor}]} \qquad (7.58)$$

A mixture of a weak base and its salt acts as a buffer in the same manner as a weak acid and its salt. When a strong acid is added, it combines with some of the base B to form the salt BH^+. Conversely, a base combines with BH^+ to form B. Since the change in the ratio will be small, the change in pH will be small. Again, the buffering capacity is maximum at a pH equal to $pK_a = 14 - pK_b$ (or at $pOH = pK_b$), with a useful range of $pK_a \pm 1$.

> $pK_a = 14 - pK_b$ for a weak base. The alkaline buffering capacity is maximum at $pOH = pK_b$ ($pH = pK_a$).

 Example 7.14

Calculate the volume of concentrated ammonia and the weight of ammonium chloride you would have to take to prepare 100 mL of a buffer at pH 10.00 if the final concentration of salt is to be 0.200 M.

Solution

We want 100 mL of 0.200 M NH_4Cl. Therefore, mmol NH_4Cl = 0.200 mmol/mL $\times$ 100 mL = 20.0 mmol

$$mg\ NH_4Cl = 20.0\ mmol \times 53.5\ mg/mmol = 1.07 \times 10^3\ mg$$

Therefore, we need 1.07 g NH_4Cl. We calculate the concentration of NH_3 by

$$pH = pK_a + \log \frac{[\text{proton acceptor}]}{[\text{proton donor}]}$$

$$= (14.00 - pK_b) - \log \frac{[NH_3]}{[NH_4^+]}$$

$$10.0 = (14.00 - 4.76) + \log \frac{[NH_3]}{0.200\ mmol/mL}$$

$$\log \frac{[NH_3]}{0.200\ mmol/mL} = 0.76$$

$$\frac{[NH_3]}{0.200\ mmol/mL} = 10^{0.76} = 5.8$$

$$[NH_3] = (0.200)(5.8) = 1.1_6\ mmol/mL$$

The molarity of concentrated ammonia is 14.8 M. Therefore,

$$100 \text{ mL} \times 1.1_6 \text{ mmol/mL} = 14.8 \text{ mmol/mL} \times \text{mL NH}_3$$
$$\text{mL NH}_3 = 7.8 \text{ mL}$$

Example 7.15

How many grams ammonium chloride and how many milliliters 3.0 M sodium hydroxide should be added to 200 mL water and diluted to 500 mL to prepare a buffer of pH 9.50 with a salt concentration of 0.10 M?

Solution

We need the ratio of $[NH_3]/[NH_4^+]$. From Example 7.14.

$$pH = pK_a + \log \frac{[NH_3]}{[NH_4^+]} = 9.24 + \log \frac{[NH_3]}{[NH_4^+]}$$

$$9.50 = 9.24 + \log \frac{[NH_3]}{[NH_4^+]}$$

$$\log \frac{[NH_3]}{[NH_4^+]} = 0.26$$

$$\frac{[NH_3]}{[NH_4^+]} = 10^{0.26} = 1.8$$

The final concentration of NH_4^+ is 0.10 M, so

$$[NH_3] = (1.8)(0.10) = 0.18 \ M$$
$$\text{mmol NH}_4^+ \text{ in final solution} = 0.10 \ M \times 500 \text{ mL} = 50 \text{ mmol}$$
$$\text{mmol NH}_3 \text{ in final solution} = 0.18 \ M \times 500 \text{ mL} = 90 \text{ mmol}$$

The NH_3 is formed from an equal number of millimoles of NH_4Cl. Therefore, a total of 50 + 90 = 140 mmol NH_4Cl must be taken:

$$\text{mg NH}_4Cl = 140 \text{ mmol} \times 53.5 \text{ mg/mmol} = 7.49 \times 10^3 \text{ mg} = 7.49 \text{ g}$$

The millimoles of NaOH needed are equal to the millimoles of NH_3:

$$3.0 \ M \times x \text{ mL} = 90 \text{ mmol}$$
$$x = 30 \text{ mL NaOH}$$

Select a buffer with a pK_a value near the desired pH.

We see that a buffer solution for a given pH is prepared by choosing a weak acid (or a weak base) and its salt, with a *pK_a value near the pH that we want*. There are a number of such acids and bases, and any pH region can be buffered by a proper choice of these. A weak acid and its salt give the best buffering in acid solution, and a weak base and its salt give the best buffering in alkaline solution. Some useful buffers for measurements in physiological solutions are described below. National Institute of Standards and Technology (NIST) buffers used for calibrating pH electrodes are described in Chapter 13.

See Chapter 13 for a list of NIST standard buffers.

You may have wondered why, in buffer mixtures, the salt does not react with water to hydrolyze as an acid or base. This is because the reaction is suppressed by the presence of the acid or base. In Equation 7.28, the presence of appreciable amounts of either HA or OH^- will suppress the ionization to a negligible amount. In Equation 7.33, the presence of either B or H_3O^+ will suppress the ionization.

Buffer salts do not hydrolyze appreciably.

7.8 Polyprotic Acids and Their Salts

Many acids or bases are polyfunctional, that is, have more than one ionizable proton or hydroxide ion. These substances ionize stepwise, and an equilibrium constant can be written for each step. Consider, for example, the ionization of phosphoric acid:

$$H_3PO_4 \rightleftharpoons H^+ + H_2PO_4^- \qquad K_{a1} = 1.1 \times 10^{-2} = \frac{[H^+][H_2PO_4^-]}{[H_3PO_4]} \qquad (7.59)$$

$$H_2PO_4^- \rightleftharpoons H^+ + HPO_4^{2-} \qquad K_{a2} = 7.5 \times 10^{-8} = \frac{[H^+][HPO_4^{2-}]}{[H_2PO_4^-]} \qquad (7.60)$$

$$HPO_4^{2-} \rightleftharpoons H^+ + PO_4^{3-} \qquad K_{a3} = 4.8 \times 10^{-13} = \frac{[H^+][PO_4^{3-}]}{[HPO_4^{2-}]} \qquad (7.61)$$

The stepwise K_a values of polyprotic acids get progressively smaller as the increased negative charge makes dissociation of the next proton more difficult.

Recall from Chapter 16 that the overall ionization is the sum of these individual steps and the overall ionization constant is the product of the individual ionization constants:

$$H_3PO_4 \rightleftharpoons 3H^+ + PO_4^{3-}$$

$$K_a = K_{a1}K_{a2}K_{a3} = 4.0 \times 10^{-22} = \frac{[H^+][PO_4^{3-}]}{[H_3PO_4]} \qquad (7.62)$$

The individual pK_a values are 1.96, 7.12, and 12.32, respectively, for pK_{a1}, pK_{a2}, and pK_{a3}. In order to make precise pH calculations, the contributions of protons from each ionization step must be taken into account. Exact calculation is difficult and requires a tedious iterative procedure since $[H^+]$ is unknown in addition to the various phosphoric acid species. See, for example, Refs. 7 and 10 for calculations.

In most cases, approximations can be made so that each ionization step can be considered individually. If the difference between successive ionization constants is at least 10^4, each proton can be differentiated in a titration, that is, each is titrated separately to give stepwise pH breaks in the titration curve. (If an ionization constant is less than about 10^{-8}, then the ionization is too small for a pH break to be exhibited in the titration curve—for example, the third proton for H_3PO_4.) Under these conditions, calculations are simplified because *the system can be considered as simply a mixture of three weak acids.*

We can titrate the first two protons of H_3PO_4 separately. The third is too weak to titrate.

BUFFER CALCULATIONS FOR POLYPROTIC ACIDS

The anion on the right side in each ionization step can be considered the salt (conjugate base) of the acid from which it is derived. That is, in Equation 7.59, $H_2PO_4^-$ is the salt of the acid H_3PO_4. In Equation 7.60, HPO_4^{2-} is the salt of the acid

We can prepare phosphate buffers with pH centered around 1.96 (pK_{a1}), 7.12 (pK_{a2}), and 12.32 (pK_{a3}).

$H_2PO_4^-$, and in Equation 7.61, PO_4^{3-} is the salt of the acid HPO_4^{2-}. So each of these pairs constitutes a buffer system, and orthophosphate buffers can be prepared over a wide pH range. The optimum buffering capacity of each pair occurs at a pH corresponding to its pK_a. The $HPO_4^{2-}/H_2PO_4^-$ couple is an effective buffer system in the blood (see below).

 Example 7.16

The pH of blood is 7.40. What is the ratio of $[HPO_4^{2-}]/[H_2PO_4^-]$ in the blood (assume 25°C)?

Solution

$$pH = pK_a + \log \frac{[\text{proton acceptor}]}{[\text{proton donor}]}$$

$$pK_{a2} = 7.12$$

Therefore,

$$pH = 7.12 + \log \frac{[HPO_4^{2-}]}{[H_2PO_4^-]}$$

$$7.40 = 7.12 + \log \frac{[HPO_4^{2-}]}{[H_2PO_4^-]}$$

$$\frac{[HPO_4^{2-}]}{[H_2PO_4^-]} = \frac{1.9}{1}$$

DISSOCIATION CALCULATIONS FOR POLYPROTIC ACIDS

Because the individual ionization constants are sufficiently different, *the pH of a solution of H_3PO_4 can be calculated by treating it just as we would any weak acid.* The H^+ from the first ionization step effectively suppresses the other two ionization steps, so that the H^+ contribution from them is negligible compared to the first ionization. The quadratic equation must, however, be solved because K_{a1} is relatively large.

Example 7.17

Calculate the pH of a 0.100 *M* H_3PO_4 solution.

Solution

$$H_3PO_4 \approx H^+ + H_2PO_4^-$$
$$0.100 - x \qquad x \qquad x$$

From Equation 7.59,

$$\frac{(x)(x)}{0.100 - x} = 1.1 \times 10^{-2}$$

In order to neglect x, C should be $\geq 100K_a$. Here, it is only 10 times as large. Therefore, use the quadratic equation to solve:

$$x^2 + 0.011x - 1.1 \times 10^{-3} = 0$$

$$x = \frac{-0.011 \pm \sqrt{(0.011)^2 - 4(-1.1 \times 10^{-3})}}{2}$$

$$x = [H^+] = 0.028 \ M$$

Treat H_3PO_4 as a monoprotic acid. But x can't be neglected compared to C.

The acid is 28% ionized:

$$pH = -\log 2.8 \times 10^{-2} = 2 - 0.45 = 1.55$$

• •

We can determine if our assumption that the only important source of protons is H_3PO_4 was a realistic one. $H_2PO_4^-$ would be the next most likely source of protons. From Equation 7.60, $[HPO_4^{2-}] = K_{a2} [H_2PO_4^-]/[H^+]$. Assuming the concentrations of $H_2PO_4^-$ and H^+ as a first approximation are 0.028 M as calculated, then $[HPO_4^{2-}] \approx K_{a2} = 7.5 \times 10^{-8} \ M$. This is very small compared to 0.028 M $H_2PO_4^-$, and so further dissociation is indeed insignificant. We were justified in our approach.

FRACTIONS OF DISSOCIATING SPECIES AT A GIVEN pH: α VALUES—HOW MUCH OF EACH SPECIES?

Often, it is of interest to know the distribution of the different species of a polyprotic acid as a function of pH, that is, at known hydrogen ion concentration as in a buffered solution.

Consider, for example, to the dissociation of phosphoric acid. The equilibria are given in Equations 7.59 to 7.61. At any given pH, all the four phosphoric acid species will coexist in equilibrium with one another, although the concentrations of some may be diminishingly small. By changing the pH, the equilibria shift, and the relative concentrations change. It is possible to derive general equations for calculating the fraction of the acid that exists in a given form, from the given hydrogen ion concentration.

For a given total **analytical concentration** of phosphoric acid, $C_{H_3PO_4}$, we can write

$$C_{H_3PO_4} = [PO_4^{3-}] + [HPO_4^{2-}] + [H_2PO_4^-] + [H_3PO_4] \quad (7.63)$$

H_3PO_4, $H_2PO_4^-$, HPO_4^{2-}, and PO_4^{3-} all exist together in equilibrium. The pH determines the fraction of each.

where the terms on the right-hand side of the equation represent the **equilibrium concentrations** of the individual species. We presumably know the initial total concentration $C_{H_3PO_4}$ and wish to find the fractions or concentrations of the individual species at equilibrium.

We define

$$\alpha_0 = \frac{[H_3PO_4]}{C_{H_3PO_4}} \qquad \alpha_1 = \frac{[H_2PO_4^-]}{C_{H_3PO_4}} \qquad \alpha_2 = \frac{[HPO_4^{2-}]}{C_{H_3PO_4}}$$

$$\alpha_3 = \frac{[PO_4^{3-}]}{C_{H_3PO_4}} \qquad \alpha_0 + \alpha_1 + \alpha_2 + \alpha_3 = 1$$

where the α's are the **fractions** of each species present at equilibrium. Note that the subscripts denote the number of dissociated protons or the charge on the species. We can use Equation 7.63 and the equilibrium constant expressions 7.59 through 7.61 to obtain an expression for $C_{H_3PO_4}$ in terms of the desired species. This is substituted into the appropriate equation to obtain α in terms of $[H^+]$ and the equilibrium constants. In order to calculate α_0, for example, we can rearrange Equations 7.59 through 7.61 to solve for all the species except $[H_3PO_4]$ and substitute into Equation 7.63:

$$[PO_4^{3-}] = \frac{K_{a3}[HPO_4^{2-}]}{[H^+]} \tag{7.64}$$

$$[HPO_4^{2-}] = \frac{K_{a2}[H_2PO_4^-]}{[H^+]} \tag{7.65}$$

$$[H_2PO_4^-] = \frac{K_{a1}[H_3PO_4]}{[H^+]} \tag{7.66}$$

We want all these to contain only $[H_3PO_4]$ (and $[H^+]$, the variable). We can substitute Equation 7.66 for $[H_2PO_4^-]$ in Equation 7.65:

$$[HPO_4^{2-}] = \frac{K_{a1}K_{a2}[H_3PO_4]}{[H^+]^2} \tag{7.67}$$

And we can substitute Equation 7.67 into Equation 7.64 for $[HPO_4^{2-}]$:

$$[PO_4^{3-}] = \frac{K_{a1}K_{a2}K_{a3}[H_3PO_4]}{[H^+]^3} \tag{7.68}$$

Finally, we can substitute 7.66 through 7.68 in Equation 7.63:

$$C_{H_3PO_4} = \frac{K_{a1}K_{a2}K_{a3}[H_3PO_4]}{[H^+]^3} + \frac{K_{a1}K_{a2}[H_3PO_4]}{[H^+]^2} + \frac{K_{a1}[H_3PO_4]}{[H^+]} + [H_3PO_4] \tag{7.69}$$

We can either divide each side of this expression by $[H_3PO_4]$ to obtain $1/\alpha_0$ or we can substitute the expression into the denominator of the above α_0 expression to obtain a value for α_0 ($[H_3PO_4]$ cancels). Doing the former, we have

$$\frac{1}{\alpha_0} = \frac{K_{a1}K_{a2}K_{a3}}{[H^+]^3} + \frac{K_{a1}K_{a2}}{[H^+]^2} + \frac{K_{a1}}{[H^+]} + 1 \tag{7.70}$$

or doing the latter, we have

$$\alpha_0 = \frac{1}{(K_{a1}K_{a2}K_{a3}/[H^+]^3) + (K_{a1}K_{a2}/[H^+]^2) + (K_{a1}/[H^+]) + 1} \tag{7.71}$$

either of which can be rearranged to

Use this equation to calculate the fraction of H_3PO_4 in solution.

$$\boxed{\alpha_0 = \frac{[H^+]^3}{[H^+]^3 + K_{a1}[H^+]^2 + K_{a1}K_{a2}[H^+] + K_{a1}K_{a2}K_{a3}}} \tag{7.72}$$

Similar approaches can be taken to obtain expressions for the other α's. For α_1, for example, the equilibrium constant expressions would be solved for all species in terms of $[H_2PO_4^-]$ and substituted into Equation 7.63 to obtain an expression for $C_{H_3PO_4}$ containing only $[H_2PO_4^-]$ and $[H^+]$, from which α_1 is calculated. The results for the other α's are

$$\alpha_1 = \frac{K_{a1}[H^+]^2}{[H^+]^3 + K_{a1}[H^+]^2 + K_{a1}K_{a2}[H^+] + K_{a1}K_{a2}K_{a3}} \qquad (7.73)$$

$$\alpha_2 = \frac{K_{a1}K_{a2}[H^+]}{[H^+]^3 + K_{a1}[H^+]^2 + K_{a1}K_{a2}[H^+] + K_{a1}K_{a2}K_{a3}} \qquad (7.74)$$

$$\alpha_3 = \frac{K_{a1}K_{a2}K_{a3}}{[H^+]^3 + K_{a1}[H^+]^2 + K_{a1}K_{a2}[H^+] + K_{a1}K_{a2}K_{a3}} \qquad (7.75)$$

Derive these equations in Problem 58.

Note that all have the *same denominator* and that *the sum of the numerators equals the denominator*. See Problem 58 for a more detailed derivation of the other α's.

Example 7.18

Calculate the equilibrium concentration of the different species in a 0.10 M phosphoric acid solution at pH 3.00 ($[H^+] = 1.0 \times 10^{-3}\ M$).

Solution

Substituting into Equation 7.72,

$$\alpha_0 = \frac{(1.0 \times 10^{-3})^3}{\begin{array}{c}(1.0 \times 10^{-3})^3 + (1.1 \times 10^{-2})(1.0 \times 10^{-3})^2 + (1.1 \times 10^{-2})(7.5 \times 10^{-8})(1.0 \times 10^{-3}) \\ + (1.1 \times 10^{-2})(7.5 \times 10^{-8})(4.8 \times 10^{-13})\end{array}}$$

$$= \frac{1.0 \times 10^{-9}}{1.2 \times 10^{-8}} = 8.3 \times 10^{-2}$$

$[H_3PO_4] = C_{H_3PO_4}\ \alpha_0 = 0.10 \times 8.3 \times 10^{-2} = 8.3 \times 10^{-3}\ M$

Similarly,

$$\alpha_1 = 0.92$$
$$[H_2PO_4^-] = C_{H_3PO_4}\ \alpha_1 = 0.10 \times 0.92 = 9.2 \times 10^{-2}\ M$$
$$\alpha_2 = 6.9 \times 10^{-5}$$
$$[HPO_4^{2-}] = C_{H_3PO_4}\ \alpha_2 = 0.10 \times 6.9 \times 10^{-5} = 6.9 \times 10^{-6}\ M$$
$$\alpha_3 = 3.3 \times 10^{-14}$$
$$[PO_4^{3-}] = C_{H_3PO_4}\ \alpha_3 = 0.10 \times 3.3 \times 10^{-14} = 3.3 \times 10^{-15}\ M$$

We see that at pH 3, the majority (91%) of the phosphoric acid exists as $H_2PO_4^-$ and 8.3% exists as H_3PO_4. Only $3.3 \times 10^{-12}\%$ exists as PO_4^{3-}!

We can prepare a spreadsheet to calculate the fraction of each species as a function of pH. Formulas and calculations are shown in the spreadsheet, and Figure 7.1 shows the corresponding α versus pH plot. The K_a values are entered in cells B4,

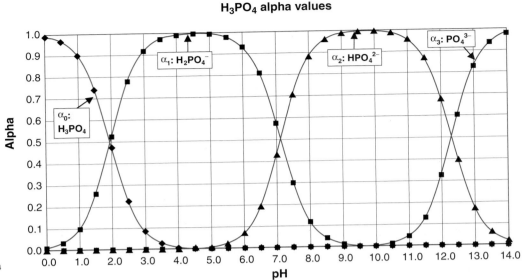

H_3PO_4 alpha values

Fig. 7.1. Fractions of H_3PO_4 species as a function of pH.

The overlapping curves represent buffer regions. The pH values where α_1 and α_2 are 1 represent the end points in titrating H_3PO_4.

See Section 7.12 for a way of representing these plots as straight lines (log–log plots).

D4, and F4. The pH values are entered in column A. All the formulas needed for each cell are listed at the bottom of the spreadsheet, and they are initially entered in the boldfaced cells. The formula for calculating the corresponding hydrogen ion concentration (used in the α calculation) is entered in cell B6. The formula for the denominator used for each α calculation is entered in cell C6. Note that the constants are entered as absolute values, while the hydrogen ion concentration is a relative value. The formulas for the three α calculations are entered in cells D6, E6, and F6. All formulas are copied down to row 34.

The curves are prepared using Chart Wizard by sequentially plotting four series of A6:A34 vs. D6:D34, E6:E34, F6:F34, and G6:G34. The spreadsheet (Sheet 1) and chart (Chart 1) are reproduced in your CD, Chapter 7. You can use the spreadsheet to reconstruct the plot of Figure 7.1. Save Sheet 1 to your desktop. Open Chart Wizard and select XY Scatter with data points connected by a smooth line (the second chart subunit down on the left). Click on Next and for Data Range, enter A6:A34,D6:D34 (Chart Wizard automatically formats it to: Sheet1!$A$6:$A$34,$D$6:$D$34). (Sheet 1 is the one from which the Chart is being plotted. The ! is a delineator that separates this from the entered cells, and when X- and Y-Values are entered in separate series, you will enter Sheet1! prior to entering the cell ranges.) Click on Series. Name series 1 alpha 0 (H3P04). Copy the X-Values to your clipboard so you can paste it in subsequent series. Click on Add to create a new Series 2 and note that you must enter X-Values and Y-Values. Paste the X-Values from the clipboard. For Y-Values, type: =Sheet1!E6:E34 (this is automatically formatted to absolute values). Name the series alpha 1 (H2P04-). Repeat for columns F and G, to create Series 3 and 4 for α_2 and α_3. Under Chart Options, the Title and the names of the X- and Y-axes are entered. Major and Minor Gridlines are added. Click on Finish, and you will see the plotted α-values. The format of the labels can be changed by clicking on Format in the header (the numbers, which may be in scientific notation, can be changed to "Number"). The Scale can be adjusted. The Series names can be moved to inside the chart by highlighting and dragging to the chart.

A plot of the fractions of each phosphoric acid species as a function of pH is given in Figure 7.1. This figure illustrates how the ratios of the four phosphoric acid species change as the pH is adjusted, for example, in titrating H_3PO_4 with NaOH. While some appear to go to zero concentration above or below certain pH

	A	B	C	D	E	F	G
1	Calculation of alpha values for H_3PO_4 vs. pH.						
2	Alpha (α_i) denominator = $[H^+]^3 + K_{a1}[H^+]^2 + K_{a1}K_{a2}[H^+] + K_{a1}K_{a2}K_{a3}$						
3	Numerators: $\alpha_0 = [H^+]^3$; $\alpha_1 = K_{a1}[H^+]^2$; $\alpha_2 = K_{a1}K_{a2}[H^+]$; $\alpha_3 = K_{a1}K_{a2}K_{a3}$						
4	$K_{a1}=$	1.10E-02		$K_{a2}=$	7.50E-08	$K_{a3}=$	4.80E-13
5	pH	$[H^+]$	Denominator	α_0	α_1	α_2	α_3
6	0.0	1	**1.01E+00**	**9.89E-01**	**1.09E-02**	**8.16E-10**	**3.92E-22**
7	0.5	0.316228	3.27E-02	9.66E-01	3.36E-02	7.97E-09	1.21E-20
8	1.0	0.1	1.11E-03	9.01E-01	9.91E-02	7.43E-08	3.57E-19
9	1.5	0.031623	4.26E-05	7.42E-01	2.58E-01	6.12E-07	9.29E-18
10	2.0	0.01	2.10E-06	4.76E-01	5.24E-01	3.93E-06	1.89E-16
11	2.5	0.003162	1.42E-07	2.23E-01	7.77E-01	1.84E-05	2.80E-15
12	3.0	0.001	1.20E-08	8.33E-02	9.17E-01	6.87E-05	3.30E-14
13	3.5	0.000316	1.13E-09	2.79E-02	9.72E-01	2.30E-04	3.50E-13
14	4.0	0.0001	1.11E-10	9.00E-03	9.90E-01	7.43E-04	3.56E-12
15	4.5	3.16E-05	1.11E-11	2.86E-03	9.95E-01	2.36E-03	3.58E-11
16	5.0	0.00001	1.11E-12	9.02E-04	9.92E-01	7.44E-03	3.57E-10
17	5.5	3.16E-06	1.13E-13	2.81E-04	9.77E-01	2.32E-02	3.52E-09
18	6.0	0.000001	1.18E-14	8.46E-05	9.30E-01	6.98E-02	3.35E-08
19	6.5	3.16E-07	1.36E-15	2.32E-05	8.08E-01	1.92E-01	2.91E-07
20	7.0	1E-07	1.93E-16	5.19E-06	5.71E-01	4.29E-01	2.06E-06
21	7.5	3.16E-08	3.71E-17	8.53E-07	2.97E-01	7.03E-01	1.07E-05
22	8.0	1E-08	9.35E-18	1.07E-07	1.18E-01	8.82E-01	4.24E-05
23	8.5	3.16E-09	2.72E-18	1.16E-08	4.05E-02	9.59E-01	1.46E-04
24	9.0	1E-09	8.36E-19	1.20E-09	1.32E-02	9.86E-01	4.73E-04
25	9.5	3.16E-10	2.62E-19	1.21E-10	4.19E-03	9.94E-01	1.51E-03
26	10.0	1E-10	8.30E-20	1.20E-11	1.33E-03	9.94E-01	4.77E-03
27	10.5	3.16E-11	2.65E-20	1.19E-12	4.15E-04	9.85E-01	1.49E-02
28	11.0	1E-11	8.65E-21	1.16E-13	1.27E-04	9.54E-01	4.58E-02
29	11.5	3.16E-12	3.00E-21	1.05E-14	3.66E-05	8.68E-01	1.32E-01
30	12.0	1E-12	1.22E-21	8.19E-16	9.01E-06	6.76E-01	3.24E-01
31	12.5	3.16E-13	6.57E-22	4.81E-17	1.67E-06	3.97E-01	6.03E-01
32	13.0	1E-13	4.79E-22	2.09E-18	2.30E-07	1.72E-01	8.28E-01
33	13.5	3.16E-14	4.22E-22	7.49E-20	2.61E-08	6.18E-02	9.38E-01
34	14.0	1E-14	4.04E-22	2.47E-21	2.72E-09	2.04E-02	9.80E-01
35	Formulas for cells in **boldface**:						
36	**Cell B6** = $[H^+]$ =	10^-A6					
37	**Cell C6**=denom.=	B6^3+$B$4*B6^2+$B$4*$D$4*B6+$B$4*$D$4*$F$4					
38	**Cell D6** = α_0 =	B6^3/C6					
39	**Cell E6** = α_1 =	($B$4*B6^2)/C6					
40	**Cell F6** = α_2 =	($B$4*$D$4*B6)/C6					
41	**Cell G6** = α_3 =	($B$4*$D$4*$F$4)/C6					
42	Copy each formula down through Cell 34						
43	Plot A6:A34 vs. D6:D34, E6:E34, F6:F34, and G6:G34 (series 1, 2, 3, and 4)						

values, they are not really zero, but diminishingly small. For example, we saw in Example 7.18 that at pH 3.00, the concentration of the PO_4^{3-} ion for 0.1 M H_3PO_4 is only 3.3×10^{-15} M, but it is indeed present in equilibrium. The pH regions where two curves overlap (with appreciable concentrations) represent regions in which buffers may be prepared using those two species. For example, mixtures of H_3PO_4 and $H_2PO_4^-$ can be used to prepare buffers around pH 2.0 ± 1, mixtures of $H_2PO_4^-$ and HPO_4^{2-} around pH 7.1 ± 1, and mixtures of HPO_4^{2-} and PO_4^{3-} around pH 12.3 ± 1. The pH values at which the fraction of a species is essentially 1.0 correspond to the end points in the titration of phosphoric acid with a strong base, that is, $H_2PO_4^{2-}$ at the first end point (pH 4.5), HPO_4^{2-} at the second end point (pH 9.7).

Equation 7.69 could be used for a rigorous calculation of the hydrogen ion concentration from dissociation of a phosphoric acid solution at a given H_3PO_4

concentration (no other added H^+), but this involves tedious iterations. As a first approximation, $[H^+]$ could be calculated from K_{a1} as in Example 7.17, assuming that only the first dissociation step of phosphoric acid was significant. (This is, in fact, what we did in that example). The first calculated $[H^+]$ could then be substituted in Equation 7.69 to calculate a second approximation of $[H_3PO_4]$, which would be used for a second iterative calculation of $[H^+]$ using K_{a1}, and so forth, until the concentration was constant.

SALTS OF POLYPROTIC ACIDS—ACID OR BASE?

Salts of acids such as H_3PO_4 may be acidic or basic. The protonated salts possess both acidic and basic properties ($H_2PO_4^-$, HPO_4^{2-}), while the unprotonated salt is simply a Brønsted base that hydrolyzes (PO_4^{3-}).

1. Amphoteric Salts. $H_2PO_4^-$ possesses both acidic and basic properties. That is, it is **amphoteric**. It ionizes as a weak acid and it also is a Brønsted base that hydrolyzes:

H₂PO₄⁻ acts as both an acid and a base. See the end of Section 7.12 for how to estimate the extent of each reaction using log–log diagrams.

$$H_2PO_4^- \rightleftharpoons H^+ + HPO_4^{2-} \qquad K_{a2} = \frac{[H^+][HPO_4^{2-}]}{[H_2PO_4^-]} = 7.5 \times 10^{-8} \quad (7.76)$$

$$H_2PO_4^- + H_2O \rightleftharpoons H_3PO_4 + OH^- \qquad K_b = \frac{K_w}{K_{a1}} = \frac{[H_3PO_4][OH^-]}{[H_2PO_4^-]}$$

$$= \frac{1.00 \times 10^{-14}}{1.1 \times 10^{-2}} = 9.1 \times 10^{-13} \quad (7.77)$$

The solution could, hence, be either alkaline or acidic, depending on which ionization is more extensive. Since K_{a2} for the first ionization is nearly 10^5 greater than K_b for the second ionization, the solution in this case will obviously be acidic.

An expression for the hydrogen ion concentration in a solution of an ion such as $H_2PO_4^-$ can be obtained as follows. The total hydrogen ion concentration is equal to the amounts produced from the ionization equilibrium in Equation 7.76 and the ionization of water, less the amount of OH^- produced from the hydrolysis in Equation 7.77. We can write, then,

$$C_{H^+} = [H^+]_{total} = [H^+]_{H_2O} + [H^+]_{H_2PO_4^-} - [OH^-]_{H_2PO_4^-} \quad (7.78)$$

or

$$[H^+] = [OH^-] + [HPO_4^{2-}] - [H_3PO_4] \quad (7.79)$$

We have included the contribution from water since it will not be negligible if the pH of the salt solution happens to be near 7—although in this particular case, the solution will be acid, making the water ionization negligible.

We can solve for $[H^+]$ by substituting expressions in the right-hand side of Equation 7.79 from the equilibrium constant expressions 7.59 and 7.60 and K_w to eliminate all but $[H_2PO_4^-]$, the experimental variable, and $[H^+]$:

$$[H^+] = \frac{K_w}{[H^+]} + \frac{K_{a2}[H_2PO_4^-]}{[H^+]} - \frac{[H_2PO_4^-][H^+]}{K_{a1}} \quad (7.80)$$

from which (by multiplying each side of the equation by $[H^+]$, collecting the terms containing $[H^+]^2$ on the left side, and solving for $[H^+]^2$)

$$[H^+]^2 = \frac{K_w + K_{a2}[H_2PO_4^-]}{1 + \dfrac{[H_2PO_4^-]}{K_{a1}}} \tag{7.81}$$

$$[H^+] = \sqrt{\frac{K_{a1}K_w + K_{a1}K_{a2}[H_2PO_4^-]}{K_{a1} + [H_2PO_4^-]}} \tag{7.82}$$

That is, for the general case HA^-,

$$[H^+] = \sqrt{\frac{K_{a1}K_w + K_{a1}K_{a2}[HA^-]}{K_{a1} + [HA^-]}} \tag{7.83}$$

For HA^{2-}, substitute $[HA^{2-}]$ for $[HA^-]$, K_{a2} for K_{a1}, and K_{a3} for K_{a2}.

This equation is valid for any salt HA^- derived from an acid H_2A (or for HA^{2-} derived from H_2A^-, etc.) where $[H_2PO_4^-]$ is replaced by $[HA^-]$.

If we assume that the equilibrium concentration $[HA^-]$ is equal to the concentration of salt added, that is, that the extent of ionization and hydrolysis is fairly small, then this value along with the constants can be used for the calculation of $[H^+]$. This assumption is reasonable if the two equilibrium constants (K_{a1} and K_b) involving the salt HA^- are small and the solution is not too dilute. In many cases, $K_{a1}K_w \ll K_{a1}K_{a2}[HA^-]$ in the numerator and can be neglected. This is the equation we would have obtained if we had neglected the dissociation of water. Furthermore, if $K_{a1} \ll [HA^-]$ in the denominator, the equation simplifies to

$$[H^+] = \sqrt{K_{a1}K_{a2}} \tag{7.84}$$

For HA^{2-}, $[H^+] = \sqrt{K_{a2}K_{a3}}$.

Therefore, if the assumptions hold, the pH of a solution of $H_2PO_4^-$ is independent of its concentration! This approximation is adequate for our purposes. The equation generally applies if there is a large difference between K_{a1} and K_{a2}. For the case of $H_2PO_4^-$, then,

$$[H^+] \approx \sqrt{K_{a1}K_{a2}} = \sqrt{1.1 \times 10^{-2} \times 7.5 \times 10^{-8}} = 2.9 \times 10^{-5}\ M \tag{7.85}$$

and the pH is approximately independent of the salt concentration (pH $\approx$ 4.54). This would be the approximate pH of a NaH_2PO_4 solution.

Similarly, HPO_4^{2-} is both an acid and a base. The K values involved here are K_{a2} and K_{a3} of H_3PO_4 ($H_2PO_4^- \equiv H_2A$ and $HPO_4^{2-} \equiv HA^-$). Since $K_{a2} \gg K_{a3}$, the pH of a Na_2HPO_4 solution can be calculated from

$$[H^+] \approx \sqrt{K_{a2}K_{a3}} = \sqrt{7.5 \times 10^{-8} \times 4.8 \times 10^{-13}} = 1.9 \times 10^{-10} \tag{7.86}$$

and the calculated pH is 9.72. Because the pH of amphoteric salts of this type is essentially independent of concentration, the salts are useful for preparing solutions of known pH for standardizing pH meters. For example, potassium acid phthalate, $KHC_8H_4O_2$, gives a solution of pH 4.0 at 25°C. However, these salts are poor buffers against acids or bases; their pH does not fall in the buffer region but occurs at the end point of a titration curve, where the pH changes markedly, that is, when a proton has just been neutralized.

KHP is a NIST "standard buffer" (Chapter 13). The pH of its solution is fixed, but it is not buffered.

2. Unprotonated Salt. Unprotonated phosphate is a fairly strong Brønsted base in solution and ionizes as follows:

$$PO_4^{3-} + H_2O \rightleftharpoons HPO_4^{2-} + OH^- \qquad K_b = \frac{K_w}{K_{a3}} \qquad (7.87)$$

The constant K_{a3} is very small, and so the equilibrium lies significantly to the right. Because $K_{a3} \ll K_{a2}$, hydrolysis of HPO_4^{2-} is suppressed by the OH^- from the first step, and the pH of PO_4^{3-} can be calculated just as for a salt of a monoprotic weak acid. However, because K_{a3} is so small, K_b is relatively large, and the amount of OH^- is not negligible compared with the initial concentration of PO_4^{3-} (C_B), and the quadratic equation must be solved, that is, PO_4^{3-} is quite a strong base.

 Example 7.19

Calculate the pH of 0.100 M Na_3PO_4.

Solution

$$PO_4^{3-} + H_2O \rightleftharpoons HPO_4^{2-} + OH^-$$
$$0.100 - x \qquad\qquad x \qquad x$$

$$\frac{[HPO_4^{2-}][OH^-]}{[PO_4^{3-}]} = K_b = \frac{K_w}{K_{a3}} = \frac{1.0 \times 10^{-14}}{4.8 \times 10^{-13}} = 0.020$$

$$\frac{(x)(x)}{0.100 - x} = \frac{1.0 \times 10^{-14}}{4.8 \times 10^{-13}} = 0.020$$

The concentration is only five times K_b, so the quadratic equation is used:

$$x^2 + 0.020x - 2.0 \times 10^{-3} = 0$$

$$x = \frac{-0.200 \pm \sqrt{(0.020)^2 - 4(-2.0 \times 10^{-3})}}{2}$$

$$x = [OH^-] = 0.036 \ M$$
$$pH = 12.56$$

The dissociation (hydrolysis) is 36% complete, and phosphate is quite a strong base. Perform this calculation using the Solver setup in your CD, Chapter 6, and see if you get the same result.

 Example 7.20

EDTA is a polyprotic acid with four protons (H_4Y). Calculate the hydrogen ion concentration of a 0.0100 M solution of Na_2EDTA (Na_2H_2Y).

Solution

The equilibria are

$$H_2Y^{2-} \rightleftharpoons H^+ + HY^{3-} \qquad K_{a3} = 6.9 \times 10^{-7}$$

and

$$H_2Y^{2-} + H_2O \rightleftharpoons H_3Y^- + OH^- \qquad K_b = \frac{K_w}{K_{a2}} = \frac{1.0 \times 10^{-14}}{2.2 \times 10^{-3}}$$

H_2Y^{2-} is the equivalent of HA^-, and H_3Y^- is the equivalent of H_2A. The equilibrium constants involved are K_{a2} and K_{a3} (the former for the conjugate acid H_3Y^- of the hydrolyzed salt). Thus,

$$[H^+] = \sqrt{K_{a2}K_{a3}} = \sqrt{(2.2 \times 10^{-3})(6.9 \times 10^{-7})}$$
$$= 3.9 \times 10^{-5} \ M$$

7.9 Physiological Buffers—They Keep You Alive

The pH of the blood in a healthy individual remains remarkably constant at 7.35 to 7.45. This is because the blood contains a number of buffers that protect against pH change due to the presence of acidic or basic metabolites. From a physiological viewpoint, a change of ± 0.3 pH unit is extreme. Acid metabolites are ordinarily produced in greater quantities than basic metabolites, and carbon dioxide is the principal one. The buffering capacity of blood for handling CO_2 is estimated to be distributed among various buffer systems as follows: hemoglobin and oxyhemoglobin, 62%; $H_2PO_4^-/HPO_4^{2-}$, 22%; plasma protein, 11%; bicarbonate, 5%. Proteins contain carboxylic and amino groups, which are weak acids and bases, respectively. They are, therefore, effective buffering agents. The combined buffering capacity of blood to neutralize acids is designated by clinicians as the "alkali reserve," and this is frequently determined in the clinical laboratory. Certain diseases cause disturbances in the acid balance of the body. For example, diabetes may give rise to "acidosis," which can be fatal.

An important diagnostic analysis is the CO_2/HCO_3^- balance in blood. This ratio is related to the pH of the blood by the Henderson–Hasselbalch equation (7.45):

$$\boxed{pH = 6.10 + \log \frac{[HCO_3^-]}{[H_2CO_3]}} \qquad (7.88)$$

where H_2CO_3 can be considered equal to the concentration of dissolved CO_2 in the blood; 6.10 is pK_{a1} of carbonic acid in blood at body temperature (37°C). Normally, the bicarbonate concentration in blood is about 26.0 mmol/L, while the concentration of carbon dioxide is 1.3 mmol/L. Accordingly, for the blood,

$$pH = 6.10 + \log \frac{26 \text{ mmol/L}}{1.3 \text{ mmol/L}} = 7.40$$

The HCO_3^- concentration may be determined by titrimetry (Experiment 8), or the total carbon dioxide content (HCO_3^- + dissolved CO_2) can be determined by acidification and measurement of the evolved gas.[4] If both analyses are performed, the

The CO_2/HCO_3^- balance can be assessed from measuring two of the parameters in Equation (7.88).

---

[4]The volume of CO_2 is measured, but from the temperature and atmospheric pressure, the number of millimoles of CO_2 and hence its concentration in mmol/L in the solution it originated from can be calculated. At standard temperature and pressure, 22.4 L contain one mole gas.

ratio of HCO_3^-/CO_2 can be calculated, and hence conclusions can be drawn concerning acidosis or alkalosis in the patient. Alternatively, if the pH is measured (at 37°C), only HCO_3^- or total CO_2 need be measured for a complete knowledge of the carbonic acid balance because the ratio of $[HCO_3^-]/[H_2CO_3]$ can be calculated from Equation 7.88.

The partial pressure, p_{CO_2}, of CO_2 may also be measured (e.g., using a CO_2 electrode), in which case $[H_2CO_3] \approx 0.30 p_{CO_2}$. Then, only pH or $[HCO_3^-]$ need be determined.

Note that these equilibria and Equation 7.88 hold although there are other buffer systems in the blood. The pH is the result of all the buffers and the $[HCO_3^-]/H_2CO_3]$ ratio is set by this pH.

The HCO_3^-/H_2CO_3 buffer system is the most important one in buffering blood in the lung (alveolar blood). As oxygen from inhaled air combines with hemoglobin, the oxygenated hemoglobin ionizes, releasing a proton. This excess acid is removed by reacting with HCO_3^-.

$$H^+ + HCO_3^- \rightarrow H_2CO_3$$

But note that the $[HCO_3^-]/[H_2CO_3]$ ratio at pH 7.4 is $26/1.3 = 20:1$. This is not a very effective buffering ratio; and as significant amounts of HCO_3^- are converted to H_2CO_3, the pH would have to decrease to maintain the new ratio. But, fortunately, the H_2CO_3 produced is rapidly decomposed to CO_2 and H_2O by the enzyme decarboxylase, and the CO_2 is exhaled by the lungs. Hence, the ratio of HCO_3^-/H_2CO_3 remains constant at $20:1$

Example 7.21

The total carbon dioxide content ($HCO_3^- + CO_2$) in a blood sample is determined by acidifying the sample and measuring the volume of CO_2 evolved with a Van Slyke manometric apparatus. The total concentration was determined to be 28.5 mmol/L. The blood pH at 37°C was determined to be 7.48. What are the concentrations of HCO_3^- and CO_2 in the blood?

Solution

$$pH = 6.10 + \log \frac{[HCO_3^-]}{[CO_2]}$$

$$7.48 = 6.10 + \log \frac{[HCO_3^-]}{[CO_2]}$$

$$\log \frac{[HCO_3^-]}{[CO_2]} = 1.38$$

$$\frac{[HCO_3^-]}{[CO_2]} = 10^{1.38} = 10^1 \times 10^{.38} = 24$$

$$[HCO_3^-] = 24[CO_2]$$

But

$$[HCO_3^-] + [CO_2] = 28.5 \text{ mmol/L}$$
$$24[CO_2] + [CO_2] = 28.5$$
$$[CO_2] = 1.1_4 \text{ mmol/L}$$
$$[HCO_3^-] = 28.5 - 1.1 = 27.4 \text{ mmol/L}$$

7.10 Buffers for Biological and Clinical Measurements

Many biological reactions of interest occur in the pH range of 6 to 8. A number, particularly specific enzyme reactions that might be used for analyses (see Chapter 22), may occur in the pH range of 4 to 10 or even greater. The proper selection of buffers for the study of biological reactions or for use in clinical analyses can be critical in determining whether or not they influence the reaction. A buffer must have the correct pK_a, near physiological pH so the ratio of $[A^-]/[HA]$ in the Henderson–Hasselbalch equation is not too far from unity, and it must be physiologically compatible.

PHOSPHATE BUFFERS

One useful series of buffers are phosphate buffers. Biological systems usually contain some phosphate already, and phosphate buffers will not interfere in many cases. By choosing appropriate mixtures of $H_3PO_4/H_2PO_4^-$, $H_2PO_4^-/HPO_4^{2-}$, or HPO_4^{2-}/PO_4^{3-}, solutions over a wide pH range can be prepared. See G. D. Christian and W. C. Purdy, *J. Electroanal. Chem.*, **3** (1962) 363 for the compositions of a series of phosphate buffers at a constant ionic strength of 0.2. Ionic strength is a measure of the total salt content of a solution (see Chapter 6), and it frequently influences reactions, particularly in kinetic studies. Hence, these buffers could be used in cases where the ionic strength must be constant. However, the buffering capacity decreases markedly as the pH approaches the values for the single salts listed, and the single salts are not buffers at all; the best buffering capacity is within ±1 pH unit of the respective pK_a value, that is, 1.96 ± 1, 7.12 ± 1, and 12.32 ± 1. Other solutions are most useful in establishing a given pH when little or no acid or base is to be added to or generated in the solution. A pH 7.40 mixture is very good for buffering at physiological pH values, although above pH 7.5 its buffering capacity begins to diminish.

Example 7.22

What weights of NaH_2PO_4 and Na_2HPO_4 would be required to prepare 1 L of a buffer solution of pH 7.45 that has an ionic strength of 0.100?

Solution

Let $x = [Na_2HPO_4]$ and $y = [NaH_2PO_4]$. There are two unknowns, and two equations are needed. (Remember there must be the same number of equations as unknowns to solve.) Our first equation is the ionic strength equation:

$$\mu = \tfrac{1}{2}\sum C_i Z_i^2$$
$$0.100 = \tfrac{1}{2}[Na^+](1)^2 + [HPO_4^{2-}](2)^2 + [H_2PO_4^-](1)^2$$
$$0.100 = \tfrac{1}{2}[(2x + y)(1)^2 + x(2)^2 + y(1)^2)] \tag{1}$$
$$0.100 = 3x + y$$

Our second equation is the Henderson–Hasselbalch equation:

$$pH = pK_{a2} + \log \frac{[HPO_4^{2-}]}{[H_2PO_4^-]}$$

$$7.45 = 7.12 + \log \frac{x}{y} \tag{2}$$

$$\frac{x}{y} = 10^{0.33} = 2.1_4$$

$$x = 2.1_4\, y \tag{3}$$

Substitute in (1):

$$0.100 = 3(2.1_4)y + y$$

$$y = 0.013_5\, M = [NaH_2PO_4]$$

Substitute in (3):

$$x = (2.1_4)(0.013_5) = 0.028_9\, M = [Na_2HPO_4]$$

$$g\ NaH_2PO_4 = 0.013_5\ mol/L \times 120\ g/mol = 1.6_2\ g/L$$

$$g\ Na_2HPO_4 = 0.028_9\ mol/L \times 142\ g/mol = 4.1_0\ g/L$$

• •

The use of phosphate buffers is limited in certain applications. Besides the limited buffering capacity at certain pH values, phosphate will precipitate or complex many polyvalent cations, and it frequently will participate in or inhibit a reaction. It should not be used, for example, when calcium is present if its precipitation would affect the reaction of interest.

TRIS BUFFERS

Tris buffers are commonly used in clinical chemistry measurements.

A buffer that is widely used in the clinical laboratory and in biochemical studies in the physiological pH range is that prepared from *tris*(hydroxymethyl)aminomethane [$(HOCH_2)_3CNH_2$—**Tris**, or **THAM**] and its conjugate acid (the amino group is protonated). It is a primary standard and has good stability, has a high solubility in physiological fluids, is nonhygroscopic, does not absorb CO_2 appreciably, does not precipitate calcium salts, does not appear to inhibit many enzyme systems, and is compatible with biological fluids. It has a pK close to physiological pH (p$K_a = 8.08$ for the conjugate acid), but its buffering capacity below pH 7.5 does begin to diminish, a disadvantage. Other disadvantages are that the primary aliphatic amine has considerable potential reactivity and it is reactive with linen fiber junctions, as found in saturated calomel reference electrodes used in pH measurements (Chapter 13); a reference electrode with a ceramic, quartz, or sleeve junction should be used. These buffers are usually prepared by adding an acid such as hydrochloric acid to a solution of Tris to adjust the pH to the desired value.

7.11 Diverse Ion Effect on Acids and Bases: K_a° and K_b°—Salts Change the pH

In Chapter 6, we discussed the thermodynamic equilibrium constant based on activities rather than on concentrations. Diverse salts affect the activities and therefore the extent of dissociation of weak electrolytes such as weak acids or bases.

The activity coefficient of the undissociated acid or base is essentially unity if it is uncharged. Then, for the acid HA,

$$K_a^\circ = \frac{a_{H^+} \cdot a_{A^-}}{a_{HA}} \approx \frac{a_{H^+} \cdot a_{A^-}}{[HA]} \tag{7.89}$$

$$K_a^\circ = \frac{[H^+] f_{H^+} \cdot [A^-] f_{A^-}}{[HA]} = K_a f_{H^+} f_{A^-} \tag{7.90}$$

$$K_a = \frac{K_a^\circ}{f_{H^+} f_{A^-}} \tag{7.91}$$

Therefore, we would predict an increase in K_a and in the dissociation with increased ionic strength, as the activity coefficients decrease. See Example 6.18, and Problem 21 in Chapter 6. A similar relationship holds for weak bases (see Problem 60 in Chapter 6).

Since the ionic strength affects the dissociation of weak acids and bases, it will have an effect on the pH of a buffer. We can write the Henderson–Hasselbalch equation in terms of either K_a or K_a°:

$$pH = pK_a + \log \frac{[A^-]}{[HA]} \tag{7.92}$$

$$pH = pK_a^\circ + \log \frac{a_{A^-}}{[HA]} \tag{7.93}$$

If the ionic strength increases, K_a will increase and pK_a will decrease. So Equation 7.92 predicts a decrease in pH. Likewise, the activity of A^- will decrease, and a similar decrease in pH is predicted by Equation 7.93. If a buffer solution is diluted, its ionic strength will decrease, and so there would be a slight *increase* in pH, even though the ratio of [A⁻]/[HA] remains constant. See Footnote 3, earlier in this chapter.

For a $HPO_4^{2-}/H_2PO_4^-$ buffer, the ratio of $a_{HPO_4^{2-}}/a_{H_2PO_4^-}$ will also decrease with increased ionic strength because the effect is greater on the multiply charged ion.

7.12 Logarithmic Concentration Diagrams—How to View Large Concentration Changes

A log concentration diagram is a log–log plot where the x axis is the same as for the α-distribution diagram, but the y axis is the logarithm of the concentration of the species of interest. Because the y axis indicates specific concentrations, we can indicate such species as H^+ and OH^-, in addition to those of the various forms of the acid.

Suppose we wish to plot a log concentration diagram for all the species present in a $1.0 \times 10^{-2}\,M$ solution of acetic acid ($pK_a = 4.76$). The species whose concentrations are the simplest to represent are H^+ and OH^-. By definition, $\log[H^+] = -pH$, so at, say pH 4, $[H^+] = 10^{-4}$, or $\log[H^+] = -4$. The curve for $[H^+]$ is a straight line with a slope of -1, and it passes through the log C point -4 at pH 4. Similarly, $\log[OH^-] = -pOH = pH - pK_w$. The curve for $[OH^-]$ is then a straight line with a slope of $+1$, and it passes through the log C point -4 at pH 10. The slopes and the one reference point for each can be used to draw the two straight lines. Figure 7.2 shows a log concentration diagram containing only the $[H^+]$ and $[OH^-]$ curves.

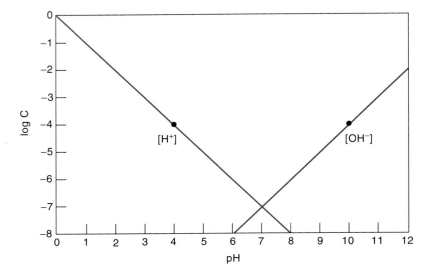

Fig. 7.2. Log concentration diagram for acid–base system. Reference points for H⁺ (pH 4, log C = −4) and OH⁻ (pH 10, log C = −4) are marked by the circled points.

Next, we need to consider the curves for [HOAc] and [OAc⁻]; you should refer to Figure 7.3 during the following discussion. In strongly acid solutions, the dissociation of HOAc is suppressed and HOAc is the major form. Since $C_{HOAc} = 1.0 \times 10^{-2}\ M$, then in acid solutions [HOAc] is essentially constant at $1.0 \times 10^{-2}\ M$. Conversely, we know that in alkaline solutions [HOAc] becomes very small due to dissociation and that $[OAc^-] \approx C_{HOAc}$. For alkaline solutions, then, we can rearrange the acid–base equilibrium expression (substituting C_{HOAc} for [OAc⁻]) to

$$[HOAc] = \frac{[H^+]\,C_{HOAc}}{K_a} \tag{7.94}$$

Taking the logarithm of both sides yields

$$\log[HOAc] = \log\frac{C_{HOAc}}{K_a} - pH \tag{7.95}$$

$$= \text{constant} - pH$$

For $1.0 \times 10^{-2}\ M$ HOAc, the constant is 2.76. The curve for HOAc in alkaline solutions is then a straight line with a slope of −1. At pH 6.76, log[HOAc] is −4.00, which could serve as a reference point on the line, along with the slope, for plotting the HOAc curve in alkaline solution.

Similarly for OAc⁻ in acid solutions, [HOAc] is approximately constant at C_{HOAc} and in strongly acid solutions,

$$[OAc^-] = \frac{K_a}{[H^+]C_{HOAc}} \tag{7.96}$$

$$\log[OAc^-] = \log K_a \cdot C_{HOAc} + pH \tag{7.97}$$

$$= \text{constant} + pH$$

and the curve for [OAc⁻] is a straight line with a slope of +1. For $1.0 \times 10^{-2}\ M$ HOAc, the constant is −6.76, and so at pH 2.76, log[OAc⁻] = −4.00. This point and the +1 slope could be used to construct the [OAc⁻] curve in acid solution.

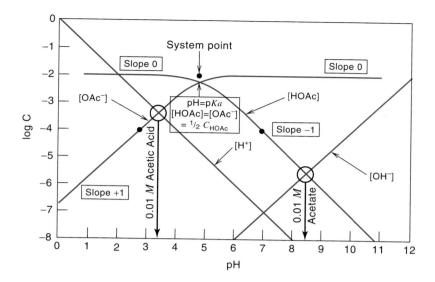

Fig. 7.3. Log concentration diagram for acetic acid system, at 0.01 M HOAc. Reference points for HOAc (pH 2.76, log $C = -4$) and OAc$^-$ (pH 6.76, log $C = -4$) are marked by circled points. Only the system point and the intersection point are needed, along with the slopes, to construct the HOAc and OAc$^-$ curves.

In solutions where pH $\approx$ pK_a, there is a smooth transition between the limiting conditions found in solutions at more acid or more alkaline conditions. Right at pH = pK_a we have the situation that [HOAc] = [OAc$^-$]; so at that pH the curves for [HOAc] and [OAc$^-$] must cross. In this example, this occurs where $C = 5.0 \times 10^{-3}$ M (one-half the total concentration), or log $C = -2.30$, and pH = 4.76.

If the linear sections of the two plots were extended, they would intersect at the point where log C = log C_{HOAc} and pH = pK_a. This point is often called the **system point** and is indicated in Figure 7.3. The system point can be used as one point along with the above example reference points for drawing the linear portions of the curves. Actually, only one reference point is needed along with the slope to draw the curves, and the system point could always serve as the reference point (the curves don't actually extend to the system point, but curve off and intersect at pH = pK_a). The position of the system point (and the curves) moves up or down the log C scale, depending on the concentration.

When a log concentration diagram is constructed, then the system point is usually the first point located; it then serves as a reference for construction of the remainder of the diagram. If the log concentration diagram were drawn for a 10^{-4} M solution instead of this 10^{-2} M solution, the curves for [OAc$^-$] and [HOAc] would have the same shape but be shifted vertically down by 2 log concentration units. The lines for [H$^+$] and [OH$^-$] would be unchanged.

Just as with the α-distribution diagram, the log concentration diagram can be used to determine which species dominate at a given pH. Because the concentration scale employed is logarithmic rather than linear, we can also use this diagram to make quite accurate predictions concerning the concentrations of species that are present, even at small concentrations. We can see at pH 2 that [H$^+$] = 10^{-2} M, [HOAc] = 10^{-2} M, and [OAc$^-$] is a bit greater than 10^{-5} M [$10^{-4.76}$, as confirmed by Equation (7.97)]. This would be too close to zero on the semilog distribution diagram of α to estimate its value. Using the acid–base equilibrium expression, we could calculate that [OAc$^-$] = 1.8×10^{-5} M, which is sufficiently close to our estimate. Given C_{HOAc} at 0.01 M, that concentration corresponds to $\alpha_1 = 0.0018$, and this would be too small to estimate from the α-distribution diagram.

Because of the resolution available on the log concentration axis, it is often possible to use a log concentration diagram to obtain an answer (at least an

Use the system point and the slope to construct a log–log diagram.

A log–log diagram allows you to estimate at a glance the concentrations of all the species in equilibrium at a given pH.

The log–log diagram can also be used to solve complex equilibrium problems.

approximate answer) to an equilibrium problem. Suppose we want to know the pH of the 1.0×10^{-2} M solution of acetic acid. The charge balance expression is

$$[H^+] = [OAc^-] + [OH^-]$$

We can assume that in an acid solution the concentration of OH^- will be negligibly small. The charge balance then simplifies to

$$[H^+] + [OAc^-]$$

In other words, the pH of the solution will be the pH indicated at the intersection of the curves for $[H^+]$ and $[OAc^-]$. That intersection point is indicated in Figure 7.3, where we see that the pH is 3.35. Also, on this plot we can see that at this pH the concentration of OH^- is very low (the point is off the bottom of the graph at about 10^{-11} M) and we were correct in our assumption that it could be ignored. Similarly, for a 1.0×10^{-2} M solution of acetate we could write

$$[OH^-] = [HOAc] + [H^+] \approx [HOAc]$$

From the intersection of the $[OH^-]$ and $[HOAc]$ lines we find that the solution pH is 8.35 and that $[H^+]$ was justifiably neglected because it is about a thousand times smaller than $[OH^-]$ or $[HOAc]$.

Log concentration diagrams can be drawn for systems of greater complexity. Figure 7.4 shows a log concentration diagram for the phosphoric acid system, at 0.001 M. See Problem 67 for derivation of the log C expressions for the different species and Problem 68 for a spreadsheet calculation of the curves using α values (as discussed below). Just as in a distribution diagram of phosphoric acid (Figure 7.1), you can deduce the major species at a given pH. The log concentration diagram also allows approximation of concentrations of minor species. Hence, at pH 1, the H_3PO_4 concentration is just under 10^{-3} M and the $H_2PO_4^-$ concentration is 10^{-4} M, while the HPO_4^2 and PO_4^{3-} concentrations are negligibly small (this means the H_3PO_4 concentration, subtracting the $H_2PO_4^-$ concentration from 10^{-3} M, is 9×10^{-4} M).

We can estimate the pH of solutions of different species of phosphoric acid from Figure 7.4. For a 10^{-3} M solution of H_3PO_4, the charge balance expression is

$$[H^+] = [H_2PO_4^-] + [HPO_4^{2-}] + [PO_4^{3-}] + [OH^-] \approx [H_2PO_4^-]$$

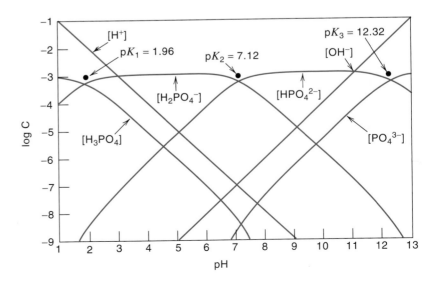

Fig. 7.4. Log concentration diagram for phosphoric acid system, at 0.001 M H_3PO_4.

We see that when $[H^+] = [H_2PO_4^-]$, the pH is 3.05. The concentrations of HPO_4^{2-}, PO_4^{3-}, and OH^- at this pH are indeed negligible and the approximation that $[H^+] = [H_2PO_4^-]$ is valid.

For a 10^{-3} M solution of $H_2PO_4^-$, we know that $[H^+] \approx \sqrt{K_{a1}K_{a2}}$, or pH $\approx$ $(pK_{a1} + pK_{a2})/2 \approx (1.96 + 7.12)/2 = 4.54$. The apparent species present at this pH are $H_2PO_4^-$ (10^{-3} M), H^+ ($10^{-4.54}$ M), and $HPO_4^{2-} \approx H_3PO_4$ ($10^{-5.6}$ M).

For a 10^{-3} M HPO_4^{2-} solution, pH $\approx (pK_{a2} + pK_{a3})/2 \approx (7.12 + 12.32)/2 \approx 9.72$. We see that at this pH, $[H_2PO_4^-] \approx [PO_4^{3-}] \approx 10^{-6.55}$ M, and $[OH^-] = 10^{-4.28}$.

For 10^{-3} M PO_4^{3-}, we can write a mass balance expression for the stepwise hydrolysis of each species:

$$[OH^-] = [HPO_4^{2-}] + [H_2PO_4^-] + [H_3PO_4] \approx [HPO_4^{2-}]$$

We can estimate the extent of the two amphoteric reactions (Equations 7.76 and 7.77).

From Figure 7.4, we see that when $[OH^-] = [HPO_4^{2-}]$, the pH is 10.96. At this pH, $[H_2PO_4^-]$ and $[H_3PO_4]$ are $\ll [HPO_4^{2-}]$, and the above approximation is valid. For this solution $[PO_4^{3-}] = 10^{-4.35}$ and $[OH^-] = 10^{-3.04}$.

SPREADSHEET EXERCISE—LOG DIAGRAM FOR HOAc

Let's prepare a spreadsheet (below) to plot the log concentration diagram for the acetic acid system at 0.01 M HOAc (Figure 7.3) and compare the plot obtained with Figure 7.3. If we construct it properly, we can use the spreadsheet to obtain log plots for different concentrations of acetic acid (hint: enter C_{HOAc} in equation form in a cell whenever it appears, rather than calculating and entering a constant value for a fixed concentration).

We need key points for each of the four species, HOAc, OAc$^-$, H$^+$, and OH$^-$. For HOAc, we know that in acid solution, [HOAc] is essentially equal to C_{HOAc}. So we assign that concentration at pH 0 and 4 (cells **B4** and **B5**). (The boldface cell is where the formula or value is entered, and then copied down.) At pH = $pK_a = 4.76$, [HOAc] = [OAc$^-$], and so each will be one-half C_{HOAc} (cells **B6** and

	A	B	C	D	E	F	G	H
1	Log curves for acetic acid system							
2		C_{HOAc}=	0.01	K_a=	1.75E-05			
3	pH	[HOAc]	log[HOAc]	[OAc$^-$]	log[OAc$^-$]	log[H$^+$]	log[OH$^-$]	
4	0.00	**0.01**	**-2.00**		**-6.76**	**0.00**	**-14.00**	
5	4.00	0.01	-2.00		-2.76	-4.00	-10.00	
6	4.76	**0.005**	-2.30	**0.005**	-2.30	-4.76	-9.24	
7	5.50		**-2.74**	**0.01**	-2.00	-5.50	-8.50	
8	12.00		-9.24	0.01	-2.00	-12.00	-2.00	
9	Formulas for cells in **boldface**:							
10	**Cell B4**: [HOAc] = C_{HOAc} =		$C$2	(Copy to Cell B5)				
11	**Cell C4**: log[HOAc] =		LOG10(B4)		(Copy to Cell C6)			
12	**Cell E4**: log[OAc$^-$] = log(K_a x C_{HOAc})+ pH =							
13	= LOG10($E$2*$C$2)+A4			(Copy to Cell E5)				
14	**Cell F4**: log[H$^+$] = -pH =		(-A4)	(Copy to end)				
15	**Cell G4**: log[OH$^-$] = 14 - log[H$^+$] =			(-14-F4)	Copy to end)			
16	**Cell B6**: [HOAc] = C_{HOAc}/2 =		$C$2/2					
17	**Cell D6**: [OAc$^-$] = [HOAc] =		B6					
18	**Cell E6**: log[OAc$^-$] =	LOG10(D6)						
19	**Cell C7**: log[HOAc] = log(C_{HOAc}/K_a) - pH =			LOG10($C$2/$E$2)-A7	(Copy to Cell C8)			
20	**Cell D7**: [OAc$^-$] = C_{HOAc} =		$C$2	Copy to Cell D8)				
21	Plot A4:A8 vs. C4:C8, D4:D8, E4:E8, and F4:F8 (Series 1, 2, 3, and 4)							
22	This spreadsheet can be used for other concentrations of acetic acid just by							
23	changing the value of Cell C2.							

D6). This is where the curves for the two species cross. Beyond pH 4.76, log[HOAc] is given by Equation 7.95 (cells **C7** and C8).

For OAc$^-$, in acid solution, log[OAc] is given by Equation 7.97 (cells **E4** and E5). Beyond pH $= K_a$, [OAc$^-$] becomes essentially equal to C_{HOAc} (cells **D7** and D8). Note that log[H$^+$] and log[OH$^-$] are obtained directly from pH throughout (cells **F4** and **G4**).

Note that by entering log $K_a \times C_{HOAc}$ and log K_a/C_{HOAc} in cells E4 and C7, instead of the actual constants for 0.01 M HOAc, the spreadsheet applies to all concentrations of HOAc entered in cell C2.

The spreadsheet and a corresponding chart showing the log C vs. pH plots are given in your CD, Chapter 7 (Chapter text spreadsheets). Save it to your desktop. Open it and click on sheet 1 to see the data and on Chart 1 to see the plot. Compare the plot to Figure 7.3.

The plots are made using Chart Wizard, selecting Scatter data connected by smooth lines. Click on Chart 1, and then on the Chart Wizard icon, and go through the steps to see how the chart was constructed. For Series 1 (log[HOAc]), X values: =Sheet1!A4:A8, and Y values: =Sheet1!C4:C8. The X values are the same for Series 2, 3, and 4, but for the Y values, E4:E8, F4:F8, and G4:G8 are entered in place of C4:C8 to give the plots for log[OAc$^-$], log[H$^+$], and log[OH$^-$]. Try reproducing the chart using a new chart number by clicking on Sheet 1, and then the Chart Wizard icon to begin a new plot. Click on Sheet 1. Change the concentration of acetic acid to see how the HOAc and OAc$^-$ curves move up or down on the chart.

USE OF α VALUES TO CONSTRUCT LOG DIAGRAMS—THIS IS EASIER

Since we have general formulas for α values, it is a simple matter to construct spreadsheets to calculate these, as we did for H_3PO_4 for Figure 7.1. Then it is straightforward to enter formulas in columns for log($C \times \alpha_i$) for each α value, and copy the formula down for calculation at each entered pH. Thus, say you have a diprotic acid, H_2A. For log[HA$^-$], if the formula for α_1 is in cell D4, and if you enter the concentration value, C, in cell B2, you could enter in cell E4: =LOG10($B$2*D4). This calculates log[HA$^-$] at the given pH. You would copy that formula down to the last pH row to calculate log[HA$^-$] at each pH. Then you would plot the pH column versus the log[HA$^-$] column. You would repeat the process for H_2A (α_0) and A^{2-} (α_2) and also prepare columns of log[H$^+$] and log[OH$^-$]. The concentration C can be changed in cell B2 to see how the plots change with concentration. Often, the calculated concentrations are very small at extreme pH values, for example, -20 or less. You can limit the range of the log C axis on the chart to say -10 to give better resolution of other plotted concentrations.

See your CD, Chapter 7, auxiliary data for such a plot for acetic acid, and Problem 64 for a similar plot for a diprotic acid, malic acid. Note that not all species in polyprotic acids give slopes of $+1$ or -1.

Learning Objectives

WHAT ARE SOME OF THE KEY THINGS WE LEARNED FROM THIS CHAPTER?

- Acid–base theories, p. 219
- Acid–base equilibria in water (key equations: 7.11, 7.13, 7.19), p. 221
- Weak acids and bases, p. 228

- Salts of weak acids and bases (key equations: 7.27, 7.29, 7.32, 7.36, 7.39), p. 230
- Buffers (key equations: 7.45, 7.56), p. 234
- Polyprotic acids—α values (key equations: 7.72–7.75), pp. 241, 243
- Using spreadsheets to prepare α vs. pH plots, p. 245
- Salts of polyprotic acids (key equations: 7.83, 7.84, 7.86, 7.87), p. 248
- Logarithmic concentration diagrams, p. 255
  - Using spreadsheets for plotting, p. 259
  - Using α values for spreadsheet plotting, p. 260

Questions

1. Explain the difference between a strong electrolyte and a weak electrolyte. Is an "insoluble" salt a weak or a strong electrolyte?
2. What is the Brønsted acid–base theory?
3. What is a conjugate acid? Conjugate base?
4. Write the ionization reaction of aniline, $C_6H_5NH_2$, in glacial acetic acid, and identify the conjugate acid of aniline. Write the ionization reaction of phenol, C_6H_5OH, in ethylene diamine, $NH_2CH_2CH_2NH_2$, and identify the conjugate base of phenol.
5. What is the Lewis acid–base theory?

Problems

STRONG ACIDS AND BASES

6. Calculate the pH and pOH of the following strong acid solutions: (a) 0.020 M $HClO_4$, (b) 1.3×10^{-4} M HNO_3, (c) 1.2 M HCl, (d) 1.2×10^{-9} M HCl, (e) 2.4×10^{-7} M HNO_3.
7. Calculate the pH and pOH of the following strong base solutions: (a) 0.050 M NaOH, (b) 0.14 M $Ba(OH)_2$, (c) 2.4 M NaOH, (d) 3.0×10^{-7} M KOH, (e) 3.7×10^{-3} M KOH.
8. Calculate the hydroxyl ion concentration of the following solutions: (a) 2.6×10^{-5} M HCl, (b) 0.20 M HNO_3, (c) 2.7×10^{-9} M $HClO_4$, (d) 1.9 M $HClO_4$.
9. Calculate the hydrogen ion concentration of the solutions with the following pH values: (a) 3.47, (b) 0.20, (c) 8.60, (d) −0.60, (e) 14.35, (f) −1.25.
10. Calculate the pH and pOH of a solution obtained by mixing equal volumes of 0.10 M H_2SO_4 and 0.30 M NaOH.
11. Calculate the pH of a solution obtained by mixing equal volumes of a strong acid solution of pH 3.00 and a strong base solution of pH 12.00.

TEMPERATURE EFFECT

12. Calculate the hydrogen ion concentration and pH of a neutral solution at 50°C ($K_w = 5.5 \times 10^{-14}$ at 50°C).
13. Calculate the pOH of a blood sample whose pH is 7.40 at 37°C.

WEAK ACIDS AND BASES

14. The pH of an acetic acid solution is 3.26. What is the concentration of acetic acid and what is the percent acid ionized?

15. The pH of a 0.20 M solution of a primary amine, RNH_2, is 8.42. What is the pK_b of the amine?

16. A monoprotic organic acid with a K_a of 6.7×10^{-4} is 3.5% ionized when 100 g of it is dissolved in 1 L. What is the formula weight of the acid?

17. Calculate the pH of a 0.25 M solution of propanoic acid.

18. Calculate the pH of a 0.10 M solution of aniline, a weak base.

19. Calculate the pH of a 0.1 M solution of iodic acid, HIO_3.

20. The first proton of sulfuric acid is completely ionized, but the second proton is only partially dissociated, with an acidity constant K_{a2} of 1.2×10^{-2}. Calculate the hydrogen ion concentration in a 0.0100 M H_2SO_4 solution.

21. Calculate the hydrogen ion concentration in a 0.100 M solution of trichloro-acetic acid.

22. An amine, RNH_2, has a pK_b of 4.20. What is the pH of a 0.20 M solution of the base?

23. What is the concentration of a solution of acetic acid if it is 3.0% ionized?

24. By how much should a 0.100 M solution of a weak acid HA be diluted in order to double its percent ionization? Assume $C > 100K_a$.

SALTS OF WEAK ACIDS AND BASES

25. If 25 mL of 0.20 M NaOH is added to 20 mL of 0.25 M boric acid, what is the pH of the resultant solution?

26. Calculate the pH of a 0.010 M solution of NaCN.

27. Calculate the pH of a 0.050 M solution of sodium benzoate.

28. Calculate the pH of a 0.25 M solution of pyridinium hydrochloride (pyridine · HCl, $C_6H_5NH^+Cl$).

29. Calculate the pH of the solution obtained by adding 12.0 mL of 0.25 M H_2SO_4 to 6.0 mL of 1.0 M NH_3.

30. Calculate the pH of the solution obtained by adding 20 mL of 0.10 M HOAc to 20 mL of 0.10 M NaOH.

31. Calculate the pH of the solution prepared by adding 0.10 mol each of hydroxylamine and hydrochloric acid to 500 mL water.

32. Calculate the pH of a 0.0010 M solution of sodium salicylate, $C_6H_4(OH)COONa$.

33. Calculate the pH of a 1.0×10^{-4} M solution of NaCN.

POLYPROTIC ACIDS AND THEIR SALTS

34. What is the pH of 0.0100 M solution of phthalic acid?

35. What is the pH of a 0.0100 M solution of potassium phthalate?

36. What is the pH of a 0.0100 M solution of potassium acid phthalate (KHP)?

37. Calculate the pH of a 0.600 M solution of Na_2S.

38. Calculate the pH of a 0.500 M solution of Na_3PO_4.

39. Calculate the pH of a 0.250 M solution of $NaHCO_3$.

40. Calculate the pH of a 0.600 M solution of NaHS.

41. Calculate the pH of a 0.050 M solution of the trisodium salt of EDTA (ethylenediaminetetraacetic acid), Na_3HY.

BUFFERS

42. Calculate the pH of a solution that is 0.050 M in formic acid and 0.10 M in sodium formate.

43. Calculate the pH of a solution prepared by mixing 5.0 mL of 0.10 M NH_3 with 10.0 mL of 0.020 M HCl.

44. An acetic acid–sodium acetate buffer of pH 5.00 is 0.100 M in NaOAc. Calculate the pH after the addition of 10 mL of 0.1 M NaOH to 100 mL of the buffer.

45. A buffer solution is prepared by adding 20 mL of 0.10 M sodium hydroxide solution to 50 mL of 0.10 M acetic acid solution. What is the pH of the buffer?

46. A buffer solution is prepared by adding 25 mL of 0.050 M sulfuric acid solution to 50 mL of 0.10 M ammonia solution. What is the pH of the buffer?

47. Aspirin (acetylsalicylic acid) is absorbed from the stomach in the free (non-ionized) acid form. If a patient takes an antacid that adjusts the pH of the stomach contents to 2.95 and then takes two 5-grain aspirin tablets (total 0.65 g), how many grams of aspirin are available for immediate absorption from the stomach, assuming immediate dissolution? Also assume that aspirin does not change the pH of the stomach contents. The pK_a of aspirin is 3.50, and its formula weight is 180.2.

48. *Tris*(hydroxymethyl)aminomethane [$(HOCH_2)_3CNH_2$—Tris, or THAM] is a weak base frequently used to prepare buffers in biochemistry. Its K_b is 1.2×10^{-6} and pK_b is 5.92. The corresponding pK_a is 8.08, which is near the pH of the physiological buffers, and so it exhibits good buffering capacity at physiological pH. What weight of THAM must be taken with 100 mL of 0.50 M HCl to prepare 1 L of a pH 7.40 buffer?

49. Calculate the hydrogen ion concentration for Problem 21 if the solution contains also 0.100 M sodium trichloroacetate.

BUFFERS FROM POLYPROTIC ACIDS

50. What is the pH of a solution that is 0.20 M in phthalic acid (H_2P) and 0.10 M in potassium acid phthalate (KHP)?

51. What is the pH of a solution that is 0.25 M each in potassium acid phthalate (KHP) and potassium phthalate (K_2P)?

52. The total phosphate concentration in a blood sample is determined by spectrophotometry to be 3.0×10^{-3} M. If the pH of the blood sample is 7.45, what are the concentrations of $H_2PO_4^-$ and HPO_4^2 in the blood?

BUFFER CAPACITY

53. A buffer solution contains 0.10 M NaH_2PO_4 and 0.070 M Na_2HPO_4. What is its buffer capacity in moles/liter per pH? By how much would the pH change if 10 μL (0.010 mL) of 1.0 M HCl or 1.0 M NaOH were added to 10 mL of the buffer?

54. You wish to prepare a pH 4.76 acetic acid–sodium acetate buffer with a buffer capacity of 1.0 M per pH. What concentrations of acetic acid and sodium acetate are needed?

CONSTANT-IONIC-STRENGTH BUFFERS

55. What weight of Na_2HPO_4 and KH_2PO_4 would be required to prepare 200 mL of a buffer solution of pH 7.40 that has an ionic strength of 0.20? (See Chapter 6 for a definition of ionic strength.)

56. What volume of 85% (wt/wt) H_3PO_4 (sp. gr. 1.69) and what weight of KH_2PO_4 are required to prepare 200 mL of a buffer of pH 3.00 that has an ionic strength of 0.20?

α CALCULATIONS

57. Calculate the equilibrium concentrations of the different species in a 0.0100 M solution of sulfurous acid, H_2SO_3, at pH 4.00 ($[H^+] = 1.0 \times 10^{-4}$ M).

58. Derive Equations 7.73, 7.74, and 7.75 for α_1, α_2, and α_3 of phosphoric acid.

DIVERSE SALT EFFECT

59. Calculate the hydrogen ion concentration for a 0.0200 M solution of HCN in 0.100 M NaCl (diverse ion effect).

60. Derive the equivalent of Equation 7.91 for the diverse salt effect on an uncharged weak base B.

LOGARITHMIC CONCENTRATION DIAGRAMS

You can use the HOAc spreadsheet exercise on your CD for Problem 61. But also try constructing the curves manually. Prepare a spreadsheet for Problem 64 using α-values—see CD auxiliary data for HOAc log plots.

61. Construct the log–log diagram for a 10^{-3} M solution of acetic acid.

62. From the diagram in Problem 61, estimate the pH of a 10^{-3} M solution of acetic acid. What is the concentration of acetate ion in this solution?

63. For Problem 61, derive the expression for log$[OAc^-]$ in acid solution and calculate the acetate concentration at pH 2.00 for a 10^{-3} M acetic acid solution. Compare with the value estimated from the log–log diagram.

64. Construct the log–log diagram for a 10^{-3} M solution of malic acid by preparing a spreadsheet using α values.

65. From the diagram in Problem 64, estimate the pH and concentrations of each species present in (a) 10^{-3} M malic acid and (b) 10^{-3} M sodium malate solution.

66. For Problem 64, derive the expressions for the HA^- curves in the acid and alkaline regions.

67. Derive expressions for (a) log$[H_3PO_4]$ between pH = pK_{a1} and pK_{a2}, (b) log $[H_2PO_4^-]$ between pH = pK_{a2} and pK_{a3}, (c) log$[HPO_4^{2-}]$ at between pH = pK_{a2} and pK_{a1}, and (d) log $[PO_4^{3-}]$ at between pH = pK_{a3} and pK_{a2}. Check with representative points on the curves.

68. Construct a log–log diagram for 0.001 M H_3PO_4 (Figure 7.4) using α values. Start with the spreadsheet for Figure 7.1 (given in your CD). Compare the chart with Figure 7.4. Vary the H_3PO_4 concentration and see how the curves change.

Recommended References

ACID–BASE THEORIES, BUFFERS

1. R. G. Bates, "Concept and Determination of pH," in I. M. Kolthoff and P. J. Elving, eds., *Treatise on Analytical Chemistry*, Part I, Vol. 1. New York: Wiley-Interscience, 1959, pp. 361–401.

2. N. W. Good, G. D. Winget, W. Winter, T. N. Connally, S. Izawa, and R. M. M. Singh, "Hydrogen Ion Buffers for Biological Research," *Biochemistry*, **5** (1966) 467.

3. D. E. Gueffroy, ed., *A Guide for the Preparation and Use of Buffers in Biological Systems*. La Jolla, CA: Calbiochem, 1975.

4. I. M. Kolthoff, "Concepts of Acids and Bases," in I. M. Kolthoff and P. J. Elving, eds., *Treatise on Analytical Chemistry*, Part I, Vol. 1. New York: Wiley-Interscience, 1959, pp. 405–420.

5. D. D. Perrin and B. Dempsey, *Buffers for pH and Metal Ion Control*. New York: Chapman and Hall, 1974.

EQUILIBRIUM CALCULATIONS

6. S. Brewer, *Solving Problems in Analytical Chemistry*. New York: Wiley, 1980. Describes iterative approach for solving equilibrium calculations.

7. J. N. Butler, *Ionic Equilibria. A Mathematical Approach*. Reading, MA: Addison-Wesley, 1964.

8. W. B. Guenther, *Unified Equilibrium Calculations*. New York: Wiley, 1991.

9. D. D. DeFord, "The Reliability of Calculations Based on the Law of Chemical Equilibrium," *J. Chem. Ed.*, **31** (1954) 460.

10. E. R. Nightingale, "The Use of Exact Expressions in Calculating H^+ Concentrations," *J. Chem. Ed.*, **34** (1957) 277.

11. R. J. Vong and R. J. Charlson, "The Equilibrium pH of a Cloud or Raindrop: A Computer-Based Solution for a Six-Component System," *J. Chem. Ed.*, **62** (1985) 141.

12. R. deLevie, *A Spreadsheet Workbook for Quantitative Chemical Analysis*. New York: McGraw-Hill, 1992.

13. H. Freiser, *Concepts and Calculations in Analytical Chemistry. A Spreadsheet Approach*. Boca Raton, FL: CRC Press, 1992.

WEB VIRTUAL CALCULATOR

14. http://hamers.chem.wisc.edu/chapman/Titrator/. Virtual Calculator 1.5, from the website of Professor Robert Hamers, University of Wisconsin. Use it to calculate alpha curves. In the Virtual Titrator window, you can select acids, graphs, and, under Window, a stoichiometric calculator for preparing solutions.

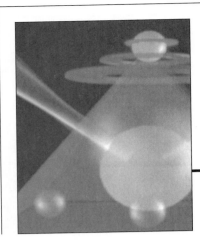

Chapter Eight

ACID—BASE TITRATIONS

In Chapter 7, we introduced the principles of acid–base equilibria. These are important for the construction and interpretation of titration curves in acid–base titrations. In this chapter, we discuss the various types of acid–base titrations, including the titration of strong acids or bases and of weak acids or bases. The shapes of titration curves obtained are illustrated. Through a description of the theory of indicators, we discuss the selection of a suitable indicator for detecting the completion of a particular titration reaction. The titrations of weak acids or bases with two or more titratable groups and of mixtures of acids or bases are presented. The important Kjeldahl analysis method is described for analyzing nitrogen in organic and biological samples.

Courtesy of Metrohm AG.

Only a strong acid or base is used as the titrant.

8.1 Strong Acid versus Strong Base—The Easy Titrations

An acid–base titration involves a **neutralization** reaction in which an acid is reacted with an equivalent amount of base. By constructing a **titration curve**, we can easily explain how the **end points** of these titrations can be detected; the end point signals the completion of the reaction. A titration curve is constructed by plotting the pH of the solution as a function of the volume of titrant added. *The titrant is always a strong acid or a strong base*. The analyte may be either a strong base or acid or a weak base or acid.

In the case of a strong acid versus a strong base, both the titrant and the analyte are completely ionized. An example is the titration of hydrochloric acid with sodium hydroxide:

$$H^+ + Cl^- + Na^+ + OH^- \rightarrow H_2O + Na^+ + Cl^- \qquad (8.1)$$

The H^+ and OH^- combine to form H_2O, and the other ions (Na^+ and Cl^-) remain unchanged, so the net result of neutralization is conversion of the HCl to a neutral solution of NaCl. The titration curve for 100 mL of 0.1 *M* HCl titrated with 0.1 *M* NaOH is shown in Figure 8.1, plotted from the spreadsheet exercise setup below.

The calculations of titration curves simply involves computation of the pH from the concentration of the particular species present at the various stages of the titration, using the procedures given in Chapter 7. The volume changes during the titration must be employed for determining the concentration of the species.

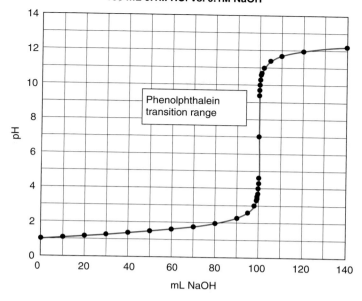

100 mL 0.1M HCl vs. 0.1M NaOH

Fig. 8.1. Titration curve for 100 mL of 0.1 *M* HCl versus 0.1 *M* NaOH.

	A	B	C	D	E	F	G
1	100.00 mL of 0.1000 M HCl vs. 0.1000 M NaOH						
2	mL$_{HCl}$=	100.00	M$_{HCl}$=	0.1000			
3	M$_{NaOH}$=	0.1000	K$_w$=	1.00E-14			
4	mL$_{NaOH}$	[H$^+$]	[OH$^-$]	pOH	pH		
5	0.00	0.1			1.00		
6	10.00	0.0818182			1.09		
7	20.00	0.0666667			1.18		
8	30.00	0.0538462			1.27		
9	40.00	0.0428571			1.37		
10	50.00	0.0333333			1.48		
11	60.00	0.025			1.60		
12	70.00	0.0176471			1.75		
13	80.00	0.0111111			1.95		
14	90.00	0.0052632			2.28		
15	95.00	0.0025641			2.59		
16	98.00	0.0010101			3.00		
17	99.00	0.0005025			3.30		
18	99.20	0.0004016			3.40		
19	99.40	0.0003009			3.52		
20	99.60	0.0002004			3.70		
21	99.80	0.0001001			4.00		
22	99.90	5.003E-05			4.30		
23	99.95	2.501E-05			4.60		
24	100.00	0.0000001			7.00		
25	100.05		2.5E-05	4.60	9.40		
26	100.10		5E-05	4.30	9.70		
27	100.20		1E-04	4.00	10.00		
28	100.40		0.0002	3.70	10.30		
29	100.80		0.0004	3.40	10.60		
30	101.00		0.0005	3.30	10.70		
31	102.00		0.00099	3.00	11.00		
32	105.00		0.00244	2.61	11.39		
33	110.00		0.00476	2.32	11.68		
34	120.00		0.00909	2.04	11.96		
35	140.00		0.01667	1.78	12.22		
36	Formulas for cells in **boldface**:						
37	**Cell B5**: [H$^+$] = (mL$_{HCl}$ x M$_{HCl}$ - mL$_{NaOH}$ x M$_{NaOH}$)/(mL$_{HCl}$ + mL$_{NaOH}$)						
38		=($B$2*$D$2-A5*$B$3)/($B$2+A5)		Copy through Cell B23			
39	**Cell E5** = pH =		(-LOG10(B5))	Copy through Cell E24			
40	**Cell B24** = [H$^+$] = K$_w^{1/2}$ =		SQRT(D3)				
41	**Cell C25** = [OH$^-$] = (mL$_{NaOH}$ x M$_{NaOH}$ - mL$_{HCl}$ x M$_{HCl}$)/(mL$_{HCl}$ + mL$_{NaOH}$)						
42		=(A25*$B$3-$B$2*$D$2)/($B$2+A25)		Copy to end			
43	**Cell D25** = pOH = -log[OH$^-$] =		(-LOG10(C25))	Copy to end			
44	**Cell E25** = pH = 14 - pH =		14-D25	Copy to end			

Table 8.1
Equations Governing a Strong-Acid (HX) or Strong-Base (BOH) Titration

Fraction F Titrated	Strong Acid		Strong Base	
	Present	Equation	Present	Equation
$F = 0$	HX	$[H^+] = [HX]$	BOH	$[OH^-] = [BOH]$
$0 < F < 1$	HX/X^-	$[H^+] = $ [remaining HX]	BOH/B^+	$[OH^-] = $ [remaining BOH]
$F = 1$	X^-	$[H^+] = \sqrt{K_w}$ (Eq. 7.13)	B^+	$[H^+] = \sqrt{K_w}$ [Eq. 7.13]
$F > 1$	OH^-/X^-	$[OH^-] = $ [excess titrant]	H^+/B^+	$[H^+] = $ [excess titrant]

Table 8.1 summarizes the equations governing the different portions of the titration curve. At the beginning of the titration, we have 0.1 M HCl, so the initial pH is 1.0. As the titration proceeds, part of the H^+ is removed from solution as H_2O. So the concentration of H^+ gradually decreases. At 90% neutralization (90 mL NaOH), only 10% of the H^+ remains. Neglecting the volume change, the H^+ concentration at this point would be 10^{-2} M, and the pH would have risen by only one pH unit. (If we correct for volume change, it will be slightly higher—see the spreadsheet below.) However, as the **equivalence point** is approached (the point at which a stoichiometric amount of base is added), the H^+ concentration is rapidly reduced until at the equivalence point, when the neutralization is complete, a neutral solution of NaCl remains and the pH is 7.0. As we continue to add NaOH, the OH^- concentration rapidly increases from 10^{-7} M at the equivalence point and levels off between 10^{-2} and 10^{-1} M; we then have a solution of NaOH plus NaCl. Thus, the pH remains fairly constant on either side of the equivalence point, but it changes markedly very near the equivalence point. This large change allows the determination of the completion of the reaction by measurement of either the pH or some property that changes with pH.

> The equivalence point is where the reaction is theoretically complete.

 Example 8.1

Calculate the pH at 0, 10, 90, 100, and 110% titration for the titration of 50.0 mL of 0.100 M HCl with 0.100 M NaOH.

Solution

> Keep track of millimoles reacted and remaining!

At 0% pH $= -\log 0.100 = 1.00$
At 10%, 5.0 mL NaOH is added. We start with 0.100 $M \times$ 50.0 mL $=$ 5.00 mmol H^+. Calculate the concentration of H^+ after adding the NaOH:

$$
\begin{aligned}
\text{mmol } H^+ \text{ at start} &= 5.00 \text{ mmol } H^+ \\
\text{mmol } OH^- \text{ added} = 0.100 \, M \times 5.0 \text{ mL} &= \underline{0.500 \text{ mmol } OH^-} \\
\text{mmol } H^+ \text{ left} &= 4.50 \text{ mmol } H^+ \text{ in } 55.0 \text{ mL}
\end{aligned}
$$

$$[H^+] = 4.50 \text{ mmol}/55.0 \text{ mL} = 0.0818 \, M$$
$$pH = -\log 0.0818 = 1.09$$

At 90%

$$
\begin{aligned}
\text{mmol } H^+ \text{ at start} &= 5.00 \text{ mmol } H^+ \\
\text{mmol } OH^- \text{ added} = 0.100 \, M \times 45.0 \text{ mL} &= \underline{4.50 \text{ mmol } OH^-} \\
\text{mmol } H^+ \text{ left} &= 0.50 \text{ mmol } H^+ \text{ in } 95.0 \text{ mL}
\end{aligned}
$$

$$[H^+] = 0.00526\ M$$
$$pH = -\log 0.00526 = 2.28$$

At 100%: All the H^+ has been reacted with OH^-, and we have a 0.0500 M solution of NaCl. Therefore, the pH is 7.00.

At 110%: We now have a solution consisting of NaCl and excess added NaOH.

$$mmol\ OH^- = 0.100\ M \times 5.00\ mL = 0.50\ mmol\ OH^-\ in\ 105\ mL$$
$$[OH^-] = 0.00476\ M$$
$$pOH = -\log 0.00476 = 2.32;\ pH = 11.68$$

• •

Note that prior to the equivalence point, when there is excess acid, the relationship is $[H^+] = (M_{acid} \times V_{acid} - M_{base} \times V_{base})/V_{total}$, where V is the volume. You can simply apply this to calculate $[H^+]$ once you understand the solution to Example 8.1. Likewise, beyond the equivalence point when there is excess base, $[OH^-] = (M_{base} \times V_{base} - M_{acid} \times V_{acid})/V_{total}$.

The magnitude of the break will depend on both the concentration of the acid and the concentration of the base. Titration curves at different concentrations are shown in Figure 8.2. The reverse titration gives the mirror image of these curves. The titration of 0.1 M NaOH with 0.1 M HCl is shown in Figure 8.3. The selection of the indicators as presented in the figure is discussed below.

> The selection of the indicator becomes more critical as the solutions become more dilute.

SPREADSHEET EXERCISE—HCl VERSUS NaOH

Let's prepare the spreadsheet for the construction of Figure 8.1. Enter the values of the HCl concentration and volume, the NaOH concentration, and K_w in specific cells (see cells B2, D2, B3, and D3). These are absolute values that will be used in the formulas. In the spreadsheet, the hydrogen ion concentration is calculated as the remaining millimoles of HCl divided by the total volume (**cell B5 formula**). At the equivalence point, it is the square root of K_w (**cell B24 formula**). And

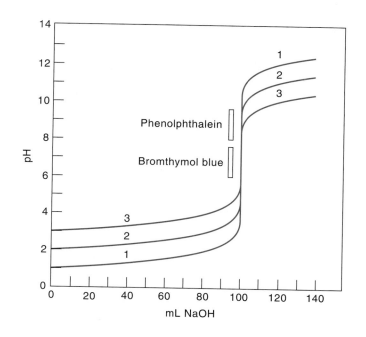

Fig. 8.2. Dependence of the magnitude of end-point break on concentration. Curve 1: 100 mL of 0.1 M HCl versus 0.1 M NaOH. Curve 2: 100 mL of 0.01 M HCl versus 0.01 M NaOH. Curve 3: 100 mL of 0.001 M HCl versus 0.0001 M NaOH.

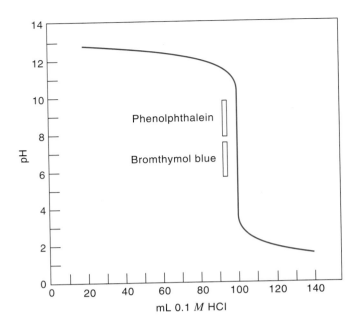

Fig. 8.3. Titration curve for 100 mL 0.1 M NaOH versus 0.1 M HCl.

beyond the equivalence point, the hydroxide concentration is the millimoles of excess NaOH divided by the total volume (**cell C25** formula). Note that the concentrations of HCl and NaOH taken, the volume of HCl, and K_w are all absolute numbers in the formulas. We enter formulas to calculate pH from [H$^+$] (**cell E5**), pOH from [OH$^-$] (**cell D25**), and pH from pOH (**cell E25**).

The titration curve is plotted using Chart Wizard by plotting A5:A35 (X-axis) versus E5:E35 (Y-axis). See the spreadsheet exercise in Chapter 7 for the α versus pH plot, Figure 7.1, for details on preparing the chart. The plot (Chart 1) is given in your CD, Chapter 8.

8.2 Detection of the End Point: Indicators

The goal is for the end point to coincide with the equivalence point.

We wish to determine when the equivalence point is reached. The point at which the reaction is *observed to be complete* is called the **end point**. A measurement is chosen such that the end point coincides with or is very close to the equivalence point. The most obvious way of determining the end point is to measure the pH at different points of the titration and make a plot of this versus milliliters of titrant. This is done with a pH meter, which is discussed in Chapter 13.

It is usually more convenient to add an **indicator** to the solution and visually detect a color change. An indicator for an acid–base titration is a weak acid or weak base that is highly colored. The color of the ionized form is markedly different from that of the nonionized form. One form may be colorless, but the other must be colored. These substances are usually composed of highly conjugated organic constituents that give rise to the color (see Chapter 16).

Assume the indicator is a weak acid, designated HIn, and assume that the nonionized form is red while the ionized form is blue:

$$\text{HIn} \rightleftharpoons \text{H}^+ + \text{In}^- \tag{8.2}$$
$$\text{(red)} \qquad\qquad \text{(blue)}$$

We can write a Henderson–Hasselbalch equation for this, just as for other weak acids:

$$pH = pK_{In} + \log\frac{[In^-]}{[HIn]} \qquad (8.3)$$

The indicator changes color over a pH **range**. The transition range depends on the ability of the observer to detect small color changes. With indicators in which both forms are colored, generally only one color is observed if the ratio of the concentration of the two forms is 10:1; only the color of the more concentrated form is seen. From this information, we can calculate the pH transition range required to go from one color to the other. When only the color of the nonionized form is seen, $[In^-]/[HIn] = \frac{1}{10}$. Therefore,

Your eyes can generally discern only one color if it is 10 times as intense as the other.

$$pH = pK_a + \log\frac{1}{10} = pK_a - 1 \qquad (8.4)$$

When only the color of the ionized form is observed, $[In^-]/[HIn] = \frac{10}{1}$, and

$$pH = pK_a + \log\frac{10}{1} = pK_a + 1 \qquad (8.5)$$

So the pH in going from one color to the other has changed from $pK_a - 1$ to $pK_a + 1$. This is a pH change of 2, and *most indicators require a transition range of about two pH units*. During this transition, the observed color is a mixture of the two colors.

Midway in the transition, the concentrations of the two forms are equal, and $pH = pK_a$. Obviously, then, *the pK$_a$ of the indicator should be close to the pH of the equivalence point.*

Choose an indicator with a pK$_a$ near the equivalence point pH.

Calculations similar to these can be made for weak base indicators, and they reveal the same transition range; the pOH midway in the transition is equal to pK_b, and the pH equals $14 - pK_b$. Hence, a weak-base indicator should be selected such that $pH = 14 - pK_b$.

See the inside back cover for a comprehensive list of indicators.

Figure 8.4 illustrates the colors and transition ranges of some commonly used indicators. The range may be somewhat less in some cases, depending on the colors; some colors are easier to see than others. The transition is easier to see if one form of the indicator is colorless. For this reason, phenolphthalein is usually used as an indicator for strong acid–base titrations when applicable (see Figure 8.1, titration of 0.1 M HCl). In dilute solutions, however, phenolphthalein falls outside the steep portion of the titration curve (Figure 8.2), and an indicator such as bromothymol blue must be used. A similar situation applies to the titration of NaOH with HCl (Figure 8.3). A more complete list of indicators is given on the inside back cover.

Since an indicator is a weak acid or base, the amount added should be kept minimal so that it does not contribute appreciably to the pH and so that only a small amount of titrant will be required to cause the color change. That is, the color change will be sharper when the concentration is lower because less acid or base is required to convert it from one form to the other. Of course, sufficient indicator must be added to impart an easily discernible color to the solution. Generally, a few tenths percent solution (wt/vol) of the indicator is prepared and two or three drops are added to the solution to be titrated.

Two drops (0.1 mL) of 0.01 M indicator (0.1% solution with f wt = 100) is equal to 0.01 mL of 0.1 M titrant.

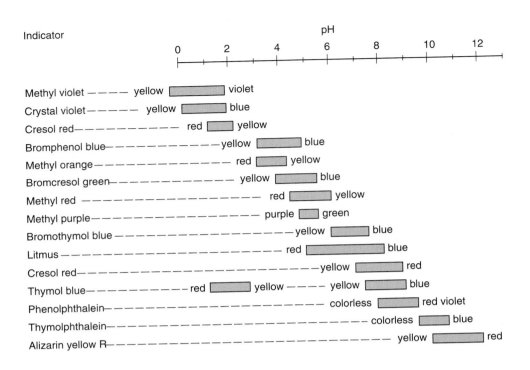

Fig. 8.4. pH transition ranges and colors of some common indicators.

8.3   Standard Acid and Base Solutions

See Chapter 2 for special procedures required to prepare and standardize acid and base solutions.

Hydrochloric acid is usually used as the strong acid titrant for the titration of bases, while sodium hydroxide is the usual titrant for acids. Most chlorides are soluble, and few side reactions are possible with HCl. It is convenient to handle.

Neither of these is a primary standard and so solutions of approximate concentrations are prepared, and then they are standardized by titrating a primary base or acid. Special precautions are required in preparing the solutions, particularly the sodium hydroxide solution. The preparation and standardization of hydrochloric acid and sodium hydroxide titrants is presented in Chapter 2.

The curve is flattest and the buffering capacity the greatest at the midpoint.

8.4   Weak Acid versus Strong Base—A Bit Less Straightforward

The titration curve for 100 mL of 0.1 M acetic acid titrated with 0.1 M sodium hydroxide is shown in Figure 8.5. The neutralization reaction is

$$\text{HOAc} + \text{Na}^+ + \text{OH}^- \rightarrow \text{H}_2\text{O} + \text{Na}^+ + \text{OAc}^- \tag{8.6}$$

The acetic acid, which is only a few percent ionized, depending on the concentration, is neutralized to water and an equivalent amount of the salt, sodium acetate. Before the titration is started, we have 0.1 M HOAc, and the pH is calculated as described for weak acids in Chapter 7. Table 8.2 summarizes the equations

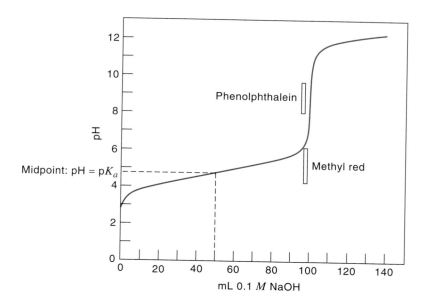

Fig. 8.5. Titration curve for 100 mL 0.1 M HOAc versus 0.1 M NaOH.

governing the different portions of the titration curve, as developed in Chapter 7. As soon as the titration is started, some of the HOAc is converted to NaOAc, and a buffer system is set up. As the titration proceeds, the pH slowly increases as the ratio [OAc$^-$]/[HOAc] changes. At the *midpoint of the titration*, [OAc$^-$] = [HOAc], and the *pH is equal to p*K_a. At the equivalence point, we have a solution of NaOAc. Since this is a Brønsted base (it hydrolyzes), *the pH at the equivalence point will be alkaline*. The pH will depend on the concentration of NaOAc (see Equation 7.32 and Figure 8.6). The greater the concentration, the higher the pH. As excess NaOH is added beyond the equivalence point, the ionization of the base OAc$^-$ is suppressed to a negligible amount (see Equation 7.23), and the pH is determined only by the concentration of excess OH$^-$. Therefore, *the titration curve beyond the equivalence point follows that for the titration of a strong acid.*

Table 8.2

Equations Governing a Weak-Acid (HA) or Weak-Base (B) Titration

Fraction F Titrated	Weak Acid		Weak Base	
	Present	Equation	Present	Equation
$F = 0$	HA	$[H^+] = \sqrt{K_a \cdot C_{HA}}$ (Eq. 7.20)	B	$[OH^-] = \sqrt{K_b \cdot C_B}$ (Example 7.8)
$0 < F < 1$	HA/A$^-$	$pH = pK_a + \log \dfrac{C_{A^-}}{C_{HA}}$ (Eq. 7.45)	B/BH$^+$	$pH = (pK_w - pK_b) + \log \dfrac{C_B}{C_{BH^+}}$ (Eq. 7.56)
$F = 1$	A$^-$	$[OH^-] = \sqrt{\dfrac{K_w}{K_a} \cdot C_{A^-}}$ (Eq. 7.32)	BH$^+$	$[H^+] = \sqrt{\dfrac{K_w}{K_b} \cdot C_{BH^+}}$ (Eq. 7.39)
$F > 1$	OH$^-$/A$^-$	$[OH^-] = $ [excess titrant]	H$^+$/BH$^+$	$[H^+] = $ [excess titrant]

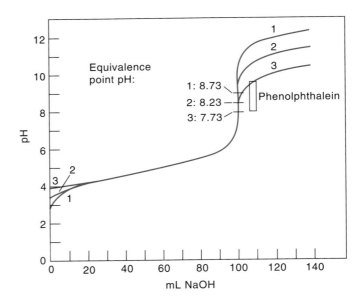

Fig. 8.6. Dependence of the titration curve of weak acid on concentration. Curve 1: 100 mL of 0.1 M HOAc versus 0.1 M NaOH. Curve 2: 100 mL of 0.01 M HOAc versus 0.01 M NaOH. Curve 3: 100 mL of 0.001 M HOAc versus 0.001 M NaOH.

· ·

Example 8.2

Calculate the pH at 0, 10.0, 25.0, 50.0, and 60.0 mL titrant in the titration of 50.0 mL of 0.100 M acetic acid with 0.100 M NaOH.

Solution

At 0 mL, we have a solution of only 0.100 M HOAc:

$$\frac{(x)(x)}{0.100 - x} = 1.75 \times 10^{-5}$$

$$[H^+] = x = 1.32 \times 10^{-3}\ M$$

$$pH = 2.88$$

Keep track of millimoles reacted and remaining.

At 10.0 mL, we started with 0.100 $M \times$ 50.0 mL = 5.00 mmol HOAc; part has reacted with OH^- and has been converted to OAc^-:

mmol HOAc at start   = 5.00 mmol HOAc
mmol OH^- added= 0.100 $M \times$ 10.0 mL = 1.00 mmol OH^-
                                                                          = mmol OAc^- formed in 60.0 mL
mmol HOAc left   = 4.00 mmol HOAc in 60.0 mL

We have a buffer. Since volumes cancel, use millimoles:

$$pH = pK_a + \log \frac{[OAc^-]}{[HOAc]}$$

$$pH = 4.76 + \log \frac{1.00}{4.00} = 4.16$$

At 25.0 mL, one-half the HOAc has been converted to OAc⁻, so pH = pK_a:

mmol HOAc at start   = 5.00 mmol HOAc

mmol OH⁻ = 0.100 M × 25.0 mL = 2.50 mmol OAc⁻ formed

mmol HOAc left   = 2.50 mmol HOAc

$$pH = 4.76 + \log \frac{2.50}{2.50} = 4.76$$

At 50.0 mL, all the HOAc has been converted to OAc⁻ (5.00 mmol in 100 mL, or 0.0500 M):

$$[OH^-] = \sqrt{\frac{K_w}{K_a}[OAc^-]}$$

$$= \sqrt{\frac{1.0 \times 10^{-14}}{1.75 \times 10^{-5}} \times 0.0500} = 5.35 \times 10^{-6} \ M$$

pOH = 5.27   pH = 8.73

At 60.0 mL, we have a solution of NaOAc and excess added NaOH. The hydrolysis of the acetate is negligible in the presence of added OH⁻. So the pH is determined by the concentration of excess OH⁻:

mmol OH⁻ = 0.100 M × 10.0 mL = 1.00 mmol in 110 mL

[OH⁻] = 0.00909 M

pOH = −2.04; pH = 11.96

• •

The slowly rising region before the equivalence point is called the **buffer region**. It is flattest at the midpoint, that is, where the ratio [OAc⁻]/[HOAc] is unity (see buffers and buffering capacity, section 7.7), and so the **buffering capacity is greatest at a pH corresponding to pK_a**. The buffering capacity also depends on the concentrations of HOAc and OAc⁻, and the **total buffering capacity** increases as the concentration increases. In other words, the distance of the flat portion on either side of pK_a will increase as [HOAc] and [OAc⁻] increase. As the pH deviates to the acid side for pK_a, the buffer will tolerate more base but less acid; the change in pH with a given small amount of added base will be greater, though, than at a pH equal to pK_a because the curve is not so flat. Conversely, on the alkaline side of pK_a, more acid but less base can be tolerated. See Chapter 7 for a discussion of buffer capacity.

See Section 7.7 for a quantitative description of buffer capacity.

You may have noticed that the corresponding region for a strong acid–strong base titration (Figures 8.1 and 8.2) is much flatter than for the weak-acid case. In this respect, a solution of a strong acid or of a strong base is much more resistant to pH change upon addition of H⁺ or OH⁻ than the buffer systems we have discussed. The problem is that they are restricted to a very narrow pH region, either very acid or very alkaline, especially if the acid or base concentration is to be strong enough to have any significant capacity against a pH change. So these are regions that are rarely of practical value for buffering. Also, solutions of strong acids and bases are not resistant to pH change upon dilution, as buffers are. Therefore, we usually use mixtures of weak acids or bases with their salts, so that the desired pH

Strong acids are actually good buffers, except their pH changes with dilution.

region can be selected. Often, a buffer is used only to give a specified pH, and no extraneous acids or bases are added. A desired pH can be obtained more easily with conventional buffers than with a strong acid or a strong base.

The transition range of the indicator for this titration of a weak acid must fall within a pH range of about 7 to 10 (Figure 8.5). Phenolphthalein fits this nicely. If an indicator such as methyl red were used, it would begin changing color shortly after the titration began and would gradually change to the alkaline color up to pH 6, before the equivalence point was even reached.

The dependence of the shape of the titration curve and of the equivalence point pH on concentration is shown in Figure 8.6 for different concentrations of HOAc and NaOH. Obviously, phenolphthalein could not be used as an indicator for solutions as dilute as 10^{-3} M (curve 3). Note that the equivalence point pH decreases as the weak-acid system becomes more dilute (which doesn't happen in the strong-acid system).

Weak-acid titrations require careful selection of the indicator.

	A	B	C	D	E	F	G	H
							K_w:	1.00E-14
1	50.00 mL 0.05000M HOAc vs. 0.05000M NaOH.							
2	mL_{HOAc}:	50.00	M_{HOAc}:	0.05000	M_{NaOH}:	0.05000	K_a:	1.75E-05
3	mL_{NaOH}	Solution	[HOAc]	[OAc⁻]	[H⁺]	[OH⁻]	pOH	pH
4	0.00	HOAc			0.0009354			3.03
5	2.00	HOAc/OAc⁻	0.046154	0.001923				3.38
6	5.00	HOAc/OAc⁻	0.040909	0.004545				3.80
7	10.00	HOAc/OAc⁻	0.033333	0.008333				4.15
8	15.00	HOAc/OAc⁻	0.026923	0.011538				4.39
9	20.00	HOAc/OAc⁻	0.021429	0.014286				4.58
10	25.00	HOAc/OAc⁻	0.016667	0.016667				4.76
11	30.00	HOAc/OAc⁻	0.0125	0.01875				4.93
12	35.00	HOAc/OAc⁻	0.008824	0.020588				5.12
13	40.00	HOAc/OAc⁻	0.005556	0.022222				5.36
14	45.00	HOAc/OAc⁻	0.002632	0.023684				5.71
15	48.00	HOAc/OAc⁻	0.00102	0.02449				6.14
16	49.00	HOAc/OAc⁻	0.000505	0.024747				6.45
17	49.50	HOAc/OAc⁻	0.000251	0.024874				6.75
18	49.95	HOAc/OAc⁻	2.5E-05	0.024987				7.76
19	50.00	OAc⁻				3.77964E-06	5.42	8.58
20	50.05	OH⁻				2.49875E-05	4.60	9.40
21	50.50	OH⁻				0.000248756	3.60	10.40
22	51.00	OH⁻				0.00049505	3.31	10.69
23	52.00	OH⁻				0.000980392	3.01	10.99
24	55.00	OH⁻				0.002380952	2.62	11.38
25	60.00	OH⁻				0.004545455	2.34	11.66
26	65.00	OH⁻				0.006521739	2.19	11.81
27	70.00	OH⁻				0.008333333	2.08	11.92
28	Formulas for cells in **boldface**:							
29	Cell E4:	[H⁺] = $(K_a[HOAc])^{1/2}$ =		SQRT($H$2*$D$2)				
30	Cell H4:	pH = -log[H⁺] =		(-LOG10(E4))				
31	Cell C5:	[HOAc] = (mL_{HOAc} x M_{HOAc} - mL_{NaOH} x M_{NaOH})/(mL_{HOAc} + mL_{NaOH})						
32		= ($B$2*$D$2-A5*$F$2)/($B$2+A5)		(Copy through Cell C18)				
33	Cell D5:	[OAc-] = (mL_{NaOH} x M_{NaOH})/(mL_{HOAc} + mL_{NaOH})						
34		= (A5*$F$2)/($B$2+A5)		(Copy through Cell D18)				
35	Cell H5:	pH = pKa + log([OAc-]/HOAc])						
36		= (-LOG10($H$2)+LOG10(D5/C5))		(Copy through Cell H18)				
37	Cell F19:	[OH⁻] = $(K_w[OAc⁻]/K_a)^{1/2}$						
38		= SQRT($H$1*($D$2/2)/$H$2)						
39	Cell G19:	pOH = -log[OH⁻] =		(-LOG10(F19))		(Copy to end)		
40	Cell H19:	pH = 14 - pOH =		14-G19	(Copy to end)			
41	Cell F20:	[OH⁻] = (mL_{NaOH} x M_{NaOH} - mL_{HOAc} x M_{HOAc})/(mL_{HOAc} + mL_{NaOH})						
42		= (A20*$F$2-$B$2*$D$2)/($B$2+A20)		(Copy to end)				

Spreadsheet to accompany discussion on the facing page.

SPREADSHEET EXERCISE—HOAc VERSUS NaOH

Prepare a spreadsheet to construct the titration curve for the titration of 50 mL 0.05 M HOAc with 0.05 M NaOH. First enter the absolute values for the volume (cell B2) and concentration (cell D2) of HOAc, the concentration of NaOH (cell F2), K_w (cell G1), and K_a (cell H2). These will be used in the following formulas to calculate the pH as a function of volume of NaOH added.

At the beginning of the titration, [H$^+$] is equal to $\sqrt{K_a[\text{HOAc}]}$ (**cell E4** formula). The pH is calculated from this (**cell H4** formula). Once the titration is begun and up to the equivalence point, we have a buffer of HOAc and OAc$^-$, and so we must calculate the concentrations of each. The former is the number of millimoles of remaining HOAc divided by the total volume (**cell C5** formula). The [OAc$^-$] is the number of millimoles NaOH added divided by the total volume (**cell D5** formula). The pH is calculated from the Henderson–Hasselbach equation (**cell H5** formula). At the equivalence point, [OH$^-$] is determined by the concentration of OAc$^-$ (**cell F19** formula), and pOH (**cell G19**) and pH (**cell H19**) are calculated from this. The pH for the remainder of the titration is calculated from the excess millimoles NaOH and total volume (**cell F20** formula), as was done for the titration of HCl.

The titration curve is plotted using Chart Wizard, by plotting A4:A27 (X-axis) versus H4:H27 (Y-axis). It is shown (Chart 1) in your CD, Chapter 8.

The equivalence point for the titration of any weak acid with a strong base will be alkaline. *The weaker the acid* (the smaller the K_a), the larger the K_b of the salt and *the more alkaline the equivalence point.*

We should note that the pH calculated from the relatively simple equations presented here break down near the equivalence point for weak acids and bases because the assumptions used in deriving them no longer apply. Even for strong acid–base titrations, we reach a point very near the equivalence point where the ionization of water becomes appreciable compared to the acid or excess base concentration, and the calculations are in error. You can satisfy yourself of these limitations by inserting in the spreadsheet examples titrant values that are, say, 99.99% or 99.999% and see where the calculated pH falls off the otherwise smooth titration curve.

Figure 8.7 shows the titration curves for 100 mL of 0.1 M solutions of weak acids of different K_a values titrated with 0.1 M NaOH. The sharpness of the end

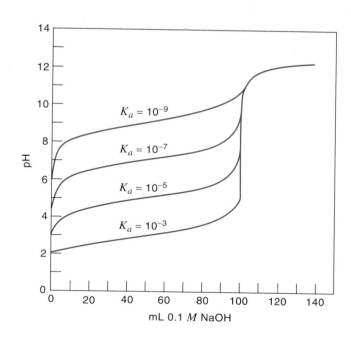

Fig. 8.7. Titration curves for 100 mL 0.1 M weak acids of different K_a values versus 0.1 M NaOH.

point decreases as K_a decreases. As in Figure 8.6, the sharpness will also decrease as the concentration decreases. Generally, for macrotitrations (ca. 0.1 M), acids with K_a values of 10^{-6} can be titrated accurately with a visual indicator; and with suitable color comparisons, those with K_a values approaching 10^{-8} can be titrated with reasonable accuracy. A pH meter can be used to obtain better precision for the very weak acids by plotting the titration curve.

8.5 Weak Base versus Strong Acid

The titration of a weak base with a strong acid is completely analogous to the above case, but the titration curves are the reverse of those for a weak acid versus a strong base. The titration curve for 100 mL of 0.1 M ammonia titrated with 0.1 M hydrochloric acid is shown in Figure 8.8. The neutralization reaction is

$$NH_3 + H^+ + Cl^- \rightarrow NH_4^+ + Cl^- \tag{8.7}$$

At the beginning of the titration, we have 0.1 M NH_3, and the pH is calculated as described for weak bases in Chapter 7. See Table 8.2. As soon as some acid is added, some of the NH_3 is converted to NH_4^+, and we are in the buffer region. At the *midpoint of the titration*, $[NH_4^+]$ equals $[NH_3]$, and the *pH is equal to (14 − pK$_b$)*. At the equivalence point, we have a solution of NH_4Cl, a weak Brønsted acid that hydrolyzes to give an **acid solution**. Again, the pH will depend on the concentration; the greater the concentration, the lower the pH (see Equation 7.39). Beyond the equivalence point, the free H^+ suppresses the ionization (see Equation 7.33), and the pH is determined by the concentration of H^+ added in excess. Therefore, the titration curve beyond the equivalence point will follow that for titration of a strong base (Figure 8.3). Because K_b for ammonia happens to be equal to K_a for acetic acid, the titration curve for ammonia versus a strong acid is just the mirror image of the titration curve for acetic acid versus a strong base.

The indicator for the titration in Figure 8.8 must have a transition range within about pH 4 to 7. Methyl red meets this requirement, as shown in the figure. If phenolphthalein had been used as the indicator, it would have gradually lost its color between pH 10 and 8, before the equivalence point was reached.

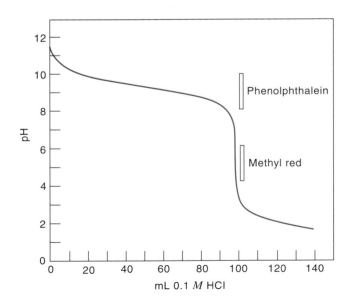

Fig. 8.8. Titration curve for 100 mL 0.1 M NH$_3$ versus 0.1 M HCl.

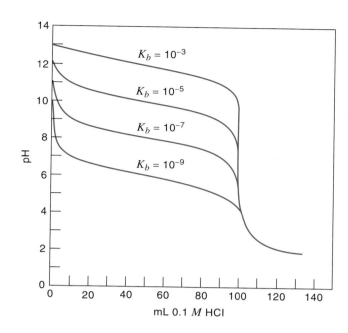

Fig. 8.9. Titration curves for 100 mL 0.1 M weak bases of different K_b values versus 0.1 M HCl.

Titration curves for different concentrations of NH_3 titrated with varying concentrations of HCl would be the mirror images of the curves in Figure 8.6. Methyl red could not be used as an indicator in dilute solutions. The titration curves for weak bases of different K_b values (100 mL, 0.1 M) versus 0.1 M HCl are shown in Figure 8.9. In macrotitrations, one can accurately titrate a base with a K_b of 10^{-6} using a visual indicator.

8.6 Titration of Sodium Carbonate—A Diprotic Base

Sodium carbonate is a Brønsted base that is used as a primary standard for the standardization of strong acids. It hydrolyzes in two steps:

$$CO_3^{2-} + H_2O \rightleftharpoons HCO_3^- + OH^- \quad K_{H1} = K_{b1} = \frac{K_w}{K_{a2}} = 2.1 \times 10^{-4} \quad (8.8)$$

$$HCO_3^- + H_2O \rightleftharpoons CO_2 + H_2O + OH^- \quad K_{H2} = K_{b2} = \frac{K_w}{K_{a1}} = 2.3 \times 10^{-8} \quad (8.9)$$

where K_{a1} and K_{a2} refer to the K_a values of H_2CO_3; HCO_3^- is the conjugate acid of CO_3^{2-} and H_2CO_3 is the conjugate acid of HCO_3^-; and the K_b values are calculated as described in Chapter 7 for salts of weak acids and bases (i.e., from $K_a K_b = K_w$).

Sodium carbonate can be titrated to give end points corresponding to the stepwise additions of protons to form HCO_3^- and CO_2. (Carbonic acid, H_2CO_3, dissociates in acid solution to CO_2—its acid anhydride—and H_2O in acid solution.) *The K_b values should differ by at least 10^4 to obtain good separation of the equivalence point breaks in a case such as this.*

A titration curve for Na_2CO_3 versus HCl is shown in Figure 8.10 (solid line). Even though K_{b1} is considerably larger than the 10^{-6} required for a sharp end point, the pH break is decreased by the formation of CO_2 beyond the first equivalence point. The second end point is not very sharp either because K_{b2} is smaller than the

The dashed line is for HCO_3^- only, for which $[H^+] = \sqrt{K_{a1}K_{a2}}$ for the horizontal portion.

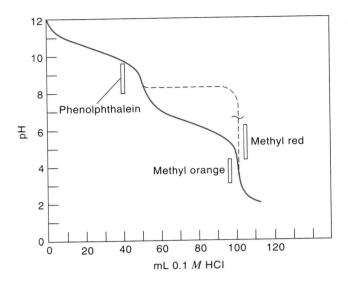

Fig. 8.10. Titration curve for 50 mL 0.1 M Na$_2$CO$_3$ versus 0.1 M HCl. Dashed line represents a boiled solution with CO$_2$ removed.

10^{-6} we would like. Fortunately, this end point can be sharpened, because the CO$_2$ produced from the neutralization of HCO$_3^-$ is volatile and can be boiled out of the solution. This is described below.

At the start of the titration, the pH is determined by the hydrolysis of the Brønsted base CO$_3^{2-}$. After the titration is begun, part of the CO$_3^{2-}$ is converted to HCO$_3^-$, and a CO$_3^{2-}$/HCO$_3^-$ buffer region is established. At the first equivalence point, there remains a solution of HCO$_3^-$, and $[H^+] \approx \sqrt{K_{a1}K_{a2}}$. Beyond the first equivalence point, the HCO$_3^-$ is partially converted to H$_2$CO$_3$(CO$_2$) and a second buffer region is established, the pH being established by [HCO$_3^-$]/[CO$_2$]. The pH at the second equivalence point is determined by the concentration of the weak acid CO$_2$.

Phenolphthalein is used to detect the first end point, and methyl orange is used to detect the second one. Neither end point, however, is very sharp. In actual practice, the phenolphthalein end point is used only to get an approximation of where the second end point will occur; phenolphthalein is colorless beyond the first end point and does not interfere. The second equivalence point, which is used for accurate titrations, is normally not very accurate with methyl orange indicator because of the gradual change in the color of the methyl orange. This is caused by the gradual decrease in the pH due to the HCO$_3^-$/CO$_2$ buffer system beyond the first end point.

If beyond the first equivalence point we were to boil the solution after each addition of HCl to remove the CO$_2$ from the solution, the buffer system of HCO$_3^-$/CO$_2$ would be removed, leaving only HCO$_3^-$ in solution. This is both a weak acid and a weak base whose pH ($\approx$8.3) is independent of concentrations ($[H^+] = \sqrt{K_{a1}K_{a2}}$ or $[OH^-] = \sqrt{K_{b1}K_{b2}}$; see Chapter 7). In effect, then, the pH would remain essentially constant until the equivalence point, when we are left with a neutral solution of water and NaCl (pH = 7). The titration curve would then follow the dashed line in Figure 8.10.

The following procedure can, therefore, be employed to sharpen the end point, as illustrated in Figure 8.11. Methyl red is used as the indicator, and the titration is continued until the solution just turns from yellow through orange to a definite red color. This will occur just before the equivalence point. The change will be very gradual because the color will start changing at about pH 6.3, well before the equivalence point. At this point, the titration is stopped and the solution is gently boiled to remove CO$_2$. The color should now revert to yellow because we have a

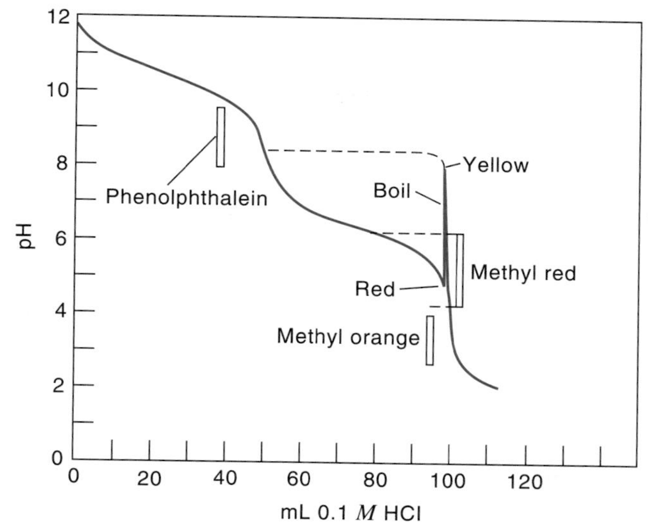

Boiling the solution removes the CO_2, raising the pH to that of HCO_3^-.

Fig. 8.11. Titration of 50 mL 0.1 *M* Na_2CO_3 with 0.1 *M* HCl using methyl red indicator.

dilute solution of only HCO_3^-. The solution is cooled and the titration is continued to a sharp color change to red or pink. The equivalence point here does not occur at pH 7, as indicated on the dashed line in Figure 8.10, because there is a small amount of HCO_3^- still remaining to be titrated after boiling. That is, there will still be a slight buffering effect throughout the remainder of the titration, and dilute CO_2 will still remain at the equivalence point.

Bromcresol green can be used in a manner similar to methyl red. Its transition range is pH 3.8 to 5.4, with a color change from blue through pale green to yellow (see Experiment 7). Similarly, methyl purple can be used (see Experiment 19).

The methyl orange end point can be used (without boiling) by adding a blue dye, xylene cyanole FF, to the indicator. This mixture is called **modified methyl orange**. The blue color is complementary to the orange color of the methyl orange at about pH 2.8. This imparts a gray color at the equivalence point, the transition range of which is smaller than that of methyl orange. This results in a sharper end point. It is still not so sharp as the methyl red end point. Methyl orange can also be used by titrating to the color of the indicator in a solution of potassium acid phthalate, which has a pH close to 4.0.

8.7 Titration of Polyprotic Acids

Diprotic acids can be titrated stepwise just as sodium carbonate was. In order to obtain good end point breaks for titration of the first proton, K_{a1} should be at least $10^4 \times K_{a2}$. If K_{a2} is in the required range of 10^{-7} to 10^{-8} for a successful titration, an end point break is obtained for titrating the second proton. Triprotic acids (e.g., H_3PO_4) can be titrated similarly, but K_{a3} is generally too small to obtain an end point break for titration of the third proton. Figure 8.12 illustrates the titration curve for a diprotic acid H_2A, and Table 8.3 summarizes the equations governing the different portions of the titration curve. The pH at the beginning of the titration is determined from the ionization of the first proton if the solution is not too dilute (see discussion of polyprotic acids in Chapter 7). If K_{a1} is not too large and the amount dissociated is ignored compared to the analytical concentration of the acid, the approximate equation given can be used to calculate $[H^+]$. Otherwise the quadratic formula must be used to solve Equation 7.20 (see Example 7.17).

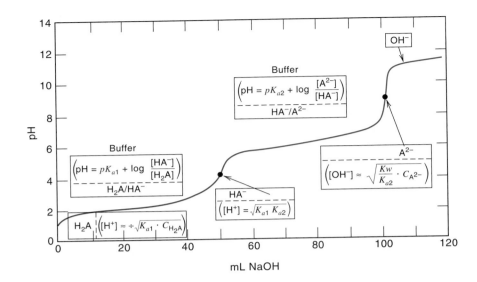

Fig. 8.12. Titration of diprotic acid, H_2A, with sodium hydroxide.

During titration up to the first equivalence point, an HA^-/H_2A buffer region is established. At the first equivalence point, a solution of HA^+ exists, and $[H^+] \approx \sqrt{K_{a1}K_{a2}}$. Beyond this point, an A^{2-}/HA^- buffer exists; and finally at the second equivalence point, the pH is determined from the hydrolysis of A^{2-}. If the salt, A^{2-}, is not too strong a base, then the approximate equation given can be used to calculate $[OH^-]$. Otherwise the quadratic equation must be used to solve Equation 7.29 (see Example 7.19).

Figure 8.12 and Table 8.3 illustrate which species are present at each part of the titration curve. As noted, if either K_{a1} or K_{b1} ($=K_w/K_{a1}$) is fairly large, we can't make the simplifying assumptions, and the quadratic equation must be used for the *beginning of the titration* or at the *second equivalence point*. In practice, there are few diprotic acids where this is not the case for one or the other. Furthermore, if K_{a1} is fairly large (e.g., chromic acid, $K_{a1} = 0.18$, $K_{a2} = 3.2 \times 10^{-7}$), then the

Table 8.3
Equations Governing Diprotic Acid (H_2A) Titration

Fraction F Titrated	Present	Equation
$F = 0$ (0%)	H_2A	$[H^+] \approx \sqrt{K_{a1}C_{H_2A}}$ (Example 7.7) (or Eq. 7.20; Ex. 7.17, quadratic, if ~strong acid)
$0 < F < 1$ (>0 to <100%)	H_2A/HA^-	$pH = pK_{a1} + \log \dfrac{C_{HA^-}}{C_{H_2A}}$ (Eq. 7.45) (or $C_{HA^-} + [H^+]$ and $C_{H_2A}-[H^+]$ if a strong acid)
$F = 1$ (100%) (1st eq. pt.)	HA^-	$[H^+] \approx \sqrt{K_{a1}K_{a2}}$ (Eq. 7.84) (or Eq. 7.83 if H_2A ~strong acid)
$1 < F < 2$ (>100 to <200%)	HA^-/A^{2-}	$pH = pK_{a2} + \log \dfrac{C_{A^{2-}}}{C_{HA^-}}$ (Eq. 7.45, Ex. 7.16, 7.22)
$F = 2$ (200%) (2nd eq. pt.)	A^{2-}	$[OH^-] \approx \sqrt{\dfrac{K_w}{K_{A2}} \cdot C_{A^{2-}}}$ (Eq. 7.32) (or Eq. 7.29, Ex. 7.19, quadratic if A^{2-} ~strong base)
$F > 2$ (>200%)	OH^-/A^{2-}	$[OH^-] = $ [excess titrant]

Henderson–Hasselbalch equation cannot be used for the *first buffer region* because the assumption in deriving that from the K_{a1} expression was that the amount of H^+ or OH^- from dissociation or hydrolysis of H_2A or HA^- was not appreciable compared to their concentrations. For a fairly strong acid, then, we can write

$$H_2A = H^+ + HA^-$$

and

$$HA^- + H_2O = H_2A + OH^-$$

The equilibrium concentration of H_2A is decreased from the calculated analytical concentration by an amount equal to $[H^+]$ and increased by an amount equal to $[OH^-]$:

$$[H_2A] = C_{H_2A} - [H^+] - [OH^-]$$

That of HA^- is increased by $[OH^-]$ and decreased by $[H^+]$:

$$[HA^-] = C_{HA^-} + [H^+] - [OH^-]$$

Since the solution is acid, we can neglect $[OH^-]$:

$$[H_2A] = C_{H_2A} - [H^+]$$
$$[HA^-] = C_{HA^-} + [H^+]$$

(For simplification, we could have written the same equations just from the H_2A dissociation above.) So, for a fairly strong acid, we must substitute in the $K_{a1} = [H^+][HA^-]/[H_2A]$ expression: $[H_2A] = C_{H_2A} - [H^+]$, and $[HA^-] = C_{HA^-} + [H^+]$, and solve a quadratic equation for the buffer region, where C_{H_2A} and C_{HA^-} are the calculated concentrations resulting from the acid–base reaction at a given point in the titration:

$$[H^+]^2 + (K_{a1} + C_{HA^-})[H^+] - K_{a1}C_{H_2A} = 0$$

Also, if K_{b1} for A^{2-} is fairly large, then just beyond the second equivalence point, OH^- from hydrolysis of A^{2-} cannot be ignored compared to the concentration of the OH^- from excess titrant. So we continue to use the K_{b1} expression, $K_{b1} = K_w/K_{a2} = [HA^-][OH^-]/[A^{2-}]$: $[A^{2-}] = C_{A^{2-}} - [OH^-]$, $[HA^-] = [OH^-]$, and solve the quadratic equation. After addition of, say, 0.5 mL, of NaOH, there is enough excess OH^- to suppress the hydrolysis of A^{2-}, and we can calculate the pH just from the excess OH^- concentration. Finally, for HA^- at the *first equivalence point*, we may have to use the more exact Equation 7.83 instead of 7.84 for calculation of $[H^+]$ to get the correct pH, since K_{a1} may not be negligible compared to $[HA^-]$ ($K_{a1}K_w$ in the numerator will probably still be negligible). So,

$$[H^+] = \sqrt{\frac{K_{a1}K_{a2}[HA^-]}{K_{a1} + [HA^-]}}$$

And very near the equivalence points, even more complicated expressions may be required.

    If the systematic approach using mass balance equations is used for calculations, then the exact mass balance expressions are used in which, similarly, simplifying assumptions are not invoked (except in acid solution, OH^- is excluded—but not H^+—and in the alkaline solution of A^{2-}, H^+ can be excluded—but not OH^-).

We will not construct a diprotic titration curve here, but if you want a good mental exercise, try it! You just can't make the simplifying assumptions that we can usually use with monoprotic acids that are sufficiently weak or not too dilute. See your CD, Chapter 8, for auxiliary data for the spreadsheet calculation of the titration curve for 50.00 mL 0.1000 M H_2CrO_4 versus 0.1000 M NaOH. You can download that and enter the K_{a1} and K_{a2} values for other diprotic acids and see what their titration curves look like. Try, for example, maleic acid. For the calculations, we used the more exact equations mentioned above for the initial pH, the first buffer zone, and the first equivalence point. We did not use the quadratic equation for the second equivalence point since CrO_4^{2-} is a quite weak base ($K_{b1} = 3.12 \times 10^{-8}$). See Ref. 8 for other examples of calculated titration curves.

8.8 Mixtures of Acids or Bases

One acid should be at least 10^4 weaker than the other to titrate separately.

Mixtures of acids (or bases) can be titrated stepwise if there is an appreciable difference in their strengths. *There must generally be a difference in* K_a *values of at least 10^4*, unless perhaps a pH meter is used to construct the titration curve. If one of the acids is a strong acid, a separate end point will be observed for the weak acid *only if K_a is about 10^{-5} or smaller.* See, for example, Figure 8.13, where only a small break is seen for the HCl. The stronger acid will titrate first and will give a pH break at its equivalence point. This will be followed by titration of the weaker acid and a pH break at its equivalence point. The titration curve for a mixture of hydrochloric acid and acetic acid versus sodium hydroxide is shown in Figure 8.13. At the equivalence point for HCl, a solution of HOAc and NaCl remains, and so the equivalence point is acidic. Beyond the equivalence point, the OAc^-/HOAc buffer region is established, and this markedly suppresses the pH break for HCl. The remainder of the titration curve is identical to Figure 8.5 for the titration of HOAc.

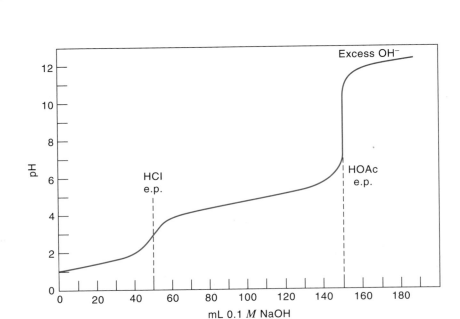

Fig. 8.13. Titration curve for 50 mL of mixture of 0.1 M HCl and 0.2 M HOAc with 0.1 M NaOH.

If two strong acids are titrated together, there will be no differentiation between them, and only one equivalence point break will occur, corresponding to the titration of both acids. The same is true for two weak acids if their K_a values are not too different. For example, a mixture of acetic acid, $K_a = 1.75 \times 10^{-5}$, and propionic acid, $K_a = 1.3 \times 10^{-5}$, would titrate together to give a single equivalence point.

With H_2SO_4, the first proton is completely dissociated and the second proton has a K_a of about 10^{-2}. Therefore, the second proton is ionized sufficiently to titrate as a strong acid, and only one equivalence point break is found. The same is true for a mixture of a strong acid and a weak acid with a K_a in the neighborhood of 10^{-2}.

The first ionization constant of sulfurous acid, H_2SO_3, is 1.3×10^{-2}, and the second ionization constant is 5×10^{-6}. Therefore, in a mixture with HCl, the first proton of H_2SO_3 would titrate along with the HCl, and the pH at the equivalence point would be determined by the HSO_3^- remaining; that is, $[H^+] = \sqrt{K_{a1}K_{a2}}$, since HSO_3^- is both an acid and a base. This would be followed by titration of the second proton to give a second equivalence point. The volume of titrant required to reach the first end point would always be greater than that in going from the first to the second since the first includes the titration of both acids. The amount of H_2SO_3 could be determined from the amount of base required for the titration of the second proton. The amount of HCl could be found by subtracting from the first end point the volume of base required to titrate the second proton of H_2SO_3, which is equal to the volume required to titrate the first proton. In practice, this titration actually would find little practical use because H_2SO_3 is volatilized as SO_2 in strong acid solution.

Phosphoric acid in mixture with a strong acid acts in a manner similar to the above example. The first proton titrates with the strong acid, followed by titration of the second proton to give a second equivalence point; the third proton is too weakly ionized to be titrated.

Example 8.3

A mixture of HCl and H_3PO_4 is titrated with 0.1000 M NaOH. The first end point (methyl red) occurs at 35.00 mL, and the second end point (bromthymol blue) occurs at a total of 50.00 mL (15.00 mL after the first end point). Calculate the millimoles HCl and H_3PO_4 present in the solution.

Solution

The second end point corresponds to that in the titration of one proton of H_3PO_4 ($H_2PO_4^- \rightarrow HPO_4^-$). Therefore, the millimoles H_3PO_4 is the same as the millimoles NaOH used in the 15.00 mL for titrating that proton:

$$mmol_{H_3PO_4} = M_{NaOH} \times mL_{NaOH} = 0.1000 \text{ mmol/mL} \times 15.00 \text{ mL}$$
$$= 1.500 \text{ mmol}$$

The HCl and the first proton of H_3PO_4 titrate together. A 15.00-mL portion of base was used to titrate the first proton of H_3PO_4 (same as for the second proton), leaving 20.00 mL used to titrate the HCl. Therefore,

$$mmol_{HCl} = 0.1000 \text{ mmol/mL} \times (35.00 - 15.00) \text{ mL} = 2.000 \text{ mmol}$$

Similarly, mixtures of bases can be titrated if their strengths are sufficiently different. Again, the difference in K_b values must be at least 10^4. Also, if one base is a strong base, the weak base must have a K_b no greater than about 10^{-5} to obtain separate end points. For example, sodium hydroxide does not give a separate end point from that for the titration of CO_3^{2-} to HCO_3^- when titrated in the presence of Na_2CO_3 ($K_{b1} = 2.1 \times 10^{-4}$).

8.9 Titration of Amino Acids—They Are Acids and Bases

Amino acids are important in pharmaceutical chemistry and biochemistry. These are amphoteric substances that contain both acidic and basic groups (i.e., they can act as acids or bases). The acid group is a carboxylic acid group ($-CO_2H$), and the basic group is an amine group ($-NH_2$). In aqueous solutions, these substances tend to undergo internal proton transfer from the carboxylic acid group to the amino group because the RNH_2 is a stronger base than RCO_2^-. The result is a **zwitterion**:

$$R-CH-CO_2^-$$
$$|$$
$$NH_3^+$$

Since they are amphoteric, these substances can be titrated with either a strong acid or a strong base. Many amino acids are too weak to be titrated in aqueous solutions, but some will give adequate end points, especially if a pH meter is used to construct a titration curve.

We can consider the conjugate acid of the zwitterion as a **diprotic acid**, which ionizes stepwise:

$$R-CH-CO_2H \overset{K_{a1}}{\rightleftharpoons} H^+ + R-CH-CO_2^- \overset{K_{a2}}{\rightleftharpoons} H^+ + R-CH-CO_2^-$$

with NH_3^+ (conjugate acid of zwitterion), NH_3^+ (zwitterion), NH_2 (conjugate base of zwitterion) below the respective structures. (8.10)

For the zwitterion, $[H^+] \approx \sqrt{K_{a1}K_{a2}}$.

K_{a1} and K_{a2} values are frequently tabulated for amino acids (see Table C.1 in Appendix C). The values listed represent the successive ionization of the protonated form (i.e., the conjugate acid of the zwitterion); it ionizes to give first the amphoteric zwitterion and second to give the conjugate base, which is the same as a salt of a weak acid that hydrolyzes. Acid–base equilibria of amino acids are therefore treated just as for any other diprotic acid. The hydrogen ion concentration of the zwitterion is calculated in the same way as for any amphoteric salt, such as HCO_3^-, as we described in Chapter 7; that is,

$$[H^+] = \sqrt{K_{a1}K_{a2}} \tag{8.11}$$

The titration of amino acids is not unlike the titration of other amphoteric substances, such as HCO_3^-. In the latter case, titration with a base gives CO_3^{2-}, with an intermediate CO_3^{2-}/HCO_3^- buffer region, and titration with an acid gives H_2CO_3, with an intermediate HCO_3^-/H_2CO_3 buffer region.

When the zwitterion of an amino acid is titrated with strong acid, a buffer region is first established, consisting of the zwitterion (the "salt") and the conjugate acid. Halfway to the equivalence point, pH = pK_{a1} (just as with $HCO_3^-/$

H_2CO_3); and at the equivalence point, the pH is determined by the conjugate acid (and K_{a1}, as with H_2CO_3). When the zwitterion is titrated with a strong base, a buffer region of conjugate base (the "salt") and zwitterion (now the "acid") is established. Halfway to the equivalence point, $pH = pK_{a2}$ (as with CO_3^{2-}/HCO_3^-); and at the equivalence point, the pH is determined by the conjugate base (whose $K_b = K_{a2}/K_w$, as with CO_3^{2-}).

Amino acids may contain more than one carboxyl or amine group; in these cases, they may yield stepwise end points like other polyprotic acids (or bases), provided the different groups differ in K's by 10^4 and are still strong enough to be titrated.

8.10 Kjeldahl Analysis: Protein Determination

An important method for accurately determining nitrogen in proteins and other nitrogen-containing compounds is the **Kjeldahl analysis**. The quantity of protein can be calculated from a knowledge of the percent nitrogen contained in it. Although other more rapid methods for determining proteins exist, the Kjeldahl method is the standard on which all other methods are based.

The material is digested with sulfuric acid to decompose it and convert the nitrogen to ammonium hydrogen sulfate:

$$C_aH_bN_c \xrightarrow[\text{catalyst}]{H_2SO_4} aCO_2 \uparrow + \tfrac{1}{2}bH_2O + cNH_4HSO_4$$

The solution is cooled, concentrated alkali is added to make the solution alkaline, and the volatile ammonia is distilled into a solution of standard acid, which is in excess. Following distillation, the excess acid is back-titrated with standard base.

$$cNH_4HSO_4 \xrightarrow{OH^-} cNH_3 \uparrow + cSO_4^{2-}$$

$$cNH_3 + (c + d)HCl \longrightarrow cNH_4Cl + dHCl$$

$$dHCl + dNaOH \longrightarrow \tfrac{1}{2}dH_2O + dNaCl$$

> mmol N(c) = mmol reacted HCl = mmol HCl taken
> $\times$ ($c + d$) − mmol NaOH(d)
> mmol $C_aH_bN_c$ = mmol N $\times$ 1/c

The digestion is speeded up by adding potassium sulfate to increase the boiling point and by a catalyst such as a selenium or copper salt. The amount of the nitrogen-containing compound is calculated from the weight of nitrogen analyzed by multiplying it by the gravimetric factor.

Example 8.4

A 0.2000-g sample containing urea,

$$\underset{NH_2 - C - NH_2}{\overset{\displaystyle O \atop \displaystyle \|}{}}$$

is analyzed by the Kjeldahl method. The ammonia is collected in 50.00 mL of 0.05000 M H_2SO_4, and the excess acid is back-titrated with 0.05000 M NaOH, a procedure requiring 3.40 mL. Calculate the percent urea in the sample.

Solution

The titration reaction is

$$H_2SO_4 + 2NaOH \rightarrow Na_2SO_4 + 2H_2O$$

So there are one-half as many millimoles excess H_2SO_4 as NaOH that reacted with it. The reaction with the NH_3 is

$$2NH_3 + H_2SO_4 \rightarrow (NH_4)_2SO_4$$

There are twice as many millimoles NH_3 as H_2SO_4 that reacted with it. And there are one-half as many millimoles urea as NH_3. Therefore,

$$\% \text{ urea} = \{[\text{mmol}_{H_2SO_4} - \text{mmol}_{NaOH} \times \tfrac{1}{2} \text{ (mmol } H_2SO_4/\text{mmol NaOH)}]$$
$$\times 2(\text{mmol } NH_3/\text{mmol } H_2SO_4) \times \tfrac{1}{2} [\text{mmol } (NH_2)_2CO/\text{mmol } NH_3]$$
$$\times (NH_2)_2CO \text{ mg/mmol urea}\}/\text{mg}_{\text{sample}} \times 100\%$$
$$= \frac{(0.0500 \ M \times 50.00 \text{ mL} - 0.0500 \ M \times 3.40 \text{ mL} \times \tfrac{1}{2}) \times 2 \times \tfrac{1}{2} \times 60.05}{200}$$
$$\times 100\% = 72.51\%$$

● ●

Instead of first calculating the millimoles of NH_3, we could have calculated the weight of $(NH_2)_2CO$ directly by multiplying the millimoles of reacted H_2SO_4 by the molecular weight of urea, 60.05; the number of millimoles of $(NH_2)_2CO$ is equal to the number of millimoles of H_2SO_4.

Many proteins contain nearly the same amounts of nitrogen.

> A large number of different **proteins** contain very nearly the same percentage of nitrogen. The gravimetric factor for conversion of weight of N to weight of protein for normal mixtures of serum proteins (globulins and albumin) and protein in feeds is 6.25, (i.e., the proteins contain 16% nitrogen). When the sample is made up almost entirely of gamma globulin, the factor 6.24 is more accurate, while if it contains mostly albumin, 6.27 is preferred.

In the conventional Kjedahl method, two standard solutions are required, the acid for collecting the ammonia and the base for back-titration. A modification can be employed that requires only standard acid for direct titration. The ammonia is collected in a solution of boric acid. In the distillation, an equivalent amount of ammonium borate is formed:

$$NH_3 + H_3BO_3 \rightleftharpoons NH_4^+ + H_2BO_3^- \tag{8.12}$$

Boric acid is too weak to be titrated, but the borate, which is equivalent to the amount of ammonia, is a fairly strong Brønsted base that can be titrated with a standard acid to a methyl red end point. The boric acid is so weak it does not interfere, and its concentration need not be known accurately.

Example 8.5

A 0.300-g feed sample is analyzed for its protein content by the modified Kjeldahl method. If 25.0 mL of 0.100 M HCl is required for titration, what is the percent protein in the sample?

Solution

Since this is a direct titration with HCl which reacts 1 : 1 with NH_3, the millimoles of NH_3 (and therefore of N) equals the millimoles of HCl. Multiplication by 6.25 gives the milligrams of protein.

$$\% \text{ protein} = \frac{0.10 \text{ mmol/mL HCl} \times 25.0 \text{ mL HCl} \times 14.01 \text{ mg N/mmol HCl} \times 6.25 \text{ mg protein/mg N}}{300 \text{ mg}}$$

$$\times 100\% = 73.0\%$$

The boric acid method (a direct method) is simpler and is usually more accurate since it requires the standardization and accurate measurement of only one solution. However, the end point break is not so sharp, and the indirect method requiring back-titration is usually preferred for **micro-Kjeldahl analysis**. A macro-Kjeldahl analysis of blood requires about 5 mL blood, while a micro-Kjeldahl analysis requires about 0.1 mL.

We have confined our discussion to those substances in which the nitrogen exists in the −3 valence state, as in ammonia. Such compounds include amines and amides. Compounds containing oxidized forms of nitrogen, such as organic nitro and azo compounds, must be treated with a reducing agent prior to digestion in order to obtain complete conversion to ammonium ion. Reducing agents such as iron(II) or thiosulfate are used.

Learning Objectives

WHAT ARE SOME OF THE KEY THINGS WE LEARNED FROM THIS CHAPTER?

Questions

1. What is the minimum pH change required for a sharp indicator color change at the end point? Why?
2. What criterion is used in selecting an indicator for a particular acid–base titration?
3. At what pH is the buffering capacity of a buffer the greatest?
4. Is the pH at the end point for the titration of a weak acid neutral, alkaline, or acidic? Why?
5. What would be a suitable indicator for the titration of ammonia with hydrochloric acid? Of acetic acid with sodium hydroxide?
6. Explain why boiling the solution near the end point in the titration of sodium carbonate increases the sharpness of the end point.
7. What is the approximate pK of the weakest acid or base that can be titrated in aqueous solution?
8. What must be the difference in the strengths of two acids in order to differentiate between them in titration?
9. Distinguish between a primary standard and a secondary standard.
10. What is a zwitterion?
11. What percent nitrogen is contained in a typical protein?
12. What is the preferred acid for titrating bases? Why?

Problems

STANDARDIZATION CALCULATIONS

13. A hydrochloric acid solution is standardized by titrating 0.4541 g of primary standard tris(hydroxymethyl)aminomethane. If 35.37 mL is required for the titration, what is the molarity of the acid?
14. A hydrochloric acid solution is standardized by titrating 0.2329 g of primary standard sodium carbonate to a methyl red end point by boiling the carbonate solution near the end point to remove carbon dioxide. If 42.87 mL acid is required for the titration, what is its molarity?
15. A sodium hydroxide solution is standardized by titrating 0.8592 g of primary standard potassium acid phthalate to a phenolphthalein end point, requiring 32.67 mL. What is the molarity of the base solution?
16. A 10.00-mL aliquot of a hydrochloric acid solution is treated with excess silver nitrate, and the silver chloride precipitate formed is determined by gravimetry. If 0.1682 g precipitate is obtained, what is the molarity of the acid?

INDICATORS

17. Write a Henderson–Hasselbalch equation for a weak-base indicator, B, and calculate the required pH change to go from one color of the indicator to the other. Around what pH is the transition?

TITRATION CURVES

You may wish to use spreadsheets for some of these calculations.

18. Calculate the pH at 0, 10.0, 25.0, and 30.0 mL of titrant in the titration of 50.0 mL of 0.100 M NaOH with 0.200 M HCl.

19. Calculate the pH at 0, 10.0, 25.0, 50.0, and 60.0 mL of titrant in the titration of 25.0 mL of 0.200 M HA with 0.100 M NaOH. $K_a = 2.0 \times 10^{-5}$.

20. Calculate the pH at 0, 10.0, 25.0, 50.0, and 60.0 mL of titrant in the titration of 50.0 mL of 0.100 M NH_3 with 0.100 M HCl.

21. Calculate the pH at 0, 25.0, 50.0, 75.0, 100, and 125% titration in the titration of both protons of the diprotic acid H_2A with 0.100 M NaOH, starting with 100 mL of 0.100 M H_2A. $K_{a1} = 1.0 \times 10^{-3}$, $K_{a2} = 1.0 \times 10^{-7}$.

22. Calculate the pH at 0, 25.0, 50.0, 75.0, 100, and 150% titration in the titration of 100 mL of 0.100 M Na_2HPO_4 with 0.100 M HCl to $H_2PO_4^{-}$.

QUANTITATIVE DETERMINATIONS

23. A 0.492-g sample of KH_2PO_4 is titrated with 0.112 M NaOH, requiring 25.6 mL:

$$H_2PO_4^{-} + OH^{-} \rightarrow HPO_4^{2-} + H_2O$$

What is the percent purity of the KH_2PO_4?

24. What volume of 0.155 M H_2SO_4 is required to titrate 0.293 g of 90.0% pure LiOH?

25. An indication of the average formula weight of a fat is its saponification number, expressed as the milligrams KOH required to hydrolyze (saponify) 1 g of the fat:

$$\begin{array}{ccc}
CH_2CO_2R & & CH_2OH \\
| & & | \\
CHCO_2R + 3KOH \rightarrow & CHOH + 3RCO_2K \\
| & & | \\
CH_2CO_2R & & CH_2OH
\end{array}$$

where R can be variable. A 1.10-g sample of butter is treated with 25.0 mL of 0.250 M KOH solution. After the saponification is complete, the unreacted KOH is back-titrated with 0.250 M HCl, requiring 9.26 mL. What is the saponification number of the fat and what is its average formula weight (assuming the butter is all fat)?

26. A sample containing the amino acid alanine, $CH_3CH(NH_2)COOH$, plus inert matter is analyzed by the Kjeldahl method. A 2.00-g sample is digested, the NH_3 is distilled and collected in 50.0 mL of 0.150 M H_2SO_4, and a volume of 9.0 mL of 0.100 M NaOH is required for back-titration. Calculate the percent alanine in the sample.

27. A 2.00-mL serum sample is analyzed for protein by the modified Kjeldahl method. The sample is digested, the ammonia is distilled into boric acid solution, and 15.0 mL of standard HCl is required for the titration of the ammonium borate. The HCl is standardized by treating 0.330 g pure $(NH_4)_2SO_4$ in the same manner. If 33.3 mL acid is required in the standardization titration, what is the concentration of protein in the serum in g% (wt/vol)?

QUANTITATIVE DETERMINATIONS OF MIXTURES

28. A 100-mL aliquot of a solution containing HCl and H_3PO_4 is titrated with 0.200 M NaOH. The methyl red end point occurs at 25.0 mL, and the bromthymol blue end point occurs at 10.0 mL later (total 35.0 mL). What are the concentrations of HCl and H_3PO_4 in the solution?

29. A 0.527-g sample of a mixture containing $NaCO_3$, $NaHCO_3$, and inert impurities is titrated with 0.109 M HCl, requiring 15.7 mL to reach the phenolphthalein end point and a total of 43.8 mL to reach the modified methyl orange end point. What is the percent each of Na_2CO_3 and $NaHCO_3$ in the mixture?

30. Sodium hydroxide and Na_2CO_3 will titrate together to a phenolphthalein end point ($OH^- \rightarrow H_2O$; $CO_3^{2-} \rightarrow HCO_3^-$). A mixture of NaOH and Na_2CO_3 is titrated with 0.250 M HCl, requiring 26.2 mL for the phenolphthalein end point and an additional 15.2 mL to reach the modified methyl orange end point. How many milligrams NaOH and Na_2CO_3 are in the mixture?

31. Sodium carbonate can coexist with either NaOH or $NaHCO_3$ but not with both simultaneously, since they would react to form Na_2CO_3. Sodium hydroxide and Na_2CO_3 will titrate together to a phenolphthalein end point ($OH^- \rightarrow H_2O$; $CO_3^{2-} \rightarrow HCO_3^-$). A mixture of either NaOH and Na_2CO_3 or of Na_2CO_3 and $NaHCO_3$ is titrated with HCl. The phenolphthalein end point occurs at 15.0 mL and the modified methyl orange end point occurs at 50.0 mL (35.0 mL beyond the first end point). The HCl was standardized by titrating 0.477 g Na_2CO_3, requiring 30.0 mL to reach the modified methyl orange end point. What mixture is present and how many millimoles of each constituent are present?

32. What would be the answers to Problem 31 if the second end point had occurred at 25.0 mL (10.0 mL beyond the first end point)?

33. A mixture containing only $BaCO_3$ and Li_2CO_3 weighs 0.150 g. If 25.0 mL of 0.120 M HCl is required for complete neutralization ($CO_3^{2-} \rightarrow H_2CO_3$), what is the percent $BaCO_3$ in the sample?

34. A sample of P_2O_5 contains some H_3PO_4 impurity. A 0.405-g sample is reacted with water ($P_2O_5 + 3H_2O \rightarrow 2H_3PO_4$), and the resulting solution is titrated with 0.250 M NaOH ($H_3PO_4 \rightarrow Na_2HPO_4$). If 42.5 mL is required for the titration, what is the percent of H_3PO_4 impurity?

SPREADSHEET PROBLEM

See CD, Sheet 1, for suggested setup, and Chart 1 for plot.

35. Prepare a spreadsheet to construct the titration curve for the titration of 100 mL 0.1 M NH_3 with 0.1 M HCl (Figure 8.8).

Recommended References

INDICATORS

1. G. Gorin, "Indicators and the Basis for Their Use," *J. Chem. Ed.*, **33** (1956) 318.
2. E. Bishop. *Indicators*. Oxford: Pergamon, 1972.

TITRATION CURVES

3. R. K. McAlpine, "Changes in pH at the Equivalence Point," *J. Chem. Ed.*, **25** (1948) 694.
4. A. K. Covington, R. N. Goldberg, and M. Sarbar, "Computer Simulation of Titration Curves with Application to Aqueous Carbonate Solutions," *Anal. Chim. Acta*, **130** (1981) 103.

TITRATIONS

5. Y.-S. Chen, S. V. Brayton, and C. C. Hach, "Accuracy in Kjeldahl Protein Analysis," *Am. Lab.*, June (1988) 62.

6. R. M. Archibald, "Nitrogen by the Kjeldahl Method," in D. Seligson, ed., *Standard Methods of Clinical Chemistry*, Vol. 2. New York: Academic, 1958, pp. 91–99.

7. M. E. Hodes, "Carbon Dioxide Content (Titrimetric)" in M. Reiner, ed., *Standard Methods of Clinical Chemistry*, Vol. 1. New York: Academic, 1953, pp. 19–22.

WEB VIRTUAL CALCULATOR

8. http://hamers.chem.wisc.edu/chapman/Titrator/. Virtual Calculator 1.5, from the website of Profesor Robert Hamers, University of Wisconsin. Use it to construct titration curves. You can select different types of acids and enter concentrations and volumes. Also construct alpha plots.

Chapter Nine

COMPLEXOMETRIC REACTIONS AND TITRATIONS

"Simple things should be simple. Complex things should be possible."
—Alan Kay

Many metal ions form slightly dissociated complexes with various ligands (complexing agents). The analytical chemist makes judicious use of complexes to mask undesired reactions. The formation of complexes can also serve as the basis of accurate and convenient titrations for metal ions in which the titrant is a complexing agent. Complexometric titrations are useful for determining a large number of metals. Selectivity can be achieved by appropriate use of *masking agents* (addition of other complexing agents that react with interfering metal ions) and by pH control, since most complexing agents are weak acids or weak bases whose equilibria are influenced by the pH. In this chapter, we discuss metal ions, their equilibria, and the influence of pH on these equilibria. We describe titrations of metal ions with the very useful complexing agent EDTA, the factors that affect them, and indicators for the titrations. The EDTA titration of calcium plus magnesium is commonly used to determine water hardness. Nearly all other metals can be accurately determined by complexometric titration. Complexing reactions are useful for gravimetry, spectrophotometry, and fluorometry, and for masking interfering ions.

9.1 Complexes and Formation Constants—How Stable Are Complexes?

Complexes play an important role in many chemical and biochemical processes. For example, the heme molecule in blood holds the iron atom tightly because the nitrogen atoms of the heme form strong ligand or complexing bonds, that is, nitrogen is a good complexer. The iron [as iron(II)] in turn bonds readily with oxygen to transport oxygen gas from the lungs to the points of oxidation in the body and then easily releases it because oxygen is a weak ligand or complexer. Cyanide kills because it is a strong complexer and displaces oxygen, as does carbon monoxide. Carbon monoxide binds heme 200 times more strongly than does oxygen, forming carboxyhemoglobin.

Many cations will form complexes in solution with a variety of substances that have a pair of unshared electrons (e.g., on N, O, S atoms in the molecule) capable of satisfying the coordination number of the metal. [The metal ion is a Lewis acid (electron pair acceptor), and the complexer is a Lewis base (electron pair donor).] The number of molecules of the complexing agent, called the **ligand**, will depend on the coordination number of the metal and on the number of complexing groups on the ligand molecule.

> Most ligands contain O, S, or N as the complexing atoms.

Ammonia is a simple complexing agent with one pair of unshared electrons that will complex copper ion:

$$Cu^{2+} + 4:NH_3 \rightleftharpoons \left[\begin{array}{c} NH_3 \\ \ddots \\ H_3N:Cu:NH_3 \\ \ddots \\ NH_3 \end{array} \right]^{2+}$$

Here, the copper ion acts as a Lewis acid, and the ammonia is a Lewis base. The Cu^{2+} (hydrated) ion is pale blue in solution, while the ammonia (the ammine) complex is deep blue. A similar reaction occurs with the green hydrated nickel ion to form a deep blue ammine complex.

Ammonia will also complex with silver ion to form a colorless complex. (The formation of this complex may be used to dissolve silver chloride precipitate remaining in a filter crucible and thereby clean the crucible.) Two ammonia molecules complex with each silver ion in a stepwise fashion, and we can write an equilibrium constant for each step, called the **formation constant K_f**:

$$Ag^+ + NH_3 \rightleftharpoons Ag(NH_3)^+ \quad K_{f1} = \frac{[Ag(NH_3)^+]}{[Ag^+][NH_3]} \tag{9.1}$$
$$= 2.5 \times 10^3$$

$$Ag(NH_3)^+ + NH_3 \rightleftharpoons Ag(NH_3)_2{}^+ \quad K_{f2} = \frac{[Ag(NH_3)_2{}^+]}{[Ag(NH_3)^+][NH_3]} \tag{9.2}$$
$$= 1.0 \times 10^4$$

The overall reaction is the sum of the two steps, and the overall formation constant is the product of the stepwise formation constants:

$$Ag^+ + 2NH_3 \rightleftharpoons Ag(NH_3)_2{}^+ \quad K_f = K_{f1} \cdot K_{f2} = \frac{[Ag(NH_3)_2{}^+]}{[Ag^+][NH_3]^2} \tag{9.3}$$
$$= 2.5 \times 10^7$$

For the formation of a simple $1:1$ complex, for example, M + L = ML, the formation constant is simply $K_f = [M][L]/[ML]$. Note that the products of the reactions are written in the numerators of the equilibrium constant expressions in the usual manner, even though we wrote the reactions as associations rather than dissociations. The formation constant is also called the **stability constant K_s, or K_{stab}**.

> $K_f = K_s = 1/K_i$ or $1/K_d$.

We could write the equilibria in the opposite direction, as dissociations. If we do this, the concentration terms are inverted in the equilibrium constant expressions. The equilibrium constants then are simply the reciprocals of the formation constants, and they are called **instability constants K_i**, or **dissociation constants K_d**:

$$Ag(NH_3)_2^+ \rightleftharpoons Ag^+ + 2NH_3 \quad K_d = \frac{1}{K_f} = \frac{[Ag^+][NH_3]^2}{[Ag(NH_3)_2^+]} = 4.0 \times 10^{-8} \quad (9.4)$$

You can use either constant in calculations, as long as you use it with the proper reaction and the correct expression.

Example 9.1

A divalent metal M^{2+} reacts with a ligand L to form a 1:1 complex:

$$M^{2+} + L \rightleftharpoons ML^{2+} \quad K_f = \frac{[ML^{2+}]}{[M^{2+}][L]}$$

Calculate the concentration of M^{2+} in a solution prepared by mixing equal volumes of 0.20 M M^{2+} and 0.20 M L. $K_f = 1.0 \times 10^8$.

Solution

We have added stoichiometrically equal amounts of M^{2+} and L. The complex is sufficiently strong that their reaction is virtually complete. Since we added equal volumes, the initial concentrations were halved. Let x represent $[M^{2+}]$. At equilibrium, we have

$$M^{2+} + L \rightleftharpoons ML^{2+}$$
$$x \qquad x \qquad 0.10 - x \approx 0.10$$

Essentially, all the M^{2+} (original concentration 0.20 M) was converted to an equal amount of ML^{2+}, with only a small amount of uncomplexed metal remaining. Substituting into the K_f expression,

$$\frac{0.10}{(x)(x)} = 1.0 \times 10^8$$

$$x = [M^{2+}] = 3.2 \times 10^{-5} M$$

Example 9.2

Silver ion forms a stable 1:1 complex with triethylenetetraamine, called "trien" $[NH_2(CH_2)_2NH(CH_2)_2NH(CH_2)_2NH_2]$. Calculate the silver ion concentration at equilibrium when 25 mL of 0.010 M silver nitrate is added to 50 mL of 0.015 M trien. $K_f = 5.0 \times 10^7$.

Solution

$$Ag^+ + trien \rightleftharpoons Ag(trien)^+ \quad K_f = \frac{[Ag(trien)^+]}{[Ag^+][trien]}$$

Calculate the millimoles Ag^+ and trien added:

$$\text{mmol } Ag^+ = 25 \text{ mL} \times 0.010 \text{ mmol/mL} = 0.25 \text{ mmol}$$
$$\text{mmol trien} = 50 \text{ mL} \times 0.015 \text{ mmol/mL} = 0.75 \text{ mmol}$$

The equilibrium lies far to the right, so you can assume that virtually all the Ag^+ reacts with 0.25 mmol of the trien (leaving 0.50 mmol trien in excess) to form 0.25 mmol complex. Calculate the molar concentrations:

$$[Ag^+] = x = \text{mol/L unreacted}$$
$$[\text{trien}] = (0.50 \text{ mmol/75 mL}) + x = 6.7 \times 10^{-3} + x$$
$$\approx 6.7 \times 10^{-3}$$
$$[Ag(\text{trien})^+] = 0.25 \text{ mmol/75 mL} - x$$
$$= 3.3 \times 10^{-3} - x \approx 3.3 \times 10^{-3}$$

Try neglecting x compared to the other concentrations:

$$\frac{3.3 \times 10^{-3}}{(x)(6.7 \times 10^{-3})} = 5.0 \times 10^{7}$$
$$x = [Ag^+] = 9.8 \times 10^{-9} \; M$$

We were justified in neglecting x.

● ●

9.2 Chelates: EDTA——The Ultimate Titrating Agent for Metals

Simple complexing agents such as ammonia are rarely used as titrating agents because a sharp end point corresponding to a stoichiometric complex is generally difficult to achieve. This is because the stepwise formation constants are frequently close together and are not very large, and a single stoichiometric complex cannot be observed. Certain complexing agents that have two or more complexing groups on the molecule, however, do form well-defined complexes and can be used as titrating agents. Schwarzenbach demonstrated that a remarkable increase in stability is achieved if a bidentate ligand (one with two complexing groups) is used. For example, replacing ammonia with the bidentate ethylenediamine, $NH_2CH_2CH_2NH_2$ (en), results in a highly stable $Cu(en)_2^{2+}$ complex. If the N hydrogens are replaced with acetic acid groups, only a single step is required to complex the copper, and titrations are straightforward.

The term *chelate* is derived from the Greek term meaning "clawlike." Chelating agents literally wrap themselves around a metal ion.

The most generally useful titrating agents are aminocarboxylic acids, in which nitrogen and carboxylate groups serve as ligands. The amino nitrogens are more basic and are protonated ($—NH_3^+$) more strongly than the carboxylate groups. When these groups bind to metal atoms, they lose their protons. The metal complexes formed with these multidentate complexing agents are often 1 : 1, regardless of the charge on the metal ion, because there are sufficient complexing groups on one molecule to satisfy the coordination sites of the metal ion.

An organic agent that has two or more groups capable of complexing with a metal ion is called a **chelating agent**. The complex formed is called a **chelate**. The chelating agent is called the *ligand*. Titration with a chelating agent is called a **chelometric titration**, a type of complexometric titration.

The protons in EDTA are displaced upon complexing with a metal ion. A negatively charged chelate results.

The most widely used chelating agent in titrations is **ethylenediaminetetra-acetic acid (EDTA)**. The formula for EDTA is

Each of the two nitrogens and each of the four carboxyl groups contains a pair of unshared electrons capable of complexing with a metal ion. Thus, EDTA contains six complexing groups. We will represent EDTA by the symbol H_4Y. It is a tetraprotic acid, and the hydrogens in H_4Y refer to the four ionizable hydrogens. It is the unprotonated ligand Y^{4-} that forms complexes with metal ions, that is, the protons are displaced by the metal ion upon complexation.

THE CHELON EFFECT—THE MORE COMPLEXING GROUPS, THE BETTER

For more discussion of the design of chelating agents, see C. N. Reilley, R. W. Schmid, and F. S. Sadek, "Chelon Approach to Analysis (I). Survey of Theory and Application," *J. Chem. Ed.*, **36** (1959) 555. Illustrated experiments are given in a second article in *J. Chem. Ed.*, **36** (1959) 619.

Multidentate chelating agents form stronger complexes with metal ions than do similar bidentate or monodentate ligands. This is the result of thermodynamic effects in complex formation. Chemical reactions are driven by decreasing enthalpy (liberation of heat, negative ΔH) and by increasing entropy (increased disorder, positive ΔS). Recall from Chapter 6, Equation 6.7, that a chemical process is spontaneous when the free energy change, ΔG, is negative, and $\Delta G = \Delta H - T\Delta S$. The enthalpy change for ligands with similar groups is often similar. For example, four ammonia molecules complexed to Cu^{2+} and four amino groups from two ethylenediamine molecules complexed to Cu^{2+} will result in about the same release of heat. However, more disorder or entropy is created by the dissociation of the $Cu(NH_3)_4^{2+}$ complex in which five species are formed than in the dissociation of the $Cu(H_2NCH_2CH_2NH_2)_2^{2+}$ complex, in which three species are formed. Hence, ΔS is greater for the former *dissociation*, creating a more negative ΔG and a greater tendency for dissociation. Thus, multidentate complexes are more stable (have larger K_f values), largely because of the entropy effect. This is known as the **chelon effect** or **chelate effect**. It is pronounced for chelating agents such as EDTA, which have sufficient ligand atoms to occupy up to six coordination sites on metal ions.

The chelon effect is an entropy effect.

EDTA EQUILIBRIA

We can represent EDTA as having four K_a values corresponding to the stepwise dissociation of the four protons[1]:

$$H_4Y \rightleftharpoons H^+ + H_3Y^- \qquad K_{a1} = 1.0 \times 10^{-2} = \frac{[H^+][H_3Y^-]}{[H_4Y]} \qquad (9.5)$$

$$H_3Y^- \rightleftharpoons H^+ + H_2Y^{2-} \qquad K_{a2} = 2.2 \times 10^{-3} = \frac{[H^+][H_2Y^{2-}]}{[H_3Y^-]} \qquad (9.6)$$

$$H_2Y^{2-} \rightleftharpoons H^+ + HY^{3-} \qquad K_{a3} = 6.9 \times 10^{-7} = \frac{[H^+][HY^{3-}]}{[H_2Y^{2-}]} \qquad (9.7)$$

---

[1]Actually all four carbonyl groups plus the nitrogens on the EDTA molecule can protonate, and so there are in reality six dissociation steps and six K_a values, the first two being 1.0 and 0.032. The two nitrogens are more basic than the carbonyl oxygens, and so protonate more readily. The nitrogen protonation does affect the solubility of EDTA in acid.

$$HY^{3-} \rightleftharpoons H^+ + Y^{4-} \qquad K_{a4} = 5.5 \times 10^{-11} = \frac{[H^+][Y^{4-}]}{[HY^{3-}]} \qquad (9.8)$$

Polyprotic acid equilibria are treated in Section 7.8, and you should review this section before proceeding with the following discussion.

Figure 9.1 illustrates the fraction of each form of EDTA as a function of pH. Since the anion Y^{4-} is the ligand species in complex formation, the complexation equilibria are affected markedly by the pH. H_4Y has a very low solubility in water, and so the disodium salt $Na_2H_2Y \cdot 2H_2O$ is generally used, in which two of the acid groups are neutralized. This salt dissociates in solution to give predominantly H_2Y^{2-}; the pH of the solution is approximately 4 to 5 (theoretically 4.4 from $[H^+] = \sqrt{K_{a2}K_{a3}}$).

FORMATION CONSTANT

Consider the formation of the EDTA chelate of Ca^{2+}. This can be represented by:

$$Ca^{2+} + Y^{4-} \rightleftharpoons CaY^2 \qquad (9.9)$$

The formation constant for this is

$$\boxed{K_f = \frac{[CaY^{2-}]}{[Ca^{2+}][Y^{4-}]}} \qquad (9.10)$$

The values of some representative EDTA formation constants are given in Appendix C.

EFFECT OF pH ON EDTA EQUILIBRIA—HOW MUCH IS Y⁴⁻?

The equilibrium in Equation 9.9 is shifted to the left as the hydrogen ion concentration is increased, due to competition for the chelating anion by the hydrogen ion.

The dissociation may be represented by:

$$CaY_2 \rightleftharpoons Ca^{2+} + \underbrace{Y^{4-} \overset{H^+}{\rightleftharpoons} HY^{3-} \overset{H^+}{\rightleftharpoons} H_2Y^{2-} \overset{H^+}{\rightleftharpoons} H_3Y^- \overset{H^+}{\rightleftharpoons} H_4Y}_{C_{H_4Y}}$$

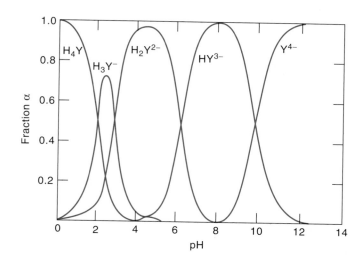

Fig. 9.1. Fraction of EDTA species as function of pH.

Protons compete with the metal ion for the EDTA ion. To apply Equation 9.10, we must replace $[Y^{4-}]$ with $\alpha_4 C_{H_4Y}$ as the equilibrium concentration of Y^{4-}.

Note that $C_{H_4Y} = [Ca^{2+}]$. Or, from the overall equilibrium:

$$Ca^{2+} + H_4Y \rightleftharpoons CaY^{2-} + 4H^+$$

According to Le Châtelier's principle, increasing the acidity will favor the competing equilibrium, that is, the protonation of Y^{4-} (all forms of EDTA are present in equilibrium, but some are diminishing small; see Figure 9.1). Decreasing the acidity will favor formation of the CaY^{2-} chelate.

From a knowledge of the pH and the equilibria involved, Equation 9.10 can be used to calculate the concentration of free Ca^{2+} under various solution conditions (e.g., in a titration to interpret a titration curve). The Y^{4-} concentration is calculated at different pH values as follows (see Chapter 7, polyprotic acids). If we let C_{H_4Y} represent the total concentration of all forms of uncomplexed EDTA, then

$$C_{H_4Y} = [Y^{4-}] + [HY^{3-}] + [H_2Y^{2-}] + [H_3Y^-] + [H_4Y] \qquad (9.11)$$

Solving for the equilibrium concentrations of H_4Y, H_3Y^-, H_2Y^{2-}, and HY^{3-} in Equations 9.5 through 9.8, substituting these in Equation 9.11 to eliminate all forms but Y^{4-}, and dividing by $[Y^{4-}]$ gives

$$\frac{C_{H_4Y}}{[Y^{4-}]} = \frac{1}{\alpha_4} = 1 + \frac{[H^+]}{K_{a4}} + \frac{[H^+]^2}{K_{a3}K_{a4}} + \frac{[H^+]^3}{K_{a2}K_{a3}K_{a4}} + \frac{[H^+]^4}{K_{a1}K_{a2}K_{a3}K_{a4}} \qquad (9.12)$$

where α_4 is the **fraction** of the total EDTA species that exists as Y^{4-} ($\alpha_4 = [Y^{4-}]/C_{H_4Y}$). Similar equations can be derived for the fraction of each of the other EDTA species α_0, α_1, α_2, and α_3, as in Chapter 7. (This is the way Figure 9.1 was constructed.)

We can use Equation 9.12, then, to calculate the fraction of the EDTA that exists as Y^{4-} at a given pH; and from a knowledge of the concentration of uncomplexed EDTA (C_{H_4Y}), we can calculate the free Ca^{2+} using Equation 9.10.

· ·

 Example 9.3

Calculate the fraction of EDTA that exists as Y^{4-} at pH 10, and from this calculate pCa in 100 mL of solution of 0.100 M Ca^{2+} at pH 10 after adding 100 mL of 0.100 M EDTA.

Solution

From Equation 9.12,

$$\frac{1}{\alpha_4} = 1 + \frac{1.0 \times 10^{-10}}{5.5 \times 10^{-11}} + \frac{(1.0 \times 10^{-10})^2}{(6.9 \times 10^{-7})(5.5 \times 10^{-11})}$$

$$+ \frac{(1.0 \times 10^{-10})^3}{(2.2 \times 10^{-3})(6.9 \times 10^{-7})(5.5 \times 10^{-11})}$$

$$+ \frac{(1.0 \times 10^{-10})^4}{(1.0 \times 10^{-2})(2.2 \times 10^{-3})(6.9 \times 10^{-7})(5.5 \times 10^{-11})}$$

$$= 1 + 1.82 + 2.6 \times 10^{-4} + 1.2 \times 10^{-11} + 1.2 \times 10^{-19} = 2.82$$

$$\alpha_4 = 0.35$$

Stoichiometric amounts of Ca^{2+} and EDTA are added to produce an equivalent amount of CaY^{2-}, less the amount dissociated:

$$mmol\ Ca^{2+} = 0.100\ M \times 100\ mL = 10.0\ mmol$$
$$mmol\ EDTA = 0.100\ M \times 100\ mL = 10.0\ mmol$$

We have formed 10.0 mmol CaY^{2-} in 200 mL, or 0.0500 M:

$$Ca^{2+} + EDTA \rightleftharpoons CaY^{2-}$$
$$x \qquad x \qquad 0.0500\ M - x$$
$$\approx 0.0500\ M\ (\text{since } K_f \text{ is large})$$

where x represents the total equilibrium EDTA concentration in all forms, C_{H_4Y}. $[Y^{4-}]$, needed to apply Equation 9.10, is equal to $\alpha_4 C_{H_4Y}$. Hence, we can write Equation 9.10 as:

$$K_f = \frac{[CaY^{2-}]}{[Ca^{2+}]\alpha_4[C_{H_4Y}]}$$

From Appendix C, $K_f = 5.0 \times 10^{10}$. Hence,

$$5.0 \times 10^{10} = \frac{0.0500}{(x)(0.35)(x)}$$
$$x = 1.7 \times 10^{-6}\ M \qquad -\log[H]$$
$$pCa = 5.77$$

CONDITIONAL FORMATION CONSTANT—USE FOR A FIXED pH

We can substitute $\alpha_4 C_{H_4Y}$ for $[Y^{4-}]$ in Equation 9.10:

$$K_f = \frac{[CaY^{2-}]}{[Ca^{2+}]\alpha_4 C_{H_4Y}} \qquad (9.13)$$

We can then rearrange the equation to yield

$$K_f\alpha_4 = K'_f = \frac{[CaY^{2-}]}{[Ca^{2+}]C_{H_4Y}} \qquad (9.14)$$

where K'_f is the **conditional formation constant** and is dependent on α_4 and, hence, the pH. We can use this equation to calculate the equilibrium concentrations of the different species at a given pH in place of Equation 9.10.

The conditional formation constant value holds for only a specified pH.

Example 9.4

The formation constant for CaY^{2-} is 5.0×10^{10}. At pH 10, α_4 is calculated (Example 9.3) to be 0.35 to give a conditional constant (from Equation 9.14) of 1.8×10^{10}. Calculate pCa in 100 mL of a solution of 0.100 M Ca^{2+} at pH 10 after addition of (a) 0 mL, (b) 50 mL, (c) 100 mL, and (d) 150 mL of 0.100 M EDTA.

Solution

(a) $$pCa = -\log[Ca^{2+}] = -\log \times 1.00 \times 10^{-1} = 1.00$$

(b) We started with $0.100\ M \times 100$ mL $= 10.0$ mmol Ca^{2+}. The millimoles of EDTA added are $0.100\ M \times 50$ mL $= 5.0$ mmol. The conditional constant is large, so the reaction 9.9 will be far to the right. Hence, we can neglect the amount of Ca^{2+} from the dissociation of CaY^{2-} and the number of millimoles of free Ca^{2+} is essentially equal to the number of unreacted millimoles:

$$mmol\ Ca^{2+} = 10.0 - 5.0 = 5.0\ mmol$$

$$[Ca^{2+}] = 5.0\ mmol/150\ mL = 0.030\ M$$

$$pCa = -\log 3.0 \times 10^{-2} = 1.48$$

(c) At the equivalence point, we have converted all the Ca^{2+} to CaY^{2-}. We must, therefore, use Equation 9.14 to calculate the equilibrium concentration of Ca^{2+}. The number of millimoles of CaY^{2-} formed is equal to the number of millimoles of Ca^{2+} started with, and $[CaY^{2-}] = 10.0$ mmol/200 mL $= 0.0500\ M$. From the dissociation of CaY^{2-}, $[Ca^{2+}] = C_{H_4Y} = x$, and the equilibrium $[CaY^{2-}] = 0.050\ M - x$. But since the dissociation is slight, we can neglect x compared to $0.050\ M$. Therefore, from Equation 9.14,

$$\frac{0.050}{(x)(x)} = 1.8 \times 10^{10}$$

$$x = 1.7 \times 10^{-6}\ M = [Ca^{2+}]$$

$$pCa = -\log 1.7 \times 10^{-6} = 5.77$$

Compare this value with that calculated using K_f in Example 9.3, instead of K_f'.

(d) The concentration C_{H_4Y} is equal to the concentration of excess EDTA added (neglecting the dissociation of CaY^{2-}, which will be even smaller in the presence of excess EDTA). The millimoles of CaY^{2-} will be as in (c). Hence,

$$[CaY^{2-}] = 10.0\ mmol/250\ mL = 0.0400\ M$$

$$mmol\ excess\ C_{H_4Y} = 0.100\ M \times 150\ mL - 0.100\ M \times 100\ mL = 5.0\ mmol$$

$$C_{H_4Y} = 5.0\ mmol/250\ mL = 0.020\ M$$

$$\frac{0.040}{[Ca^{2+}](0.020)} = 1.8 \times 10^{10}$$

$$[Ca^{2+}] = 1.1 \times 10^{-10}\ M$$

$$pCa = -\log 1.1 \times 10^{-10} = 9.95$$

• •

The pH can affect stability of the complex (i.e., K_f') by affecting not only the form of the EDTA but also that of the metal ion. For example, hydroxy species may form ($M^{2+} + OH^- \rightarrow MOH^+$). That is, OH^- competes for the metal ion just as H^+ competes for the Y^{4-}. Figure 9.2 (prepared from a spreadsheet—see Problem 20) shows how K_f' changes with pH for three metal–EDTA chelates with moderate (Ca) to strong (Hg) formation constants. The calcium chelate is obviously too weak to be titrated in acid solution ($K_f' < 1$), while the mercury chelate is strong enough

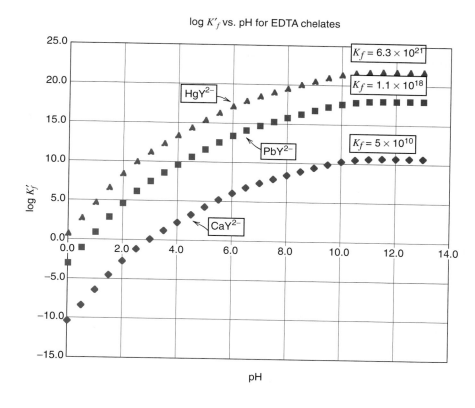

log K'_f vs. pH for EDTA chelates

$K_f = 6.3 \times 10^{21}$

$K_f = 1.1 \times 10^{18}$

HgY^{2-}

PbY^{2-}

$K_f = 5 \times 10^{10}$

CaY^{2-}

pH

Fig. 9.2. Effect of pH on K'_f values for EDTA chelates.

to be titrated in acid. At pH 13, all K'_f values are virtually equal to the K_f values because α_4 is essentially unity; that is, the EDTA is completely dissociated to Y^{4-}. The curves parallel one another because at each pH, each K_f is multiplied by the same α_4 value to obtain K'_f.

9.3 Metal–EDTA Titration Curves

A titration is performed by adding the chelating agent to the sample; the reaction occurs as in Equation 9.9. Figure 9.3 shows the titration curve for Ca^{2+} titrated with EDTA at pH 10. Before the equivalence point, the Ca^{2+} concentration is nearly equal to the amount of unchelated (unreacted) calcium since the dissociation of the chelate is slight (analogous to the amount of an unprecipitated ion). At the equivalence point and beyond, pCa is determined from the dissociation of the chelate at the given pH as described in Example 9.3 or 9.4, using K_f or K'_f. The effect of pH on the titration is apparent from the curve in Figure 9.3 for titration at pH 7.

See Problem 21 for a spreadsheet calculation of the Ca–EDTA titration curve in Figure 9.3 at pH 10. As with calculated acid–base titration curves, the calculations here break down very near the equivalence point due to simplifying assumptions we have made.

The more stable the chelate (the larger K_f), the farther to the right will be the equilibrium of the reaction (Equation 9.9), and the larger will be the end-point break. Also, the more stable the chelate, the lower the pH at which the titration can be performed (Figure 9.2). This is important because it allows the titration of some metals in the presence of others whose EDTA chelates are too weak to titrate at the lower pH.

Only some metal chelates are stable enough to allow titrations in acid solution; others require alkaline solution.

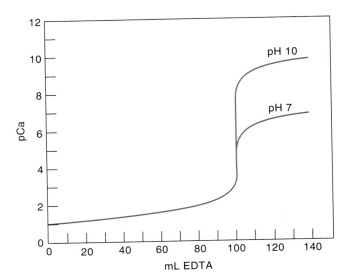

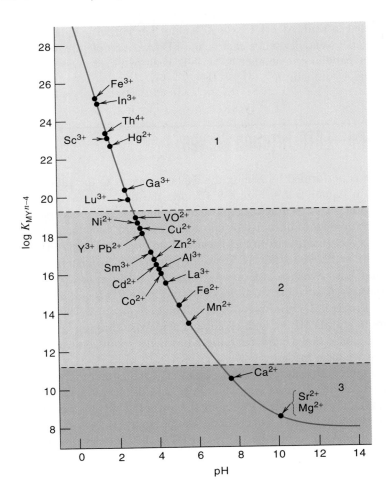

Fig. 9.3. Titration curves for 100 mL 0.1 M Ca^{2+} versus 0.1 M Na_2EDTA at pH 7 and pH 10.

Figure 9.4 shows the minimum pH at which different metals can be titrated with EDTA. The points on the curve represent the pH at which the *conditional formation constant* K_f' for each metal is 10^6 (log $K_f' = 6$), which was arbitrarily chosen as the minimum needed for a sharp end point. Note that the smaller K_f, the more alkaline the solution must be to obtain a K_f' of 10^6 (i.e., the larger α_4 must be). Thus, Ca^{2+} with K_f only about 10^{10} requires a pH of about 8 or above. The

Fig. 9.4. Minimum pH for effective titration of various metal ions with EDTA. (Reprinted with permission from C. N. Reilley and R. W. Schmid, *Anal. Chem.*, **30** (1958) 947. Copyright by the American Chemical Society.)

dashed lines in the figure divide the metals into separate groups according to their formation constants. One group is titrated in a highly acidic (pH < 4) solution, a second group at pH 4 to 7, and a third group at pH > 7. At the highest pH range, all the metals will react, but not all can be titrated directly due to precipitation of hydroxides. For example, titration of Fe^{3+} or Th^{4+} is not possible without the use of back-titration or auxiliary complexing agents to prevent hydrolysis. At the intermediate pH range, the third group will not titrate, and the second group of metals can be titrated in the presence of the third group. And finally, in the most acidic pH range, only the first group will titrate and can be determined in the presence of the others.

Masking can be achieved by precipitation, complex formation, oxidation–reduction, and kinetically. A combination of these techniques may be employed. For example, Cu^{2+} can be masked by reduction to Cu(I) with ascorbic acid and by complexation with I^-. Lead can be precipitated with sulfate when bismuth is to be titrated. Most masking is accomplished by selectively forming a stable, soluble complex. Hydroxide ion complexes aluminum ion [$Al(OH)_4^-$ or AlO_2^-] so calcium can be titrated. Fluoride masks Sn(IV) in the titration of Sn(II). Ammonia complexes copper so it cannot be titrated with EDTA using murexide indicator. Metals can be titrated in the presence of Cr(III) because its EDTA chelate, although very stable, forms only slowly.

9.4 Detection of the End Point: Indicators—They Are Chelating Agents

We can measure the pM potentiometrically if a suitable electrode is available, for example, an ion-selective electrode (see Chapter 13), but it is simpler if an indicator can be used. Indicators used for chelometric titrations are themselves chelating agents. They are usually dyes of the o,o'-dihydroxy azo type.

Eriochrome Black T is a typical indicator. It contains three ionizable protons, so we will represent it by H_3In. This indicator can be used for the titration of Mg^{2+} with EDTA. A small amount of indicator is added to the sample solution, and it forms a red complex with part of the Mg^{2+}; the color of the uncomplexed indicator is blue. As soon as all the free Mg^{2+} is titrated, the EDTA displaces the indicator from the magnesium, causing a change in the color from red to blue:

$$\underset{\text{(red)}}{MgIn^-} + \underset{\text{(colorless)}}{H_2Y^{2-}} \rightarrow \underset{\text{(colorless)}}{MgY^{2-}} + \underset{\text{(blue)}}{HIn^{2-}} + H^+ \qquad (9.15)$$

This will occur over a pMg range, and the change will be sharper if the indicator is kept as dilute as possible and will still give a good color.

Of course, the metal–indicator complex must be less stable than the metal–EDTA complex, or else the EDTA will not displace it from the metal. On the other hand, it must not be too weak, or the EDTA will start replacing it at the beginning of the titration, and a diffuse end point will result. In general, *the metal–indicator complex should be 10 to 100 times less stable than the metal–titrant complex.*

The formation constants of the EDTA complexes of calcium and magnesium are too close to differentiate between them in an EDTA titration, even by adjusting pH (see Figure 9.4). So they will titrate together, and the Eriochrome Black T end point can be used as above. This titration is used to determine **total hardness of water**, (Ca^{2+} plus Mg^{2+}—see Experiment 9). Eriochrome Black T cannot be

Water hardness is expressed as ppm $CaCO_3$ and represents the sum of calcium and magnesium.

used to indicate the direct titration of calcium in the absence of magnesium, with EDTA, however, because the indicator forms too weak a complex with calcium to give a sharp end point. Therefore, a small measured amount of Mg^{2+} is added to the Ca^{2+} solution; and as soon as the Ca^{2+} and the small amount of free Mg^{2+} are titrated, the end-point color change occurs as above. (The Ca^{2+} titrates first since its EDTA chelate is more stable.) A correction is made for the amount of EDTA used for titration of the Mg^{2+} by performing a "blank" titration on the same amount of Mg^{2+} added to the buffer.

It is more convenient to add, instead, about 2 mL of 0.005 M Mg–EDTA instead of $MgCl_2$. This is prepared by adding together equal volumes of 0.01 M $MgCl_2$ and 0.01 M EDTA solutions and adjusting the ratio with dropwise additions until a portion of the reagent turns a dull violet when treated with pH 10 buffer and Eriochrome Black T indicator. When this occurs, one drop of 0.01 M EDTA will turn the solution blue, and one drop of 0.01 M $MgCl_2$ will turn it red.

If we add Mg–EDTA to the sample, the Ca^{2+} in the sample displaces the EDTA from the Mg^{2+} (since the Ca–EDTA is more stable) so that the Mg^{2+} is free to react with the indicator. At the end point, an equivalent amount of EDTA displaces the indicator from the Mg^{2+}, causing the color change, and no correction is required for the added Mg–EDTA. This procedure is used in Experiment 9.

An alternative method is to add a small amount of Mg^{2+} to the EDTA solution. This immediately reacts with EDTA to form MgY^{2-} with very little free Mg^{2+} in equilibrium. This, in effect, reduces the molarity of the EDTA. So the EDTA is standardized *after* adding the Mg^{2+} by titrating primary standard calcium carbonate (dissolved in HCl and pH adjusted). When the indicator is added to the calcium solution, it is pale red. But as soon as the titration is started, the indicator is complexed by the magnesium and turns wine red. At the end point, it changes to blue, as the indicator is displaced from the magnesium. No correction is required for the Mg^{2+} added because it is accounted for in the standardization. *This solution should not be used to titrate metals other than calcium.*

High-purity EDTA can be prepared from $Na_2H_2Y \cdot 2H_2O$ by drying at 80°C for 2 h. The waters of hydration remained intact, and this can be used as a primary standard for preparing a standard EDTA solution.

The titration of calcium and magnesium with EDTA is done at pH 10, using an ammonia–ammonium chloride buffer. The pH must not be too high or the metal hydroxide may precipitate, causing the reaction with EDTA to be very slow. Calcium can actually be titrated in the presence of magnesium by raising the pH to 12 with strong alkali; $Mg(OH)_2$ precipitates and does not titrate.

Since Eriochrome Black T and other indicators are weak acids, their color will depend on the pH because the different ionized species of the indicator have different colors. For example, with Eriochrome Black T, H_2In^- is red (pH < 6), HIn^{2-} is blue (pH 6 to 12), and In^{3-} is yellow orange (pH > 12). Thus, indicators can be used over definite pH ranges. The pH also affects the stability of the complex formed between the indicator and the metal ion, as well as that formed between EDTA and the metal ion.

An indicator is useful for indication of titrations of only those metals that form a more stable complex with the titrant than with the indicator at the given pH. This all sounds rather complex but, fortunately, suitable indicators have been described for many titrations with several different chelating agents.

Calmagite indicator gives a somewhat improved end point over Eriochrome Black T for the titration of calcium and magnesium with EDTA. It also has a longer shelf life. Xylenol orange is useful for titration of metal ions that form very strong EDTA complexes and are titrated at pH 1.5 to 3.0. Examples are the direct titration of thorium(IV) and bismuth(III), and the indirect determination of zirconium(IV) and iron(III) by back-titration with one of the former two metals. There are many other indicators for EDTA titrations. The treatise by Wilson and Wilson (Ref. 6) gives many examples of EDTA titrations, and you are referred to this excellent source for detailed descriptions of procedures for different metals.

There are a number of other useful reagents for complexometric titrations. A notable example is ethyleneglycol bis(β-aminoethyl ether)-N,N,N′,N′-tetraacetic acid (**EGTA**). This is an ether analog of EDTA that will selectively titrate calcium in the presence of magnesium:

EGTA allows the titration of calcium in the presence of magnesium.

$$\text{HOOCCH}_2 \quad \diagdown \text{H}^+ \qquad\qquad\qquad \text{H}^+ \diagup \quad \text{CH}_2\text{COOH}$$
$$\text{NCH}_2\text{CH}_2\text{OCH}_2\text{CH}_2\text{OCH}_2\text{CH}_2\text{N}$$
$$^-\text{OOCCH}_2 \diagup \qquad\qquad\qquad\qquad\qquad \diagdown \text{CH}_2\text{COO}^-$$

Complexing agents having ether linkages have a strong tendency to complex the alkaline earths heavier than magnesium. Log K_f for calcium–EGTA is 11.0, while that for magnesium–EGTA is only 5.2. For other chelating agents and their applications, see Refs. 4 and 6.

With the exception of the alkali metals, *nearly every metal can be determined with high precision and accuracy by chelometric titration.* These methods are much more rapid and convenient than gravimetric procedures and have largely replaced them, except in those few instances when greater accuracy may be offered and required.

Complexometric titrations in the clinical laboratory are limited to those substances that occur in fairly high concentrations since volumetric methods are generally not too sensitive. The most important complexometric titration is the determination of calcium in blood (see Ref. 8). Chelating agents such as EDTA are used in the treatment of heavy-metal poisoning, for example, when children ingest chipped paint that contains lead. The calcium chelate (as Na_2CaY) is administered to prevent complexation and removal of calcium in the bones. Heavy metals such as lead form more stable EDTA chelates than calcium does and will displace the calcium from the EDTA. The chelated lead is then excreted via the kidneys.

A table of formation constants of some EDTA chelates appears in Appendix C.

9.5 Other Uses of Complexes

Analytical chemists can use the formation of complexes to advantage in ways other than in titrations. For example, metal ion chelates may be formed and extracted into a water-immiscible solvent for separation by **solvent extraction**. Complexes of metal ions with the chelating agent dithizone, for example, are useful for extractions. The chelates are often highly colored. Their formation can then serve as the basis for **spectrophotometric determination** of metal ions. Complexes that fluoresce may also be formed. Even metal chelate precipitates may be formed. The nickel–dimethylglyoxime precipitate is an example used in gravimetric analysis. Table 10.2 lists several other metal chelate precipitates. Complex equilibria may influence separations by chromatographic separations, and we have mentioned the use of complexing agents as masking agents to prevent interfering reactions. For example, in the solvent extraction of vanadium with the chelating agent oxine, copper extraction is avoided by complexing the copper ion with EDTA, preventing formation of its oxine chelate.

Chelates are used in gravimetry, spectrophotometry, fluorometry, solvent extraction, and chromatography.

All of these complexing reactions are pH dependent, and pH adjustment and control (with buffers) is always necessary to optimize the desired reaction or to minimize undesired reactions.

9.6 Fraction of Dissociating Species in Polyligand Complexes: β Values—How Much of Each Species?

Complexes such as $Ag(NH_3)_2^+$ dissociate stepwise, just as polyprotic acids do. As with the acids, calculation of the equilibrium concentrations of the individual complex species is difficult unless excess ligand is present. Otherwise, an iterative procedure must be used. When excess ligand is present, treatment of the equilibria is similar to the calculation of the equilibrium concentration of the various species of a polyprotic acid at a given hydrogen ion concentration. (See Chapter 7, Equations 7.63 through 7.75.) Suppose, for example, that we wish to calculate the equilibrium concentration of the various silver species in Equations 9.1 and 9.2 at a given ammonia concentration. We can define the fractions of each species as follows:

$$\beta_0 = \frac{[Ag^+]}{C_{Ag}} \qquad \beta_1 = \frac{[Ag(NH_3)^+]}{C_{Ag}}$$

$$\beta_2 = \frac{[Ag(NH_3)_2^+]}{C_{Ag}} \qquad \beta_0 + \beta_1 + \beta_2 = 1$$

where C_{Ag} is the total silver ion concentration in all forms. We use β to represent the various fractions for complexes to avoid confusion with a used for the fractions of acid species. (The symbol β is also sometimes used to denote stepwise equilibrium constants, and these should not be confused with the β terms here.) The subscript denotes the number of ligands associated with the metal ion. The total concentration of silver species is given by:

$$C_{Ag} = [Ag(NH_3)_2^+] + [Ag(NH_3)^+] + [Ag^+] \qquad (9.16)$$

If we wish to calculate β_0, the fraction of Ag^+, then we can use the equilibrium constant expressions in Equation 9.1 and 9.2 to substitute in Equation 9.16 to obtain an expression containing only $[Ag^+]$ as the silver species. From Equation 9.1,

$$[Ag(NH_3)^+] = K_{f1}[Ag^+][NH_3] \qquad (9.17)$$

From Equations 9.2 and 9.17,

$$[Ag(NH_3)_2^+] = K_{f2}[Ag(NH_3)^+][NH_3] = K_{f1}K_{f2}[Ag^+][NH_3]^2 \qquad (9.18)$$

Substituting Equations 9.17 and 9.18 in 9.16,

$$C_{Ag} = K_{f1}K_{f2}[Ag^+][NH_3]^2 + K_{f1}[Ag^+][NH_3] + [Ag^+] \qquad (9.19)$$

Substituting in the denominator of the β_0 expression, we have

Use this equation to calculate the fraction of Ag^+ in solution.

$$\boxed{\beta_0 = \frac{1}{K_{f1}K_{f2}[NH_3]^2 + K_{f1}[NH_3] + 1}} \qquad (9.20)$$

We can derive similar equations for the other β's by substituting from Equations 9.1 and 9.2 to obtain expressions in terms of either $[Ag(NH_3)^+]$ for β_1 or $[Ag(NH_3)_2^+]$ for β_2. Or, since we already have an expression for the denominators of β_1 and β_2, we can substitute 9.17 and 9.18 in the numerators of β_1 and β_2, respectively. The result is

$$\beta_1 = \frac{K_{f1}[NH_3]}{K_{f1}K_{f2}[NH_3]^2 + K_{f1}[NH_3] + 1} \tag{9.21}$$

$$\beta_2 = \frac{K_{f1}K_{f2}[NH_3]^2}{K_{f1}K_{f2}[NH_3]^2 + K_{f1}[NH_3] + 1} \tag{9.22}$$

These equations may be compared with Equations 7.72 through 7.75 in Chapter 7 for α's. They are somewhat different in form because the equilibria were written as associations rather than dissociations. If dissociation constants were used in place of formation constants, the equations would be identical to the α equations for a diprotic acid, except that $[H^+]$ would be replaced by $[NH_3]$, the K_a's would be replaced by the K_d's, and the order of β's would be reversed from the order of α's ($\beta_0 \approx \alpha_2$) since they were defined in reverse order.

 **Example 9.5**

Calculate the equilibrium concentrations of the different silver ion species for 0.010 M silver(I) in 0.10 M NH$_3$.

Solution

From Equations 9.20 through 9.22,

$$\beta_0 = \frac{1}{(2.5 \times 10^3)(1.0 \times 10^4)(0.10)^2 + (2.5 \times 10^3)(0.10) + 1} = 4.0 \times 10^{-6}$$

Similarly,

$$\beta_1 = 1.0 \times 10^{-3} \qquad \beta_2 = 1.0$$
$$[Ag^+] = C_{Ag}\beta_0 = (0.010)(4.0 \times 10^{-6}) = 4.0 \times 10^{-8} \, M$$
$$[Ag(NH_3)^+] = C_{Ag}\beta_1 = (0.010)(1.0 \times 10^{-3}) = 1.0 \times 10^{-5} \, M$$
$$[Ag(NH_3)_2^+] = C_{Ag}\beta_2 = (0.010)(1.0) = 0.010 \, M$$

Essentially, all the silver exists as the diammine complex in 0.10 M ammonia.

We neglected in the above calculation any consumption of ammonia in forming the complexes. We see that 20% of it was consumed, as a first approximation. If we were to recalculate the β's at 0.08 M NH$_3$, β_2 would still be equal to 1.0, and most of the silver would still exist as Ag(NH$_3$)$^+$. The relative values of β_0 and β_1 would change, however. This is an **iterative procedure** or **method of successive approximations**. It can be used in any equilibrium calculation in which assumptions are made to simplify the calculations, including simple acid–base equilibria where the amount of acid dissociated is assumed negligible compared to the

Perform an iterative calculation to correct for the NH$_3$ consumed in complexation.

initial concentration (see Chapter 6). Usually, two or at most three iterative calculations are adequate.

See Problem 22 for spreadsheet calculation of the β values for $Ag(NH_3)_2^+$ species as a function of $[NH_3]$.

Learning Objectives

WHAT ARE SOME OF THE KEY THINGS WE LEARNED FROM THIS CHAPTER?

- Formation constants, p. 294
- EDTA equilibria (key equations: 9.12, 9.13), p. 298
- Indicators for EDTA titrations, p. 305
- β values (key equations: 9.20–9.22), p. 308

Questions

1. Distinguish between a complexing agent and a chelating agent.
2. Explain the principles of chelation titration indicators.
3. Why is a small amount of magnesium salt added to the EDTA solution used for the titration of calcium with an Eriochrome Black T indicator?

Problems

COMPLEX EQUILIBRIUM CALCULATIONS (K_f)

4. Calcium ion forms a weak 1:1 complex with nitrate ion with a formation constant of 2.0. What would be the equilibrium concentrations of Ca^{2+} and $Ca(NO_3)^+$ in a solution prepared by adding 10 mL each of 0.010 M $CaCl_2$ and 2.0 M $NaNO_3$? Neglect diverse ion effects.

5. The formation constant of the silver–ethylenediamine complex, $Ag(NH_2CH_2CH_2NH_2)^+$, is 5.0×10^4. Calculate the concentration of Ag^+ in equilibrium with a 0.10 M solution of the complex. (Assume no higher order complexes.)

6. What would be the concentration of Ag^+ in Problem 5 if the solution contained also 0.10 M ethylenediamine, $NH_2CH_2CH_2NH_2$?

7. Silver ion forms stepwise complexes with thiosulfate ion, $S_2O_3^{2-}$, with $K_{f1} = 6.6 \times 10^8$ and $K_{f2} = 4.4 \times 10^4$. Calculate the equilibrium concentrations of all silver species for 0.0100 M $AgNO_3$ in 1.00 M $Na_2S_2O_3$. Neglect diverse ion effects.

CONDITIONAL FORMATION CONSTANTS

8. The formation constant for the lead–EDTA chelate (PbY^{2-}) is 1.10×10^{18}. Calculate the conditional formation constant (a) at pH 3 and (b) at pH 10.

9. Using the conditional constants calculated in Problem 8 calculate the pPb ($-\log[Pb^{2+}]$) for 50.0 mL of a solution of 0.0250 M Pb^{2+} (a) at pH 3 and (b) at pH 10 after the addition of (1) 0 mL, (2) 50 mL, (3) 125 mL, and (4) 200 mL of 0.0100 M EDTA.

10. The conditional formation constant for the calcium–EDTA chelate was calculated for pH 10 in Example 9.4 to be 1.8×10^{10}. Calculate the conditional formation constant at pH 3. Compare this with that calculated for lead at pH 3 in Problem 8. Could lead be titrated with EDTA at pH 3 in the presence of calcium?

STANDARD SOLUTIONS

11. Calculate the weight of $Na_2H_2Y \cdot 2H_2O$ required to prepare 500.0 mL of 0.05000 *M* EDTA.

12. An EDTA solution is standardized against high-purity $CaCO_3$ by dissolving 0.3982 g $CaCO_3$ in hydrochloric acid, adjusting the pH to 10 with ammoniacal buffer, and titrating. If 38.26 mL was required for the titration, what is the molarity of the EDTA?

13. Calculate the titer of 0.1000 *M* EDTA in mg $CaCO_3$/mL.

14. If 100.0 mL of a water sample is titrated with 0.01000 *M* EDTA, what is the titer of the EDTA in terms of water hardness/mL?

QUANTITATIVE COMPLEXOMETRIC DETERMINATIONS

15. Calcium in powdered milk is determined by ashing a 1.50-g sample and then titrating the calcium with EDTA solution, 12.1 mL being required. The EDTA was standardized by titrating 10.0 mL of a zinc solution prepared by dissolving 0.632 g zinc metal in acid and diluting to 1 L (10.8 mL EDTA required for titration). What is the concentration of calcium in the powdered milk in parts per million?

16. Calcium is determined in serum by microtitration with EDTA. A 100-μL sample is treated with two drops of 2 *M* KOH, Cal-Red indicator is added, and the titration is performed with 0.00122 *M* EDTA, using a microburet. If 0.203 mL EDTA is required for titration, what is the level of calcium in the serum in mg/dL and in meq/L?

17. In the Liebig titration of cyanide ion, a soluble complex is formed; and at the equivalence point, solid silver cyanide is formed, signaling the end point:

$$2CN^- + Ag^+ \rightarrow Ag(CN)_2^- \quad \text{(titration)}$$
$$Ag(CN)_2^- + Ag^+ \rightarrow Ag[Ag(CN)_2] \quad \text{(end point)}$$

A 0.4723-g sample of KCN was titrated with 0.1025 *M* $AgNO_3$, requiring 34.95 mL. What is the percent purity of the KCN?

18. Copper in saltwater near the discharge of a sewage treatment plant is determined by first separating and concentrating it by solvent extraction of its dithizone chelate at pH 3 into methylene chloride and then evaporating the solvent, ashing the chelate to destroy the organic portion, and titrating the copper with EDTA. Three 1-L portions of the sample are each extracted with 25-mL portions of methylene chloride, and the extracts are combined in a 100-mL volumetric flask and diluted to volume. A 50-mL aliquot is evaporated, ashed, and titrated. If the EDTA solution has a $CaCO_3$ titer of 2.69 mg/mL and 2.67 mL is required for titration of the copper, what is the concentration of copper in the seawater in parts per million?

19. Chloride in serum is determined by titration with $Hg(NO_3)_2$; $2Cl^- + Hg^{2+} \rightleftharpoons HgCl_2$. The $Hg(NO_3)_2$ is standardized by titrating 2.00 mL of a 0.0108 *M* NaCl solution, requiring 1.12 mL to reach the diphenylcarbazone end point. A 0.500-mL serum sample is treated with 3.50 mL water, 0.50 mL 10% sodium tungstate solution, and 0.50 mL of 0.33 M H_2SO_4 solution to precipitate proteins. After the proteins are precipitated, the sample is filtered through a dry filter into a dry flask. A 2.00-mL aliquot of the filtrate is titrated with the $Hg(NO_3)_2$ solution, requiring 1.23 mL. Calculate the meg/L chloride in the serum. (Note: mercury is rarely used today due to its toxicity. The problem is illustrative.)

SPREADSHEET PROBLEMS

See your CD, Chapter 9, for suggested setups.

20. Prepare a spreadsheet for Figure 9.2, log K_f' vs. pH for the EDTA chelates of calcium, lead, and mercury. This will require calculating α_4 for EDTA and the K_f values for the chelates of calcium, lead, and mercury. Calculate at 0.5 pH intervals. Compare your plot with Figure 9.2.

21. Prepare a spreadsheet for the titration of 100.00 mL of 0.1000 M Ca^{2+} with 0.1000 M Na_2EDTA at pH 10 (pCa vs. mL EDTA—Figure 9.3). Start out at 10 mL and then 20-mL intervals of titrant, and gradually decrease to 0.05 mL just before and after the equivalue point. See Example 9.4.

22. Prepare a spreadsheet to plot the three β values for $Ag(NH_3)_2^+$ as a function of $[NH_3]$. Plot from 0 to 0.01 M NH_3, beginning with increments of 0.0005 M, and after four intervals, use 0.001 M.

Recommended References

1. *Stability Constants of Metal-Ion Complexes. Part A: Inorganic Ligands*, E. Hogfeldt, ed., *Part B: Organic Ligands*, D. D. Perrin, ed., Oxford: Pergamon, 1979, 1981.

2. A Martell and R. J. Motekaitis, *The Determination and Use of Stability Constants.* New York: VCH, 1989.

3. J. Kragten, *Atlas of Metal-Ligand Equilibria in Aqueous Solution.* London: Ellis Horwood, 1978.

4. G. Schwarzenbach, *Complexometric Titrations.* New York: Interscience, 1957.

5. H. Flaschka, *EDTA Titrations.* New York: Pergamon, 1959.

6. H. A. Flaschka and A. J. Barnard, Jr., "Titrations with EDTA and Related Compounds," in C. L. Wilson and D. W. Wilson, eds. *Comprehensive Analytical Chemistry*, Vol. 1B. New York: Elsevier, 1960.

7. F. J. Welcher, *The Analytical Uses of Ethylenediaminetetraacetic Acid.* Princeton: Van Nostrand, 1958.

8. F. W. Fales, "Calcium (Complexometric)," in *Standard Methods of Clinical Chemistry*, Vol. 2. D. Seligson, ed. New York: Academic, 1958, pp. 1–11.

9. J. Stary, ed., *Critical Evaluation of Equilibrium Constants Involving 8-Hydroxyquinoline and Its Metal Chelates.* Oxford: Pergamon, 1979.

10. H. A. Flaschka, *Chelates in Analytical Chemistry*, Vols. 1–5. New York: Dekker, 1967–1976.

Chapter Ten

GRAVIMETRIC ANALYSIS AND PRECIPITATION EQUILIBRIA

*"Some loads are light, some heavy.
Some people prefer the light to the heavy . . ."*
—Mao Tse-tung

Gravimetric analysis is one of the most accurate and precise methods of macroquantitative analysis. In this process the analyte is selectively converted to an insoluble form. The separated precipitate is dried or ignited, possibly to another form, and is accurately weighed. From the weight of the precipitate and a knowledge of its chemical composition, we can calculate the weight of analyte in the desired form.

This chapter describes the specific steps of gravimetric analysis, including preparing the solution in proper form for precipitation, the precipitation process and how to obtain the precipitate in pure and filterable form, the filtration and washing of the precipitate to prevent losses and impurities, and heating the precipitate to convert it to a weighable form. It gives calculation procedures for computing the quantity of analyte from the weight of precipitate, following the principles introduced in Chapter 5. It also provides some common examples of gravimetric analysis. Finally, it discusses the solubility product and associated precipitation equilibria.

Gravimetry is among the most accurate analytical techniques (but it is tedious!). T. W. Richards used it to determine atomic weights! He received the Nobel Prize in 1914 for his work. See *Z. Anorg. Chem.*, **8** (1895), 413, 419, and 421 for some of his careful studies on contamination. See also http://nobelprizes.com.

10.1 How to Perform a Successful Gravimetric Analysis

A successful gravimetric analysis consists of a number of important operations designed to obtain a pure and filterable precipitate suitable for weighing. You may wish to precipitate silver chloride from a solution of chloride by adding silver nitrate. There is more to the procedure than simply pouring in silver nitrate solution and then filtering.

Accurate gravimetric analysis requires careful manipulation in forming and treating the precipitate.

WHAT STEPS ARE NEEDED?

The steps required in a gravimetric analysis, after the sample has been dissolved, can be summarized as follows:

1. Preparation of the solution
2. Precipitation
3. Digestion
4. Filtration
5. Washing
6. Drying or igniting
7. Weighing
8. Calculation

These operations and the reasons for them are described below.

FIRST PREPARE THE SOLUTION

The first step in performing gravimetric analysis is to prepare the solution. Some form of preliminary separation may be necessary to eliminate interfering materials. Also, we must adjust the solution conditions to maintain low solubility of the precipitate and to obtain it in a form suitable for filtration. Proper adjustment of the solution conditions prior to precipitation may also mask potential interferences. Factors that must be considered include the volume of the solution during precipitation, the concentration range of the test substance, the presence and concentrations of other constituents, the temperature, and the pH.

Usually, the precipitation reaction is selective for the analyte.

Although preliminary separations may be required, in other instances the precipitation step in gravimetric analysis is sufficiently selective that other separations are not required. The pH is important because it often influences both the solubility of the analytical precipitate and the possibility of interferences from other substances. For example, calcium oxalate is insoluble in basic medium, but at low pH the oxalate ion combines with the hydrogen ions to form a weak acid. 8-Hydroxyquinoline (oxine) can be used to precipitate a large number of elements, but by controlling pH, we can precipitate elements selectively. Aluminum ion can be precipitated at pH 4, but the concentration of the anion form of oxine is too low at this pH to precipitate magnesium ion.

A higher pH is required to shift the ionization step to the right in order to precipitate magnesium. If the pH is too high, however, magnesium hydroxide will precipitate, causing interference.

The effects of the other factors mentioned above will become apparent as we discuss the precipitation step.

THEN DO THE PRECIPITATION—BUT UNDER THE RIGHT CONDITIONS

After preparing the solution, the next step is to do the precipitation. Again, certain conditions are important. The precipitate should first be *sufficiently insoluble* that the amount lost due to solubility will be negligible. It should consist of *large crystals* that can be easily filtered. All precipitates tend to carry some of the other constituents of the solution with them. This contamination should be negligible. Keeping the crystals large can minimize this contamination.

We can achieve an appreciation of the proper conditions for precipitation by first looking at the **precipitation process**: When a solution of precipitating agent is added to a test solution to form a precipitate, such as in the addition of $AgNO_3$ to a chloride solution to precipitate AgCl, the actual precipitation occurs in a series of steps. The precipitation process involves heterogeneous equilibria and, as such, is not instantaneous (see Chapter 6). The equilibrium condition is described by the solubility product, discussed at the end of the chapter. First, **supersaturation** occurs, that is, the solution phase contains more of the dissolved salt than occurs at equilibrium. This is a metastable condition, and the driving force will be for the system to approach equilibrium (saturation). This is started by **nucleation**. For nucleation to occur, a minimum number of particles must come together to produce microscopic nuclei of the solid phase. The higher the degree of supersaturation, the greater the rate of nucleation. The formation of a greater number of nuclei per unit time will ultimately produce more total crystals of smaller size. The total crystal surface area will be larger, and there will be more danger that impurities will be adsorbed (see below).

Although nucleation should theoretically occur spontaneously, it is usually induced, for example, on dust particles, scratches on the vessel surface, or added seed crystals of the precipitate (not in quantitative analysis).

Following nucleation, the initial nucleus will grow by depositing other precipitate particles to form a crystal of a certain geometric shape. Again, the greater the supersaturation, the more rapid the crystal growth rate. An increased growth rate increases the chances of imperfections in the crystal and trapping of impurities.

Von Weimarn discovered that the particle size of precipitates is inversely proportional to the relative supersaturation of the solution during the precipitation process:

$$\boxed{\text{Relative supersaturation} = \frac{Q - S}{S}}$$

where Q is the concentration of the mixed reagents *before* precipitation occurs and is the **degree of supersaturation**, and S is the **solubility** of the precipitate at equilibrium. This ratio is also called the **von Weimarn ratio**.

As previously mentioned, when a solution is supersaturated, it is in a state

During the precipitation process, supersaturation occurs (this should be minimized!), followed by nucleation and precipitation.

of metastable equilibrium, and this favors rapid nucleation to form a large number of small particles. That is,

---

High relative supersaturation → many small crystals
(high surface area)

Low relative supersaturation → fewer, larger crystals
(low surface area)

Obviously, then, we want to keep Q low and S high during precipitation. Several steps are commonly taken to maintain *favorable conditions for precipitation*:

> Here is how to minimize supersaturation and obtain large crystals.

1. Precipitate from *dilute solution*. This keeps Q low.
2. Add dilute precipitating reagents *slowly*, with effective *stirring*. This also keeps Q low. Stirring prevents local excesses of the reagent.
3. Precipitate from *hot solution*. This increases S. The solubility should not be too great or the precipitation will not be quantitative (with less than 1 part per thousand remaining). The bulk of the precipitation may be performed in the hot solution, and then the solution may be cooled to make the precipitation quantitative.
4. Precipitate at as *low* a *pH* as is possible to maintain quantitative precipitation. As we have seen, many precipitates are more soluble in acid medium, and this slows the rate of precipitation. They are more soluble because the anion of the precipitate combines with protons in the solution.

> Very insoluble precipitates are not the best candidates for gravimetric analysis! They supersaturate too easily.

Most of these operations can also decrease the degree of contamination. The concentration of impurities is kept lower and their solubility is increased, and the slower rate of precipitation decreases their chance of being trapped. The larger crystals have a smaller specific surface area (i.e., a smaller surface area relative to the mass) and so have less chance of adsorbing impurities. Note that the most insoluble precipitates do not make the best candidates for pure and readily filterable precipitates. An example is hydrous iron oxide (or iron hydroxide), which forms a gelatinous precipitate of large surface area.

> Don't add too much excess precipitating agent. This will increase adsorption.

When the precipitation is performed, a slight excess of precipitating reagent is added to decrease the solubility by mass action (common ion effect) and to assure complete precipitation. A large excess of precipitating agent should be avoided because this increases chances of adsorption on the surface of the precipitate, in addition to being wasteful. If the approximate amount of analyte is known, a 10% excess of reagent is generally added. Completeness of precipitation is checked by waiting until the precipitate has settled and then adding a few drops of precipitating reagent to the clear solution above it. If no new precipitate forms, precipitation is complete.

> Check for completeness of precipitation!

DIGEST THE PRECIPITATE TO MAKE LARGER AND MORE PURE CRYSTALS

We know that very small crystals with a large specific surface area have a higher surface energy and a higher apparent solubility than large crystals. This is an initial rate phenomenon and does not represent the equilibrium condition, and it is

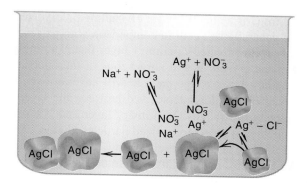

Fig. 10.1. Ostwald ripening.

one consequence of heterogeneous equilibria. When a precipitate is allowed to stand in the presence of the **mother liquor** (the solution from which it was precipitated), the large crystals grow at the expense of the small ones. This process is called **digestion**, or **Ostwald ripening**, and is illustrated in Figure 10.1. The small particles tend to dissolve and reprecipitate on the surfaces of the larger crystals. In addition, individual particles **agglomerate** to effectively share a common counterion layer (see below), and the agglomerated particles finally *cement* together by forming connecting bridges. This noticeably decreases surface area.

Also, imperfections of the crystals tend to disappear, and adsorbed or trapped impurities tend to go into solution. Digestion is usually done at elevated temperatures to speed the process, although in some cases it is done at room temperature. It improves both the filterability of the precipitate and its purity.

Many precipitates do not give a favorable von Weimarn ratio, especially very insoluble ones. Hence, it is impossible to yield a crystalline precipitate (small number of large particles), and the precipitate is first **colloidal** (large number of small particles).

Colloidal particles are very small (1 to 100 μm) and have a very large surface-to-mass ratio, which promotes surface adsorption. They are formed by virtue of the precipitation mechanism. As a precipitate forms, the ions are arranged in a fixed pattern. In AgCl, for example, there will be alternating Ag^+ and Cl^- ions on the surface (see Figure 10.2). While there are localized $+$ and $-$ charges on the surface, the net surface charge is zero. However, the surface does tend to adsorb the ion of the precipitate particle that is in excess in the solution, for example, Cl^-

Ostwald ripening improves the purity and crystallinity of the precipitate.

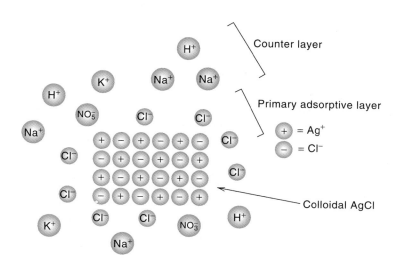

Fig. 10.2. Representation of silver chloride colloidal particle and adsorptive layers when Cl^- is in excess.

if precipitating Cl^- with Ag^+; this imparts a charge. (With crystalline precipitates, the degree of such adsorption will generally be small in comparison with particles that tend to form colloids.) The adsorption creates a **primary layer** that is strongly adsorbed and is an integral part of the crystal. It will attract ions of the opposite charge in a **counterlayer** or secondary layer so the particle will have an overall neutral charge. There will be solvent molecules interspersed between the layers. Normally, the counterlayer completely neutralizes the primary layer and is close to it, so the particles will collect together to form larger-sized particles; that is, they will **coagulate**. However, if the secondary layer is loosely bound, the primary surface charge will tend to repel like particles, maintaining a colloidal state.

When coagulated particles are filtered, they retain the adsorbed primary and secondary ion layers along with solvent. Washing the particles with water increases the extent of solvent (water) molecules between the layers, causing the secondary layer to be loosely bound, and the particles revert to the colloidal state. This process is called **peptization** and is discussed in more detail below where we describe washing precipitate. Adding an electrolyte will result in a closer secondary layer and will promote coagulation. Heating tends to decrease adsorption and the effective charge in the adsorbed layers, thereby aiding coagulation. Stirring will also help.

> Peptization is the reverse of coagulation (the precipitate reverts to a colloidal state and is lost). It is avoided by washing with an electrolyte that can be volatized by heating.

While all colloidal precipitates cause difficulties in analytical precipitates, some are worse than others. There are two types of colloids, **hydrophilic** and **hydrophobic**. Hydrophilic means "water loving," and these colloids have a strong affinity for water. A solution of a hydrophilic colloid is therefore viscous. A hydrophobic colloid has little attraction for water. A solution of this type of colloid is called a **sol**.

Coagulation of a hydrophobic colloid is fairly easy and results in a curdy precipitate. An example is silver chloride. Coagulation of a hydrophilic colloid, such as hydrous ferric oxide, is more difficult, and it produces a gelatinous precipitate that is difficult to filter because it tends to clog the pores of the filter. In addition, gelatinous precipitates adsorb impurities readily because of their very large surface area. Sometimes a *reprecipitation* of the filtered precipitate is required. During the reprecipitation, the concentration of impurities in solution (from the original sample matrix) has been reduced to a low level, and adsorption will be very small.

> AgCl forms a hydrophobic colloid (a sol), which readily coagulates. $Fe_2O_3 \cdot xH_2O$ forms a hydrophilic colloid (a gel) with large surface area.

Despite the colloidal nature of silver chloride, the gravimetric determination of chloride is one of the most accurate determinations.

IMPURITIES IN PRECIPITATES

Precipitates tend to carry down from the solution other constituents that are normally soluble, causing the precipitate to become contaminated. This process is called **coprecipitation**. The process may be equilibrium based or kinetically controlled. There are a number of ways in which a foreign material may be coprecipitated.

1. Occlusion and Inclusion. In the process of **occlusion**, material that is not part of the crystal structure is trapped within a crystal. For example, water may be trapped in pockets when $AgNO_3$ crystals are formed, and this can be driven off by melting. If such mechanical trapping occurs during a precipitation process, the water will contain dissolved impurities. **Inclusion** occurs when ions, generally of similar size and charge, are trapped within the crystal lattice (isomorphous inclusion, as with K^+ in NH_4MgPO_4 precipitation). These are not equilibrium processes.

> Occlusion is the trapping of impurities inside the precipitate.

Occluded or included impurities are difficult to remove. Digestion may help some but is not completely effective. The impurities cannot be removed by washing. Purification by dissolving and reprecipitating may be helpful.

2. Surface Adsorption. As we have already mentioned, the surface of the precipitate will have a primary adsorbed layer of the lattice ions in excess. This results in **surface adsorption**, the most common form of contamination. After the barium sulfate is completely precipitated, the lattice ion in excess will be barium, and this will form the primary layer. The counterion will be a foreign anion, say, a nitrate anion, two for each barium ion. The net effect then is an adsorbed layer of barium nitrate, an equilibrium-based process. These adsorbed layers can often be removed by washing, or they can be replaced by ions that are readily volatilized. Gelatinous precipitates are especially troublesome, though. Digestion reduces the surface area and the amount of adsorption.

Surface adsorption of impurities is the most common source of error in gravimetry. It is reduced by proper precipitation technique, digestion, and washing.

3. Isomorphous Replacement. Two compounds are said to be **isomorphous** if they have the same type of formula and crystallize in similar geometric forms. When their lattice dimensions are about the same, one ion can replace another in a crystal, resulting in a **mixed crystal**. This process is called **isomorphous replacement**. For example, in the precipitation of Mg^{2+} as magnesium ammonium phosphate, K^+ has nearly the same ionic size as NH_4^+ and can replace it to form magnesium potassium phosphate. Isomorphous replacement, when it occurs, is very serious, and little can be done about it. Precipitates in which it occurs are seldom used analytically. Mixed crystal formation is a form of equilibrium precipitates formation, although it may be influenced by the rate of precipitation.

4. Postprecipitation. Sometimes, when the precipitate is allowed to stand in contact with the mother liquor, a second substance will slowly form a precipitate with the precipitating reagent. This is called **postprecipitation**. For example, when calcium oxalate is precipitated in the presence of magnesium ions, magnesium oxalate does not immediately precipitate because it tends to form supersaturated solutions. But it will come down if the solution is allowed to stand too long before being filtered. Similarly, copper sulfide will precipitate in acid solution in the presence of zinc ions, but eventually zinc sulfide will precipitate. Postprecipitation is a slow equilibrium process.

WASHING AND FILTERING THE PRECIPITATES—TAKE CARE OR YOU MAY LOSE SOME

Coprecipitated impurities, especially those on the surface, can be removed by washing the precipitate after filtering. The precipitate will be wet with the mother liquor, which is also removed by washing. Many precipitates cannot be washed with pure water, because **peptization** occurs. This is the reverse of coagulation, as previously mentioned.

The process of coagulation discussed above is at least partially reversible. As we have seen, coagulated particles have a neutral layer of adsorbed primary and counterions. We also saw that the presence of another electrolyte will cause the counterions to be forced into closer contact with the primary layer, thus promoting coagulation. These foreign ions are carried along in the coagulation. Washing with water will dilute and remove foreign ions, and the counterion will occupy a larger volume, with more solvent molecules between it and the primary layer. The result is that the repulsive forces between particles become strong again, and the particles partially revert to the colloidal state and pass through the filter. This can be prevented by adding an electrolyte to the wash liquid, for example, HNO_3 or NH_4NO_3 for AgCl precipitate (but not KNO_3 since it is nonvolatile—see below).

The electrolyte must be one that is volatile at the temperature to be used for drying or ignition, and it must not dissolve the precipitate. For example, dilute

Test for completeness of washing.

nitric acid is used as the wash solution for silver chloride. The nitric acid replaces the adsorbed layer of $Ag^+|anion^-$, and it is volatilized when dried at 110°C. Ammonium nitrate is used as the wash electrolyte for hydrous ferric oxide. It is decomposed to NH_3, HNO_3, N_2, and oxides of nitrogen when the precipitate is dried by ignition at a high temperature.

When you wash a precipitate, you should conduct a test to determine when the washing is complete. This is usually done by testing the filtrate for the presence of an ion of the precipitating reagent. After several washings with small volumes of the wash liquid, a few drops of the filtrate are collected in a test tube for the testing. For example, if chloride ion is determined by precipitating with silver nitrate reagent, the filtrate is tested for silver ion by adding sodium chloride or dilute HCl. We describe the technique of filtering in Chapter 2.

DRYING OR IGNITING THE PRECIPITATE

Drying removes the solvent and wash electrolytes.

If the collected precipitate is in a form suitable for weighing, it must be heated to remove water and to remove the adsorbed electrolyte from the wash liquid. This drying can usually be done by heating at 110 to 120°C for 1 to 2 h. **Ignition** at a much higher temperature is usually required if a precipitate must be converted to a more suitable form for weighing. For example, magnesium ammonium phosphate, $MgNH_4PO_4$, is decomposed to the pyrophosphate, $Mg_2P_2O_7$, by heating at 900°C. Hydrous ferric oxide, $Fe_2O_3 \cdot xH_2O$, is ignited to the anhydrous ferric oxide. Many metals that are precipitated by organic reagents (e.g., 8-hydroxyquinoline) or by sulfide can be ignited to their oxides. The technique of igniting a precipitate is also described in Chapter 2.

10.2 Gravimetric Calculations—How Much Analyte Is There?

The precipitate we weigh is usually in a different form than the analyte whose weight we wish to report. The principles of converting the weight of one substance to that of another are given in Chapter 5 (Section 5.8), using stoichiometric mole relationships. We introduced the **gravimetric factor** (GF), which represents the weight of analyte per unit weight of precipitate. It is obtained from the ratio of the formula weight of the analyte to that of the precipitate, multiplied by the moles of analyte per mole of precipitate obtained from each mole of analyte, that is,

$$GF = \frac{\text{f wt analyte (g/mol)}}{\text{f wt precipitate (g/mol)}} \times \frac{a}{b} \text{ (mol analyte/mol precipitate)} \qquad (10.1)$$
$$= \text{g analyte/g precipitate}$$

Grams precipitate $\times$ GF gives grams analyte.

So, if Cl_2 in a sample is converted to chloride and precipitated as AgCl, the weight of Cl_2 that gives 1 g of AgCl is

$$\text{g Cl}_2 = \text{g AgCl} \times \frac{\text{f wt Cl}_2\text{ (g Cl}_2\text{/mol Cl}_2)}{\text{f wt AgCl (g AgCl/mol AgCl)}} \times \frac{1}{2} \text{ (mol Cl}_2\text{/mol AgCl)}$$
$$= \text{g AgCl} \times \text{GF (g Cl}_2\text{/g AgCl)}$$
$$= \text{g AgCl} \times 0.2473_7 \text{ (g Cl}_2\text{/g AgCl)}$$

Example 10.1

Calculate the grams of analyte per gram of precipitate for the following conversions:

Analyte	*Precipitate*
P	Ag_3PO_4
K_2HPO_4	Ag_3PO_4
Bi_2S_3	$BaSO_4$

Solution

$$\text{g P/g } Ag_3PO_4 = \frac{\text{at wt P (g/mol)}}{\text{g wt } Ag_3PO_4 \text{ (g/mol)}} = \frac{1}{1} \text{ (mol P/mol } Ag_3PO_4)$$

$$= \frac{30.97 \text{ (g P/mol)}}{418.58 \text{ (g } Ag_3PO_4/\text{mol)}} \times \frac{1}{1} = 0.07399 \text{ g P/g } Ag_3PO_4 = GF$$

$$\text{g } K_2HPO_4/\text{g } Ag_3PO_4 = \frac{\text{f wt } K_2HPO_4 \text{ (g/mol)}}{\text{f wt } Ag_3PO_4 \text{ (g/mol)}} \times \frac{1}{1} \text{ (mol } K_2HPO_4/\text{mol } Ag_3PO_4)$$

$$= \frac{174.18 \text{ (g } K_2HPO_4/\text{mol)}}{418.58 \text{ (g } Ag_3PO_4/\text{mol)}} \times \frac{1}{1} = 0.41612 \text{ g } K_2HPO_4/\text{g } Ag_3PO_4$$

$$= GF$$

$$\text{g } Bi_2S_3/\text{g } BaSO_4 = \frac{\text{f wt } Bi_2S_3 \text{ (g/mol)}}{\text{f wt } BaSO_4 \text{ (g/mol)}} \times \frac{1}{3} \text{ (mol } Bi_2S_3/\text{mol } BaSO_4)$$

$$= \frac{514.15 \text{ (g } Bi_2S_3/\text{mol)}}{233.40 \text{ (g } BaSO_4/\text{mol)}} \times \frac{1}{3} = 0.73429 \text{ g } Bi_2S_3/\text{g } BaSO_4$$

$$= GF$$

In gravimetric analysis, we are generally interested in the percent composition by weight of the analyte in the sample, that is,

$$\% \text{ substance sought} = \frac{\text{weight of substance sought (g)}}{\text{weight of sample (g)}} \times 100\% \qquad (10.2)$$

We obtain the weight of substance sought from the weight of the precipitate and the corresponding weight/mole relationship (Equation 10.1):

Weight of substance sought (g) = weight of precipitate (g)

$$\times \frac{\text{f wt substance sought (g/mol)}}{\text{f wt precipitate (g/mol)}}$$

$$\times \frac{a}{b} \text{ (mol substance sought/mol precipitate)}$$

= weight of precipitate (g)

$$\times GF \text{ (g sought/g precipitate)}$$

$$(10.3)$$

Calculations are usually made on a percentage basis:

$$\% \ A = \frac{g_A}{g_{sample}} \times 100\% \qquad (10.4)$$

where g_A represents the grams of analyte (the desired test substance) and g_{sample} represents the grams of sample taken for analysis.

Check units!

We can write a general formula for calculating the percentage composition of the substance sought:

$$\% \ \text{sought} = \frac{\text{weight of precipitate (g)} \times \text{GF (g sought/g precipitate)}}{\text{weight of sample (g)}} \times 100\%$$

$$(10.5)$$

Example 10.2

Orthophosphate (PO_4^{3-}) is determined by weighing as ammonium phosphomolybdate, $(NH_4)PO_4 \cdot 12MoO_3$. Calculate the percent P in the sample and the percent P_2O_5 if 1.1682 g precipitate (ppt) were obtained from a 0.2711-g sample. Perform the % P calculation using the gravimetric factor and just using dimensional analysis.

Solution

$$\% \ P = \frac{1.1682 \ \text{g ppt} \times \dfrac{P}{(NH_4)_3PO_4 \cdot 12MoO_3} \ (\text{g P/g ppt})}{0.2711 \ \text{g sample}} \times 100\%$$

$$= \frac{1.1682 \ \text{g} \times (30.97/1876.5)}{0.2711 \ \text{g}} \times 100\% = 7.111\%$$

$$\% \ P_2O_5 = \frac{1.1682 \ \text{g ppt} \times \dfrac{P_2O_5}{2(NH_4)_3PO_4 \cdot 12MoO_3} \ (\text{g} \ P_2O_5/\text{g ppt})}{0.2711 \ \text{g sample}} \times 100\%$$

$$= \frac{1.1682 \ \text{g} \times [141.95/(2 \times 1876.5)]}{0.2711 \ \text{g}} \times 100\%$$

$$= 16.30\%$$

Let's do the same calculation using dimensional analysis for the % P setup.

$$\% \ P = \frac{1.982 \ \text{g} \ \cancel{(NH_4)_2PO_4 \cdot MoO_4} \times (30.97/1867.5) \text{g P/g} \ \cancel{(NH_4)_2PO_4 \cdot 12MoO_4}}{0.2771 \ \text{g sample}}$$

$$\times 100\%$$

$$= (7.111 \ \text{g P/g sample}) \times 100\% = 7.111\% \ P$$

Note that the $(NH_4)_2PO_4 \cdot MoO_4$ species cancel one another (dimensional analysis), leaving only g $\cdot$P in the numerator.

When we compare this approach with the gravimetric factor calculation, we see that the setups are really identical. However, this approach better shows which units cancel and which remain.

Example 10.3

An ore is analyzed for the manganese content by converting the manganese to Mn_3O_4 and weighing it. If a 1.52-g sample yields Mn_3O_4 weighing 0.126 g, what would be the percent Mn_2O_3 in the sample? The percent Mn?

Solution

$$\% \ Mn_2O_3 = \frac{0.126 \text{ g } Mn_3O_4 \times \dfrac{3Mn_2O_3}{2Mn_3O_4} \ (\text{g } Mn_2O_3/\text{g } Mn_3O_4)}{1.52 \text{ g sample}} \times 100\%$$

$$= \frac{0.126 \text{ g} \times [3(157.9)/2(228.8)]}{1.52 \text{ g}} \times 100\% = 8.58\%$$

$$\% \ Mn = \frac{0.126 \text{ g } Mn_3O_4 \times \dfrac{3Mn}{Mn_3O_4}(\text{g } Mn/\text{g } Mn_3O_4)}{1.52 \text{ g sample}} \times 100\%$$

$$= \frac{0.126 \text{ g} \times [3(54.94)/228.8]}{1.52 \text{ g}} \times 100\% = 5.97\%$$

The following two examples illustrate some special additional capabilities of gravimetric computations.

Example 10.4

What weight of pyrite ore (impure FeS_2) must be taken for analysis so that the $BaSO_4$ precipitate weight obtained will be equal to one-half that of the percent S in the sample?

Solution

If we have A% of S, then we will obtain $\frac{1}{2}$A g of $BaSO_4$. Therefore,

$$A\% \ S = \frac{\dfrac{1}{2}A(\text{g } BaSO_4) \times \dfrac{S}{BaSO_4}(\text{g } S/\text{g } BaSO_4)}{\text{g sample}} \times 100\%$$

or

$$1\% \ S = \frac{\dfrac{1}{2} \times \dfrac{32.064}{233.40}}{\text{g sample}} \times 100\%$$

$$\text{g sample} = 6.869 \text{ g}$$

PRECIPITATE MIXTURES—WE NEED TWO WEIGHTS

 Example 10.5

A mixture containing only $FeCl_3$ and $AlCl_3$ weighs 5.95 g. The chlorides are converted to the hydrous oxides and ignited to Fe_2O_3 and Al_2O_3. The oxide mixture weighs 2.62 g. Calculate the percent Fe and Al in the original mixture.

Solution

There are two unknowns, so two simultaneous equations must be set up and solved. Let x = g Fe and y = g Al. Then, for the first equation,

$$\text{g } FeCl_3 + \text{g } AlCl_3 = 5.95 \text{ g} \tag{1}$$

$$x\left(\frac{FeCl_3}{Fe}\right) + y\left(\frac{AlCl_3}{Al}\right) = 5.95 \text{ g} \tag{2}$$

$$x\left(\frac{162.21}{55.85}\right) + y\left(\frac{133.34}{26.98}\right) = 5.95 \text{ g} \tag{3}$$

$$2.90x + 4.94y = 5.95 \text{ g} \tag{4}$$

$$\text{g } Fe_2O_3 + \text{g } Al_2O_3 = 2.62 \text{ g} \tag{5}$$

$$x\left(\frac{Fe_2O_3}{2Fe}\right) + y\left(\frac{Al_2O_3}{2Al}\right) = 2.62 \text{ g} \tag{6}$$

$$x\left(\frac{159.69}{2 \times 55.85}\right) + y\left(\frac{101.96}{2 \times 26.98}\right) = 2.62 \text{ g} \tag{7}$$

$$1.43x + 1.89y = 2.62 \text{ g} \tag{8}$$

Solving (4) and (8) simultaneously for x and y:

$$x = 1.07 \text{ g}$$
$$y = 0.58 \text{ g}$$

$$\% \text{ Fe} = \frac{1.07 \text{ g}}{5.95 \text{ g}} \times 100\% = 18.0\%$$

$$\% \text{ Al} = \frac{0.58 \text{ g}}{5.95 \text{ g}} \times 100\% = 9.8\%$$

10.3 Examples of Gravimetric Analysis

Some of the most precise and also accurate analyses are gravimetric analyses. There are many examples, and you should be familiar with some of the more common ones. These are summarized in Table 10.1, which lists the substance sought, the precipitate formed, the form in which it is weighed, and the common elements that will interfere and must be absent. You should consult more advanced texts and comprehensive analytical reference books for details on these and other determinations.

Table 10.1

Some Commonly Employed Gravimetric Analyses

Substance Analyzed	Precipitate Formed	Precipitate Weighed	Interferences
Fe	$Fe(OH)_3$	Fe_2O_3	Many. Al, Ti, Cr, etc.
	Fe cupferrate	Fe_2O_3	Tetravalent metals
Al	$Al(OH)_3$	Al_2O_3	Many. Fe, Ti, Cr, etc.
	$Al(ox)_3{}^a$	$Al(ox)_3$	Many. Mg does not interfere in acidic solution
Ca	CaC_2O_4	$CaCO_3$ or CaO	All metals except alkalis and Mg
Mg	$MgNH_4PO_4$	$Mg_2P_2O_7$	All metals except alkalis
Zn	$ZnNH_4PO_4$	$Zn_2P_2O_7$	All metals except Mg
Ba	$BaCrO_4$	$BaCrO_4$	Pb
$SO_4{}^{2-}$	$BaSO_4$	$BaSO_4$	$NO_3{}^-$, $PO_4{}^{3-}$, $ClO_3{}^-$
Cl^-	$AgCl$	$AgCl$	Br^-, I^-, SCN^-, CN^-, S^{2-}, $S_2O_3{}^{2-}$
Ag	$AgCl$	$AgCl$	Hg(I)
$PO_4{}^{3-}$	$MgNH_4PO_4$	$Mg_2P_2O_7$	$MoO_4{}^{2-}$, $C_2O_4{}^{2-}$, K^+
Ni	$Ni(dmg)_2{}^b$	$Ni(dmg)_2$	Pd

^{a}ox = Oxine (8-hydroxyquinoline) with 1 H^+ removed.
^{b}dmg = Dimethylglyoxime with 1 H^+ removed.

10.4 Organic Precipitates

All the precipitating agents we have talked about so far, except for oxine, cupferrate, and dimethylglyoxime (Table 10.1), have been inorganic in nature. There are also a large number of organic compounds that are very useful precipitating agents for metals. Some of these are very selective, and others are very broad in the number of elements they will precipitate.

Organic precipitating agents have the advantages of giving precipitates with very low solubility in water and a favorable gravimetric factor. Most of them are **chelating agents** that form slightly soluble, uncharged **chelates** with the metal ions. A chelating agent is a type of complexing agent that has two or more groups capable of complexing with a metal ion. The complex formed is called a chelate. See Chapter 9 for a more thorough discussion of chelates.

Chelates are described in Chapter 9.

Since chelating agents are weak acids, the number of elements precipitated, and thus the selectivity, can usually be regulated by adjustment of the pH. The reactions can be generalized as

$$M^{n+} + nHX \rightleftharpoons \underline{MX_n} + nH^+$$

There may be more than one ionizable proton on the organic reagent. The weaker the metal chelate, the higher the pH needed to achieve precipitation. Some of the commonly used organic precipitants are listed in Table 10.2. Some of these precipitates are not stoichiometric, and more accurate results are obtained by igniting them to form the metal oxides. Some, such as sodium diethyldithiocarbamate, can be used to perform group separations, as is done with hydrogen sulfide. You should consult treatises on analytical chemistry for applications of these and other organic

Metal chelate precipitates (which give selectivity) are sometimes ignited to metal oxides for improved stoichiometry.

Table 10.2
Some Organic Precipitating Agents

Reagent	Structure	Metals Precipitated
Dimethylglyoxime	$CH_3 — C = NOH$ $\mid$ $CH_3 — C = NOH$	Ni(II) in NH_3 or buffered HOAc; Pd(II) in HCl ($M^{2+} + 2HR \rightarrow \underline{MR_2} + 2H^+$)
α-Benzoinozime (cupron)	OH NOH CH—C (with two phenyl rings)	Cu(II) in NH_3 and tartrate; Mo(VI) and W(VI) in H^+ ($M^{2+} + H_2R \rightarrow \underline{MR} + 2H^+$; $M^{2+} = Cu^{2+}$, MoO_2^+, WO_2^{2+}) Metal oxide weighed
Ammonium nitrosophenylhydroxylamine (cupferron)	$N=O$ $N—O—NH_4$ (with phenyl ring)	Fe(III), V(V), Ti(IV), Zr(IV), Sn(IV), U(IV) ($M^{n+} + nNH_4R \rightarrow \underline{MR_n} + nNH_4^+$) Metal oxide weighed
8-Hydroxyquinoline (oxine)	OH (quinoline ring with N)	Many metals. Useful for Al(III) and Mg(II) ($M^{n+} + nHR \rightarrow \underline{MR_n} + nH^+$)
Sodium diethyldithiocarbamate	S $\parallel$ $(C_2H_5)_2N — C — S^-Na^+$	Many metals from acid solution ($M^{n+} + nNaR \rightarrow \underline{MRn} + nNa^+$)
Sodium tetraphenylboron	$NaB(C_6H_5)_4$	K^+, Rb^+, Cs^+, Tl^+, Ag^+, Hg(I), Cu(I), NH_4^+, RNH_3^+, $R_2NH_2^+$, R_3NH^+, R_4N^+. Acidic solution ($M^+ + NaR \rightarrow \underline{MR} + Na^+$)
Tetraphenylarsonium chloride	$(C_6H_5)_4AsCl$	$Cr_2O_7^{2-}$, MnO_4^-, ReO_4^-, MoO_4^{2-}, WO_4^{2-}, ClO_4^-, I_3^-. Acidic solution ($A^{n-} + nRCl \rightarrow R_nA + nCl^-$)

precipitating reagents. The multivolume treatise by Hollingshead on the uses of oxine and its derivatives is very helpful for applications of this versatile reagent. (See Ref. 4 at the end of the chapter.)

10.5 Precipitation Equilibria: The Solubility Product

When substances have limited solubility and their solubility is exceeded, the ions of the dissolved portion exist in equilibrium with the solid material. So-called insoluble compounds generally exhibit this property.

"Insoluble" substances still have slight solubility.

When a compound is referred to as insoluble, it is actually not completely insoluble but is **slightly soluble**. For example, if solid AgCl is added to water, a small portion of it will dissolve:

$$\underline{AgCl} \rightleftharpoons (AgCl)_{aq} \rightleftharpoons Ag^+ + Cl^- \tag{10.6}$$

The precipitate will have a definite solubility (i.e., a definite amount that will dissolve) in g/L, or mol/L, at a given temperature (a saturated solution). A small amount of undissociated compound usually exists in equilibrium in the aqueous phase (e.g., on the order of 0.1%), and its concentration is constant. It is difficult to measure the undissociated molecule, and we are interested in the ionized form as a measure of a compound's solubility and chemical availability. Hence, we can generally neglect the presence of any undissociated species.

We can write an overall equilibrium constant for the above stepwise equilibrium, called the **solubility product** K_{sp}. $(AgCl)_{aq}$ cancels when the two stepwise equilibrium constants are multiplied together.

$$K_{sp} = [Ag^+][Cl^-] \tag{10.7}$$

The solid does not appear in K_{sp}.

The "concentration" of any solid such as AgCl is constant and is combined in the equilibrium constant to give K_{sp}. The above relationship holds regardless of the presence of any undissociated intermediate; that is, the concentrations of free ions are rigorously defined by Equation 10.7, and we will take these as a measure of a compound's solubility. From a knowledge of the value of the solubility product at a specified temperature, we can calculate the equilibrium solubility of the compounds. (The solubility product is determined in the reverse order, by measuring the solubility.)

The amount of a slightly soluble salt that dissolves does *not* depend on the amount of the solid in equilibrium with the solution, so long as there is enough to saturate the solution. Instead, the amount that dissolves depends on the *volume* of the solvent. A nonsymmetric salt such as Ag_2CrO_4 would have a K_{sp} as follows:

The concentration of solute in a saturated solution is the same whether the solution fills a beaker or a swimming pool, so long as there is solid in equilibrium with it. But much more solid will dissolve in the swimming pool!

$$Ag_2CrO_4 \rightleftharpoons 2\,Ag^+ + CrO_4^{2-} \tag{10.8}$$
$$K_{sp} = [Ag^+]^2[CrO_4^{2-}] \tag{10.9}$$

Such electrolytes do not dissolve or dissociate in steps because they are really strong electrolytes. That portion that dissolves ionizes completely. *Therefore, we do not have stepwise K_{sp} values.* As with any equilibrium constant, the K_{sp} product holds under all equilibrium conditions at the specified temperature. Since we are dealing with heterogeneous equilibria, the equilibrium state will be achieved more slowly than with homogeneous solution equilibria.

THE SATURATED SOLUTION

 **Example 10.6**

The K_{sp} of AgCl at 25°C is 1.0×10^{-10}. Calculate the concentrations of Ag^+ and Cl^- in a saturated solution of AgCl, and the molar solubility of AgCl.

Solution

When AgCl ionizes, equal amounts of Ag^+ and Cl^- are formed; $AgCl \rightleftharpoons Ag^+ + Cl^-$ and $K_{sp} = [Ag^+][Cl^-]$. Let s represent the molar solubility of AgCl. Since each mole of AgCl that dissolves gives one mole of either Ag^+ or Cl^-, then

$$[Ag^+] = [Cl^-] = s$$
$$s^2 = 1.0 \times 10^{-10}$$
$$s = 1.0 \times 10^{-5}\ M$$

The solubility of AgCl is $1.0 \times 10^{-5}\ M$.

DECREASING THE SOLUBILITY—THE COMMON ION EFFECT

If there is an excess of one ion over the other, the concentration of the other is suppressed (**common ion effect**), and the solubility of the precipitate is decreased. We can still calculate the concentration from the solubility product.

 Example 10.7

Adding a common ion decreases the solubility.

Ten milliliters of 0.20 M AgNO$_3$ is added to 10 mL of 0.10 M NaCl. Calculate the concentration of Cl$^-$ remaining in solution at equilibrium, and the solubility of the AgCl.

Solution

The final volume is 20 mL. The millimoles Ag$^+$ added equals $0.20 \times 10 = 2.0$ mmol. The millimoles Cl$^-$ taken equals $0.10 \times 10 = 1.0$ mmol. Therefore, the millimoles excess Ag$^+$ equals $(2.0 - 1.0) = 1.0$ mmol. From Example 10.6, we see that the Ag$^+$ concentration contributed from the precipitate is small, that is, on the order of 10^{-5} mmol/mL in the absence of a common ion. This will be even smaller in the presence of excess Ag$^+$ since the solubility is suppressed. Therefore, we can neglect the amount of Ag$^+$ contributed from the precipitate compared to the excess Ag$^+$. Hence, the final concentration of Ag$^+$ is 1.0 mmol/20 mL = 0.050 M, and

$$(0.050)[\text{Cl}^-] = 1.0 \times 10^{-10}$$
$$[\text{Cl}^-] = 2.0 \times 10^{-9} \, M$$

The Cl$^-$ concentration again equals the solubility of the AgCl, and so the solubility is 2.0×10^{-9} M.

The solubility product must be exceeded for precipitation to occur.

Because the K_{sp} product always holds, *precipitation will not take place unless the product of [Ag$^+$] and [Cl$^-$] exceeds the* K$_{sp}$. If the product is just equal to K_{sp}, all the Ag$^+$ and Cl$^-$ remains in solution.

SOLUBILITY DEPENDS ON THE STOICHIOMETRY

Table 10.3 lists some solubility products along with the corresponding calculated molar solubilities for some slightly soluble salts. The molar solubility is not necessary directly proportional to the K_{sp} value since it depends on the stoichiometry

Table 10.3
Solubility Product Constants of Selected Slightly Soluble Salts

Salt	K_{sp}	Solubility, s (mol/L)
PbSO$_4$	1.6×10^{-8}	1.3×10^{-4}
AgCl	1.0×10^{-6}	1.0×10^{-5}
AgBr	4×10^{-13}	6×10^{-7}
AgI	1×10^{-16}	1×10^{-8}
Al(OH)$_3$	2×10^{-32}	5×10^{-9}
Fe(OH)$_3$	4×10^{-38}	2×10^{-10}
Ag$_2$S	2×10^{-49}	4×10^{-17}
HgS	4×10^{-53}	6×10^{-27}

of the salt. The K_{sp} of AgI is 5×10^{15} larger than that of $Al(OH)_3$, but its molar solubility is only twice that of $Al(OH)_3$. That is, a $1:1$ salt has a lower solubility than a nonsymmetric salt for a given K_{sp}. Note that HgS has a solubility product of only 4×10^{-53}, with a molar solubility of 6×10^{-27} M! This corresponds to less than one ion each of Hg^{2+} and S^{2-} in a liter in equilibrium with the precipitate, and it would take some 280 L for two ions to exist together (can you calculate this using Avogadro's number?). So it is like two ions finding each other in a good size bathtub! (Actually, they find the precipitate). A more complete list of solubility products appears in Appendix C.

Example 10.8

What must be the concentration of added Ag^+ to just start precipitation of AgCl in a 1.0×10^{-3} M solution of NaCl?

Solution

$$[Ag^+](1.0 \times 10^{-3}) = 1.0 \times 10^{-10}$$
$$[Ag^+] = 1.0 \times 10^{-7} \ M$$

The concentration of Ag^+ must, therefore, just exceed 10^{-7} M to begin precipitation.

Example 10.9

What is the solubility of PbI_2, in g/L, if the solubility product is 7.1×10^{-9}?

Solution

The equilibrium is $PbI_2 \rightleftharpoons Pb^{2+} + 2I^-$, and $K_{sp} = [Pb^{2+}][I^-]^2 = 7.1 \times 10^{-9}$. Let s represent the molar solubility of PbI_2. Then

$$[Pb^{2+}] = s \quad and \quad [I^-] = 2s$$
$$(s)(2s)^2 = 7.1 \times 10^{-9}$$
$$s = \sqrt[3]{\frac{7.1 \times 10^{-9}}{4}} = 1.2 \times 10^{-3} \ M$$

Therefore, the solubility, in g/L, is

$$1.2 \times 10^{-3} \ mol/L \times 461.0 \ g/mol = 0.55 \ g/L$$

Note that the concentration of I^- was *not* doubled before squaring; $2s$ represented its actual equilibrium concentration, not twice its concentration. We could have let s represent the concentration of I^-, instead of the molar solubility of PbI_2, in which case $[Pb^{2+}]$ and the solubility of PbI_2 would have been $\frac{1}{2}s$. The calculated s would have been twice as great, but the concentrations of each species would have been the same. You try this calculation!

Example 10.10

A smaller K_{sp} with a nonsymmetrical precipitate does not necessarily mean a smaller solubility compared to a symmetrical one.

Calculate the molar solubility of $PbSO_4$ and compare it with that of PbI_2.

Solution

$$PbSO_4 \rightleftharpoons Pb^{2+} + SO_4^{2-}$$
$$[Pb^{2+}][SO_4^{2-}] = 1.6 \times 10^{-8}$$
$$(s)(s) = 1.6 \times 10^{-8}$$
$$s = 1.3 \times 10^{-4} \ M$$

Although the K_{sp} of PbI_2 is smaller than that of $PbSO_4$, the solubility of PbI_2 is greater (see Example 10.9), due to the nonsymmetrical nature of the precipitate.

For electrolytes of the same valence type, the order of solubility will be the same as the order of the corresponding solubility products. But when we compare salts of difference valence type, the order may be different. Compound AB will have a smaller molar solubility than compound AC_2 when both have identical K_{sp} values.

We take advantage of the common ion effect to decrease the solubility of a precipitate in gravimetric analysis. For example, sulfate ion is determined by precipitating $BaSO_4$ with added barium chloride solution. Figure 10.3 illustrates the effect of excess barium ion on the solubility of $BaSO_4$.

Example 10.11

Fe(OH)$_3$ actually precipitates in acid solution due to the small K_{sp}!

What pH is required to just precipitate iron(III) hydroxide form from a 0.10 M FeCl$_3$ solution?

Fig. 10.3. Predicted effect of excess barium ion on solubility of BaSO$_4$. Sulfate concentration is amount in equilibrium and is equal to BaSO$_4$ solubility. In absence of excess barium ion, solubility is 10^{-5} M.

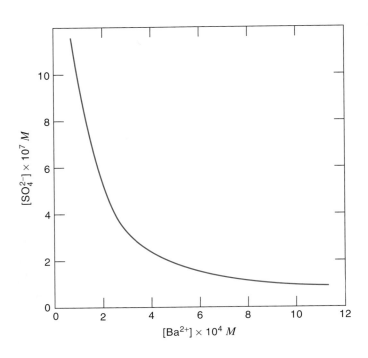

Solution

$$Fe(OH)_3 \rightleftharpoons Fe^{3+} + 3OH^-$$
$$[Fe^{3+}][OH^-]^3 = 4 \times 10^{-38}$$
$$(0.1)[OH^-]^3 = 4 \times 10^{-38}$$
$$[OH^-] = \sqrt[3]{\frac{4 \times 10^{-38}}{0.1}} = 7 \times 10^{-13} \text{ M}$$
$$pOH = -\log 7 \times 10^{-13} = 12.2$$
$$pH = 14.0 - 12.2 = 1.8$$

Hence, we see that iron hydroxide precipitate in acid solution, when the pH just exceeds 1.8! When you prepare a solution of $FeCl_3$ in water, it will slowly hydrolyze to form ion hydroxide (hydrous ferric oxide), a rust-colored gelatinous precipitate. To stabilize the iron(III) solution, you must acidify the solution with, for example, hydrochloric acid.

•••

Example 10.12

Twenty-five milliliters of 0.100 M $AgNO_3$ is mixed with 35.0 mL of 0.0500 M K_2CrO_4 solution. (a) Calculate the concentrations of each ionic species at equilibrium. (b) Is the precipitation of silver quantitative (>99.9%)?

Solution

(a) The reaction is

$$2Ag^+ + CrO_4^{2-} \rightleftharpoons \underline{Ag_2CrO_4}$$

We mix

$$25.0 \text{ mL} = 0.100 \text{ mmol/mL} = 2.50 \text{ mmol } AgNO_3$$

and

$$35.0 \text{ mL} \times 0.0500 \text{ mmol/mL} = 1.75 \text{ mmol } K_2CrO_4$$

Hence, 1.25 mmol of CrO_4^{2-} will react with 2.50 mmol Ag^+, leaving an excess of 0.50 mmol CrO_4^{2-}. The final volume is 60.0 mL. If we let s be the molar solubility of Ag_2CrO_4, then at equilibrium:

$$[CrO_4^{2-}] = 0.50/60.0 + s = 0.0083 + s \approx 0.0083 \text{ } M$$

s will be very small due to the excess CrO_4^{2-} and may be neglected compared to 0.0083.

$$[Ag^+] = 2s$$
$$[K^+] = 3.50/60.0 = 0.0583 \text{ } M$$
$$[NO_3^-] = 2.50/60.0 = 0.0417 \text{ } M$$
$$[Ag^+]^2 [CrO_4^{2-}] = 1.1 \times 10^{-12}$$
$$(2s)^2 (8.3 \times 10^{-3}) = 1.1 \times 10^{-12}$$

$$s = \sqrt{\frac{1.1 \times 10^{-12}}{4 \times 8.3 \times 10^{-3}}} = 5.8 \times 10^{-6} \, M$$

$$[\text{Ag}^+] = 2(5.8 \times 10^{-6}) = 1.1_6 \times 10^{-5} \, M$$

(b) The percentage of silver precipitated is

$$\frac{2.50 \, \text{mmol} - 60.0 \, \text{mL} \times 1.1_6 \times 10^{-5} \, \text{mmol/mL}}{2.50 \, \text{mmol}} \times 100\% = 99.97\%$$

Or the percent remaining in solution is

$$\frac{60.0 \, \text{mL} \times 1.1_6 \times 10^{-5} \, \text{mmol/L}}{2.50 \, \text{mmol}} \times 100\% = 0.028\%$$

Hence, the precipitation is quantitative.

• •

10.6   Diverse Ion Effect on Solubility: K_{sp}° and Activity Coefficients

In Chapter 6 we defined the thermodynamic equilibrium constant written in terms of activities to account for the effects of inert electrolytes on equilibria. The presence of diverse salts will generally increase the solubility of precipitates due to the shielding of the dissociated ion species. (Their activity is decreased.) Consider the solubility of AgCl. The thermodynamic solubility product K_{sp}° is

$$K_{sp}^{\circ} = a_{\text{Ag}^+} \cdot a_{\text{Cl}^-} = [\text{Ag}^+] f_{\text{Ag}^+} [\text{Cl}^-] f_{\text{Cl}^-} \qquad (10.10)$$

K_{sp}° holds at all ionic strengths. K_{sp} must be corrected for ionic strength.

Since the *concentration* solubility product K_{sp} is $[\text{Ag}^+][\text{Cl}^-]$, then

$$K_{sp}^{\circ} = K_{sp} f_{\text{Ag}^+} f_{\text{Cl}^-} \qquad (10.11)$$

or

$$K_{sp} = \frac{K_{sp}^{\circ}}{f_{\text{Ag}^+} f_{\text{Cl}^-}} \qquad (10.12)$$

The numerical value of K_{sp}° holds at all activities. K_{sp} equals K_{sp}° at zero ionic strength, but at appreciable ionic strengths, a value must be calculated for each ionic strength using Equation 10.12. Note that this equation predicts, as we predicted qualitatively, that decreased activity of the ions will result in an increased K_{sp} and, therefore, increased molar solubility.

• •

Example 10.13

Calculate the solubility of silver chloride in 0.10 M NaNO$_3$.

Solution

The equilibrium constants listed in the Appendix C are for zero ionic strength; that is, they are really thermodynamic equilibrium constants.[1] Therefore, from Table C.3, $K_{sp}^{\circ} = 1.0 \times 10^{-10}$.

We need the activity coefficients of Ag^+ and Cl^-. The ionic strength is 0.10. From Ref. 9 in Chapter 6, we find that $f_{Ag^+} = 0.75$ and $f_{Cl^-} = 0.76$. (You could also have used the values of α_{Ag^+} and α_{Cl^-} in the reference to calculate the activity coefficients using Equation 6.19.) From Equation 10.12

$$K_{sp} = \frac{1.0 \times 10^{-10}}{(0.75)(0.76)} = 1.8 \times 10^{-10} = [Ag^+][Cl^-] = s^2$$

$$s = \sqrt{1.8 \times 10^{-10}} = 1.3 \times 10^{-5} \, M$$

This is 30% greater than at zero ionic strength ($s = 1.0 \times 10^{-5} \, M$).

• •

Diverse salts increase the solubility of precipitates and have more effect on precipitates with multiply charged ions.

Figure 10.4 illustrates the increase in solubility of $BaSO_4$ in the presence of $NaNO_3$ due to the diverse ion effect.

The increase in solubility is greater with precipitates containing multiply charged ions. At very high ionic strengths, where activity coefficients may become greater than unity, the solubility is decreased. In gravimetric analysis, a sufficiently large excess of precipitating agent is added so that the solubility is reduced to such a small value that we do not need to worry about this effect.

Acids frequently affect the solubility of a precipitate. As the H^+ concentration increases, it competes more effectively with the metal ion of interest for the

---

[1]Experimental K_{sp} values are available at different ionic strengths and can be used to calculate molar solubilities at the listed ionic strengths without having to calculate activity coefficients.

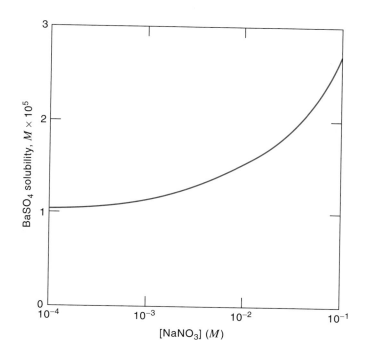

Fig. 10.4. Predicted effect of increased ionic strength on solubility of $BaSO_4$. Solubility at zero ionic strength is $1.0 \times 10^{-5} \, M$.

precipitating agent (which may be the anion of a weak acid). With less free reagent available, and a constant K_{sp}, the solubility of the salt must increase:

$$M^{n+} + nR^- \rightleftharpoons MR_n \qquad \text{(desired reaction)}$$

$$R^- + H^+ \rightleftharpoons HR \qquad \text{(competing reaction)}$$

$$MR_n + nH^+ \rightleftharpoons M^{n+} + nHR \qquad \text{(overall reaction)}$$

Similarly, a complexing agent that reacts with the metal ion of the precipitate will increase the solubility, for example, when ammonia reacts with silver chloride:

$$AgCl + 2NH_3 \rightleftharpoons Ag(NH_3)_2^+ + Cl^-$$

The quantitative treatment of these effects in solubility calculations will be treated in Chapter 11.

SPREADSHEET EXAMPLES

The iron content in a 2.287-g sample of iron ore is determined by precipitating as $Fe(OH)_3$, igniting to Fe_2O_3, and weighing. The result is a net weight of 0.8792 g. Prepare a spreadsheet to calculate the % Fe in the ore.

	A	B	C	D	E	F	G	H
1	Calculation of % Fe.							
2	g. sample:	2.287	g. Fe$_2$O$_3$:	0.8792				
3	% Fe:	**26.88797**						
4								
5	%Fe = {[g Fe$_2$O$_3$ x 2Fe/Fe$_2$O$_3$ (g Fe/g Fe$_2$O$_3$)]/g sample] x 100%							
6	=			{[0.8792 g Fe$_2$O$_3$ x 2(55.845/159.69)g Fe/g Fe$_2$O$_3$]/2.287 g sample} x 100%				
7	**B3 =**		(D2*2*(55.845/159.69))/B2)*100					
8								
9	The answer is 26.89% Fe.							

The solubility product for AgCl is 1.0×10^{-10}. Use Excel Solver to calculate the solubility.

	A	B	C	D	E	F
1	Calculation of the Solubility of AgCl Using Solver.					
2	K$_{sp}$:	1.00E-10		formula = s^2/K$_{sp}$ = 1		
3						
4		s =		1E-05	formula:	1.000001
5						
6				formula Cell E4: = C4^2/B2		
7				**Solver:**		
8				**E4 = Target Cell**		
9				**Set value to 1**		
10				**Changing Cell = C4**		
11	The solubility is 1.0×10^{-5} M.					

_____ Learning Objectives

WHAT ARE SOME OF THE KEY THINGS WE LEARNED FROM THIS CHAPTER?

- Steps of a gravimetric analysis: precipitation, digestion, filtration, washing, drying, weighing, calculation, p. 313
- Gravimetric calculations (key equations: 10.1, 10.3, 10.5), p. 320
- The solubility product, the common ion effect, p. 326
- The diverse ion effect (key equation: 10.10), p. 332

_____ Questions

1. Describe the unit operations commonly employed in gravimetric analysis, and briefly indicate the purpose of each.

2. What is the von Weimarn ratio? Define the terms in it.

3. What information concerning optimum conditions of precipitation does the von Weimarn ratio give us?

4. What is digestion of a precipitate, and why is it necessary?

5. Outline the optimum conditions for precipitation that will obtain a pure and filterable precipitate.

6. What is coprecipitation? List the different types of coprecipitation, and indicate how they may be minimized or treated.

7. Why must a filtered precipitate be washed?

8. Why must a wash liquid generally contain an electrolyte? What are the requirements for this electrolyte?

9. What advantages do organic precipitating agents have?

_____ Problems

GRAVIMETRIC FACTOR

10. Calculate the weight of sodium present in 50.0 g Na_2SO_4.

11. If the salt in Problem 10 is analyzed by precipitating and weighing $BaSO_4$, what weight of precipitate would be obtained?

12. Calculate the gravimetric factors for:

Substance Sought	Substance Weighed
As_2O_3	Ag_3AsO_4
$FeSO_4$	Fe_2O_3
K_2O	$KB(C_6H_5)_4$
SiO_2	$KAlSi_3O_8$

13. How many grams CuO would 1.00 g Paris green, $Cu_3(AsO_3)_2 \cdot 2As_2O_3 \cdot Cu(C_2H_3O_2)_2$, give? Of As_2O_3?

QUANTITATIVE CALCULATIONS

14. A 523.1-mg sample of impure KBr is treated with excess $AgNO_3$ and 814.5 mg AgBr is obtained. What is the purity of the KBr?

15. What weight of Fe_2O_3 precipitate would be obtained from a 0.4823-g sample of iron wire that is 99.89% pure?

16. The aluminum content of an alloy is determined gravimetrically by precipitating it with 8-hydroxyquinoline (oxine) to give $Al(C_9H_6ON)_3$. If a 1.021-g sample yielded 0.1862 g of precipitate. what is the percent aluminum in the alloy?

17. Iron in an ore is to be analyzed gravimetrically by weighing as Fe_2O_3. It is desired that the results be obtained to four significant figures. If the iron content ranges between 11 and 15%, what is the minimum size sample that must be taken to obtain 100.0 mg of precipitate?

18. The chloride in a 0.12-g sample of 95% pure $MgCl_2$ is to be precipitated as AgCl. Calculate the volume of 0.100 M $AgNO_3$ solution required to precipitate the chloride and give a 10% excess.

19. Ammonium ions can be analyzed by precipitating with H_2PtCl_6 as $(NH_4)_2PtCl_6$ and then igniting the precipitate to platinum metal, which is weighed $[(NH_4)_2PtCl_6 \xrightarrow{\text{heat}} Pt + 2NH_4Cl \uparrow + 2Cl_2 \uparrow]$. Calculate the percent ammonia in a 1.00-g sample that yields 0.100 g Pt by this method.

20. A sample is to be analyzed for its chloride content by precipitating and weighing silver chloride. What weight of sample would have to be taken so that the weight of precipitate is equal to the percent chloride in the sample?

21. Pyrite ore (impure FeS_2) is analyzed by converting the sulfur to sulfate and precipitating $BaSO_4$. What weight of ore should be taken for analysis so that the grams of precipitate will be equal to 0.1000 times the percent of FeS_2?

22. A mixture containing only BaO and CaO weighs 2.00 g. The oxides are converted to the corresponding mixed sulfates, which weigh 4.00 g. Calculate the percent Ba and Ca in the original mixture.

23. A mixture containing only $BaSO_4$ and $CaSO_4$ contains one-half as much Ba^{2+} as Ca^{2+} by weight. What is the percentage of $CaSO_4$ in the mixture?

24. A mixture containing only AgCl and AgBr weighs 2.000 g. It is quantitatively reduced to silver metal, which weighs 1.300 g. Calculate the weight of AgCl and AgBr in the original mixture.

SOLUBILITY PRODUCT CALCULATIONS

25. Write solubility product expressions for the following: (a) AgSCN, (b) $La(IO_3)_3$, (c) Hg_2Br_2, (d) $Ag[Ag(CN)_2]$; (e) $Zn_2Fe(CN)_6$, (f) Bi_2S_3.

26. Bismuth iodide, BiI_3, has a solubility of 7.76 mg/L. What is its K_{sp}?

27. What is the concentration of Ag^+ and CrO_4^{2-} in a saturated solution of Ag_2CrO_4?

28. Calculate the concentration of barium in the solution at equilibrium when 15.0 mL of 0.200 M K_2CrO_4 is added to 25.0 mL of 0.100 M $BaCl_2$.

29. What must be the concentration of PO_4^{3-} to just start precipitation of Ag_3PO_4 in a 0.10 M $AgNO_3$ solution?

30. What must be the concentration of Ag^+ to just start precipitating 0.10 M PO_4^{3-}? 0.10 M Cl^-?

31. At what pH will $Al(OH)_3$ begin to precipitate from 0.10 M $AlCl_3$?

32. What weight of Ag_3AsO_4 will dissolve in 250 mL water?

33. What is the solubility of Ag_2CrO_4 in 0.10 M K_2CrO_4?

34. Compounds AB and AC_2 each have solubility products equal to 4×10^{-18}. Which is more soluble, as expressed in moles per liter?

35. The solubility product of Bi_2S_3 is 1×10^{-97} and that of HgS is 4×10^{-53}. Which is the least soluble?

36. A student proposes to analyze barium gravimetrically by precipitating BaF_2 with NaF. Assuming a 200-mg sample of Ba^{2+} in 100 mL is to be precipitated and that the precipitation must be 99.9% complete for quantitative results, comment on the feasibility of the analysis.

DIVERSE ION EFFECT ON SOLUBILITY

37. Write the thermodynamic solubility product expressions for the following:

    (a) $\underline{BaSO_4} \rightleftharpoons Ba^{2+} + SO_4^{2-}$

    (b) $\underline{Ag_2CrO_4} \rightleftharpoons 2Ag^+ + CrO_4^{2-}$

38. Calculate the solubility of $BaSO_4$ in 0.0125 M $BaCl_2$. Take into account the diverse ion effect.

39. You are to determine fluoride ion gravimetrically by precipitating CaF_2. $Ca(NO_3)_2$ is added to give an excess of 0.015 M calcium ion after precipitation. The solution also contains 0.25 M $NaNO_3$. How many grams fluoride will be in solution at equilibrium if the volume is 250 mL?

EXCEL EXERCISES

40. Prepare a spreadsheet to calculate the percent P_2O_5 in Example 10.2. Use it to calculate the % P_2O_5 for the sample given in the example. Do a second calculation for a 0.5267 g sample that gives a precipitate of 2.0267 g.

41. Prepare a spreadsheet to calculate the solubility of $BaSO_4$ as a function of concentration of excess Ba^{2+} concentration, as in Figure 10.3. Prepare a graph of solubility versus Ba^{2+} concentration, using the Chart function of Excel, and compare it with Figure 10.3.

42. Prepare a spreadsheet to calculate the solubility of $BaSO_4$ as a function of ionic strength, as in Figure 10.4. Prepare a graph of solubility versus ionic strength using the Chart function of Excel, and compare it with Figure 10.4.

43. Solve Example 10.9, using Solver to calculate the solubility, s, of PbI_2.

Recommended References

GENERAL AND INORGANIC

1. F. E. Beamish and W. A. E. McBryde, "Inorganic Gravimetric Analysis," in C. L. Wilson and D. W. Wilson, eds., *Comprehensive Analytical Chemistry*, Vol. 1A. New York: Elsevier, 1959, Chapter VI.

2. C. L. Wilson and D. W. Wilson, eds., *Comprehensive Analytical Chemistry*, Vol. 1C, *Classical Analysis: Gravimetric and Titrimetric Determination of the Elements*, New York: Elsevier, 1962.

ORGANIC REAGENTS

3. K. L. Cheng, K. Ueno, and T. Imamura, eds. *Handbook of Organic Analytical Reagents*. Boca Raton, FL: CRC Press, 1982.

4. R. G. W. Hollingshead, *Oxine and Its Derivatives.* London: Butterworth Scientific, 1954–56.

5. F. Holmes, "Organic Reagents in Inorganic Analysis," in C. L. Wilson and D. W. Wilson, eds., *Comprehensive Analytical Chemistry*, Vol. 1A. New York: Elsevier, 1959, Chapter II.8.

PRECIPITATION FROM HOMOGENEOUS SOLUTION

6. L. Gordon, M. L. Salulsky, and H. H. Willard, *Precipitation from Homogeneous Solution.* New York: Wiley, 1959.

Chapter Eleven

PRECIPITATION REACTIONS AND TITRATIONS

"If you're not part of the solution, then you're part of the precipitate."
—Anonymous

A number of anions form slightly soluble precipitates with certain metal ions and can be titrated with the metal solutions; for example, chloride can be titrated with silver ion and sulfate with barium ion. The precipitation equilibrium may be affected by pH or by the presence of complexing agents. The anion of the precipitate may be derived from a weak acid and therefore combine with protons in acid solution to cause the precipitate to dissolve. On the other hand, the metal ion may complex with a ligand (the complexing agent) to shift the equilibrium toward dissolution. Silver ion will complex with ammonia and cause silver chloride to dissolve.

In this chapter, we describe the quantitative effects of acidity and complexation in precipitation equilibria and discuss precipitation titrations using silver nitrate and barium nitrate titrants with different kinds of indicators and their theory. You should review fundamental precipitation equilibria described in Chapter 10. Most ionic analytes, especially inorganic anions, are conveniently determined using ion chromatography (Chapter 21), but for high concentrations more precise determinations can be made by precipitation titration when applicable.

11.1 Effect of Acidity on Solubility of Precipitates: Conditional Solubility Product

Before describing precipitation titrations, we shall consider the effects of competing equilibria on the solubility of precipitates. Before you read any further, you might want to review the discussion of polyprotic acid equilibria and the calculation of α's, the fractions of each acid species in equilibrium at a given pH, in Chapter 7.

The solubility of a precipitate whose anion is derived from a weak acid will increase in the presence of added acid because the acid will tend to combine with the anion and thus remove the anion from solution. For example, the precipitate MA that partially dissolves to give M^+ and A^- ions will exhibit the following equilibria:

$$\left. \begin{array}{c} MA \rightleftharpoons M^+ + A^- \\ + \\ H^+ \\ \uparrow\downarrow \\ HA \end{array} \right\} C_{HA}$$

The anion A^- can combine with protons to increase the solubility of the precipitates. The combined equilibrium concentrations of A^- and HA make up the total analytical concentration, C_{HA}, which will be equal to $[M^+]$ from the dissolved precipitate (if neither M^+ or A^- is in excess). By applying the equilibrium constants for the equilibria involved, we can calculate the solubility of the precipitate at a given acidity.

Consider, for example, the solubility of CaC_2O_4 in the presence of a strong acid. The equilibria are

Protons compete with calcium ion for the oxalate ion.

$$CaC_2O_4 \rightleftharpoons Ca^{2+} + C_2O_4^{2-} \quad K_{sp} = [Ca^{2+}][C_2O_4^{2-}] = 2.6 \times 10^{-9} \tag{11.1}$$

$$C_2O_4^{2-} + H^+ \rightleftharpoons HC_2O_4^- \quad K_{a2} = \frac{[H^+][C_2O_4^{2-}]}{[HC_2O_4^-]} = 6.1 \times 10^{-5} \tag{11.2}$$

$$HC_2O_4^- + H^+ \rightleftharpoons H_2C_2O_4 \quad K_{a1} = \frac{[H^+][HC_2O_4^-]}{[H_2C_2O_4]} = 6.5 \times 10^{-2} \tag{11.3}$$

The solubility s of CaC_2O_4 is equal to $[Ca^{2+}] = C_{H_2C_2O_4}$, where $C_{H_2C_2O_4}$ represents the concentrations of all the oxalate species in equilibrium ($= [H_2C_2O_4] + [HC_2O_4^-] + [C_2O_4^{2-}]$). We can substitute $C_{H_2C_2O_4}\alpha_2$ for $[C_2O_4^{2-}]$ in the K_{sp} expression:

$$\boxed{K_{sp} = [Ca^{2+}]C_{H_2C_2O_4}\alpha_2} \tag{11.4}$$

where α_2 is the fraction of the oxalate species present as $C_2O_4^{2-}$ ($\alpha_2 = [C_2O_4^{2-}]/C_{H_2C_2O_4}$). Using the approach described in Chapter 7 for H_3PO_4 to calculate α's, we find that

$$\alpha_2 = \frac{K_{a1}K_{a2}}{[H^+]^2 + K_{a1}[H^+] + K_{a1}K_{a2}} \tag{11.5}$$

We can write, then, that

$$\boxed{\frac{K_{sp}}{\alpha_2} = K'_{sp} = [Ca^{2+}]\, C_{H_2C_2O_4} = s^2} \tag{11.6}$$

The conditional solubility product value holds for only a specified pH.

where K'_{sp} is the **conditional solubility product**, similar to the conditional formation constant described in Chapter 9.

Example 11.1

Calculate the solubility of CaC_2O_4 in a solution containing 0.0010 M hydrochloric acid.

Solution

$$\alpha_2 = \frac{(6.5 \times 10^{-2})(6.1 \times 10^{-5})}{(1.0 \times 10^{-3})^2 + (6.5 \times 10^{-2})(1.0 \times 10^{-3}) + (6.5 \times 10^{-2})(6.1 \times 10^{-5})}$$

$$= 5.7 \times 10^{-2}$$

$$s = \sqrt{K_{sp}/\alpha_2} = \sqrt{2.6 \times 10^{-9}/5.7 \times 10^{-2}} = 2.1 \times 10^{-4} \ M$$

This compares with a calculated solubility in water using Equation 11.1 of 5.1×10^{-5} M (a 400% increase in solubility). Note that both $[Ca^{2+}]$ and $C_{H_2C_2O_4} = 2.1 \times 10^{-4}$ M. We can obtain the concentrations of the other oxalate species in equilibrium by multiplying this number by α_0, α_1, and α_2 for oxalic acid at 0.0010 M H^+ to obtain $[H_2C_2O_4]$, $[HC_2O_4^-]$, and $[C_2O_4^{2-}]$, respectively. We will not derive α_0 and α_1 here, but the results would be $[C_2O_4^{2-}] = 1.2 \times 10^{-5}$ M, $[HC_2O_4^-] = 2.0 \times 10^{-4}$ M, and $[H_2C_2O_4] = 3.1 \times 10^{-6}$ M. (You can try the calculations.)

Since the solution is unbuffered, perform an iterative calculation to correct for the protons consumed.

In the above calculations we neglected the fact that some of the acid was consumed by reaction with oxalate. We see that one-fifth of it reacted to form $HC_2O_4^{2-}$. The amount reacted to form $H_2C_2O_4$ is negligible. If we desire a more exact solution, then we can subtract the amount of acid reacted, as calculated above, from the initial acid concentration and then repeat the calculation using the new acid concentration. We then repeat this process until the change in the final answer is within the desired accuracy, an iterative procedure. Recalculation using 0.8×10^{-3} M acid gives a calcium concentration of 1.9×10^{-4} M, 10% less.

We should emphasize that, when dealing with multiple equilibria, the validity of a given equilibrium expression is in no way compromised by the existence of additional competing equilibria. Thus, in the above example, the solubility product expression for CaC_2O_4 describes the relationship between Ca^{2+} and $C_2O_4^{2-}$ ions, whether or not acid is added. In other words, the product $[Ca^{2+}][C_2O_4^{2-}]$ is always a constant as long as there is solid CaC_2O_4 present. The quantity of CaC_2O_4 that dissolves is increased, however, because part of the $C_2O_4^{2-}$ in solution is converted to $HC_2O_4^-$ and $H_2C_2O_4$.

11.2 Mass Balance Approach for Multiple Equilibria

We may solve the multiple equilibrium problem as well by using the systematic approaches described in Chapter 6, using the equilibrium constant expressions, the mass balance expressions, and the charge balance expression.

The systematic approach is well suited for competing equilibria calculations.

Example 11.2

How many moles of MA will dissolve in 1 L of 0.10 M HCl if K_{sp} for MA is 1.0×10^{-8} and K_a for HA is 1.0×10^{-6}?

Solution

The equilibria are

$$MA \rightleftharpoons M^+ + A^-$$
$$A^- + H^+ \rightleftharpoons HA$$
$$H_2O \rightleftharpoons H^+ + OH^-$$
$$HCl \rightarrow H^+ + Cl^-$$

The equilibrium expressions are

$$K_{sp} = [M^+][A^-] = 1.0 \times 10^{-8} \tag{1}$$

$$K_a = \frac{[H^+][A^-]}{[HA]} = 1.0 \times 10^{-6} \tag{2}$$

$$K_w = [H^+][OH^-] = 1.0 \times 10^{-14} \tag{3}$$

The mass balance expressions are

$$[M^+] = [A^-] + [HA] - C_{HA} \tag{4}$$
$$[H^+] = [Cl^-] + [OH^-] = [HA] \tag{5}$$
$$[Cl^-] = 0.10 \ M \tag{6}$$

The charge balance expression is

$$[H^+] + [M^+] = [A^-] + [Cl^-] + [OH^-] \tag{7}$$

Number of expressions versus number of unknowns:

There are six unknowns ($[H^+]$, $[OH^-]$, $[Cl^-]$, $[HA]$, $[M^+]$, and $[A^-]$) and six independent equations (the charge balance equation can be generated as a linear combination of the others, and is not used here).

Simplifying assumptions:

(1) In an acid solution, dissociation of HA is suppressed, making $[A^-] \ll [HA]$, so from (4):

$$[M^+] = [A^-] + [HA] \approx [HA]$$

(2) In an acid solution $[OH^-]$ is very small, so from (5) and (6):

$$[H^+] = 0.10 + [OH^-] - [HA] \approx 0.10 - [HA]$$

Calculate:

We need to calculate $[M^+]$ in order to obtain the moles of MA dissolved in a liter.

The number of equations must equal the number of unknowns. Make assumptions to simplify calculations.

From (1)

$$[M^+] = \frac{K_{sp}}{[A^-]} \qquad (8)$$

From (2)

$$[A^-] = \frac{K_a[HA]}{[H^+]} \qquad (9)$$

So, dividing (8) by (9):

$$[M^+] = \frac{K_{sp}[H^+]}{K_a[HA]} = 1.0 \times 10^{-2} \frac{[H^+]}{[HA]} \qquad (10)$$

From assumption (1),

$$[M^+] \approx [HA]$$

From assumption (2),

$$[H^+] \approx 0.10 - [HA] \approx 0.10 - [M^+]$$

$$[M^+] = \frac{(1.0 \times 10^{-2})(0.10 - [M^+])}{[M^+]}$$

$$\frac{[M^+]^2}{0.10 - [M^+]} = 1.0 \times 10^{-2}$$

Use of the quadratic equation gives $[M] = 0.027\ M$.

So, in 1 L, 0.027 mol of MA will dissolve. This compares with 0.00010 mol in water. Check

The validity of the assumptions can be checked.

(1)
$$[HA] \approx [M^+] = 0.027\ M$$

$$[A^-] = \frac{K_{sp}}{[M^+]} = \frac{1.0 \times 10^{-8}}{0.027} = 3.7 \times 10^{-7}\ M$$

Assumption (1) is acceptable because $[A^-] \ll [HA]$.

(2)
$$[H^+] \approx 0.10 - [M^+] = 0.073\ M$$

$$[OH^-] = \frac{K_w}{[H^+]} = \frac{1.0 \times 10^{-14}}{0.073} = 1.4 \times 10^{-13}$$

Assumption (2) is acceptable because $[OH^-] \ll [Cl^-]$ or $[HA]$.

• •

Example 11.3

Calculate the solubility of CaC_2O_4 in a solution of 0.0010 M hydrochloric acid, using the systematic approach.

Solution

The equilibria are

$$CaC_2O_4 \rightleftharpoons Ca^{2+} + C_2O_4^{2-}$$
$$C_2O_4^{2-} + H^+ \rightleftharpoons HC_2O_4^-$$
$$HC_2O_4^- + H^+ \rightleftharpoons H_2C_2O_4$$
$$H_2O \rightleftharpoons H^+ + OH^-$$
$$HCl \rightarrow H^+ + Cl^-$$

The equilibrium constant expressions are

$$K_{sp} = [Ca^{2+}][C_2O_4^{2-}] = 2.6 \times 10^{-9} \tag{1}$$

$$K_{a1} = \frac{[H^+][HC_2O_4^-]}{[H_2C_2O_4]} = 6.5 \times 10^{-2} \tag{2}$$

$$K_{a2} = \frac{[H^+][C_2O_4^{2-}]}{[HC_2O_4^-]} = 6.1 \times 10^{-5} \tag{3}$$

$$K_w = [H^+][OH^-] = 1.00 \times 10^{-14} \tag{4}$$

The mass balance expressions are

$$[Ca^{2+}] = [C_2O_4^{2-}] + [HC_2O_4^-] + [H_2C_2O_4] = C_{H_2C_2O_4} \tag{5}$$
$$[H^+] = [Cl^-] + [OH^-] - [HC_2O_4^-] - [H_2C_2O_4] \tag{6}$$
$$[Cl^-] = 0.0010 \; M \tag{7}$$

The charge balance expression is

$$[H^+] + 2[Ca^{2+}] = 2[C_2O_4^{2-}] + [HC_2O_4^-] + [Cl^-] + [OH^-] \tag{8}$$

There are seven unknowns ($[H^+]$, $[OH^-]$, $[Cl^-]$, $[Ca^{2+}]$, $[C_2O_4^{2-}]$, $[HC_2O_4^-]$, and $[H_2C_2O_4]$) and seven independent equations.

Simplifying assumptions:

(1) K_{a1} is rather large and K_{a2} is rather small, so assume $[HC_2O_4^-] \gg [H_2C_2O_4]$, $[C_2O_4^{2-}]$.

(2) In an acid solution, $[OH^-]$ is very small, so from (6) and (7):

$$[H^+] = 0.0010 + [OH^-] - [H_2C_2O_4^-] - [H_2C_2O_4] \approx 0.0010 - [HC_2O_4^-] \tag{9}$$

Calculate:

We need to calculate $[Ca^{2+}]$ in order to obtain the moles of CaC_2O_4 dissolved in a liter.

From (1)

$$[Ca^{2+}] = \frac{K_{sp}}{[C_2O_4^{2-}]} \tag{10}$$

From (3)

$$[C_2O_4^{2-}] = \frac{K_{a2}[HC_2O_4^-]}{[H^+]} \tag{11}$$

So,

$$[Ca^{2+}] = \frac{K_{sp}[H^+]}{K_{a2}[HC_2O_4{}^{2-}]} \qquad (12)$$

From assumption (1),

$$[Ca^{2+}] = [HC_2O_4{}^-] \qquad (13)$$

From assumption (2),

$$[H^+] \approx 0.0010 - [HC_2O_4{}^-] \approx 0.0010 - [Ca^{2+}] \qquad (14)$$

Substitute (13) and (14) in (12):

$$[Ca^{2+}] = \frac{K_{sp}(0.0010 - [Ca^{2+}])}{K_{a2}[Ca^{2+}]} = \frac{(2.6 \times 10^{-9})(0.0010 - [Ca^{2+}])}{(6.1 \times 10^{-5})[Ca^{2+}]}$$

$$[Ca^{2+}] = \frac{(4.6 \times 10^{-5})(0.0010 - [Ca^{2+}])}{[Ca^{2+}]}$$

Solving the quadratic equation gives $[Ca^{2+}] = 1.9 \times 10^{-4}$ *M*. This is the same as that calculated in Example 11.1, using the conditional solubility product approach, *after correcting for the H$^+$ consumed.* In the present example, we corrected for the H$^+$ consumed in the calculation. Note that in Example 11.1 we calculated $HC_2O_4{}^-$ to be 95% of the $[Ca^{2+}]$ value, so our assumption (1) was reasonable.

The answer is the same as when using K'_{sp} (Example 11.1).

• •

11.3 Effect of Complexation on Solubility: Conditional Solubility Product

Complexing agents can compete for the metal ion in a precipitate, just as acids compete for the anion. A precipitate MA that dissociates to give M$^+$ and A$^-$ and whose metal complexes with the ligand L to form ML$^+$ would have the following equilibria:

$$\begin{array}{c} MA \rightleftharpoons M^+ \\ + \\ L \\ \updownarrow \\ ML^+ \end{array} \Biggr\} + A^- \quad \Biggr\} C_M$$

The sum of $[M^+]$ and $[ML^+]$ is the analytical concentration C_M in equilibrium, which is equal to $[A^-]$. Calculations for such a situation are handled in a manner completely analogous to those for the effects of acids on solubility.

Consider the solubility of AgBr in the presence of NH_3. The equilibria are

$$AgBr \rightleftharpoons Ag^+ + Br^- \qquad (11.7)$$

$$Ag^+ + NH_3 \rightleftharpoons Ag(NH_3)^+ \qquad (11.8)$$

$$Ag(NH_3)^+ + NH_3 \rightleftharpoons Ag(NH_3)_2{}^+ \qquad (11.9)$$

The solubility s of AgBr is equal to $[Br^-] = C_{Ag}$, where C_{Ag} represents the concentrations of all the silver species in equilibrium $[= [Ag^+] + [Ag(NH_3)^+] + [Ag(NH_3)_2^+]$. As before, we can substitute $C_{Ag}\beta_0$ for $[Ag^+]$ in the K_{sp} expression, where β_0 is the fraction of silver species present as Ag^+ (Equation 9.20):

$$K_{sp} = [Ag^+][Br^-] = C_{Ag}\beta_0[Br^-] = 4 \times 10^{-13} \qquad (11.10)$$

Then,

$$\boxed{\frac{K_{sp}}{\beta_0} = K'_{sp} + C_{Ag}[Br^-] = s^2} \qquad (11.11)$$

The K'_{sp} value holds for only a given NH_3 concentration.

where K'_{sp} is again the **conditional solubility product**, whose value depends on the concentration of ammonia.

 Example 11.4

Calculate the molar solubility of silver bromide in a 0.10 M ammonia solution.

Solution

From Example 9.5, β_0 for 0.10 M $NH_3 = 4.0 \times 10^{-6}$.

$$s = \sqrt{\frac{K_{sp}}{\beta_0}} = \sqrt{4 \times 10^{-13}/4.0 \times 10^{-6}} = 3.2 \times 10^{-4} \ M$$

This compares with a solubility in water of 6×10^{-7} M (530 times more soluble). Note again that both $[Br^-]$ and $C_{Ag} = 3.2 \times 10^{-4}$ M. The concentrations of the other silver species in equilibrium can be obtained by multiplying this number by β_0, β_1, and β_2 at 0.10 M NH_3 to obtain $[Ag^+]$, $[Ag(NH_3)^+]$, and $[Ag(NH_3)_2^+]$, respectively. Taking the β values from Example 9.5, the results are $[Ag^+] = 1.3 \times 10^{-9}$ M, $[Ag(NH_3)^+] = 3.2 \times 10^{-7}$ M, and $[Ag(NH_3)_2^+] = 3.2 \times 10^{-4}$ M. Note that the majority of the dissolved silver exists in the $Ag(NH_3)_2^+$ form.

Check that the equilibrium ammonia concentration assumed was correct.

We neglected the amount of ammonia consumed in the reaction with the silver. We see that it was indeed negligible compared to 0.10 M [6×10^{-4} M was used in forming $Ag(NH_3)_2^+$, even less in forming $Ag(NH_3)^+$]. Had the amount of ammonia consumed been appreciable, we could have used an iterative procedure to obtain a more exact solution; that is, we could have subtracted the amount of ammonia consumed from the original concentration and then used the new concentration to calculate new β's and a new solubility.

11.4 Precipitation Titrations

Titrations with precipitating agents are useful for determining certain analytes, provided the equilibria are rapid and a suitable means of detecting the end point is available. A consideration of titration curves will increase our understanding of indicator selection, precision, and the titration of mixtures.

TITRATION CURVES—CALCULATING pX

Consider the titration of Cl^- with a standard solution of $AgNO_3$. A titration curve can be prepared by plotting pCl ($-\log[Cl^-]$) against the volume of $AgNO_3$, in a manner similar to that used for acid–base titrations. A typical titration curve is illustrated in Figure 11.1; pX in the figure refers to the negative logarithm of the halide concentration. At the beginning of the titration, we have 0.1 M Cl^-, and pCl is 1. As the titration continues, part of the Cl^- is removed from solution by precipitation as AgCl, and the pCl is determined by the concentration of the remaining Cl^-; the contribution of Cl^- from dissociation of the precipitate is negligible, except near the equivalence point. At the equivalence point, we have a saturated solution of AgCl, pCl = 5, and $[Cl^-] = \sqrt{K_{sp}} = 10^{-5}$ M (see Chapter 10). Beyond the equivalence point, there is excess Ag^+, and the Cl^- concentration is determined from the concentration of Ag^+ and K_{sp} as in Example 10.7 in Chapter 10 ($[Cl^-] = K_{sp}/[Ag^+]$).

Example 11.5

Calculate pCl for the titration of 100.0 mL of 0.1000 M Cl^- with 0.1000 M $AgNO_3$ for the addition of 0.00, 20.00, 99.00, 99.50, 100.00, 100.50, and 110.00 mL $AgNO_3$.

Solution
At 0.00 mL:

$$pCl = -\log 0.1000 = 1.000$$

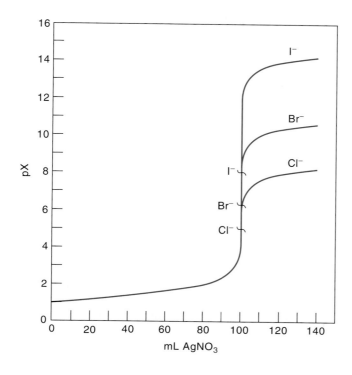

Agl has the lowest solubility, so $[I^-]$ beyond the equivalence point is smaller and pI is larger.

Fig. 11.1. Titration curves for 100 mL 0.1 M chloride, bromide, and iodide solutions versus 0.1 M $AgNO_3$.

At 20.00 mL:

$$\text{mmol } Cl^- = 100.0 \text{ mL} \times 0.1000 \text{ mmol/mL} = 10.00 \text{ mmol}$$
$$\text{mmol } Ag^+ = 20.00 \text{ mL} \times 0.1000 \text{ mmol/mL} = 2.000 \text{ mmol}$$
$$Cl^- \text{ left} = 10.00 - 2.00 = 8.00 \text{ mmol/120.0 mL} = 0.0667 \ M$$
$$pCl = -\log 0.0667 = 1.18$$

At 99.00 mL:

$$\text{mmol } Ag^+ = 99.00 \text{ mL} \times 0.1000 \text{ mmol/mL} = 9.900 \text{ mmol}$$
$$Cl^- \text{ left} = 10.00 - 9.90 = 0.10 \text{ mmol/199.0 mL} = 5.0 \times 10^{-4} \ M$$
$$pCl = -\log 5.0 \times 10^{-4} = 3.26$$

At 99.50 mL:

$$\text{mmol } Ag^+ = 99.50 \text{ mL} \times 0.1000 \text{ mmol/mL} = 9.950 \text{ mmol}$$
$$Cl^- \text{ left} = 10.00 - 9.95 = 0.05 \text{ mmol/199.5 mL} = 2_5 \times 10^{-4} \ M$$
$$pCl = -\log 2_5 \times 10^{-4} = 3.60$$

At 100.00 mL, all the Cl^- is reacted with Ag^+:

$$[Cl^-] = \sqrt{K_{sp}} = \sqrt{1.0 \times 10^{-10}} = 1.0 \times 10^{-5} \ M$$
$$pCl = -\log 1.0 \times 10^{-5} = 5.00$$

At 100.50 mL:

$$\text{mmol } Ag^+ = 100.50 \text{ mL} \times 0.1000 \text{ mmol/mL} = 10.05 \text{ mmol}$$
$$Ag^+ \text{ left} = 10.05 - 10.00 = 0.05 \text{ mmol/200.5 mL} = 0.02_4 \ M$$
$$[Cl^-] = K_{sp}/[Ag^+] = 1.0 \times 10^{-10}/2_4 \times 10^{-2} = 4_2 \times 10^{-9} \ M$$
$$pCl = -\log 4_2 \times 10^{-9} = 8.38$$

At 110.00 mL:

$$\text{mmol } Ag^+ = 110.00 \text{ mL} \times 0.1000 \text{ mmol/mL} = 11.00 \text{ mmol}$$
$$Ag^+ \text{ left} = 11.00 - 10.00 = 1.00 \text{ mmol/210 mL} = 4.76 \times 10^{-3} \ M$$
$$[Cl^-] = 1.0 \times 10^{-10}/4.76 \times 10^{-3} = 2.1 \times 10^{-8} \ M$$
$$pCl = -\log 2.1 \times 10^{-8} = 7.67$$

• •

The smaller the K_{sp}, the sharper the end point.

The smaller the K_{sp}, the larger the break at the equivalence point. We can illustrate this point by comparing the titration curves for Cl^-, Br^-, and I^- versus Ag^+ in Figure 11.1. The K_{sp} values of AgCl, AgBr, and AgI are 1×10^{-10}, 4×10^{-13}, and 1×10^{-16}, respectively. The concentration of each anion has been chosen to be the same at the beginning of the titration, and so up to near the equivalence point the concentration of each remains the same, since the same fraction is removed from solution. At the equivalence point, $[X^-]$ is smaller for the smaller K_{sp} values; hence pX is larger for a saturated solution of the salt. Beyond the equivalence point, $[X^-]$ is smaller when K_{sp} is smaller; also resulting in a larger jump in pX. So the overall effect is a larger pX break at the equivalence point when the compound is more insoluble.

If the chloride titration were performed in reverse, that is, Ag^+ titrated with Cl^-, the titration curve would be the reverse of the curve in Figure 11.1 if pCl were plotted against the volume of Cl^-. Before the equivalence point, the Cl^- concentration would be governed by the concentration of excess Ag^+ and K_{sp}; and beyond the equivalence point, it would be governed merely by the excess Cl^-. The pAg value could be plotted, instead, against the volume of chloride solution, and the curve would look the same as that of Figure 11.1.

DETECTION OF THE END POINT: INDICATORS

We can detect the end point by measuring either pCl or pAg with an appropriate electrode and a potentiometer. We discuss this in Chapter 13. It is more convenient if an indicator can be employed. The indicator theory for these titrations is different from that for acid–base indicators. The properties of the indicators do not necessarily depend on the concentration of some ion in solution, that is, on pCl or pAg.

Chemists commonly employ two types of indicators. The first type forms a colored compound with the titrant when the titrant is in excess. The second type, called an **adsorption indicator**, suddenly becomes adsorbed on the precipitate at the equivalence point owing to a property of the precipitate at the equivalence point, and the color of the indicator changes when it is adsorbed. Both mechanisms are discussed below.

1. Indicators Reacting with the Titrant. There are several examples of an indicator forming a colored compound with a titrant. The **Mohr method** for determining chloride serves as an example. The chloride is titrated with standard silver nitrate solution. A soluble chromate salt is added as the indicator. This produces a yellow solution. When the precipitation of the chloride is complete, the first excess of Ag^+ reacts with the indicator to precipitate red silver chromate:

$$\boxed{\underset{\text{(yellow)}}{CrO_4^{2-}} + 2Ag^+ \rightarrow \underset{\text{(red)}}{Ag_2CrO_4}} \qquad (11.12)$$

The concentration of the indicator is important. The Ag_2CrO_4 should just start precipitating at the equivalence point, where we have a saturated solution of AgCl. From K_{sp}, the concentration of Ag^+ at the equivalence point is 10^{-5} M. (It is less than this before the equivalence point.) So, Ag_2CrO_4 should precipitate just when $[Ag^+] = 10^{-5}$ M. The solubility product of Ag_2CrO_4 is 1.1×10^{-12}. By inserting the Ag^+ concentration in the K_{sp} equation for Ag_2CrO_4, we calculate that, for this to occur, $[CrO_4^{2-}]$ should be 0.011 M:

$$(10^{-5})^2[CrO_4^{2-}] = 1.1 \times 10^{-12}$$
$$[CrO_4^{2-}] = 1.1 \times 10^{-2} \, M$$

If the concentration is greater than this, Ag_2CrO_4 will begin to precipitate when $[Ag^+]$ is less than 10^{-5} M (before the equivalence point). If it is less than 0.011 M, then the $[Ag^+]$ will have to exceed 10^{-5} M (beyond the equivalence point) before precipitation of Ag_2CrO_4 begins.

In actual practice, the indicator concentration is kept at 0.002 to 0.005 M. If it is much higher than this, the intense yellow color of the chromate ion obscures the red Ag_2CrO_4 precipitate color, and an excess of Ag^+ is required to produce enough precipitate to be seen. An indicator blank should always be run and subtracted from the titration to correct for errors.

The Mohr titration is performed in slightly alkaline solution.

The Mohr titration must be performed at a pH of about 8. If the solution is too acid (pH < 6), then part of the indicator is present as $HCrO_4^-$, and more Ag^+ will be required to form the Ag_2CrO_4 precipitate. Above pH 8, silver hydroxide may be precipitated (at pH > 10). The pH is properly maintained by adding solid calcium carbonate to the solution. (While the carbonate ion is a fairly strong Brønsted base, the concentration in a saturated calcium carbonate solution is just sufficient to give a pH about 8.) The Mohr titration is useful for determining chloride in neutral or unbuffered solutions, such as drinking water.

The Volhard titration is performed in acid solution.

A second example of this type of indicator is illustrated in the **Volhard titration**. This is an indirect titration procedure for determining anions that precipitate with silver (Cl^-, Br^-, SCN^-), and it is performed in acid (HNO_3) solution. In this procedure, we add a measured excess of $AgNO_3$ to precipitate the anion and then determine the excess Ag^+ by back-titration with standard potassium thiocyanate solution:

$$\boxed{\begin{array}{l} X^- + Ag^+ \rightarrow \underline{AgX} + \text{excess } Ag^+ \\ \text{excess } Ag^+ + SCN^- \rightarrow \underline{AgSCN} \end{array}} \qquad (11.13)$$

We detect the end point by adding iron(III) as a ferric alum (ferric ammonium sulfate), which forms a soluble red complex with the first excess of titrant:

$$\boxed{Fe^{3+} + SCN^- \rightarrow Fe(SCN)^{2+}} \qquad (11.14)$$

If the precipitate, AgX, is less soluble than AgSCN, we do not have to remove the precipitate before titrating. Such is the case with I^-, Br^-, and SCN^-. In the case of I^-, we do not add the indicator until all the I^- is precipitated, since it would be oxidized by the iron(III). If the precipitate is more soluble than AgSCN, it will react with the titrant to give a high and diffuse end point. Such is the case with AgCl:

$$\underline{AgCl} + SCN^- \rightarrow \underline{AgSCN} + Cl^- \qquad (11.15)$$

Therefore, we remove the precipitate by filtration before titrating.

Obviously, these indicators must not form a compound with the titrant that is more stable than the precipitate, or the color reaction would occur when the first drop of titrant is added.

2. Adsorption Indicators. With adsorption indicators, the indicator reaction takes place on the surface of the precipitate. The indicator, which is a dye, exists in solution as the ionized form, usually an anion, In^-. To explain the mechanism of the indicator action, we must invoke the mechanism occurring during precipitation (see Chapter 10 for more detail).

Consider the titration of Cl^- with Ag^+. Before the equivalence point, Cl^- is in excess, and the *primary adsorbed layer* is Cl^-. This repulses the indicator anion, and the more loosely held *secondary (counter) layer* of adsorbed ions is cations, such as Na^+:

$$AgCl:Cl^-::Na^+$$

Beyond the equivalence point, Ag^+ is in excess, and the surface of the precipitate becomes positively charged, with the primary layer being Ag^+. This will now attract the indicator anion and adsorb it in the counterlayer:

$$AgCl:Ag^+::In^-$$

The color of the adsorbed indicator is different from that of the unadsorbed indicator, and this difference signals the completion of the titration. A possible explanation for this color change is that the indicator forms a colored complex with Ag^+, which is too weak to exist in solution, but whose formation is facilitated by adsorption on the surface of the precipitate (it becomes "insoluble").

The pH is important. If it is too low, the indicator, which is usually a weak acid, will dissociate too little to allow it to be adsorbed as the anion. Also, the indicator must not be too strongly adsorbed at the given pH, or it will displace the anion of the precipitate (e.g., Cl^-) in the primary layer before the equivalence point is reached. This will, of course, depend on the degree of adsorption of the anion of the precipitate. For example, Br^- forms a less soluble precipitate with Ag^+ and is more strongly adsorbed. A more strongly adsorbed indicator can therefore be used.

> The more insoluble precipitates can be titrated in more acid solutions, using more strongly adsorbed indicators.

The degree of adsorption of the indicator can be decreased by increasing the acidity. The stronger an acid the indicator is, the wider the pH range over which it can be adsorbed. In the case of Br^-, since a more acidic (more strongly adsorbed) indicator can be used, the pH of the titration can be more acidic than with Cl^-.

Table 11.1 lists some adsorption indicators. Fluorescein can be used as an indicator for any of the halides at pH 7 because it will not displace any of them. Dichlorofluoroscein will displace Cl^- at pH 7 but not at pH 4. Hence, results tend to be low when titrations are performed at pH 7. Titration of chloride using these indicators is called **Fajans' method**. Fluorescein was the original indicator described by Fajans, but dichlorofluoroscein is now preferred. Eosin cannot be used for the titration of chloride at any pH because it is too strongly adsorbed.

Because most of these end points do not coincide with the equivalence point, *the titrant should be standardized by the same titration as used for the sample.* In this way, the errors will nearly cancel if about the same amount of titrant is used for both the standardization and analysis.

A chief source of error in titrations involving silver is photodecomposition of AgX, which is catalyzed by the adsorption indicator. By proper standardization, however, accuracies of one part per thousand can be achieved.

The precipitate is uncharged at the equivalence point (neither ion is in excess). Colloidal precipitates, such as silver chloride, therefore tend to coagulate at this point, especially if the solution is shaken. This is just what we want for gravimetry, but the opposite of what we want here. Coagulation decreases the surface area for absorption of the indicator, which in turn decreases the sharpness of the end point. We can prevent coagulation of silver chloride by adding some dextrin to the solution.

> For adsorption indicators, we want the maximum surface area for adsorption, in contrast to gravimetry.

Visual indicators are convenient for rapid precipitation titrimetry with silver ion. Potentiometric end-point detection is also widely used, particularly for dilute solutions, for example, millimolar (see Chapter 14).

Table 11.1
Adsorption Indicators

Indicator	Titration	Solution
Fluorescein	Cl^- with Ag^+	pH 7–8
Dichlorofluorescein	Cl^- with Ag^+	pH 4
Bromcresol green	SCN^- with Ag^+	pH 4–5
Eosin	Br^-, I^-, SCN^- with Ag^+	pH 2
Methyl violet	Ag^+ with Cl^-	Acid solution
Rhodamine 6 G	Ag^+ with Br^-	HNO_3 ($\leq 0.3\ M$)
Thorin	SO_4^{2-} with Ba^{2+}	pH 1.5–3.5
Bromphenol blue	Hg^{2+} with Cl^-	$0.1\ M$ solution
Orthochrome T	Pb^{2+} with CrO_4^{2-}	Neutral, $0.02\ M$ solution

TITRATION OF SULFATE WITH BARIUM

Sulfate can be determined by titrating with barium ion, to precipitate $BaSO_4$. As in the gravimetric determination of sulfate by precipitation of barium sulfate, this titration is subject to errors by coprecipitation. Cations such as K^+, Na^+, and NH_4^+ (especially the first) coprecipitate as sulfates:

$$BaSO_4 : SO_4^{2-} : 2M^+$$

As a result, less barium ion is required to complete the precipitation of the sulfate ion, and the calculated results are low. Some metal ions will complex the indicator and interfere. Foreign anions may coprecipitate as the barium salts to cause high results. Errors from chloride, bromide, and perchlorate are small, but nitrate causes large errors and must be absent.

Cation interferences are readily removed with a strong cation exchange resin in the hydrogen form:

$$2Rz^-H^+ + (M)_2SO_4 \rightarrow 2Rz^-M^+ + H_2SO_4$$

The cations are replaced by protons. We discuss the principles of ion exchange chromatography in Chapter 21.

The titration is carried out in an aqueous–nonaqueous solvent mixture. The organic solvent decreases the dissociation of the indicator and thereby hinders formation of a barium–indicator complex. It also results in a more flocculant precipitate with better adsorption properties for the indicator.

Learning Objectives

WHAT ARE SOME OF THE KEY THINGS WE LEARNED FROM THIS CHAPTER?

- Effects of acids on solubility (key equations: 11.4, 11.6), p. 339
- Mass balance calculations, p. 341
- Effect of complexation on solubility (key equations: 11.10, 11.11), p. 345
- Calculating precipitation titration curves, p. 346
- Indicators for precipitation titrations, p. 349

Questions

1. Explain the Volhard titration of chloride. The Fajan titration. Which is used for acid solutions? Why?
2. Explain the principles of adsorption indicators.

Problems

EFFECT OF ACIDITY ON SOLUBILITY

3. Calculate the solubility of $AgIO_3$ in 0.100 M HNO_3. Also calculate the equilibrium concentrations of IO_3^- and HIO_3.
4. Calculate the solubility of CaF_2 in 0.100 M HCl. Also calculate the equilibrium concentrations of F^- and HF.

5. Calculate the solubility of PbS in 0.0100 M HCl. Also calculate the equilibrium concentrations of S^{2-}, HS^-, and H_2S.

EFFECT OF COMPLEXATION ON SOLUBILITY

6. Silver ion forms a stepwise 1 : 2 complex with ethylenediamine (en) with formation constants of $K_{f1} = 5.0 \times 10^4$ and $K_{f2} = 1.4 \times 10^3$. Calculate the solubility of silver chloride in 0.100 M ethylenediamine. Also calculate the equilibrium concentrations of $Ag(en)^+$ and $Ag(en)_2^+$.

MASS BALANCE CALCULATIONS

7. Calculate the solubility of $AgIO_3$ in 0.100 M HNO_3, using the mass balance approach. Compare with Problem 3.

8. Calculate the solubility of PbS in 0.0100 M HCl, using the mass balance approach. Compare with Problem 5.

9. Calculate the solubility of AgCl in 0.100 M ethylenediamine. Compare with Problem 6. The formation constant is given in Problem 6.

QUANTITATIVE PRECIPITATION DETERMINATIONS

10. Chloride in a brine solution is determined by the Volhard method. A 10.00-mL aliquot of the solution is treated with 15.00 mL of standard 0.1182 M $AgNO_3$ solution. The excess silver is titrated with standard 0.101 M KSCN solution, requiring 2.38 mL to reach the red $Fe(SCN)^{2+}$ end point. Calculate the concentration of chloride in the brine solution, in g/L.

11. In a Mohr titration of chloride with silver nitrate, an error is made in the preparation of the indicator. Instead of 0.011 M chromate indicator in the titration flask at the end point, there is only 0.0011 M. If the flask contains 100 mL at the end point, what is the error in the titration in milliliters of 0.100 M titrant? Neglect errors due to the color of the solution.

SPREADSHEET PROBLEM

12. Prepare a spreadsheet to plot the titration curve of 100 mL 0.1 M chloride titrated with 0.1 M silver nitrate (Figure 11.1). See your CD for a suggested setup. Use the spreadsheet to change the concentrations of chloride and silver (e.g., 0.2 M each, 0.05 M each), and notice how the titration curve changes. Note that there is a limit to how low the concentrations can go in these calculated plots because eventually the solubility of the AgCl at 99.9 and 100.1 mL titrant becomes appreciable.

Recommended Reference

1. O. Schales, "Chloride," in M. Reiner, ed., *Standard Methods of Clinical Chemistry*, Vol. 1, New York: Academic, 1953, pp. 37–42.

Chapter Twelve

ELECTROCHEMICAL CELLS AND ELECTRODE POTENTIALS

"I know of nothing sublime which is not some modification of power."
—Edmund Burke

Oxidation is a loss of electrons.
Reduction is a gain of electrons.

An important class of titrations is reduction–oxidation or "redox" titrations, in which an oxidizing agent and a reducing agent react (see Equation 12.1). We define an **oxidation** as a loss of electrons to an oxidizing agent (which gets reduced) to give a higher or more positive oxidation state, and we define **reduction** as a gain of electrons from a reducing agent (which gets oxidized) to give a lower or more negative oxidation state. We can gain an understanding of these reactions from a knowledge of electrochemical cells and electrode potentials. In this chapter, we discuss electrochemical cells, standard electrode potentials, the Nernst equation (which describes electrode potentials), and limitations of those potentials. Chapter 13 discusses potentiometry, the use of potential measurements for determining concentration, including the glass pH electrode and ion-selective electrodes. In Chapter 14, we describe redox titrations and potentiometric titrations in which potentiometric measurements are used to detect the end point. We review in that chapter the balancing of redox reactions since this is required for volumetric calculations. You may wish to review that material now.

12.1 What Are Redox Reactions?

A reduction–oxidation reaction—commonly called a **redox** reaction—is one that occurs between a reducing and an oxidizing agent:

$$\boxed{Ox_1 + Red_2 \rightleftharpoons Red_1 + Ox_2} \tag{12.1}$$

The oxidizing agent is reduced.
The reducing agent is oxidized.

Ox_1 is reduced to Red_1, and Red_2 is oxidized to Ox_2. Ox_1 is the oxidizing agent, and Red_2 is the reducing agent. The reducing or oxidizing tendency of a substance will depend on its reduction potential, described below. An oxidizing substance will tend to take on an electron or electrons and be reduced to a lower oxidation state:

$$M^{a+} + ne^- \rightarrow M^{(a-n)+} \qquad (12.2)$$

for example, $Fe^{3+} + e^- \rightarrow Fe^{2+}$. Conversely, a reducing substance will tend to give up an electron or electrons and be oxidized:

$$M^{a+} \rightarrow M^{(a+n)+} + ne^- \qquad (12.3)$$

for example, $2I^- \rightarrow I_2 + 2e^-$. If the oxidized form of a metal ion is complexed, it is more stable and will be more difficult to reduce; so its tendency to take on electrons will be decreased if the reduced form is not also complexed to make it more stable and easier to form.

We can better understand the oxidizing or reducing tendencies of substances by studying electrochemical cells and electrode potentials.

12.2 Electrochemical Cells—What Electroanalytical Chemists Use

There are two kinds of electrochemical cells, *voltaic* (galvanic) and **electrolytic**. In voltaic cells, a chemical reaction spontaneously occurs to *produce electrical energy*. The lead storage battery and the ordinary flashlight battery are common examples of voltaic cells. In electrolytic cells, on the other hand, *electrical energy is used* to force a nonspontaneous chemical reaction to occur, that is, to go in the reverse direction it would in a voltaic cell. An example is the electrolysis of water. In both types of these cells, the electrode at which oxidation occurs is the **anode**, and that at which reduction occurs is the **cathode**. Voltaic cells will be of importance in our discussions in the next two chapters, dealing with potentiometry. Electrolytic cells are important in electrochemical methods such as voltammetry, in which electroactive substances like metal ions are reduced at an electrode to produce a measurable current by applying an appropriate potential to get the nonspontaneous reaction to occur (Chapter 15). The current that results from the forced electrolysis is proportional to the concentration of the electroactive substance.

In a voltaic cell, a spontaneous chemical reaction produces electricity. This occurs only when the cell circuit is closed, as when you turn on a flashlight. The cell voltage (e.g., in a battery) is determined by the potential difference of the two half reactions. When the reaction has gone to completion, the cell runs down, and the voltage is zero (the battery is "dead"). In an electrolytic cell, the reaction is forced the other way by applying an external voltage greater than and opposite to the spontaneous voltage.

VOLTAIC CELL AND SPONTANEOUS REACTIONS—WHAT IS THE CELL POTENTIAL?

Consider the following redox reaction in a voltaic cell:

$$Fe^{2+} + Ce^{4+} \rightleftharpoons Fe^{3+} + Ce^{3+} \qquad (12.4)$$

If we mix a solution containing Fe^{2+} with one containing Ce^{4+}, there is a certain tendency for the ions to transfer electrons. Assume the Fe^{2+} and Ce^{4+} are in separate beakers connected by a **salt bridge**, as shown in Figure 12.1. (A salt bridge allows charge transfer through the solutions but prevents mixing of the solutions.) No reaction can occur since the solutions do not make contact. A salt bridge is not always needed—only when the reactants or products at the anode or cathode react with each other so that it is necessary to keep them from mixing freely. Now put an inert platinum wire in each solution and connect the two wires. The setup now

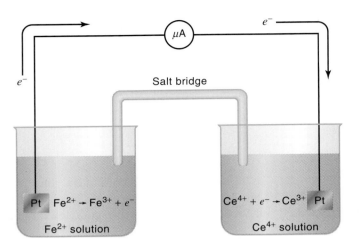

Fig. 12.1. Voltaic cell.

constitutes a voltaic cell. If a microammeter is connected in series, it indicates that a current is flowing. The Fe^{2+} is being oxidized at the platinum wire (the anode):

$$Fe^{2+} \rightarrow Fe^{3+} + e^- \qquad (12.5)$$

The released electrons flow through the wire to the other beaker where the Ce^{4+} is reduced (at the cathode):

$$Ce^{4+} + e^- \rightarrow Ce^{3+} \qquad (12.6)$$

This process occurs because of the tendency of these ions to transfer electrons. The net result is the reaction written in Equation 12.4, which would occur if Fe^{2+} and Ce^{4+} were added together in a single beaker. The platinum wires can be considered **electrodes**. Each will adopt an electrical **potential** that is determined by the tendency of the ions to give off or take on electrons, and this is called the **electrode potential**. A voltmeter placed between the electrodes will indicate the *difference* in the potentials between the two electrodes. The larger the potential difference, the greater the tendency for the reaction between Fe^{2+} and Ce^{4+}. The driving force of the chemical reaction (the potential difference) can be used to perform work such as lighting a light bulb or running a motor, as is done with a battery.

HALF-REACTIONS—GIVING AND ACCEPTING ELECTRONS

Equations 12.5 and 12.6 are **half-reactions**. No half-reaction can occur by itself. There must be an **electron donor** (a reducing agent) and an **electron acceptor** (an oxidizing agent). In this case, Fe^{2+} is the reducing agent and Ce^{4+} is the oxidizing agent. Each half-reaction will generate a definite potential that would be adopted by an inert electrode dipped in the solution.

HALF-REACTION POTENTIALS—THEY ARE MEASURED RELATIVE TO EACH OTHER

If the potentials of all half-reactions could be measured, then we could determine which oxidizing and reducing agents will react. Unfortunately, there is no way to measure individual electrode potentials. But, as we just saw, the *difference* between two electrode potentials can be measured. The electrode potential of the half-reaction.[1]

$$2H^+ + 2e^- \rightleftharpoons H_2$$
(12.7)

has arbitrarily been assigned a value of 0.000 V. This is called the **normal hydrogen electrode** (NHE), or the **standard hydrogen electrode** (SHE). This consists of a platinized platinum electrode (one coated with fine "platinum black" by electroplating platinum on the electrode) contained in a glass tube, over which hydrogen gas is bubbled. The platinum black catalyzes Reaction 12.7. The potential differences between this half-reaction and other half-reactions have been measured using voltaic cells and arranged in decreasing order. Some of these are listed in Table 12.1. Potentials are dependent on concentrations, and all standard potentials refer to conditions of unit activity for all species (or 1 atmosphere partial pressure in the case of gases, as for hydrogen in the NHE). The effects of concentrations on potentials are described below. A more complete listing of potentials appears in Appendix C.

We arbitrarily define the potential of this half-reaction as zero (at standard conditions). All others are measured relative to this.

The potentials are for the half-reaction written as a *reduction*, and so they represent **reduction potentials**. We will use the Gibbs–Stockholm electrode potential convention, adopted at the 17th Conference of the International Union of Pure and Applied Chemistry in Stockholm, 1953. In this convention, the half-reaction is written as a reduction, and the potential increases as the tendency for reduction (of the oxidized form of the half-reaction) increases.

In the Gibbs–Stockholm convention, we always write the half-reaction as a reduction.

---

[1] The reaction could have been written $H^+ + e^- = \frac{1}{2}H_2$. The way it is written does not affect its potential.

Table 12.1
Some Standard Potentials

Half-Reaction	E^0(V)
$H_2O_2 + 2H^+ + 2e^- = 2H_2O$	1.77
$MnO_4^- + 4H^+ + 3e^- = MnO_2 + 2H_2O$	1.695
$Ce^{4+} + e^- = Ce^{3+}$	1.61
$MnO_4^- + 8H^+ + 5e^- = Mn^{2+} + 4H_2O$	1.51
$Cr_2O_7^{2-} + 14H^+ + 6e^- = 2Cr^{3+} + 7H_2O$	1.33
$MnO_2 + 4H^+ + 2e^- = Mn^{2+} + 2H_2O$	1.23
$2IO_3^- + 12H^+ + 10e^- = I_2 + 6H_2O$	1.20
$H_2O_2 + 2e^- = 2OH^-$	0.88
$Cu^2 + I^- + e^- = CuI$	0.86
$Fe^{3+} + e^- = Fe^{2+}$	0.771
$O_2 + 2H^+ + 2e^- = H_2O_2$	0.682
$I_2(aq) + 2e^- = 2I^-$	0.6197
$H_3AsO_4 + 2H^+ + 2e^- = H_3AsO_3 + H_2O$	0.559
$I_3^- + 2e^- = 3I^-$	0.5355
$Sn^{4+} + 2e^- = Sn^{2+}$	0.154
$S_4O_6^{2-} + 2e^- = 2S_2O_3^{2-}$	0.08
$2H^+ + 2e^- = H_2$	0.000
$Zn^{2+} + 2e^- = Zn$	−0.763
$2H_2O + 2e^- = H_2 + 2OH^-$	−0.828

The electrode potential for $Sn^{4+} + 2e^- \rightleftharpoons Sn^{2+}$ is $+0.15$ V. In other words, the potential of this half-reaction relative to the NHE in a cell like that in Figure 12.1 would be 0.15 V. Since the above couple has a larger (more positive) reduction potential than the NHE, Sn^{4+} has a stronger tendency to be reduced than H^+ has. We can draw some general conclusions from the electrode potentials:

1. The more *positive* the electrode potential, the greater the tendency of the oxidized form to be reduced. In other words, **the more positive the electrode potential, the stronger an oxidizing agent the oxidized form is and the weaker a reducing agent the reduced form is.**

2. The more *negative* the electrode potential, the greater the tendency of the reduced form to be oxidized. In other words, **the more negative the reduction potential, the weaker an oxidizing agent is the oxidized form is and the stronger a reducing agent the reduced form is.**

Ce^{4+} is a good oxidizing agent because of the high reduction potential. (But Ce^{3+} is a poor reducing agent.) Zn is a good reducing agent because of the low reduction potential. (But Zn^{2+} is a poor oxidizing agent.)

The reduction potential for $Ce^{4+} + e^- \rightleftharpoons Ce^{3+}$ is very positive, so Ce^{4+} is a strong oxidizing agent, while Ce^{3+} is a very weak reducing agent. On the other hand, the potential for $Zn^{2+} + 2e^- \rightleftharpoons Zn$ is very negative, and so Zn^{2+} is a very weak oxidizing agent, while metallic zinc is a very strong reducing agent.

WHAT SUBSTANCES REACT?

> The oxidized form of a species in a half-reaction is capable of oxidizing the reduced form of a species in a half-reaction whose reduction potential is more *negative* than its own, and vice versa: The reduced form in a half-reaction is capable of reducing the oxidized form in a half-reaction with a more *positive* potential.

For example, consider the two half-reactions

$$Fe^{3+} + e^- = Fe^{2+} \quad E^0 = 0.771 \text{ V} \tag{12.8}$$
$$Sn^{4+} + 2e^- = Sn^{2+} \quad E^0 = 0.154 \text{ V} \tag{12.9}$$

There are two combinations for possible reaction between an oxidizing and a reducing agent in these two half-reactions, which we arrive at by *subtracting* one from the other (multiplying the first half-reaction by 2 so the electrons cancel):

$$2Fe^{3+} + Sn^{2+} = 2Fe^{2+} + Sn^{4+} \tag{12.10}$$

and

$$Sn^{4+} + 2Fe^{2+} = Sn^{2+} + 2Fe^{3+} \tag{12.11}$$

[There is no possibility of reaction between Fe^{3+} and Sn^{4+} (both oxidizing agents) or between Fe^{2+} and Sn^{2+} (both reducing agents).] Perusal of the potentials tells us that Reaction 12.10 will take place; that is, the reduced form Sn^{2+} of Reaction 12.9 (with the more negative potential) will react with the oxidized form of Reaction 12.8 (with the more positive potential). Note that the number of electrons donated and accepted must be equal (see Chapter 14 on balancing redox reactions).

Example 12.1

For the following substances, list the oxidizing agents in decreasing order of oxidizing capability, and the reducing agents in decreasing order of reducing capability: MnO_4^-, Ce^{3+}, Cr^{3+}, IO_3^-, Fe^{3+}, I^-, H^+, Zn^{2+}.

Solution

Looking at Table 12.1, the following must be oxidizing agents (are in the oxidized forms) and are listed from the most positive E^0 to the least positive: MnO_4^-, IO_3^-, Fe^{3+}, H^+, Zn^{2+}. MnO_4^- is a very good oxidizing agent, Zn^{2+} is very poor. The remainder are in the reduced form, and their reducing power is in the order I^-, Cr^{3+}, and Ce^{3+}. I^- is a reasonably good reducing agent; Ce^{3+} is poor.

If the potentials are subtracted in the same manner as the half-reactions to give the net reaction, the result is the **cell voltage** that would be observed in a voltaic cell (Equation 12.8 minus Equation 12.9, or 0.771 V − 0.154 V = +0.617 V in the above).[2] *If this calculated cell voltage is positive, the reaction goes as written.* If it is negative, the reaction will occur in the reverse direction. This is the result of the convention that, for a spontaneous reaction, the free energy is negative. The free energy at standard conditions is given by

$$\Delta G^\circ = -nF\,\Delta E^0 \tag{12.12}$$

and so a positive potential difference provides the necessary negative free energy. Hence, we can tell from the relative standard potentials for two reactions, and from their signs, which reaction combination will produce a negative free-energy change and thus be spontaneous. For example, for the Ce^{4+}/Ce^{3+} half-reaction, E^0 is +1.61 V (Table 12.1); and for the Fe^{3+}/Fe^{2+} half-reaction, E^0 is +0.771 V. ΔG° for the former is more negative than for the latter, and subtraction of the iron half-reaction from the cerium one will provide the spontaneous reaction that would occur to give a negative free energy. That is, Ce^{4+} would spontaneously oxidize Fe^{2+}.

> The spontaneous cell reaction is the one that gives a positive cell voltage when subtracting one half-reaction from the other.

WHICH IS THE ANODE? THE CATHODE?

By convention, a cell is written with the anode on the left:

$$\boxed{\text{anode/solution/cathode}} \tag{12.13}$$

The single lines represent a boundary between either an electrode phase and a solution phase or two solution phases. In Figure 12.1, the cell would be written as

$$Pt/Fe^{2+}(C_1),\ Fe^{3+}(C_2)//Ce^{4+}(C_3),\ Ce^{3+}(C_4)/Pt \tag{12.14}$$

---

[2] We refer to electrode **potentials** and cell **voltages** to distinguish between half-reactions and complete reactions.

where C_1, C_2, C_3, and C_4 represent the concentrations of the different species. The double line represents the salt bridge. If a voltaic cell were constructed for the above iron and tin half-reactions, it would be written as

$$Pt/Sn^{2+}(C_1), Sn^{4+}(C_2)//Fe^{3+}(C_3), Fe^{2+}(C_4)/Pt \qquad (12.15)$$

> The anode is the electrode where oxidation occurs, i.e., the more negative half-reaction.

Since oxidation occurs at the anode and reduction occurs at the cathode, *the stronger reducing agent is placed on the left* and *the stronger oxidizing agent is placed on the right*. The potential of the voltaic cell is given by

$$\boxed{E_{cell} = E_{right} - E_{left} = E_{cathode} - E_{anode} = E_+ - E_-} \qquad (12.16)$$

where E_+ is the more positive electrode potential and E_- is the more negative of the two electrodes.

When the cell is set up properly, *the calculated voltage will always be positive*, and the cell reaction is written correctly, that is, the correct cathode half-reaction is written as a reduction and the correct anode half-reaction is written as an oxidation. In cell (12.15), we would have at standard conditions

$$E^0_{cell} = E^0_{Fe^{3+},Fe^{2+}} - E^0_{Sn^{4+},Sn^{2+}} = 0.771 - 0.154 = 0.617 \text{ V}$$

> The potential difference between the titrant and the analyte half-reaction should be 0.2–0.3 V for a sharp end point.

To take some more examples of possible redox reactions, Fe^{3+} will not oxidize Mn^{2+}. Quite the contrary, MnO_4^- will oxidize Fe^{2+}. I_2 is a moderate oxidizing agent and will oxidize Sn^{2+}. On the other hand, I^- is a fairly good reducing agent and will reduce Fe^{3+}, $Cr_2O_7^{2-}$, and so on. *For a reaction to be complete enough to obtain a sharp end point in a titration, there should be at least 0.2 to 0.3 V difference between the two electrode potentials.*

Example 12.2

From the potentials listed in Table 12.1, determine the reaction between the following half-reactions, and calculate the corresponding cell voltage:

$$Fe^{3+} + e^- = Fe^{2+} \qquad E^0 = 0.771 \text{ V}$$
$$I_3^- + 2e^- = 3I^- \qquad E^0 = 0.5355 \text{ V}$$

Solution

Since the Fe^{3+}/Fe^{2+} potential is the more positive, Fe^{3+} is a better oxidizing agent than I_3^-. Hence, Fe^{3+} will oxidize I^- and $E^0_{cell} = E_{cathode} - E_{anode} = E^0_{Fe^{3+},Fe^{2+}} - E^0_{I_3^-,I^-}$. In the same fashion, the second half-reaction must be subtracted from the first (multiplied by 2) to give the overall cell reaction:

$$2Fe^{3+} + 3I^- = 2Fe^{2+} + I_3^- \qquad E^0_{cell} = 0.771 \text{ V} - 0.536 \text{ V} = +0.235 \text{ V}$$

Note that multiplying a half-reaction by any number does not change its potential.

12.3 Nernst Equation—Effects of Concentrations on Potentials

The potentials listed in Table 12.1 were determined for the case when the concentrations of both the oxidized and reduced forms (and all other species) were at **unit activity**, and they are called the **standard potentials**, designated by E^0. Volta originally set up empirical E^0 tables under very controlled and defined conditions. Nernst made them practical by establishing quantitative relationships between potential and concentrations. This potential is dependent on the concentrations of the species and varies from the standard potential. This potential dependence is described by the **Nernst equation**[3]:

Activities should be used in the Nernst equation. We will use concentrations here because titrations deal with large potential changes, and the errors are small by doing so.

$$a\text{Ox} + ne^- \rightleftharpoons b\text{Red} \tag{12.17}$$

$$E = E^0 - \frac{2.3026RT}{nF} \log \frac{[\text{Red}]^b}{[\text{Ox}]^a} \tag{12.18}$$

where E is the reduction potential at the specific concentrations, n is the number of electrons involved in the half-reaction (equivalents per mole), R is the gas constant (8.3143 V coul $\deg^{-1}$ mol^{-1}), T is the absolute temperature, and F is the Faraday constant (96,487 coul eq^{-1}). At 25°C (298.16 K), the value of $2.3026RT/F$ is 0.05916, or 1.9842×10^{-4} (°C + 273.16). *The concentration of pure substances such as precipitates and liquids (H_2O) is taken as unity. Note that the log term of the reduction half-reaction is the ratio of the right-side concentration(s) over the left-side concentration(s).*

Example 12.3

A solution if 10^{-3} M in $Cr_2O_7^{2-}$ and 10^{-2} M in Cr^{3+}. If the pH is 2.0 what is the potential of the half-reaction?

Solution

$$Cr_2O_7^{2-} + 14H^+ + 6e^- \rightleftharpoons 2Cr^{3+} + 7H_2O$$

$$E = E^0_{Cr_2O_7^{2-},Cr^{3+}} - \frac{0.059}{6} \log \frac{[Cr^{3+}]^2}{[Cr_2O_7^{2-}][H^+]^{14}}$$

$$= 1.33 - \frac{0.059}{6} \log \frac{(10^{-2})^2}{(10^{-3})(10^{-2})^{14}}$$

$$= 1.33 - \frac{0.059}{6} \log 10^{27} = 1.33 - 27\left(\frac{0.059}{6}\right)$$

$$= 1.06 \text{ V}$$

---

[3]More correctly, activities, rather than concentrations, should be used; but we will use concentrations for this discussion. In the next chapter, involving potential measurements for direct calculation of concentrations, we will use activities.

This calculated potential is the potential an electrode would adopt, relative to the NHE, if it were placed in the solution, and it is a measure of the oxidizing or reducing power of that solution. Theoretically, the potential would be infinite if there were no Cr^{3+} at all in solution. In actual practice, the potential is always finite (but impossible to calculate from the simple Nernst equation). Either there will be a small amount of impurity of the oxidized or reduced form present or, more probably, the potential will be limited by another half-reaction, such as the oxidation or reduction of water, that prevents it from going to infinity.

$\bullet$

EQUILIBRIUM POTENTIAL— AFTER THE REACTION HAS OCCURRED

To construct a titration curve, we are interested in the equilibrium *electrode* potential (i.e., when the *cell* potential is zero—after the titrant and analyte have reacted). The two electrodes have identical potentials then, as determined by the Nernst equation for each half-reaction.

The potential of an inert electrode in a solution containing the ions of two half-reactions at equilibrium (e.g., at different points in a titration) can be calculated relative to the NHE using the Nernst equation for *either* half-reaction. This is because when the reaction comes to equilibrium, the potentials for the two half-reactions become identical; otherwise, the reaction would still be going on. An electrode dipped in the solution will adopt the **equilibrium potential**. The equilibrium potential is dictated by the equilibrium concentrations of either half-reaction and the Nernst equation.

$\bullet\bullet\bullet\bullet\bullet$ $\bullet$

Example 12.4

5.0 mL of 0.10 M Ce^{4+} solution is added to 5.0 mL of 0.30 M Fe^{2+} solution. Calculate the potential of a platinum electrode dipping in the solution (relative to the NHE).

Solution

We start with $0.30 \times 5.0 = 1.5$ mmol Fe^{2+} and add $0.10 \times 5.0 = 0.50$ mmol Ce^{4+}. So we form 0.50 mmol each of Fe^{3+} and Ce^{3+} and have 1.0 mmol Fe^{2+} remaining. The reaction lies far to the right at equilibrium if there is at least 0.2 V difference between the standard electrode potentials of two half-reactions. But a small amount of Ce^{4+} ($= x$) will exist at equilibrium, and an equal amount of Fe^{2+} will be formed:

$$Fe^{2+} + Ce^{4+} \rightleftharpoons Fe^{3} + Ce^{3+}$$
$$1.0 + x \qquad x \qquad 0.50 - x \quad 0.50 - x$$

These are the equilibrium concentrations, following reaction.

where the numbers and x represent millimoles. This is analogous to "ionization" of the product in precipitation or acid–base reactions; a slight shift of the equilibrium here to the left would be the "ionization." The quantity x is very small compared with 0.50 or 1.0 and can be neglected. Either half-reaction can be used to calculate the potential. Since the concentrations of both species in the Fe^{3+}/Fe^{2+} couple are known, we will use this:

$$Fe^{3+} + e^- \rightleftharpoons Fe^{2+}$$
$$0.50 \qquad\qquad 1.0$$
$$E = 0.771 - 0.059 \log \frac{[Fe^{2+}]}{[Fe^{3+}]}$$

The final volume is 10 mL, so

$$E = 0.771 - 0.059 \log \frac{1.0 \text{ mmol/10 mL}}{0.50 \text{ mmol/10 mL}} = 0.771 - 0.059 \log 2.0$$

$$= 0.771 - 0.059(0.30)$$

$$= 0.753 \text{ V}$$

• •

CELL VOLTAGE—BEFORE REACTION

The voltage of a cell can be calculated by taking the difference in potentials of the two half-reactions, to give a positive potential, calculated using the Nernst equation,

$$E_{cell} = E_+ - E_- \tag{12.19}$$

as given in Equation (12.16).

In Example 12.2 for $2Fe^{3+} + 3I^- = 2Fe^{2+} + I_3^-$,

$$E_{cell} = E_{Fe^{3+},Fe^{2+}} - E_{I_3^-,I^-}$$

$$= \left(E^0_{Fe^{3+},Fe^{2+}} - \frac{0.059}{2} \log \frac{[Fe^{2+}]^2}{[Fe^{3+}]^2} \right) - \left(E^0_{I_3^-,I^-} - \frac{0.059}{2} \log \frac{[I^-]^3}{[I_3^-]} \right) \tag{12.20}$$

$$= E^0_{Fe^{3+},Fe^{2+}} - E^0_{I_3^-,I^-} - \frac{0.059}{2} \log \frac{[Fe^{2+}]^2[I_3^-]}{[Fe^{3+}]^2[I^-]^3}$$

Note that the log term for the cell potential of a spontaneous reaction is always the *ratio of the product concentration(s) over the reactant concentration(s), that is, right side over left side* (as for a reduction half-reaction). Notice it was necessary to multiply the Fe^{3+}/Fe^{2+} half-reaction by 2 (as when subtracting the two half-reactions) in order to combine the two log terms (with $n = 2$), and the final equation is the same as we would have written from the cell reaction. Note also that $E^0_{Fe^{3+},Fe^{2+}} - E^0_{I_3^-,I^-}$ is the cell standard potential, E^0_{cell}.

The term on the right of the log sign is the **equilibrium constant expression** for the reaction:

$$2Fe^{3+} + 3I^- \rightleftharpoons 2Fe^{2+} + I_3^- \tag{12.21}$$

> The cell voltage represents the tendency of a reaction to occur when the reacting species are put together (just as it does in a battery; that is, it represents the potential for work). After the reaction has reached equilibrium, the cell voltage necessarily becomes zero and the reaction is complete (i.e., no more work can be derived from the cell). That is, the potentials of the two half-reactions are equal at equilibrium. This is what happens when a battery runs down.

Example 12.5

One beaker contains a solution of 0.0200 M KMnO$_4$, 0.00500 M MnSO$_4$, and 0.500 M H$_2$SO$_4$; and a second beaker contains 0.150 M FeSO$_4$ and 0.00150 M Fe$_2$(SO$_4$)$_3$. The two beakers are connected by a salt bridge, and platinum electrodes are placed in each. The electrodes are connected via a wire with a voltmeter in between. What would be the potential of each half-cell (a) before reaction and (b) after reaction? What would be the measured cell voltage (c) at the start of the reaction and (d) after the reaction reaches equilibrium? Assume H$_2$SO$_4$ to be completely ionized and in equal volumes in each beaker.

Solution

The cell reaction is

$$5Fe^{2+} + MnO_4^- + 8H^+ \rightleftharpoons 5Fe^+ + Mn^{2+} + 4H_2O$$

and the cell is

Pt/Fe^{2+} (0.150 M), Fe^{3+} (0.00300 M)//MnO$_4^-$(0.0200 M),

$$Mn^{2+}(0.00500\ M),\ H^+(1.00\ M)/Pt$$

(a)
$$E_{Fe} = E^0_{Fe^{3+},Fe^{2+}} - 0.059 \log \frac{[Fe^{2+}]}{[Fe^{3+}]}$$

$$= 0.771 - 0.059 \log \frac{0.150}{0.00300} = 0.671\ V$$

$$E_{Mn} = E^0_{MnO_4^-Mn^{2+}} - \frac{0.059}{5} \log \frac{[Mn^{2+}]}{[MnO_4^-][H^+]^8}$$

$$= 1.51 - \frac{0.059}{5} \log \frac{0.00500}{(0.0200)(1.00)^8} = 1.52\ V$$

(b) At equilibrium, $E_{Fe} = E_{Mn}$. We can calculate E from either half-reaction. First, calculate the equilibrium concentrations. Five moles of Fe^{2+} will react with each mole of MnO$_4^-$. The Fe^{2+} is in excess. It will be decreased by 5 × 0.0200 = 0.100 M, so 0.050 M Fe^{2+} remains and 0.100 M Fe^{3+} is formed (total now is 0.100 + 0.003 = 0.103 M). Virtually all the MnO$_4^-$ is converted to Mn^{2+} (0.0200 M) to give a total of 0.0250 M. A small unknown amount of MnO$_4^-$ remains at equilibrium, and we would need the equilibrium constant to calculate it; this can be obtained from $E_{cell} = 0$ at equilibrium—as in Equation 12.20—and will be treated in Chapter 14. But we need not go to this trouble since [Fe^{2+}] and [Fe^{3+}] are known:

$$E_{Mn} = E_{Fe} = 0.771 - 0.059 \log \frac{0.050}{0.103} = 0.790\ V$$

Note that the half-cell potentials at equilibrium are in between the values for the two half-cells before reaction.

(c) $E_{cell} = E_{Mn} - E_{Fe} = 1.52 - 0.671 = 0.85\ V$

(d) At equilibrium, $E_{Mn} = E_{Fe}$, and so E_{cell} is zero volts.

Note that if one of the species had not been initially present in a half-reaction, we could not have calculated an initial potential for that half-reaction.

12.4 Formal Potential—Use It for Defined Nonstandard Solution Conditions

The E^0 values listed in Table 12.1 are for the case where *all* species are at an activity of 1 M. However, the potential of a half-reaction may depend on the conditions of the solution. For example, the E^0 value for $Ce^{4+} + e^- \rightleftharpoons Ce^{3+}$ is 1.61 V. However, we can change this potential by changing the acid used to acidify the solution. (See Table C.5 in Appendix C.) This change in potential happens because the anion of the acid complexes with the cerium, and the concentration of free cerium ion is thereby reduced.

> The formal potential is used when not all species are known.

If we know the form of the complex, we could write a new half-reaction involving the acid anion and determine an E^0 value for this reaction, keeping the acid and all other species at unit activity. However, the complexes are frequently of unknown composition. So we define the **formal potential** and designate this as $E^{0\prime}$. This is the standard potential of a redox couple with the oxidized and reduced forms at 1 M concentrations and *with the solution conditions specified*. For example, the formal potential of the Ce^{4+}/Ce^{3+} couple in 1 M HCl is 1.28 V. The Nernst equation is written as usual, using the formal potential in place of the standard potential. Table C.5 lists some formal potentials.

DEPENDENCE OF POTENTIAL ON pH

Hydrogen or hydroxyl ions are involved in many redox half-reactions. We can change the potential of these redox couples by changing the pH of the solution. Consider the As(V)/As(III) couple:

> Many redox reactions involve protons, and their potentials are influenced greatly by pH.

$$H_3AsO_4 + 2H^+ + 2e^- \rightleftharpoons H_3AsO_3 + H_2O \qquad (12.22)$$

$$E = E^0 - \frac{0.059}{2} \log \frac{[H_3AsO_3]}{[H_3AsO_4][H^+]^2} \qquad (12.23)$$

This can be rearranged to[4]

$$E = E^0 + 0.059 \log[H^+] - \frac{0.059}{2} \log \frac{[H_3AsO_3]}{[H_3AsO_4]} \qquad (12.24)$$

or $$E = E^0 - 0.059 \text{ pH} - \frac{0.059}{2} \log \frac{[H_3AsO_3]}{[H_3AsO_4]} \qquad (12.25)$$

The term $E^0 - 0.059$ pH, where E^0 is the standard potential for the half-reaction, can be considered as equal to a formal potential $E^{0\prime}$, which can be calculated from the pH of the solution.[5] In 0.1 M HCl (pH 1), $E^{0\prime} = E^0 - 0.059$. In neutral condition, it is $E^0 - 0.059(7) = E^0 - 0.41$.

In strongly acid solution, H_3AsO_4 will oxidize I^- to I_2. But in neutral solution, the potential of the As(V)/As(III) couple ($E^{0\prime} = 0.146$ V) is less than that for I_2/I^-, and the reaction goes in the reverse; that is, I_2 will oxidize H_3AsO_3.

---

[4] The H^+ term in the log term can be separated as $(-0.059/2) \log (1/[H^+]^2) = (+0.059/2) \log [H^+]^2$. The squared term can be brought to the front of the log term to give $0.059 \log [H^+]$.

[5] Actually, this is an oversimplification of the effect of pH in this particular case because H_3AsO_4 and H_3AsO_3 are also weak acids, and the effect of their ionization, that is, their K_a values, should be taken into account as well.

DEPENDENCE OF POTENTIAL ON COMPLEXATION

Complexing one ion reduces its effective concentration, which changes the potential.

If an ion in a redox couple is complexed, the concentration of the free ion is reduced. This causes the potential of the couple to change. For example, E^0 for the Fe^{3+}/Fe^{3+} couple is 0.771 V. In HCl solution, the Fe^{3+} is complexed with the chloride ion, probably to a variety of species. This reduces the concentration of Fe^{3+}, and so the potential is decreased. In 1 M HCl, the formal potential is 0.70 V. If we assume that the complex is $FeCl_4^-$, then the half-reaction would be

$$FeCl_4^- + e^- \rightleftharpoons Fe^{2+} + 4Cl^- \qquad (12.26)$$

and if we assume that [HCl] is constant at 1 M,

$$E = 0.70 - 0.059 \log \frac{[Fe^{2+}]}{[FeCl_4^-]} \qquad (12.27)$$

In effect, we have stabilized the Fe^{3+} by complexing it, making it more difficult to reduce. So the reduction potential is decreased. If we complexed the Fe^{2+}, the reverse effect would be observed.

12.5 Limitations of Electrode Potentials

Electrode potentials predict whether a reaction *can* occur. They say nothing about the kinetics or rate of the reaction.

Electrode potentials (E^0 or $E^{0\prime}$) will predict whether a given reaction will occur, but they indicate nothing about the **rate** of the reaction. If a reaction is reversible, it will occur fast enough for a titration. But if the rate of the electron transfer step is slow, the reaction may be so slow that equilibrium will be reached only after a very long time. We say that such a reaction is **irreversible**.

Some reactions in which one half-reaction is irreversible do occur rapidly. Several oxidizing and reducing agents containing oxygen are reduced or oxidized irreversibly but may be speeded up by addition of an appropriate catalyst. The oxidation of arsenic(III) by cerium(IV) is slow, but it is catalyzed by a small amount of osmium tetroxide, OsO_4.

So, while electrode potentials are useful for predicting many reactions, they do not assure the success of a given reaction. They are useful in that they will predict that a reaction will *not* occur if the potential differences are not sufficient.

Learning Objectives

WHAT ARE SOME OF THE KEY THINGS WE LEARNED FROM THIS CHAPTER?

- Voltaic cells, p. 355
- Using standard potentials to predict reactions, p. 358
- Anodes, cathodes, and cell voltages (key equation: 12.16), p. 359
- The Nernst equation (key equation: 12.18), p. 361
- Calculating electrode potentials before and after reaction, p. 362
- The formal potentials, p. 365

1. What is an oxidizing agent? A reducing agent?

2. What is the Nernst equation?

3. What is the standard potential? The formal potential?

4. What is the function of a salt bridge in an electrochemical cell?

5. What is the NHE? SHE?

6. The standard potential for the half-reaction $M^{4+} + 2e^- = M^{2+}$ is +0.98 V. Is M^{2+} a good or a poor reducing agent?

7. What should be the minimum potential difference between two half-reactions so that a sharp end point will be obtained in a titration involving the two half-reactions?

8. Why cannot standard or formal electrode potentials always be used to predict whether a given titration will work?

REDOX STRENGTHS

9. Arrange the following substances in decreasing order of oxidizing strengths: H_2SeO_3, H_3AsO_4, Hg^{2+}, Cu^{2+}, Zn^{2+}, O_3, $HClO$, K^+, Co^{2+}.

10. Arrange the following substances in decreasing order of reducing strengths: I^-, V^{3+}, Sn^{2+}, Co^{2+}, Cl^-, Ag, H_2S, Ni, HF.

11. Which of the following pairs would be expected to give the largest end-point break in a titration of one component with the other in each pair?

   (a) $Fe^{2+} - MnO_4^-$ or $Fe^{2+} - Cr_2O_7^{2-}$

   (b) $Fe^{2+} - Ce^{4+}$ (H_2SO_4) or $Fe^{2+} - Ce^{4+}$ ($HClO_4$)

   (c) $H_3AsO_3 - MnO_4^-$ or $Fe^{2+} - MnO_4^-$

   (d) $Fe^{3+} - Ti^{2+}$ or $Sn^{2+} - I_3^-$

VOLTAIC CELLS

12. Write the equivalent voltaic cells for the following reactions (assume all concentrations are 1 M):

   (a) $6Fe^{2+} + Cr_2O_7^{2-} + 14H^+ \rightleftharpoons 6Fe^{3+} + 2Cr^{3+} + 7H_2O$

   (b) $IO_3^- + 5I^- + 6H^+ \rightleftharpoons 3I_2 + 3H_2O$

   (c) $Zn + Cu^{2+} \rightleftharpoons Zn^{2+} + Cu$

   (d) $Cl_2 + H_2SeO_3 + H_2O \rightleftharpoons 2Cl^- + SeO_4^{2-} + 4H^+$

13. For each of the following cells, write the cell reactions:

   (a) $Pt/V^{2+}, V^{3+}//PtCl_4^{2-}, PtCl_6^{2+}, Cl^-/Pt$

   (b) $Ag/AgCl(s)/Cl^-//Fe^{3+}, Fe^{2+}/Pt$

   (c) $Cd/Cd^{2+}//ClO_3^-, Cl^-, H^+/Pt$

   (d) $Pt/I^-, I_2//H_2O_2, H^+/Pt$

POTENTIAL CALCULATIONS

14. What is the electrode potential (vs. NHE) in a solution containing 0.50 M $KBrO_3$ and 0.20 M Br_2 at pH 2.5?

15. What is the electrode potential (vs. NHE) in the solution prepared by adding 90 mL of 5.0 M KI to 10 mL of 0.10 M H_2O_2 buffered at pH 2.0?

16. A solution of a mixture of Pt^{4+} and Pt^{2+} is 3.0 M in HCl, which produces the chloro complexes of the Pt ions (see Problem 18). If the solution is 0.015 M in Pt^{4+} and 0.025 M in Pt^{2+}, what is the potential of the half-reaction?

17. Equal volumes of 0.100 M UO_2^{2+} and 0.100 M V^{2+} in 0.10 M H_2SO_4 are mixed. What would the potential of a platinum electrode (vs. NHE) dipped in the solution be at equilibrium? Assume H_2SO_4 is completely ionized.

CELL VOLTAGES

18. From the standard potentials of the following half-reactions, determine the reaction that will occur, and calculate the cell voltage from the reaction:

$$PtCl_6^{2-} + 2e^- = PtCl_4^{2-} + 2Cl^-$$
$$E^0 = 0.68 \text{ V}$$
$$V^{3+} + e^- = V^{2+}$$
$$E^0 = -0.255 \text{ V}$$

19. Calculate the voltages of the following cells:

(a) Pt/I^- (0.100 M), I_3^- (0.0100 M)//IO_3^- (0.100 M), I_2 (0.0100 M), H^+ (0.100 M)/Pt

(b) Ag/AgCl(s)/Cl^- (0.100 M)//UO_2^{2+} (0.200 M), U^{4+} (0.050 M), H^+ (1.00 M)/Pt

(c) Pt/Tl^+ (0.100 M), Tl^{3+} (0.0100 M)//MnO_4^- (0.0100 M), Mn^{2+} (0.100 M), H^+ (pH 2.00)/Pt

20. From the standard potentials, determine the reaction between the following half-reactions, and calculate the corresponding standard cell voltage:

$$VO_2^+ + 2H^+ + e^- = VO^{2+} + H_2O$$
$$UO_2^{2+} + 4H^+ + 2e^- = U^{4+} + 2H_2O$$

Recommended References

NERNST EQUATION

1. L. Meites, "A 'Derivation' of the Nernst Equation for Elementary Quantitative Analysis," *J. Chem. Ed.*, **29** (1952) 142.

ELECTRODE SIGN CONVENTIONS

2. F. C. Anson, "Electrode Sign Convention," *J. Chem. Ed.*, **36** (1959) 394.

3. T. S. Light and A. J. de Bethune, "Recent Developments Concerning the Signs of Electrode Conventions," *J. Chem. Ed.*, **34** (1957) 433.

STANDARD POTENTIALS

4. A. J. Bard, R. Parsons, and J. Jordan, eds., *Standard Potentials in Aqueous Solution*. New York: Marcel Dekker, 1985.

5. W. M. Latimer, *The Oxidation States of the Elements and Their Potentials in Aqueous Solutions*, 2nd ed. New York: Prentice Hall, 1952.

Chapter Thirteen

POTENTIOMETRIC ELECTRODES AND POTENTIOMETRY

In Chapter 12, we mentioned measurement of the potential of a solution and described a platinum electrode whose potential was determined by the half-reaction of interest. This was a special case, and there are a number of electrodes available for measuring solution potentials. In this chapter, we list the various types of electrodes that can be used for measuring solution potentials and how to select the proper one for measuring a given analyte. The apparatus for making potentiometric measurements is described along with limitations and accuracies of potentiometric measurements. The important glass pH electrode is described, as well as standard buffers required for its calibration. The various kinds of ion-selective electrodes are discussed. The use of electrodes in potentiometric titrations is described in Chapter 14.

Potentiometric electrodes measure activity rather than concentration, a unique feature, and we will use activities in this chapter in describing electrode potentials. An understanding of activity and the factors that affect it are important for direct potentiometric measurements, as in pH or ion-selective electrode measurements. You should, therefore, review the material on activity and activity coefficients in Chapter 6.

Review activities in Chapter 6, for an understanding of potentiometric measurements.

13.1 Metal Electrodes for Measuring the Metal's Cation

An electrode of this type is a metal in contact with a solution containing its cation. An example is a silver metal electrode dipping in a solution of silver nitrate.

For all electrode systems, an electrode half-reaction can be written from which the potential of the electrode is described. The electrode system can be represented by M/M^{n+}, in which the line represents an electrode–solution interface. For the silver electrode, we have

$$Ag|Ag^+ \qquad (13.1)$$

and the half-reaction is

$$Ag^+ + e^- \rightleftharpoons Ag \qquad (13.2)$$

The potential of the electrode is described by the Nernst equation:

$$E = E^0_{Ag^+,Ag} - \frac{2.303RT}{F} \log \frac{1}{a_{Ag^+}} \qquad (13.3)$$

where a_{Ag^+} represents the **activity** of the silver ion (see Chapter 6). We will use the more correct unit of activity in discussions in this chapter because, in the interpretation of direct potentiometric measurements, significant errors would result if concentrations were used in calculations.

Increasing cation activity always causes the electrode potential to become more positive (if you write the Nernst equation properly).

The potential calculated from Equation 13.3 is the potential *relative to the normal hydrogen electrode* (NHE—see Section 13.3). The potential becomes increasingly positive with increasing Ag^+ (the case for any electrode measuring a cation). That is, in a cell measurement using the NHE as the second half-cell, the voltage is

$$E_{measd.} = E_{cell} = E_{ind\ vs.\ NHE} = E_{ind} - E_{NHE} \qquad (13.4)$$

The indicator electrode is the one that measures the analyte.

where E_{ind} is the potential of the **indicator electrode** (the one that responds to the test solution, Ag^+ ions in this case). Since E_{NHE} is zero,

$$E_{cell} = E_{ind} \qquad (13.5)$$

corresponds to writing the cells as

$$E_{ref}|solution|E_{ind} \qquad (13.6)$$

and

$$E_{cell} = E_{right} - E_{left} = E_{ind} - E_{ref} = E_{ind} - constant \qquad (13.7)$$

The reference electrode completes the cell but does not respond to the analyte. It is usually separated frm the test solution by a salt bridge.

where E_{ref} is the potential of the **reference electrode**, whose potential is constant. Note that E_{cell} (or E_{ind}) may be positive or negative, depending on the activity of the silver ion or the relative potentials of the two electrodes. This is in contrast to the convention used in Chapter 12 for a voltaic cell, in which a cell was always set up to give a positive voltage and thereby indicate what the spontaneous cell reaction would be. In potentiometric measurements, we, in principle, measure the potential at zero current so as not to disturb the equilibrium and, therefore, the relative concentrations of the species being measured at the indicating electrode surface—which establishes the potential (see measurement of potential, below). We are interested in how the potential of the test electrode (indicating electrode) changes with analyte concentration, as measured against some constant reference electrode. Equation 13.7 is arranged so that changes in E_{cell} reflect the same changes in E_{ind}, *including sign*. This point is discussed further when we talk about cells and measurement of electrode potentials.

Any pure substance does not appear in the Nernst equation (e.g., Cu, H_2O).

The activity of silver metal above, as with other pure substances, is taken as unity. So an electrode of this kind can be used to monitor the activity of a metal ion in solution. There are few reliable electrodes of this type because many metals tend to form an oxide coating that changes the potential.

13.2 Metal–Metal Salt Electrodes for Measuring the Salt's Anion

The general form of this type of electrode is $M|MX|X^{n-}$, where MX is a slightly soluble salt. An example is the silver–silver chloride electrode:

$$Ag|AgCl(s)|Cl^- \tag{13.8}$$

The (s) indicates a solid, (g) is used to indicate a gas, and (l) is used to indicate a pure liquid. A vertical line denotes a phase boundary between two different solids or a solid and a solution. The half-reaction is

$$\underline{AgCl} + e^- \rightleftharpoons \underline{Ag} + Cl^- \tag{13.9}$$

and the potential is defined by

$$E = E^0_{AgCl,Ag} - \frac{2.303RT}{F} \log a_{Cl^-} \tag{13.10}$$

This electrode, then, can be used to measure the activity of chloride ion in solution. Note that, as the activity of chloride increases, the potential *decreases*. This is true of any electrode measuring an anion—the opposite for a cation electrode. A silver wire is coated with silver chloride precipitate (e.g., by electrically oxidizing it in a solution containing chloride ion, the reverse reaction of Equation 13.9). Actually, as soon as a silver wire is dipped in a chloride solution, it adopts a thin layer of silver chloride and pretreatment is usually not required.

Increasing anion activity always causes the electrode potential to decrease.

Note that this electrode can be used to monitor either a_{Cl^-} or a_{Ag^+}. It really measures ("sees") only silver ion, and the activity of this is determined by the solubility of a slightly soluble salt. Since $a_{Cl^-} = K_{sp}/a_{Ag^-}$, Equation 13.10 can be rewritten:

The Ag metal really responds to Ag^+, whose activity is determined by K°_{sp} and a_{Cl^-}.

$$E = E^0_{AgCl,Ag} - \frac{2.303RT}{F} \log \frac{K_{sp}}{a_{Ag^+}} \tag{13.11}$$

$$E = E^0_{AgCl,Ag} - \frac{2.303RT}{F} \log K_{sp} - \frac{2.303RT}{F} \log \frac{1}{a_{Ag^+}} \tag{13.12}$$

Comparing this with Equation 13.3, we see that

$$E^0_{Ag^+,Ag} = E^0_{AgCl,Ag} - \frac{2.303RT}{F} \log K_{sp} \tag{13.13}$$

K_{sp} here really represents the thermodynamic solubility product K°_{sp} (see Chapter 6), since activities were used in arriving at it in these equations. We could have arrived at an alternative form of Equation 13.10 by substituting K_{sp}/a_{Cl^-} for a_{Ag^+} in Equation 13.3 (see Example 13.1).

In a solution containing a mixture of Ag^+ and Cl^- (e.g., a titration of Cl^- with Ag^+), the concentrations of each *at equilibrium* will be such that the potential of a silver wire dipping in the solution can be calculated by either Equation 13.3 or Equation 13.10. This is completely analogous to the statement in Chapter 12 that the potential of one half-reaction must be equal to the potential of the other in a chemical reaction at equilibrium. Equations 13.2 and 13.9 are the two half-reactions in this case, and when one is subtracted from the other, the result is the *overall chemical reaction*.

$$Ag^+ + Cl^- \rightleftharpoons AgCl \tag{13.14}$$

Note that as Cl^- is titrated with Ag^+, the former decreases and the latter increases. Equation 13.10 predicts an increase in potential as Cl^- decreases; and similarly, Equation 13.12 predicts the same increase as Ag^+ increases.

The silver electrode can also be used to monitor other anions that form slightly soluble salts with silver, such as I^-, Br^-, and S^{2-}. The E^0 in each case would be that for the particular half-reaction $AgX + e^- \rightleftharpoons Ag + X^-$.

Another widely used electrode of this type is the **calomel electrode**, Hg, $Hg_2Cl_2(s)|Cl^-$. This will be described in more detail when we talk about reference electrodes.

Example 13.1

Given that the standard potential of the calomel electrode is 0.268 V and that of the Hg/Hg_2^{2+} electrode is 0.789 V, calculate K_{sp} for calomel (Hg_2Cl_2).

Solution

For $Hg_2^{2+} + 2e^- \rightleftharpoons Hg$,

$$E = 0.789 - \frac{0.05916}{2} \log \frac{1}{a_{Hg_2^{2+}}} \tag{1}$$

For $Hg_2Cl_2 + 2e^- \rightleftharpoons 2Hg + 2Cl^-$,

$$E = 0.268 - \frac{0.05916}{2} \log(a_{Cl^-})^2 \tag{2}$$

Since $K_{sp} = a_{Hg_2^{2+}} \cdot (a_{Cl^-})^2$,

$$E = 0.268 - \frac{0.05916}{2} \log \frac{K_{sp}}{a_{Hg_2^{2+}}} \tag{3}$$

$$E = 0.268 - \frac{0.05916}{2} \log K_{sp} - \frac{0.05916}{2} \log \frac{1}{a_{Hg_2^{2+}}} \tag{4}$$

From (1) and (4),

$$0.789 = 0.268 - \frac{0.05916}{2} \log K_{sp}$$

$$K_{sp} = 2.4_4 \times 10^{-18}$$

13.3 Redox Electrodes—Inert Metals

In the redox electrode, an inert metal is in contact with a solution containing the soluble oxidized and reduced forms of the redox half-reaction. This type of electrode was mentioned in Chapter 12.

The inert metal used is usually platinum. The potential of such an inert electrode is determined by the ratio at the electrode surface of the reduced and oxidized species in the half-reaction:

$$M^{a+} + ne^- \rightleftharpoons M^{(a-n)+} \tag{13.15}$$

$$E = E^0_{M^{a+}, M^{(a-n)+}} - \frac{2.303RT}{nF} \log \frac{a_{M^{(a-n)+}}}{a_{M^{a+}}} \tag{13.16}$$

An example is the measurement of the ratio of MnO_4^-/Mn^{2+}:

$$MnO_4^- + 8H^+ + 5e^- \rightleftharpoons Mn^{2+} + 4H_2O \tag{13.17}$$

$$E = E^0_{MnO_4^-, Mn^{2+}} - \frac{2.303RT}{5F} \log \frac{a_{Mn^{2+}}}{a_{MnO_4^-} \cdot (a_{H^+})^8} \tag{13.18}$$

The pH is usually held constant, and so the ratio $a_{Mn^{2+}}/a_{MnO_4^-}$ is measured, as in a redox titration.

A very important example of this type of electrode is the **hydrogen electrode**, $Pt|H_2, H^+$:

$$H^+ + e^- \rightleftharpoons \tfrac{1}{2}H_2 \tag{13.19}$$

$$E = E^0_{H^+, H_2} - \frac{2.303RT}{F} \log \frac{(p_{H_2})^{1/2}}{a_{H^+}} \tag{13.20}$$

The construction of the hydrogen electrode is shown in Figure 13.1. A layer of platinum black must be placed on the surface of the platinum electrode by

For gases, we use pressures, p (in atmospheres), in place of activities.

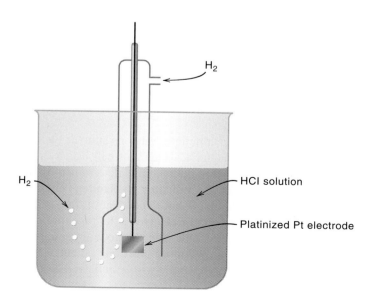

Fig. 13.1. Hydrogen electrode.

cathodically electrolyzing in a H_2PtCl_6 solution. The platinum black provides a larger surface area for adsorption of hydrogen molecules and catalyzes their oxidation. Too much platinum black, however, can adsorb traces of other substances such as organic molecules or H_2S, causing erratic behavior of the electrode.

The pressure of gases, in atmospheres, is used in place of activities. If the hydrogen pressure is held at 1 atm, then, since E^0 for Equation 13.19 is defined as zero,

$$E = \frac{2.303RT}{F} \log \frac{1}{a_{H^+}} = -\frac{2.303RT}{F} \text{pH}$$ (13.21)

 Example 13.2

The vapor pressure of water above the solution must be subtracted from the measured gas pressure.

Calculate the pH of a solution whose potential at 25°C measured with a hydrogen electrode at an atmospheric pressure of 1.012 atm (corrected for the vapor pressure of water at 25°C) is −0.324 V (relative to the NHE).

Solution
From Equation 13.20,

$$-0.324 = -0.05916 \log \frac{(1.012)^{1/2}}{a_{H^+}}$$

$$= -0.05916 \log(1.012)^{1/2} - 0.05916 \text{ pH}$$

$$\text{pH} = 5.48$$

While the hydrogen electrode is very important for specific applications (e.g., establishing standard potentials or the pH of standard buffers—see below), its use for routine pH measurements is limited. First, it is inconvenient to prepare and use. The partial pressure of hydrogen must be established at the measurement temperature. The solution should not contain other oxidizing or reducing agents since these will alter the potential of the electrode.

13.4 Voltaic Cells without Liquid Junction—For Maximum Accuracy

To make potential measurements, a complete cell consisting of two half-cells must be set up, as was described in Chapter 12. One half-cell usually is comprised of the test solution and an electrode whose potential is determined by the analyte we wish to measure. This electrode is the **indicator electrode**. The other half-cell is any arbitrary half-cell whose potential is not dependent on the analyte. This half-cell electrode is designated the **reference electrode**. Its potential is constant, and the measured cell voltage reflects the indicator electrode potential relative to that of the reference electrode. Since the reference electrode potential is constant, any *changes* in potential of the indicator electrode will be reflected by an equal change in the cell voltage.

There are two basic ways a cell may be set up, either without or with a salt bridge. The first is called a *cell without liquid junction*. An example of a cell of this type would be

$$Pt \mid H_2(g), HCl(solution) \mid AgCl(s) \mid Ag \qquad (13.22)$$

The solid line represents an electrode—solution interface. An electrical cell such as this is a voltaic cell, and the cell is written for the *spontaneous cell reaction* by convention (positive E_{cell}—although we may actually measure a negative cell voltage if the indicator electrode potential is the more negative one; we haven't specified which of the half-reactions represents the indicator electrode). The hydrogen electrode is the anode, since its potential is the more negative (see Chapter 12 for a review of cell voltage conventions for voltaic cells). The potential of the left electrode would be given by Equation 13.20, and that for the right electrode would be given by Equation 13.10, and the cell voltage would be equal to the difference in these two potentials:

$$E_{cell} = \left(E_{AgCl,Ag}^0 - \frac{2.303RT}{F} \log a_{Cl^-} \right) - \left(E_{H^+,H_2}^0 - \frac{2.303RT}{F} \log \frac{(p_{H_2})^{\frac{1}{2}}}{a_{H^+}} \right) \qquad (13.23)$$

$$E_{cell} = E_{AgCl,Ag}^0 - E_{H^+,H_2}^0 - \frac{2.303RT}{F} \log \frac{a_{H^+} a_{Cl^-}}{(p_{H_2})^{\frac{1}{2}}} \qquad (13.24)$$

The **cell reaction** would be (half-reaction)$_{right}$ − (half-reaction)$_{left}$ (to give a positive E_{cell} and the spontaneous reaction), or

$$\underline{AgCl} + e^- \rightleftharpoons \underline{Ag} + Cl^- \qquad (13.25)$$

$$\frac{- (H^+ + e^- \rightleftharpoons \frac{1}{2}H_2)}{\underline{AgCl} + \frac{1}{2}H_2 \rightleftharpoons \underline{Ag} + Cl^- + H^+} \qquad (13.26)$$

$$\underline{AgCl} + \frac{1}{2}H_2 \rightleftharpoons \underline{Ag} + Cl^- + H^+ \qquad (13.27)$$

Equation 13.23 would also represent the cell voltage if the right half-cell were used as an indicating electrode in a potentiometric measurement of chloride ion and the left cell were the reference electrode (see Equations 13.6 and 13.7). That is, the voltage (and hence the indicator electrode potential) would decrease with increasing chloride ion. If we were to use the hydrogen electrode as the indicating electrode to measure hydrogen ion activity or pH, we would reverse the cell setup to represent the *measurement*, Cell 13.22, and Equation 13.23, and the voltage (and indicator electrode potential) would increase with increasing acidity or decreasing pH ($E_{cell} = E_{ind} - E_{ref}$, Equation 13.7).

Cells without liquid junction are always used for the most accurate measurements because there are no uncertain potentials to account for and were used for measuring the pH of NIST standard buffers (see below). However, there are few examples of cells without liquid junction (sometimes called **cells without transference**), and these are inconvenient to use. Therefore, the more convenient but less accurate cells with liquid junction are most commonly used.

It is possible to construct a cell without a salt bridge. For practical purposes, this is rare because of the tendency of the reference electrode potential to be influenced by the test solution.

This cell is used to accurately measure the pH of "standard buffers." See Section 13.12.

13.5 Voltaic Cells with Liquid Junction—The Practical Kind

An example of this type of cell is

$$Hg \mid Hg_2Cl_2(s) \mid KCl(saturated) \parallel HCl(solution), H_2(g) \mid Pt \qquad (13.28)$$

The double line represents the **liquid junction** between two dissimilar solutions and is usually in the form of a **salt bridge**. The purpose of this is to prevent mixing of the two solutions. In this way, the potential of one of the electrodes will be constant, independent of the composition of the test solution, and determined by the solution in which it dips. The electrode on the left of cell 13.28 is the **saturated calomel electrode**, which is a commonly used reference electrode (see below). The cell is set up using the hydrogen electrode as the indicating electrode to measure pH.

LIQUID-JUNCTION POTENTIAL—WE CAN'T IGNORE THIS

The presence of a liquid-junction potential limits the accuracy of potentiometric measurements.

The disadvantage of a cell of this type is that there is a potential associated with the liquid junction, called the **liquid-junction potential**. The potential of the above cells is

$$E_{cell} = (E_{right} - E_{left}) + E_j \qquad (13.29)$$

where E_j is the liquid-junction potential; E_j may be positive or negative. The liquid-junction potential results from the unequal diffusion of the ions on each side of the boundary. A careful choice of salt bridge (or reference electrode containing a suitable electrolyte) can minimize the liquid-junction potential and make it reasonably constant so that a calibration will account for it. The basis for such a selection is discussed as follows.

A typical boundary might be a fine-porosity sintered-glass frit with two different solutions on either side of it; the frit prevents appreciable mixing of the two solutions. The simplest type of liquid junction occurs between two solutions containing the same electrolyte at different concentrations. An example is HCl $(0.1\ M)\|HCl\ (0.01\ M)$, illustrated in Figure 13.2. Both hydrogen ions and chloride ions will migrate across the boundary in both directions, but the net migration will be from the more concentrated to the less concentrated side of the boundary, the driving force for this migration being proportional to the concentration difference. Hydrogen ions migrate about five times faster than chloride ions. Therefore, a net positive charge is built up on the right side of the boundary, leaving a net negative charge on the left side; that is, there is a separation of charge, and this represents a potential. A steady state is rapidly achieved by the action of this built-up positive charge in repulsing the further migration of hydrogen ions; the converse applies to the negative charge on the left-hand side. Hence, a constant potential difference is quickly attained between the two solutions.

The E_j for this junction is +40 mV, and $E_{cell} = (E_{right} - E_{left}) + 40$ mV. This E_j is very large, owing to the rapid mobility of the hydrogen ion. As the concentration of HCl on the left side of the boundary is decreased, the net charge built up

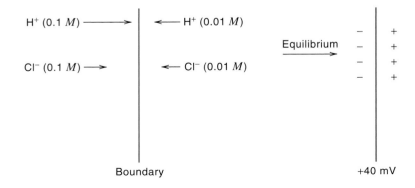

Fig. 13.2. Representation of liquid-junction potential.

will be less, and the liquid-junction potential will be decreased until, at equal concentration, it will be zero, because equal amounts of HCl diffuse in each direction.

A second example of this type of liquid junction is 0.1 M KCl/0.01 M KCl. This situation is completely analogous to that above, except that in this case the K^+ and Cl^- ions migrate at nearly the same rate, with the chloride ion moving only about 4% faster. So a net *negative* charge is built up on the right side of the junction, but it will be relatively small. Thus, E_j will be negative and is equal to -1.0 mV.

We minimize the liquid-junction potential by using a high concentration of a salt whose ions have nearly equal mobility, for example, KCl.

HOW DO WE MINIMIZE THE LIQUID-JUNCTION POTENTIAL?

The nearly equal migration of potassium and chloride ions makes it possible to decrease significantly the liquid-junction potential. Decreasing is possible because, if an electrolyte on one side of a boundary is in large excess over that on the other side, the flux of the migration of the ions of this electrolyte will be much greater than that of the more dilute electrolyte, and the liquid-junction potential will be determined largely by the migration of this more concentrated electrolyte. Thus, E_j of the junction KCl (3.5 M)‖H_2SO_4 (0.05 M) is only -4 mV, even though the hydrogen ions diffuse at a much more rapid rate than sulfate.

Some examples of different liquid-junction potentials are given in Table 13.1. (The signs are for those as set up, and they would be the signs in a potentiometric measurement if the solution on the left were used for the salt bridge and the one on the right were the test solution. If solutions on each side of the junction were reversed, the signs of the junction potentials would be reversed.) It is apparent that the liquid junction potential can be minimized by keeping on one side of the boundary a high concentration of a salt whose ions have nearly the same mobility, such as KCl. Ideally, the same high concentration of such a salt should be on both sides of the junction. This is generally not possible for the test solution side of a salt bridge. However, the solution in the other half-cell in which the other end of the salt bridge forms a junction can often be made high in KCl to minimize that junction potential. As noted before, this half-cell, which is connected via the salt bridge to form a complete cell, is the reference electrode. See the discussion of the saturated calomel electrode below.

As the concentration of the (dissimilar) electrolyte on the other side of the boundary (in the test solution) increases, or as the ions are made different, the liquid-junction potential will get larger. Very rarely can the liquid-junction potential be considered to be negligible. The liquid-junction potential with neutral salts

Table 13.1
Some Liquid-Junction Potentials at 25°C[a]

Boundary	E_j(mV)
0.1 M KCl‖0.1 M NaCl	+6.4
3.5 M KCl‖0.1 M NaCl	+0.2
3.5 M KCl‖1 M NaCl	+1.9
0.01 M KCl‖0.01 M HCl	−26
0.1 M KCl‖0.1 M HCl	−27
3.5 M KCl‖0.1 M HCl	+3.1
0.1 M KCl‖0.1 M NaOH	+18.9
3.5 M KCl‖0.1 M NaOH	+2.1
3.5 M KCl‖1 M NaOH	+10.5

[a]Adapted from G. Milazzo, *Electrochemie*. Vienna: Springer, 1952; and D. A. MacInnes and Y. L. Yeh, *J. Am. Chem. Soc.*, **43** (1921) 2563.

Liquid-junction potentials are highly pH dependent because of the high mobilities of the proton and hydroxide ions.

is less than when a strong acid or base is involved. The variation is due to the unusually high mobilities of the hydrogen ion and the hydroxyl ion. Therefore, *the liquid-junction potential will vary with the pH of the solution*, an important fact to remember in potentiometric pH measurements. A potassium chloride salt bridge, at or near saturation, is usually employed, except when these ions may interfere in a determination. Ammonium chloride or potassium nitrate may be used if the potassium or chloride ion interferes.

A commonly used salt bridge is a 3% agar-saturated potassium chloride salt bridge, which is prepared by adding 100 mL cold water to 3 g granulated agar. The mixture is heated on a steam bath and shaken until a homogeneous solution is obtained. Then, 25 g solid potassium chloride is added and the solution is stirred until the salt dissolves. The mixture will gel on cooling. Various other types of electrolyte junctions or salt bridges have been designed, such as a ground-glass joint, a wick of asbestos sealed into glass, a porous glass or ceramic plug, or a fine capillary drip. The reference electrode solution then contains saturated KCl solution, which slowly leaks through the bridge to create the liquid junction with the test solution.

13.6 Reference Electrodes: The Saturated Calomel Electrode

A requirement of a reference electrode is that its potential be fixed and stable, unaffected by the passage of small amounts of current required in making potentiometric measurements (ideally, the current in the measurement is zero, but in practice some small current must be passed—see below). Metal–metal salt electrodes generally possess the needed properties.

Reference electrodes are usually metal–metal salt types. The two most common are the Hg/Hg$_2$Cl$_2$ (calomel) and the Ag/AgCl electrodes.

A commonly used reference electrode is the **saturated calomel electrode** (SCE). The term "saturated" refers to the concentration of potassium chloride; and at 25°C, the potential of the SCE is 0.242 V versus NHE. An SCE consists of a small amount of mercury mixed with some solid Hg$_2$Cl$_2$ (calomel), solid KCl, and enough saturated KCl solution to moisten the mixture. This is contacted with a saturated KCl solution containing some solid KCl to maintain saturation. A platinum electrode is immersed in the paste to make contact with the small mercury pool formed, and the lead from the wire goes to one terminal of the potential measuring device. A salt bridge serves as the contact between the KCl solution and the test solution and is usually a fiber or frit wetted with the saturated KCl solution. If a different salt bridge is needed to prevent contamination of the test solution (you can't use the SCE for chloride measurements!), then a double-junction reference electrode is used in which the KCl junction contacts a different salt solution that in turn contacts the test solution. This, of course, creates a second liquid-junction potential, but it is constant.

A commercial probe-type SCE is shown in Figure 13.3. This contains a porous fiber or frit as the salt bridge in the tip that allows very slow leakage of the saturated potassium chloride solution. It has a small mercury pool area and so the current it can pass without its potential being affected is limited (as will be seen below, a small current is usually drawn during potential measurements). The fiber salt bridge *has a high resistance* (about 2500 Ω). This limits the sensitivity in measurements with a potentiometer, in which the reading is made as close as possible at zero current flow; with increased external resistance, a given current deflection from zero will result in an increased error in the potential reading. This is no serious problem in many potential measurements (e.g., in titrations), but a lower-resistance electrode is still preferred. The fiber SCE is perfectly satisfactory for

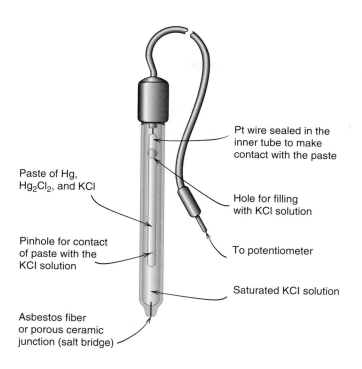

Pt wire sealed in the inner tube to make contact with the paste

Paste of Hg, Hg_2Cl_2, and KCl

Hole for filling with KCl solution

To potentiometer

Pinhole for contact of paste with the KCl solution

Saturated KCl solution

Asbestos fiber or porous ceramic junction (salt bridge)

Fig. 13.3. Commercial saturated calomel electrode. (Courtesy of Arthur H. Thomas Company.)

use with a pH meter, though, which is designed to make measurements with high-resistance electrodes.

Example 13.3

Calculate the potential of the cell consisting of a silver electrode dipping in a silver nitrate solution with $a_{Ag^+} = 0.0100\ M$ and an SCE reference electrode.

Solution

Neglecting the liquid-junction potential,

$$E_{cell} = E_{ind} - E_{ref}$$

$$E_{cell} = \left(E^0_{Ag^+,Ag} - 0.0592 \log \frac{1}{a_{Ag^+}} \right) - E_{SCE}$$

$$= 0.799 - 0.0592 \log \frac{1}{0.0100} - 0.242$$

$$= 0.439\ V$$

Example 13.4

A cell voltage measured using an SCE reference electrode is -0.774 V. (The indicating electrode is the more negative half-cell.) What would the cell voltage be with a silver/silver chloride reference electrode (1 M KCl; $E = 0.228$ V) or with an NHE?

Solution

The potential of the Ag/AgCl electrode is more negative than that of the SCE by $0.242 - 0.228 = 0.014$ V. Hence, the cell voltage using the former electrode is less negative by this amount:

$$E_{vs.\ Ag/AgCl} = E_{vs.\ SCE} + 0.014$$
$$= -0.774 + 0.014 = -0.760\ V$$

Similarly, the cell voltage using the NHE is 0.242 V less negative:

$$E_{vs.\ NHE} = E_{vs.\ SCE} + 0.242\ V$$
$$= -0.774 + 0.242 = -0.532\ V$$

• •

Reference electrode potentials are all relative. The measured cell potential depends on which one is used.

Potentials relative to different reference electrodes may be represented schematically on a scale on which the different electrode potentials are placed (see Ref. 2). Figure 13.4 illustrates this for Example 13.4.

13.7 Measurement of Potential

We create a voltaic cell with the indicator and reference electrodes. We measure the voltage of the cell, giving a reading of the indicator electrode potential relative to the reference electrode. We can relate this to the analyte activity or concentration using the Nernst equation.

THE POTENTIOMETER AND pH METER

There are two commonly used instruments for making potential measurements. One is the **potentiometer** and the other is the **pH meter** (a voltmeter), with the latter almost always used today. pH measurements with a glass (or other) electrode involve the measurement of potentials (see below).

The potentiometer can be used for measurements of low-resistance circuits. The pH meter is a voltage measuring device designed for use with high-resistance glass electrodes and can be used with both low- and high-resistance circuits. **Electrometers** can also be used with high-resistance circuits.

A pH meter is a voltmeter that converts the unknown voltage to a current that is amplified and read out. These are "high-input impedance" devices. (Impedance in an ac circuit is comparable to resistance in a dc circuit. These devices convert the signal to an ac signal for amplification.) Because of their high-input resistance, very little current is drawn, typically 10^{-13} to 10^{-15} A, and so chemical equilibrium is not greatly disturbed. A voltmeter must be used for irreversible reactions that do not return to the prior state when disturbed by appreciable drawing of current.

A pH meter or electrometer draws very small currents and is best suited for irreversible reactions that are slow to reestablish equilibrium. They are also required for high-resistance electrodes, like glass pH or ion-selective electrodes.

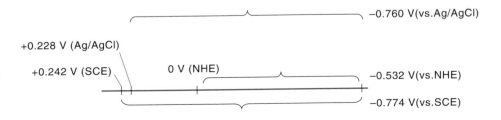

Fig. 13.4. Schematic representation of electrode potential relative to different reference electrodes.

High-input impedance circuits must be used with high-resistance electrodes (e.g., several megohms—10^6 Ω). Also, the current drawn must be very small in order for the voltage drop across the cell ($=iR$ or current $\times$ cell resistance) to be low enough not to cause error in the measurement; the cell resistance is high since it includes the glass electrode.

Expanded-scale pH meters are available that will measure the potential to a few tenths of a millivolt, about 10 times more closely than conventional pH meters. They are well suited for direct potentiometric measurements with ion-selective electrodes.

THE CELL FOR POTENTIAL MEASUREMENTS

In potentiometric measurements, a cell of the type shown in Figure 13.5 is set up. For direct potentiometric measurements in which the activity of one ion is to be calculated from the potential of the indicating electrode, the potential of the reference electrode will have to be known or determined. The voltage of the cell is described by Equation 13.7, and when a salt bridge is employed, the liquid-junction potential must be included. Then,

$$E_{\text{cell}} = (E_{\text{ind}} - E_{\text{ref}}) + E_j \tag{13.30}$$

The E_j can be combined with the other constants in Equation 13.30 into a single constant, assuming that the liquid-junction potential does not differ significantly from one solution to the next. We are forced to accept this assumption since E_j cannot be evaluated under most circumstances. E_{ref}, E_j, and E_{ind}^0 are lumped together into a constant k:

$$k = E_{\text{ind}}^0 - E_{\text{ref}} + E_j \tag{13.31}$$

Then (for a 1:1 reaction),

$$\boxed{E_{\text{cell}} = k - \frac{2.303RT}{nF} \log \frac{a_{\text{red}}}{a_{\text{ox}}}} \tag{13.32}$$

The constant k is determined by measuring the potential of a standard solution in which the activities are known.

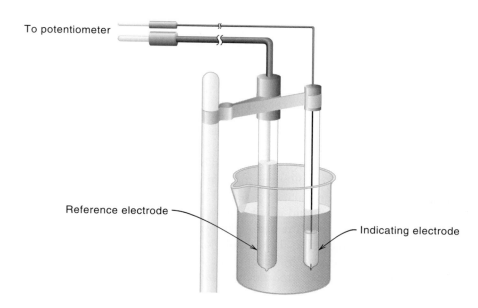

To potentiometer

Reference electrode

Indicating electrode

Fig. 13.5. Cell for potentiometric measurements.

13.8 Determination of Concentrations from Potential Measurements

Usually, we are interested in determining the concentration of a test substance rather than its activity. Activity coefficients are not generally available, and it is inconvenient to calculate activities of solutions used to standardize the electrode.

If the ionic strength of all solutions is held constant at the same value, the activity coefficient of the test substance remains nearly constant for all concentrations of the substance. We can then write for the log term in the Nernst equation:

$$-\frac{2.303RT}{nF} \log f_i C_i = -\frac{2.303RT}{nF} \log f_i - \frac{2.303RT}{nF} \log C_i \qquad (13.33)$$

> If the ionic strength is maintained constant, activity coefficients are constant and can be included in k. So *concentrations* can be determined from measured cell potentials.

Under the prescribed conditions, the first term on the right-hand side of this equation is constant and can be included in k, so that at constant ionic strength,

$$\boxed{E_{\text{cell}} = k - \frac{2.303RT}{nF} \log \frac{C_{\text{red}}}{C_{\text{ox}}}} \qquad (13.34)$$

In other words, the electrode potential changes by $\pm 2.303RT/nF$ volts for each 10-fold change in *concentration* of the oxidized or reduced form.

It is best to determine a **calibration curve** of potential versus log concentration; this should have a slope of $\pm 2.303RT/nF$. In this way, any deviation from this theoretical response will be accounted for in the calibration curve.

Since the ionic strength of an unknown solution is usually not known, a high concentration of an electrolyte is added both to the standards and to the samples to maintain about the same ionic strength. The standard solutions should contain any species in the test solution that will change the activity of the analyte, such as complexing agents.

13.9 Residual Liquid-Junction Potential—It Should Be Minimum

> If the liquid-junction potentials of the calibrating and test solutions are identical, no error results (the residual $E_j = 0$). Our goal is to keep residual E_j as small as possible.

We have assumed above in Equations 13.32 and 13.34 that k is the same in measurements of both standards and samples. This is so only if the liquid-junction potential at the reference electrode is the same in both solutions. But the test solution will usually have a somewhat different composition from the standard solution, and the magnitude of the liquid-junction potential will vary from solution to solution. The difference in the two liquid-junction potentials is called the **residual liquid-junction potential**, and it will remain unknown. The difference can be kept to a minimum by keeping the pH of the test solution and the pH of the standard solution as close as possible, and by keeping the ionic strength of both solutions as close as possible. *The former is particularly important.*

13.10 Accuracy of Direct Potentiometric Measurements—Voltage Error versus Activity Error

We can get an idea of the accuracy required in potentiometric measurements from the percent error caused by a 1-mV error in the reading at 25°C. For an electrode responsive to a monovalent ion such as silver,

$$E_{cell} = k - 0.05916 \log \frac{1}{a_{Ag^+}} \tag{13.35}$$

and

$$a_{Ag^+} = \text{antilog} \frac{E_{cell} - k}{0.05916} \tag{13.36}$$

A ± 1-mV error results in an error in a_{Ag^+} of $\pm 4\%$. This is quite significant in direct potentiometric measurements. The same percent error in activity will result for all activities of silver ion with a 1-mV error in the measurement. *The error is doubled when* n *is doubled* to 2. So, a 1-mV error for a copper/copper(II) electrode would result in an 8% error in the activity of copper(II). It is obvious, then, that the residual liquid junction potential can have an appreciable effect on the accuracy.

The accuracy and precision of potentiometric measurements are also limited by the **poising capacity** of the redox couple being measured. This is analogous to the buffering capacity in pH measurements. If the solution is very dilute, the solution is poorly poised and potential readings will be sluggish. That is, the solution has such a low ion concentration that it takes longer for the solution around the electrode to rearrange its ions and reach a steady state, when the equilibrium is disturbed during the measurement process. This is why an electrometer or pH meter that draws very small current is preferred for potentiometric measurements in such solutions. To help correct this problem and to maintain a constant ionic strength, a relatively high concentration of an inert salt (ionic strength "buffer") can be added. Stirring helps speed up the equilibrium response.

In very dilute solutions, the potential of the electrode may be governed by other electrode reactions. In a very dilute silver solution, for example, $-\log(1/a_{Ag^+})$ becomes very negative and the potential of the electrode is very reducing. Under these conditions, an oxidizing agent in solution (such as oxygen) may be reduced *at the electrode surface*, setting up a second redox couple (O_2/OH^-); the potential will be a **mixed potential**.

Usually, the lower limit of concentration that can be measured with a degree of certainty is 10^{-5} to 10^{-6} M, although the actual range should be determined experimentally. As the solution becomes more dilute, a longer time should be taken to establish the equilibrium potential reading because of the sluggishness. An exception to this limit is in pH measurements in which the hydrogen ion concentration of the solution is well poised, either by a buffer or by excess acid or base. At pH 10, the hydrogen ion concentration is 10^{-10} M, and this can be measured with a glass pH electrode (see Section 13.11). A neutral, unbuffered solution is poorly poised, however, and pH readings are sluggish.

For dilute or poorly poised solution, stirring the solution helps achieve an equilibrium reading.

13.11 Glass pH Electrode—Workhorse of Chemists

The glass electrode, because of its convenience, is used almost universally for pH measurements today. Its potential is essentially not affected by the presence of oxidizing or reducing agents, and it is operative over a wide pH range. It is fast responding and functions well in physiological systems. No other pH-measuring electrode possesses all these properties.

PRINCIPLE OF THE GLASS ELECTRODE

A typical construction of a pH glass electrode is shown in Figure 13.6. For measurement, only the bulb need be submerged. There is an internal reference electrode and electrolyte ($Ag|AgCl|Cl^-$) for making electrical contact with the glass membrane; its potential is necessarily constant and is set by the concentration of HCl. A complete cell, then, can be represented by

$$
\begin{array}{c|c|c|c|c}
\text{reference} & & & & \text{reference} \\
\text{electrode} \parallel & H^+ \text{ (unknown)} & \text{glass} & H^+ \text{ (internal)} & \text{electrode} \\
\text{(external)} & & \text{membrane} & & \text{(internal)}
\end{array}
$$

The double line represents the salt bridge of the reference electrode. The glass electrode is attached to the indicating electrode terminal of the pH meter while the external reference electrode (e.g., SCE) is attached to the reference terminal.

The potential of the glass membrane is given by

$$
E_{\text{glass}} = \text{constant} - \frac{2.303RT}{F} \log \frac{a_{H^+ \text{ int}}}{a_{H^+ \text{ unk}}} \tag{13.37}
$$

and the voltage of the cell is given by

$$
E_{\text{cell}} = k + \frac{2.303RT}{F} \log a_{H^+ \text{ unk}} \tag{13.38}
$$

where k is a constant that includes the potentials of the two reference electrodes, the liquid-junction potential, a potential at the glass membrane due to H^+ (internal), and a term known as the **asymmetry potential**.

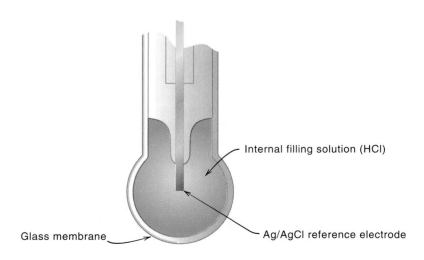

Internal filling solution (HCl)

Glass membrane   Ag/AgCl reference electrode

Fig. 13.6. Glass pH electrode.

The asymmetry potential is a small potential across the membrane that is present even when the solutions on both sides of the membrane are identical. It is associated with factors such as nonuniform composition of the membrane, strains within the membrane, mechanical and chemical attack of the external surface, and the degree of hydration of the membrane. It slowly changes with time, especially if the membrane is allowed to dry out, and is unknown. For this reason, *a glass pH electrode should be calibrated from day to day*. The asymmetry potential will vary from one electrode to another, owing to differences in construction of the membrane.

Since pH $= -\log a_{H^+}$, Equation 13.38 can be rewritten[i]

> The glass pH electrode must be calibrated using "standard buffers." See Section 13.12.

$$E_{cell} = k - \frac{2.303RT}{F} \, pH_{unk} \qquad (13.39)$$

or

$$pH_{unk} = \frac{k - E_{cell}}{2.303RT/F} \qquad (13.40)$$

It is apparent that the glass electrode will undergo a $2.303RT/F$-volt response for each change of 1 pH unit (10-fold change in a_{H^+}); k must be determined by calibration with a **standard buffer** (see below) of known pH:

$$k = E_{cell} + \frac{2.303RT}{F} \, pH_{std} \qquad (13.41)$$

Substitution of Equation 13.41 into Equation 13.39 yields

$$pH_{unk} = pH_{std} + \frac{E_{cell \, std} - E_{cell \, unk}}{2.303RT/F} \qquad (13.42)$$

> We usually don't resort to this calculation in pH measurements. Rather, the potential scale of the pH meter is calibrated in pH units (see Section 13.14).

Note that since the determination involves potential measurements with a very high-resistance membrane electrode (1 to 100 MΩ), it is very important to minimize the iR drop by using a pH meter that draws very little current (see before, measurement of potential).

Example 13.5

A glass electrode–SCE pair is calibrated at 25°C with a pH 4.01 standard buffer, the measured voltage being 0.814 V. What voltage would be measured in a 1.00×10^{-3} *M* acetic acid solution? Assume $a_{H^+} = [H^+]$.

---

[i]We will assume the proper definition of pH as $-\log a_{H^+}$ in this chapter since this is what the glass electrode measures.

Solution

From Example 6.7 in Chapter 6, the pH of a 1.00×10^{-3} M acetic solution is 3.88;

$$\therefore 3.88 = 4.01 + \frac{0.814 - E_{\text{cell unk}}}{0.0592}$$

$$E_{\text{cell unk}} = 0.822 \text{ V}$$

Note that the potential increases as the H^+ (a cation) increases, as expected.

• •

COMBINATION pH ELECTRODES—A COMPLETE CELL

A combination electrode is a complete cell when dipped in a test solution.

Both an indicating and a reference electrode (with salt bridge) are required to make a complete cell so that potentiometric measurements can be made. It is convenient to combine the two electrodes into a single probe, so that only small volumes are needed for measurements. A typical construction of a combination pH–reference electrode is shown in Figure 13.7. It consists of a tube within a tube, the inner one housing the pH indicator electrode and the outer one housing the reference electrode (e.g., a Ag/AgCl electrode) and its salt bridge. There is one lead from the combination electrode, but it is split into two connectors at the end, one (the largest) going to the pH terminal and the other going to the reference electrode terminal. It is important that the salt bridge be immersed in the test solution in order to complete the cell. The salt bridge may be a small plug in the outer ring rather than a complete ring as illustrated here. Combination electrodes are the most commonly used because of their convenience.

WHAT DETERMINES THE GLASS MEMBRANE POTENTIAL?

The pH glass electrode functions as a result of ion exchange on the surface of a hydrated layer. The membrane of a pH glass electrode consists of chemically

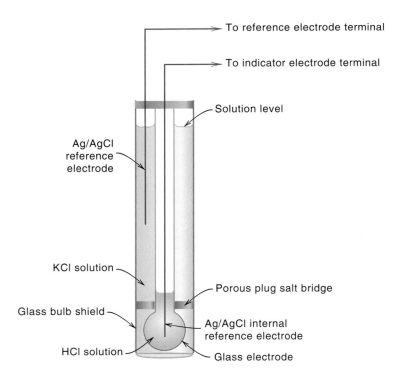

Fig. 13.7. Combination pH–reference electrode.

bonded Na_2O and SiO_2. The surface of a new glass electrode contains fixed silicate groups associated with sodium ions, $—SiO^-Na^+$. It is known that for the electrode to become operative, it must be soaked in water. During this process, the outer surface of the membrane becomes *hydrated*. The inner surface is already hydrated. The glass membrane is usually 0.03 to 0.1 mm thick, and the hydrated layers are 10^{-5} to 10^{-4} mm thick.

When the outer layer becomes hydrated, the sodium ions are exchanged for protons in the solution:

$$\underset{\text{solid}}{—SiO^-Na^+} + \underset{\text{solution}}{H^+} \rightleftharpoons \underset{\text{solid}}{—SiO^-H^+} + \underset{\text{solution}}{Na^+} \qquad (13.43)$$

Other ions in the solution can exchange for the Na^+ (or H^+) ions, but the equilibrium constant for the above exchange is very large because of the large affinity of the glass for protons. Thus, the surface of the glass is made up almost entirely of silicic acid, except in very alkaline solution, where the proton concentration is small. The $—SiO^-$ sites are fixed, but the protons are free to move and exchange with other ions. (By varying the glass composition, the exchange for other ions becomes more favorable, and this forms the basis of electrodes selective for other ions—see below.)

The potential of the membrane consists of two components, the boundary potential and the diffusion potential. The former is almost the sole hydrogen ion activity-determining potential. The **boundary potential** resides at the surface of the glass membrane, that is, at the interface between the hydrated gel layer and the external solution. When the electrode is dipped in an aqueous solution, a boundary potential is built up, which is determined by the activity of hydrogen ions in the external solution and the activity of hydrogen ions on the surface of the gel. One explanation of the potential is that the ions will tend to migrate in the direction of lesser activity, much as at a liquid junction. The result is a microscopic layer of charge built up on the surface of the membrane, which represents a potential. Hence, as the solution becomes more acidic (the pH decreases), protons migrate to the surface of the gel, building up a positive charge, and the potential of the electrode increases, as indicated by Equations 13.37 and 13.38. The reverse is true as the solution becomes more alkaline.

The **diffusion potential** results from a tendency of the protons in the inner part of the gel layer to diffuse toward the dry membrane, which contains $—SiO^-Na^+$, and a tendency of the sodium ions in the dry membrane to diffuse to the hydrated layer. The ions migrate at a different rate, creating a type of liquid-junction potential. But a similar phenomenon occurs on the other side of the membrane, only in the opposite direction. These in effect cancel each other, and so the potential of the membrane is determined largely by the boundary potential. (Small differences in boundary potentials may occur due to differences in the glass across the membrane—these represent a part of the asymmetry potential.)

Pungor has presented evidence that the establishment of an electrode potential is caused by charge separation, due to chemisorption of the primary ion (H^+) from the solution phase onto the electrode surface, that is, a surface chemical reaction. Counter ions of the opposite charge accumulate in the solution phase, and this charge separation represents a potential. A similar mechanism applies to other ion-selective electrodes (below). [See E. Pungor, "The New Theory of Ion-Selective Electrodes," *Sensors*, **1** (2001) 1–12 (this is an electronic journal: www.mdpi.net/sensors).]

K. L. Cheng has proposed a theory of glass electrodes based on capacitor theory in which the electrode senses the hydroxide ion in alkaline solution (where a_{H^+} is very small), rather than sensing protons. [K. L. Cheng, "Capacitor Theory

The pH of the test solution determines the external boundary potential.

Does the glass electrode sense H^+ or OH^- in alkaline solutions?

for Nonfaradaic Potentiometry," *Microchem. J.*, **42** (1990) 5.] Nonfaradaic refers to nonredox reaction. Cheng has performed isotope experiments that suggest the generally accepted ion exchange reaction between H^+ and Na^+ does not occur. Cheng argues that the electrode actually responds to OH^- ions in alkaline solution (remember, $[H^+]$ at pH 14 is only 10^{-14} M!) [C.-M. Huang et al., "Isotope Evidence Disproving Ion Exchange Reaction Between H^+ and Na^+ in pH Glass Electrode," *J. Electrochem. Soc.*, **142** (1995) L175]. His capacitor theory is not generally acknowledged, but he and his coworkers present some compelling arguments and experimental results that make this an interesting theory to contemplate. It has some commonality with Pungor's double-layer theory.

ALKALINE ERROR

Two types of error do occur that result in non-Nernstian behavior (deviation from the theoretical response). The first is called the **alkaline error**. Such error is due to the capability of the membrane for responding to other cations besides the hydrogen ion. As the hydrogen ion activity becomes very small, these other ions can compete successfully in the potential-determining mechanism. Although the hydrated gel layer prefers protons, sodium ions will exchange with the protons in the layer when the hydrogen ion activity in the external solution is very low (reverse of Equation 13.43). The potential then depends partially on the ratio of $a_{Na^+\ external}/a_{Na^+\ gel}$; that is, the electrode becomes a sodium ion electrode.

The error is negligible at pH less than about 9; but at pH values above this, the H^+ concentration is very small relative to that of other ions, and the electrode response to the other ions such as Na^+, K^+, and so on, becomes appreciable. In effect, the electrode appears to "see" more hydrogen ions than are present, and the pH reading is too low. The magnitude of this negative error is illustrated in Figure 13.8. Sodium ion causes the largest errors, which is unfortunate, because many

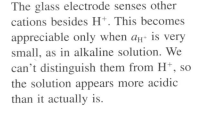

The glass electrode senses other cations besides H^+. This becomes appreciable only when a_{H^+} is very small, as in alkaline solution. We can't distinguish them from H^+, so the solution appears more acidic than it actually is.

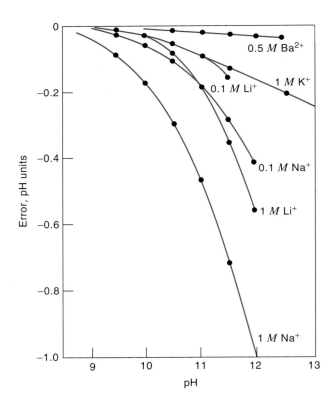

Fig. 13.8. Error of Corning 015 glass electrode in strongly alkaline solutions containing various cations. (From L. Meites and L. C. Thomas, *Advanced Analytical Chemistry.* Copyright © 1958, McGraw-Hill, New York. Used with permission of McGraw-Hill Book Company.)

analytical solutions, especially alkaline ones, contain significant amounts of sodium. Commercial general-purpose glass electrodes will usually be supplied with a nomogram for correcting the alkaline error if the sodium ion concentration is known, and these electrodes are useful up to pH about 11.

By a change in the composition of the glass, the affinity of the glass for sodium ion can be reduced. If the Na_2O in the glass membrane is largely replaced by Li_2O, then the error due to sodium ions is markedly decreased. This is the so-called lithium glass electrode, high-pH electrode, or full-range electrode (0 to 14 pH range). A general-purpose electrode is preferred for use below pH 11 because it will provide faster response and greater stability due to its lower-resistance glass. As mentioned before, it was the discovery that variation in the glass composition could change its affinity for different ions that led to the development of glasses selective for ions other than protons, that is, of ion-selective electrode.

ACID ERROR

The **acid error**, which is more aptly described as the **water activity error**, is the second type causing non-Nernstian response. Such error occurs because the potential of the membrane depends on the activity of the water with which it is in contact. If the activity is unity, the response is Nernstian. In very acid solutions, the activity of water is less than unity (an appreciable amount is used in solvating the protons), and a positive error in the pH reading results (Figure 13.9). A similar type of error will result if the activity of the water is decreased by a high concentration of dissolved salt or by addition of nonaqueous solvent such as ethanol. In these cases, a large liquid-junction potential may also be introduced and another error will thereby result, although this is not very large with small amounts of ethanol.

13.12 Standard Buffers—Reference for pH Measurements

The National Institute of Standards and Technology (NIST) has developed a series of certified standard buffers for use in calibrating pH measurements. The pH values of the buffers were determined by measuring their pH using a hydrogen-indicating electrode in a cell without liquid junction (similar to the cell given by Equation 13.22). A silver/silver chloride reference electrode was used. From Equation 13.24, we see that the activity of the chloride ion must be calculated (to calculate the potential of the reference electrode) using the Debye–Hückel theory; *this ultimately limits the accuracy of the pH of the buffers to about ±0.01 pH unit*. The partial pressure of

The pH of NIST buffers is determined with a cell without liquid junction (Cell 13.22) and is calculated using Equation 13.24. The activity of Cl^- must be calculated from the Debye–Hückel equation, which limits the accuracy of the a_{H^+} calculated from the measured potential.

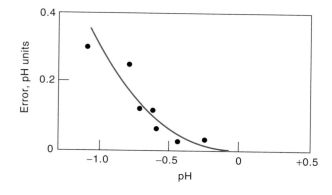

Fig. 13.9. Error of glass electrode in hydrochloric acid solutions. (From L. Meites and L. C. Thomas, *Advanced Analytical Chemistry.* Copyright © 1958, McGraw-Hill, New York. Used with permission of McGraw-Hill Company.)

Table 13.2
pH Values of NIST Buffer Solutions[a]

t (°C)	Tetroxalate[b]	Tartrate[c]	Phthalate[d]	Phosphate[e]	Phosphate[f]	Borax[g]	Calcium Hydroxide[h]
0	1.666	—	4.003	6.984	7.534	9.464	13.423
5	1.668	—	3.999	6.951	7.500	9.395	13.207
10	1.670	—	3.998	6.923	7.472	9.332	13.003
15	1.672	—	3.999	6.900	7.448	9.276	12.810
20	1.675	—	4.002	6.881	7.429	9.225	12.627
25	1.679	3.557	4.008	6.865	7.413	9.180	12.454
30	1.683	3.552	4.015	6.853	7.400	9.139	12.289
35	1.688	3.549	4.024	6.844	7.389	9.102	12.133
38	1.691	3.549	4.030	6.840	7.384	9.081	12.043
40	1.694	3.547	4.035	6.838	7.380	9.068	11.984
45	1.700	3.547	4.047	6.834	7.373	9.038	11.841
50	1.707	3.549	4.060	6.833	7.367	9.011	11.705
55	1.715	3.554	4.075	6.834	—	8.985	11.574
60	1.723	3.560	4.091	6.836	—	8.962	11.449
70	1.743	3.580	4.126	6.845	—	8.921	—
80	1.766	3.609	4.164	6.859	—	8.885	—
90	1.792	3.650	4.205	6.877	—	8.850	—
95	1.806	3.674	4.227	6.886	—	8.833	—

[a] From R. G. Bates, *J. Res. Natl. Bur. Std.*, **A66** (1962) 179. (Reprinted by permission of the U.S. Government Printing Office.)
[b] 0.05 *m* potassium tetroxalate (*m* refers to molality, but only small errors result if molarity is used instead).
[c] Satd. (25°C) potassium hydrogen tartrate.
[d] 0.05 *m* potassium hydrogen phthalate.
[e] 0.025 *m* potassium dihydrogen phosphate, 0.025 *m* disodium monohydrogen phosphate.
[f] 0.008695 *m* potassium dihydrogen phosphate, 0.03043 *m* disodium hydrogen phosphate.
[g] 0.01 *m* borax.
[h] Satd. (25°C) calcium hydroxide.

Only the phosphate mixtures are really buffers. The pH values change with temperature due to the temperature dependence of the K_a values.

hydrogen is determined from the atmospheric pressure at the time of the measurement (minus the vapor pressure of the water at the temperature of the solution).

The compositions and pH of NIST standard buffers are given in Table 13.2. Although the absolute value of the pH accuracy is no better than 0.01 unit, the buffers have been measured *relative to one another* to 0.001 pH. The potentials used in calculating the pH can be measured reproducibly this closely, and the discrimination of differences of thousandths of pH units is sometimes important (i.e., an electrode may have to be calibrated to a thousandth of a pH unit). The pH of the buffers is temperature dependent because of the dependence of the ionization constants of the parent acids or bases on temperature.

Note that several of these solution ions are not really buffers, and they are actually standard pH solutions whose pH is stable since we do not add acid or base. They are resistant to pH change with minor dilutions (e.g., $H^+ \approx \sqrt{K_{a1}K_{a2}}$). Only the two phosphate solutions are actually buffers.

It should be pointed out that if a glass electrode–SCE cell is calibrated with one standard buffer and is used to measure the pH of another, the new reading will not correspond exactly to the standard value of the second because of the residual liquid-junction potential.

The KH_2PO_4–Na_2HPO_4 buffer (pH 7.384 at 38°C) is particularly suited for calibration for blood pH measurements. Many blood pH measurements are made at 38°C, which is near body temperature; thus, the pH of the blood in the body is indicated.

For a discussion of the above NIST pH standard and other proposed definitions of pH, see the letters by W. F. Koch (*Anal. Chem.*, December 1, 1997, 700A; *Chem. & Eng. News*, October 20, 1997, 6).

13.13 Accuracy of pH Measurements

The accuracy of pH measurements is governed by the accuracy to which the hydrogen ion activity of the standard buffer is known. As mentioned above, this accuracy is not better than ±0.01 pH unit because of limitations in calculating the activity coefficient of a single ion.

A second limitation in the accuracy is the residual liquid-junction potential. The cell is standardized in one solution, and then the unknown pH is measured in a solution of a different composition. We have mentioned that this residual liquid-junction potential is minimized by keeping the pH and compositions of the solutions as near as possible. Because of this, *the cell should be standardized at a pH close to that of the unknown.* The error in standardizing at a pH far removed from that of the test solution is generally within 0.01 to 0.02 pH unit but can be as large as 0.05 pH unit for very alkaline solutions.

The residual liquid-junction potential, combined with the uncertainty in the standard buffers, limits the *absolute accuracy of measurement of pH of an unknown solution to about ±0.02 pH unit.* It may be possible, however, to *discriminate* between the pH of two similar solutions with differences as small as ±0.004 or even ±0.002 pH units, although their accuracy is no better than ±0.02 pH units. Such discrimination is possible because the liquid-junction potentials of the two solutions will be virtually identical in terms of true a_{H^+}. For example, if the pH values of two blood solutions are close, we can measure the difference between them accurately to ±0.004 pH. If the pH difference is fairly large, however, then the residual liquid-junction potential will increase and the difference cannot be measured as accurately. For discrimination of 0.02 pH unit, large changes in the ionic strength may not be serious, but they are important for smaller changes than this.

An error of ±0.02 pH unit corresponds to an error in a_{H^+} of ±4.8% (±1.2 mV),[2] and a discrimination of ±0.004 pH unit would correspond to a discrimination of ±1.0% in a_{H^+} (±0.2 mV).

If pH measurements are made at a temperature other than that at which the standardization is made, other factors being equal, the liquid-junction potential will change with temperature. For example, in a rise from 25° to 38°C, a change of +0.76 mV has been reported for blood and −0.55 mV for buffer solutions. Thus, for very accurate work, the cell should be standardized at the same temperature as the test solution.

The residual liquid-junction potential limits the accuracy of pH measurement. Always calibrate at a pH close to that of the test solution.

Potentiometric measurements of a_{H^+} are only about 5% accurate.

13.14 Using the pH Meter—How Does It Work?

We have already mentioned that owing to the high resistance of the glass electrode, an electrometer or pH meter must be used to make the potential measurements. If voltage is measured directly, Equation 13.40 or 13.42 is applied to calculate the pH.

Check out www.ph-measurement. co.uk for a useful tutorial on the basics of pH measurement and information on pH measurements in different media.

---

[2]The electrode response is 59 mV/pH at 25°C.

The temperature knob on the pH meter adjusts T in the RT/nF value, which determines the slope of the potential versus pH buffers.

The value of $2.303RT/F$ at 298.16 K (25°C) is 0.05916; if a different temperature is used, this value should be corrected in direct proportion to the temperature.

A digital pH meter is shown in Figure 13.10. The potential scale is calibrated in pH units, with each pH unit equal to 59.16 mV at 25°C (Equation 13.39). The pH meter is adjusted with the calibrate knob to indicate the pH of the standard buffer. Then, the standard buffer is replaced by the unknown solution and the pH is read from the scale. This procedure, in effect, sets the constant k in Equation 13.40 and adjusts for the asymmetry potential as well as the other constants included in k.

The pH meter contains a temperature adjustment dial, which changes the sensitivity response (mV/pH) so that it will be equal to $2.303RT/F$. For example, it is 54.1 mV at 0°C and 66.0 mV at 60°C. Note that this does *not* compensate for the change in the pH of standard buffers with temperature, and the pH value of the buffer at the given temperature is used.

Electrodes and meters are designed to have a point in calibrations lines, in the midrange of activity measurements, where the potential essentially has no variation with temperature. For pH glass electrodes, this is set at pH 7 (Figure 13.11). This is called the **isopotential point**, and the potential is zero. (pH meters actually measure potential which is converted to pH reading, and potentials can be recorded directly.) Any potential reading different from 0 mV for a pH 7.0 standard buffer is called the **offset** of that electrode. When the temperature is changed, the calibration slope changes, and the intersection of the curves establishes the actual isopotential point. If the isopotential point of the electrode differs from pH 7, then the temperature of the calibration buffer and the test solution should be the same for highest accuracy because a slight error will occur in the slope adjustment at different temperatures. For more details of the isopotential point and its quantitative interpretation, see A. A. S. C. Machado, *Analyst*, **19** (1994) 2263.

In calibrating the pH meter, the electrodes are inserted in a pH 7.0 standard buffer. The temperature of the buffer is checked and the temperature adjustment knob is adjusted to that temperature. Using the standardized or calibration knob, the meter is adjusted to read 7.00. Many pH meters have microprocessors that

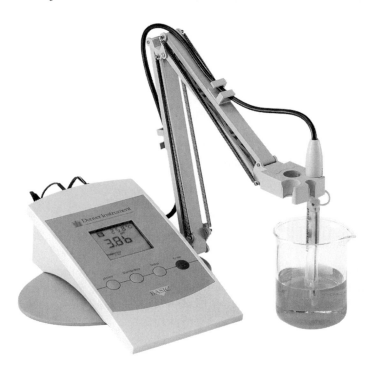

Fig. 13.10. Typical pH meter. (Courtesy of Denver Instrument Company.)

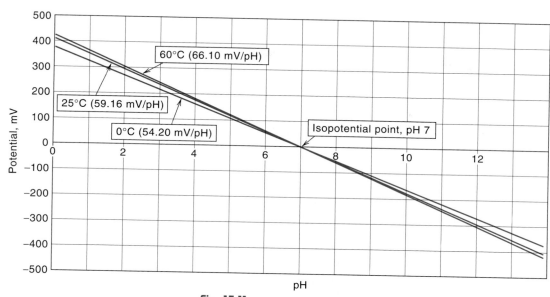

Fig. 13.11. Isopotential point.

recognize specific pH values (e.g., 4.0, 7.0, and 10.0), and you can simply depress the stand, cal, or set key, and the meter will automatically read the buffer value. Next the slope is set by repeating with either a pH 4.0 or 10.0 standard buffer, depending on the pH range for the sample measurements. Here, though, the Slope Control is used to adjust the meter to the correct pH reading. Most pH meters now include a separate temperature probe so that temperature compensation can be automatic. The temperature probe may be incorporated in the electrode probe.

Ordinary pH meters are precise to ± 0.1 to ± 0.01 pH unit (± 6 to ± 0.6 mV) with a full-meter scale of 14 pH units (about 840 mV). The meters can be set to read millivolts directly (usually with a sensitivity of 1400 mV full scale). Expanded-scale pH meters, which amplify the potential signal, are capable of reading to ± 0.001 pH unit with a full-scale reading typically of 1.4 pH units (140 mV on the millivolt-scale setting); to accomplish this, the potential must be read to closer than 0.1 mV.

When the pH of an unbuffered solution near neutrality is measured, readings will be sluggish because the solution is poorly poised and a longer time will be required to reach a stable reading. The solution should be stirred because a small amount of the glass tends to dissolve, making the solution at the electrode surface alkaline (Equation 13.43, where H_2O—the source of H^+—is replaced by NaOH solution).

13.15 pH Measurement of Blood—Temperature Is Important

Recall from Chapter 7 that, because the equilibrium constants of the blood buffer systems change with temperature, the pH of blood at the body temperature of 37°C is different than at room temperature. Hence, to obtain meaningful blood pH measurements that can be related to actual physiological conditions, the measurements should be made at 37°C and the samples should not be exposed to the atmosphere. (Also recall that the pH of a neutral aqueous solution at 37°C is 6.80, and so the acidity scale is changed by 0.20 pH unit.)

The pH measurement of blood samples must be made at body temperature to be meaningful.

Some useful rules in making blood pH measurements are as follows:

1. Calibrate the electrodes using a standard buffer at 37°C, making sure to select the proper pH of the buffer at 37°C and to set the temperature knob on the pH meter at 37°C (slope = 61.5 mV/pH). It is a good idea to use two standards for calibration, narrowly bracketing the sample pH; this assures that the electrode is functioning properly. Also, the electrodes must be equilibrated at 37°C before calibration and measurement. The potential of the internal reference electrode inside the glass electrode is temperature dependent, as may be the potential-determining mechanism at the glass membrane interface; and the potentials of the SCE reference electrode and the liquid junction are temperature dependent. (We should note here that if pH or other potential measurements are made at less than room temperature, the salt bridge or the reference electrode should not contain saturated KCl, but somewhat less concentrated KCl, because solid KCl crystals will precipitate in the bridge and increase its resistance.)

2. Blood samples must be kept anaerobically to prevent loss or absorption of CO_2. Make pH measurements within 15 min after sample collection, if possible, or else keep the sample on ice and make the measurements within 2 h. The sample is equilibrated to 37°C before measuring. (If a pCO_2 measurement is to be performed also, do this within 30 min.)

3. To prevent coating of the electrode, flush the sample from the electrode with saline solution after each measurement. A residual blood film can be removed by dipping for *only* a few minutes in 0.1 *M* NaOH, followed by 0.1 *M* HCl and water or saline.

Generally, venous blood is taken for pH measurement, although arterial blood may be required for special applications. The 95% confidence limit range (see Chapter 3) for arterial blood pH is 7.31 to 7.45 (mean 7.40) for all ages and sexes. A range of 7.37 to 7.42 has been suggested for subjects at rest. Venous blood may differ from arterial blood by up to 0.03 pH unit and may vary with the vein sampled. Intracellular erythrocyte pH is about 0.15 to 0.23 unit lower than that of the plasma.

13.16 pH Measurements in Nonaqueous Solvents

Measurement of pH in a nonaqueous solvent when the electrode is standardized with an aqueous solution has little significance in terms of possible hydrogen ion activity because of the unknown liquid-junction potential, which can be rather large, depending on the solvent. Measurements made in this way are usually referred to as "apparent pH." pH scales and standards for nonaqueous solvents have been suggested using an approach similar to the one for aqueous solutions. These scales have no rigorous relation to the aqueous pH scale, however. You are referred to the book by Bates (Ref. 3) for a discussion of this topic. See also M. S. Frant, "How to Measure pH in Mixed & Nonaqueous Solutions," *Today's Chemist at Work*, American Chemical Society, June, 1995, p. 39.

13.17 Ion-Selective Electrodes

Various types of membrane electrodes have been developed in which the membrane potential is selective toward a given ion or ions, just as the potential of the glass membrane of a conventional glass electrode is selective toward hydrogen ions. These electrodes are important in the measurement of ions, especially in small concentrations. Generally, they are not "poisoned" by the presence of proteins, as some other electrodes are, and so they are ideally suited to measurements in biological media. This is especially true for the glass membrane ion-selective electrodes.

See www.nico2000.net for an excellent tutorial (12,000-word beginners guide) on principles of pH and ion-selective electrodes, calibration, and measuring procedures.

None of these electrodes is *specific* for a given ion, but each will possess a certain *selectivity* toward a given ion or ions. So they are properly referred to as **ion-selective electrodes** (ISEs).

GLASS MEMBRANE ELECTRODES

These are similar in construction to the pH glass electrode. Varying the composition of the glass membrane can cause the hydrated glass to acquire an increased affinity for various monovalent cations, with a much lower affinity for protons than the pH glass electrode has. The membrane potential becomes dependent on these cations, probably through an ion exchange mechanism similar to that presented for the glass pH electrode; that is, a boundary potential is produced, determined by the relative activities of the cations on the surface of the gel and in the external solution. Increased cation activity results in increased positive charge on the membrane and a positive increase in electrode potential.

The construction is similar to Figure 13.6. The internal filling solution will usually be the chloride salt of the cation to which the electrode is most responsive. The composition of the membrane will vary from manufacturer to manufacturer, but we can classify three general types of glass electrodes:

The glass membrane pH electrode is the ultimate ion-selective electrode.

1. pH type. This is the conventional pH glass electrode, and it has a selectivity order of $H^+ >>> Na^+ > K^+, Rb^+, Cs^+ \ldots \gg Ca^{2+}$. The response to ions other than H^+ is the "alkaline error" we talked about above.

2. Cation-sensitive type. This responds in general to monovalent cations, and the order of selectivity is $H^+ > K^+ > Na^+ > NH_4^+, Li^+ \ldots \gg Ca^{2+}$.

3. "Sodium-sensitive" type. The selectivity order is $Ag^+ > H^+ > Na^+ \gg K^+, Li^+, \ldots \gg Ca^{2+}$.

Note that all the electrodes are responsive to hydrogen ion, but the second two are much less so than the first type. Because of this response, the electrodes must be used at a pH sufficiently high that the hydrogen ion activity is so low that the response will be determined primarily from the ion of interest. This lower pH limit will vary from electrode to electrode and ion to ion. (Refer to What Determines the Glass Membrane Potential? above regarding K. L. Cheng's theory of response to OH^- at high pH.)

H^+ is a common interferent with other ISEs, and so the pH must be above a limiting value, depending on the concentration of the primary ion (the one being measured).

The sodium-sensitive type of electrode can be used to determine the activity of sodium ion in the presence of appreciable amounts of potassium ion. Its selectivity for sodium over potassium is on the order of 3000 or more. Glass electrodes can be obtained that show a selectivity ratio for silver over sodium of greater than 1000. Glass electrodes can be used in liquid ammonia and molten salt media.

SOLID-STATE ELECTRODES

The fluoride ion-selective electrode is one of the most successful and useful since the determination of fluoride is rather difficult by most other methods.

The construction of these electrodes is shown in Figure 13.12. The most successful example is the fluoride electrode. The membrane consists of a single crystal of lanthanum fluoride doped with some europium(II) to increase the conductivity of the crystal. Lanthanum fluoride is very insoluble, and this electrode exhibits Nerstian response to fluoride down to 10^{-5} M and non-Nerstian response down to 10^{-6} M (19 ppb!). This electrode has at least a 1000-fold selectivity for fluoride ion over chloride, bromide, iodide, nitrate, sulfate, monohydrogen phosphate, and bicarbonate anions and a 10-fold selectivity over hydroxide ion. Hydroxide ion appears to be the only serious interference. The pH range is limited by the formation of hydrofluoric acid at the acid end and by hydroxide ion response at the alkaline end; a pH range of 4 to 9 is claimed

A useful solution for minimizing interferences with the fluoride electrode consists of a mixture of an acetate buffer at pH 5.0 to 5.5, 1 M NaCl, and cyclohexylenedinitrilo tetraacetic acid (CDTA). This solution is commercially available as TISAB (total ionic-strength adjustment buffer). A 1:1 dilution of samples and standards with the solution provides a high ionic-strength background, swamping out moderate variations in ionic strength between solutions. This keeps both the junction potential and the activity coefficient of the fluoride ion constant from solution to solution. The buffer provides a pH at which appreciable HF formation is avoided and hydroxide response is not present. CDTA is a chelating agent, similar to EDTA, that complexes with polyvalent cations such as Al^{3+}, Fe^{3+}, and Si^{4+}, which would otherwise complex with F^- and change the fluoride activity.

TISAB serves to adjust the ionic strength and the pH, and to prevent Al^{3+}, Fe^{3+}, and Si^{4+} from complexing the fluoride ion.

A useful solid-state electrode is based on an Ag_2S membrane. By itself, this electrode responds to either Ag^+ or S^{2-} ions, down to about 10^{-8} M. This lower limit of detection is too high to be caused by solubility of Ag_2S ($K_{sp} = 10^{-51}$). It probably reflects difficulties in preparing extremely dilute solutions and ionic adsorption on, and desorption from, the surfaces of the electrodes and the vessel containing the solution. This membrane is a good ionic conductor with low resistance; and by mixing the Ag_2S with other silver or sulfide salts (whose resistances might be high), it will become responsive to other ions. For example, a mixed membrane of AgI/Ag_2S will respond to I^- ion in addition to Ag^+ and S^{2-}. A mixed CuS/Ag_2S membrane will respond to Cu^{2+} in addition to the other membrane ions.

The chief restriction of these mixed-salt electrodes is that the solubility of the second salt must be much larger than that of Ag_2S; but, on the other hand, it must be sufficiently insoluble that its dissolution does not limit to relatively high values the test ion concentration that can be detected. As long as the membrane contains sufficient silver sulfide to provide silver ion conducting pathways through the membrane, it will function as a silver electrode. The potential, then, is related

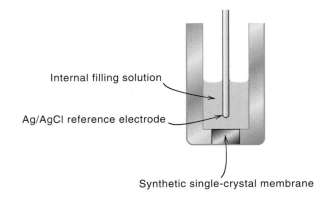

Internal filling solution

Ag/AgCl reference electrode

Synthetic single-crystal membrane

Fig. 13.12. Crystal membrane electrode. (Reproduced by permission of Orion Research, Inc.)

to the test ion through a series of equilibria similar to those described for a metal–metal salt electrode; that is, the available Ag^+ is governed by the solubility equilibria. Other Ag_2S mixed-crystal electrodes are available for Cl^-, Br^-, SCN^-, CN^-, Pb^{2+}, and Cd^{2+}. These electrodes, of course, are subject to interference from other ions that react with Ag^+ (in the case of anion electrodes) or with S^{2-} (in the case of cation electrodes).

LIQUID–LIQUID ELECTRODES

The basic construction of these electrodes is shown in Figure 13.13. Here, the potential-determining "membrane" is a layer of a water-immiscible liquid ion exchanger held in place by an inert, porous membrane. The porous membrane allows contact between the test solution and the ion exchanger but minimizes mixing. It is either a synthetic, flexible membrane or a porous glass frit, depending on the manufacturer. The internal filling solution contains the ion for which the ion exchanger is specific plus a halide ion for the internal reference electrode.

An example of this electrode is the calcium-selective electrode. The ion exchanger is a calcium organophosphorus compound. The sensitivity of the electrode is governed by the solubility of the ion exchanger in the test solution. A Nernstian response is obtained down to about 5×10^{-5} M. The selectivity of this electrode is about 3000 for calcium over sodium or potassium, 200 over magnesium, and 70 over strontium. It can be used over the pH range of 5.5 to 11. Above pH 11, calcium hydroxide precipitates. A phosphate buffer should not be used for calcium measurements because the calcium activity will be lowered by complexation or precipitation. Experience has shown that these and other liquid-membrane electrodes are often subject to poisoning, for example, in biological fluids.

A "divalent cation" ion exchange electrode that responds to several cations is available. Its response is nearly equal for calcium and magnesium, and it is useful for measuring water hardness. A copper and a lead electrode are also available. Anion-selective electrodes of this type are available for nitrate, perchlorate, and chloride. They are the same in principle, except that a liquid anion exchanger is used instead of a cation exchanger.

Table 13.3 summarizes the characteristics of some commercial ion-selective electrodes.

The filling solution for ISEs usually contains a chloride salt of the primary ion, for example, $CaCl_2$ for a Ca^{2+} electrode or KCl for a K^+ electrode. The chloride establishes the potential of the internal Ag/AgCl electrode.

PLASTIC MEMBRANE–IONOPHORE ELECTRODES

A very versatile and relatively easy to prepare type of electrode is that in which a neutral lipophilic (organic loving) **ionophore** that selectively complexes with the ion of interest is dissolved in a soft plastic membrane. The ionophore should

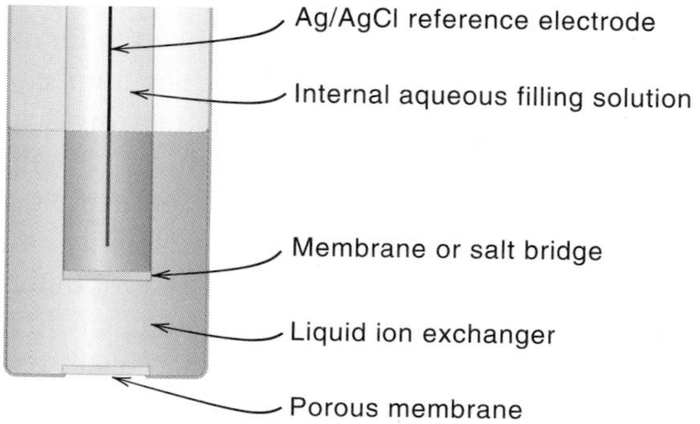

- Ag/AgCl reference electrode
- Internal aqueous filling solution
- Membrane or salt bridge
- Liquid ion exchanger
- Porous membrane

Fig. 13.13. Liquid–membrane electrode.

Table 13.3

Typical Properties of Some Commercial Ion-Selective Electrodes

Electrode	Concentration Range (M)	Principal Interferences[a]
Liquid–liquid ion exchange electrodes		
Ca^{2+}	10^0–10^{-5}	$Zn^{2+}(3)$; $Fe^{2+}(0.8)$; $Pb^{2+}(0.6)$; $Mg^{2+}(0.1)$; $Na^+(0.003)$
Cl^-	10^{-1}–10^{-5}	$I^-(17)$; $NO_3^-(4)$; $Br^-(2)$; $HCO_3^-(0.2)$; SO_4^{2-}, $F^-(0.1)$
Divalent cation	10^0–10^{-8}	Fe^{2+}, $Zn^{2+}(3.5)$; $Cu^{2+}(3.1)$; $Ni^{2+}(1.3)$; Ca^{2+}, $Mg^{2+}(1)$; $Ba^{2+}(0.94)$; $Sr^{2+}(0.54)$; $Na^+(0.015)$
BF_4^-	10^{-1}–10^{-5}	$NO_3^-(0.1)$; $Br^-(0.04)$; OAc^-, $HCO_3^-(0.004)$; $Cl^-(0.001)$
NO_3^-	10^{-1}–10^{-5}	$ClO_4^-(1000)$; $I^-(20)$; $Br^-(0.1)$; $NO_2^-(0.04)$; $Cl^-(0.004)$; $CO_3^{2-}(0.0002)$; $F^-(0.00006)$; $SO_4^{2-}(0.00003)$
ClO_4^-	10^{-1}–10^{-5}	$I^-(0.01)$; NO_3^-, $OH^-(0.0015)$; $Br^-(0.0006)$; F^-, $Cl^-(0.0002)$
K^+	10^0–10^{-5}	$Cs^+(1)$; $NH_4^+(0.03)$; $H^+(0.01)$; $Na^+(0.002)$; Ag^+, $Li^+(0.001)$
Solid–state electrodes[b]		
F^-	10^0–10^{-6}	Maximum level: $OH^- < 0.1\ F^-$
Ag^+ or S^{2-}	10^0–10^{-7}	$Hg^{2+} < 10^{-7}\ M$

[a] Number in parentheses is the relative selectivity for the interfering ion over the test ion (see The Selectivity Coefficient below).
[b] Interference concentrations given represent maximum tolerable concentrations.

be lipophilic (as opposed to hydrophilic) so that it is not leached from the membrane upon exposure to aqueous solutions. The plastic membrane is usually polyvinylchloride (PVC) based and consists of about 33% PVC; about 65% plasticizer, for example, o-nitrophenyl ether (o-NPOE); about 1.5% ionophore; and about 0.5% potassium tetrakis(p-chlorophenyl)borate (KT$_p$ClB) to increase the conductivity and minimize interference from lipophilic anions such as SCN. The $(\phi Cl)_4 B^-$ ion is itself lipophilic and repels lipophilic anions that would otherwise penetrate the membrane and counter the metal ion response. A solution of these components is prepared in a solvent such as tetrahydrofuran (THF) and then is poured onto a glass plate to allow the THF to evaporate. The pliable membrane that results can then be mounted at the end of an electrode body.

Perhaps the most successful example of this type of electrode is the potassium ion-selective electrode incorporating the ionophore, valinomycin. Valinomycin is a naturally occurring antibiotic with a cyclic polyether ring that has a cage of oxygens in the ring of just the right size for selectively complexing the potassium ion. Its selectivity for potassium is about 10^4 that for sodium.

Useful ionophores for a number of metal ions, especially alkali and alkaline earth ions, are **crown ethers**. These are synthetic neutral cyclic ether compounds that can be tailor made to provide cages of the right size to selectively complex a given ion. A long hydrocarbon chain or phenyl group is usually attached to make the compound lipophilic. An example is the 14-crown-4-ether illustrated in Figure 13.14, which is selective for lithium ion in the presence of sodium. The number 4 refers to the number of oxygens in the ring and the number 14 is the ring size. 14-Crown-4 compounds have the proper cage size to complex lithium. Placing bulky phenyl groups on the compound causes steric hindrance in the formation of the 2:1 crown ether:sodium complex and enhances the lithium selectivity (the lithium:crown complex is 1:1). The result is about 800-fold selectivity for lithium. Crown ether-based electrodes have been prepared for sodium, potassium, calcium, and other ions. Amide-based ionophores have been synthesized that selectively complex certain ions. Figure 13.15 shows some ionophores that have been used in PVC-based electrodes.

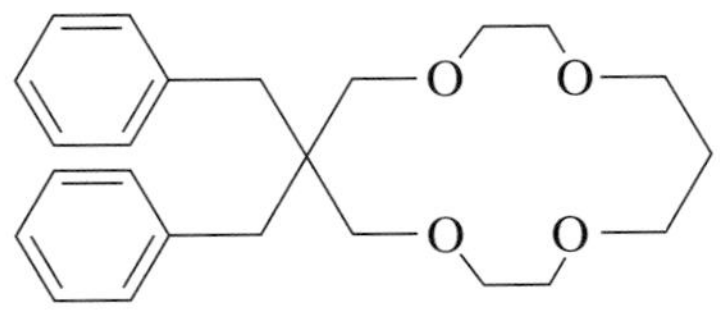

Pederson received the 1987 Nobel Prize for his pioneering work on crown ethers. See http://almaz.com/nobel/chemistry/1987c.html.

Fig. 13.14. 14-Crown-4 ether that selectively binds lithium ion.

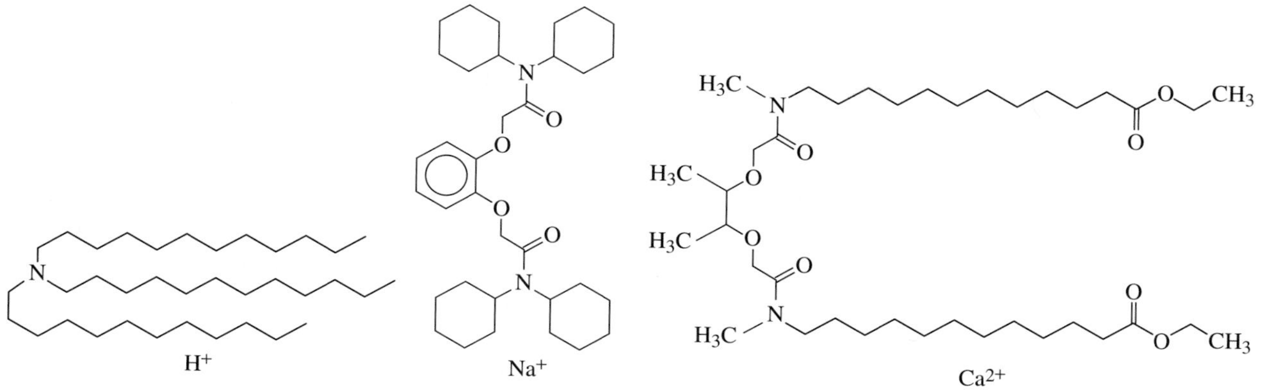

Fig. 13.15. Ionophores for H^+, Na^+, and Ca^{2+}.

COATED-WIRE ELECTRODES

Cattrall and Freiser reported that ion-selective electrodes could be prepared by simply coating a wire with the above PVC membranes, to make electrical contact. (See *Anal. Chem.*, **43** (1971) 1905.) While thermodynamically the contact wire should be potentiometrically well poised (e.g., a metal–metal salt electrode), bare metal electrodes do work satisfactorily (e.g., Pt, Cu, Ag). These electrodes are very convenient to prepare and use. A THF solution of the membrane ingredients is coated on the wire, and the solvent is allowed to evaporate.

POTENTIOMETRIC ENZYME-BASED ELECTRODES FOR MEASUREMENT OF SUBSTRATES

Ion-selective electrodes used in conjunction with immobilized enzymes can serve as the basis of electrodes that are selective for specific enzyme substrates. Enzymes are proteins that catalyze specific reactions to a high degree of specificity. The reactants are the substrates. Enzymes and their properties are discussed in more detail in Chapter 24.

Consider the hydrolysis of the substrate urea in the presence of the enzyme urease:

$$NH_2CONH_2 + 2H_2O + H^+ \xrightarrow{\text{urease}} 2NH_4^+ + HCO_3^-$$

A urea electrode can be prepared by immobilizing urease in a gel and coating it on the surface of a cation-sensitive-type glass electrode (that responds to monovalent cations). When the electrode is dipped in a solution containing urea, the urea diffuses into the gel layer and the enzyme catalyzes its hydrolysis to form ammonium ions. The ammonium ions diffuse to the surface of the electrode where they are sensed by the cation-sensitive glass to give a potential reading. After about 30 to 60 s, a steady-state reading is reached which, over a certain working range, is a linear function of the logarithm of the urea concentration. By appropriate choice of immobilized enzyme and electrode, a number of other substrate-selective enzyme electrodes have been described.

"Enzyme" electrodes incorporate enzymes to measure their substrates.

MECHANISM OF MEMBRANE RESPONSE

The mechanisms of ion-selective electrode membrane response have not been as extensively studied as the glass pH electrode, and even less is known about how their potentials are determined. Undoubtedly, there is a similarity of mechanisms.

The active membrane generally contains the ion of interest selectively bound to a reagent in the membrane, either as a precipitate or a complex. Or else the electrode must be equilibrated in solution of the test ion, in which case the ion also binds selectively to the membrane reagent. This can be compared to the —SiO⁻H⁺ sites on the glass pH electrode. When the ion-selective electrode is immersed in a solution of the test ion, a boundary potential is established at the interface of the membrane and the external solution. The possible mechanism again is due to the tendency of the ions to migrate in the direction of lesser activity to produce a liquid-junction-type potential. Positive ions will result in a positive charge and a change in the potential in the positive direction, while negative ions will result in a negative charge and a change in the potential in the negative direction.

The secret in constructing ion-selective electrodes, then, is to find a material with sites that show strong affinity for the ion of interest. Thus, the calcium liquid ion exchange electrode exhibits high selectivity for calcium over magnesium and sodium ions because the organic phosphate cation exchanger (in the calcium form) has a high chemical affinity for calcium ions. The ion exchange equilibrium at the membrane–solution interface involves calcium ions, and the potential depends on the ratio of the activity of calcium ions in the external solution to that of calcium ion in the membrane phase.

SELECTIVITY COEFFICIENT

The potential of an ion-selective electrode in the presence of a single ion follows an equation similar to Equation 13.38 for the pH glass membrane electrode:

Don't forget the sign of z.

$$E_{\mathrm{ISE}} = k + \frac{S}{z} \log a_{\mathrm{ion}} \qquad (13.44)$$

where S represents the slope (theoretically $2.303RT/F$) and z is the ion charge, *including sign*. Often, the slope is less than Nernstian; but for monovalent ion electrodes, it is usually close. The constant k depends on the nature of the internal reference electrode, the filling solution, and the construction of the membrane. It is determined by measuring the potential of a solution of the ion of known activity.

Example 13.6

A fluoride electrode is used to determine fluoride in a water sample. Standards and samples are diluted 1:10 with TISAB solution. For a 1.00×10^{-3} M (before dilution) standard, the potential reading relative to the reference electrode is -211.3 mV; and for a 4.00×10^{-3} M standard, it is -238.6 mV. The reading with the unknown is -226.5 mV. What is the concentration of fluoride in the sample?

Solution

Since the ionic strength remains constant due to dilution with the ionic-strength adjustment solution, the response is proportional to $\log[\mathrm{F}^-]$:

$$E = k + \frac{S}{z} \log[\mathrm{F}^-] = k - S \log[\mathrm{F}^-]$$

where z is -1. First calculate S:

$$-211.3 = k - S \log(1.00 \times 10^{-3}) \qquad (1)$$
$$-238.6 = k - S \log(4.00 \times 10^{-3}) \qquad (2)$$

Subtract (2) from (1):

$$27.3 = S \log(4.00 \times 10^{-3}) - S \log(1.00 \times 10^{-3}) = S \log \frac{4.00 \times 10^{-3}}{1.00 \times 10^{-3}}$$

$$27.3 = S \log 4.00$$

$$S = 45.3 \text{ mV (somewhat sub-Nernstian)}$$

Calculate k:

$$-211.3 = k - 45.3 \log(1.00 \times 10^{-3})$$

$$k = -347.2 \text{ mV}$$

For the unknown:

$$-226.5 = -347.2 - 45.3 \log[F^-]$$

$$[F^-] = 2.16 \times 10^{-3} \ M$$

• •

If the electrode is in a solution containing a mixture of cations (or anions, if it is an anion-responsive electrode), it may respond to the other cations. Suppose, for example, we have a mixture of sodium and potassium ions and an electrode that responds to both. The Nernst equation must include an additive term for the potassium activity:

No electrode is totally specific. Ideally, we can keep the product $K_{NaK}a_{K^+}$ negligible compared to a_{Na^+}.

$$E_{NaK} = k_{Na} + S \log(a_{Na^+} + K_{NaK}a_{K^+}) \qquad (13.45)$$

The constant k_{Na} corresponds to that in the Nerst equation for the primary ion, sodium, alone. E_{NaK} is the potential of the electrode in a mixture of sodium and potassium. K_{NaK} is the **selectivity coefficient** of the electrode for *potassium over sodium*. It is equal to the reciprocal of K_{KNa}, which is the selectivity coefficient of *sodium over potassium*. Obviously, we want K_{NaK} to be small, or else the product $a_{K^+}K_{NaK}$ to be small.

$K_{NaK} = 1/K_{KNa}$.

K_{NaK} and k_{Na} are determined by measuring the potential of two different standard solutions containing sodium and potassium and then solving the two simultaneous equations for the two constants. Alternatively, one of the solutions may contain only sodium, and k_{Na} is determined from Equation 13.44.

A general equation, called the *Nicolsky* equation, can be written for mixtures of two ions of different charges:

$$E_{AB} = k_A + \frac{S}{z_A} \log (a_A + K_{AB}a_B^{z_A/z_B}) \qquad (13.46)$$

where z_A is the charge on ion A (the primary ion) and z_B is the charge on ion B. Thus, measurement of sodium in the presence of calcium using a sodium ion electrode would follow the expression:

$$E_{NaCa} = k_{Na} + S \log(a_{Na^+} + K_{NaCa}a_{Ca^{2+}}^{1/2}) \qquad (13.47)$$

Since all electrodes respond more or less to hydrogen ions, the practice is to keep the activity of the hydrogen ion low enough that the product $K_{AH}a_{H^+}^{z_A}$ in the Nikolsky equation is negligible.

$K_{NaH}a_{H^+}$ compared to a_{Na^+} determines the lower pH limit of the electrode.

Selectivity coefficients are generally not sufficiently constant to use in quantitative calculations.

One problem with selectivity coefficients should be mentioned. They often vary with the relative concentrations of ions and are not constant. For this reason, it is difficult to use them in calculations involving mixtures of ions. They are useful for predicting conditions under which interfering ions can be neglected. That is, in practice, conditions are generally adjusted so the product $K_{AB}a_B^{z_A/z_B}$ is negligible and a simple Nernst equation applies for the test ion. Usually, a calibration curve is prepared, and if an interfering ion is present, this can be added to standards at the same concentrations as in the unknowns; the result would be a nonlinear but corrected calibration curve. This technique can obviously only be used if the concentration of the interfering ion remains nearly constant in the samples.

An example of how the selectivity coefficient might be used in a calculation is given in the following example.

 Example 13.7

A cation-sensitive electrode is used to determine the activity of calcium in the presence of sodium. The potential of the electrode in 0.0100 M CaCl$_2$ measured against an SCE is +195.5 mV. In a solution containing 0.0100 M CaCl$_2$ and 0.0100 M NaCl, the potential is 201.8 mV. What is the activity of calcium ion in an unknown solution if the potential of the electrode in this is 215.6 mV versus SCE and the sodium ion activity has been determined with a sodium ion-selective electrode to be 0.0120 M? Assume Nernstian response.

Solution

The ionic strength of 0.0100 M CaCl$_2$ is 0.0300, and that of the mixture is 0.0400. Therefore, from Equation 6.20, the activity coefficient of calcium ion in the pure CaCl$_2$ solution is 0.55, and for calcium and sodium ions in the mixture, it is 0.51 and 0.83, respectively. Therefore,

$$k_{Ca} = E_{Ca} - 29.58 \log a_{Ca^{2+}}$$
$$= 195.5 - 29.58 \log(0.55 \times 0.0100)$$
$$= 262.3 \text{ mV}$$
$$E_{CaNa} = k_{Ca} + 29.58 \log(a_{Ca^{2+}} + K_{CaNa}a_{Na+}^2)$$
$$201.8 = 262.3 + 29.58 \log[0.51 \times 0.0100 + K_{CaNa}(0.83 \times 0.0100)^2]$$
$$K_{CaNa} = 47$$
$$215.6 = 262.3 + 29.58 \log(a_{Ca^{2+}} + 47 \times 0.0120^2)$$
$$a_{Ca^{2+}} = 0.0196 \text{ M}$$

Note that although the selectivity coefficient for Ca^{2+} is not very good (the electrode is a better sodium sensor!), the sodium contribution (0.0068) in the mixture is only about 0.3 that of the calcium (0.0196), due to the squared term for sodium.

EXPERIMENTAL METHODS FOR DETERMINING SELECTIVITY COEFFICIENTS

The above discussions illustrate the ideal application of the Nicolsky equation. As mentioned, selectivity coefficients may not be really constant, and the values determined will depend on the method of evaluation. One of two approaches is generally used, the **separate solution method** and the **mixed solution method**. An empirical variation of the latter is the matched potential method. These methods are briefly discussed.

1. Separate Solution Method. In this method, calibration curves are prepared for each ion being tested. Parallel curves should result, with the potentials for the primary ion (if a cation) being more positive. The selectivity ratio can be related to the difference in potential between the two curves (see Problem 21). For two monovalent ions:

$$-\log K_{AB} = \frac{E_A - E_B}{S} \qquad (13.48)$$

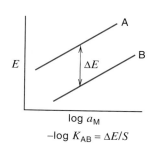

$-\log K_{AB} = \Delta E/S$

See Problem 21 for a derivation of a related equation.

where E_A and E_B correspond to the potentials at a fixed concentration of the two ions and will ideally be the potential difference between two parallel curves. Obviously, the more selective the electrode is for ion A, the larger will be the potential difference (the smaller will be the potential for ion B).

2. Mixed Solution Method. There are various measurement methods using mixed solutions of the two ions. The *fixed interference method* is commonly used. Consider, for example, the testing of a lithium ion-selective electrode in the presence of sodium ion. A lithium calibration curve is prepared in the presence of a fixed concentration of sodium, for example, 140 mM as found in blood. A plot such as that given in Figure 13.16 results. In the upper portion of the curve, the electrode responds in a Nernstian manner to the lithium ion. As the lithium concentration decreases, the electrode potential is increasingly affected by the constant background of sodium ions, and in the lower portion of the curve the electrode exhibits a mixed response to both the lithium and the sodium. When the lithium concentration is very small, the response is due solely to sodium (the baseline potential).

Two methods can be employed to estimate the selectivity ratio based on this curve. The first is based on finding graphically the point at which the electrode is responding equally to both ions. This corresponds to the activity of A from the extrapolated linear portion of the curve at which the potential is equal to the background potential due to B (this is the concentration of A that would give that potential if there were no B present, that is, if the curve follows the Nernst equation for A).

Mixed solution methods better represent conditions of actual samples.

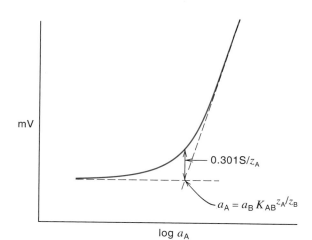

mV

$0.301S/z_A$

$a_A = a_B K_{AB}{}^{z_A/z_B}$

log a_A

Fig. 13.16. Fixed interference calibration curve.

Since at that point each ion is contributing equally to the potential, we can write (from Equation 13.46)

$$a_A = K_{AB} a_B^{z_A/z_B} \qquad (13.49)$$

or

$$K_{AB} = \frac{a_A}{a_B^{z_A/z_B}} \qquad (13.50)$$

where a_B is the fixed concentration of the secondary ion and a_A is the activity of A at the intersection. If, for example, the lithium concentration were 1.0 mM at this point in the presence of 150 mM sodium, K_{LiNa} would be 1.00/150 = 6.7×10^{-3}, and the electrode is 150 times more selective to lithium ion.

A second method is based on the theoretical point of equal contributions of each ion to the potential. The linear portion of the upper curve, which is extrapolated, is represented at this point by Equation 13.44, that is,

$$E_A = k_A + \frac{S}{z_A} \log 2a_A \qquad (13.51)$$

The nonlinear portion of the curve is expressed by Equation 13.46. When both ions contribute equally to the potential, $a_A = K_{AB} a_B^{z_A/z_B}$, and Equation 13.46 becomes

$$E_{AB} = k_A + \frac{S}{z_A} \log 2a_A \qquad (13.52)$$

$$= k_A + \frac{S}{z_A} \log a_A + \frac{0.301 S}{z_A}$$

For a Nernstian slope at 25°C

$$E_{AB} = k_A + \frac{S}{z_A} \log a_A + \frac{17.8}{z_Z} \text{ mV} \qquad (13.53)$$

The a_A at which the theoretical line deviates from the experimental line by 17.8 mV (for $z_A = 1$) or 8.9 mV (for $z_A = 2$) gives the same response as the background a_B. Then $K_{AB} = a_A/a_B^{z_A/z_B}$.

Equation 13.52 implies that the activity of A that gives the same response as that from background B can be found graphically from the point at which the extrapolated line and the experimental line differ by 0.301 S/z_A mV, that is, 17.8/z_A mV. For a monovalent ion electrode, this is 17.8 mV for a Nernstian response. Then Equation 13.50 applies again.

Although these two methods of calculation rarely give identical values of the selectivity ratio, the principle of the calculation is the same, and calculated values are comparable.

3. Matched Potential Method. This is a strictly empirical variation of the fixed interference method. Numerical values of selectivity coefficients may vary with solution conditions, for example, relative concentrations of the ions. The matched potential method allows the analyst to obtain an empirical value under the experimental conditions of the analysis. Suppose you wish to know the relative interference of sodium ion in blood in the measurement of lithium ion (ca. 1 mM) in the serum of a bipolar (manic depressive) patient taking Li_2CO_3. A reference potential is established for 140 m*M* sodium (the concentration in serum). Then the potential of a 140 m*M* sodium chloride solution containing 1 m*M* lithium ion is measured. The increased potential is due to lithium. Finally, increasing concentrations of sodium chloride solutions are measured to establish what additional sodium concentration corresponds to the 1 m*M* lithium response. That ratio of Li$^+$ (1 m*M*) to increased Na$^+$ gives the empirical selectivity coefficient. If, for example, 240 m*M* Na$^+$ (an increase of 100 m*M*) gives the same response, then the selectivity coefficient is 1.0×10^{-2} (100 times more selective to lithium).[3] See Refs. 16 to 18 for further details.

The matched potential method has been recommended as the preferred method by the International Union of Pure and Applied Chemistry (IUPAC; Ref. 17). Horvai has examined the procedure in more detail for applicability to all sensors (Ref. 18).

This empirical method gives relative responses of ions under the analytical solution conditions. The ion charge is not considered.

HOW VALID IS THE NICOLSKY EQUATION?

We mentioned that selectivity coefficients determined using the Nicolsky equation often vary, depending on relative concentrations and measurement conditions, and the equation does not apply if the response is non-Nernstian (often the case with secondary ions). The matched potential method is an attempt to deal with this. Bakker and co-workers introduced a new formulism that provides a clear interpretation of the matched potential method, elegantly deriving an equation based on ion exchange in which the last term in Equation 13.49 becomes $k_{AB}{}^{Psel}a_B$, where $k_{AB}{}^{Psel}$ is defined as he *selectivity factor* (see Ref. 19). Note that the charge on the secondary ion relative to the primary ion is not taken into account. They point out that since for ions of different charge, the slopes of the calibration curves differ, selectivity factor values (and consequently matched potential values) are not constants and comparisons are difficult unless the exact experimental conditions are known. For this reason, the Nicolsky coefficients may still be preferred. The authors also derive a formula for calculating required K_{AB} values for given target samples containing a_A and a_B for a specified maximum tolerable error, p_{AB}, in percent, which is valid as long as p_{AB} is less than about 10%:

$$K_{AB} = a_A/a_B^{z_A/z_B}(p_{AB}/100)^{z_A/z_B} \qquad (13.54)$$

This equation can be used to assess the feasibility of a specific measurement with a given ion-selective electrode.

One problem in assessing true selectivity coefficients for neutral carrier-based ISEs is that although an electrode may exhibit Nernstian response to a secondary ion, once it has been exposed to a solution of the primary ion, it loses that Nernstian response to the secondary ion. Bakker therefore developed a procedure for determining unbiased selectivity coefficients for newly prepared electrodes by first making measurements with the secondary ion, before exposing the electrode to the primary ion (see Refs. 20 and 21). In this manner, data can be obtained about which ionophores are really the most selective for the primary ion. See Ref. 22 for a review of selectivity of ion-selective electrodes. For an excellent review of ionophore-based electrodes, see Ref. 14 by Bakker et al. (a comprehensive 50-page review).

---

[3] See V. P. Y. Gadzekpo and G. D. Christian, *Anal. Chim. Acta*, **164** (1984) 279 for further details.

HOW SENSITIVE ARE ISES?

The detection limits for ion-selective electrodes are often taken as the activity or concentration where the extrapolated linear response curve intersects the horizontal part, as in Figure 13.16. Detection limits are typically on the order of 10^{-5} to 10^{-6} M. Ion-selective electrodes usually have a fairly high concentration of a salt of the primary ion (the one being measured) added to the internal solution to maintain a stable potential at the membrane. An example is 10 mM $CaCl_2$ in a calcium–ISE internal solution. Bakker and co-workers hypothesized that this internal primary ion flows through the sensor membrane to the analyte solution side and pollutes the thin layer of sample solution in contact with the membrane surface. This leakage results in micromolar concentrations at the membrane surface. Since the electrodes respond to phase boundary activities, this limits the analyte concentration that can be detected to that range. They demonstrated that by lowering the inner electrolytes to micromolar levels, detection limits could be lowered by 100-fold or more, to the 10-nM range. See Refs. 23 and 24.

Pretsch and co-workers in Switzerland were able to lower detections even further by also setting up an ion exchange process in the membrane that prevents the primary ion from flowing out at all. They do this by adding a relatively high concentration of another salt in the inner solution, for example, NaCl for the calcium electrode. This creates a concentration gradient that causes the calcium ions to diffuse inward from the membrane, and so none diffuses out. In this manner, they were able to achieve detection limits in the picomolar range (10^{-12} M!). See Ref. 25. A practical limit is that if the induced flow of primary ions toward the inner solution is too rapid, there may be depletion of the primary ion at the analyte membrane surface, causing a negative error. Bakker and Pretsch have collaborated to understand the mechanisms and develop systems that are stable and robust and do not exhibit bias. See, for example, Refs. 26 and 27.

MEASUREMENT WITH ION-SELECTIVE ELECTRODES

Ion-selective electrodes are subject to the same accuracy limitations as pH electrodes. For $z_A = 2$, the errors per millivolt are doubled.

As with pH glass electrodes, most ion-selective electrodes have high resistance, and an electrometer or pH meter must be used to make measurements. An expanded-scale pH meter is generally used. It is often necessary to pretreat ion-selective electrodes by soaking in a solution of the ion to be determined.

The problems and accuracy limitations discussed under pH and other direct potentiometric measurements apply to ion-selective electrodes.

Ion-selective electrodes measure only the *free* ion.

A calibration curve of potential versus log activity is usually prepared. If concentrations are to be measured, then the technique of maintaining a constant ionic strength as described earlier is used (Equation 13.34). For example, the concentration of unbound calcium ion in serum is determined by diluting samples and standards with 0.15 M NaCl. Only the *unbound* calcium is measured and not the fraction that is complexed.

The activity coefficient of sodium ion in normal human serum has been estimated, using ion-selective electrodes, to be 0.780 ± 0.001, and in serum water to be 0.747 (serum contains about 96% water by volume). Standard solutions of sodium chloride and potassium chloride are usually used to calibrate electrodes for the determination of sodium and potassium in serum. Concentrations of 1.0, 10.0, and 100.0 mmol/L can be prepared with respective activities of 0.965, 9.03, and 77.8 mmol/L for sodium ion in pure sodium chloride solution and 0.965, 9.02, and 77.0 mmol/L for potassium ion in pure potassium chloride solution.

The response of ion-selective electrodes is frequently slow, and considerable time must be taken to establish an equilibrium reading. The response becomes more sluggish as the concentration is decreased. Some electrodes, on the other hand, respond sufficiently rapidly that they can be used to monitor reaction rates.

We can summarize some of the advantages and disadvantages of ion-selective electrodes and some precautions and limitations in their use as follows:

---

1. They measure activities rather than concentrations, a unique advantage but a factor that must be considered in obtaining concentrations from measurements. Interference can occur from ionic-strength effects.

2. They measure "free" ions (i.e., the portion that is not associated with other species). Chemical interference can occur from complexation, protonation, and the like.

3. They are not specific but merely more selective toward a particular ion. Hence, they are subject to interference from other ions. They respond to hydrogen ions and are, therefore, pH-limited.

4. They function in turbid or colored solutions, where photometric methods fail.

5. They have a logarithmic response, which results in a wide dynamic working range, generally from four to six orders of magnitude. This logarithmic response also results in an essentially constant, albeit relatively large, error over the working range where the Nernst relation holds.

6. Except in dilute solution, their response is fairly rapid, often requiring less than 1 min for a measurement. Electrode response is frequently rapid enough to allow process stream monitoring.

7. The response is temperature dependent, by RT/nF.

8. The required measuring equipment can be made portable for field operations, and small samples (e.g., 1 mL) can be measured.

9. The sample is not destroyed in measurement.

10. While certain electrodes may operate down to 10^{-6} M, many will not, and electrodes are not ultrasensitive tools as are some other techniques that may approach 10^{-9} M sensitivities or lower.

11. Frequent calibration is generally required.

12. Few primary activity standards are available, as there are for pH measurements,[4] and the standard solutions that are used are not "buffered" in the ion being tested. Impurities, especially in dilute standards, may cause erroneous results.

The logarithmic response of potentiometric electrodes gives a wide dynamic range, but at some expense in precision.

In spite of some limitations, ion-selective electrodes have become important because they represent an approach to the analytical chemist's dream of a probe that is specific for the test substance and, therefore, requires essentially no sample preparation for fluids.

---

[4] Activity standards similar to pH standards are available from the National Institute of Standards and Technology for some salts, such as sodium chloride.

13.18 Solid-State ISFET Electrodes

Small solid-state chemical sensors can be prepared by using semiconductors as the base for electrical contact. Such sensors are termed ion-selective field-effect transistors (ISFETs). The transistor is coated with an insulating layer of SiO_2 plus a layer of Si_3N_4, which enhances the electrical stability and gives a layer that is impervious to ions in the sample solution. The layer is coated with an ion-sensing membrane, for example, a PVC membrane-sensing layer. A potential applied to the transistor causes a small current to flow, and this varies as the potential of the ion-sensing membrane changes. The applied voltage is varied to bring the current back to its original value, and this represents the potentiometric response to the analyte ion.

A nonglass ISFET pH electrode is shown in Figure 13.17. The reference electrode and temperature sensor are incorporated in the small probe. For information on pH ISFETS, see www.phmeters.com (IQ Scientific Instruments, Inc.), www.sentronph.com (Sentron, Inc.), and www.servonics.com/Jenco.htm (Jenco Instruments, Inc.).

Learning Objectives

WHAT ARE SOME OF THE KEY THINGS WE LEARNED FROM THIS CHAPTER?

- Types of electrodes and electrode potentials from the Nernst equation (key equations: 13.3, 13.10, 13.16), p. 369
- Liquid junctions and junction potentials, p. 375
- Reference electrodes, p. 378
- Accuracy of potentiometric measurements (key equation: 13.36), p. 383
- The pH glass electrode (key equation: 13.42), p. 384
- Standard buffers and the accuracy of pH measurements, pp. 389, 391
- The pH meter, p. 391
- Ion-selective electrodes, p. 395
- The selectivity coefficient (key equation: 13.46), p. 400

Fast response silicon chip sensor

Built-in reference and medical grade temperature sensor

Fig. 13.17. Solid-state ISFET electrode. (Courtesy of IQ Scientific Instruments, Inc.)

1. What is the liquid-junction potential? Residual liquid-junction potential? How can these be minimized?

2. Discuss the mechanism of the glass membrane electrode response for pH measurements.

3. What is the alkaline error and the acid error of a glass membrane pH electrode?

4. Describe the different types of ion-selective electrodes. Include in your discussion the construction of the electrodes, differences in membranes, and their usefulness.

5. What is the selectivity coefficient? Discuss its significance and how you would determine its value.

6. What is a crown ether? What would a 16-crown-6 ether represent?

7. What is the Nicolsky equation?

STANDARD POTENTIALS

8. The standard potential of the silver/silver bromide electrode is 0.073 V. Calculate the solubility product of silver bromide.

9. A sample of thiocyanate is titrated with silver solution. The potential at the end point of 0.202 V versus SCE. Calculate the standard potential for $Ag^+ + e^- = Ag$. The K_{sp} for silver thiocyanate is 1.00×10^{-12}.

VOLTAIC CELLS

10. For each of the following reactions, (1) separate the reaction into its component half-reactions; (2) write a schematic representation of a cell in which the reaction would occur in the direction as written; (3) calculate the standard potential of the cell; (4) assign the polarity of each electrode under conditions that the reaction would occur as written.

   (a) $Ag + Fe^{3+} = Ag^+ + Fe^{2+}$

   (b) $VO_2^+ + V^{3+} = 2VO^{2+}$

   (c) $Ce^{4+} + Fe^{2+} = Ce^{3+} + Fe^{3+}$

11. For the following cells, write the half-reactions occurring at each electrode and the complete cell reaction, and calculate the cell potential:

   (a) Pt, H_2(0.2 atm)$|$HCl(0.5 M)$|$Cl$_2$(0.2 atm), Pt

   (b) Pt$|$Fe^{2+}(0.005 M), Fe^{3+}(0.05 M), HClO$_4$(0.1 M)$\|$HClO$_4$(0.1 M), VO$_2^+$ (0.001 M), VO^{2+}(0.002) $M|$Pt

REDOX POTENTIOMETRIC MEASUREMENTS

12. What would the potentials of the following half-cells at standard conditions be versus a saturated calomel electrode? (a) Pt/Br$_2$(aq), Br$^-$; (b) Ag/AgCl/Cl$^-$; (c) Pt/V^{3+}, V^{2+}.

13. What would be the observed potential if the following half-cell were connected with a saturated calomel electrode?

   Fe^{3+}(0.00200 M), Fe^{2+}(0.0500 M)$|$Pt (numbers represent activities)

14. A 50-mL solution that is 0.10 M in chloride and iodide ions is titrated with 0.10 M silver nitrate. (a) Calculate the percent iodide remaining unprecipitated when silver chloride begins to precipitate. (b) Calculate the potential of a silver electrode versus the SCE when silver chloride begins to precipitate and compare this with the theoretical potential corresponding to end point for the titration of iodide. (c) Calculate the potential at the end point for chloride. Use concentrations in calculations.

15. The potential of the electrode Hg|Hg–EDTA, N–EDTA, N^{n+} is a function of the metal in N^{n+} and can be shown as

$$E = E^0_{Hg^{2+},Hg} - \frac{2.303RT}{2F} \log \frac{K_{f(HgEDTA)}}{K_{f(NEDTA)}}$$
$$- \frac{2.303RT}{2F} \log \frac{a_{NEDTA}}{a_{HgEDTA}} - \frac{2.303RT}{2F} \log \frac{1}{a_{N^{n+}}}$$

The stability of N–EDTA must be less than that of Hg–EDTA (a very stable chelate; $K_{f(Hg–EDTA)} = 10^{22}$). A Hg|Hg–EDTA electrode can be used to monitor N^{n+} during the course of a titration with EDTA. Starting with the Hg|Hg^{2+} electrode, derive the above equation. This represents a metal–metal chelate–metal ion electrode.

16. The potential of a hydrogen electrode in an acid solution is -0.465 V when measured against an SCE reference electrode. What would the potential be measured against a normal calomel electrode (1 M KCl)?

pH POTENTIOMETRIC MEASUREMENTS

17. (a) How accurately can the pH of an unknown solution generally be measured? What limits this? What is this (calculate it) in terms of millivolts measured? In terms of percent error in the hydrogen ion activity? (b) How precisely can the pH of a solution be measured? How much is this in terms of millivolts measured? In terms of percent variation in the hydrogen ion activity?

18. A glass electrode was determined to have a potential of 0.395 V when measured against the SCE in a standard pH 7.00 buffer solution. Calculate the pH of the unknown solution for which the following potential readings were obtained (the potential decreases with increasing pH):

(a) 0.467 V

(b) 0.209 V

(c) 0.080 V

(d) -0.013 V

19. Calculate the potential of the cell consisting of a hydrogen electrode ($P_{H_2} = 1$ atm) and a saturated calomel reference electrode (a) in a solution of 0.00100 M HCl, (b) in a solution of 0.00100 M acetic acid, and (c) in a solution containing equal volumes of 0.100 M acetic acid and 0.100 M sodium acetate. Assume that activities are the same as concentrations.

20. The quinhydrone electrode can be used for the potentiometric determination of pH. The solution to be measured is saturated with quinhydrone, an equimolar mixture of quinone (Q) and hydroquinone (HQ), and the potential of the solution is measured with a platinum electrode. The half-reaction and its standard potential are as follows:

quinone (Q)   hydroquinone (HQ)

What is the pH of a solution saturated with quinhydrone if the potential of a platinum electrode in the solution, measured against a saturated calomel electrode, is -0.205 V? Assume the liquid-junction potential to be zero.

ION-SELECTIVE ELECTRODE MEASUREMENTS

21. It can be shown from Equations 13.44 and 13.45 that, for monovalent ions, $\log K_{AB} = (k_B - k_A)/S$. Derive this equation.

22. A potassium ion-selective electrode is used to measure the concentration of potassium ion in a solution that contains $6.0 \times 10^{-3} M$ cesium (activity). From Table 13.3, the electrode responds equally to either ion ($K_{KCs} = 1$). If the potential versus a reference electrode is -18.3 mV for a $5.0 \times 10^{-} M$ KCl solution and $+20.9$ mV in the sample solution, what is the activity of K^+ in the sample? Assume Nernstian response.

23. The nitrate concentration in an industrial effluent is determined using a nitrate ion-selective electrode. Standards and samples are diluted 20-fold with 0.1 M K_2SO_4 to maintain constant ionic strength. Nitrate standards of 0.0050 and 0.0100 M give potential readings of -108.6 and -125.2 mV, respectively. The sample gives a reading of -119.6 mV. What is the concentration of nitrate in the sample?

24. The perchlorate concentration in a sample containing 0.015 M iodide is determined using a perchlorate ion-selective electrode. All samples and standards are diluted 1:10 with 0.2 M KCl to maintain constant ionic strength. A 0.00100 M $KClO_4$ standard gives a reading of -27.2 mV, and a 0.0100 M KI standard gives a reading of $+32.8$ mV. The sample solution gives a reading of -15.5 mV. Assuming Nernstian response, what is the concentration of perchlorate in the sample?

25. The potential of a glass cation-sensitive electrode is measured against an SCE. In a sodium chloride solution of activity 0.100 M, this potential is 113.0 mV, and in a potassium chloride solution of the same activity, it is 67.0 mV. (a) Calculate the selectivity coefficient of this electrode for potassium over sodium, using the relationship derived in Problem 21. (b) What would be the expected potential in a mixture of sodium ($a = 1.00 \times 10^{-3} M$) and potassium ($a = 1.00 \times 10^{-2} M$) chlorides? Assume Nernstian response, 59.2 mV/decade.

26. The selectivity coefficient for a cation-selective electrode for B^+ with respect to A^+ is determined from measurements of two solutions containing different activities of the two ions. The following potential readings were obtained: (1) $2.00 \times 10^{-4} M$ $A^+ + 1.00 \times 10^{-3} M$ B^+, $+237.8$ mV; and (2) $4.00 \times 10^{-4} M$ $A^+ + 1.00 \times 10^{-3} M$ B^+, $+248.2$ mV. Calculate K_{AB}. The electrode response is 56.7 mV/decade.

27. A sodium glass ion-selective electrode is calibrated using the separate solution method, for sodium response and potassium response. The two calibration curves have slopes of 58.1 mV per decade, and the sodium curve is 175.5 mV more positive than the potassium curve. What is K_{NaK} for the electrode?

28. A valinomycin-based potassium ion-selective electrode is evaluated for sodium interference using the fixed interference method. A potassium calibration curve is prepared in the presence of 140 mM sodium. The straight line obtained from extrapolation of the linear portion deviates from the experimental curve by 17.4 mV at a potassium concentration corresponding to 1.5×10^{-5} M. If the linear slope is 57.8 mV per decade, what is K_{NaK} for the electrode?

Recommended References

ACTIVITY COEFFICIENTS

1. J. Kielland, "Individual Activity Coefficients of Ions in Aqueous Solutions," *J. Am. Chem. Soc.*, **59** (1937) 1675.

RELATIVE POTENTIALS

2. W.-Y. Ng, "Conversion of Potentials in Voltammetry and Potentiometry," *J. Chem. Ed.*, **65** (1988) 726.

pH ELECTRODES AND MEASUREMENTS

3. R. G. Bates, *Determination of pH*. New York: Wiley, 1964.

4. H. B. Kristensen, A. Salomon, and G. Kokholm, "International pH Scales and Certification of pH," *Anal. Chem.*, **63** (1991) 885A.

5. J. V. Straumford, Jr., "Determination of Blood pH," in D. Seligson, ed., *Standard Methods of Clinical Chemistry*, Vol. 2. New York: Academic, 1958, pp. 107–121.

6. C. C. Westcott, *pH Measurements*. New York: Academic, 1978.

7. H. Galster, *pH Measurement. Fundamentals, Methods, Applications, Instrumentation*. New York: VCH, 1991.

ION-SELECTIVE ELECTRODES AND MEASUREMENTS

8. R. G. Bates, "Approach to Conventional Scales of Ionic Activity for the Standardization of Ion-Selective Electrodes," *Pure Appl. Chem.*, **37** (1974) 573.

9. R. A. Durst, ed., *Ion Selective Electrodes*. Washington, DC: National Bureau of Standards Special Publication 314, U.S. Government Printing Office, 1969.

10. E. Pungor, "Theory and Applications of Anion Selective Membrane Electrodes," *Anal. Chem.*, **39**(13) (1967) 28A.

11. G. A. Rechnitz, "New Directions for Ion Selective Electrodes," *Anal. Chem.*, **41**(12) (1969) 109A. Dated but gives good description of principles.

12. G. A. Rechnitz, "Ion Selective Electrodes," *Chem. Eng. News*, **45** (1967) (June 12) 146. Also good description of principles.

13. B. Fu, E. Bakker, M. E. Mehyerhoff, V. C. Yhang, and J. H. Yoon, "Polyion Sensitive Membrane Electrodes; Principles and Biomedical Applications," *Anal. Chem.*, **68** (1996) 168A.

14. E. Bakker. P. Buhlmann, and E. Pretsch, "Carrier-Based Ion-Selective Electrodes and Bulk Optodes. 1. General Characteristics," *Chem. Rev.*, **97** (1997) 3083.

SELECTIVITY COEFFICIENTS

15. Y. Umezawa, *CRC Handbook of Ion Selective Electrodes: Selectivity Coefficients.* Boca Raton, FL: CRC Press, 1990.

16. V. P. Y. Gadzekpo and G. D. Christian, "Determination of Selectivity Coefficients of Ion-Selective Electrodes by a Matched-Potential Method." *Anal. Chim. Acta,* **166** (1984) 279.

17. Y. Umezawa, K. Umezawa, and H. Ito, "Selectivity Coefficients for Ion-Selective Electrodes: Recommended Methods for Reporting k_{ij}^{pot} Values," *Pure Appl. Chem.,* **67** (1995) 507.

18. G. Horvai, "The Matched Potential Method, a Generic Approach to Characterize the Differential Selectivity of Chemical Sensors," *Sensors and Actuators B,* **43** (1997) 94. See also G. Horvai, *Trends in Anal. Chem.,* **16** (1997) 260.

19. E. Bakker, R. K. Meruwa, E. Pretsch, and M. E. Meyerhoff "Selectivity of Polymer Membrane-Based Ion-Selective Electrodes: Self-Consistent Model Describing the Potentiometric Response in Mixed Ion Solutions of Different Charge," *Anal. Chem.,* **66** (1994) 3021.

20. E. Bakker, "Determination of Improved Selectivity Coefficients of Polymer Membrane Ion-Electrodes by Conditioning with a Discriminating Ion," *J. Electrochem. Soc.,* **143**(4) (1996) L83.

21. E. Bakker, "Determination of Unbiased Selectivity Coefficients of Neutral Carrier-Based Cation-Selective Electrodes, *Anal. Chem.,* **69** (1997) 1061.

22. E. Bakker. "Review: Selectivity of Liquid Membrane Ion-Selective Electrodes," *Electroanalysis,* **9** (1997) 7.

ULTRASENSITIVE ION-SELECTIVE ELECTRODES

23. S. Mathison and E. Bakker, "Effect of Transmembrane Electrolyte Diffusion on the Detection Limit of Carrier-Based Potentiometric Sensors," *Anal. Chem.,* **70** (198) 303.

24. Y. Mi, S. Mathison, R. Goines, A. Logue, and E. Bakker, "Detection Limit of Polymeric Membrane Potentiometric Ion Sensors: How Low Can We Go Down to Trace Levels?" *Anal. Chim. Acta,* **397** (1999) 103.

25. T. Sokalski, A. Ceresa, T. Zwickl, and E. Pretsch, "Large Improvement of the Lower Detection Limit of Ion-Selective Polymer Membrane Electrodes," *J. Am. Chem. Soc.,* **119** (1997) 11,347.

26. T. Sokalski, T. Zwickl, E. Bakker, and E. Pretsch, "Lowering the Detection Limit of Solvent Polymeric Ion-Selective Electrodes. 1. Modeling the Influence of Steady-State Ion Fluxes," *Anal. Chem.,* **71** (1999) 1204.

27. T. Sokalski, A. Cera, M. Fibbioli, T. Zwickl, E. Bakker, and E. Pretsch, "Lowering the Detection Limit of Solvent Polymeric Ion-Selective Electrodes. 2. Influence of Composition of Sample and Internal Electrolyte Solution," *Anal. Chem.* **71** (1999) 1210.

Chapter Fourteen

REDOX AND POTENTIOMETRIC TITRATIONS

Volumetric analyses based on titrations with reducing or oxidizing agents are very useful for many determinations. They may be performed using visual indicators or by measuring the potential with an appropriate indicating electrode to construct a potentiometric titration curve. In this chapter, we discuss redox titration curves based on half-reaction potentials and describe representative redox titrations and the necessary procedures to obtain the sample analyte in the correct oxidation state for titration. The construction of potentiometric titration curves is described, including derivative titration curves and Gran plots. You should first review the balancing of redox reactions since balanced reactions are required for volumetric calculations.

14.1 First: Balance the Reduction—Oxidation Reaction

The calculations in volumetric analysis require that the balanced reaction be known. The balancing of redox reactions is reviewed in your CD.

Various methods are used to balance redox reactions, and we shall use the **half-reaction method**. In this technique, the reaction is broken down into two parts: the oxidizing part and the reducing part. In every redox reaction, an oxidizing agent reacts with a reducing agent. The oxidizing agent is reduced in the reaction while the reducing agent is oxidized. Each of these constitutes a *half-reaction*, and the overall reaction can be broken down into these two half-reactions. Thus, in the reaction

$$Fe^{2+} + Ce^{4+} \rightarrow Fe^{3+} + Ce^{3+}$$

Fe^{2+} is the reducing agent and Ce^{4+} is the oxidizing agent. The corresponding half-reactions are:

$$Fe^{2+} \rightarrow Fe^{3+} + e^-$$

and

$$Ce^{4+} + e^- \rightarrow Ce^{3+}$$

There are various ways of balancing redox reactions. Use the method you are most comfortable with.

To balance a reduction–oxidation reaction, each half-reaction is first balanced. There must be a net gain or loss of zero electrons in the overall reaction, and so the second step is multiplication of one or both of the half-reactions by an appropriate factor or factors so that, when they are added, the electrons cancel. The final step is addition of the half-reactions. Review the procedures posted on the Wiley website for your text.

14.2 Calculation of the Equilibrium Constant of a Reaction— Needed to Calculate Equivalence Point Potentials

Before we discuss redox titration curves based on reduction–oxidation potentials, we need to learn how to calculate equilibrium constants for redox reactions from the half-reaction potentials. The reaction equilibrium constant is used in calculating equilibrium concentrations at the equivalence point, in order to calculate the equivalence point potential. Recall from Chapter 12 that since a cell voltage is zero at reaction equilibrium, the difference between the two half-reaction potentials is zero (or the two potentials are equal), and the Nernst equations for the half-reactions can be equated. When the equations are combined, the log term is that of the equilibrium constant expression for the reaction (see Equation 12.20), and a numerical value can be calculated for the equilibrium constant. This is a consequence of the relationship between the free energy and the equilibrium constant of a reaction. Recall from Equation 6.10 that $\Delta G° = -RT \ln K$. Since $\Delta G° = -nFE^0$ for the reaction, then

At the equivalence point, we have unknown concentrations that must be calculated from K_{eq}. This is calculated by equating the two Nernst equations, combining the concentration terms to give K_{eq}, and then solving for K_{eq} from ΔE^0.

$$-RT \ln K = -nFE^0 \tag{14.1}$$

or

$$E^0 = \frac{RT}{nF} \ln K$$

For the spontaneous reaction, $\Delta G°$ is negative and E^0 is positive.

Example 14.1

Calculate the potential in a solution (vs. NHE) when 5.0 mL of 0.10 M Ce^{4+} solution is added to 5.0 mL of 0.30 M Fe^{2+} solution, using the cerium half-reaction. Compare with Example 12.4.

Solution

This is the same as Example 12.4 where we used the iron half-reaction to calculate the potential since both [Fe^{2+}] and [Fe^{3+}] were known. We begin with 0.30 × 5.0 = 1.5 mmol Fe^{2+} and add 0.10 × 5.0 = 0.50 mmol Ce^{4+}. So we form 0.50 mmol each of Fe^{3+} and Ce^{3+}, leaving 1.0 mmol Fe^{2+}:

$$
\begin{array}{ccccc}
\text{Fe}^{2+} & + \text{Ce}^{4+} \rightleftharpoons & \text{Fe}^{3+} & + & \text{Ce}^{3+} \\
1.0 + x & x & 0.50 - x & & 0.50 - x
\end{array}
$$

where the numbers and x represent millimoles. In order to use the cerium half-reaction, we need to solve for x. This can only be done using the equilibrium constant, which is obtained by equating the two half-reaction potentials. The Ce^{4+}/Ce^{3+} half-reaction is

$$Ce^{4+} + e^- \rightleftharpoons Ce^{3+}$$

$$E = 1.61 - 0.059 \log \frac{[Ce^{3+}]}{[Ce^{4+}]}$$

Therefore,

$$1.61 - 0.059 \log \frac{[Ce^{3+}]}{[Ce^{4+}]} = 0.771 - 0.059 \log \frac{[Fe^{2+}]}{[Fe^{3+}]}$$

$$0.84 = 0.059 \log \frac{[Ce^{3+}][Fe^{3+}]}{[Ce^{4+}][Fe^{2+}]} = 0.059 \log K_{eq}$$

$$\frac{[Ce^{3+}][Fe^{3+}]}{[Ce^{4+}][Fe^{2+}]} = 10^{0.84/0.059} = 10^{14.2} = 1.6 \times 10^{14} = K_{eq}$$

Note that the large magnitude of K_{eq} indicates the reaction lies far to the right at equilibrium. Now, since the volumes cancel, we can use millimoles instead of millimoles/milliliter (molarity) and

$$[Ce^{3+}] = 0.50 - x \approx 0.50 \text{ mmol}$$
$$[Ce^{4+}] = x \text{ mmol}$$
$$[Fe^{3+}] = 0.50 - x \approx 0.50 \text{ mmol}$$
$$[Fe^{2+}] = 1.0 + x \cong 1.0 \text{ mmol}$$

Therefore,

$$\frac{(0.50 \text{ mmol})(0.50 \text{ mmol})}{(x \text{ mmol})(1.0 \text{ mmol})} = 1.6 \times 10^{14}$$

$$x = 1.6 \times 10^{-15} \text{ mmol} (= 1.6 \times 10^{-16} M)$$

We see how very small $[Ce^{4+}]$ is. Nevertheless, it is finite, and knowing its concentration, we calculate the potential from the Nernst equation, using millimoles:

$$E = 1.61 - 0.059 \log \frac{[Ce^{3+}]}{[Ce^{4+}]} = 1.61 - 0.059 \log \frac{0.50 \text{ mmol}}{1.6 \times 10^{-15} \text{ mmol}}$$

$$= 0.75 \text{ V}$$

This compares with 0.753 V calculated in Example 12.4.

• •

Obviously, it is easier to make the calculations using the half-reaction we have the most information about; in essence, the potential of that half-reaction must be calculated anyway during the calculation using the other half-reaction. The calculations illustrate that in a mixture, the concentrations of all species *at equilibrium* are such that the potential of each half-reaction is the same. Note that the potential will be close to the standard potential (E^0) of the half-reaction in which there is an excess of the reactant; in this case, there is an excess of Fe^{2+}.

It should be pointed out here that the n values in the two half-reactions do not have to be equal in order to equate the Nernst equations. For convenience, the two half-reactions are generally adjusted to the same n value before the Nernst equations are equated.

When there are stoichiometric amounts of reactants, for example, at the equivalence point of a titration, the equilibrium concentrations of the species in neither half-reaction is known, and an approach similar to the calculation in Example 14.1 is required.

 Example 14.2

Calculate the potential of a solution obtained by reacting 10 mL each of 0.20 M Fe^{2+} and 0.20 M Ce^{4+}.

Solution

The reactants are essentially quantitatively converted to equivalent quantities of Fe^{3+} and Ce^{3+} and the concentration of each of the products is 0.10 M (neglecting the amount of the reverse reaction):

$$Fe^{2+} + Ce^{4+} \rightleftharpoons Fe^{3+} + Ce^{3+}$$
$$x \qquad x \qquad 0.10 - x \qquad 0.10 - x$$

where x represents the molar concentration of Fe^{2+} and Ce^{4+}. We can solve for x as in Example 14.1 and then plug it in the Nernst equation for either half-reaction to calculate the potential (do this for practice). Another approach follows.

The potential is given by either Nernst equation:

$$E = E^0_{Fe^{3+},Fe^{2+}} - \frac{0.059}{n_{Fe}} \log \frac{[Fe^{2+}]}{[Fe^{3+}]}; \qquad n_{Fe}E = n_{Fe}E^0_{Fe^{3+},Fe^{2+}} - 0.059 \log \frac{x \text{ mmol/mL}}{0.10 \text{ mmol/mL}}$$

$$E = E^0_{Ce^{4+},Ce^{3+}} - \frac{0.059}{n_{Ce}} \log \frac{[Ce^{3+}]}{[Ce^{4+}]}; \qquad n_{Ce}E = n_{Ce}E^0_{Ce^{4+},Ce^{3+}} - 0.059 \log \frac{0.10 \text{ mmol/mL}}{x \text{ mmol/mL}}$$

Note that Nernst equations for both species are written for reductions even though one of the species, here Fe^{2+}, is actually being oxidized in the reaction. We can add these equations together and solve for E, the potential of each half-reaction, and hence the potential of the solution at equilibrium:

$$n_{Fe}E + n_{Ce}E = n_{Fe}E^0_{Fe^{3+},Fe^{2+}} + n_{Ce}E^0_{Ce^{4+},Ce^{3+}} - 0.059 \log \frac{x \text{ mmol/mL}}{0.10 \text{ mmol/mL}} \cdot \frac{0.10 \text{ mmol/mL}}{x \text{ mmol/mL}}$$

$$E = \frac{n_{Fe}E^0_{Fe^{3+},Fe^{2+}} + n_{Ce}E^0_{Ce^{4+},Ce^{3+}}}{n_{Fe} + n_{Ce}} = \frac{(1)0.77 + (1)1.61}{1 + 1} = 1.19 \text{ V}$$

The above approach is general, that is, E for stoichiometric quantities of reactants (E at the equivalence point of a titration) is given by

$$E = \frac{n_1 E^0_1 + n_2 E^0_2}{n_1 + n_2} \qquad (14.2)$$

Use this equation to calculate the equivalence point potential, *if* there is no polyatomic species or proton dependence. See Problem 13 for those cases.

where n_1 and E_1^0 are the n value and standard potential for one half-reaction and n_2 and E_2^0 are the values for the other half-reaction. In other words, E is the weighted average of the E^0 values. In the above example, it was the simple average since the n values were each unity. *This equation holds only for reactions in which there are no polyatomic species (e.g., $Cr_2O_7^{2-}$) and no hydrogen ion dependence (or when the pH is zero).* The equation contains additional terms if pH and concentration factors must be considered (see Problem 13). It can be applied if **formal potentials** are used, that is, for the specified conditions of acidity (see Chapter 12).

14.3 Calculating Redox Titration Curves

The potential change at the end point will be approximately ΔE^0 for the reactant and titrant half-reactions.

We can use our understanding of redox equilibria to describe titration curves for redox titrations. The shape of a titration curve can be predicted from the E^0 values of the analyte half-reaction and the titrant half-reaction. Roughly, the potential change in going from one side of the equivalence point to the other will be equal to the difference in the two E^0 values; the potential will be near E^0 for the analyte half-reaction before the equivalence point and near that of the titrant half-reaction beyond the equivalence point.

Consider the titration of 100 mL of 0.1 M Fe^{2+} with 0.1 M Ce^{4+} in 1 M HNO_3. Each millimole Ce^{4+} will oxidize one millimole Fe^{2+}, and so the end point will occur at 100 mL. The titration curve is shown in Figure 14.1. This is actually a plot of the potential of the titration solution relative to the NHE, whose potential by definition is zero.

The indicator electrode monitors potential changes throughout the titration.

In Chapter 13, we saw that the potential difference between redox half-cells can be measured with inert electrodes such as platinum in a cell similar to that in Figure 12.1. The electrode dipped in the titration or test solution is called the **indicator electrode**, and the other is called the **reference electrode**, whose potential remains constant. Hence, the potential of the indicator electrode will change relative to that of the reference electrode as indicated in Figure 14.1. The potential relative to the NHE is plotted against the volume of titrant. This is analogous to plotting the pH of a solution versus the volume of titrant in an acid–base titration, or pM against volume of titrant in a precipitation or complexometric titration. In a redox titration, it is the potential rather than the pH that changes with con-

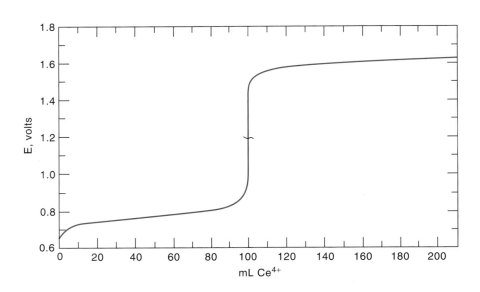

Fig. 14.1. Titration curve for 100 mL 0.1 M Fe^{2+} versus 0.1 M Ce^{4+}.

centration. At the beginning of the titration, we have only a solution of Fe^{2+}, and so we cannot calculate the potential. As soon as the first drop of titrant is added, a known amount of Fe^{2+} is converted to Fe^{3+}, and we know the ratio of $[Fe^{2+}]/[Fe^{3+}]$. So the potential can be determined from the Nernst equation of this couple. It will be near the E^0 value for this couple (the sample) *before* the end point.

Note that, since $[Fe^{2+}]/[Fe^{3+}]$ is equal to unity at the midpoint of the titration and $\log 1 = 0$, the potential is equal to E^0 at this point in the titration. This will only be true if the half-reaction is symmetrical. For example, in the half-reaction $I_2 + 2e^- \rightleftharpoons 2I^-$, the $[I^-]$ would be twice $[I_2]$ at midway in a titration, and the ratio would be $[I^-]^2/[I_2] = (2)^2/(1) = 4$. So the potential would be less than E^0 by $(-0.059/2)\log 4$, or -0.018 V.

At the equivalence point of our titration, we have the following conditions:

$$Fe^{2+} + Ce^{4+} \rightleftharpoons Fe^{3+} + Ce^{3+}$$
$$\phantom{Fe^{2+} +}x x C-x C-x$$

where C is the concentration of Fe^{3+}, which we know since all the Fe^{2+} is converted to Fe^{3+} (x is negligible compared to C). Now, we have an unknown quantity in both half-reactions, and so we must solve for x by equating the two Nernst equations, as was done in Example 14.1. Then, we can calculate the potential from either half-reaction. Alternatively, Equation 14.2 could be used since this is a symmetrical reaction (no polyatomic species).

Beyond the equivalence point, we have an excess of Ce^{4+} and an unknown amount of Fe^{2+}. Since we now have more information about the Ce^{4+}/Ce^{3+} half-reaction, it is easier to calculate the potential from its Nernst equation. Note that here, with an excess of titrant, the potential is near the E^0 value of the *titrant*. At 200% of the titration, $[Ce^{4+}]/[Ce^{3+}] = 1$, and E is E^0 of the cerium couple.

Example 14.3 illustrates that the magnitude of the end-point break is directly related to the difference in the E^0 values of the sample and the titrant half-reactions. At least a 0.2-V difference is required for a sharp end point.

A potential change of 0.2 V is needed for a sharp end point.

The equivalence point for this titration is indicated in Figure 14.1. Because the reaction is symmetrical, the equivalence point (*inflection point* of the curve—that point at which it is steepest) occurs at the midpoint of the rising part of the curve. In nonsymmetrical titrations, the inflection point will not occur at the midpoint. For example, in the titration of Fe^{2+} with MnO_4^-, the steepest portion occurs near the top of the break because of the consumption of protons in the reaction, causing it to be nonsymmetrical. See Problem 21 for a spreadsheet calculation of the curve in Figure 14.1.

Example 14.3

Calculate the potential as a function of titrant volume in the above titration of 100 mL of 0.100 M Fe^{2+} at 10.0, 50.0, 100, and 200 mL of 0.100 M Ce^{4+}.

Solution

The reaction is

$$Fe^{2+} + Ce^{4+} \rightleftharpoons Fe^{3+} + Ce^{3+}$$

10.0 mL: mmol Ce^{4+} added $= 0.100\ M \times 10.0$ mL $= 1.00$ mmol

mmol Fe^{2+} reacted $= 1.00$ mmol $=$ mmol Fe^{3+} formed

mmol Fe^{2+} left $= 0.100\ M \times 100$ mL $- 1.00$ mmol $= 9.0$ mmol Fe^{2+}

$$E = 0.771 - 0.059 \log \frac{9.0}{1.00} = 0.715 \text{ V}$$

50.0 mL: One-half the Fe^{2+} is converted to Fe^{3+} (5.00 mmol each)

$$E = 0.771 - 0.059 \log \frac{5.00}{5.00} = 0.771 \text{ V}$$

100 mL: mmol $Fe^{3+} = 10.0 - x \approx 10.0$

mmol $Fe^{2+} = x$

mmol $Ce^{3+} = 10.0 - x \approx 10.0$

mmol $Ce^{4+} = x$

We must solve for x. Since our calculations are for when equilibrium is achieved between the two half-reactions, the two Nernst equations are equal:

$$0.771 - \frac{0.059}{1} \log \frac{[Fe^{2+}]}{[Fe^{3+}]} = 1.61 - \frac{0.059}{1} \log \frac{[Ce^{3+}]}{[Ce^{4+}]}$$

$$-0.84 = -0.059 \log \frac{[Fe^{3+}][Ce^{3+}]}{[Fe^{2+}][Ce^{4+}]} = -0.059 \log K_{eq}$$

$$K_{eq} = 1.7 \times 10^{14}$$

Substituting into K_{eq} and solving for x (use millimoles, since volumes cancel):

$$\frac{(10.0)(10.00)}{(x)(x)} = 1.7 \times 10^{14}$$

$$x = 7.7 \times 10^{-7} \text{ mmol } Fe^{2+} = \text{mmol } Ce^{4+}$$

Use either half-reaction to calculate the potential:

$$E = 0.771 - 0.059 \log \frac{7.7 \times 10^{-7}}{10.0} = 1.19 \text{ V}$$

Compare this with the potential calculated in Example 14.2. Try calculating this potential using the Ce^{4+}/Ce^{3+} Nernst equation. Note that this potential is halfway between the two E^0 potentials.

200 mL: We have 100 mL excess titrant (Ce^{4+}). It is now easier to use the Ce^{4+}/Ce^{3+} half-reaction:

mmol $Ce^{3+} = 10.0 - x \approx 10.0$

mmol $Ce^{4+} = 0.100 \ M \times 100 \text{ mL} + x \approx 10.0 \text{ mmol}$

$$E = 1.61 - 0.059 \log \frac{10.0}{10.0} = 1.61 \text{ V}$$

[We could have used the Fe^{3+}/Fe^{2+} half-reaction to calculate this potential by calculating x ($[Fe^{2+}]$) as above from K_{eq}.]

• •

For nonsymmetrical reactions, we must keep track of the ratio in which the chemicals react. Also, if protons are consumed or produced in the reaction, the change in $[H^+]$ must be calculated.

Example 14.4

Calculate the potential at the equivalence point in the titration of 100 mL of $0.100\ M\ Fe^{2+}$ in $0.500\ M\ H_2SO_4$ with 100 mL of $0.0200\ M\ MnO_4^-$.[1]

Solution

The reaction is

$$5Fe^{2+} + MnO_4^- + 8H^+ \rightleftharpoons 5Fe^{3+} + Mn^{2+} + 4H_2O$$

$$x \qquad \tfrac{1}{5}x \qquad\qquad C-x \quad \tfrac{1}{5}C - \tfrac{1}{5}x$$

> Keep track of millimoles and the ratio in which things react. One millimole Fe^{2+} reacts with $\tfrac{1}{5}$ mmol MnO_4^-.

mmol $Fe^{3+} = 0.100\ M \times 100\ mL - x \approx 10.0$

mmol $Fe^{2+} = x$

mmol $Mn^{2+} = \tfrac{1}{5}(10.0) - \tfrac{1}{5}x \approx 2.00$

mmol $MnO_4^- = \tfrac{1}{5}x$

Solve for x by equating the two Nernst equations (since we are at equilibrium, they are equal). Multiply the Fe^{2+}/Fe^{3+} half-reaction by 5 to equate the electrons:

$$0.771 - \frac{0.059}{5} \log \frac{[Fe^{2+}]^5}{[Fe^{3+}]^5} = 1.51 - \frac{0.059}{5} \log \frac{[Mn^{2+}]}{[MnO_4^-][H^+]^8}$$

$$-0.74 = \frac{0.059}{5} \log \frac{[Mn^{2+}][Fe^{3+}]^5}{[MnO_4^-][Fe^{2+}]^5[H^+]^8} = -\frac{0.059}{5} \log K_{eq}$$

$$K_{eq} = 5_{.0} \times 10^{62}$$

We started with $1.00 \times 100 = 100$ mmol H^+. We consumed 8 mmol per 5 mmol Fe^{2+} reacted ($= \tfrac{8}{5} \times 10.0 = 16.0$ mmol). Therefore, we have 84 mmol H^+ left in 200 mL, or 0.42 M. For the other species, we can use millimoles since the volumes all cancel:

> We must calculate the concentration of H^+ after reaction.

$$\frac{(2.00)(10.0)^5}{(1/5x)(x)^5(0.42)^8} = 5_{.0} \times 10^{62}$$

$$x = 1.1 \times 10^{-9}\ \text{mmol}\ Fe^{2+}; \qquad \text{mmol}\ MnO_4^- = \tfrac{1}{5}(1.1 \times 10^{-9})$$
$$= 2.2 \times 10^{-10}$$

Use either half-reaction to calculate the potential:

$$E = 0.771 - 0.059 \log \frac{1.1 \times 10^{-9}}{10} = 1.35_9\ V$$

E^0 for the Mn^{2+}/MnO_4^- couple is 1.51 V. Note that the potential halfway between the two E^0 potentials is 1.14 V. The equivalence point (inflection point) for this unsymmetrical titration reaction is therefore closer to the titrant couple and the titration curve is unsymmetrical.

---

[1]In 0.5 $M\ H_2SO_4$, the second proton is only about 2% dissociated (calculate it from the ionization constant and see!); but for simplicity, we will assume that it is completely ionized to give a H^+ concentration of 1 M.

14.4 Visual Detection of the End Point

Obviously, the end point can be determined by measuring potential with an indicating electrode (Chapter 13) relative to a reference and plotting this against the volume of titrant. But as in other titrations, it is usually more convenient to use a visual indicator. There are three methods used for visual indication.

SELF-INDICATION

If the titrant is highly colored, this color may be used to detect the end point. For example, a 0.02 M solution of potassium permanganate is deep purple. A dilute solution of potassium permanganate is pink. The product of its reduction, Mn^{2+}, is nearly colorless, being a very faint pink. During a titration with potassium permanganate, the purple color of the MnO_4^- is removed as soon as it is added because it is reduced to Mn^{2+}. As soon as the titration is complete, a fraction of a drop of excess MnO_4^- solution imparts a definite pink color to the solution, indicating that the reaction is complete. Obviously, the end point does not occur at the equivalence point, but at a fraction of a drop beyond. The error is small and can be corrected for by running a blank titration, or it is accounted for in standardization.

STARCH INDICATOR

This indicator is used for titrations involving iodine. Starch forms a not very reversible complex with I_2 that is a very dark-blue color. The color reaction is sensitive to very small amounts of iodine. In titrations of reducing agents with iodine, the solution remains colorless up to the equivalence point. A fraction of a drop of excess titrant turns the solution a definite blue.

REDOX INDICATORS

Compare redox indicators with acid–base indicators. Here, the potential determines the ratio of the two colors, rather than the pH.

The above two methods of indication do not depend on the half-reaction potentials, although the completeness of the titration reaction and hence the sharpness of the end point do. Examples of these first two methods of visual indication are few, and most types of redox titrations are detected using **redox indicators**. These are highly colored dyes that are weak reducing or oxidizing agents that can be oxidized or reduced; the colors of the oxidized and reduced forms are different. The oxidation state of the indicator and hence its color will depend on the potential at a given point in the titration. A half-reaction and Nernst equation can be written for the indicator:

$$Ox_{ind} + ne^- \rightleftharpoons Red_{ind} \tag{14.3}$$

$$E_{ind} = E_{ind}^0 - \frac{0.059}{n} \log \frac{[Red_{ind}]}{[Ox_{ind}]} \tag{14.4}$$

E_{In}^0 must be near the equivalence point potential. A potential change of 120 mV is needed for a color change for $n = 1$ (of the indicator half-reaction) and 60 mV for $n = 2$.

The half-reaction potentials during the titration determine E_{In} and hence the ratio of $[Red_{ind}]/[Ox_{ind}]$. This is analogous to the ratio of the different forms of a pH indicator being determined by the pH of the solution. So the ratio, and therefore the color, will change as the potential during the titration changes. If we assume, as with acid–base indicators, that the ratio must change from 10/1 to 1/10 in order that a sharp color change can be seen, then *a potential equal to 2 × (0.059/n) V is required*. If n for the indicator is equal to 1, then a 0.12 V change is required. If

Table 14.1
Redox Indicators

Indicator	Color		Solution	E^0 (V)
	Reduced Form	Oxidized Form		
Nitroferroin	Red	Pale blue	1 M H$_2$SO$_4$	1.25
Ferroin	Red	Pale blue	1 M H$_2$SO$_4$	1.06
Diphenylaminesulfonic acid	Colorless	Purple	Diluted acid	0.84
Diphenylamine	Colorless	Violet	1 M H$_2$SO$_4$	0.76
Methylene blue	Blue	Colorless	1 M acid	0.53
Indigo tetrasulfonate	Colorless	Blue	1 M acid	0.36

E_{In}^0 *is near the equivalence point potential* of the titration, where there is a rapid change in potential in excess of 0.12 V, then the color change occurs at the equivalence point. Again, this is analogous to the requirement that the pK_a value of an acid–base indicator be near the pH of the equivalence point.

If there is a hydrogen ion dependence in the indicator reaction, Equation 14.3, then this will appear in the corresponding Nernst equation, Equation 14.4, and the potential at which the indicator changes color will be displaced from E_{In}^0 by the hydrogen ion term.

So, redox indicators will have a transition range over a certain potential, and this transition range must fall within the steep equivalence point break of the titration curve. The redox indicator reaction must be *rapid*, and to use terms of the electrochemist, it must be *reversible*. If the reaction is slow or is *irreversible* (slow rate of electron transfer), the color change will be gradual and a sharp end point will not be detected.

There are not many good redox indicators. Table 14.1 lists some of the common indicators arranged in order of decreasing standard potentials. Ferroin [tris(1,10-phenanthroline)iron(II) sulfate] is one of the best indicators. It is useful for many titrations with cerium(IV). It is oxidized from the red color to a pale blue at the equivalence point. Other phenanthroline-type indicators are listed in Table 14.1. Diphenylaminesulfonic acid is used as an indicator for titrations with dichromate in acid solution. The potential of the Cr$_2$O$_7^{2-}$/Cr^{3+} couple is lower than that of the cerium couple, and so this indicator with a lower E^0 is required, but care must be taken in disposing of Cr(VI) since it is an environmental pollutant that is carcinogenic. The color at the end point is purple. The indicator used may depend on the sample titrated since the magnitude of the end-point break is also dependent on the potential of the sample half-reaction.

14.5 Titrations Involving Iodine: Iodimetry and Iodometry

Redox titrations are among the most important types of analyses performed in many areas of application, for example, in food analyses, industrial analyses, and pharmaceutical analyses. Titration of sulfite in wine using iodine is a common example. Alcohol can be determined by reacting with potassium dichromate. Examples in clinical laboratories are rare since most analyses are for traces, but these titrations are still extremely useful for standardizing reagents. You should be familiar with some of the more commonly used titrants.

Iodine is an oxidizing agent that can be used to titrate fairly strong reducing agents. On the other hand, iodide ion is a mild reducing agent and serves as the basis for determining strong oxidizing agents.

IODIMETRY

In iodimetry, the titrant is I_2 and the analyte is a reducing agent. The end point is detected by the appearance of the blue starch–iodine color.

Iodine is a moderately strong oxidizing agent and can be used to titrate reducing agents. Titrations with I_2 are called **iodimetric methods**. These titrations are usually performed in neutral or mildly alkaline (pH 8) to weakly acid solutions. If the pH is too alkaline. I_2 will disproportionate to hypoiodate and iodide:

$$I_2 + 2OH^- = IO^- + I^- + H_2O \tag{14.5}$$

There are three reasons for keeping the solution from becoming strongly acidic. First, the starch used for the end-point detection tends to hydrolyze or decompose in strong acid, and so the end point may be affected. Second, the reducing power of several reducing agents is increased in neutral solution. For example, consider the reaction of I_2 with As(III):

$$H_3AsO_3 + I_2 + H_2O = H_3AsO_4 + 2I^- + 2H^+ \tag{14.6}$$

This equilibrium is affected by the hydrogen ion concentration. At low hydrogen ion concentration, the equilibrium is shifted to the right. We have already seen in Equation 12.25 that in neutral solution the potential of the As(V)/As(III) couple is decreased sufficiently that arsenic(III) will reduce I_2. But in acid solution, the equilibrium is shifted the other way, and the reverse reaction occurs. The third reason for avoiding acid solutions is that the I^- produced in the reaction tends to be oxidized by dissolved oxygen in acid solution:

$$4I^- + O_2 + 4H^+ \rightarrow 2I_2 + 2H_2O \tag{14.7}$$

The pH for the titration of arsenic(III) with I_2 can be maintained neutral by adding $NaHCO_3$. The bubbling action of the CO_2 formed also removes the dissolved oxygen and maintains a blanket of CO_2 over the solution to prevent air oxidation of the I^-.

The I_2 is a more selective oxidizing titrant than stronger ones.

Because I_2 is not a strong oxidizing agent, the number of reducing agents that can be titrated is limited. Nevertheless, several examples exist, and the moderate oxidizing power of I_2 makes it a more selective titrant than the strong oxidizing agents. Some commonly determined substances are listed in Table 14.2. Antimony

Table 14.2
Some Substances Determined by Iodimetry

Substance Determined	Reaction with Iodine	Solution Conditions
H_2S	$H_2S + I_2 \rightarrow S + 2I^- + 2H^+$	Acid solution
SO_3^{2-}	$SO_3^{2-} + I_2 + H_2O \rightarrow SO_4^{2-} + 2I^- + 2H^+$	
Sn^{2+}	$Sn^{2+} + I_2 \rightarrow Sn^{4+} + 2I^-$	Acid solution
As(III)	$H_2AsO_3^- + I_2 + H_2O \rightarrow HAsO_4^{2-}$	pH 8
	$+ 2I^- + 3H^+$	
N_2H_4	$N_2H_4 + 2I_2 \rightarrow N_2 + 4H^+ + 4I^-$	

behaves similarly to arsenic, and the pH is critical for the same reasons. Tartrate is added to complex the antimony and keep it in solution to prevent hydrolysis.

Although high-purity I_2 can be obtained by sublimation, iodine solutions are usually standardized against a primary standard reducing agent such as As_2O_3 (As_4O_6). Arsenious oxide is not soluble in acid, and so it is dissolved in sodium hydroxide. The solution is neutralized after dissolution is complete. If arsenic(III) solutions are to be kept for any length of time, they should be neutralized or acidified because arsenic(III) is slowly oxidized in alkaline solution.

Iodine has a low solubility in water but the complex I_3^- is very soluble. So iodine solutions are prepared by dissolving I_2 in a concentrated solution of potassium iodide:

$$I_2 + I^- \rightarrow I_3^- \tag{14.8}$$

Therefore, I_3^- is the actual species used in the titration.

Example 14.5

The purity of a hydrazine, N_2H_4 (note: a violent poison!), sample is determined by titration with iodine. A sample of the oily liquid weighing 1.4286 g is dissolved in water and diluted to 1 L in a volumetric flask. A 50.00-mL aliquot is taken with a pipet and titrated with standard iodine solution, requiring 42.41 mL. The iodine was standardized against 0.4123 g primary standard As_2O_3 by dissolving the As_2O_3 in a small amount of NaOH solution, adjusting the pH to 8, and titrating, requiring 40.28 mL iodine solution. What is the percent purity by weight of the hydrazine?

Solution

Standardization:

$$H_2AsO_3^- + I_2 + H_2O \rightarrow HAsO_4^{2-} + 2I^- + 3H^+$$

Each As_2O_3 gives $2H_2AsO_3^-$, so mmol $I_2 = 2 \times$ mmol As_2O_3.

> With molarity, keep track of millimoles and the ratios in which things react.

$$M_{I_2} \times 40.28 \text{ mL } I_2 = \frac{412.3 \text{ mg } As_2O_3}{197.85 \text{ mg } As_2O_3/\text{mmol}} \times 2 \text{ mmol } I_2/\text{mmol } As_2O_3$$

$$M_{I_2} = 0.1034_7 \text{ mmol/mL}$$

Analysis:

$$N_2H_4 + 2I_2 \rightarrow N_2 + 4H^+ + 4I^-$$

$$\text{mmol } N_2H_4 = \tfrac{1}{2} \times \text{mmol } I_2$$

$$\text{weight of } N_2H_4 \text{ titrated} = 1.4286 \text{ g} \times \frac{50.00}{1000.0} = 0.07143 \text{ g}$$

$$\% N_2H_4 = [0.1034_7 \; M \; I_2 \times 42.41 \text{ mL } I_2 \times \tfrac{1}{2} \text{ (mmol } N_2H_4/\text{mmol } I_2)$$
$$\times 32.045 \text{ mg } N_2H_4/\text{mmol}]/71.43 \text{ mg} \times 100\% = 98.43\%$$

The problem can, of course, also be worked using equivalents and normality. To illustrate this approach, the calculation is given. The equivalent weight of

> With normality, keep track of milliequivalents and milliequivalent weights.

As_2O_3 is one-fourth its formula weight, since each arsenic is oxidized from +3 to +5 valence and there are two arsenics per molecule. Therefore,

$$N_{I_2} \times 40.28 \text{ mL I}_2 = \frac{412.3 \text{ mg As}_2\text{O}_3}{197.85/4 \text{ (mg As}_2\text{O}_3\text{/meq)}}$$

$$N_{I_2} = 0.2069_4 \text{ meq/mL}$$

Each nitrogen in hydrazine is oxidized from −2 to 0 valence for a total valence change of 4 electrons per molecule. Therefore, its equivalent weight is one-fourth its formula weight. Hence,

$$\% \text{ N}_2\text{H}_4 = \frac{0.2069_4 \ N \times 42.41 \text{ mL} \times 32.045/4 \text{ (mg N}_2\text{H}_4\text{/meq)}}{71.43 \text{ mg}} \times 100\% = 98.43\%$$

Note that because of the low molecular weight of hydrazine, it would have been difficult to weigh out the required sample to four significant figures, and by titrating an accurately measured aliquot, a larger sample can be weighed.

• •

IODOMETRY

In iodometry, the analyte is an oxidizing agent that reacts with I^- to form I_2. The I_2 is titrated with thiosulfate, using disappearance of the starch–iodine color for the end point.

Iodide ion is a weak reducing agent and will reduce strong oxidizing agents. It is not used, however, as a titrant mainly because of the lack of a convenient visual indicator system, as well as other factors such as speed of the reaction.

When an excess of iodide is added to a solution of an oxidizing agent, *I_2 is produced in an amount equivalent to the oxidizing agent present.* This I_2 can, therefore, be titrated with a reducing agent, and the result will be the same as if the oxidizing agent were titrated directly. The titrating agent used is sodium thiosulfate.

Analysis of an oxidizing agent in this way is called an **iodometric method**. Consider, for example, the determination of dichromate:

$$\text{Cr}_2\text{O}_7{}^{2-} + 6\text{I}^- \text{ (excess)} + 14\text{H}^+ \rightarrow 2\text{Cr}^{3+} + 3\text{I}_2 + 7\text{H}_2\text{O} \qquad (14.9)$$

$$\boxed{\text{I}_2 + 2\text{S}_2\text{O}_3{}^{2-} \rightarrow 2\text{I}^- + \text{S}_4\text{O}_6{}^{2-}} \qquad (14.10)$$

The millimoles thiosulfate per millimole analyte is needed for calculations. There are 2 mmol for each mmol I_2 produced.

Each $\text{Cr}_2\text{O}_7{}^{2-}$ produces 3I_2, which in turn react with $6\text{S}_2\text{O}_3{}^{2-}$. The millimoles of $\text{Cr}_2\text{O}_7{}^{2-}$ are equal to one-sixth the millimoles of $\text{S}_2\text{O}_3{}^{2-}$ used in the titration.

Iodate can be determined iodometrically:

$$\text{IO}_3{}^- + 5\text{I}^- + 6\text{H}^+ \rightarrow 3\text{I}_2 + 3\text{H}_2\text{O} \qquad (14.11)$$

Each $\text{IO}_3{}^-$ produces 3I_2, which again react with $6\text{S}_2\text{O}_3{}^{2-}$, and the millimoles of $\text{IO}_3{}^-$ are obtained by multiplying the millimoles of $\text{S}_2\text{O}_3{}^{2-}$ used in the titration by $\frac{1}{6}$.

• •

 Example 14.6

A 0.200-g sample containing copper is analyzed iodometrically. Copper(II) is reduced to copper(I) by iodide:

$$2\text{Cu}^{2+} + 4\text{I}^- \rightarrow \underline{2\text{CuI}} + \text{I}_2$$

What is the percent copper in the sample if 20.0 mL of 0.100 M $Na_2S_2O_3$ is required for titration of the liberated I_2?

Solution

One-half mole of I_2 is liberated per mole of Cu^{2+}, and since each I_2 reacts with $2S_2O_3^{2-}$, each Cu^{2+} is equivalent to one $S_2O_3^{2-}$, and mmol Cu^{2+} = mmol $S_2O_3^{2-}$.

$$\% \ Cu = \frac{0.100 \ \text{mmol} \ S_2O_3^{2-}/\text{mL} \times 20.0 \ \text{mL} \ S_2O_3^{2-} \times Cu}{200 \ \text{mg sample}} \times 100\%$$

$$= \frac{0.100 \ \text{mmol/mL} \times 20.0 \ \text{mL} \times 63.54 \ \text{mg Cu/mmol}}{200 \ \text{mg sample}} \times 100\% = 63.5\%$$

• •

Why not titrate the oxidizing agents directly with the thiosulfate? Because strong oxidizing agents oxidize thiosulfate to oxidation states higher than that of tetrathionate (e.g., to SO_4^{2-}), but the reaction is generally not stoichiometric. Also, several oxidizing agents form mixed complexes with thiosulfate (e.g., Fe^{3+}). By reaction with iodide, the strong oxidizing agent is destroyed and an equivalent amount of I_2 is produced, which will react stoichiometrically with thiosulfate and for which a satisfactory indicator exists. The titration can be considered a direct titration.

The end point for iodometric titrations is detected with starch. The disappearance of the blue starch–I_2 color indicates the end of the titration. The starch is not added at the beginning of the titration when the iodine concentration is high. Instead, it is added just before the end point when the dilute iodine color becomes pale yellow. There are two reasons for such timing. One is that the iodine–starch complex is only slowly dissociated, and a diffuse end point would result if a large amount of the iodine were adsorbed on the starch. The second reason is that most iodometric titrations are performed in strongly acid medium and the starch has a tendency to hydrolyze in acid solution. The reason for using acid solutions is that reactions between many oxidizing agents and iodide are promoted by high acidity. Thus,

The starch is added near the end point.

$$2MnO_4^- + 10I^- + 16H^+ \rightarrow 5I_2 + 2Mn^{2+} + 8H_2O \tag{14.12}$$

$$H_2O_2 + 2I^- + 2H^+ \rightarrow I_2 + 2H_2O \tag{14.13}$$

as examples.

The titration should be performed rapidly to minimize air oxidation of the iodide. Stirring should be efficient to prevent local excesses of thiosulfate because it is decomposed in acid solution:

$$S_2O_3^{2-} + 2H^+ \rightarrow H_2SO_3 + S \tag{14.14}$$

Indications of such excess is the presence of colloidal sulfur, which makes the solution cloudy. In iodometric methods, a large excess of iodide is added to promote the reaction (common ion effect). The unreacted iodide does not interfere, but it may be air-oxidized if the titration is not performed immediately.

Sodium thiosulfate solution is standardized iodometrically against a pure oxidizing agent such as $K_2Cr_2O_7$, KIO_3, $KBrO_3$, or metallic copper (dissolved to give Cu^{2+}). With potassium dichromate, the deep green color of the resulting chromic ion makes it a little more difficult to determine the iodine–starch end point. When copper(II) is titrated iodometrically, the end point is diffuse unless thiocyanate ion

Table 14.3
Iodometric Determinations

Substance Determined	Reaction with Iodide
MnO_4^-	$2MnO_4^- + 10I^- + 16H^+ \rightleftharpoons 2Mn^{2+} + 5I_2 + 8H_2O$
$Cr_2O_7^{2-}$	$Cr_2O_7^{2-} + 6I^- + 14H^+ \rightleftharpoons 2Cr^{3+} + 3I_2 + 7H_2O$
IO_3^-	$IO_3^- + 5I^- + 6H^+ \rightleftharpoons 3I_2 + 3H_2O$
BrO_3^-	$BrO_3^- + 6I^- + 6H^+ \rightleftharpoons Br^- + 3I_2 + 3H_2O$
Ce^{4+}	$2Ce^{4+} + 2I^- \rightleftharpoons 2Ce^{3+} + I_2$
Fe^{3+}	$2Fe^{3+} + 2I^- \rightleftharpoons 2Fe^{2+} + I_2$
H_2O_2	$H_2O_2 + 2I^- + 2H^+ \xrightleftharpoons{\text{[Mo(VI) catalyst]}} 2H_2O + I_2$
As(V)	$H_3AsO_4 + 2I^- + 2H^+ \rightleftharpoons H_3AsO_3 + I_2 + H_2O$
Cu^{2+}	$2Cu^{2+} + 4I^- \rightleftharpoons 2CuI + I_2$
HNO_2	$2HNO_2 + 2I^- \rightleftharpoons I_2 + 2NO + H_2O$
SeO_3^{2-}	$SeO_3^{2-} + 4I^- + 6H^+ \rightleftharpoons Se + 2I_2 + 3H_2O$
O_3	$O_3 + 2I^- + 2H^+ \rightleftharpoons O_2 + I_2 + H_2O$
	(can determine in presence of O_2 above pH 7)
Cl_2	$Cl_2 + 2I^- \rightleftharpoons 2Cl^- + I_2$
Br_2	$Br_2 + 2I^- \rightleftharpoons 2Br^- + I_2$
HClO	$HClO + 2I^- + H^+ \rightleftharpoons Cl^- + I_2 + H_2O$

is added. The primary reaction is given in Example 14.6. But iodine is adsorbed on the surface of the cuprous iodide precipitate and only slowly reacts with the thiosulfate titrant. The thiocyanate coats the precipitate with CuSCN and displaces the iodine from the surface. The potassium thiocyanate should be added near the end point since it is slowly oxidized by iodine to sulfate. The pH must be buffered to around 3. If it is too high, copper(II) hydrolyzes and cupric hydroxide will precipitate. If it is too low, air oxidation of iodide becomes appreciable because it is catalyzed in the presence of copper. Copper metal is dissolved in nitric acid, with oxides of nitrogen being produced. These oxides will oxidize iodide, and they are removed by addition of urea. Some examples of iodometric determinations are listed in Table 14.3.

Example 14.7

A solution of $Na_2S_2O_3$ is standardized iodometrically against 0.1262 g of high-purity $KBrO_3$, requiring 44.97 mL $Na_2S_2O_3$. What is the molarity of the $Na_2S_2O_3$?

Solution

The reactions are

$$BrO_3^- + 6I^- + 6H^+ \rightarrow Br^- + 3I_2 + 3H_2O$$
$$3I_2 + 6S_2O_3^{2-} \rightarrow 6I^- + 3S_4O_6^{2-}$$

So mmol $S_2O_3^{2-} = 6 \times$ mmol BrO_3^-:

$$M_{S_2O_3^{2-}} \times 44.97 \text{ mL} = \frac{126.2 \text{ mg KBrO}_3}{167.01 \text{ (mg/mmol KBrO}_3)} \times 6 \text{ (mmol } S_2O_3^{2-}/\text{mmol } BrO_3^-)$$

$$M_{S_2O_3^{2-}} = 0.1008_2 \text{ mmol/mL}$$

14.6 Titrations with Other Oxidizing Agents

We have already mentioned some oxidizing agents that can be used as titrants. The titrant should be fairly stable and should be convenient to prepare and to handle. If is too strong an oxidizing agent, it will be so reactive that its stability will not be great. Thus, fluorine is one of the strongest oxidizing agents known, but it is certainly not convenient to use in the analytical laboratory ($E^0 = 3.06$ V). Chlorine would make a good titrant, except that it is volatile from aqueous solution, and to prepare and maintain a standard solution would be difficult.

Potassium permanganate is a widely used oxidizing titrant. It acts as a self-indicator for end-point detection and is a very strong oxidizing agent ($E^0 = 1.51$ V). The solution is stable if precautions are taken in its preparation. When the solution is first prepared, small amounts of reducing impurities in the solution reduce a small amount of the MnO_4^-. In neutral solution, the reduction product of this permanganate is MnO_2, rather than Mn^{2+} produced in acid medium. The MnO_2 acts as a catalyst for further decomposition of the permanganate, which produces more MnO_2, and so on. This is called **autocatalytic decomposition**. The solution can be stabilized by removing the MnO_2. So, before standardizing, the solution is boiled to hasten oxidation of all impurities and is allowed to stand overnight. The MnO_2 is then removed by filtering through a sintered-glass filter. Potassium permanganate can be standardized by titrating primary standard sodium oxalate, $Na_2C_2O_4$, which, dissolved in acid, forms oxalic acid:

$$5H_2C_2O_4 + 2MnO_4^- + 6H^+ = 10CO_2 + 2Mn^{2+} + 8H_2O \qquad (14.15)$$

The solution must be heated for rapid reaction. The reaction is catalyzed by the Mn^{2+} product and it goes very slowly at first until some Mn^{2+} is formed. Pure electrolytic iron metal can also be used as the primary standard. It is dissolved in acid and reduced to Fe^{2+} for titration (see Section 14.8).

A difficulty arises when permanganate titrations of iron(II) are performed in the presence of chloride ion. The oxidation of chloride ion to chlorine by permanganate at room temperature is normally slow. However, the oxidation is catalyzed by the presence of iron. If an iron sample has been dissolved in hydrochloric acid, or if stannous chloride has been used to reduce it to iron(II) (see below), the titration can be performed by adding the **Zimmermann–Reinhardt reagent**. This contains manganese(II) and phosphoric acid. The manganese(II) reduces the potential of the MnO_4^-/Mn^{2+} couple sufficiently so that permanganate will not oxidize chloride ion; the formal potential is less than E^0, due to the large concentration of Mn^{2+}. This decrease in the potential decreases the magnitude of the end-point break. Therefore, phosphoric acid is added to complex the iron(III) and decrease the potential of the Fe^{3+}/Fe^{2+} couple also; the iron(II) is not complexed. In other words, iron(III) is removed from the solution as it is formed to shift the equilibrium of the titration reaction to the right and give a sharp end point. The overall effect is still a large potential break in the titration curve, but the entire curve has been shifted to a lower potential.

An added effect of complexing the iron(III) is that the phosphate complex is nearly colorless, while the chloro complex (normally present in chloride medium) is deep yellow. A sharper end-point color change results.

Potassium dichromate, $K_2Cr_2O_7$, is a slightly weaker oxidizing agent than potassium permanganate. The great advantage of this reagent is its availability as a primary standard, and the solution need not be standardized in most cases. In the titration of iron(II), standardizing potassium dichromate against electrolytic iron is preferable, however, because the green color of the chromic ion introduces a small

The Z–R reagent prevents oxidation of Cl^- by MnO_4^- and sharpens the end point.

error in the end point (diphenylamine sulfonate indicator). Standardization is necessary only for the most accurate work.

Oxidation of chloride ion is not a problem with dichromate. However, the formal potential of the $Cr_2O_7^{2-}/Cr^{3+}$ couple is reduced from 1.33 to 1.00 V in 1 M hydrochloric acid, and phosphoric acid must be added to reduce the potential of the Fe^{3+}/Fe^{2+} couple. Such addition is also necessary because it decreases the equivalence point potential to near the standard potential for the diphenylamine sulfonate indicator (0.84 V). Otherwise, the end point would occur too soon.

Cerium(IV) is a powerful oxidizing agent. Its formal potential depends on the acid used to keep it in solution (it hydrolyzes to form ceric hydroxide if the solution is not acid). Titrations are usually performed in sulfuric acid or perchloric acid. In the former acid, the formal potential is 1.44 V, and in the latter acid, it is 1.70 V. So cerium(IV) is a stronger oxidizing agent in perchloric acid. Cerium(IV) can be used for most titrations in which permanganate is used, and it possesses a number of advantages. It is a very strong oxidizing agent and its potential can be varied by choice of the acid used. The rate of oxidation of chloride ion is slow, even in the presence of iron, and titrations can be carried out in the presence of moderate amounts of chloride without the use of a Zimmermann–Reinhardt type of preventive solution. The solution can be heated but should not be boiled, or chloride ion will be oxidized. Sulfuric acid solutions of cerium(IV) are stable indefinitely. Nitric acid and perchloric acid solutions, however, do decompose, but only slowly. An added advantage of cerium is that a salt of cerium(IV), ammonium hexanitratocerate, $(NH_4)_2Ce(NO_3)_6$, can be obtained as a primary standard, and the solution does not have to be standardized. The main disadvantage of cerium(IV) is its increased cost over potassium permanganate, although this should not be a serious factor if a saving in time is achieved. Ferroin is a suitable indicator for many cerate titrations.

Cerium(IV) solutions can be standardized against primary standard As_2O_3, $Na_2C_2O_4$, or electrolytic iron. The reaction with arsenic(III) is slow, and it must be catalyzed by adding either osmium tetroxide (OsO_4) or iodine monochloride (ICl). Ferroin is used as the indicator. The reaction with oxalate is also slow at room temperature, and the same catalyst can be used. The reaction is rapid, however, at room temperature in the presence of 2 M perchloric acid. Nitroferroin is used as the indicator.

Cerium(IV) solutions to be standardized are usually prepared from ammonium sulfatocerate, $(NH_4)_4Ce(SO_4)_4 \cdot 2H_2O$; ammonium nitratocerate, $(NH_4)_2Ce(NO_3)_6$ (*not* the high-purity primary standard variety, though); or hydrous ceric oxide, $CeO_2 \cdot 4H_2O$. Primary standard ammonium nitratocerate is used only if the solution is not to be standardized, because of its increased expense.

14.7 Titrations with Other Reducing Agents

Standard solutions of reducing agents are not used as widely as oxidizing agents are because most of them are oxidized by dissolved oxygen. They are, therefore, less convenient to prepare and use. **Thiosulfate** is the only common reducing agent that is stable to air oxidation and that can be kept for long periods of time. This is the reason that iodometric titrations are so popular for determining oxidizing agents. However, stronger reducing agents than iodide ion are sometimes required.

Iron(II) is only slowly oxidized by air in sulfuric acid solution and is a common titrating agent. It is not a strong reducing agent ($E^0 = 0.771$ V) and can be used to titrate strong oxidizing agents such as cerium(IV), chromium(VI) (dichromate), and vanadium(V) (vanadate). Ferroin is a good indicator for the first two titrations, and oxidized diphenylamine sulfonate is used for the last titration. The iron(II) standardization should be checked daily.

Chromium(II) and **titanium(III)** are very powerful reducing agents, but they are readily air-oxidized and difficult to handle. The standard potential of the former is -0.41 V (Cr^{3+}/Cr^{2+}) and that of the latter is 0.04 V (TiO^{2+}/Ti^{3+}). The oxidized forms of copper, iron, silver, gold, bismuth, uranium, tungsten, and other metals have been titrated with chromium(II). The principal use of Ti^{3+} is in the titration of iron(III) as well as copper(II), tin(IV), chromate, vanadate, and chlorate.

14.8 Preparing the Solution—Getting the Analyte in the Right Oxidation State before Titration

When samples are dissolved, the element to be analyzed is usually in a mixed oxidation state or is in an oxidation state other than that required for titration. There are various oxidizing and reducing agents that can be used to convert different metals to certain oxidation states prior to titration. The excess preoxidant or prereductant must generally be removed before the metal ion is titrated.

REDUCTION OF THE SAMPLE PRIOR TO TITRATION

The reducing agent should not interfere in the titration or, if it does, unreacted reagent should be readily removable. Most reducing agents will, of course, react with oxidizing titrants, and they must be removable. **Sodium sulfite**, Na_2SO_3, and **sulfur dioxide** are good reducing agents in acid solution ($E^0 = 0.17$ V), and the excess can be removed by bubbling with CO_2 or in some cases by boiling. If SO_2 is not available, sodium sulfite or bisulfite can be added to an acidified solution. Thallium(III) is reduced to the $+1$ state, arsenic(V) and antimony(V) to the $+3$ state, vanadium(V) to the $+4$ state, and selenium and tellurium to the elements. Iron(III) and copper(II) can be reduced to the $+2$ and $+1$ states, respectively, if thiocyanate is added to catalyze the reaction.

> Reducing agents that can be readily removed are used to reduce the analyte, prior to titration with an oxidizing agent.

Stannous chloride, $SnCl_2$, is usually used for the reduction of iron(III) to iron(II) for titrating with cerium(IV) or dichromate. The reaction is rapid in the presence of chloride (hot HCl). When iron samples (e.g., ores) are dissolved (usually in hydrochloric acid), part or all of the iron is in the $+3$ oxidation state and must be reduced. The reaction with stannous chloride is

$$2Fe^{3+} + SnCl_2 + 2Cl^- \rightarrow 2Fe^{2+} + SnCl_4 \qquad (14.16)$$

The reaction is complete when the yellow color of the iron(III)–chloro complex disappears. The excess tin(II) is removed by addition of mercuric chloride:

$$SnCl_2 + 2HgCl_2 \text{ (excess)} \rightarrow SnCl_4 + \underline{Hg_2Cl_2} \text{ (calomel)} \qquad (14.17)$$

A large excess of cold $HgCl_2$ must be added rapidly with stirring. If too little is added, or if it is added slowly, some of the mercury will be reduced, by local excesses of $SnCl_2$, to elemental mercury, a gray precipitate. The calomel, Hg_2Cl_2, which is a milky-white precipitate, does not react at an appreciable rate with dichromate or cerate, but mercury will. In order to prevent a large excess of tin(II) and subsequent danger of formation of mercury, the stannous chloride is added dropwise until the yellow color of iron(III) just disappears. If a gray precipitate is noted after the $HgCl_2$ is added, the sample must be discarded. Stannous chloride can also be used to reduce As(V) to As(III), Mo(VI) to Mo(V), and, with $FeCl_3$ catalyst, U(VI) to U(IV).

Metallic reductors are widely used for preparing samples. These are usually used in a granular form in a column through which the sample solution is passed. The sample is eluted from the column by slowly passing dilute acid through it. The oxidized metal ion product does not interfere in the titration and no excess reductant is present since the metal is insoluble. For example, lead can be used to reduce tin(IV):

$$Pb + Sn^{4+} \rightarrow Pb^{2+} + Sn^{2+} \tag{14.18}$$

The solution eluted from the column will contain Pb^{2+} and Sn^{2+}, but no Pb. Table 14.4 lists several commonly used metallic reductors and some elements they will reduce. The reductions are carried out in acid solution. In the case of zinc, metallic zinc is amalgamated with mercury to prevent attack by acid to form hydrogen:

$$Zn + 2H^+ \rightarrow Zn^{2+} + H_2 \tag{14.19}$$

Sometimes, the reduced sample is rapidly air-oxidized and the sample must be titrated under an atmosphere of CO_2, by the addition of sodium bicarbonate to an acid solution. Air must be excluded from tin(II) and titanium(III) solutions. Sometimes, elements rapidly air-oxidized are eluted from the column into an iron(III) solution, with the end of the column immersed in the solution. The iron(III) is reduced by the sample to give an equivalent amount of iron(II), which can be titrated with dichromate. Molybdenum(III), which is oxidized to molybdenum(VI) by the iron, and copper(I) are determined in this way.

OXIDATION OF THE SAMPLE PRIOR TO TITRATION

Oxidizing agents that can be readily removed are used to oxidize the analyte, prior to titration with a reducing agent.

Very strong oxidizing agents are required to oxidize most elements. Hot anhydrous **perchloric acid** is a strong oxidizing agent. It can be used to oxidize chromium(III) to dichromate. The mixture must be diluted and cooled very quickly to prevent reduction. Dilute perchloric acid is not a strong oxidizing agent, and the solution needs only to be diluted following the oxidation. Chlorine is a product of perchloric acid reduction, and this must be removed by boiling the diluted solution. See Chapter 2 for precautions in using perchloric acid.

Table 14.4
Metallic Reductors

Reductor	Element Reduced
Zn (Hg) (Jones reductor)	Fe(III) → Fe(II), Cr(VI) → Cr(II), Cr(III) → Cr(II), Ti(IV) → Ti(III), V(V) → V(II), Mo(VI) → Mo(III), Ce(IV) → Ce(III), Cu(II) → Cu
Ag (1 *M* HCl) (Walden reductor)	Fe(III) → Fe(II), U(VI) → U(IV), Mo(VI) → Mo(V) (2 *M* HCl), Mo(VI) → Mo(III) (4 *M* HCl), V(V) → V(IV), Cu(II) → Cu(I)
Al	Ti(IV) → Ti(III)
Pb	Sn(IV) → Sn(II), U(VI) → U(IV)
Cd	$ClO_3^- \rightarrow Cl^-$

Potassium persulfate, $K_2S_2O_8$, is a powerful oxidizing agent that can be used to oxidize chromium(III) to dichromate, vanadium(IV) to vanadium(V), cerium(III) to cerium(IV), and manganese(II) to permanganate. The oxidations are carried out in hot acid solution, and a small amount of silver(I) catalyst must be added. The excess persulfate is destroyed by boiling. This boiling will always reduce some permanganate.

Bromine can be used to oxidize several elements, such as Tl(I) to Tl(III) and iodide to iodate. The excess is removed by adding phenol, which is brominated. **Chlorine** is an even stronger oxidizing agent. **Permanganate** oxidizes V(IV) to V(V) and Cr(III) to Cr(VI). The latter reaction is rapid only in alkaline solution. It has been used to oxidize trace quantities of Cr(III) in acid solution, however, by heating. Excess permanganate is destroyed by adding hydrazine, the excess of which is destroyed by boiling. **Hydrogen peroxide** will oxidize Fe(II) to Fe(III), Co(II) to Co(III) in mildly alkaline solution, and Cr(II) to Cr(VI) in strongly alkaline solution.

For most redox determinations, specified procedures have been described for the preparation of different elements in various types of samples. You should be able to recognize the reasoning behind the operations from the discussions in this chapter.

The only common redox titration applied in the clinical laboratory is for the analysis of calcium in biological fluids. Calcium oxalate is precipitated and filtered, the precipitate is dissolved in acid, and the oxalate, which is equivalent to the calcium present, is titrated with standard potassium permanganate solution. This method is largely replaced now by more convenient techniques such as complexometric titration with EDTA (Chapter 9) or measurement by atomic absorption spectrophotometry (Chapter 17).

14.9 Potentiometric Titrations [Indirect Potentiometry]

Volumetric titrations are usually most conveniently performed with a visual indicator. In cases where a visual indicator is unavailable, potentiometric indication of the end point can often be used. Potentiometric titrations are among the most accurate known because the potential follows the actual change in activity and, therefore, the end point will often coincide directly with the equivalence point. And, as we have mentioned in our discussions of volumetric titrations, they are more sensitive than visual indicators. Potentiometry is, therefore, often employed for dilute solutions.

Potentiometric titrations are straightforward. They involve measurement of an indicating electrode potential against a convenient reference electrode and plotting the change of this potential difference against volume of titrant. See Figure 13.5 for a potentiometric titration setup. A large potential break will occur at the equivalence point. Since we are interested only in the potential *change*, the *correct* potential of the indicating electrode need not be known. For example, in pH titrations, the glass electrode does not have to be calibrated with a standard buffer; it will still give the same *shape* of titration curve that may be shifted up or down on the potential axis. It is a good idea, however, to have some indication of the correct value so the end point can be anticipated and any anomalous difficulties can be detected.

Because we are not interested in "absolute" potentials, the liquid-junction potential becomes unimportant. It will remain somewhat constant throughout the titration, and small changes will be negligible compared to the change in potential at

Potentiometric indication is more sensitive and accurate than visual indication.

the end point. Also, the potential need not be read very closely, and so a conventional pH meter, the scale of which is divided to the nearest 10 mV and can be estimated to the nearest 1 mV, can be used for most titrations.

pH TITRATIONS—USING pH ELECTRODES

A glass pH electrode is used to follow acid–base titrations.

We have shown in Chapter 8 that in acid–base titrations the pH of the solution exhibits a large break at the equivalence point. This pH change can easily be monitored with a glass pH electrode. By plotting the measured pH against volume of titrant, one can obtain titration curves similar to those shown in Chapter 8. The end point is taken as the **inflection point** of the large pH break occurring at the equivalence point; this is the steepest part of the curve.

PRECIPITATION TITRATIONS—USING SILVER ELECTRODES

A silver electrode is used to follow titrations with silver ion.

The indicating electrode in precipitation titrations is used to follow the change in pM or pA, where M is the cation of the precipitate and A is the anion. In the titration of chloride ion with silver ion, for example, either Equation 13.3 or 13.10 will hold. In the former equation, the term $\log (1/a_{Ag^+})$ is equal to pAg; and in the latter equation, the term $\log a_{Cl^-}$ is equal to $-$pCl. Therefore, the potential of the silver electrode will vary in direct proportion to pAg or pCl, changing 2.30 RT/F V (ca. 59 mV) for each 10-fold change in a_{Ag^+} or a_{Cl^-}. A plot of the potential versus volume of titrant will give a curve identical in shape to that in Figure 11.1. (Note that since $a_{Ag^+}a_{Cl^-}$ = constant, a_{Cl^-} is proportional to $1/a_{Ag^+}$ and pCl is proportional to $-$pAg, so the same shape curve results if we plot or measure either pCl or pAg.)

REDOX TITRATIONS—USING PLATINUM ELECTRODES

An inert electrode (e.g., Pt) is used to follow redox titrations.

Because there is generally no difficulty in finding a suitable indicator electrode, redox titrations are widely used; an inert metal such as platinum is usually satisfactory for the electrode. Both the oxidized and reduced forms are usually soluble and their ratio varies throughout the titration. The potential of the indicating electrode will vary in direct proportion to $\log (a_{red}/a_{ox})$, as in the calculated potential for the titration curves shown in Figure 14.1 for the titration of Fe^{2+} with Ce^{4+}. As pointed out, the potential is determined by either half-reaction. Generally, the pH in these titrations is held nearly constant, and any H^+ term in the Nernst equation will drop out of the log term.

A potentiometric titration curve is used to select the appropriate redox indicator (with E^0_{In} near $E_{eq. pt.}$).

A potentiometric plot, as in Figure 14.1, is useful for evaluating or selecting a suitable visual indicator for the titration, particularly for a new titration. From a knowledge of the transition potential, it is often possible to select an indicator whose color transition occurs within this potential range. Or the potential can actually be measured during the visual titration and the color transition range on the potentiometric curve noted to see whether it corresponds to the equivalence point.

ION-SELECTIVE ELECTRODES IN TITRATIONS—MEASURING pM

The term $\log a_{ion}$ in Equation 13.44 is equal to $-$pIon, and so ion-selective electrodes (ISE) can be used to monitor changes in pM during a titration. For example, a cation-selective glass electrode that is sensitive to silver ion can be used to follow changes in pAg in titrations with silver nitrate. A calcium-sensitive electrode can be used for the titration of calcium with EDTA. The electrode should not respond to sodium ion since the disodium salt of EDTA is usually used. If the electrode responds to a second ion in the solution whose activity remains approximately constant throughout the titration, then Equation 13.46 will hold, and the titration curve will be distorted; this is so because the electrode potential is determined by $\log(a_{ion} + constant)$ and not $\log a_{ion}$. If the contribution from the second ion is not too large, then the distortion will not be too great and a good break will still occur at the end point. Titrations involving anions can also be monitored with anion-selective electrodes. For example, fluoride ion can be precipitated with lanthanum(III), and a fluoride electrode can be used to mark the end point of the titration.

Potentiometric titrations are always more accurate than direct potentiometry because of the uncertainties involved in potential measurements. Whereas accuracies of better than a few percent are rarely possible in direct potentiometry, accuracies of a few tenths of a percent are common by potentiometric titration. We can make some general statements concerning potentiometric titrations:

> Potentiometric titrations are more accurate than direct ISE measurements because the liquid-junction potential is not important.

1. The potential readings are usually sluggish in dilute solutions and near the end point because the solution is poorly poised.

2. It is necessary to plot the potential only near the end point. Small increments of titrant are added near the end point, 0.1 or 0.05 mL, for example. The exact end point volume need not be added, but it is determined by interpolation of the E versus volume plot.

3. The polarity of the indicating electrode relative to the reference electrode may change during the titration. That is, the potential difference may go from one polarity to zero and then to the reverse polarity; hence, the polarity of the potential-measuring device may have to be changed.

DERIVATIVE TITRATIONS

Plotting or recording the first or second derivative of a titration curve can more accurately pinpoint the end point.

1. First Derivative Plot. We noted above that at the end point the slope of the titration curve was maximum. In other words, the rate of change of potential with addition of titrant is maximum at the end point. So, if we could plot the rate of change of potential with change in volume ($\Delta E/\Delta V$) against volume, then a "spiked" curve should result, and the peak of this spike should occur at the end point. This is conveniently done by adding equal increments of titrant near the end point. Consider the data collected near the end point during a titration, as shown in the spreadsheet below, used to calculate and plot the derivative curves. Disregard the last four columns for the time being. We want to plot $\Delta E/\Delta V$ against the volume to get the first derivative. Such a plot is shown in Figure 14.2. The volume used is the *average* of the two volumes used to calculate ΔE (column III in the spreadsheet). So, the volume for $\Delta E/\Delta V = 0.4$ is 35.475 mL, and so on. The end point is theoretically the maximum of this plot, which in the figure occurs at 35.577 mL (we would record this as 35.58 mL). In a manual plot, you would extrapolate the points on each side of the curve to obtain the intersection. The extrapolation of a first derivative plot may lead to an uncertainty that can be partially avoided by a second derivative plot (see below).

Note that we have used equal volume increments here, and so ΔE could have been plotted in place of $\Delta E/\Delta V$. These equal increments are not necessary but do shorten the calculations. Although the average volume may be calculated to 0.001 mL for plotting, experimentally, we are not justified in reporting the end point to more than 0.01 mL.

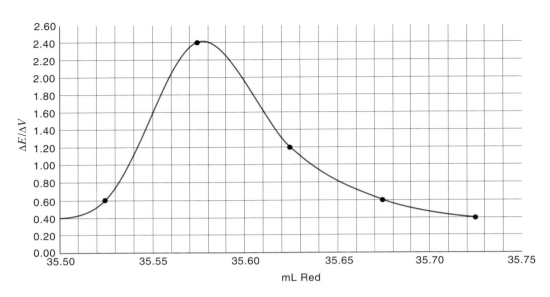

Fig. 14.2. First derivative plot of a potentiometric titration curve.

2. Second Derivative Plot. Mathematically, the second derivative of a titration curve should pass through zero at the equivalence point. The last four columns in the spreadsheet illustrate how such a plot can be accomplished. The second derivative is the rate of change of the first (column VIII) with respect to the change in the average volume (column IX). Division of column VIII by column IX, then, gives the second derivative $[\Delta^2 E/\Delta V^2$ or $\Delta(\Delta E/\Delta V_1)/\Delta V_2$—column X]. The average of the two successive volumes used for the first derivative plot (column III) is used for the second derivative plot (column VII). See Figure 14.3. Again, there is some extrapolation, but it is less significant than in the first derivative plot. The curve passes through zero at 35.579 mL, and so the end point is taken as 35.58 mL. As

	A	B	C	D	E	F	G	H	I	J
1	**Derivative Titration: Red vs. Ox**									
2	I	II	III	IV	V	VI	VII	VIII	IX	X
3			**First derivative**				**Second derivative**			
4	**V, mL**	**E**	**$V_{ave\,1}$**	**ΔE**	**ΔV_1**	**$\Delta E/\Delta V_1$**	**$V_{ave\,2}$**	**$\Delta(\Delta E/\Delta V_1)$**	**ΔV_2**	**$\Delta^2 E/\Delta V^2$**
5	35.45	0.630								$(\Delta(\Delta E/\Delta V_1)/\Delta V_2)$
6	35.50	0.650	**35.475**	**0.020**	**0.05**	**0.400**				
7	35.55	0.680	35.525	0.030	0.05	0.600	**35.500**	**0.20**	**0.050**	**4.00**
8	35.60	0.800	35.575	0.120	0.05	2.400	35.550	1.80	0.050	36.00
9	35.65	0.860	35.625	0.060	0.05	1.200	35.600	-1.20	0.050	-24.00
10	35.70	0.890	35.675	0.030	0.05	0.600	35.650	-0.60	0.050	-12.00
11	35.75	0.910	35.725	0.020	0.05	0.400	35.700	-0.20	0.050	-4.00
12	**Cell C6 = V_{ave1} =**	(A5+A6)/2	Copy all formulas down to end							
13	**Cell D6 = ΔE =**	B6-B5								
14	**Cell E6 = ΔV_1 =**	A6-A5								
15	**Cell F6 = $\Delta E/\Delta V_1$ =**	D6/E6								
16	**Cell G7 = V_{ave2} =**	(C6+C7)/2								
17	**Cell H7 = $\Delta(\Delta E/\Delta V_1)$ =**	F7-F6								
18	**Cell I7 = ΔV_2 =**	C7-C6								
19	**Cell J7 = $\Delta^2 E/\Delta V^2$ =**	H7/I7								
20	1st Derivative: Plot C6:C11 vs. F6:F11									
21	2nd Derivative: Plot G7:G11 vs. J7:J11									

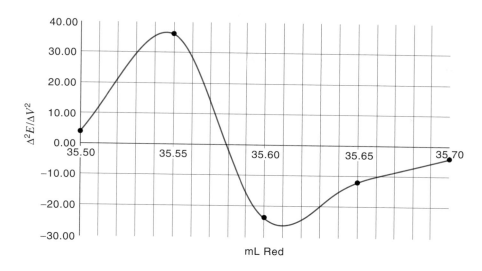

Fig. 14.3. Second derivative plot of a potentiometric titration curve.

before, we are experimentally justified in reporting it to only the nearest 0.01 mL. Again, since equal volume increments were added, V_{2nd} was constant, and we could have plotted column VIII rather than column X to save calculations. This will not work, however, if equal volume increments are not added. See your CD, Chapter 14, which contains the above spreadsheet, to compare the derivative plots with the conventional plot near the end point (Charts 1, 2, and 3). See Ref. 4 by Carter and Huff for a discussion of second derivative plots.

In both these methods, the volume increment should not be too large or there will not be sufficient points near the end point. If the increments are small enough, then extrapolation of the second derivative plots may not be necessary at all because there will be two or more points on the straight-line portion of the plot that passes through zero. On the other hand, the increments should not be small enough to be tedious and to fall within experimental error of the volume measurement. Usually, more points are taken than we have illustrated here. Of course, these small volume increments are taken only near the end point. In some titrations, the potential break is sufficiently large that the magnitude of potential change can be noted with equal added volume increments and the end point taken as that point where the change is largest. Also, it is convenient sometimes merely to titrate to an end-point potential, which has been determined by calculations or empirically from a measured titration curve.

A word of caution should be mentioned with respect to derivative methods. The derivatives tend to emphasize noise or scatter in the data points, being worse for the second derivative. Hence, if a particular titration is subject to noise or potential drift, a direct plot may be preferred.

Each time a derivative is taken, the noise is amplified.

GRAN'S PLOTS FOR END-POINT DETECTION

Assume that instead of plotting the electrode potential (which is a logarithmic function of concentration) against volume of titrant, we plotted the concentration of analyte remaining at each point in the titration. A straight-line plot would in principle result (neglecting volume changes) in which the concentration would decrease to zero at the equivalence point (assuming the equilibrium for the titration reaction lies far to the right). This is because at 20% titrated, 80% of the sample will remain, at 50% titrated, 50% will remain, at 80% titrated, 20% will remain, and so on. (In practice, a plot in the region of the end point is made.) Similarly, a plot of titrant concentration beyond the equivalence point would be a linear plot of increasing concentration that would extrapolate to zero concentration at the end point.

Consider the titration of chloride ion with silver nitrate solution. Except near the equivalence point where the solubility becomes appreciable compared to the unreacted chloride, the concentration of chloride in solution at any point in the titration is calculated from the initial moles less the moles reacted with $AgNO_3$:

$$[Cl^-] = \frac{M_{Cl}mL_{Cl} - M_{Ag}mL_{Ag}}{mL_{Cl} + mL_{Ag}} \qquad (14.20)$$

The potential of a chloride ion-selective electrode (neglecting activity coefficients) is

$$E_{cell} = k - S \log[Cl^-] \qquad (14.21)$$

or

$$\log[Cl^-] = \frac{k - E_{cell}}{S} \qquad (14.22)$$

where S is the empirical potentiometric slope in the Nernst equation (theoretically 0.059) and k the empirical potentiometric cell constant (theoretically the difference in E^0 values for the indicating and reference electrodes). Substituting (14.20) in (14.22):

$$\log\left(\frac{M_{Cl}mL_{Cl} - M_{Ag}mL_{Ag}}{mL_{Cl} + mL_{Ag}}\right) = \frac{k - E_{cell}}{S} \qquad (14.23)$$

$$(mL_{Cl} + mL_{Ag}) \text{ antilog}\left(\frac{k - E_{cell}}{S}\right) = M_{Cl}mL_{Cl} - M_{Ag}mL_{Ag} \qquad (14.24)$$

A plot of mL_{Ag} (the variable) versus the left-hand side of the equation will give a straight line (the readings are corrected for volume changes in the above calculations). This is called a **Gran plot** (see Refs. 5–7). The equivalence point occurs when mmol Cl = mmol Ag; that is, when the left-hand term (y axis) is zero. The plot would be as illustrated in Figure 14.4. There is curvature near the end point because of the finite solubility of the silver chloride; that is, the antilog term does not go to zero (the potential would have to go to infinity), so extrapolation is made over several points slightly before the end point.

The application of Equation (14.24) to a Gran plot implies knowledge of the constant k in the Nernst equation in order to construct the zero intercept on the y axis. This (and the slope) can be determined from standards.

The Gran plot can also be performed empirically in a number of ways. A calibration curve of potential versus analyte concentration can be constructed and used to convert potential readings directly into concentration readings: the end-point intercept would then correspond to zero concentration on the y axis. Or the log scale on the potential measuring device (e.g., pH meter) can be used to read concentration values directly, after the scale is calibrated with one or more standards (each $59/n$ mV being equal to a 10-fold change in concentration). Alternatively, the antilogarithm of the potential or pH reading can be calculated and plotted against volume of titrant ($E \propto \log C$, antilog $E \propto C$). The intercept then would correspond to the potential determined for zero analyte concentration.

A Gran plot can also be obtained for the titrant beyond the end point (in which the antilog term increases linearly from zero at the end point). In this case, the intercept potential is best determined from a blank titration and extrapolation of the linear portion of the y axis to zero milliliters.

A Gran plot converts a logarithmic response to a linear plot.

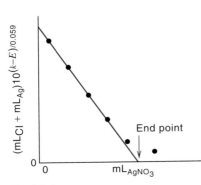

Fig. 14.4. Gran plot as given for Equation 14.24.

The antilog values, which are directly proportional to concentration, must be corrected for volume changes. Corrected values are obtained by multiplying the observed values by $(V + v)/V$, where V is the initial volume and v is the added volume.

In addition to the advantage of a linear plot, Gran plots do not require measurements around the end point, where the potential tends to drift because of the low level of the ion being sensed and where very small increments of titrant must be added. Only a few points are needed on the straight line at a distance away from the end point.

A typical Gran plot is shown in Figure 14.5 for the titration of small amounts of chloride with silver ion. The excess titrant is monitored with a Ag/Ag$_2$S electrode. A plot proportional to the titrant concentration is shown (right-hand ordinate) along with the usual S-shaped potentiometric plot (left-hand ordinate); a small potentiometric inflection point occurs due to the small concentrations involved. The straight-line plot is extrapolated back to the horizontal axis to determine the end point (a blank titration is performed and the linear blank plot is extrapolated to zero milliliters to accurately determine the horizontal axis). Curvature of the straight line around the end point generally indicates appreciable solubility of a precipitate, dissociation of a complex, and so on.

Several advantages accrue from the linear plots. It is only necessary to obtain a few points to define the straight line, and the end point is easily identified by extrapolating the line to the horizontal axis. Points only need be accurately determined a bit away from the equivalence point, where the titrant is in sufficient excess to suppress dissociation of the titration product and where electrode response is rapid because one of the ions is at relatively high levels compared to the levels at the equivalence point. In case of small inflection points (Figure 14.5), the end point is more readily defined by a Gran plot.

A Gran-type plot can also be obtained by plotting the reciprocal of a first derivative curve, that is, $\Delta E/\Delta V$ versus V. Since in a derivative titration $\Delta V/\Delta E$ goes to infinity at the equivalence point, the reciprocal will go to zero where the intersection of the two lines occurs, and a V-shaped plot results. In this application, the average volume between the two increments is plotted, as in the first derivative plot. The $\Delta E/\Delta V$ values must be corrected for volume changes to obtain straight lines ($\Delta E/\Delta V$ is linearly dependent on volume changes).

> With a Gran plot, we do not have to hit the end point in the titration.

> A first derivative titration can be used to prepare a Gran plot.

Electrode potential (mV) vs **mL of 2×10^{-3} AgNO$_3$**

Potentionmetric S-titration curve

Linear titration curve

Antilog electrode potential (volume corrected)

Fig. 14.5. Gran plot for titration of 100 mL 5×10^{-5} M Cl$^-$ with AgNO$_3$ using Ag$_2$S electrode. (Courtesy of Orion Research, Inc.)

Standard additions calibration corrects for sample matrix effects. The standard is added to the sample.

A Gran-type plot is convenient in **standard additions** or **known additions** procedures. Standard additions methods are useful ways of calibration when the sample matrix affects the analyte signal. In these methods, a signal is recorded for the sample, and then a known amount of standard is added to the sample and the change in signal is measured. This latter measurement provides calibration in the same matrix as the unknown analyte, and the matrix should have the same effect on both unknown and standard. In this case, it is the electrode response that is calibrated. Most analytical methods give a linear response to analyte; but in potentiometry, it is a logarithmic response. By employing a Gran-type plot, a linear graph can be obtained, simplifying calculation. Here, the potential of the sample is initially recorded and then known amounts of standard are added to the sample. The antilog values are plotted as a function of the amount of standard added, and the best straight line is drawn through them (e.g., least-squares analysis). Extrapolation to the horizontal axis (determined from similar measurements on a "blank" with extrapolation to zero concentration) gives the equivalent amount of analyte in the sample (Figure 14.6).

In applying the standard additions method, it is most convenient to add small volumes of concentrated standard to the sample solution in order to minimize volume change and thereby make volume corrections unnecessary. For example, 100 μL of a 1000-ppm standard might be added to 10 mL of sample to increase the concentration by 10 ppm. The volume change is only 1% and can probably be ignored. The concentration increments should be close to the unknown concentration.

The standard additions method can also be applied mathematically, as illustrated in Example 14.8.

Example 14.8

The calcium ion concentration in serum is determined using an ion-selective electrode. The potential measured with the electrode in the sample is +217.6 mV. Addition of 100 μL of a 2000-ppm standard to 2.00 mL of sample and measurement of the potential gives +226.8 mV. Assuming a Nernstian response (59.2/2 mV per 10-fold change in activity), what is the concentration of calcium in the sample?

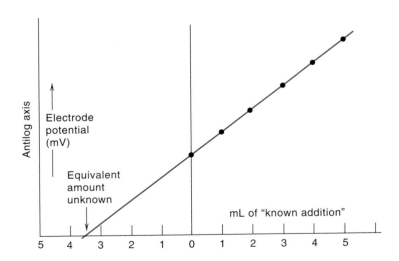

Fig. 14.6. Known addition using Gran plot paper. (Courtesy of Orion Research, Inc.)

Solution

Since the analyte and the standard are subjected to the same matrix and ionic strength, the electrode responds in a Nernstian fashion to concentration (see Section 13.8). We can write

$$E = k + 29.6 \log[Ca^{2+}]$$

The standard (0.100 mL) is diluted in the sample (2.00 mL) about 1:20 to give an added concentration of 100 ppm or, more precisely, correcting for the 5% volume change:

$$C = 2000 \text{ ppm} \times \frac{0.100 \text{ mL}}{2.10 \text{ mL}} = 95.2 \text{ ppm}$$

Let x equal the unknown concentration in parts per million:

$$217.6 \text{ mV} = k + 29.6 \log x \tag{1}$$
$$\underline{226.8 \text{ mV} = k + 29.6 \log(x + 95.2)} \tag{2}$$

Subtracting (2) from (1):

$$-9.2 \text{ mV} = 29.6 \log x - 29.6 \log(x + 95.2)$$

$$-9.2 \text{ mV} = 29.6 \log \frac{x}{x + 95.2}$$

$$\log \frac{x}{x + 95.2} = -0.31_1$$

$$\frac{x}{x + 95.2} = 0.467$$

$$x = 83.5 \text{ ppm}$$

If the actual slope for the electrode is not known, then multiple additions of the standard should be made to determine the actual shape.

● ●

AUTOMATIC TITRATORS

There are numerous automatic titrators that employ potentiometric end-point detection. They usually can automatically record the first or second derivative of the titration curve and read out the end-point volume. The sample is placed in the titration vessel, and the titrant, drawn from a reservoir, is placed in a syringe-driven buret. The volume is digitally read from the displacement of the syringe plunger by the electronic driver. Titrators may also employ photometric detection of indicator color changes. An automatic titrator is shown in Figure 14.7. Automatic titrators make volumetric analyses rapid, reproducible, and convenient. While instrumental methods provide many advantages, classical volumetric analyses are still widely used and are very useful, especially for major constituents, for example, in the pharmaceutical industry.

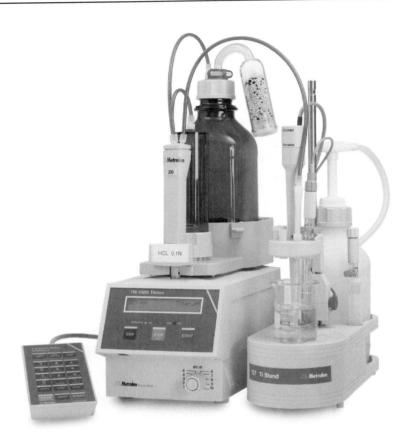

Fig. 14.7. Automatic potentiometric titrator. (Courtesy of Brinkmann Instruments, Inc.)

Learning Objectives

WHAT ARE SOME OF THE KEY THINGS WE LEARNED FROM THIS CHAPTER?

- Balancing redox reactions, p. 414
- Calculating the reaction equilibrium constant from standard potentials (key equation: 14.1; Example 14.3), p. 415
- Calculating redox titration curves, p. 418
- Redox indicators, p. 422
- Iodimetry and iodometry, pp. 424, 426
- Preparing the analyte for titration, p. 431
- Potentiometric titrations, p. 433
- Derivative titrations—using spreadsheets for plotting, p. 435
- Gran plots, p. 437

Questions

1. Describe the ways in which the end points of redox titrations may be detected visually.
2. Distinguish between iodimetry and iodometry.
3. Why are iodimetric titrations usually done in neutral solution and iodometric titrations in acid solution?

4. Does the end point in a permanganate titration coincide with the equivalence point? Explain and suggest how any discrepancies might be corrected.

5. Explain the function of the Zimmermann–Reinhardt reagent in the titration of iron(II) with permanganate.

Problems

BALANCING REDOX REACTIONS

6. Balance the following aqueous reactions:

   (a) $IO_3^- + I^- \rightarrow I_2$ (acid solution)

   (b) $Se_2Cl_2 \rightarrow H_2SeO_3 + Se + HCl$

   (c) $H_3PO_3 + HgCl_2 \rightarrow \underline{Hg_2Cl_2} + H_3PO_4 + HCl$

7. Balance the following aqueous reactions:

   (a) $MnO_4^{2-} \rightarrow MnO_2 + MnO_4^-$ (alkaline solution)

   (b) $MnO_4^- + H_2S \rightarrow Mn^{2+} + S$

   (c) $SbH_3 + Cl_2O \rightarrow H_4Sb_2O_7 + HCl$

   (d) $FeS + NO_3^- \rightarrow Fe^{3+} + NO_2 + S$ (acid solution)

   (e) $Al + NO_3^- \rightarrow AlO_2^- + NH_3$

   (f) $FeAsS + ClO_2 \rightarrow Fe^{3+} + AsO_4^{3-} + SO_4^{2-} + Cl^-$ (acid solution)

   (g) $K_2NaCo(NO_2)_6 + MnO_4^- \rightarrow K^+ + Na^+ + Co^{3+} + NO_3^- + Mn^{2+}$ (acid solution)

8. Equal volumes of 0.10 M $TlNO_3$ and 0.20 M $Co(NO_3)_3$ are mixed. What is the potential in the solution versus the normal hydrogen electrode (NHE)?

9. Calculate the potential in the solution (vs. NHE) in the titration of 50.0 mL of 0.100 M Fe^{2+} in 1.00 M $HClO_4$ with 0.0167 M $Cr_2O_7^{2-}$ at 10, 25, 50, and 60 mL titrant added.

10. Calculate the potential of the solution (vs. NHE) in the titration of 100 mL of 0.100 M Fe^{2+} in 0.500 M H_2SO_4 with 0.0200 M $KMnO_4$ at 10.0, 50.0, 100, and 200 mL of 0.020 M $KMnO_4$ titrant. Assume the H_2SO_4 is completely ionized.

EQUIVALENCE POINT POTENTIALS

11. What would be the potential at the equivalence point in the titration of Fe^{3+} with Sn^{2+}?

12. Equation 14.2 was derived using two half-reactions with equal n values. Derive a similar equation for the following reaction (used in Problem 11), using the numerical n values:

$$2Fe^{3+} + Sn^{2+} \rightleftharpoons 2Fe^{2+} + Sn^{4+}$$

13. Derive an equation similar to Equation 14.2 for the following reaction, remembering to include a hydrogen ion term:

$$5Fe^{2+} + MnO_4^- + 8H^+ \rightleftharpoons 5Fe^{3+} + Mn^{2+} + 4H_2O$$

Use the derived equation to calculate the end-point potential in the titration in Example 14.4 and compare with the value obtained in that example by calculating equilibrium concentrations.

14. For the following cells, calculate the cell voltage before reaction and each half-cell potential after reaction. Also calculate the equilibrium constants of the reactions:

   (a) $Zn|Zn^{2+}$ (0.250 M)$\|Cd^{2+}$ (0.0100 M)$|Cd$

   (b) $Pb|Pb^{2+}$ (0.0100 M)$\|I_3^-$ (0.100 M), I^- (1.00 M)$|Pt$

QUANTITATIVE CALCULATIONS

15. Selenium in a 10.0-g soil sample is distilled as the tetrabromide, which is collected in aqueous solution where it is hydrolyzed to SeO_3^{2-}. The SeO_3^{2-} is determined iodometrically, requiring 4.5 mL of standard thiosulfate solution for the titration. If the thiosulfate titer is 0.049 mg $K_2Cr_2O_7$/mL, what is the concentration of selenium in the soil in ppm?

16. The calcium in a 5.00-mL serum sample is precipitated as CaC_2O_4 with ammonium oxalate. The filtered precipitate is dissolved in acid, the solution is heated, and the oxalate is titrated with 0.00100 M $KMnO_4$, requiring 4.94 mL. Calculate the concentration of calcium in the serum in meq/L (equivalents based on charge).

17. A 2.50-g sample containing As_2O_5, Na_2HAsO_3 and inert material is dissolved and the pH is adjusted to neutral with excess $NaHCO_3$. The As(III) is titrated with 0.150 M I_2 solution, requiring 11.3 mL to just reach the end point. Then, the solution (all the arsenic in the +5 state now) is acidified with HCl, excess KI is added, and the liberated I_2 is titrated with 0.120 M $Na_2S_2O_3$, requiring 41.2 mL. Calculate the percent As_2O_5 and Na_2HAsO_3 in the sample.

18. If 1.00 mL $KMnO_4$ solution will react with 0.125 g Fe^{2+} and if 1.00 mL $KHC_2O_4 \cdot H_2C_2O_4$ solution will react with 0.175 mL of the $KMnO_4$ solution, how many milliliters of 0.200 M NaOH will react with 1.00 mL of the tetroxalate solution? (All three protons on the tetroxalate are titratable.)

19. The sulfide content in a pulp plant effluent is determined with a sulfide ion-selective electrode using the method of standard additions for calibration. A 10.0-mL sample is diluted to 25.0 mL with water and gives a potential reading of −216.4 mV. A similar 10.0-mL sample plus 1.00 mL of 0.030 M sulfide standard diluted to 25.0 mL gives a reading of −224.0 mV. Calculate the concentration of sulfide in the sample.

GRAN PLOTS

20. Starting with the K_a expression for a weak acid HA and substituting the titrant volumes in [HA] and [A^-], show that the following expression holds up to the equivalence point for the titration of HA with a strong base B:

$$V_B[H^+] = K_a(V_{eq.\ pt.} - V_B) = V_B 10^{-pH}$$

where V_B is the volume of added base and $V_{eq.\ pt.}$ the volume added at the equivalence point. A plot of V_B versus $V_B 10^{-pH}$ gives a straight line with a slope of $-K_a$ and an intercept corresponding to the equivalence point.

SPREADSHEET PROBLEM

21. Prepare a spreadsheet to plot the titration curve in Figure 14.1. Take K_{eq} to be 1.7×10^{14}. (See your CD, Chapter 14, for suggested setup after you have prepared the spreadsheet.)

Recommended References

REDOX EQUATIONS

1. C. A. Vanderhoff, "A Consistent Treatment of Oxidation–Reduction," *J. Chem. Ed.*, **25** (1948) 547.

2. R. G. Yolman, "Writing Oxidation–Reduction Equations," *J. Chem. Ed.*, **36** (1959) 215.

EQUIVALENCE POINT POTENTIAL

3. A. J. Bard and S. H. Simpsonsen, "The General Equation for the Equivalence Point Potential in Oxidation–Reduction Titrations," *J. Chem. Educ.*, **37** (1960) 364.

DERIVATIVE TITRATIONS

4. K. N. Carter and R. B. Huff, "Second Derivative Curves and End-Point Determination," *J. Chem. Ed.*, **56** (1979) 26.

GRAN'S PLOTS

5. G. Gran, "Determination of Equivalent Point in Potentiometric Titrations," *Acta Chem. Scand.*, **4** (1950) 559.

6. G. Gran, "Determination of the Equivalence Point in Potentiometric Titrations. Part II," *Analyst*, **77** (1952) 661.

7. C. C. Westcott, "Ion-Selective Measurements by Gran Plots with a Gran Ruler," *Anal. Chim. Acta*, **86** (1976) 269.

8. H. Li, "Improvement of Gran's Plot Method in Standard Addition and Subtraction Methods by a New Plot Method," *Anal. Lett.*, **24** (1991) 473.

Chapter Fifteen

VOLTAMMETRY AND ELECTROCHEMICAL SENSORS

Electrolytic methods include some of the most accurate, as well as most sensitive, instrumental techniques. In these methods, an analyte is oxidized or reduced at an appropriate electrode in an electrolytic cell by application of a voltage (Chapter 12), and the amount of electricity (quantity or current) involved in the electrolysis is related to the amount of analyte. The fraction of analyte electrolyzed may be very small, in fact negligible, in the current–voltage techniques of voltammetry. Micromolar or smaller concentrations can be measured. Since the potential at which a given analyte will be oxidized or reduced is dependent on the particular substance, selectivity can be achieved in electrolytic methods by appropriate choice of the electrolysis potential. Owing to the specificity of the methods, prior separations are often unnecessary. These methods can therefore be rapid.

In this chapter, we discuss voltammetric methods and associated electrochemical sensors, including chemically modified electrodes. Voltammetric techniques use a microelectrode for microelectrolysis. Here, the potential is scanned and a dilute solution of the analyte produces, at a given potential, a limiting current (microampere range or less), which is proportional to the analyte concentration. Amperometry is the application of voltammetry at a fixed potential to follow, via the current, changes in concentration of a given species, for example, during a titration. Amperometric measurements also form the bases of electrochemical sensors.

We describe in detail each of these techniques in this chapter. It will be helpful to review Chapter 13 on potentiometry before reading this material.

15.1 Voltammetry

In voltammetry, the potential is scanned at a microelectrode, and at a certain potential, the analyte will be reduced or oxidized. The current increases in proportion to the analyte concentration.

Voltammetry is essentially an electrolysis on a microscale, using a micro working electrode (e.g., a platinum wire). As the name implies, it is a current–voltage technique. The potential of the micro working electrode is varied (scanned slowly) and the resulting current is recorded as a function of applied potential. The recording is called a **voltammogram**. If an electroactive (reducible or oxidizable) species is, present, a current will be recorded when the applied potential becomes sufficiently negative or positive for it to be electrolyzed. [By convention, a cathodic (reduction) current is + and an anodic (oxidation) current is −]. If the solution is dilute, the current will reach a limiting value because the analyte can only diffuse to the electrode and be electrolyzed at a finite rate, depending on its concentration. We will see below that the limiting current is proportional to the concentration of the species. The microelectrode restricts the current to a few microamperes or less, and so in most applications the concentration of the test substance in solution remains essentially unchanged after the voltammogram is recorded.

THE VOLTAMMETRIC CELL—AN ELECTROLYTIC CELL

A voltammetric cell consists of the micro **working electrode**, the **auxiliary electrode**, and a **reference electrode**, usually a saturated calomel electrode (SCE). A **potentiostat** is employed to control the potential. The current of the working electrode is recorded as a function of its potential measured against the reference electrode, but the voltage is applied between and the current passes between the working and auxiliary electrodes, as in Figure 15.1. In this manner, the current–voltage curve is not disturbed by an appreciable solution resistance, which creates an iR drop (voltage drop) between the working and auxiliary electrodes, as in nonaqueous solvents. Ohm's law states that voltage is equal to the product of current and resistance: $E = iR$, where i is the current in amperes and R the resistance in ohms. When current flows, the recorded potential is distorted—shifted—by an amount equal to $i \times R$, where R is the solution resistance. If appreciable, this causes the current–potential curve (see Fig. 15.2) to be distorted and drawn out over a larger potential range. With a three-electrode system, the recorded potential is that between the working electrode and the reference electrode, with essentially no flow of current and no distorting iR drop.

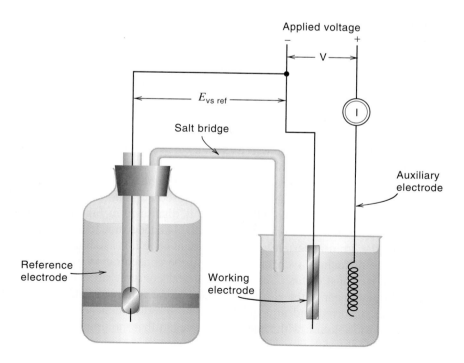

Fig. 15.1. Setup for voltammetric measurements.

THE CURRENT–VOLTAGE CURVE—THE BASIS OF VOLTAMMETRY

See Example 13.4 and Figure 13.4 for conversion of potentials from one reference electrode to another.

Potentials in voltammetry are by convention referred to the SCE. The following relationship can be used to convert potentials versus SCE to the corresponding potentials versus normal hydrogen electrode (NHE), and vice versa:

$$E_{\text{vs. SCE}} = E_{\text{vs. NHE}} - 0.242 \qquad (15.1)$$

This relationship can be used to calculate the applied potential required for the electrolysis of the test ion at the microelectrode. Suppose, for example, we place a $10^{-3}\, M$ solution of cadmium nitrate in the test cell with a carbon microelectrode and impress a voltage difference between the working and auxiliary electrodes, making the microelectrode negative relative to the SCE. The electrode reaction will be

$$\text{Cd}^{2+} + 2e^- \rightarrow \text{Cd} \qquad E^0 = -0.403 \text{ V} \qquad (15.2)$$

The voltammetric cell is really an electrolytic cell in which the electrochemical reaction as a result of the applied potential is the reverse of the spontaneous reaction (as in a voltaic cell). See Section 12.2.

The minimum working electrode potential to begin reducing cadmium [back-emf (electromotive force) required to force the reaction] can be calculated from the Nernst equation (Chapter 13):

$$E_{\text{vs. SCE}} = -0.403 - \frac{0.0592}{2} \log \frac{1}{10^{-3}} - 0.242 = -0.556 \text{ V} \qquad (15.3)$$

This is called the **decomposition potential**. As the potential is increased beyond the decomposition potential, the current will increase linearly in accordance with Ohm's law,

$$\frac{E}{R_{\text{circuit}}} = kE \qquad (15.4)$$

A limiting current is reached because the analyte is being electrolyzed as fast as it can diffuse to the electrode.

As the electrolysis proceeds, the ions in the vicinity of the electrode are depleted by being reduced, creating a concentration gradient between the surface of the electrode and the bulk solution. As long as the applied potential is small, the ions from the bulk of the solution can diffuse rapidly enough to the electrode surface to maintain the electrolysis current. But as the potential is increased, the current is increased, creating a larger concentration gradient. Hence, the ions must diffuse at a more rapid rate in order to maintain the current. The concentration gradient, and hence the rate of diffusion, is proportional to the bulk concentration. If the solution is dilute, a potential will eventually be reached at which the rate of diffusion reaches a maximum and all the ions are reduced as fast as they can diffuse to the electrode surface. Hence, a **limiting current** value, i_l, is reached, and further increased potential will not result in increased current.

The solution can be recovered unchanged following voltammetric measurement due to the small currents passed.

A typical recorded current–voltage curve is illustrated in Figure 15.2. If the solution is stirred or the electrode is rotated, an S-shaped plot [curve (*a*)] is obtained; that is, the limiting current remains constant once it is established. This occurs because the **diffusion layer**, or thickness of the concentration gradient across which the analyte must diffuse, remains small and constant since the analyte is continually brought near to the electrode surface by mass transfer (stirring). But if the electrode is unstirred and in a quiet solution, the diffusion layer will extend farther out into the solution with time, with the result that the limiting current decreases exponentially with time and a "peaked" wave is recorded [curve (*b*)]. For this reason and others, the voltage scan using stationary microelectrodes is usually

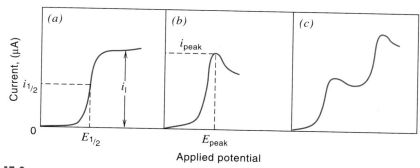

Fig. 15.2. Different types of voltammetric curves. (*a*) Stirred solution or rotated electrode, (*b*) unstirred solution, and (*c*) stepwise reduction (or oxidation) of analyte or of a mixture of two electroactive substances (unstirred solution).

fairly rapid, for example, 50 mV seconds. (Actually, even with stirred solutions, the waves tend to be "peaked" to some extent.)

Although the decomposition potential required to initiate the electrolysis will vary slightly with concentration, the potential at which the current is one-half the limiting current is independent of concentration. This is called the **half-wave potential** $E_{1/2}$. It is a constant related to the standard or formal potential of the redox couple, and so voltammetry serves as a qualitative tool for identifying the reducible or oxidizable species.

An electrode whose potential is dependent on the current flowing is called a **polarizable electrode**. If the electrode area is small and a limiting current is reached, then the electrode is said to be **depolarized**. Therefore, a substance that is reduced or oxidized at a microelectrode is referred to as a **depolarizer**.

If a depolarizer is reduced at the working electrode, a **cathodic current** is recorded at potentials more negative than the decomposition potential. If the depolarizer is oxidized, then an **anodic current** is recorded at potentials more positive than the decomposition potential.

The potential at which the analyte is electrolyzed is a qualitative measure of the analyte.

STEPWISE REDUCTION OR OXIDATION

An electroactive substance may be reduced to a lower oxidation state at a certain potential and then be reduced to a still lower oxidation state when the potential reaches another more negative value. For example, copper(II) in ammonia solution is reduced at a graphite electrode to a stable Cu(I)–ammine complex at -0.2 V versus SCE, which is then reduced to the metal at -0.5 V, each a one-electron reduction step. In such cases, two successive voltammetric waves will be recorded as in curve (*c*) in Figure 15.2. The relative heights of the waves will be proportional to the number of electrons involved in the reduction or oxidation. In this case, the two waves would be of equal height.

When a solution contains two or more electroactive substances that are reduced at different potentials, then a similar stepwise reduction will occur. For example, lead is reduced at potentials more negative than -0.4 V versus SCE ($Pb^{2+} + 2e^- \rightarrow Pb$), and cadmium is reduced at potentials more negative than -0.6 V ($Cd^{2+} + 2e^- \rightarrow Cd$). So a solution containing a mixture of these would exhibit two voltammetric waves at a graphite electrode, one for lead at -0.4 V, followed by another stepwise wave for cadmium at -0.6 V. The relative heights will be proportional to the relative concentrations of the two substances, as well as the relative n values in their reduction or oxidation.

The height of a voltammetric wave is proportional to the number of electrons in the electrolysis reaction.

Mixtures of electroactive substances can be determined by their stepwise voltammetric waves. There should be at least 0.2 V between the $E_{1/2}$ values for good resolution. If the $E_{1/2}$ values are equal, then a single composite wave will be seen, equal in height to the sum of the individual waves. If a major component in much higher concentration is reduced (or oxidized for anodic scans) before the test substance(s) of interest, its wave will mask succeeding waves and it may not even reach a limiting current. In such cases, most of the interfering substance will have to be removed before the analysis can be performed. A common procedure is to preelectrolyze it at a macroelectrode using a setup similar to Figure 15.1, at a potential corresponding to its limiting current plateau but not sufficient to electrolyze the test substance.

Stepwise oxidations may also occur to give stepwise anodic waves.

THE SUPPORTING ELECTROLYTE—NEEDED FOR VOLTAMMETRIC MEASUREMENTS

It was assumed above that when a concentration gradient existed in a quiescent solution, the only way the reducible ion could get to the electrode surface was by diffusion. It can also get to the electrode surface by electrical (coulombic) attraction or repulsion.

The supporting electrolyte is an "inert" electrolyte in high concentration that "swamps out" the attraction or repulsion of the analyte ion at the charged electrode.

The electrode surface will be either positively or negatively charged, depending on the applied potential, and this surface charge will either repulse or attract the ion diffusing to the electrode surface. This will cause an increase or decrease in the limiting current, which is called the **migration current**. The migration current can be prevented by adding a high concentration, at least 100-fold greater than the test substance, of an inert **supporting electrolyte** such as potassium nitrate. Potassium ion is reduced only at a very negative potential and will not interfere. The high concentration of inert ions essentially eliminates the attraction or repulsion forces between the electrode and the analyte, and the inert ions are attracted or repulsed instead. The inert ions are not electrolyzed, however.

A second reason for adding a supporting electrolyte is to decrease the iR drop of the cell. For this reason, at least 0.1 M supporting electrolyte is commonly added. This is true for nearly all electrochemical techniques, potentiometry being an exception. The supporting electrolyte is frequently chosen to provide optimum conditions for the particular analysis, such as buffering at the proper pH or elimination of interferences by selective complexation of some species. When a metal ion is complexed, it is generally stabilized against electrolysis and its voltammetric half-wave is shifted to more negative reduction potentials, or it may even become non-electroactive. Commonly used complexing agents include tartrate, citrate, cyanide, ammonia, and EDTA.

IRREVERSIBLE REDUCTION OR OXIDATION

If a substance is reduced or oxidized reversibly, then its half-wave potential will be near the standard potential for the redox reaction. If it is reduced or oxidized irreversibly, the mechanism of electron transfer at the electrode surface involves a slow step with a high energy of activation. Therefore, extra energy must be applied to the electrode for the electrolysis to occur at an appreciable rate. This is in the form of increased applied potential and is called the **activation overpotential**. Therefore, $E_{1/2}$ will be more negative than the standard potential in the case of a reduction, or if will be more positive in the case of an oxidation. An irreversible wave is more drawn out than a reversible wave. Nevertheless, an S-shaped wave is still obtained, and its diffusion current will be the same as if it were reversible because i_l is limited only by the rate of diffusion of the substance to the electrode surface.

THE WORKING POTENTIAL RANGE—IT DEPENDS ON THE ELECTRODE

The potential range over which voltammetric techniques can be used will depend on the electrode material, the solvent, the supporting electrolyte, and the acidity of the solution. If a platinum electrode is used in aqueous solution, the limiting positive potential would be oxidation of water ($H_2O \rightarrow \frac{1}{2} O_2 + 2H^+ + 2e^-$), unless the supporting electrolyte contains a more easily oxidizable ion (e.g., Cl^-). E^0 for the water half-reaction is $+ 1.0$ V versus SCE, and so the limiting positive potential is about $+1$ V versus SCE, depending on the pH. The negative limiting potential will be from the reduction of hydrogen ions. Platinum has a low hydrogen overvoltage at low current densities, and so this will occur at about -0.1 V versus SCE. Because oxygen is not reduced at these potentials, it need not be removed from the solution, unless oxygen interferes chemically.

Carbon electrodes are frequently used for voltammetry. Their positive potential limit will be essentially the same as with platinum electrodes, but more negative potentials can be reached because hydrogen has a rather high overvoltage on carbon. Potentials of about -1 V versus SCE or more can be used, again depending on the pH. With potentials more negative than -0.1 versus SCE, oxygen must be removed from solutions because it is electrochemically reduced. This is conveniently done by bubbling nitrogen through the solution for 10 to 15 min through a small tube. The nitrogen is passed through water first to saturate it with water vapor so the test solution is not evaporated. Following deaeration, the tube is withdrawn and nitrogen is passed above the surface of the solution to prevent air from being absorbed.

An advantage of carbon electrodes is that they are not troubled by oxide formation on the surface, as platinum electrodes are. While carbon electrodes can be used at fairly negative potentials, a dropping mercury electrode (DME) is often preferred because better reproducibility can be achieved. This is because the electrode surface is constantly renewed (small mercury drops fall from a capillary attached to a mercury reservoir). Voltammetric techniques using a dropping mercury electrode are called **polarography**.

Solid electrode voltammetry is used largely for the oxidation of substances at fairly positive potentials, although for very easily reducible substances, it is also useful. However, reproducibility frequently suffers because the surface characteristics of the electrodes are not reproducible and the surface becomes contaminated. For this reason, the related technique of polarography is preferred when applicable.

Water or protons are easily reduced at a platinum electrode, limiting the available negative potential range to about -0.1 V versus SCE.

Potentials of -1 V versus SCE can be reached with a carbon electrode and -2 V with a mercury electrode. Oxygen must be removed for measurements more negative than -0.1 V. This is done by bubbling with nitrogen.

15.2 Amperometric Electrodes—Measurement of Oxygen

Amperometry is the application of voltammetric measurements at a fixed potential to detect changes in currents as a function of concentration of electroactive species. Electrochemical sensors can be designed based on amperometric measurement. An important example is the oxygen electrode.

The oxygen electrode consists of a thin plastic film such as Teflon stretched over a platinum or gold cathode that allows diffusion of gases but is impermeable to ions in solution (Figure 15.3). Oxygen diffuses through the membrane and is reduced at the cathode, producing an amperometric current. A potential suitable to

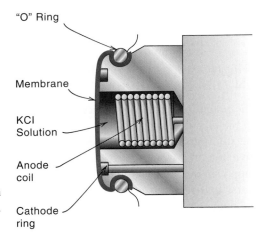

"O" Ring

Membrane

KCl
Solution

Anode
coil

Cathode
ring

Fig. 15.3. Construction of an oxygen electrode. (Courtesy of Arthur H. Thomas Company.)

cause reduction of the oxygen is applied between the oxygen-indicating electrode and the reference electrode, usually a silver–silver chloride electrode built into the probe. An electrolyte in solution or a gel is usually placed between the membrane and the glass insulator to provide electrical contact between the reference electrode and the indicating electrode.

The rate of diffusion of oxygen to the cathode is proportional to the partial pressure of the oxygen in the sample to which the electrode is exposed, and the amperometric current is proportional to this. Measurements are read at atmospheric pressure. Halogens and other gases (e.g., SO_2) that are also reduced at the fixed polarization potential interfere. Hydrogen sulfide poisons the electrode.

The meter is precalibrated by exposure of the probe to samples with known oxygen content, for example, air with assumed 20.9% O_2 content or water saturated with either oxygen or air. At 37°C and sea level (P_{O_2} of 159 torr), air-saturated water contains 5.6 μL O_2 per milliliter and oxygen-saturated water contains 28 μL O_2 per milliliter. See Ref. 7 for a discussion of the calibration of electrodes and the calculation of P_{O_2} and oxygen concentration.

A polarographic oxygen analyzer is often used by biochemists to follow the consumption or release of oxygen in biochemical and enzymatic reactions in order to determine the kinetics of the reactions, and in the clinical laboratory for analytical measurements of enzymes or substrates whose reactions involve consumption of oxygen.

15.3 Electrochemical Sensors: Chemically Modified Electrodes

Amperometric electrodes are a type of electrochemical sensor, as are potentiometric electrodes discussed in Chapter 13. In recent years there has been a great deal of interest in the development of various types of electrochemical sensors that exhibit increased selectivity or sensitivity. These enhanced measurement capabilities of amperometric sensors are achieved by chemical modification of the electrode surface to produce **chemically modified electrodes** (CMEs).

All chemical sensors consist of a **transducer**, which transforms the response into a signal that can be detected (a current in the case of amperometric sensors) and a **chemically selective layer**. The transducer may be optical (e.g., a fiber-optic cable sensor), electrical (potentiometric, amperometric), thermal, and so on. We are concerned here with amperometric transducers.

An enzyme layer imparts chemical selectivity to the electrode.

ENZYME-BASED ELECTRODES FOR MEASURING SUBSTRATES

Enzymes are often employed in the chemical layer to impart the selectivity needed. We saw an example of this in Chapter 13 when discussing potentiometric enzyme electrodes. An example of an amperometric enzyme electrode is the glucose electrode, illustrated in Figure 15.4. The enzyme glucose oxidase is immobilized in a gel (e.g., acrylamide) and coated on the surface of a platinum wire cathode. The gel also contains a chloride salt and makes contact with silver–silver chloride ring to complete the electrochemical cell. Glucose oxidase enzyme catalyzes the aerobic oxidation of glucose as follows:

$$\text{glucose} + O_2 + H_2O \xrightarrow{\text{glucose oxidase}} \text{gluconic acid} + H_2O_2 \qquad (15.5)$$

(see Chapter 22). A potential (ca. +0.6 V vs. Ag/AgCl) is applied to the platinum electrode at which H_2O_2 is electrochemically oxidized:

$$H_2O_2 \rightarrow O_2 + 2H^+ + 2e^- \qquad (15.6)$$

Glucose and oxygen from the test solution diffuse into the gel where their reaction is catalyzed to produce H_2O_2; part of this diffuses to the platinum cathode where it is oxidized to give a current in proportion to the glucose concentration. The remainder eventually diffuses back out of the membrane. An alternative design of a glucose electrode is to coat the membrane of a Clark oxygen electrode with the glucose oxidase gel. Then the depletion of oxygen due to the reaction is measured.

Other examples of amperometric enzyme electrodes based on the measurement of oxygen or hydrogen peroxide include electrodes for the measurement of galactose in blood (galactose oxidase enzyme), oxalate in urine (oxalate oxidase), and cholesterol in blood serum (cholesterol oxidase). Ethanol is determined by reacting with a cofactor, nicotinamide adenine dinucleotide (NAD^+) in the presence of the enzyme alcohol dehydrogenase to produce the reduced form of NAD^+, NADH, which is electrochemically oxidized. Lactate in blood is similarly determined (lactate dehydrogenase enzyme).

CATALYTIC ELECTRODES—REDOX MEDIATORS

Often analytes are irreversibly (slowly) oxidized or reduced at an electrode, that is, require a substantial overpotential to be applied beyond the thermodynamic redox potential (E^0) for electrolysis to occur. This problem of slow electron transfer kinetics has spawned much research in the development of *electrocatalysts*, which may be covalently attached to the electrode, chemisorbed, or trapped in a polymer layer. The basis of electrocatalytic CMEs is illustrated in Figure 15.5. Red is the analyte in the reduced form, which is irreversibly oxidized, and Ox is its oxidized form. The redox mediator is electrochemically reversible and is oxidized at a lower

A redox mediator layer catalyzes the electrochemical reaction, so smaller potentials are required.

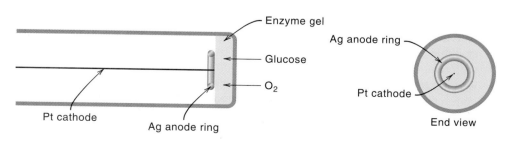

Fig. 15.4. Amperometric glucose electrode.

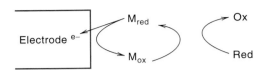

Fig. 15.5. Redox mediator chemically modified elecrode. Red is the analyte being measured, in the reduced form.

potential. The analyte reacts rapidly with the oxidized form of the mediator, M_{ox}, to produce M_{red}, which is immediately oxidized at the electrode surface. The electrochemical reaction takes place near the thermodynamic E^0 value of the mediator. If a lower potential is applied, there is reduced chance for interference from other electrochemically active species (in addition to providing a signal for the analyte!). Electrochemical mediators include ruthenium complexes, ferrocene derivatives, and o-hydroxybenzene derivatives. Mediators such as methylene blue catalyze the oxidation of H_2O_2 so that a potential of only about $+0.2$ V versus Ag/AgCl needs to be applied, instead of the usual $+0.6$ V.

Electrodes are sometimes coated with protective layers to prevent fouling from larger molecules (e.g., proteins). A layer of cellulose acetate, for example, will allow the small H_2O_2 molecule to pass but not the larger ascorbic acid molecule present in biological fluids, which is oxidized at the same potential. Anionic Nafion membranes repel anions and allow cations to pass.

> A protective layer enhances selectivity and reduces chemical fouling of the electrode.

> Ultramicroelectrode responses are independent of the diffusion layer thickness and of flow. They exhibit increased signal-to-noise ratio.

15.4 Ultramicroelectrodes

Amperometric electrodes made on a microscale, on the order of 5 to 30 μm diameter possess a number of advantages. The electrode is smaller than the diffusion layer thickness. This results in enhanced mass transport that is independent of flow, and an increased signal-to-noise ratio, and electrochemical measurements can be made in high-resistance media, such as nonaqueous solvents. An S-shaped sigmoid current–voltage curve is recorded in a quiet solution instead of a peak shaped curve because of the independence on the diffusion layer. The limiting current, i_l, of such microelectrodes is given by

$$i_l = 2nFDCd \qquad (15.7)$$

where n is the electron change, F is the Faraday constant, D is the diffusion coefficient, C is the concentration, and d is the electrode diameter.

There are various ways of constructing ultramicroelectrodes. See Ref. 14. A typical construction is shown in Figure 15.6. The microdisk is the electrode. Electrodes are typically made from carbon fiber, and they can be used for electrooxidation measurements at positive potentials in various microenvironments. An example is the detection of neurotransmitter release in the extracellular space of the brain. [See T. Edmonds, *Anal. Chim. Acta*, **175** (1985) 1, for a review of various electroanalytical applications of carbon fiber electrodes.] These electrodes generate currents that are only a few nanoamperes, and sensitive potentiostats are required for measurements.

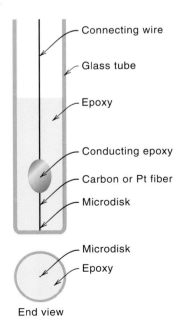

Fig. 15.6. Ultramicroelectrode construction.

_____ Questions

1. Define back-emf, overpotential, and iR drop.

2. Define half-wave potential, depolarizer, DME, residual current, and voltammetry.

3. Give two reasons for using a supporting electrolyte in voltammetry.

4. A solution contains about $10^{-2} M$ Fe^{3+} and $10^{-5} M$ Pb^{2+}. It is desired to analyze for the lead content polarographically. Fe^{3+} is reduced to Fe^{2+} at all potentials accessible with the DME up to -1.5 V versus SCE, and is reduced along with Fe^{2+} to the metal at potentials more negative than -1.5 V. Pb^{2+} is reduced at -0.4 V. Suggest a scheme for measuring the lead polarographically.

5. What effect does complexation have on the voltammetric reduction of a metal ion?

6. What is a chemically modified electrode?

7. What is the function of an electrocatalyst in a chemically modified electrode?

8. What are the advantages of an ultramicroelectrode?

_____ Problems

VOLTAMMETRY/AMPEROMETRY

9. The limiting current of lead in an unknown solution is 5.60 μA. One milliliter of a 1.00×10^{-3} M lead solution is added to 10.0 mL of the unknown solution and the limiting current of the lead is increased to 12.2 μA. What is the concentration of lead in the unknown solution?

10. Iron(III) is polarographically reduced to iron(II) at potentials more negative than about $+0.4$ V versus SCE and is further reduced to iron(0) at -1.5 V versus SCE. Iron(II) is also reduced to the metal at -1.5 V. A polarogram is run (using a DME) on a solution containing Fe^{3+} and/or Fe^{2+}. A current is recorded at zero applied volts, and its magnitude is 12.5 μA. A wave is also recorded with $E_{1/2}$ equal to -1.5 V versus SCE, and its height is 30.0 μA. Identify the iron species in solution (3+ and/or 2+) and calculate the relative concentration of each.

Recommended References

GENERAL

1. J. J. Lingane, *Electroanalytical Chemistry*, 2nd ed. New York: Interscience, 1958. A classic. Excellent general text.

2. J. Wang, *Analytical Electrochemistry*, 2nd ed. New York: John Wiley, 2000. Good introduction to modern electroanalytical techniques.

3. A. J. Bard and L. R. Faulkner, *Electrochemical Methods: Fundamentals and Applications*, 2nd ed. New York: Wiley, 2001

4. A. J. Bard and M. Stratmann, editors-in-chief, *Encyclopedia of Electrochemistry*, 10 volumes + index volume, New York: Wiley. Vol. 3, *Instrumentation and Electroanalytical Chemistry*, P. Unwin, ed., 2003; Vol. 9, *Bioelectrochemistry*, G. S. Wilson, ed., 2002; Vol. 10, *Modified Electrodes*, I. Rubinstein and M. Fujihira, eds., 2003.

VOLTAMMETRY/AMPEROMETRY

5. M. R. Smyth and J. G. Vos, *Analytical Voltammetry*. Amsterdam: Elsevier Science, 1992.

6. R. N. Adams, *Electrochemistry at Solid Electrodes*, New York: Marcel Dekker, 1969. A classic, in its 5th printing.

7. M. A. Lessler and G. P. Brierley, in D. Glick, ed., *Methods of Biochemical Analysis*, Vol. 17. New York: Interscience, 1969, p. 1. Describes the oxygen electrode.

8. Web: www.pineinst.com, link: Educator's Reference Guide for Electrochemistry. An excellent 70-page tutorial on the principles of voltammetry and instrumentation. You can download a sample voltammetry experiment from the Pine Instrument Company website.

ELECTROCHEMICAL SENSORS

9. R. W. Murray, A. E. Ewing, and R. A. Durst, "Chemically Modified Electrodes. Molecular Design for Electroanalysis," *Anal. Chem.*, **59** (1987) 379A.

10. E. R. Reynolds and A. M. Yacynych, "Miniaturized Electrochemical Biosensors," *Am. Lab.*, March (1991) 19.

11. R. I. Stefan, J. F. van Staden, and H. Y. Aboul-Enein, *Electrochemical Sensors in Bioanalysis*, New York: Marcel Dekker, 2001.

12. D. Diamond, ed., *Principles of Chemical and Biological Sensors*, New York: Wiley, 1998.

13. P. G. Edelman and J. Wang, eds., *Biosensors and Chemical Sensors: Optimizing Performance Through Polymeric Materials*, ACS Symposium Series No. 487, Washington, DC: American Chemical Society, 1992.

ULTRAMICROELECTRODES

14. M. Koppenol, J. B. Cooper, and A. M. Bond, "Tools and Procedures for Semiautomated Production of Robust Microdisk Electrodes," *Am. Lab.*, July (1994) 25.

LABORATORY TECHNIQUES

15. P. T. Kissinger and W. R. Heineman, *Laboratory Techniques in Electroanalytical Chemistry*, 2nd ed. New York: Marcel Dekker, 1996.

Chapter Sixteen

SPECTROCHEMICAL METHODS

*"I cannot pretend to be impartial about colours.
I rejoice with the brilliant ones and am
genuinely sorry for the poor browns."*
—Sir Winston Churchill

Spectrometry, particularly in the visible region of the electromagnetic spectrum, is one of the most widely used methods of analysis. It is very widely used in clinical chemistry and environmental laboratories because many substances can be selectively converted to a colored derivative. The instrumentation is readily available and generally fairly easy to operate. In this chapter, we (1) describe the absorption of radiation by molecules and its relation to molecular structure; (2) make quantitative calculations, relating the amount of radiation absorbed to the concentration of an absorbing analyte; and (3) describe the instrumentation required for making measurements. Measurements can be made in the infrared, visible, and ultraviolet regions of the spectrum. The wavelength region of choice will depend upon factors such as availability of instruments, whether the analyte is colored or can be converted to a colored derivative, whether it contains functional groups that absorb in the ultraviolet or infrared regions, and whether other absorbing species are present in the solution. Infrared spectrometry is generally less suited for quantitative measurements but better suited for qualitative or fingerprinting information than are ultraviolet (UV) and visible spectrometry. Visible spectrometers are generally less expensive and more available than UV spectrometers.

We also describe a related technique, fluorescence spectrometry, in which the amount of light emitted upon excitation is related to the concentration. This is an extremely sensitive analytical technique.

Visible spectrometry is probably the most widely used analytical technique.

Who Was the First Spectroscopist?

Johannes Marcus Marci of Kronland (1595–1667) in Eastern Bohemia can probably be considered the first spectroscopic scientist. He was interested in the phenomenon of the rainbow and performed experiments to explain it. He published a book about 1648 whose title, roughly translated, is *The Book of*

Thaumas, about the Heavenly Rainbow and the Nature of the Colors That Appear and Also about Its Origin and the Causes Thereof. He described the conditions responsible for the production of rainbows and wrote about the production of a spectrum by passing a beam of light through a prism. The phenomenon (and the rainbow) was correctly explained as being due to the diffraction of light. Newton, over 20 years later, performed experiments similar to Marci's and provided a more rigorous explanation of the colors of the rainbow, and he received more notoriety; but Marci was the first!

16.1 Interaction of Electromagnetic Radiation with Matter

Spectrometry is based on the absorption of photons by the analyte.

In spectrometric methods, the sample solution absorbs electromagnetic radiation from an appropriate source, and the amount absorbed is related to the concentration of the analyte in the solution. A solution containing copper ions is blue because it absorbs the *complementary color* yellow from white light and transmits the remaining blue light (see Table 16.1). The more concentrated the copper solution, the more yellow light is absorbed and the deeper the resulting blue color of the solution. In a spectrometric method, the amount of this yellow light absorbed would be measured and related to the concentration. We can obtain a better understanding of absorption spectrometry from a consideration of the electromagnetic spectrum and how molecules absorb radiation.

THE ELECTROMAGNETIC SPECTRUM

Wavelength, frequency, and wavenumber are interrelated.

Electromagnetic radiation, for our purposes, can be considered a form of radiant energy that is propagated as a transverse wave. It vibrates perpendicular to the direction of propagation, and this imparts a wave motion to the radiation, as illustrated in Figure 16.1. The wave is described either in terms of its **wavelength**, the distance of one complete cycle, or in terms of the **frequency**, the number of cycles passing a fixed point per unit time. The reciprocal of the wavelength is called the **wavenumber** and is the number of waves in a unit length or distance per cycle. See http://science.csustan.edu/tutorial/color/index.htm for an excellent tutorial on the basics of light and color. It also describes the additive and subtractive primary colors.

Table 16.1
Colors of Different Wavelength Regions

Wavelength Absorbed (nm)	Absorbed Color	Transmitted Color (Complement)
380–450	Violet	Yellow-green
450–495	Blue	Yellow
495–570	Green	Violet
570–590	Yellow	Blue
590–620	Orange	Green-blue
620–750	Red	Blue-green

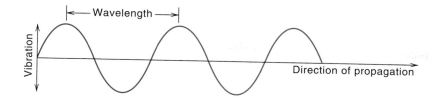

Fig. 16.1. Wave motion of electromagnetic radiation.

The relationship between the wavelength and frequency is

$$\lambda = \frac{c}{\nu} \qquad (16.1)$$

where λ is the wavelength in centimeters (cm),[1] ν is the frequency in reciprocal seconds (s^{-1}), or hertz (Hz), and c is the velocity of light (3×10^{10} cm/s). The wavenumber is represented by $\bar{\nu}$, in cm^{-1}:

$$\bar{\nu} = \frac{1}{\lambda} = \frac{\nu}{c} \qquad (16.2)$$

The wavelength of electromagnetic radiation varies from a few angstroms to several meters. The units used to describe the wavelength are as follows:

$\text{Å} = \text{angstrom} = 10^{-10} \text{ meter} = 10^{-8} \text{ centimeter} = 10^{-4} \text{ micrometer}$

$\text{nm} = \text{nanometer} = 10^{-9} \text{ meter} = 10 \text{ angstroms} = 10^{-3} \text{ micrometer}$

$\mu\text{m} = \text{micrometer} = 10^{-6} \text{ meter} = 10^{4} \text{ angstroms}$

The wavelength unit preferred for the *ultraviolet* and *visible* regions of the spectrum is nanometer, while the unit micrometer is preferred for the *infrared* region.[2] In this last case, wavenumbers are often used in place of wavelength, and the unit is cm^{-1}. See below for a definition of the ultraviolet, visible, and infrared regions of the spectrum.

Electromagnetic radiation possesses a certain amount of energy. The energy of a unit of radiation, called the **photon**, is related to the frequency or wavelength by

$$E = h\nu = \frac{hc}{\lambda} \qquad (16.3)$$

where E is the energy of the photon in ergs and h is Planck's constant, 6.62×10^{-34} joule-second (J-s). It is apparent, then, that *the shorter the wavelength or the greater the frequency, the greater the energy.*

Wavelengths in the ultraviolet and visible regions are on the order of nanometers. In the infrared region, they are micrometers, but the reciprocal of wavelength is often used (wavenumbers, in cm^{-1}).

Shorter wavelengths have greater energy. That is why ultraviolet radiation from the sun burns you!

---

[1] More correctly, the units are centimeters per cycle for wavelength and cycles per second for frequency, but the cycles unit is often assumed. In place of cycles/s, the unit **hertz** (Hz) is now commonly used.

[2] Nanometer (nm) is the preferred term over millimicron (mμ), the unit used extensively prior to this. In the infrared region, micrometer (μm) is the preferred term in place of the previously used term micron (μ).

We should note that the birth of the quantum theory came about in trying to explain the electronic structure of atoms and the properties of light. It became apparent toward the end of the nineteenth century that the classical laws of physics (classical mechanics as proposed by Isaac Newton in the seventeenth century) could not be used to describe electronic structure. The new theory of quantum mechanics, developed at the beginning of the twentieth century, was a scientific breakthrough that changed the way we view atoms.

As indicated above, the electromagnetic spectrum is arbitrarily broken down into different regions according to wavelength. The various regions of the spectrum are shown in Figure 16.2. We will not be concerned with the gamma-ray and X-ray regions in this chapter, although these high-energy radiations can be used in principle in the same manner as lower-energy radiations. The *ultraviolet* region extends from about 10 to 380 nm, but the most analytically useful region is from 200 to 380 nm, called the **near-ultraviolet or quartz UV region**. Below 200 nm, the air absorbs appreciably and so the instruments are operated under a vacuum; hence, this wavelength region is called the **vacuum-ultraviolet region**. The **visible (Vis) region** is actually a very small part of the electromagnetic spectrum, and it is the region of wavelengths that can be seen by the human eye, that is, where the light appears as a color. The visible region extends from the near-ultraviolet region (380 nm) to about 780 nm. The **infrared (IR) region** extends from about 0.78 μm (780 nm) to 300 μm, but the range from 2.5 to 15 μm is the most frequently used for analysis, corresponding to a wavenumber range of 4000 to 667 cm^{-1}. The 0.8- to 2.5-μm range is known as the **near-infrared region**, the 2.5- to 16-μm region as the **mid- or NaCl-infrared region**, and longer wavelengths as the **far-infrared region**. We shall not be concerned with lower-energy radiation (radio or microwave) in this chapter. Nuclear magnetic resonance spectroscopy involves the interaction of low-energy microwave radiation with the nuclei of atoms.

Working ranges of the UV/Vis and IR spectra.

UV	200–380 nm
Vis	380–780 nm
Near-IR	0.78–2.5 μm
Mid-IR	2.5–15 μm

HOW DOES MATTER ABSORB RADIATION?

A qualitative picture of the absorption of radiation can be obtained by considering the absorption of light in the visible region. We "see" objects as colored because they transmit or reflect only a portion of the light in this region. When polychromatic light (white light), which contains the whole spectrum of wavelengths in the visible region, is passed through an object, the object will absorb certain of the wavelengths, leaving the unabsorbed wavelengths to be transmitted. These residual transmitted wavelengths will be seen as a color. This color is **complementary** to the absorbed colors. In a similar manner, opaque objects will absorb certain wavelengths, leaving a residual color to be reflected and "seen."

The color of an object we see is due to the wavelengths transmitted or reflected. The other wavelengths are absorbed.

We see only a very small portion of electromagnetic radiation.

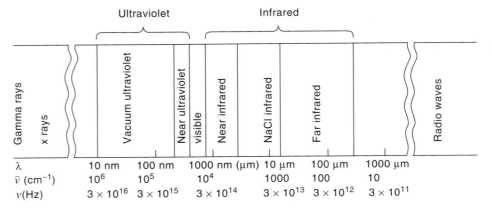

Fig. 16.2. Electromagnetic spectrum.

Table 16.1 summarizes the approximate colors associated with different wavelengths in the visible spectrum. As an example, a solution of potassium permanganate absorbs light in the green region of the spectrum with an absorption maximum of 525 nm, and the solution is purple.

There are three basic processes by which a molecule can absorb radiation; all involve raising the molecule to a higher internal energy level, the increase in energy being equal to the energy of the absorbed radiation (hv). The three types of internal energy are **quantized**; that is, they exist at discrete levels. First, the molecule rotates about various axes, the energy of rotation being at definite energy levels, so the molecule may absorb radiation and be raised to a higher rotational energy level, in a **rotational transition**. Second, the atoms or groups of atoms within a molecule vibrate relative to each other, and the energy of this vibration occurs at definite quantized levels. The molecule may then absorb a discrete amount of energy and be raised to a higher vibrational energy level, in a **vibrational transition**. Third, the electrons of a molecule may be raised to a higher electron energy, corresponding to an **electronic transition**.

Since each of these internal energy transitions is quantized, they will occur only at *definite wavelengths* corresponding to an energy hv equal to the quantized jump in the internal energy. There are, however, many *different* possible energy levels for each type of transition, and several wavelengths may be absorbed. The transitions can be illustrated by an energy level diagram like that in Figure 16.3. The relative energy levels of the three transition processes are in the order electronic > vibrational > rotational, each being about an order of magnitude different in its energy level. Rotational transitions thus can take place at very low energies (long wavelengths, that is, the microwave or far-infrared region), but vibrational transitions require higher energies in the near-infrared region, while electronic transitions require still higher energies (in the visible and ultraviolet regions).

A molecule absorbs a photon by undergoing an energy transition exactly equal to the energy of the photon. The photon must have the right energy for this quantitized transition.

ROTATIONAL TRANSITIONS

Purely rotational transitions can occur in the **far-infrared** and **microwave** regions (ca. 100 μm to 10 cm), where the energy is insufficient to cause vibrational or electronic transitions. The molecule, at room temperature, is usually in its lowest

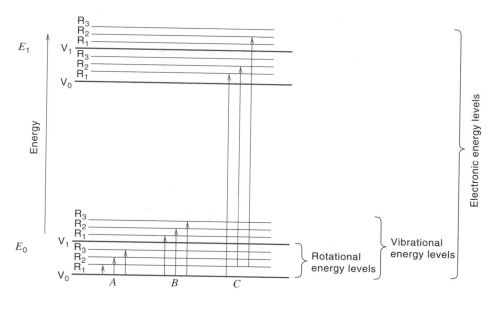

Fig. 16.3. Energy level diagram illustrating energy changes associated with absorption of electromagnetic radiation: A, pure rotational changes (far infrared); B, rotational–vibrational changes (near infrared); C, rotational–vibrational–electronic transitions (visible and ultraviolet). E_0 is electronic ground state and E_1 is first electronic excited state.

Rotational transitions occur at very long wavelengths (low energy, far infrared). Sharp line spectra are recorded.

electronic energy state, called the **ground state** (E_0). Thus, the pure rotational transition will occur at the ground-state electronic level (*A* in Figure 16.3), although it is also possible to have an appreciable population of **excited states** of the molecule. When only rotational transitions occur, discrete absorption *lines* will occur in the spectrum, the wavelength of each line corresponding to a particular transition. Hence, fundamental information can be obtained about rotational energy levels of molecules. This region has been of little use analytically, however.

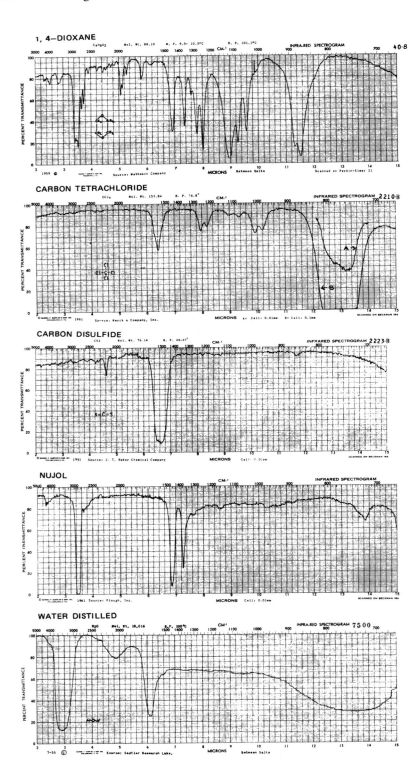

Fig. 16.4. Typical infrared spectra. (From *26 Frequently Used Spectra for the Infrared Spectroscopist*, Standard Spectra-Midget Edition. Copyright © Sadtler Research Laboratories, Inc. Permission for the publication herein of Sadtler Standard Spectra © has been granted, and all rights are reserved by Sadtler Research Laboratories, Inc.)

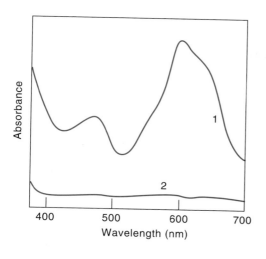

Fig. 16.5. Typical visible absorption spectrum. Tartaric acid reacted with β-naphthol in sulfuric acid. 1, Sample; 2, blank. [From G. D. Christian, *Talanta*, **16** (1969) 255. Reproduced by permission of Pergamon Press, Ltd.]

VIBRATIONAL TRANSITIONS

As the energy is increased (the wavelength decreased), vibrational transitions occur *in addition* to the rotational transitions, with different combinations of vibrational–rotational transitions. *Each* rotational level of the lowest vibrational level can be excited to different rotational levels of the excited vibrational level (*B* in Figure 16.3). In addition, there may be several different excited vibrational levels, each with a number of rotational levels. This leads to numerous discrete transitions. The result is a spectrum of *peaks* or "envelopes" of unresolved fine structure. The wavelengths at which these peaks occur can be related to vibrational modes within the molecule. These occur in the mid- and far-infrared regions. Some typical infrared spectra are shown in Figure 16.4.

Vibrational transitions are also discrete. But the overlayed rotational transitions result in a "smeared" spectrum of unresolved lines.

ELECTRONIC TRANSITIONS

At still higher energies (visible and ultraviolet wavelengths), different levels of electronic transition take place, and rotational and vibrational transitions are superimposed on these (*C* in Figure 16.3). This results in an even larger number of possible transitions. Although all the transitions occur in quantized steps corresponding to discrete wavelengths, these individual wavelengths are too numerous and too close to resolve into the individual lines or vibrational peaks, and the net result is a spectrum of broad *bands* of absorbed wavelengths. Typical visible and ultraviolet spectra are shown in Figure 16.5 and 16.6.

Discrete electronic transitions (visible and ultraviolet regions) are superimposed on vibrational and rotational transitions. The spectra are even more "smeared."

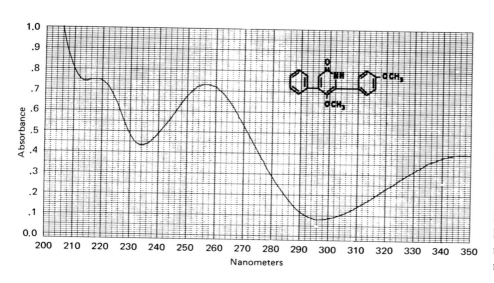

Fig. 16.6. Typical ultraviolet spectrum. 5-Methoxy-6-(*p*-methoxyphenyl)-4-phenyl-2(1*H*)-pyridone in methanol. (From *Sadtler Standard Spectra-u.v.* Copyright © Sadtler Research Laboratories, Inc., 1963. Permission for the publication herein of Sadtler Standard Spectra® has been granted and all rights are reserved by Sadtler Research Laboratories, Inc.)

WHAT HAPPENS TO ABSORBED RADIATION?

Molecules lose most of the energy from absorbing radiation as heat, via collisional processes, that is, by increasing the kinetic energy of the collided molecules.

The lifetimes of excited states of molecules are rather short, and the molecules will lose their energy of excitation and drop back down to the ground state. However, rather than emitting this energy as a photon of the same wavelength as absorbed, most of them will be deactivated by collisional processes in which the energy is lost as heat; the heat will be too small to be detected in most cases. This is the reason for a solution or a substance being colored. If the light were reemitted, then it would appear colorless.[3] In some cases, light will be emitted, usually at longer wavelengths; we discuss this more in Section 16.15 where we discuss fluorescence.

16.2 Electronic Spectra and Molecular Structure

The electronic transitions that take place in the visible and ultraviolet regions of the spectrum are due to the absorption of radiation by specific types of groups, bonds, and functional groups within the molecule. The wavelength of absorption and the intensity are dependent on the type. The wavelength of absorption is a measure of the energy required for the transition. Its intensity is dependent on the probability of the transition occurring when the electronic system and the radiation interact and on the polarity of the excited state.

Radiation	Transition Type
Microwave	Rotational
Infrared	Rotational/ vibrational
Near IR	Vibrational
Visible	Outer electrons
UV	Electronic transitions

KINDS OF TRANSITIONS

π (double or triple bond) and n (outer-shell) electrons are responsible for most UV and visible electron transitions.

Electrons in a molecule can be classified into four different types. (1) Closed-shell electrons that are not involved in bonding. These have very high excitation energies and do not contribute to absorption in the visible or UV regions. (2) Covalent single-bond electrons (σ, or sigma, electrons). These also possess too high an excitation energy to contribute to absorption of visible or UV radiation (e.g., single-valence bonds in saturated hydrocarbons, $-CH_2-CH_2-$). (3) Paired nonbonding outer-shell electrons (n electrons), such as those on N, O, S, and halogens. These are less tightly held than σ electrons and can be excited by visible or UV radiation. (4) Electrons in π (pi) orbitals, for example, in double or triple bonds. These are the most readily excited and are responsible for a majority of electronic spectra in the visible and UV regions.

Excited electrons go into antibonding (π^* or σ^*) orbitals. Most transitions above 200 nm are $\pi \rightarrow \pi^*$ or $n \rightarrow \pi^*$.

Electrons reside in orbitals. A molecule also possesses normally *unoccupied orbitals* called **antibonding orbitals**; these correspond to excited-state energy levels and are either σ^* or π^* orbitals. Hence, absorption of radiation results in an electronic transition to an antibonding orbital. The most common transitions are from π or n orbitals to antibonding π^* orbitals, and these are represented by $\pi \rightarrow \pi^*$ and $n \rightarrow \pi^*$ transitions, indicating a transition to an excited π^* state. The nonbonding n electron can also be promoted, at very short wavelengths, to an antibonding σ^* state: $n \rightarrow \sigma^*$. These occur at wavelengths less than 200 nm.

Examples of $\pi \rightarrow \pi^*$ and $n \rightarrow \pi^*$ transitions occur in ketones ($R-\overset{\overset{\displaystyle O}{\|}}{C}-R'$). Representing the electronic transitions by valence bond structures, we can write

$$\overset{\diagdown}{\underset{\diagup}{C}}=\overset{\frown}{O} \longrightarrow \overset{\diagdown}{\underset{\diagup}{C}}\!\!\!\overset{+}{}-\overset{-}{O}$$

$\pi \rightarrow \pi^*$ transition

---

[3]With unidirectional parallel radiation, the solution should still appear colored, however, because the emitted light would be emitted as a point source in all directions.

$n \to \pi^*$ transition

Acetone, for example, exhibits a high-intensity $\pi \to \pi^*$ transition and a low-intensity $n \to \pi^*$ transition in its absorption spectrum. An example of $n \to \pi^*$ transition occurs in ethers (R—O—R'). Since this occurs below 200 nm, ethers as well as thioethers (R—S—R'), disulfides (R—S—S—R), alkyl amines (R—NH$_2$), and alkyl halides (R—X) are transparent in the visible and UV regions; that is, they have no absorption bands in these regions.

The probability of $\pi \to \pi^*$ transitions is greater than for $n \to \pi^*$ transitions, and so the intensities of the absorption bands are greater for the former. Molar absorptivities, ϵ, at the band maximum for $\pi \to \pi^*$ transitions are typically 1000 to 100,000, while for $n \to \pi^*$ transitions they are less than 1000; ϵ is a direct measure of the intensities of the bands.

ABSORPTION BY ISOLATED CHROMOPHORES

The absorbing groups in a molecule are called **chromophores**. A molecule containing a chromophore is called a **chromogen**. An **auxochrome** does not itself absorb radiation, but, if present in a molecule, it can enhance the absorption by a chromophore or shift the wavelength of absorption when attached to the chromophore. Examples are hydroxyl groups, amino groups, and halogens. These possess unshared (n) electrons that can interact with the π electrons in the chromophore (n–π conjugation).

Spectral changes can be classed as follows: (1) **bathochromic shift**—absorption maximum shifted to longer wavelength, (2) **hypsochromic shift**—absorption maximum shifted to shorter wavelength, (3) **hyperchromism**—an increase in molar absorptivity, and (4) **hypochromism**—a decrease in molar absorptivity.

In principle, the spectrum due to a chromophore is not markedly affected by minor structural changes elsewhere in the molecule. For example, acetone.

$$CH_3 - \overset{\overset{\displaystyle O}{\|}}{C} - CH_3$$

and 2-butanone,

$$CH_3 - \overset{\overset{\displaystyle O}{\|}}{C} - CH_2CH_3$$

give spectra similar in shape and intensity. If the alteration is major or is very close to the chromophore, then changes can be expected.

Similarly, the spectral effects of two isolated chromophores in a molecule (separated by at least two single bonds) are, in principle, independent and are additive. Hence, in the molecule CH_3CH_2CNS, an absorption maximum due to the CNS group occurs at 245 nm with an ϵ of 800. In the molecule $SNCCH_2CH_2CH_2CNS$, an absorption maximum occurs at 247 nm, with approximately double the intensity ($\epsilon = 2000$). Interaction between chromophores may perturb the electronic energy levels and alter the spectrum.

Table 16.2 lists some common chromophores and their approximate wavelengths of maximum absorption.

Table 16.2

Electronic Absorption Bands for Representative Chromophores[a]

Chromophore	System	λ_{max}	ϵ_{max}
Amine	—NH$_2$	195	2,800
Ethylene	—C=C—	190	8,000
		195	1,000
Ketone	C=O	270–285	18–30
Aldehyde	—CHO	210	Strong
		280–300	11–18
Nitro	—NO$_2$	210	Strong
Nitrite	—ONO	220–230	1,000–2,000
		300–400	10
Azo	—N=N—	285–400	3–25
Benzene		184	46,700
		202	6,900
		255	170
Naphthalene		220	112,000
		275	5,600
		312	175
Anthracene		252	199,000
		375	7,900

[a]From M. M. Willard, L. L. Merritt, and J. A. Dean, *Instrumental Methods of Analysis*, 4th ed. Copyright © 1948, 1951, 1958, 1965, by Litton Educational Publishing, Inc., by permission of Van Nostrand Reinhold Company.

It should be noted that exact wavelengths of an absorption band and the probability of absorption (intensity) cannot be calculated, and the analyst always runs standards under carefully specified conditions (temperature, solvent, concentration, instrument type, etc.). Modern instruments may have databases of standard spectra, and standard catalogs of spectra are available for reference.

ABSORPTION BY CONJUGATED CHROMOPHORES

Where multiple (e.g., double, triple) bonds are separated by just one single bond each, they are said to be conjugated. The π orbitals overlap, which decreases the energy gap between adjacent orbitals. The result is a bathochromic shift in the absorption spectrum and generally an increase in the intensity. The greater the degree of conjugation (i.e., several alternating double, or triple, and single bonds), the greater the shift. Conjugation of multiple bonds with nonbonding electrons

(n–π conjugation) also results in spectral changes, for example,   C=CH—NO$_2$.

ABSORPTION BY AROMATIC COMPOUNDS

Aromatic compounds are good absorbers of UV radiation.

Aromatic systems (containing phenyl or benzene groups) exhibit conjugation The spectra are somewhat different, however, than in other conjugated systems, being more complex. Benzene, ⬡, absorbs strongly at 200 nm ($\epsilon_{max} = 6900$) with a weaker band at 230 to 270 nm ($\epsilon_{max} = 170$); see Figure 16.7. The weaker band exhibits considerable fine structure, each peak being due to the influence of vibrational sublevels on the electronic transitions.

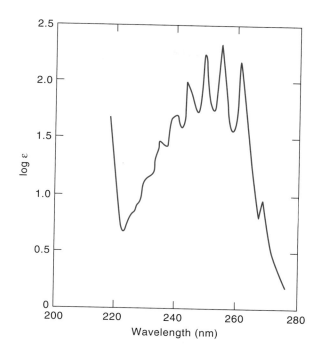

Fig. 16.7. Ultraviolet spectrum of benzene.

As substituted groups are added to the benzene ring, a smoothing of the fine structure generally results, with a bathochromic shift and an increase in intensity. Hydroxy (—OH), methoxy (—OCH$_3$), amino (—NH$_2$), nitro (—NO$_2$), and aldehydic (—CHO) groups, for example, increase the absorption about 10-fold; this large effect is due to n–π conjugation. Halogens and methyl (—CH$_3$) groups act as auxochromes.

Polynuclear aromatic compounds (fused benzene rings), for example, naphthalene, ⬡⬡, have increased conjugation and so absorb at longer wavelengths. Naphthacene (four rings) has an absorption maximum at 470 nm (visible) and is yellow, and pentacene (five rings) has an absorption maximum at 575 nm and is blue (see Table 16.1).

In polyphenyl compounds, ⬡(⬡)$_n$⬡, para-linked molecules (1,6 positions, as shown) are capable of resonance interactions (conjugation) over the entire system, and increased numbers of para-linked rings result in bathochromic shifts (e.g., from 250 to 320 nm in going from $n = 0$ to $n = 4$). In meta-linked molecules (1,3 positions), however, such conjugation is not possible and no appreciable shift occurs up to $n = 16$. The intensity of absorption increases, however, due to the additive effects of the identical chromophores.

Many heterocyclic aromatic compounds, for example, pyridine, ⬡N, absorb in the UV region, and added substituents will cause spectral changes as for phenyl compounds.

Indicator dyes used for acid–base titrations and redox titrations (Chapters 8 and 14) are extensively conjugated systems and therefore absorb in the visible region. Loss or addition of a proton or an electron will markedly change the electron distribution and hence the color.

WHAT IF A MOLECULE DOES NOT ABSORB RADIATION?

An absorbing derivative of a non-absorbing analyte can often be prepared.

If a compound (organic or inorganic) does not absorb in the ultraviolet or visible region, it may be possible to prepare a derivative of it that does. For example, proteins will form a colored complex with copper(II) (biuret reagent). Metals form highly colored chelates with many of the organic precipitating reagents listed in Table 10.2, as well as with others. These may be dissolved or extracted (Chapter 18) in an organic solvent such as ethylene chloride, and the color of the solution measured spectrometrically. The mechanism of absorption of radiation by inorganic compounds is described below.

Spectrometric measurements in the visible region or the ultraviolet region (particularly the former) are widely employed in clinical chemistry, frequently by forming a derivative or reaction product that is colored and can be related to the test substance. For example, creatinine in blood is reacted with picrate ion in alkaline solution to form a colored product that absorbs at 490 nm. Iron is reacted with bathophenanthroline and measured at 535 nm; inorganic phosphate is reacted with molybdenum(VI) and the complex formed is reduced to form "molybdenum blue" (a +5 species) that absorbs at 660 nm; and uric acid is oxidized with alkaline phosphotungstate, and the blue reduction product of phosphotungstate is measured at 680 nm. Ultraviolet measurements include the determination of barbiturates in alkaline solution at 252 nm, and the monitoring of many enzyme reactions by following the change in absorbance at 340 nm due to changes in the reduced form of nicotinamide adenine dinucleotide (NADH), a common reactant or product in enzyme reactions. Clinical measurements are discussed in more detail in Chapter 24.

INORGANIC CHELATES: HOW DO THEY ABSORB SO INTENSELY?

The absorption of ultraviolet or visible radiation by a metal complex can be ascribed to one or more of the following transitions: (1) *excitation of the metal ion*, (2) *excitation of the ligand*, or (3) *charge transfer transition*. Excitation of the metal ion in a complex usually has a very low molar absorptivity (ϵ), on the order of 1 to 100, and is not useful for quantitative analysis. Most ligands used are organic chelating agents that exhibit the absorption properties discussed above, that is, can undergo $\pi \rightarrow \pi^*$ and $n \rightarrow \pi^*$ transitions. Complexation with a metal ion is similar to protonation of the molecule and will result in a change in the wavelength and intensity of absorption. These changes are slight in most cases.

The intense color of metal chelates is frequently due to charge transfer transitions. This is simply the movement of electrons from the metal ion to the ligand, or vice versa. Such transitions include promotion of electrons from π levels in the ligand or from σ bonding orbitals to the unoccupied orbitals of the metal ion, or promotion of σ-bonded electrons to unoccupied π orbitals of the ligand.

When such transitions occur, a redox reaction actually occurs between the metal ion and the ligand. Usually, the metal ion is reduced and the ligand is oxidized, and the wavelength (energy) of maximum absorption is related to the ease with which the exchange occurs. A metal ion in a lower oxidation state, complexed with a high electron affinity ligand, may be oxidized without destroying the complex. An important example is the 1,10-phenanthroline chelate of iron(II).

Charge transfer transitions between a metal ion and complexing ligand are very intense.

Charge transfer transitions are extremely intense, with ϵ values typically 10,000 to 100,000; they occur in either the visible or UV regions. The intensity (ease of charge transfer) is increased by increasing the extent of conjugation in the ligand. Metal complexes of this type are intensely colored due to their high absorption and are well suited for the detection and measurement of trace concentration of metals.

16.3 Infrared Absorption and Molecular Structure

Infrared spectroscopy is very useful for obtaining qualitative information about molecules. But molecules must possess certain properties in order to undergo absorption.

ABSORPTION OF INFRARED RADIATION

Not all molecules can absorb in the infrared region. For absorption to occur, there must be a *change in the dipole moment (polarity) of the molecule.* A diatomic molecule must have a permanent dipole (polar covalent bond in which a pair of electrons is shared unequally) in order to absorb, but larger molecules do not. For example, nitrogen, $N{\equiv}N$, cannot exhibit a dipole and will not absorb in the infrared region. An unsymmetrical diatomic molecule such as carbon monoxide does have a permanent dipole and hence will absorb. Carbon dioxide, $O{=}C{=}O$, does not have a permanent dipole, but by vibration it may exhibit a dipole moment. Thus, in the vibration mode $O{\Rightarrow}C{\Leftarrow}O$, there is symmetry and no dipole moment. But in the mode $O{\Leftarrow}C{\Leftarrow}O$, there is a dipole moment and the molecule can absorb infrared radiation, that is, via an induced dipole. The types of absorbing groups and molecules for the infrared and other wavelength regions will be discussed below.

> The molecule must undergo a change in dipole moment in order to absorb infrared radiation.

Our discussions have been confined to molecules since nearly all absorbing species in solution are molecular in nature. In the case of single atoms (which occur in a flame or an electric arc) that do not vibrate or rotate, only electronic transitions occur. These occur as sharp lines corresponding to definite transitions and will be the subject of discussion in the next chapter.

> Single atoms only undergo electronic transitions. So the spectra are sharp lines.

INFRARED SPECTRA

Absorbing (vibrating) groups in the infrared region absorb within a certain wavelength region, and the exact wavelength will be influenced by neighboring groups. The absorption peaks are much sharper than in the ultraviolet or visible regions, however, and easier to identify. In addition, each molecule will have a complete absorption spectrum unique to that molecule, and so a "fingerprint" of the molecule is obtained. See, for example, the top spectrum in Figure 16.4. Catalogs of infrared spectra are available for a large number of compounds for comparison purposes. See the references at the end of the chapter and the Web address at the end of this section. Mixtures of absorbing compounds will, of course, exhibit the combined spectra of compounds. Even so, it is often possible to identify the individual compounds from the absorption peaks of specific groups on the molecules. Typical functional groups that can be identified include alcohol, hydroxyl, ester carbonyl, olefin, and aromatic unsaturated hydrocarbon groups. Figure 16.8 summarizes regions where certain types of groups absorb. Absorption in the 6- to 15-μm region is very dependent on the molecular environment, and this is called the **fingerprint region**. A molecule can be identified by a comparison of its unique absorption in this region with cataloged known spectra.

> The IR region is the "fingerprint" region.

Although the most important use of infrared spectroscopy is in identification and structure analysis, it is useful for quantitative analysis of complex mixtures of similar compounds because some absorption peaks for each compound will occur at a definite and selective wavelength, with intensities proportional to the concentration of absorbing species. See http://science.csustaniedu/tutorial/ir/index.htm for a quick summary of common absorption bands and compare with Figure 16.8. This gives a nice brief description of how to identify different types of compounds from combinations of bands, including noting which ones are absent.

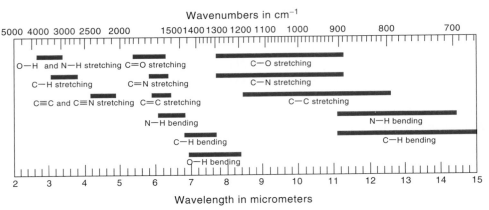

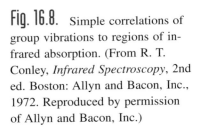

Fig. 16.8. Simple correlations of group vibrations to regions of infrared absorption. (From R. T. Conley, *Infrared Spectroscopy*, 2nd ed. Boston: Allyn and Bacon, Inc., 1972. Reproduced by permission of Allyn and Bacon, Inc.)

INFRARED SPECTROMETRY IN EVERYDAY LIFE

Infrared spectrometry is used for quantitative analysis in many applications, such as industrial hygiene and air quality monitoring. When you have your car emission tested (most or all states require this), an infrared probe is inserted in the exhaust tailpipe to measure CO, CO_2, and hydrocarbons (based on an average molar absorptivity for hydrocarbons).

If you are arrested for driving under the influence of alcohol, chances are you will be tested for blood alcohol content by measurement of the alcohol in your breath using a "breathalyzer" infrared instrument. You are directed to blow into a tube to collect aveolar (deep lung) air in a sample chamber (the aveolar air is in equilibrium with the capillary blood in the lungs). Alcohol is measured using an absorption band at 3.44 μm. This absorption is not specific for alcohol, though, and acetone in the breath is the most likely substance to interfere, which occurs in cases of acidosis (ketosis) if a person is diabetic or has not eaten for a period of time. To correct for this, the absorbance is also measured at 3.37 μm where acetone has a strong absorption band and alcohol weaker. The two absorbances are used to calculate a corrected breath alcohol content. Calibration is done by blowing air through a standard aqueous alcohol solution, at a set temperature, into the sample cell. Breath alcohol content is converted to (and read out as) percent blood alcohol, using an average conversion factor of 2100:1. Many states have adopted a blood alcohol content of under 0.08% as the legal limit, that is, at this level or above, you are presumed to be under the influence. In many European countries, it is even less. And if you are underage, the limit may be 0% and you will lose your license until you are of legal age.

16.4 Near-Infrared Spectrometry for Nondestructive Testing

The mid-infrared region (mid-IR) (1.5 to 25 μm) is widely used for qualitative purposes because of the fine structure information of the spectra. Quantitative analysis is more limited because of the necessity of diluting samples to make measurements and the difficulty in finding solvents that do not absorb in the regions of interest. The region of the spectrum just beyond the visible end of the electromagnetic spectrum, from 0.75 to 2.5 μm (750 to 2500 nm), is called the **near-infrared** region (NIR region). Absorption bands in this region are weak and rather featureless but are useful for nondestructive quantitative measurements, for example, for analysis of solid samples.

OVERTONES AND BANDS—THE BASIS OF NIR ABSORPTION

NIR absorption is due to vibrational **overtones** and **combination bands**, which are *forbidden transitions* of low probability and hence the reason they are weak. These are related to fundamental vibrations and the mid-IR. Excitation of a molecule from the ground vibrational state to a higher vibrational state, where the vibrational quantum number v is ≥ 2, results in overtone absorptions. Thus, the first overtone band results from a $v = 0$ to $v = 2$ transition, while the second and third overtones result from a $v = 0$ to $v = 3$ and a $v = 0$ to $v = 4$ transition, respectively. Combination absorption bands arise when two different molecular vibrations are excited simultaneously. The intensity of overtone bands decreases by approximately one order of magnitude for each successive overtone. Absorption in the NIR is due mainly to C—H, O—H, and N—H bond stretching and bending motions.

SHORT- AND LONG-WAVELENGTH NIR

The NIR region can be further subdivided into the short-wavelength NIR (750 to 1100 nm) and the long-wavelength NIR (1100 to 2500 nm). These subdivisions are based solely on the types of detectors used for the two regions (silicon detectors for the former and PbS, InGaAs, or germanium detectors for the latter). Absorbances are generally weaker in the short-wavelength NIR region. So a 1- to 10-cm pathlength may be used for this, while a shorter 1- to 10-mm cell may be required for the long-wavelength NIR. This is an important distinction because the longer pathlength will give a more representative measurement of the sample. NIR absorption, in general, is 10 to 1000 times less intense than in the mid-IR region, and so samples are usually run "neat" as powders, slurries, or solutions, with no dilution. In the mid-IR, samples are usually diluted, in the form of KBr pellets, thin films, mulls, or solutions, and cell pathlengths are limited to between 15 μm and 1 mm.

NIR FOR NONDESTRUCTIVE TESTING—HOW DO WE CALIBRATE?

While near-IR spectra are rather featureless and have low absorption, the signal-to-noise ratio is high due to intense radiation sources, high radiation throughput, and sensitive detectors in near-IR spectrometers. The operating noise range for the mid-IR is typically in the milliabsorbance range, while near-IR detectors operate at microabsorbance noise levels, 1000 times lower (see definition of absorbance, which follows). Hence, with proper calibration, excellent quantitative results can be achieved.

NIR absorption is useful for nondestructive quantitative measurements. For example, the protein content of grains can be rapidly measured.

Because of its penetration of undiluted samples and the ability to use relatively long pathlength cells, NIR is useful for nondestructive and rapid measurements of more representative samples. However, the low resolution of the technique limited its use for many years until the advent of laboratory computers and the development of statistical (chemometric) techniques to "train" instruments to recognize and resolve analyte spectra in a complex sample matrix, for example, using principal components regression analysis. Chemometric techniques utilize multivariate mathematical procedures for multicomponent measurements, measuring all at once, rather than one parameter or component at a time. Sophisticated software packages are available for automatic calibration and determination. In essence, calibrating standards containing the analyte at different concentrations in the sample matrix are used as training sets from whose spectra the instrument's computer software is able to extract the analyte spectrum and prepare a calibration curve. Generally, the entire spectrum is measured simultaneously (see Section 16.8) and hundreds or thousands of wavelengths are used to extract the spectrum. For quantitative analysis (the main use of NIR spectroscopy), the composition of standards must be known or determined by a standard method.

SOME USES OF NIR SPECTROMETRY

A major use of NIR is the bulk determination of nutrient values in milled grains such as wheat, corn, rice, and oats. The classical ways for analyzing these samples include Kjeldahl analysis for protein, Soxhlet extraction for fat, air drying for moisture, and refractometry for sugars. With proper calibration using mixtures of these consituents, it is possible to put a powdered grain sample in a cup and obtain a complete analysis in a few minutes. But each sample matrix (wheat, corn, etc.) requires its own set of calibrants since matrix matching is required for accurate analysis. Even different geographical sources of grains may require a different calibration model for each source. And usually hundreds of standard mixtures are needed. So the speed and flexibility of the technique is balanced by more time and effort needed to prepare standards and calibrate the instrument, and the technique is really limited to measurements in which thousands of samples are routinely analyzed. Another example of the use of NIR is the petrochemical industry, for measuring octane number, vapor pressure, aromatic content, and the like in refineries. These properties are linked to the composition of the hydrocarbons whose spectra are measured.

16.5 Spectral Databases—Identifying Unknowns

The references at the end of the chapter list a number of useful catalogued spectral sources for UV–Vis and IR spectra, for compound identification. Bourassa et al. have compiled an excellent spectral interpretation bibliography [*Spectroscopy*, **12**(1) (January) (1997) 10].

Powerful and versatile commercial spectral databases available. Also, there are some basic free databases. We list some here. Details of each are given on the text website.

1. www.galactic.com. Free Spectra Online search database, over 6000 spectra (www.galactic.com/spconline).

2. www.acdlabs.com. Lists several free downloadable databases.

3. http://webbook.nist.gov. Gateway to NIST data collection of UV–Vis and IR spectra.

4. www2.chemie.uni-erlang.en.de/services/telespec/. Simulation of infrared spectra.

5. www.ftirsearch.com. Pay-per-use spectral libraries for smaller laboratories, 71,000 spectra.

6. www.chemicalconcepts.com. "World's largest spectra collection," 660,000 spectra.

7. www.mattsonir.com. Includes an inorganic library of FTIR spectra.

8. www.sadtler.com. Integrated system of UV–Vis, and IR and other spectral data.

16.6 Solvents for Spectrometry

Obviously, the solvent used to prepare the sample must not absorb appreciably in the wavelength region where the measurement is being made. In the visible region, this is no problem. There are many colorless solvents and, of course, water is used for inorganic substances. Water can be used in the ultraviolet region. Many substances measured in the ultraviolet region are organic compounds that are insoluble in water and so an organic solvent must be used. Table 16.3 lists a number of solvents for use in the ultraviolet region. The cutoff point is the lowest wavelength at which the absorbance (see below) approaches unity, using a 1-cm cell with water as the reference. These solvents can all be used at least up to the visible region.

The choice of solvent will sometimes affect the spectrum in the ultraviolet region due to solvent–solute interactions. In going from a nonpolar to a polar solvent, loss of fine structure may occur and the wavelength of maximum absorption may shift (either bathochromic or hypsochromic, depending on the nature of the transition and the type of solute–solvent interactions).

The problem of finding a suitable solvent is more serious in the infrared region, where it is difficult to find one that is completely transparent. The use of either carbon tetrachloride or carbon disulfide (health effects aside) will cover the most widely used region of 2.5 to 15 μm (see Figure 16.4). Water exhibits strong absorption bands in the infrared region, and it can be employed only for certain portions of the spectrum. Also, special cell materials compatible with water must be used; rock salt is usually used in cells for infrared measurements because glass absorbs the radiation, but rock salt would dissolve in water. The solvents must be moisture-free if rock salt cells are used.

Transparent solvents in the IR region are limited. Rather concentrated solutions of the sample must often be used.

Table 16.3
Lower Transparency Limit of Solvents in the Ultraviolet Region

Solvent	Cutoff Point (nm)	Solvent	Cutoff Point (nm)[a]
Water	200	Dichloromethane	233
Ethanol (95%)	205	Butyl ether	235
Acetonitrile	210	Chloroform	245
Cyclohexane	210	Ethyl proprionate	255
Cyclopentane	210	Methyl formate	260
Heptane	210	Carbon tetrachloride	265
Hexane	210	N,N-Dimethylformamide	270
Methanol	210	Benzene	280
Pentane	210	Toluene	285
Isopropyl alcohol	210	m-Xylene	290
Isooctane	215	Pyridine	305
Dioxane	220	Acetone	330
Diethyl ether	220	Bromoform	360
Glycerol	220	Carbon disulfide	380
1,2-Dichloroethane	230	Nitromethane	380

[a]Wavelength at which the absorbance is unity for a 1-cm cell, with water as the reference.

16.7 Quantitative Calculations

The fraction of radiation absorbed by a solution of an absorbing analyte can be quantitatively related to its concentration. Here, we present calculations for single species and for mixtures of absorbing species.

BEER'S LAW—RELATING THE AMOUNT OF RADIATION ABSORBED TO CONCENTRATION

This is not a beverage law, although it applies to the absorption of radiation by beer (to make it yellow)!

The amount of monochromatic radiation absorbed by a sample is described by the Beer–Bouguer–Lambert law, commonly called **Beer's law**. Consider the absorption of monochromatic radiation as in Figure 16.9. Incident radiation of radiant power P_0 passes through a solution of an absorbing species at concentrations c and pathlength b, and the emergent (transmitted) radiation has radiant power P. This radiant power is the quantity measured by spectrometric detectors. Bouguer in 1729 (P. Bouguer, *Essai d'otique sur la gradation de la lumier*, Paris, 1729) and Lambert in 1760 (J. H. Lambert, *Photometria*, Ausburg, 1760) recognized that when electromagnetic energy is absorbed, the power of the transmitted energy decreases geometrically (exponentially). Assume, for example, that 25% of the radiant energy in Figure 16.9 is absorbed in a pathlength of b. Twenty-five percent of the remaining energy (25% of $0.75P_0$) will be absorbed in the next pathlength b, leaving 56.25% as the emergent radiation. Twenty-five percent of this would be absorbed in another pathlength of b, and so on, so that an infinite pathlength would be required to absorb all the radiant energy. Since the fraction of radiant energy transmitted decays exponentially with pathlength, we can write it in exponential form:

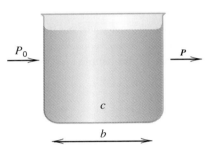

$$T = \frac{P}{P_0} = 10^{-kb} \tag{16.4}$$

Fig. 16.9. Absorption of radiation. P_0 = power of incident radiation, P = power of transmitted radiation, c = concentration, b = pathlength.

where k is a constant and T is called the **transmittance**, the fraction of radiant energy transmitted. Putting this in logarithmic form,

$$\log T = \log \frac{P}{P_0} = -kb \tag{16.5}$$

In 1852, Beer [A. Beer, *Ann. Physik Chem.*, **86** (1852) 78] and Bernard [F. Bernard, *Ann. Chim. et Phys.*, **35** (1853) 385] each stated that a similar law holds for the dependence of T on the concentration, c:

$$T = \frac{P}{P_0} = 10^{-k'c} \tag{16.6}$$

where k' is a new constant, or

$$\log T = \log \frac{P}{P_0} = -k'c \tag{16.7}$$

Combining these two laws, we have Beer's law, which describes the dependence of T on both the pathlength and the concentration:

$$T = \frac{P}{P_0} = 10^{-abc} \tag{16.8}$$

where a is a combined constant of k and k', and

$$\log T = \log \frac{P}{P_0} = -abc \qquad (16.9)$$

It is a more convenient to omit the negative sign on the right-hand side of the equation and to define a new term, **absorbance**:

$$A = -\log T = \log \frac{1}{T} = \log \frac{P_0}{P} = abc \qquad (16.10)$$

where A is the absorbance. This is the common form of Beer's law. Note that it is the *absorbance* that is directly proportional to the concentration.

Beer's law is as simple as abc!

The **percent transmittance** is given by

$$\% \, T = \frac{P}{P_0} \times 100 \qquad (16.11)$$

Equation 16.10 can be rearranged. Since $T = \% \, T/100$,

$$A = \log \frac{100}{\% \, T} = \log 100 - \log \% T$$

Or

$$A = 2.00 - \log \% \, T$$

and

$$\% \, T = \text{antilog}(2.00 - A) \qquad (16.12)$$

The following spreadsheet calculation and plot of absorbance and transmittance illustrates the exponential change of absorbance as a function of a linear change in transmittance.

	A	B	C	D	E	F	G	H	I
1	Comparison of absorbance and transmittance								
2		T	A						
3	0.001	0.001	**3.0000**						
4	0.010	0.010	2.0000						
5	0.050	0.050	1.3010						
6	0.100	0.100	1.0000						
7	0.200	0.200	0.6990						
8	0.300	0.300	0.5229						
9	0.400	0.400	0.3979						
10	0.500	0.500	0.3010						
11	0.600	0.600	0.2218						
12	0.700	0.700	0.1549						
13	0.800	0.800	0.0969						
14	0.900	0.900	0.0458						
15	1.000	1.000	0.0000						
16									
17									
18									
19									
20	**Cell C3**: A = -logT = -LOG10(B3)								
21	Copy formula down.								
22	Plot B3:B15 vs. A3:A15, and C3:C15 vs. A3:A15								

Table 16.4
Spectrometry Nomenclature

Recommended Name	Older Names or Symbols
Absorbance (A)	Optical density (OD), extinction, absorbancy
Absorptivity (a)	Extinction coefficient, absorbancy index, absorbing index
Pathlength (b)	l or d
Transmittance (T)	Transmittancy, transmission
Wavelength (nm)	mμ (millicron)

There are many Beer's law symbols and terms in the literature. Here are some of them.

The absorptivity varies with wavelength and represents the absorption spectrum.

The pathlength b in Equation 16.10 is expressed in centimeters and the concentration c in grams per liter. The constant a is called the **absorptivity** and is dependent on the wavelength and the nature of the absorbing material. In an absorption spectrum, the absorbance varies with wavelength in direct proportion to a (b and c are held constant). The product of the absorptivity and the molecular weight of the absorbing species is called the **molar absorptivity ϵ**. Thus,

$$A = \epsilon bc \qquad (16.13)$$

$a = $ cm^{-1} g^{-1} L
$\epsilon = a \times$ f.wt. $=$ cm^{-1} mol^{-1} L

where c is now in *moles per liter*. The cell pathlength in ultraviolet and visible spectrophotometry is often 1 cm; ϵ has the units cm^{-1} mol^{-1} L, while a has the units cm^{-1} g^{-1} L. The absorptivity a may be used with units other than g/L and, for example, concentrations may be expressed in ppm. But the recommended units for publication are as just described. Beer's law holds strictly for monochromatic radiation since the absorptivity varies with wavelength.

We have used the symbols and terminology recommended by the journal *Analytical Chemistry*. Other terms—such as optical density (OD) in place of absorbance, and extinction coefficient in place of absorptivity—may appear, especially in the older literature, but their use is not now recommended. Table 16.4 lists some of the older nomenclature.

 Example 16.1

A sample in a 1.0-cm cell is determined with a spectrometer to transmit 80% light at a certain wavelength. If the absorptivity of this substance at this wavelength is 2.0, what is the concentration of the substance?

Solution

T is unitless. Check dimensional units.

The percent transmittance is 80%, and so $T = 0.80$:

$$\log \frac{1}{0.80} = 2.0 \text{ cm}^{-1} \text{ g}^{-1} \text{ L} \times 1.0 \text{ cm} \times c$$

$$\log 1.2_5 = 2.0 \text{ g}^{-1} \text{ L} \times c$$

$$c = \frac{0.10}{2.0} = 0.050 \text{ g/L}$$

Example 16.2

A solution containing 1.00 mg ion (as the thiocyanate complex) in 100 mL was observed to transmit 70.0% of the incident light compared to an appropriate blank. (a) What is the absorbance of the solution at this wavelength? (b) What fraction of light would be transmitted by a solution of iron four times as concentrated?

Solution

(a)
$$T = 0.700$$

$$A = \log \frac{1}{0.700} = \log 1.43 = 0.155$$

(b)
$$0.155 = ab(0.0100 \text{ g/L})$$
$$ab = 15.5 \text{ L/g}$$

Therefore,
$$A = 15.5 \text{ L/g } (4 \times 0.0100 \text{ g/L}) = 0.620$$

$$\log \frac{1}{T} = 0.620$$

$$T = 0.240$$

The absorbance of the new solution could have been calculated more directly:

$$\frac{A_1}{A_2} = \frac{abc_1}{abc_2} = \frac{c_1}{c_2}$$

$$A_2 = A_1 \times \frac{c_2}{c_1} = 0.155 \times \frac{4}{1} = 0.620$$

Example 16.3

Amines, RNH_2, react with picric acid to form amine picrates, which absorb strongly at 359 nm ($\epsilon = 1.25 \times 10^4$). An unknown amine (0.1155 g) is dissolved in water and diluted to 100 mL. A 1-mL aliquot of this is diluted to 250 mL for measurement. If this final solution exhibits an absorbance of 0.454 at 359 nm using a 1.00-cm cell, what is the formula weight of the amine? What is a probable formula?

Solution

$$A = \epsilon bc$$
$$0.454 = 1.25 \times 10^4 \text{ cm}^{-1} \text{ mol}^{-1} \text{ L} \times 1.00 \text{ cm} \times c$$
$$c = 3.63 \times 10^{-5} \text{ mol/L}$$

$$\frac{(3.63 \times 10^{-5} \text{ mol/L})(0.250 \text{ L})}{1.00 \text{ mL}} \times 100 \text{ mL} = 9.08 \times 10^{-4} \text{ mol in original flask}$$

$$\frac{0.1155 \text{ g}}{9.08 \times 10^{-4} \text{ mol}} = 127_{.2} \text{ g/mol}$$

The formula weight of chloroaniline, $ClC_6H_4NH_2$, is 127.6, and so this is the probable amine.

Example 16.4

Chloroaniline in a sample is determined as the amine picrate as described in Example 16.3. A 0.0265-g sample is reacted with picric acid and diluted to 1 L. The solution exhibits an absorbance of 0.368 in a 1-cm cell. What is the percentage chloroaniline in the sample?

Solution

$$A = \epsilon bc$$

$$0.368 = 1.25 \times 10^4 \text{ cm}^{-1} \text{ mol}^{-1} \text{ L} \times 1.00 \text{ cm} \times c$$

$$c = 2.94 \times 10^{-5} \text{ mol/L}$$

$$(2.94 \times 10^{-5} \text{ mol/L})(127.6 \text{ g/mol}) = 3.75 \times 10^{-3} \text{ g chloroaniline}$$

$$\frac{3.75 \times 10^{-3} \text{ g chloroaniline}}{2.65 \times 10^{-2} \text{ g sample}} \times 100\% = 14.2\%$$

MIXTURES OF ABSORBING SPECIES

It is possible to make quantitative calculations when two absorbing species in solution have overlapping spectra. It is apparent from Beer's law that the total absorbance A at a given wavelength will be equal to the sum of the absorbances of all absorbing species. For two absorbing species, then, if c is in grams per liter,

$$A = a_x bc_x + a_y bc_y \tag{16.14}$$

> The absorbances of individual absorbing species are additive.

or if c is in moles per liter,

$$A = \epsilon_x bc_x + \epsilon_y bc_y \tag{16.15}$$

where the subscripts refer to substances x and y, respectively.

Consider, for example, the determination of substances x and y whose individual absorption spectra at their given concentration would appear as the solid curves in Figure 16.10, and the combined spectrum of the mixture is the dashed curve. Since there are two unknowns, two measurements will have to be made. The technique is to choose two wavelengths for measurement, one occurring at the absorption maximum for x (λ_1 in the figure) and the other at the maximum for y (λ_2 in the figure). We can write, then,

$$A_1 = A_{x1} + A_{y1} = \epsilon_{x1} bc_x + \epsilon_{y1} bc_y \tag{16.16}$$

$$A_2 = A_{x2} + A_{y2} = \epsilon_{x2} bc_x + \epsilon_{y2} bc_y \tag{16.17}$$

> We have two unknowns (c_x and c_y). We need to write two equations that can be solved simultaneously.

where A_1 and A_2 are the absorbances at wavelengths 1 and 2, respectively (for the mixture); A_{x1} and A_{y1} are the absorbances contributed by x and y, respectively, at wavelength 1; and A_{x2} and A_{y2} are the absorbances contributed by x and y, respectively, at wavelength 2. Similarly, ϵ_{x1} and ϵ_{y1} are the molar absorptivities of x and y, respectively, at wavelength 1; while ϵ_{x2} and ϵ_{y2} are the molar absorptivities of x and y, respectively, at wavelength 2. These molar absorptivities are determined by making absorbance measurements on pure solutions (known molar concentrations)

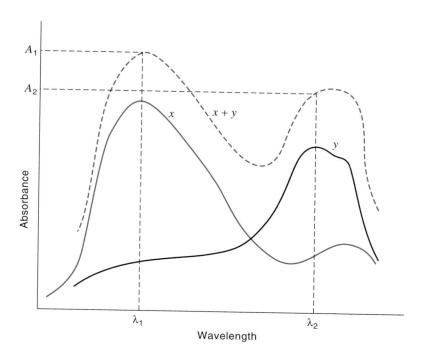

Fig. 16.10. Absorption spectra of pure substances x and y and of mixture of x and y at same concentrations.

of x and y at wavelengths 1 and 2. So c_x and c_y become the only two unknowns in Equations 16.16 and 16.17, and they can be calculated from the solution of the two simultaneous equations.

 **Example 16.5**

Potassium dichromate and potassium permanganate have overlapping absorption spectra in 1 M H_2SO_4. $K_2Cr_2O_7$ has an absorption maximum at 440 nm, and $KMnO_4$ has a band at 545 nm (the maximum is actually at 525 nm, but the longer wavelength is generally used where interference from $K_2Cr_2O_7$ is less). A mixture is analyzed by measuring the absorbance at these two wavelengths with the following results: $A_{440} = 0.405$, $A_{545} = 0.712$ in a 1-cm cell (approximate; exact length not known). The absorbances of pure solutions of $K_2Cr_2O_7$ (1.00×10^{-3} M) and $KMnO_4$ (2.00×10^{-4} M) in 1 M H_2SO_4, using the same cell gave the following results: $A_{Cr,440} = 0.374$, $A_{Cr,545} = 0.009$, $A_{Mn,440} = 0.019$, $A_{Mn,545} = 0.475$. Calculate the concentrations of dichromate and permanganate in the sample solution.

Solution

The pathlength b is not known precisely; but since the same cell is used in all measurements, it is constant. We can calculate the product ϵb from the calibration measurements and use this constant in calculations (call the constant k):

If the pathlength is held fixed, it becomes part of the constant.

$$0.374 = k_{Cr,440} \times 1.00 \times 10^{-3} \qquad k_{Cr,440} = 374$$
$$0.009 = k_{Cr,545} \times 1.00 \times 10^{-3} \qquad k_{Cr,545} = 9$$
$$0.019 = k_{Mn,440} \times 2.00 \times 10^{-4} \qquad k_{Mn,440} = 95$$
$$0.475 = k_{Mn,545} \times 2.00 \times 10^{-4} \qquad k_{Mn,545} = 2.38 \times 10^3$$
$$A_{440} = k_{Cr,440}[Cr_2O_7^{2-}] + k_{Mn,440}[MnO_4^-]$$
$$A_{545} = k_{Cr,545}[Cr_2O_7^{2-}] + k_{Mn,545}[MnO_4^-]$$
$$0.405 = 374[Cr_2O_7^{2-}] + 95[MnO_4^-]$$
$$0.712 = 9[Cr_2O_7^{2-}] + 2.38 \times 10^3[MnO_4^-]$$

Solving simultaneously,

$$[Cr_2O_7^{2-}] = 1.01 \times 10^{-3} \, M \qquad [MnO_4^-] = 2.95 \times 10^{-4} \, M$$

Note that for Cr at 545 nm, where it overlaps the main Mn peak, the absorbance was measured to only one figure since it was so small. This is fine. The smaller the necessary correction, the better. Ideally, it should be zero.

• •

SPREADSHEET EXERCISE: MIXTURE CALCULATION

Let's use Excel Solver to do the calculation for this problem. Denoting the two concentrations X and Y, we will use it to solve for Y, and then substitute this value in the appropriate formula for X to calculate the value of X. We write

$$A_1 = k_{x1}X + k_{y1}Y \tag{1}$$

$$A_2 = k_{x2}X + k_{y2}Y \tag{2}$$

From (1):

$$X = (A_1 - k_{y1}Y)/k_{x1} \tag{3}$$

Substituting in (2):

$$A_2 = k_{x2}(A_1 - k_{y1}Y)/k_{x1} + k_{y2}Y \tag{4}$$

Rearranging:

$$k_{x2}(A_1 - k_{y1}Y)/k_{x1} + k_{y2}Y - A_2 = 0 \tag{5}$$

We will use (5) to calculate Y using Solver. We enter formula (3) in the spreadsheet to automatically calculate X as well. Compare the X and Y values calculated in the spreadsheet with those in the problem above.

The spreadsheet below is included in your CD, Chapter 16. Check the above calculations. You can use the spreadsheet to solve any two-mixture problem by substituting the appropriate values for A_1, A_2, k_{x1}, k_{y1}, k_{x2}, and k_{y2}. (Be sure to save the spreadsheet to your desktop before using it.)

If the two spectral curves overlap only at one of the wavelengths, the solution becomes simpler. For example, if the spectrum of x does not overlap with that of y at wavelength 2, the concentration of y can be determined from a single measurement at wavelength 2, just as if it were not in a mixture. The concentration of x can then be calculated from the absorbance at wavelength 1 by subtracting the contribution of y to the absorbance at that wavelength, that is, from Equation 16.16. The molar absorptivity of y must, of course, be determined at wavelength 1. If there is no overlap of either spectrum at the wavelength of measurement (usually at maximum absorbance), then each substance can be determined in the usual manner.

In making these difference measurements, we have assumed that Beer's law holds over the concentration ranges encountered. If one substance is much more concentrated than the other, then its absorbance may be large at both wavelengths compared to that of the other substance, with the result that the determination of this other substance will not be very accurate.

	A	B	C	D	E	F	G	H	I
1	**Using Excel solver for solving simultaneous equations for mixture absorbances (Example 16.5)**								
2	$A_1 = k_{x1}X + k_{y1}Y$		**$A_1 =$**	0.405	**$k_{x1} =$**	374	**$k_{y1} =$**	95	
3	$A_2 = k_{x2}X + k_{y2}Y$		**$A_2 =$**	0.712	**$k_{x2} =$**	9	**$k_{y2} =$**	2.38E+03	
4	We will use Solver to calculate Y (Cell C9). X is then calculated from the Y value.								
5									
6	$X = (A_1-k_{y1}Y)/k_{x1}$ = Cell C8.			formula in C8 =		($D$2-$H$2*C9)/$F$2			
7									
8			**X =**	0.001008					
9			**Y =**	0.000295					
10			**Solver:**	**C9 = Changing Cell**					
11									
12	**formula:**	-1.57E-14		**Solver:**					
13				**B12 = target cell**					
14				**Set value at 0**					
15	formula:	$k_{x2}((A_1-k_{y1}Y)/k_{x1} + k_{y2}Y - A_2 = 0$ (the 0 is not entered)							
16	formula in B12 =		$F$3*(($D$2-$H$2*C9)/$F$2)+$H$3*C9-$D$3						
17									
18	We have used Solver to calculate Y (C9). That value is substituted in C8 to calculate X.								
19	We could have calculated X with Solver and used that to calculate Y.								
20	Note: We can not subtract the formula in Cell A3 from that in Cell A2 and equate it to								
21	$A_1 - A_2$, and use Solver to change X and Y to arrive at an answer, because there is								
22	an infinite number of iterations.								
23									
24	You can use this for solving other mixture problems by substituting the values indicated								
25	by boldface (A_1, A_2, k_{x1}, k_{y1}, k_{x2}, k_{y2})								

Modern digital instruments that record the entire spectrum of a solution often incorporate mathematical algorithms that will compute the concentrations of several different analytes with overlapping spectra, by utilizing the absorbance values at many different wavelengths (to overestimate the data and improve the confidence) and perform the simultaneous equation calculations by computer. See diode array spectrometers in Section 16.10.

With multiple wavelength measurements, we may analyze for a half dozen or more components! See Section 16.9 and Figure 16.24.

SPREADSHEET EXERCISE: CALCULATION OF UNKNOWN FROM CALIBRATION CURVE

Iron is determined spectrophotometrically by reacting with 1,10-phenanthroline to produce a complex that absorbs at 510 nm. A series of standard solutions give the following absorbances: 0.100 ppm: 0.081; 0.200 ppm: 0.171; 0.500 ppm: 0.432; and 1.0 ppm: 0.857. The sample solution gives an absorbance of 0.463. Prepare a spreadsheet to construct the calibration curve and calculate the concentration of the sample solution.

The linear least-squares line gives a slope of 0.861 and an intercept of -0.002 (using Options under Chart, Add Trendline, when highlighting the chart or line). Hence, the concentration of the unknown is equal to $(0.463 - 0.002)/0.861$, as given by the formula in the spreadsheet (below). The sample concentration is 0.540 ppm. We will now perform the same calculation without charting the calibration curve, and including the standard deviation of the sample concentration.

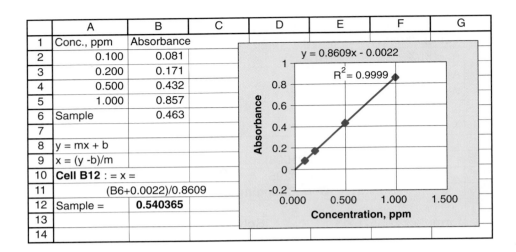

	A	B	C	D	E	F	G
1	Conc., ppm	Absorbance					
2	0.100	0.081					
3	0.200	0.171					
4	0.500	0.432					
5	1.000	0.857					
6	Sample	0.463					
7							
8	y = mx + b						
9	x = (y -b)/m						
10	**Cell B12** : = x =						
11	(B6+0.0022)/0.8609						
12	Sample =	**0.540365**					
13							
14							

STANDARD DEVIATION OF THE SAMPLE CONCENTRATION

We will now introduce some new statistics that allow us to calculate the standard deviation (S_c) of the unknown concentration determined from the calibration curve, based on the number of measurements (N), the slope (m) of the calibration curve, the average absorbance (y_{ave}), the sum of the squares of deviations (S_{xx}) of the individual concentrations (x_i) from the mean of x (x_{ave}), and the standard deviation about regression (S_r). First review the use of the Excel LINEST statistical functions in Section 3.20. We will use some new syntaxes (statistical functions) to help perform the calculations (review useful syntaxes in Section 3.8).

We define the following terms:

N = number of measurements (use COUNT to calculate)

M = number of replicate analyses of unknown ($=1$, if no repeats)

y_{ave} = mean of the y values for N calibration data (use AVERAGE to calculate)

S_{xx} = sum of deviations from the mean x_{ave} value

$$S_{xx} = \sum(x_i - x_{ave})^2 \quad \text{(use N*VARP to calculate)} \tag{16.18}$$

S_{yy} = sum of squares of deviations from the y_{ave} value

$$S_{yy} = \sum(y_i - y_{ave})^2 \quad \text{(use N*VARP to calculate)} \tag{16.19}$$

S_r = standard deviation about regression = standard error of estimate = standard error in y (LINEST calculated this for a calibration curve in Figure 3.10)

$$S_r = \sqrt{\frac{S_{yy} - m^2 S_{xx}}{N - 2}} \quad \text{(use STEYX to calculate)} \tag{16.20}$$

We wish to calculate the standard deviation of the concentration obtained from the calibration curve:

$$S_c = \frac{S_r}{m}\sqrt{\frac{1}{M} + \frac{1}{N} + \frac{y_c - y_{ave}}{m^2 S_{xx}}} \tag{16.21}$$

As indicated, the values of S_r and S_{xx} can be calculated from the Excel statistical functions (we don't have to calculate S_{yy} needed for S_r, since Excel does that for us). The following spreadsheet is set up to perform the calculations.

	A	B	C	D	E
1	Conc., ppm	Absorbance			
2	0.100	0.081		N:	4
3	0.200	0.171		M:	1
4	0.500	0.432		m:	0.860918
5	1.000	0.857		y_{ave}:	0.38525
6	Sample (y_c)	0.463		S_{xx}:	0.49
7				S_r:	0.003629
8	**Cell E2** = N = COUNT(B2:B5)			**S_c:**	0.0051
9	**Cell E3** = M = sample replicates = 1				
10	**Cell E4** =slope = m = SLOPE(B2:B5,A2:A5)				
11	**Cell E5** = y_{ave} = AVERAGE(B2:B5)				
12	**Cell E6** = S_{xx} = N*VARP(A2:A5) = E2*VARP(A2:A5)				
13	**Cell E7** = S_r = STEYX(B2:B5,A2:A5)				
14	**Cell E8** = S_c = $S_r/m((1/M + 1/N + (y_c\text{-}y_{ave})/m^2{*}S_{xx}))^{1/2}$				
15	= E7/E4*((1/E3+1/E2+(B6-E5)/(E4^2*E6))^0.5)				
16					
17	**Cell B18** = b = INTERCEPT(B2:B7,A2:A7)				
18	=	**-0.0021633**			
19	**Cell B20**: Sample = x = (y_c-b)/m =		(B6-B18)/E4		
20	=	**0.54031054**			
21	**Sample = 0.540+/-0.005 ppm**				

The standard deviation of the unknown concentration, S_c is 0.0051, for a sample concentration of 0.5403 ppm, so we report 0.540 ± 0.005 ppm.

QUANTITATIVE MEASUREMENTS FROM INFRARED SPECTRA

Infrared instruments usually record the percent transmittance as a function of wavelength. The presence of scattered radiation, especially at higher concentrations in infrared work, makes direct application of Beer's law difficult. Also, due to rather weak sources, it is necessary to use relatively wide slits (which give rise to apparent deviations from Beer's law—see below). Therefore, empirical methods are often employed in quantitative infrared analysis, keeping experimental conditions constant. The **baseline** or **ratio method** is often used, and this is illustrated in Figure 16.11. A peak is chosen that does not fall too close to others of the test substance or of other substances. A straight line is drawn at the base of the band, and P and P_0 are measured at the absorption peak. (The curve is upside down from the usual absorption spectrum because transmittance is recorded against wavelength.) Log P_0/P is plotted against concentration in the usual manner. Unknowns are compared against standards run under the same instrumental conditions. This technique minimizes relative errors that are in proportion to the sample size, but it does not eliminate simple additive errors, such as those that offset the baseline.

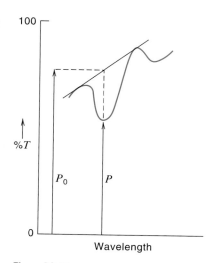

Fig. 16.11. Baseline method for quantitative determination in infrared region of spectrum.

16.8 Spectrometric Instrumentation

A **spectrometer** or **spectrophotometer** is an instrument that will resolve polychromatic radiation into different wavelengths. A block diagram of a spectrometer is shown in Figure 16.12. All spectrometers require (1) a **source** of continuous radiation over the wavelengths of interest, (2) a **monochromator** for selecting a

The types of instrument components will depend on the wavelength region.

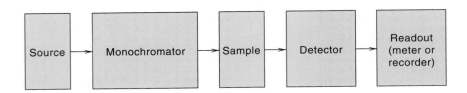

Fig. 16.12. Block diagram of spectrometer.

narrow band of wavelengths from the source spectrum, (3) a sample cell, (4) a **detector**, or transducer, for converting radiant energy into electrical energy, and (5) a device to read out the response of the detector. The sample may precede or follow the monochromator. Each of these, except the readout device, will vary depending on the wavelength region.

SOURCES

Sources for:
  Vis—incandescent lamp
  UV—H_2 or D_2 discharge tube
  IR—rare earth oxide or silicon
  carbide glowers

The source should have a readily detectable output of radiation over the wavelength region for which the instrument is designed to operate. No source, however, has a constant spectral output. The most commonly employed source for the **visible** region is a *tungsten filament incandescent lamp*. The spectral output of a typical filament bulb is illustrated in Figure 16.13. The useful wavelength range is from about 325 or 350 nm to 3 μm, so it can also be used in the near-ultraviolet and near-infrared regions. The wavelength of maximum emission can be shifted to shorter wavelengths by increasing the voltage to the lamp and hence the temperature of the filament, but its lifetime is shortened. For this reason, a stable, regulated power supply is required to power the lamp. This is true for sources for other regions of the spectrum also. Sometimes, a 6-V storage battery is used as the voltage source.

Working ranges of common UV/Vis sources:
  Pulsed xenon arc: 180–2500 nm
  dc deuterium: 185–2500 nm
  dc arc: 200–2500 nm
  quartz tungsten–halogen filament:
  220–2200 nm

For the **ultraviolet** region, a low-pressure *hydrogen* or *deuterium discharge tube* is generally used as the source. Each of these can be used from 185 to about 375 nm, but the deuterium source has about three times the spectral output of the hydrogen source. Ultraviolet sources must have a quartz window, because glass is not transparent to ultraviolet radiation. They are frequently water-cooled to dissipate the heat generated.

Infrared radiation is essentially heat, and so hot wires, light bulbs, or glowing ceramics are used as sources. The energy distribution from the black-body sources tends to peak at about 100 to 2000 nm (near-IR) and then tails off in the mid-IR. Infrared spectrometers usually operate from about 2 to 15 μm, and because of the relatively low-intensity radiation in this region, relatively large slits

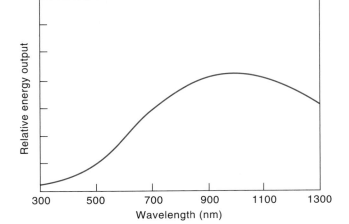

Fig. 16.13. Intensity of radiation as function of wavelength for typical tungsten bulb at 3000 K.

are used to increase the light throughput. But this degrades the wavelength resolution. For this reason, an interferometer is preferred for its increased throughput (see discussion of Fourier transform infrared instrument in Section 16.11). A typical infrared source is the *Nernst glower*. This is a rod consisting of a mixture of rare-earth oxides. It has a negative temperature coefficient of resistance and is nonconducting at room temperature. Therefore, it must be heated to excite the element to emit radiation, but once in operation it becomes conducting and furnishes maximum radiation at about 1.4 μm, or 7100 cm^{-1} (1500 to 2000°C). Another infrared source is the *Globar*. This is a rod of sintered silicon carbide heated to about 1300 to 1700°C. Its maximum radiation occurs at about 1.9 μm (5200 cm^{-1}), and it must be water-cooled. The Globar is a less intense source than the Nernst glower, but it is more satisfactory for wavelengths longer than 15 μm because its intensity decreases less rapidly. IR sources have no protection from the atmosphere, as no satisfactory envelope material exists.

In **fluorescence spectrometry**, the intensity of fluorescence is proportional to the intensity of the radiation source (see Section 16.15). Various continuum UV sources are used to excite fluorescence (see below). But the use of lasers has gained in importance because these monochromatic radiation sources can have high relative intensities. Table 16.5 lists the wavelength and power characteristics of some common laser sources. Only those that lase in the ultraviolet region are generally useful for exciting fluorescence. The nitrogen laser (337.1 nm), which can only be operated in a pulsed mode (rather than continuous wave, or CW, mode), is useful

> Lasers are intense monochromatic sources, good as fluorescence sources.

Table 16.5
Characteristics of Common Lasers

Laser	Wavelength (nm)	Power (W)
Ionic crystal		
Ruby[a]	694.3	1–10 MW
Nd: YAG[a]	1064.0	25 MW (8–9 ns)
Gas		
He–Ne	632.8	0.001–0.05
He–Cd	441.6	0.05
	325.0	0.01
Ar$^+$	514.5	7.5
	496.6	2.5
	488.0	6.0
	476.5	2.5
	465.8	7.0
	457.9	1.3
	333.6–363.8 (4 lines)	3.0
Kr$^+$	752.5	1.2
	647.1	3.5
	530.9	1.5
	482.5	0.4
	468.0	0.5
	413.1	1.8
	406.7	0.9
	337.5–356.4 (3 lines)	2.0
Nitrogen[a]	337.1	200 kW

[a]Operated in pulsed mode; values given are peak power (pulse width).
From G. D. Christian and J. E. O'Reilly, *Instrumental Analysis*, 2nd ed. Boston: Allyn and Bacon, Inc., 1986. Reproduced by permission of Allyn and Bacon, Inc.

for pumping tunable dye lasers. Dye lasers contain solutions of organic compounds that exhibit fluorescence in the UV, visible, or infrared regions. They can generally be tuned over a range of wavelengths of 20 to 50 nm. Tuned lasers are also useful as sources in absorption spectrometry because they provide good resolution (about 1 nm) and high throughput, although they tend to be less stable than continuum sources. Tunable lasers are available from about 265 to 800 nm. Several dyes are needed to cover a wide wavelength range.

We shall see below how spectrometric instruments can be adjusted to account for the variations in source intensity with wavelength as well as for the variation in detector sensitivity with wavelength.

MONOCHROMATORS

A monochromator consists chiefly of lenses or mirrors to focus the radiation, entrance and exit slits to restrict unwanted radiation and help control the spectral purity of the radiation emitted from the monochromator, and a dispersing medium to "separate" the wavelengths of the polychromatic radiation from the source. There are two basic types of dispersing elements, the prism and the diffraction grating. Various types of optical filters may also be used to select specific wavelengths.

Dispersion by prisms is good at short wavelengths, poor at long wavelengths (IR).

1. Prisms. When electromagnetic radiation passes through a prism, it is refracted because the index of refraction of the prism material is different from that in air. The index of refraction depends on the wavelength and, therefore, so does the degree of refraction. Shorter wavelengths are refracted more than longer wavelengths. The effect of refraction is to "spread" the wavelengths apart into different wavelengths (Figure 16.14). By rotation of the prism, different wavelengths of the spectrum can be made to pass through an exit slit and through the sample. A prism works satisfactorily in the ultraviolet and visible regions and can also be used in the infrared region. However, because of its **nonlinear dispersion**, it works more effectively for the shorter wavelengths. Glass prisms and lenses can be used in the visible region, but quartz or fused silica must be used in the ultraviolet region. The latter can also be used in the visible region.

In the infrared region, glass and fused silica transmit very little, and the prisms and other optics must be made from large crystals of alkali or alkaline earth halides, which are transparent to infrared radiation. Sodium chloride (rock salt) is used in most instruments and is useful for the entire region from 2.5 to 15.4 μm (4000 to 650 cm^{-1}). For longer wavelengths, KBr (10 to 25 μm) or CsI (10 to 38 μm) can be used. These (and the monochromator compartment) must be kept dry.

Dispersion by gratings is independent of wavelength, but the intensity varies with wavelength.

2. Diffraction Gratings. These consist of a large number of parallel lines (grooves) ruled on a highly polished surface such as aluminum, about 15,000 to 30,000 per inch for the ultraviolet and visible regions and 1500 to 2500 per inch for the infrared region. The grooves act as scattering centers for rays impinging on the grating. The result is equal dispersion of all wavelengths of a given order, that is, **linear dispersion** (Figure 16.15). The resolving power depends on the number

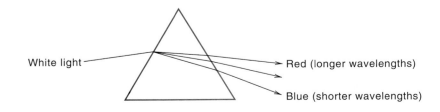

Fig. 16.14. Dispersion of polychromatic light by prism.

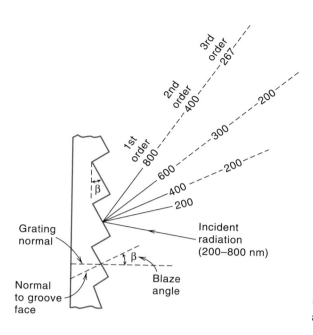

Fig. 16.15. Diffraction of radiation from grating.

of ruled grooves, but generally the resolving power of gratings is better than that of prisms, and they can be used in all regions of the spectrum. They are particularly well suited for the infrared region because of their equal dispersion of the long wavelengths. Gratings are difficult to prepare and original gratings are expensive. However, many **replica gratings** can be prepared from an original grating. This is done by coating the grating with a film of an epoxy resin that, after setting, is stripped off to yield the replica. It is made reflective by aluminizing the surface. These replica gratings are much less expensive and are even used in small inexpensive instruments.

An incident beam of radiation strikes the grating face at an angle i relative to the grating normal (Figure 16.15) and is reflected at an angle θ on the other side of the grating normal. The distance between grooves is d. The path difference between two incoming rays at angle i is $d \sin i$, and the path difference between the corresponding outgoing rays is $d \sin \theta$. The path difference for an incident and reflected ray is $d \sin i - d \sin \theta$. When this difference is equal to one or more wavelengths, fully constructive interference, and no destructive interference, occurs and a bright image results. The corresponding grating equation is

$$n\lambda = d(\sin i - \sin \theta) \tag{16.22}$$

where n is the *diffraction order*, and is an integer. It is apparent that if n is increased and the wavelength decreased by the same multiple, these shorter (higher order) wavelengths will be reflected at the same angle, θ. These have to be filtered before they reach the detector (see below). To disperse light of many wavelengths, the grating is rotated so that the angle i changes.

The *dispersion* of a grating for a given incident angle, i, is given by

$$\frac{d\theta}{d\lambda} = \frac{n}{d \cos \theta} \tag{16.23}$$

that is, it equals the order divided by the product of the grating spacing and cosine of the angle of reflection. The *resolving power* of a grating is the product of the number of rulings and the order. So a large grating has greater resolving power than a small one.

The intensity of radiation reflected by a grating varies with the wavelength, the wavelength of maximum intensity being dependent on the angle from which the radiation is reflected from the surface of the groove in the blazed grating. Hence, gratings are blazed at specific angles for specific wavelength regions, and one blazed for the blue region would be poor for an infrared spectrometer. As mentioned, gratings also will produce radiation at *multiples* of the incident radiation (see Figure 16.15). These multiples are called **higher orders** of the radiation. The primary order is called the first order, twice the wavelength is the second order, three times the wavelength is the third order, and so on. So a grating produces first-order spectra, second-order spectra, and so on. The higher order spectra are more greatly dispersed and the resolution increased. Because of the occurrence of higher orders, radiation at wavelengths less than the spectral region must be filtered out, or else its higher orders will overlap the radiation of interest. This can be accomplished with various types of optical filters (see below) that pass radiation only above a certain wavelength. For example, if incident radiation from a radiating sample (replaces the source on a spectrophotometer) in the 400- to 700-nm range is being dispersed and measured (e.g., fluorescence), any radiation by the sample at, for example, 325 nm, would have a second order at 650 nm, which would overlap first-order radiation at 650 nm. This can be filtered out by placing a filter between the radiating sample and the grating that blocks radiation of $\leq$400 nm in the path of the incident beam; then the 325-nm radiation will not reach the grating.

Ruled gratings have a problem of "ghosting" associated with periodic errors in the ruling engine drive screws, particularly if the gratings are used with high-intensity radiation sources (e.g., in fluorescence instruments—see below). This stray light is greatly reduced with **holographic gratings**. These are manufactured by exposing a photoresist layer, on a suitable substrate, to the interference pattern produced by two monochromatic laser beams, followed by photographic development to produce grooves, and then a reflective coating process. The smoother line profile results in reduced light scatter. Also, these gratings can be produced on curved surfaces and used to collimate light, eliminating mirrors or lenses that result in loss of light. While the cost of these gratings is higher than that of the more conventional type, they are commonly used in spectrometers today, particularly for measurement of radiating samples such as in fluorescence analysis. They have by-and-large replaced prisms in most instruments today.

3. Optical Filters. Various types of optical filters may be used to isolate certain wavelengths of light. There are narrow-bandpass filters, sharp-cut filters, and interference filters. The first two are usually made of glass and contain chemicals (dyes) that absorb all radiation except that desired to be passed. The sharp-cut filters absorb all radiation up to a specified wavelength, and pass radiation at longer wavelengths.

Interference filters consist of two layers of glass on whose inner surfaces a thin semitransparent film of metal is deposited and an inner layer of a transparent material such as quartz or calcium fluoride. Radiation striking the filter exhibits destructive interference, except for a narrow band of radiation for which the filter is designed to transmit. The bandwidth of the filters decreases as the transmitted radiation increases.

SAMPLE CELLS

The cell holding the sample (usually a solution) must, of course, be transparent in the wavelength region being measured. The materials described above for the

Higher orders are better dispersed.

In fluorescence, higher order radiation from a shorter emitting (primary) wavelength may overlap a longer primary wavelength that is being measured. The shorter primary radiation must be filtered before reaching the grating. See also Section 16.9, single-beam spectrometers.

Fig. 16.16. Some typical UV and visible absorption cells.

optics are used for the cell material in instruments designed for the various regions of the spectrum.

The cells for use in **visible** and **ultraviolet** spectrometers are usually cuvets 1 cm thick (*internal* distance between parallel walls), although cells of different pathlengths and volumes can be used. These are illustrated in Figure 16.16. For **infrared** instruments, various assorted cells are used. The most common is a cell of sodium chloride windows. Fixed-thickness cells are available for these purposes and are the most commonly used. The solvent, of course, must not attack the windows of the cell. Sodium chloride cells must be protected from atmospheric moisture (stored in desiccators) and moist solvents. They require periodic polishing to remove "fogging" due to moisture contamination. Silver chloride windows are often used for wet samples or aqueous solutions. These are soft and will gradually darken due to exposure to visible light.

Table 16.6 lists the properties of several infrared transmitting materials. The short pathlengths required in infrared spectrometry are difficult to reproduce, especially when the windows must be repolished, and so quantitative analysis is not

Cells for:
UV—quartz
Vis—glass, quartz
IR—salt crystals

Recommended pathlengths:
UV/Vis: 0.1–1 cm
Near-IR (800–1100 nm): 5–10 cm
(1100–3000 nm): 0.1–2 cm

Table 16.6
Properties of Infrared Materials

Material	Useful Range (cm^{-1})	General Properties
NaCl	40,000–625	Hygroscopic, water soluble, low cost, most commonly used material.
KCl	40,000–500	Hygroscopic, water soluble.
KBr	40,000–400	Hygroscopic, water soluble, slightly higher in cost than NaCl and more hygroscopic.
CsBr	40,000–250	Hygroscopic, water soluble.
CsI	40,000–200	Very hygroscopic, water soluble, good for lower wavenumber studies.
LiF	83,333–1425	Slightly soluble in water, good UV material.
CaF$_2$	77,000–1110	Insoluble in water, resists most acids and alkalis.
BaF$_2$	67,000–870	Insoluble in water, brittle, soluble in acids and NH$_4$Cl.
AgCl	10,000–400	Insoluble in water, corrosive to metals. Darkens upon exposure to short-wavelength visible light. Store in dark.
AgBr	22,000–333	Insoluble in water, corrosive to metals. Darkens upon exposure to short-wavelength visible light. Store in dark.
KRS-5	16,600–285	Insoluble in water, highly toxic, soluble in bases, soft, good for ATR work.
ZnS	50,000–760	Insoluble in water, normal acids and bases, brittle.
ZnSe	20,000–500	Insoluble in water, normal acids and bases, brittle.
Ge	5000–560	Brittle, high index of refraction.
Si	83,333–1430 400–30	Insoluble in most acids and bases.
UV quartz	56,800–3700	Unaffected by water and most solvents.
IR quartz	40,000–3000	Unaffected by water and most solvents.
Polyethylene	625–10	Low-cost material for far-IR work.

Adapted from McCarthy Scientific Co. Catalogue 489, with permission.

as accurate in this region. Use of an internal standard helps. The pathlength of the empty cell can be measured from the interference fringe patterns. Variable pathlength cells are also available in thicknesses from about 0.002 to 3 mm.

When samples exist as pure liquids, they are usually run without dilution ("neat") in the infrared region, as is often the case when an organic chemist is trying to identify or confirm the structure of an unknown or new compound. For this purpose, the cell length must be short in order to keep the absorbance within the optimum region, generally pathlengths of 0.01 to 0.05 mm. If a solution of the sample is to be prepared, a fairly high concentration is usually run, because no solvent is completely transparent in the infrared region, and this will keep the solvent absorbance minimal. So again, short pathlengths are required, generally 0.1 mm or less.

Solids are often not sufficiently soluble in the available solvents to give a high enough concentration to measure in the infrared region. However, powders may be run as a suspension or thick slurry (mull) in a viscous liquid having about the same index of refraction in order to reduce light scattering. The sample is ground in the liquid, which is often Nujol, a mineral oil (see Figure 16.4). Chlorofluorocarbon greases are useful when the Nujol masks any C—H bands present. The mull technique is useful for qualitative analysis, but it is difficult to reproduce for quantitative work. Samples may also be ground with KBr (which is transparent in the infrared region) and pressed into a pellet for mounting for measurement.

Gases may be analyzed by infrared spectrometry, and for this purpose a long-path cell is used, usually 10 cm in length, although cells as long as 20 m and up have been used in special applications. Some typical infrared cells are shown in Figure 16.17.

DETECTORS

Detectors for:
   UV—phototube, PM tube, diode array
   Vis—phototube, PM tube, diode array
   IR—thermocouples, bolometers, thermistors

UV–Vis Detectors. Again, the detectors will also vary with the wavelength region to be measured. A **phototube** (or photocell) is commonly used in the *ultraviolet* and *visible regions*. This consists of a photoemissive cathode and an anode. A high voltage is impressed between the anode and cathode. When a photon enters the window of the tube and strikes the cathode, an electron is emitted and attracted to the anode, causing current to flow that can be amplified and measured. The response of the photoemissive material is wavelength dependent, and different

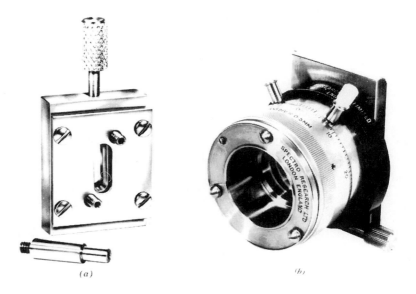

Fig. 16.17. Typical infrared cells. (*a*) Fixed-path cell. (Courtesy of Barnes Engineering Co.) (*b*) Variable-pathlength cell. (Courtesy of Wilks Scientific Corporation.)

(*a*)   (*b*)

phototubes are available for different regions of the spectrum. For example, one may be used for the blue and ultraviolet portions and a second for the red portion of the spectrum.

A **photomultiplier (PM) tube** is more sensitive than a phototube for the *visible* and *ultraviolet regions*. It consists of a photoemissive cathode, which the photon strikes, and a series of electrodes (dynodes), each at a more positive potential (50 to 90 V) than the one before it. When an electron strikes the photoemissive surface, a primary electron is emitted (this is the photoelectric effect—Albert Einstein received the 1921 Nobel Prize in Physics for its discovery in 1905, not for the special theory of relativity which he also introduced in 1905—see www.lucidcafe.com/lucidcafe/library/96mar/einstein.html). The primary electron released from the photoemissive surface is accelerated toward the first dynode. The impact of the electron on the dynode surface causes the release of many secondary electrons, which in turn are accelerated to the next electrode where each secondary electron releases more electrons, and so on, up to about 10 stages of amplication. The electrons are finally collected by the anode. The final output of the photomultiplier tube may, in turn, be electronically amplified.

Again, different photomultiplier tubes have different response characteristics, depending on the wavelength. Figure 16.18 illustrates the response characteristics of some typical photomultiplier tubes with different photoemissive cathode surfaces. The 1P28 (S-5 surface) tube is popular because it can be used in both the ultraviolet and visible regions (e.g., in a UV–visible spectrometer). A 1-S surface is needed for the red region. Because of the greater sensitivity of photomultiplier tubes, less intense radiation is required and narrower slit widths can be used for better resolution of the wavelengths.

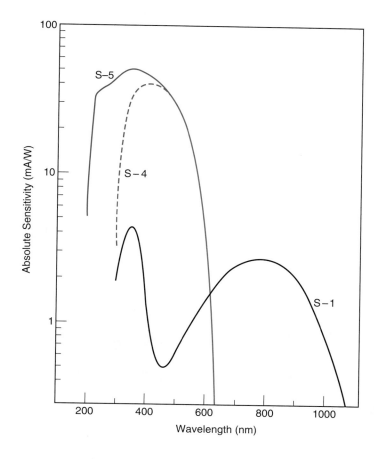

Fig. 16.18. Some spectral responses of photomultipliers. S-5 = RCA 1P28; S-4 = RCA 1P21; S-1 = RCA 7102. (From G. D. Christian and J. E. O'Reilly, *Instrumental Analysis*, 2nd ed. Boston: Allyn and Bacon, Inc., 1986. Reproduced by permission of Allyn and Bacon, Inc.)

Diode arrays can record an entire spectrum at once, from UV to near-IR.

Photomultiplier tubes have also been developed with response limited to the ultraviolet region (160 to 320 nm), the so-called **solar-blind photomultipliers**. They are helpful in reducing stray light effects from visible radiation and are useful as *UV detectors* in nondispersive systems.

Diode array detectors are used in spectrometers that record an entire spectrum simultaneously (see Section 16.10). A **diode array** consists of a series of hundreds of silicon photodiodes positioned side by side on a single silicon crystal or chip. Each has an associated storage capacitor that collects and integrates the photocurrent generated when photons strike the photodiode. They are read by periodical discharging, taking from 5 to 100 ms to read an entire array. If radiation dispersed into its different wavelengths falls on the surface area of the diode array, a spectrum can be recorded. A photograph of diode arrays is shown in Figure 16.19. They consist of 1024 diode elements in a space of a couple of centimeters. The spectral response of a silicon diode array is that of silicon, about 180 to 1100 nm; that is, ultraviolet to near infrared. See Figure 16.20. This range is wider than for photomultiplier tubes and the quantum efficiency is higher. The design of a diode array spectrometer is described in Section 16.10.

Another type of array detector is the charge-coupled device (CCD) detector. Two-dimensional CCDs are used in digital cameras.

Inexpensive silicon diodes or photocells are often used in lower priced instruments. These consist of a chip of pure silicon "doped" with a specific element in which a photon striking it causes an electric impulse (current). which is amplified for readout. Diodes are prepared to be sensitive to specific colors of light.

A spectrophotometer is a double-beam spectrometer that measures absorbance directly.

Spectrometers that use phototubes or photomultiplier tubes (or diode arrays) as detectors are generally called **spectrophotometers**, and the corresponding measurement is called **spectrophotometry**. More strictly speaking, the journal *Analytical Chemistry* defines a spectrophotometer as a spectrometer that measures the *ratio* of the radiant power of two beams, that is, P/P_0, and so it can record absorbance. The two beams may be measured simultaneously or separately, as in a double-beam or a single-beam instrument—see below. Phototube and photomultiplier instruments in practice are almost always used in this manner. An exception is when the radiation source is replaced by a radiating sample whose spectrum and intensity are to be measured, as in fluorescence spectrometry—see below. If the prism or grating monochromator in a spectrophotometer is replaced by an optical filter that passes a narrow band of wavelengths, the instrument may be called a photometer.

Common detectors:
 Photomultiplier tubes: 160–1100 nm
 Silicon-based photodiode arrays:
  180–1100 nm
 Charge-coupled devices (CCDs):
  180–1100 nm
 Silicon photodiodes: 350–1100 nm
 Indium gallium arsenide (InGaAs):
  800–1700 nm
 Lead sulfide (PbS): 1000–3000 nm

IR Detector. As with sources, detectors used in the ultraviolet and visible regions do not work in the infrared region. But *infrared* radiation possesses the property of heat, and heat detectors that transduce heat into an electrical signal can be used. Thermocouples and bolometers are used as detectors. A **thermocouple** consists of

Fig. 16.19. Photo of 1024-element diode arrays. (Courtesy of Hamatsu Photonics, K. K.)

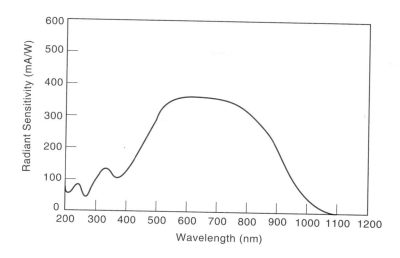

Fig. 16.20. Typical spectral response of diode array. (From M. Kendall-Tobias, *Am Lab.*, March, 1989, p. 102. Reproduced by permission of International Scientific Communications, Inc.)

two dissimilar metal wires, for example, antimony and bismuth, connected at two points. When a temperature difference exists between the two points, a potential difference is developed, which can be measured. One of the junctions, then, is placed in the path of the light from the monochromator. A **thermopile** consists of up to six thermocouples in series, mounted in a vacuum to minimize heat loss by conduction. Half are sensing and half are bonded to a substrate. Thermopiles have response times of about 30 ms. **Balometers** and **thermistors** are materials whose *resistance* is temperature dependent. The thermal resistors are made of sintered oxides of cobalt, manganese, and nickel. Their change in resistance is measured in a Wheatstone bridge circuit. The advantage of these over thermocouples is the more rapid response time (4 ms, compared with 60 ms), and thus improved resolution and faster scanning rates can be accomplished, but sensitivity is compromised. The response of thermal detectors is essentially independent of the wavelengths measured.

For rapid measurements required with FTIR instruments, and for high-sensitivity measurements, photon detectors are used. Examples are the solid-state lead sulfide (PbS), lead selenide (PbSe), or indium gallium arsenide (InGaAs) photoconductive detectors. Photovoltaic detectors are even faster (as fast as 20-ns response) and more sensitive, but they require liquid nitrogen cooling. The InSe detector is limited to 5.5 μm, while the PbSnTe detector works in the 5- to 13-μm range. InGaAs offers the highest sensitivity in the near-IR and has become the detector of choice.

SLIT WIDTH—PHYSICAL VS. SPECTRAL

We mentioned above that it is impossible to obtain spectrally pure wavelengths from a monochromator. Instead, a **band** of wavelengths emanates from the monochromator and the width of this band will depend on both the dispersion of the grating or prism and the exit slit width. The dispersive power of a prism depends on the wavelength and on the material from which it is made, as well as on its geometrical design, while that of a grating depends on the number of grooves per inch. Dispersion is also increased as the distance to the slit is increased.

After the radiation has been dispersed, a certain portion of it will fall on the exit slit, and the width of this slit determines how broad a band of wavelengths the sample and detector will see. Figure 16.21 depicts the distribution of wavelengths leaving the slit. The **nominal wavelength** is that set on the instrument and is the wavelength of maximum intensity passed by the slit. The intensity of radiation at

The radiation passed by a slit is not monochromatic.

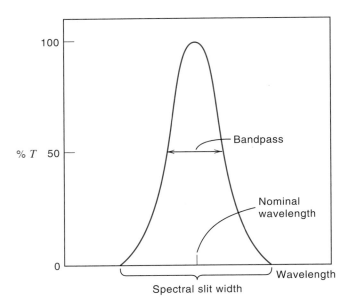

Fig. 16.21. Distribution of wavelengths leaving the slit of monochromator.

wavelengths on each side of this decreases, and the width of the band of wavelengths passed at one-half the intensity of the nominal wavelength is the **spectral bandwidth**, or **bandpass**. The **spectral slit width** is theoretically twice the spectral bandwidth (Figure 16.21 is theoretically an isosceles triangle), and this is a measure of the total wavelength spread that is passed by the slit. Note that the spectral slit width is not the same as the mechanical slit width, which may vary from a few micrometers to a millimeter or more (the spectral slit width is the band of radiation passed by the mechanical slit and is measured in units of wavelength). Seventy-five percent of the radiation intensity is theoretically contained within the wavelengths of the spectral bandwidth.

If the intensity of the source and the sensitivity of the detector are sufficient, the spectral purity can be improved (the bandpass decreased) by decreasing the slit width. The decrease may not be linear, however, and a limit is reached due to aberrations in the optics and diffraction effects caused by the slit at very narrow widths. The diffraction effectively increases the spectral slit width. In actual practice, the sensitivity limit of the instrument is usually reached before diffraction effects become too serious.

> The bandwidth varies with wavelength with a prism, but is constant with a grating.

The bandwidth or the spectral slit width is essentially constant with a grating dispersing element for all wavelengths of a given spectral order at a constant slit width setting. This is not so with a prism because of the variation of dispersion with changing wavelength. The bandwidth will be smaller at shorter wavelengths and larger at longer wavelengths.

INSTRUMENTAL WAVELENGTH AND ABSORBANCE CALIBRATION

The wavelength reading of a spectrophotometer can be checked using solutions of known absorbance maxima and minima. Potassium dichromate at pH 2.9 has maximum absorbances at 257 and 350 nm, and minima at 235 and 313 nm. A holmium oxide glass filter absorbs sharply at 279.2, 222.8, 385.8, 446.0, 536.4, and 637.5 nm.

The National Institute of Standards and Technology (NIST) provides standard reference materials (SRMs) to verify the wavelength accuracy and accuracy of absorbance (transmittance) readings. SRM 930E for UV–Vis analysis consists of a set of three neutral density glass filters of standard thickness with nominal transmittances of 10, 20, and 30%. Other SRMs consist of standard solutions of, for example, potassium dichromate or potassium acid phthalate in perchloric acid. See the NIST Web page (www.nist.gov) for details and listings. Under Subject, go to Standard Reference Materials, Catalogue, and check by SRM number, for example, 931F, 935A, and 2031A for examples. SRM 1921A is a polystyrene film for infrared calibration. See R. A. Spragg and M. Billingham, *Spectroscopy* **10**(1) (1995) 41 for routine use corrections to apply (correcting for effects of resolution, peak-finding algorithm, and temperature on the band positions).

There are commercial sources of reference materials for spectral calibration that are traceable to the NIST standards. See, for example Starna Cells, Inc. (www.starnacells.com).

16.9 Types of Instruments

Although all spectrometric instruments have the basic design presented in Figure 16.12, there are many variations depending on the manufacturer, the wavelength regions for which the instrument is designed, the resolution required, and so on. It is beyond the scope of our discussion to go into these, but we will indicate a few of the important general types of design and the general operation of a spectrometer.

SINGLE-BEAM SPECTROMETERS

These are the most common student spectrometers, since they are less expensive than more sophisticated instruments, and excellent results can be obtained with them. A diagram of the popular Bausch and Lomb Spectronic 20 spectrophotometer (phototube instrument) is shown in Figure 16.22. It consists of a tungsten lamp visible-light source and an inexpensive replica grating of 600 grooves per millimeter to disperse the radiation, ranging in wavelength from 330 to 950 nm. The exit slit allows a band of 20 nm of radiation to pass. If the wavelength is set at 480 nm, for example, radiation from 470 to 490 nm passes through the exit slit. By turning the wavelength cam, the grating is rotated to change the band of 20 nm of wavelengths passing through the exist slit (the path of only one 20-nm band is shown after reflection from the grating in the figure). The filter removes second-order and higher orders of diffraction from the grating that may pass the slit (stray

Higher order radiation from the grating must be filtered.

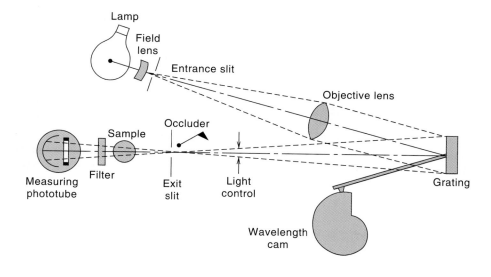

Fig. 16.22. Optical diagram of Bausch and Lomb Spectronic 20 spectrophotometer (top view). (Courtesy of Bausch and Lomb, Inc.)

light). The selection of the filter depends on what radiation must be restricted. For most applications, a cutoff-type filter is used that passes radiation below a certain wavelength where measurements are to be made, but not longer wavelengths where higher orders may appear. Narrower-range filters may be better for some applications, for example, *a red filter to remove any nonred light so the detector sees essentially pure red* (see below).

Any radiation not absorbed by the sample falls on the detector, where the intensity is converted to an electrical signal that is amplified and read on a meter. The measuring phototube for the visible region has maximum response at 400 nm, with only 5% of this response at 625 nm. Measurements above 625 nm are best made by substituting a red-sensitive phototube (RCA 6953) along with a red filter to remove second-order diffraction from the grating (it passes the desired red radiation but not undesired higher orders).

The Spectronic 20 models are available with either analog (Model 20+) or digital (Model 20D+) readout. The analog model features wavelength selection, a power/zeroing knob and *T/A* control. The absorbance is read on an analog meter. Analog output is adjustable from 0 to 1.0 V dc, and can be recorded on a chart recorder or a digital readout device. The digital instrument digitally displays the wavelength and either % *T*, *A*, or concentration. In the FACTOR mode, it electronically converts absorbance values to concentration units by multiplying the absorbance by this factor that has been established from calibration.

We have illustrated that the spectral intensity of the sources and the spectral response of the detectors are dependent on the wavelength. Therefore, some means must be employed to adjust the electrical output of the detector to the same magnitude at all wavelengths. This can be accomplished by one of two ways: by adjusting the slit width to allow more or less light to fall on the detector, or by adjusting the gain on the detector (the amount of amplification of the signal).

A single-beam instrument will have a shutter that can be placed in front of the detector so that no light reaches it. This is the occluder in the Spectronic 20, and it drops into place whenever there is no measuring cell placed in the instrument. With the shutter in position, a "dark current" adjusting knob is used to set the scale reading to zero percent transmittance (infinite absorbance). The **dark current** is a small current that may flow in the absence of light, owing to thermal emission of electrons from the cathode of the phototube. In the above operation, the dark current is set to zero scale reading by effectively changing the voltage on the tube. Now, the cell filled with solvent is placed in the beam path and the shutter is opened. By means of a slit width control to adjust the amount of radiation passed or a "sensitivity" knob (gain control), the output of the detector is adjusted so that the scale reading is 100% transmittance (zero absorbance). The dark current and 100% transmittance adjustments are usually repeated to make certain the adjustment of one has not changed the other. The instrument scale is now calibrated and it is ready to read an unknown absorbance. *The above operation must be repeated at each wavelength.*

Each time a series of samples is run, the absorbance of one or more blank solutions[4] is read versus pure solvent; and, if appreciable (≥ 0.01 A with a Spectronic 20), this is subtracted from all analyte solution readings. Actually, if the blank solution is essentially colorless (i.e., its absorbance is small), this solution is often used in place of the solvent for adjusting the 100% transmittance reading. Any blank absorbance is then automatically corrected for (subtracted). This method

Some current flows in the detector, even when no radiation falls on it. This is the dark current.

---

[4] This contains all reagents used in the sample, but no analyte.

should only be used if the blank reading is small and has been demonstrated to be constant. A large blank reading would be more likely to be variable, and it would require a large gain on the detector, causing an increase in the noise level. An advantage of zeroing the instrument with the blank is that one reading, which always contains some experimental error, is eliminated. If this technique is used, it would be a good practice to check the zero with all the blank solutions to make sure the blank is constant.

DOUBLE-BEAM SPECTROMETERS

These are in practice rather complex instruments, but they have a number of advantages. They are used largely as recording instruments, that is, instruments that automatically vary the wavelength and record the absorbance as a function of wavelength. The instrument has two light paths, one for the sample and one for the blank or reference. In a typical setup, the beam from the source strikes a vibrating or rotating mirror that alternately passes the beam through the reference cell and the sample cell and, from each, to the detector. In effect, the detector alternately sees the reference and the sample beam and the output of the detector is proportional to the ratio of the intensities of the two beams (P/P_0).

The output is an alternating signal whose frequency is equal to that of the vibrating or rotating mirror. An ac amplifier is used to amplify this signal, and stray dc signals are not recorded. The wavelength is changed by a motor that drives the dispersing element at a constant rate, and the slit is continually adjusted by a servomotor to keep the energy from the reference beam at a constant value; that is, it automatically adjusts to 100% transmittance through the reference cell (which usually contains the blank or the solvent).

This is a simplified discussion of a double-beam instrument. There are variations on this design and operation, but it illustrates the utility of these instruments. They are very useful for qualitative work in which the entire spectrum is required, and they automatically compensate for absorbance by the blank, as well as for drifts in source intensity.

Double-beam spectrometers can automatically scan the wavelength and record the spectrum.

Single Beam or Double Beam?

Early UV–Vis and IR spectrophotometers, back in the 1950s, were big clunkers that usually had double-beam monochrometers to compensate for optical drift and electronic noise. They were slow and only moderately sensitive. Improvements in optical and electronic technology have reduced the necessity for double-beam optical systems that reduce the energy of the transmitted beam. Modern single-beam instruments are smaller, faster, more sensitive, and more economical than the older versions. But double-beam instruments still provide the optimal stability, and the choice depends on your need. All modern dispersive IR instruments are single beam.

The choice of resolution of instruments ranges from low-resolution student instruments, such as the Spectronic 20 with 20 nm resolution to 0.05 nm double-grating research instruments. The typical instrument will have a resolution of about 2 nm, and built-in software that allows calibration with multiple standards, polynomial curves, and statistical calculations.

16.10   Diode Array Spectrometers—Getting the Entire Spectrum at Once

In diode array spectrometers, there is no exit slit, and all dispersed wavelengths that fall on the array are recorded simultaneously.

In discussing detectors, we mentioned the use of photodiode array detectors for recording an entire spectrum in a few milliseconds. The basic design of a diode array-based spectrometer is shown in Figure 16.23. In this instance, polychromatic light passes through the sample, and the dispersing element is placed after the sample. The use of an exit slit to isolate a given wavelength is eliminated, and the dispersed light is allowed to fall on the face of the diode array detector. Each diode, in effect, acts as an exit slit of a monochromator. Resolution is limited by the element size of the diode array, but generally, the spatial resolution is about twice the size of a single element.

Diode spectrometers are very useful for the analysis of mixtures of absorbing species whose spectra overlap. The conventional simultaneous equation approach for analyzing mixtures is limited to two or three components (absorbance is measured at two or three wavelengths) in which the spectra are substantially different. With the diode array spectrometer, the absorbance at many points can be measured, using data on the sides of absorption bands as well as at absorption maxima. This method of "overdetermination," in which more measurement points than analytes are obtained, improves the reliability of quantitative measurements, allowing six or more constituents to be determined, or simple mixtures of components with similar spectra. An example of a multicomponent analysis is shown in Figure 16.24 for the simultaneous measurement of five hemoglobins. The five spectra were quantitatively resolved by comparing against standard spectra of each compound stored in the computer memory. Full-spectrum analysis can be performed in a variety of software packages. Mixtures of standards may be used for calibration, and this can compensate for possible interactions between components.

Modern-day instruments have done away with analog strip chart recorders. Instead, spectral scans are displayed on video monitors and printed by the computer.

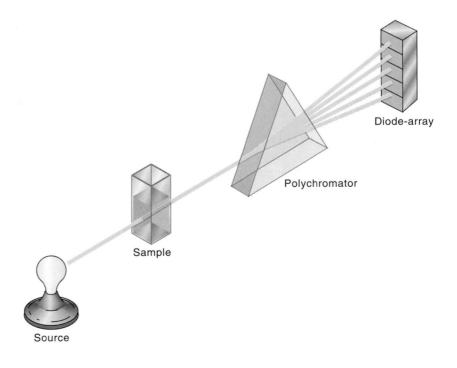

Fig. 16.23.   Schematic of diode array spectrometer.

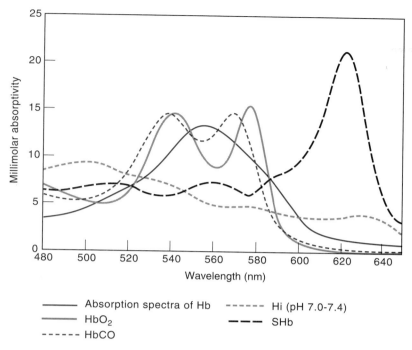

———— Absorption spectra of Hb   - - - - - Hi (pH 7.0–7.4)
———— HbO₂   – – – SHb
- - - - - - HbCO

Note: Hb = hemoglobin, HbO₂ = oxyhemoglobin, HbCO = carboxyhemo-globin, Hi = methemoglobin, SHb = sulfhemoglobin.

Fig. 16.24. Millimolar absorptivities in mmol⁻¹ L cm⁻¹. [From A. Zwart, A. Buursma, E. J. van Kampen, and W. G. Zijlstra, *Clin. Chem.*, **30** (1984) 373. Reproduced by permission.]

The ability of diode array spectrometers to acquire data rapidly also allows the use of measurement statistics to improve the quantitative data. For example, 10 measurements can be made at each point in one second, from which the standard deviation of each point is obtained. The instrument's computer then weights the data points in a least-squares fit, based on their precisions. This "maximum-likelihood" method minimizes the effect of bad data points on the quantitative calculations.

The measurement precision is improved by averaging many measurements.

16.11 Fourier Transform Infrared Spectrometers

Conventional infrared spectrometers are known as **dispersive instruments**. With the advent of computer- and microprocessor-based instruments, these have been largely replaced by Fourier transform infrared (FTIR) spectrometers, which possess a number of advantages. Rather than a grating monochromator, an FTIR instrument employs an interferometer to obtain a spectrum.

The basis of an interferometer instrument is illustrated in Figure 16.25. Radiation from a conventional IR source is split into two paths by a beam splitter, one path going to a fixed position mirror, and the other to a moving mirror. When the beams are reflected, one is slightly displaced (out of phase) from the other since it travels a smaller (or greater) distance due to the moving mirror, and they recombine to produce an interference pattern (of all wavelengths in the beam) before passing through the sample. The sample sees all wavelengths simultaneously, and the interference pattern changes with time as the mirror is continuously scanned at a linear velocity. The result of absorption of the radiation by the sample is a spectrum in the **time domain**, called an **interferogram**, that is, absorption intensity as a function of the optical path difference between the two beams.

FTIR spectrometers have largely replaced dispersive IR spectrometers.

An interferogram is a spectrum in the time domain. Fourier transformation converts it to the frequency domain.

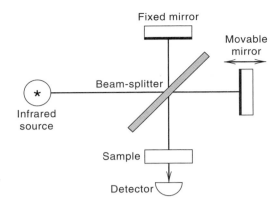

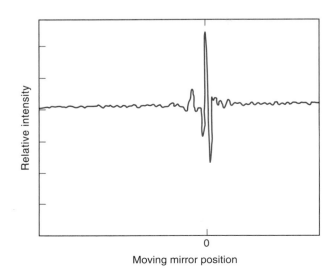

Fig. 16.25. Schematic of interferometer for FTIR spectrometry.

A typical interferogram is shown in Figure 16.26. The tall part of the signal corresponds to when the two mirrors are equidistant from the beam splitter, when destructive interference between the two beams is zero, and is called the centerburst. The intensity drops off rapidly away from this, due to destructive interference. This is converted, using a computer, into the frequency domain via a mathematical operation known as a **Fourier transformation** (hence the name **Fourier transform infrared spectrometer**). A conventional appearing infrared spectrum results.

Advantages of FTIR spectrometers: greater throughput, increased signal-to-noise ratio, simultaneous measurement of all wavelengths.

The advantages of an interferometer instrument is that there is greater throughput (Jacquinot's advantage) since all the radiation is passed. That is, the sample sees all wavelengths at all times, instead of a small portion at a time. This results in increased signal-to-noise ratio. In addition, a *multiplex advantage* (Fellget's advantage) results because the interferometer measures all IR frequencies simultaneously, and so a spectrum with resolution comparable to or better than that with a grating is obtained in a few seconds.

In order to take many interferograms and average them to increase the signal-to-noise level, the computer must average the centerburst at exactly the same position along the mirror's path every time. To achieve this, interferometers have a small red helium–neon (He–Ne) laser whose monochromatic beam passes through the interferometer, the same as the infrared source. Upon recombining, it produces interference fringes separated by the exact wavelength of the laser, 632.8 nm. These fringes serve as a calibration for the moving mirror position, allowing the computer to synchronize all the spectra.

Fig. 16.26. Typical interferogram. Point marked "0" is where both mirrors of the interferometer are the same distance from beam splitter. [From D. W. Ball, *Spectroscopy*, **9**(8) (1994) 24. Reproduced by permission.]

The principles of interferometers and Fourier transformation have been known for over a century, but practical applications had to await the advent of high-speed digital computer techniques. FTIR instruments are available as field portable units and more sophisticated laboratory instruments. They all have a salt beam splitter, typically germanium or potassium bromide, a moving mirror on either a precision mechanical or air bearing, a solid-state or cryogenic detector, and a computer to process the time-domain interferogram into a frequency-domain spectrum. Also, a monochromatic laser with a diode detector is incorporated for calibration of the wavelength.

Although less common, fast scanning single-beam dispersive instruments are available, replacing the complexity of the interferometer with the economy of the monochromator. These instruments can provide speed and resolution comparable to the FTIR instruments but lack the multiplex advantage that decreases noise for low-level measurements.

Modern IR instruments often have reflectance or other sampling capabilities for obtaining IR spectra that eliminate the necessity of salt plate cells and simplify sample handling. The most useful is an internal reflectance method called attenuated total reflectance (ATR). The sample is pressed on a diamond substrate and the infrared radiation penetrates the sample, being reflected internally, and then exits for detection.

16.12 Near-IR Instruments

Radiation sources for near-IR instruments are operated at typically 2500 to 3000 K, compared to 1700 K in the mid-IR region, resulting in about 10 times more intense radiation and improved signal-to-noise ratios. This is possible because the IR radiation of typical sources tails off in the mid-IR region and the maximum intensity shifts further into the near-IR region as the temperature is increased. The higher temperature results in weaker mid-IR radiation, but is beneficial in the near-IR region. A tungsten–halogen lamp provides intense radiation in the 800- to 1100-nm range.

NIR sources are more intense and detectors more sensitive than for the mid-IR region, so noise levels are 1000-fold lower.

A gallium indium arsenide (GaInAs) detector is most commonly used in the near-IR and is roughly 100 times more sensitive than mid-IR detectors. The combination of intense radiation sources and sensitive detectors results in very low noise levels, on the order of microabsorbance units. Glass and quartz are transparent to near-IR radiation, and so the optics and cells are easier to design and use than for the mid-IR region. Near-IR radiation can be sent for long distances over fiber optics, and commercial instruments for process or field (portable) testing often use fiber-optic probes (see below) for nondestructive sample testing.

16.13 Spectrometric Error in Measurements

There will always be a certain amount of error or irreproducibility in reading an absorbance or transmittance scale. Uncertainty in the reading will depend on a number of instrumental factors and on the region of the scale being read, and hence on the concentration.

Because of the logarithmic relationship between transmittance and concentration, small errors in measuring transmittance cause large relative errors in the calculated concentration at low and high transmittances. It is probably obvious to

It is difficult to precisely measure either very small or very large decreases in absorbance.

you that if the sample absorbs only a very small amount of the light, an appreciable *relative* error may result in reading the small decrease in transmittance. At the other extreme, if the sample absorbs nearly all the light, an extremely stable instrument would be required to read the large decrease in the transmittance accurately. There is, therefore, some optimum transmittance or absorbance where the relative error in making the reading will be minimal.

The transmittance for minimum relative error can be derived from Beer's law by calculus, assuming that the error results essentially from the uncertainty in reading the instrument scale and also that the *absolute* error in reading the transmittance is constant, independent of the value of the transmittance. The result is the prediction that the minimum relative error in the concentration theoretically occurs when $T = 0.368$ or $A = 0.434$.

Figure 16.27 illustrates the dependence of the relative error on the transmittance, calculated for a constant error of $0.01T$ in reading the scale. It is evident from the figure that, while the minimum occurs at 36.8% T, a nearly constant minimum error occurs over the range of 20 to 65% T (0.7 to 0.2 A). The percent transmittance should fall within 10 to 80% T ($A = 1$ to 0.1) in order to prevent large errors in spectrophotometric readings. Hence, samples should be diluted (or concentrated), and standard solutions prepared, so that the absorbance falls within the optimal range.

Figure 16.27 in practice approximates the error only for instruments with **Johnson** or **thermal noise-limited detectors**, such as photoconductive detectors like CdS or PbS detectors (400 to 3500 nm) or thermocouples, bolometers, and Golay detectors in the infrared region. Johnson noise is produced by random thermal motion in resistance circuit elements.

With phototubes and photomultiplier-type detectors (photoemissive detectors, ultraviolet to visible range), thermal noise becomes insignificant as compared to shot noise. **Shot noise** is the random fluctuation of the electron current from an electron-emitting surface (i.e., across a junction from cathode to anode), and in PM tubes that is amplified and becomes the noise-limiting fluctuation. In instruments with these detectors, the absolute error is not constant at all values of T, and the expressions for the spectrophotometric error become more complicated. It has been calculated that, for these cases, the minimal error should occur at $T = 0.136$ or $A = 0.87$. These instruments have a working range of about 0.1 to 1.5 A.

The absorbance should fall in the 0.1 to 1 range.

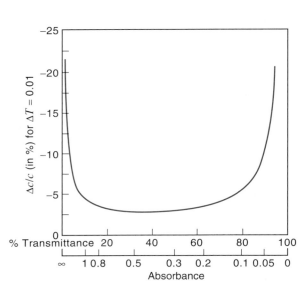

Fig. 16.27. Relative concentration error as function of transmittance for 1% uncertainty in % T.

16.14 Deviation from Beer's Law

It cannot always be assumed that Beer's law will apply, that is, that a linear plot of absorbance versus concentration will occur. Deviations from Beer's law occur as the result of chemical and instrumental factors. Most "deviations" from Beer's law are really only "apparent" deviations because if the factors causing nonlinearity are accounted for, the true or corrected absorbance-versus-concentration curve will be linear. True deviations from Beer's law will occur when the concentration is so high that the index of refraction of the solution is changed from that of the blank. A similar situation would apply for mixtures of organic solvents with water, and so the blank solvent composition should closely match that of the sample. The solvent may also have an effect on the absorptivity of the analyte.

Deviations from Beer's law result in nonlinear calibration curves, especially at higher concentrations.

CHEMICAL DEVIATIONS

Chemical causes for nonlinearity occur when nonsymmetrical chemical equilibria exist. An example is a weak acid that absorbs at a particular wavelength but has an anion that does not:

$$\underset{\text{(absorbs)}}{HA} \rightleftharpoons H^+ + \underset{\text{(transparent)}}{A^-}$$

The ratio of the acid form to the salt form will, of course, depend on the pH (Chapter 7). So long as the solution is buffered or is very acid, this ratio will remain constant at all concentrations of the acid. However, in unbuffered solution, the degree of ionization will increase as the acid is diluted, that is, the above equilibrium will shift to the right. Thus, a smaller fraction of the species exists in the acid form available for absorption for dilute solutions of the acid, causing apparent deviations from Beer's law. The result will be a positive deviation from linearity at higher concentrations (where the fraction dissociated is smaller). If the anion form were the absorbing species, then the deviation would be negative. Similar arguments apply to colored (absorbing) metal ion complexes or chelates in the absence of a large excess of the complexing agent. That is, in the absence of excess complexing agent, the degree of dissociation of the complex will increase as the complex is diluted. Here, the situation may be extremely complicated because the complex may dissociate stepwise into successive complexes that may or may not absorb at the wavelength of measurement. pH also becomes a consideration in these equilibria.

Apparent deviations may also occur when the substance can exist as a dimer as well as a monomer. Again, the equilibrium depends on the concentration. An example is the absorbance by methylene blue, which exhibits a negative deviation at higher concentrations due to association of the methylene blue.

The best way to minimize chemical deviations from Beer's law is by adequate buffering of the pH, adding a large excess of complexing agent, ionic strength adjustment, and so forth. Preparation of a calibration curve over the measurement range will correct for most deviations.

If both species of a chemical equilibrium absorb, and if there is some overlap of their absorption curves, the wavelength at which this occurs is called the **isosbestic point**, and the molar absorptivity of both species is the same. Such a point is illustrated in Figure 16.28. The spectra are plotted at different pH values since the pH generally causes the shift in the equilibrium. Obviously, the effect of pH could be eliminated by making measurements at the isosbestic point, but the sensitivity is decreased. By making the solution either very acid or very alkaline,

The absorptivity of all species is the same at the isosbestic point.

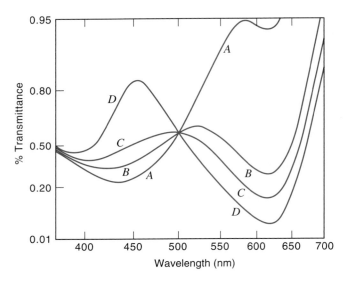

Fig. 16.28. Illustration of isosbestic point of bromthymol blue (501 nm): (*A*) pH 5.45, (*B*) pH 6.95, (*C*) pH 7.50, (*D*) pH 11.60.

one species predominates and the sensitivity is increased by measuring at this condition.

For a two-component system in which the two absorbing species are in equilibrium, all curves intersect at the isosbestic point where they have the same ϵ value. The existence of an isosbestic point is a necessary (although not sufficient) condition to prove that there are only two absorbing substances in equilibrium with overlapping absorption bands. If both of the absorbing species follow Beer's law, then the absorption spectra of all equilibrium mixtures will intersect at a fixed wavelength. For example, the different colored forms of indicators in equilibrium (e.g., the red and yellow forms of methyl orange) often exhibit an isosbestic point, supporting evidence that two and only two colored species participate in the equilibrium.

The existence of an isosbestic point is not proof of the presence of only two components. There may be a third component with $\epsilon = 0$ at this particular wavelength. The absence of an isosbestic point, however, is definite proof of the presence of a third component, provided the possibility of deviation from Beer's law in the two-component system can be dismissed. For a two-component system, the isosbestic point is a unique wavelength for quantitative determination of the total amount of two absorbing species in mutual equilibrium.

INSTRUMENTAL DEVIATIONS

The basic assumption in applying Beer's law is that monochromatic light is used. We have seen in the discussions above that it is impossible to extract monochromatic radiation from a continuum source. Instead, a band of radiation is passed, the width of which depends on the dispersing element and the slit width. In an absorption spectrum, different wavelengths are absorbed to a different degree; that is, the absorptivity changes with wavelength. At a wavelength corresponding to a fairly broad maximum on the spectrum, the band of wavelengths will all be absorbed to nearly the same extent. However, on a steep portion of the spectrum, they will be absorbed to different degrees. The slope of the spectrum increases as the concentration is increased, with the result that the fractions of the amounts of each wavelength absorbed may change, particularly if the instrument setting should drift over the period of the measurement. So a negative deviation in the absorbance-versus-concentration plot will be observed. The greater the slope of the spectrum, the greater is the deviation.

Obviously, it is advantageous to make the measurement on an absorption peak whenever possible, in order to minimize this curvature, as well as to obtain maximum sensitivity. Because a band of wavelengths is passed, the absorptivity at a given wavelength may vary somewhat from one instrument to another, depending on the resolution, slit width, and sharpness of the absorption maximum. Therefore, you should check the absorptivity and linearity on your instrument rather than relying on reported absorptivities. It is common practice to prepare calibration curves of absorbance versus concentration rather than to rely on direct calculations of concentration from Beer's law.

The absorptivity at a given wavelength may vary from instrument to instrument. Therefore, always run a standard.

If there is a second (interfering) absorbing species whose spectrum overlaps with that of the test substance, nonlinearity of the total absorbance as a function of the test substance concentration will result. It may be possible to account for this in preparation of the calibration curve by adding the interfering compound to standards at the same concentration as in the samples. This will obviously work only if the concentration of the interfering compound is essentially constant, and the concentration should be relatively small. Otherwise, simultaneous analysis as described earlier will be required.

Other instrumental factors that may contribute to deviations from Beer's law include stray radiation entering the monochromator and being detected, internal reflections of the radiation within the monochromator, and mismatched cells (in terms of pathlength) used for different analyte solutions or used in double-beam instruments (when there is appreciable absorbance by the blank or solvent in the reference cell). **Stray light** (any detected light that is not absorbed by the sample or is outside the bandwidth of the selected wavelength) becomes especially limiting at high absorbances and eventually causes deviation from linearity. Noise resulting from stray light also becomes a major contributor to the spectrometric error or imprecision at high absorbances. Radiation that does not interact with the sample can originate from light leaks in the instrument, from scattering of light from the optical components, or scattered light through the sample itself. A stray light component equivalent to 0.1% transmittance results in an error of 0.4% for a sample with 1 absorbance unit.

Stray light is the most common cause of negative deviation from Beer's law. For Beer's law, the light falling on the detector goes to zero at infinite concentration (all the light is absorbed). But this is impossible when stray light falls on the detector.

Other chemical and instrumental sources of nonlinearity in absorbance measurements may include hydrogen bonding, interaction with the solvent, nonlinear detector, nonlinear electronics, noncollimated radiation, and high signal levels (saturation).

Nonuniform cell thickness can affect a quantitative analysis. This is potentially a problem, especially in infrared spectrometry, where cell spacers are used. Air bubbles can affect the pathlength and stray light, and it is important to eliminate these bubbles, again especially in the infrared cells.

16.15 Fluorometry

Fluorometric analysis is extremely sensitive and is used widely by biochemists, clinical chemists, and analytical chemists in general.

PRINCIPLES OF FLUORESCENCE

When a molecule absorbs electromagnetic energy, this energy is usually lost as heat, as the molecule is deactivated via collisional processes. With certain molecules (ca. 5 to 10%), however, particularly when absorbing high-energy radiation such as UV radiation, only part of the energy is lost via collisions, and then the

Some molecules that absorb UV radiation lose only part of the absorbed energy by collisions. The rest is reemitted as radiation at longer wavelengths.

electron drops back to the ground state by emitting a photon of lower energy (longer wavelength) than was absorbed. Refer to Figure 16.29.

A molecule at room temperature normally resides in the ground state. The ground state is usually a **singlet state** (S_0), with all electrons paired. Electrons that occupy the same molecular orbital must be "paired," that is, have opposite spins. In a singlet state, the electrons are paired. If electrons have the same spin, they are "unpaired" and the molecule is in a **triplet state**. Singlet and triplet states refer to the **multiplicity** of the molecule. The process leading to the emission of a fluorescent photon begins with the absorption of a photon (a process that takes 10^{-15} s) by the fluorophore, resulting in an electronic transition to a higher-energy (excited) state. In most organic molecules at room temperature, this absorption corresponds to a transition from the lowest vibrational level of the ground state to one of the vibrational levels of the first or second electronic excited state of the same multiplicity (S_1, S_2). The spacing of the vibrational levels and rotational levels in these higher electronic states gives rise to the absorption spectrum of the molecule.

If the transition is to an electronic state higher than S_1, a process of **internal conversion** rapidly takes place. It is thought that the excited molecule passes from the vibrational level of this higher electronic state to a high vibrational level of S_1 that is isoenergetic with the original excited state. Collision with solvent molecules at this point rapidly removes the excess energy from the higher vibrational level of S_1; this process is called **vibrational relaxation**. These energy degradation processes (internal conversion and vibrational relaxation) occur rapidly ($\sim 10^{-12}$ s). Because of this rapid energy loss, emission fluorescence from higher states than the first excited state is rare.

The wavelengths of emitted radiation are independent of the wavelength of excitation. Their intensities are dependent, though.

Once the molecule reaches the first excited singlet, internal conversion to the ground state is a relatively slow process. Thus, decay of the first excited state by emission of a photon can effectively compete with other decay processes. This emission process is **fluorescence**. Generally, fluorescence emission occurs very rapidly after excitation (10^{-6} to 10^{-9} s). Consequently, it is not possible for the eye to perceive fluorescence emission after removal of the excitation source. Because fluorescence occurs from the lowest excited state, the fluorescence spectrum, that is, the wavelengths of emitted radiation, is independent of the wavelength of

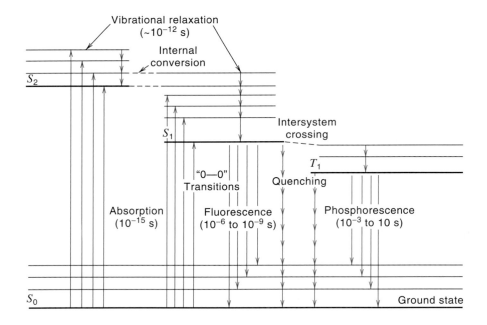

Fig. 16.29. Energy level diagram showing absorption processes, relaxation processes, and their rates.

excitation. The intensity of emitted radiation, however, will be proportional to the intensity of incident radiation (i.e., the number of photons absorbed).

Another feature of excitation and emission transitions is that the longest wavelength of excitation corresponds to the shortest wavelength of emission. This is the "0–0" band, corresponding to the transitions between the 0 vibrational level of S_0 and the 0 vibrational level of S_1 (Figure 16.29).

While the molecule is in the excited state, it is possible for one electron to reverse its spin, and the molecule transfers to a lower-energy triplet state by a process called **intersystem crossing**. Through the processes of internal conversion and vibrational relaxation, the molecule then rapidly attains the lowest vibrational level of the first excited triplet (T_1). From here, the molecule can return to the ground state S_0 by emission of a photon. This emission is referred to as **phosphorescence**. Since transitions between states of different multiplicity are "forbidden," T_1 has a much longer lifetime than S_1 and phosphorence is much longer-lived than fluorescence ($>10^{-4}$ s). Consequently, one can quite often perceive an "afterglow" in phosphorescence when the excitation source is removed. In addition, because of its relatively long life, radiationless processes can compete more effectively with phosphorescence than fluorescence. For this reason, phosphorescence is not normally observed from solutions due to collisions with the solvent or with oxygen. Phosphorescence measurements are made by cooling samples to liquid nitrogen temperature ($-196°C$) to freeze them and minimize collision with other molecules. Solid samples will also phosphoresce, and many inorganic minerals exhibit long-lived phosphorescence. Studies have been made in which molecules in solution are adsorbed on a solid support from which they can phosphoresce. Phosphorescence may be observed with minerals.

A typical excitation and emission spectrum of a fluorescing molecule is shown in Figure 16.30. The excitation spectrum usually corresponds closely in shape to the absorption spectrum of the molecule. There is frequently (but not necessarily) a close relationship between the structure of the excitation spectrum and the structure of the emission spectrum. In many relatively large molecules, the vibrational spacings of the excited states, especially S_1, are very similar to those in S_0. Thus, the form of the emission spectrum resulting from decay to the various S_0 vibrational levels tends to be a "mirror image" of the excitation spectrum arising from excitation

Phosphorescence is longer lived than fluorescence, and it may continue after the excitation source is turned off.

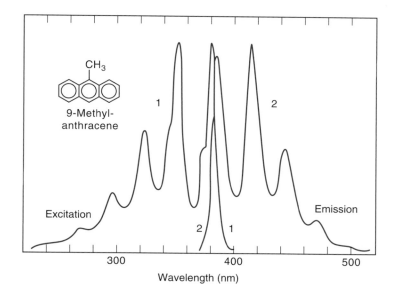

Fig. 16.30. Excitation and emission spectra of fluorescing molecule.

to the various vibrational levels in the excited state, such as S_1. Substructure, of course, results also from different rotational levels at each vibrational level.

The longest wavelength of absorption and the shortest wavelength of fluorescence tend to be the same (the 0–0 transition in Figure 16.29). More typically, however, this is not the case due to solvation differences between the excited molecule and the ground-state molecule. The heats of solvation of each are different, which results in a decrease in the energy of the emitted photon by an amount equal to these two heats of solvation.

Only those molecules that will absorb radiation, usually ultraviolet radiation, can fluoresce, and of those that do absorb, only about 5 to 10% fluoresce. This is an advantage when considering possible interference in fluorescence. The emitted radiation may be in the ultraviolet region, especially if the compound absorbs at less than 300 nm, but it is usually in the visible region. It is the emitted radiation that is measured and related to concentration.

CHEMICAL STRUCTURE AND FLUORESCENCE

In principle, any molecule that absorbs ultraviolet radiation could fluoresce. There are many reasons why they do not, but we will not go into these, other than to point out, in general, what types of substances may be expected to fluoresce.

First of all, the greater the absorption by a molecule, the greater its fluorescence intensity. Many aromatic and heterocyclic compounds fluoresce, particularly if they contain certain substituted groups. Compounds with multiple conjugated double bonds are favorable to fluorescence. One or more electron-donating groups such as —OH, —NH_2, and —OCH_3 enhances the fluorescence. Polycyclic compounds such as vitamin K, purines, and nucleosides and conjugated polyenes such as vitamin A are fluorescent. Groups such as —NO_2, —COOH, —CH_2COOH, —Br, —I, and azo groups tend to *inhibit* fluorescence. The nature of other substituents may alter the degree of fluorescence. The fluorescence of many molecules is greatly pH dependent because only the ionized or un-ionized form may be fluorescent. For example, phenol, C_6H_5OH, is fluorescent but its anion, $C_6H_5O^-$, is not.

If a compound is nonfluorescent, it may be converted to a fluorescent derivative. For example, nonfluorescent steroids may be converted to fluorescent compounds by dehydration with concentrated sulfuric acid. These cyclic alcohols are converted to phenols. Similarly, dibasic acids, such as malic acid, may be reacted with β-naphthol in concentrated sulfuric acid to form a fluorescing derivative. White and Argauer have developed fluorometric methods for many metals by forming chelates with organic compounds (see Ref. 23). Antibodies may be made to fluoresce by condensing them with fluorescein isocyanate, which reacts with the free amino groups of the proteins. NADH, the reduced form of nicotinamide adenine dinucleotide, fluoresces. It is a product or reactant (cofactor) in many enzyme reactions (see Chapter 24), and its fluorescence serves as the basis of the sensitive assay of enzymes and their substrates. Most amino acids do not fluoresce, but fluorescent derivatives are formed by reaction with dansyl chloride.

FLUORESCENCE QUENCHING

Quenching of fluorescence is often a problem in quantitative measurements.

One difficulty frequently encountered in fluorescence is that of **fluorescence quenching** by many substances. These are substances that, in effect, compete for the electronic excitation energy and decrease the quantum yield (the efficiency of conversion of absorbed radiation to fluorescent radiation—see below). Iodide ion is an extremely effective quencher. Iodide and bromide substituent groups decrease the quantum yield. Substances such as this may be determined indirectly by measuring the extent of fluorescence quenching. Some molecules do not fluoresce because they may have a bond whose dissociation energy is less than that of the radiation. In other words, a molecular bond may be broken, preventing fluorescence.

A colored species in solution with the fluorescing species may interfere by absorbing the fluorescent radiation. This is the so-called **inner-filter effect**. For example, in sodium carbonate solution, potassium dichromate exhibits absorption peaks at 245 and 348 nm. These overlap with the excitation (275 nm) and emission (350 nm) peaks for tryptophan and would interfere. The inner-filter effect can also arise from too high a concentration of the fluorophore itself. Some of the analyte molecules will reabsorb the emitted radiation of others (see the discussion of fluorescence intensity and concentration below).

RELATIONSHIP BETWEEN CONCENTRATION AND FLUORESCENCE INTENSITY

It can be readily derived from Beer's law (Problem 48) that the fluorescence intensity F is given by

Fluorescence intensity is proportional to the intensity of the source. Absorbance, on the other hand, is independent of it.

$$F = \phi P_0(1 - 10^{-abc}) \qquad (16.24)$$

where ϕ is the **quantum yield**, a proportionality constant and a measure of the fraction of absorbed photons that are converted into fluorescent photons. The quantum yield is, therefore, less than or equal to unity. The other terms in the equation are the same as for Beer's law. It is evident from the equation that if the product abc is large, the term 10^{-abc} becomes negligible compared to 1, and F becomes constant:

$$F = \phi P_0 \qquad (16.25)$$

On the other hand, if abc is small (≤ 0.01), it can be shown[5] by expanding Equation 16.18 that as a good approximation,

$$F = 2.303\phi P_0 abc \qquad (16.26)$$

Thus, for low concentrations, the fluorescence intensity is directly proportional to the concentration. Also, it is proportional to the intensity of the incident radiation.

This equation generally holds for concentrations up to a few parts per million, depending on the substance. At higher concentrations, the fluorescence intensity may decrease with increasing concentration. The reason can be visualized as follows. In dilute solutions, the absorbed radiation is distributed equally through the entire depth of the solution. But at higher concentrations, the first part of the solution in the path will absorb more of the radiation. So the equation holds only when most of the radiation goes through the solution, when more than about 92% is transmitted.

For low concentrations, fluorescence intensity becomes directly proportional to the concentration.

---

[5]It is known that $e^{-x} = 1 - x + x^2/2! \ldots$ and that $10^{-x} = e^{-2.303x}$. Therefore, $1 - e^{-2.303abc} = 1 - [1 - 2.303abc + (2.303abc)^2/2! \ldots]$. The squared term and higher-order terms can be neglected if $abc \leq 0.01$, and so the expanded term reduces to $2.303abc$. This is a Taylor expansion series.

FLUORESCENCE INSTRUMENTATION

For fluorescence measurements, it is necessary to separate the emitted radiation from the incident radiation. This is most easily done by measuring the fluorescence at right angles to the incident radiation. The fluorescence radiation is emitted in all directions, but the incident radiation passes straight through the solution.

A simple fluorometer design is illustrated in Figure 16.31. An ultraviolet source is required. Most fluorescing molecules absorb ultraviolet radiation over a band of wavelengths, and so a simple line source is sufficient for many applications. Such a source is a mercury vapor lamp. A spark is passed through mercury vapor at low pressure, and principal lines are emitted at 2537, 3650, 5200 (green), 5800 (yellow), and 7800 (red) Å. *Wavelengths shorter than 3000 Å are harmful to the eyes*, and one must never look directly at a short-wavelength UV source. The mercury vapor itself absorbs most of the 2537-Å radiation (self-absorption), and a blue filter in the envelope of the lamp may be added to remove most of the visible light. The 3650-Å line is thus the one used primarily for the activation. A high-pressure xenon arc (a continuum source) is usually used as the source in more sophisticated instruments that will scan the spectrum (spectrofluorometers). The lamp pressure is 7 atm at 25°C and 35 atm at operating temperatures. Take care!

In the simple instrument in Figure 16.31, a primary filter (filter 1) is used to filter out the wavelengths close to the wavelength of the emission because, in practice, some radiation is scattered. The primary filter allows the passage of only the wavelength of excitation. The secondary filter (filter 2) passes the wavelength of emission but not the wavelength of excitation (which may be scattered). Glass will pass appreciable amounts of the 3650-Å line, and so some instruments employ glass cuvets and filters. However, it is better to use quartz (special nonfluorescing grades are available). This simple setup is satisfactory for many purposes.

We can see why fluorometric methods are so sensitive if we compare them with absorption spectrometry. In absorption methods, the difference between two finite signals, P_0 and P, is measured. The sensitivity is then governed by the ability to distinguish between these two, which is dependent on the stability of the instrument, among other factors. In fluorescence, however, we measure the difference between zero and a finite number, and so, in principle, the limit of detection is governed by the intensity of the source and the sensitivity and stability of the detector (the "shot noise"). Also, in fluorescence, the signal depends linearly on concentration, and a much wider dynamic range of concentration can be measured; a dynamic range of 10^3 to 10^4 is not uncommon. In addition to the enhanced

Fluorescence measurements are 1000-fold more sensitive than absorbance measurements. Absorbance is like weighing a ship and captain and subtracting the ship's weight to get the captain's weight ($P = P_0 - p$). In fluorescence, we measure only the captain.

Filter 1 removes wavelengths that would pass filter 2 and appear as fluorescence. Filter 2 removes scattered excitation wavelengths and passes the fluorescence.

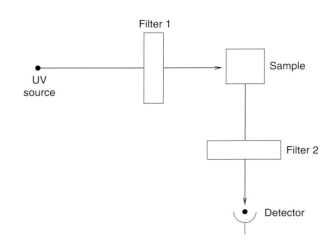

Fig. 16.31. Simple fluorometer design.

sensitivity, much wider ranges of concentrations can be measured; a 1000-fold or greater range is not uncommon.

In a **spectrofluorometer**, the measurement is also made at right angles to the direction of the incident radiation. But instead of using filters, the instrument incorporates two monochromators, one to select the wavelength of excitation and one to select the wavelength of fluorescence. The wavelength of excitation from a continuum source can be scanned and the fluorescence measured at a set wavelength to give a spectrum of the excitation wavelengths. This allows the establishment of the wavelength of maximum excitation. Then, by setting the excitation wavelength for maximum excitation, the emission wavelength can be scanned to establish the wavelength of maximum emission. When this spectrum is scanned, there is usually a "scatter peak" corresponding to the wavelength of excitation.

In spectrofluorometers, it is difficult to correct for variations in intensity from the source or response of the detector at different wavelengths, and calibration curves are generally prepared under a given set of conditions. Since the source intensity or detector response may vary from day to day, the instrument is usually calibrated by measuring the fluorescence of a standard solution and adjusting the gain to bring the instrument reading to the same value. A dilute solution of quinine in dilute sulfuric acid is usually used as the calibrating standard.

Sometimes it is desirable to obtain "absolute" spectra of a fluorescing compound to calculate quantum efficiencies for different transitions. This would require point-by-point correction for variations in the recorded signal due to variations in the instrumental parameters. Commercial instruments are available that will provide "corrected spectra." These adjust for variation in the source intensity with wavelength, so the sample is irradiated with constant energy, and they also correct for variations of the detector response. The recorded emission spectrum is presented directly in quanta of photons emitted per unit bandwidth.

> In a spectrofluorometer, the filters are replaced with scanning monochromators. Either the excitation spectrum (similar to the absorbance spectrum) or the emission spectrum may be recorded.

16.16 Optical Sensors: Fiber Optics

There has been a great deal of interest in recent years in developing optically based sensors that function much as electrochemical sensors (Chapter 15) do. These have been made possible with the advent of fiber-optic cables that transmit light along a flexible cable (waveguide) or "light pipe." Optical fibers were developed for the communications industry and are capable of transmitting light over long distances, but they have proven valuable for transmitting light to spectrometers and for developing analyte-selective sensors by coupling appropriate chemistries to the fibers. Through the use of optical fibers, a sample need not be brought to the spectrometer because light can be transmitted to and returned from the sample via the cables.

> Fiber-optic cables allow the sample to be far removed from the spectrometer.

FIBER-OPTIC PROPERTIES

The construction of a fiber-optic cable is illustrated in Figure 16.32. It consists of a cylindrical *core* that acts as the waveguide, surrounded by a *cladding* material of higher index of refraction, and a protective buffer layer. Light is transmitted along the core by total internal reflection at the core–cladding interface. The angle of acceptance, θ_a, is the greatest angle of radiation that will be totally reflected for a given core–cladding refractive index difference. Any light entering at an angle greater than θ will not be transmitted, and θ_a defines the fiber's numerical aperture (NA):

$$\text{NA} = n_{\text{ext}} \sin \theta_a = \sqrt{n_1^2 - n_2^2} \tag{16.27}$$

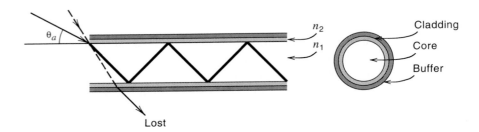

Fig. 16.32.   Fiber-optic structure.

where n_2 is the refractive index of the cladding, n_1, is that of the core, and n_{ext} is that of the external medium. The greater NA, the greater the light-gathering ability.

    Manufacturers typically provide numerical aperture data for different fibers. Another property usually provided is the light loss per unit length for different wavelengths. A spectral curve is given that shows attenuation versus wavelength. Attenuation is usually expressed in decibels per kilometer (dB/km), and is given by

$$dB = 10 \log \frac{P_0}{P} \qquad (16.28)$$

where P_0 is the input intensity and P the output intensity. Thus, the attenuation for silica-based fibers at 850 nm is in the order of 10 dB/km. Note that $dB = 10 \times$ absorbance. So a 10-m (0.01-km) fiber would exhibit an absorbance of 0.01 (0.1 dB attenuation), corresponding to 97.7% transmittance.

    Fiber optics may be purchased that transmit radiation from the ultraviolet (190 nm) to the infrared (≥ 5 μm), but each has a limited range. Plastic and compound glass materials are used for short distances in the visible region, while silica fibers can be used from the UV through the near-IR (2.3 μm) regions, but they are very costly. Fluoride and calcogenide glasses extend farther into the infrared.

    In coupling fiber optics to spectrometers, there is a trade-off between increased numerical aperture to collect more light and the collection angle of the spectrometer itself, which is usually limiting. That is, light collected with a numerical aperture greater than that for the spectrometer limit will not be seen by the spectrometer. See Ref. 23 for a discussion of design considerations for fiber optic/spectrometer coupling.

With bifurcated cables, one is used to transmit the source radiation and the other is used to receive the absorbed or fluorescent radiation.

    Fiber optics may be used as probes for conventional spectrophotometric and fluorescence measurements. Light must be transmitted from a radiation source to the sample and back to the spectrometer. While there are couplers and designs that allow light to be both transmitted and received by a single fiber, usually a **bifurcated fiber** cable is used. This consists of two fibers in one casing, split at the end that goes to the radiation source and the spectrometer. Often, the cables consist of a bundle of several dozen small fibers, and half are randomly separated from the other at one end. For absorbance measurements, a small mirror is mounted (attached to the cable) a few millimeters from the end of the fiber. The source radiation penetrates the sample solution and is reflected back to the fiber for collection and transmission to the spectrometer. The radiation path length is twice the distance between the fiber and the mirror.

    Fluorescence measurements are made in a similar fashion, but without the mirror. Radiation emitting from the end of the fiber in the shape of a cone excites fluorescence in the sample solution, which is collected by the return cable (the amount depends on the numerical aperture) and sent to the spectrometer. Often, a laser radiation source is used to provide good fluorescence intensity.

FIBER-OPTIC SENSORS

We can convert fiber-optic probes into selective absorbance- or fluorescence-based sensors by immobilizing appropriate reagents on the end of a fiber-optic cable. These possess the advantage over electrochemical sensors in that a reference electrode (and salt bridge) is not needed, and electromagnetic radiation will not influence the response. For example, a fluorometric pH sensor may be prepared by chemically immobilizing the indicator fluorescein isothiocyanate (FITC) on a porous glass bead and attaching this to the end of the fiber with epoxy. The FITC fluorescence spectrum changes with pH (Figure 16.33) over the range of about pH 3 to 7, centered around pK of the indicator. The fluorescence intensity measured at the fluorescence maximum is related to the pH via a calibration curve. The calibration curve will be sigmoid-shaped since it in effect represents a titration of the indicator. See Refs. 31 and 32 for a discussion of the limitations of fiber-optic sensors for measuring pH and ionic activity.

If an enzyme, for example, penicillinase, is immobilized along with an appropriate indicator, then the sensor is converted into a biosensor for measuring penicillin. The enzyme catalyzes the hydrolysis of penicillin to produce penicilloic acid, which causes a pH decrease.

Fiber-optic sensors have been developed for oxygen, CO_2, alkali metals, and other analytes. In order for these to function, the indicator chemistry must be reversible.

> Optical sensors do not have the requirement and associated difficulties of a reference electrode.

MINIATURE FIBER-OPTIC SPECTROMETERS

There are inexpensive spectrometers that utilize fiber optics for light transmission, either through a cuvet cell or as fiber-optic probes. An example is the Ocean Optics S2000 spectrometer (www.oceanoptics.com). It has a 2048-element charge-coupled device (CCD)—an array silicon detector that accepts light energy transmitted through a single-strand optical fiber, and disperses it via a fixed grating across the array. Figure 16.34 illustrates this compact spectrometer. It can be

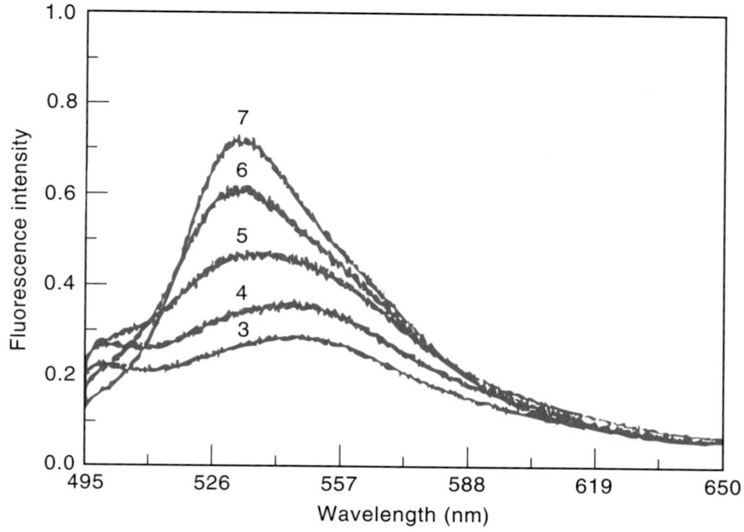

Fig. 16.33. Fluorescence spectra of FITC immobilized on porous glass bead at pH 3, 4, 5, 6, and 7. [From M.-R. S. Fuh, L. W. Burgess, T. Hirschfeld, G. D. Christian, and F. Wang, *Analyst*, **112** (1987) 1159. Reproduced by permission.]

Fig. 16.34. Miniature fiber-optic spectrometer. Box is the spectrometer. Light source is to right, and fiber-optic cable guides light to cuvet. Second cable takes transmitted radiation to spectrometer. (Photo courtesy of Ocean Optics, Inc.)

configured for UV, visible, and NIR applications, from 200 to 1100 nm, using different sources, gratings, fiber optics, and detectors. The instrument provides the entire spectrum over the wavelength of operation and has software for analyzing the spectra.

Learning Objectives

WHAT ARE SOME OF THE KEY THINGS WE LEARNED FROM THIS CHAPTER?

- Wavelength, frequencies, and photon energy (key equations: 16.1 to 16.3), p. 458
- How molecules absorb electromagnetic radiation, p. 460
- UV–visible absorption and molecular structure, p. 464
- IR absorption and molecular structure, p. 469
- Near-IR spectrometry, p. 470
- Spectral databases, p. 472
- Beer's law calculations (key equations: 16.10, 16.13), p. 474
- Mixture calculations (key equations: 16.16 and 16.17)—use of spreadsheets for calculations, pp. 478, 480
- Using spreadsheets to calculate unknown concentrations, and their standard deviations from the calibration curve, p. 481
- Spectrometers (components) for UV, visible, and IR regions, p. 483
- FTIR spectrometers, p. 499
- Spectrometric error, p. 501
- Fluorometry, p. 505
- Optical sensors and fiber optics, p. 511

ABSORPTION OF RADIATION

1. Describe the absorption phenomena taking place in the far-infrared, mid-infrared, and visible–ultraviolet regions of the spectrum.

2. What types of electrons in a molecule are generally involved in the absorption of UV or visible radiation?

3. What are the most frequent electronic transitions during absorption of electromagnetic radiation? Which results in more intense absorption? Give examples of compounds that exhibit each.

4. What is a necessary criterion for absorption to occur in the infrared region?

5. What types of molecular vibration are associated with infrared absorption?

6. What distinguishes near-infrared absorption from mid-infrared absorption? What are its primary advantages?

7. Define the following terms: chromophore, auxochrome, bathochromic shift, hypsochromic shift, hyperchromism, and hypochromism.

8. Which of the following pairs of compounds is likely to absorb radiation at the longer wavelength and with greater intensity?

    (a) $CH_3CH_2CO_2H$ or $CH_2=CHCO_2H$

    (b) $CH_3CH=CHCH=CHCH_3$ or $CH_3C\equiv C-C\equiv CCH_3$

    (c)

9. In the following pairs of compounds, describe whether there should be an increase in the wavelength of maximum absorption and whether there should be an increase in absorption intensity in going from the first compound to the second:

    (a)

    (b)

    (c)

10. Why do acid-base base indicators change color in going from acid to alkaline solution?

11. What are the mechanisms by which a metal complex can absorb radiation?

QUANTITATIVE RELATIONSHIPS

12. Define absorption, absorbance, percent transmittance, and transmittance.

13. Define absorptivity and molar absorptivity.

14. Why is a calibration curve likely to be linear over a wider range of concentrations at the wavelength of maximum absorption compared to a wavelength on a shoulder of the absorption curve?

15. List some solvents that can be used in the ultraviolet, visible, and infrared regions, respectively. Give any wavelength restrictions.

16. What is an isosbestic point?

17. Describe and compare different causes for deviations from Beer's law. Distinguish between real and apparent deviations.

INSTRUMENTATION

18. Describe radiation sources and detectors for the ultraviolet, visible, and infrared regions of the spectrum.

19. Distinguish between the two types of monochromators (light dispersers) used in spectrophotometers and list the advantages and disadvantages of each.

20. Discuss the effect of the slit width on the resolution of a spectrophotometer and the adherence to Beer's law. Compare it with the spectral slit width.

21. Compare the operations of a single-beam spectrophotometer and a double-beam spectrophotometer.

22. Given the weak absorption in the near-infrared region, why do near-infrared instruments function with reasonable sensitivity?

23. Describe the operation of a diode array spectrometer.

24. Describe the operation of an interferometer. What are its advantages?

25. Referring to Figure 16.28, what would be the color of an acid solution and an alkaline solution at maximum absorption? What color filter would be most applicable for the analysis of each in a filter colorimeter? (A filter replaces the prism and slit arrangement).

FLUORESCENCE

26. Describe the principles of fluorescence. Why is fluorescence generally more sensitive than absorption measurements?

27. Under what conditions is fluorescence intensity proportional to concentration?

28. Describe the instrumentation required for fluorescence analysis. What is a primary filter? A secondary filter?

29. Suggest an experiment by which you could determine iodide ion by fluorescence.

Problems

WAVELENGTH/FREQUENCY/ENERGY

30. Express the wavelength 2500 Å in micrometers and nanometers.

31. Convert the wavelength 4000 Å into frequency (Hz) and into wavenumbers (cm^{-1}).

32. The most widely used wavelength region for infrared analysis is about 2 to 15 μm. Express this range in angstroms and in wavenumbers.

33. One mole of photons (Avogadro's number of photons) is called an *einstein* of radiation. Calculate the energy, in calories, of one einstein of radiation at 3000 Å.

BEER'S LAW

34. Several spectrophotometers have scales that are read either in absorbance or in percent transmittance. What would be the absorbance reading at 20% T? At 80% T? What would the transmittance reading be at 0.25 absorbance? At 1.00 absorbance?

35. A 20-ppm solution of a DNA molecule (unknown molecular weight) isolated from *Escherichia coli* was found to give an absorbance of 0.80 in a 2-cm cell. Calculate the absorptivity of the molecule.

36. A compound of formula weight 280 absorbed 65.0% of the radiation at a certain wavelength in a 2-cm cell at a concentration of 15.0 μg/mL. Calculate its molar absorptivity at the wavelength.

37. Titanium is reacted with hydrogen peroxide in 1 M sulfuric acid to form a colored complex. If a 2.00×10^{-5} M solution absorbs 31.5% of the radiation at 415 nm, what would be (a) the absorbance and (b) the transmittance and percent absorption for a 6.00×10^{-5} M solution?

38. A compound of formula weight 180 has an absorptivity of 286 cm^{-1} g^{-1} L. What is its molar absorptivity'?

39. Aniline, $C_6H_5NH_2$, when reacted with picric acid gives a derivative with an absorptivity of 134 cm^{-1} g^{-1} L at 359 nm. What would be the absorbance of a 1.00×10^{-4} M solution of reacted aniline in a 1.00-cm cell?

QUANTITATIVE MEASUREMENTS

40. The drug tolbutamine (f wt = 270) has a molar absorptivity of 703 at 262 nm. One tablet is dissolved in water and diluted to a volume of 2 L. If the solution exhibits an absorbance in the UV region at 262 nm equal to 0.687 in a 1-cm cell, how many grams tolbutamine are contained in the tablet?

41. Amines (weak base) form salts with picric acid (trinitrophenol), and all amine picrates exhibit an absorption maximum at 359 nm with a molar absorptivity of 1.25×10^4. A 0.200-g sample of aniline, $C_6H_5NH_2$, is dissolved in 500 mL water. A 25.0-mL aliquot is reacted with picric acid in a 250-mL volumetric flask and diluted to volume. A 10.0-mL aliquot of this is diluted to 100 mL and the absorbance read at 359 nm in a 1-cm cell. If the absorbance is 0.425, what is the percent purity of the aniline?

42. Phosphorus in urine can be determined by treating with molybdenum(VI) and then reducing the phosphomolybdo complex with aminonaphtholsulfonic acid to give the characteristic molybdenum blue color. This absorbs at 690 nm. A patient excreted 1270 mL urine in 24 h, and the pH of the urine was 6.5. A 1.00-mL aliquot of the urine was treated with molybdate reagent and aminonaphtholsulfonic acid and was diluted to a volume of 50.0 mL. A series of phosphate standards was similarly treated. The absorbance of the solutions at 690 nm, measured against a blank, were as follows:

Solution	Absorbance
1.00 ppm P	0.205
2.00 ppm P	0.410
3.00 ppm P	0.615
4.00 ppm P	0.820
Urine sample	0.625

(a) Calculate the number of grams of phosphorus excreted per day.

(b) Calculate the phosphate concentration in the urine as millimoles per liter.

(c) Calculate the ratio of HPO_4^{2-} to $H_2PO_4^-$ in the sample:

$$K_1 = 1.1 \times 10^{-2} \qquad K_2 = 7.5 \times 10^{-8} \qquad K_3 = 4.8 \times 10^{-13}$$

43. Iron(II) is determined spectrophotometrically by reacting with 1,10-phenanthroline to produce a complex that absorbs strongly at 510 nm. A stock standard iron(II) solution is prepared by dissolving 0.0702 g ferrous ammonium sulfate, $Fe(NH_4)_2SO_4 \cdot 6H_2O$, in water in a 1-L volumetric flask, adding 2.5 mL H_2SO_4, and diluting to volume. A series of working standards is prepared by transferring 1.00-, 2.00-, 5.00-, and 10.00-mL aliquots of the stock solution to separate 100-mL volumetric flasks and adding hydroxylammonium chloride solution to reduce any iron(III) to iron(II), followed by phenanthroline solution and then dilution to volume with water. A sample is added to a 100-mL volumetric flask and treated in the same way. A blank is prepared by adding the same amount of reagents to a 100-mL volumetric flask and diluting to volume. If the following absorbance readings measured against the blank are obtained at 510 nm, how many milligrams iron are in the sample?

Solution	A
Standard 1	0.081
Standard 2	0.171
Standard 3	0.432
Standard 4	0.857
Sample	0.463

44. Nitrate nitrogen in water is determined by reacting with phenoldisulfonic acid to give a yellow color with an absorption maximum at 410 nm. A 100-mL sample that has been stabilized by adding 0.8 mL H_2SO_4/L is treated with silver sulfate to precipitate chloride ion, which interferes. The precipitate is filtered and washed (washings added to filtered sample). The sample solution is adjusted to pH 7 with dilute NaOH and evaporated just to dryness. The residue is treated with 2.0 mL phenol disulfonic acid solution and heated in a hot-water bath to aid dissolution. Twenty milliliters distilled water and 6 mL ammonia are added to develop the maximum color, and the clear solution is transferred to a 50-mL volumetric flask and diluted to volume with distilled water. A blank is prepared using the same volume of reagents, starting with the disulfonic acid step. A standard nitrate solution is prepared by dissolving 0.722 g anhydrous KNO_3 and diluting to 1 L. A standard addition calibration is performed by spiking a separate 100-mL portion of sample with 1.00 mL of the standard solution and carrying through the entire procedure. The following absorbance readings were obtained: blank, 0.032; sample, 0.270; sample plus standard, 0.854. What is the concentration of nitrate nitrogen in the sample in parts per million?

45. Two colorless species, A and B, react to form a colored complex AB that absorbs at 550 nm with a molar absorptivity of 450. The dissociation constant for the complex is 6.00×10^{-4}. What would the absorbance of a solution, prepared by mixing equal volumes of 0.0100 M solutions of A and B in a 1.00-cm cell, be at 550 nm?

MIXTURES

(You can use the Solver spreadsheet in your CD to perform the simultaneous equation calculations.)

46. Compounds A and B absorb in the ultraviolet region. Compound A exhibits an absorption maximum at 267 nm ($a = 157$) and a trailing shoulder at 312 nm ($a = 12.6$). Compound B has an absorption maximum at 312 nm ($a = 186$) and does not absorb at 267 m. A solution containing the two compounds exhibits absorbances (using a 1-cm cell) of 0.726 and 0.544 at 267 and 312 nm, respectively. What are the concentrations of A and B in mg/L?

47. Titanium(IV) and vanadium(V) form colored complexes when treated with hydrogen peroxide in 1 M sulfuric acid. The titanium complex has an absorption maximum at 415 nm, and the vanadium complex has an absorption maximum at 455 nm. A 1.00×10^{-3} M solution of the titanium complex exhibits an absorbance of 0.805 at 415 nm and of 0.465 at 455 nm, while a 1.00×10^{-2} M solution of the vanadium complex exhibits absorbances of 0.400 and 0.600 at 415 and 455 nm, respectively. A 1.000-g sample of an alloy containing titanium and vanadium was dissolved, treated with excess hydrogen peroxide, and diluted to a final volume of 100 mL. The absorbance of the solution was 0.685 at 415 nm and 0.513 at 455 nm. What were the percentages of titanium and vanadium in the alloy?

FLUORESCENCE

48. Derive Equation 16.24 relating fluorescence intensity to concentration.

Recommended References

GENERAL

1. D. F. Swinehart, "The Beer-Lambert Law," *J. Chem. Ed.*, **39** (1962) 333.

2. J. C. Lindon, *Encyclopedia of Spectroscopy and Spectrometry*, G. E. Tranter and J. L. Holmes, eds. San Diego: Academic, 1999 (3-volume set).

3. J. Workman and A. Springsteen, *Applied Spectroscopy. A Compact Reference for Practitioners*. San Diego: Academic, 1998. Covers UV/Vis and near-IR spectrometry.

4. J. M. Chalmers, ed., *Spectroscopy in Process Analysis*. Boca Raton, FL: CRC Press, 2000.

5. J. Workman, *Handbook of Organic Compounds. NIR, IR, Raman, and UV-Vis Spectra Featuring Polymers and Surfactants*. San Diego: Academic, 2000 (3-volume set).

6. R. M. Silverstein and F. X. Webster, *Spectrometric Identification of Organic Compounds*, 6th ed. New York: Wiley, 1997.

7. G. D. Christian and J. B. Callis, eds., *Trace Analysis: Spectroscopic Methods for Molecules*. New York: Wiley, 1986.

8. G. H. Morrison and H. Freiser, *Solvent Extraction in Analytical Chemistry*. New York: Wiley, 1975, pp. 189–247. Colorimetric determination of metals.

CATALOGED SPECTRA

9. *Catalogue of Infrared Spectral Data.* Washington, DC: American Petroleum Institute, Research Project 44. Multivolume series started in 1943.

10. *Catalogue of Ultraviolet Spectral Data.* Washington, DC: American Petroleum Institute, Research Project 44. Multivolume series started in 1945.

11. "Infrared Prism Spectra," in *The Sadtler Standard Spectra*, Vols. 1–36; "Standard Infrared Grating Spectra," in *The Sadtler Standard Spectra*, Vols. 1–16. Philadelphia: Sadtler Research Laboratories.

12. L. Lang, ed., *Absorption Spectra in the Ultraviolet and Visible Regions*, Vols. 1–23. New York: Academic, 1961–1979.

13. *U.V. Atlas of Organic Compounds*, Vols. I–V. London: Butterworths, 1966–1971.

14. "Ultraviolet Spectra," in *The Sadtler Standard Spectra*, Vol. 1–62. Philadelphia: Sadtler Research Laboratories. A comprehensive catalog of ultraviolet spectra of organic compounds.

15. D. L. Hansen, *The Spouse Collection of Spectra. I. Polymers, II. Solvents by Cylindrical Internal Reflectance, III. Surface Active Agents, IV. Common Solvents—Condensed Phase, Vapor Phase and Mass Spectra.* Amsterdam: Elsevier Science, 1987–1988. Peak table search software available for each.

INFRARED

16. P. R. Griffiths, *Fourier Transform Infrared Spectrometry*, 2nd ed. New York: Wiley, 1986.

17. B. C. Smith, *Fundamentals of Fourier Transform Infrared Spectroscopy.* Boca Raton, FL: CRC Press, 1996.

18. B. C. Smith, *Infrared Spectral Interpretation. A Systematic Approach.* Boca Raton, FL: CRC Press, 1999.

19. B. R. Singh, *Infrared Analysis of Peptides and Proteins.* Washington, DC: American Chemical Society, 1999.

20. D. A. Burns and E. W. Ciurczak, eds. *Handbook of Near-Infrared Analysis*, 2nd ed. New York: Marcel Dekker, 2001.

21. R. Raghavachari, ed., *Near-Infrared Applications in Biotechnology.* New York: Marcel Dekker, 2000.

FLUOROMETRY

22. A. Sharma and S. G. Schulman, *An Introduction to Fluorescence Spectroscopy.* New York: Wiley, 1999.

23. C. E. White and R. J. Argauer, *Fluorescence Analysis: A Practical Approach.* New York: Marcel Dekker, 1970.

24. R. J. Hurtubise, *Phosphorimetry. Theory, Instrumentation and Applications.* New York: VCH, 1990.

FIBER OPTICS

25. K. T. Grattan and B. T. Meggitt, *Optical Fibre Sensor Technology. Chemical and Environmental Sensing, Vol. 4.* Boca Raton, FL: CRC Press/Chapman Hall, 1999.

26. O. S. Wolfbeis, ed., *Fiber Optic Chemical Sensors and Biosensors*, Vols. 1 and 2. Boca Raton, FL: CRC Press, 1991.

27. L. W. Burgess, M.-R. S. Fuh, and G. D. Christian, "Use of Analytical Fluorescence with Fiber Optics," in P. Eastwood and L. J. Cline-Love, eds. *Progress in Analytical Luminescence*, ASTM STP 1009, Philadelphia: American Society for Testing and Materials, 1988.

28. M. J. Webb, "Practical Consideration When Using Fiber Optics with Spectrometers," *Spectroscopy*, **4** (1989) 9.

29. W. R. Seitz, "Chemical Sensors Based on Fiber Optics," *Anal. Chem.* **56** (1984) 16A.

30. W. R. Seitz, "Chemical Sensors Based on Immobilized Indicators and Fiber Optics," *CRC Crit. Rev. Anal. Chem.*, **19** (1988) 135.

31. J. Janata, "Do Optical Fibers Really Measure pH?" *Anal. Chem.*, **59** (1987) 1351.

32. J. Janata, "Ion Optrodes," *Anal. Chem.*, **64** (1992) 921A.

33. M. A. Arnold, "Fiber-Optic Chemical Sensors," *Anal. Chem.*, **64** (1992) 1015A.

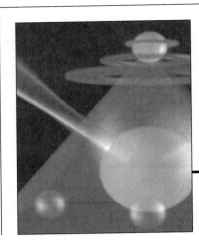

Chapter Seventeen

ATOMIC SPECTROMETRIC METHODS

Chapter 16 dealt with the spectrometric determination of substances in solution, that is, the absorption of energy by molecules, either organic or inorganic. This chapter deals with the spectroscopy of atoms. Since atoms are the simplest and purest form of matter and cannot rotate or vibrate as a molecule does, only electronic transitions within the atom can take place when energy is absorbed. Because the transitions are discrete (quantized), line spectra are observed. There are various ways to obtain free atoms (atomic vapor) and to measure the absorption or emission of radiation by these.

The principal techniques described in this chapter include flame emission spectrometry in which atoms in the form of atomic vapor are created in a flame; a portion of them is thermally and collisionally excited to a higher electronic energy level and then returned to their ground energy state by emitting photons, to create sharp-line emission spectra. Atomic absorption spectrometry is described in which the amount of radiation absorbed by ground-state atoms created in a flame or a minifurnace is measured; the absorption spectrum is sharp-line. Also included is a discussion of the types of flames used for emission or absorption, interferences in flames, and the use of nonflame (electrothermal, minifurnace) atomizers for extremely sensitive atomic absorption measurements.

Atomic spectrometry is widely used in many laboratories, particularly whenever trace element analyses are required. Environmental samples are analyzed for heavy-metal contamination, and pharmaceutical samples may be analyzed for metal impurities. The steel industry needs to determine minor components, as well as major ones. The particular technique used will depend on the sensitivity required, the number of samples to be analyzed, and whether single-element or multielement measurements are needed. The following discussion gives the capabilities of the techniques.

17.1 Flame Emission Spectrometry[1]

In this technique, formerly called flame photometry, the source of excitation energy is a flame. This is a low-energy source, and so the emission spectrum is simple and

---

[1] See Chapter 16 for the distinction between spectrometry and spectrophotometry.

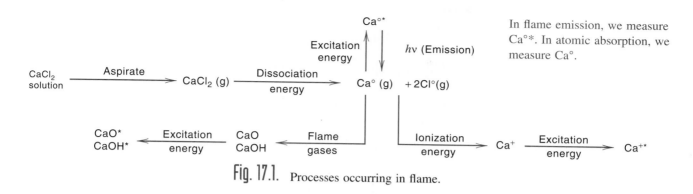

Fig. 17.1. Processes occurring in flame.

In flame emission, we measure $Ca^{\circ *}$. In atomic absorption, we measure Ca°.

there are few emission lines. The sample is introduced into the flame in the form of a solution, and so the technique is very easy to quantify.

There are numerous types of aspirator burners used. Basically, the solution is introduced into the flame as a fine spray. The mechanism of obtaining atomic vapor is complex, but an attempt at explaining the basic processes is illustrated in Figure 17.1. The solvent evaporates, leaving the dehydrated salt. The salt is dissociated into free gaseous atoms in the ground state. A certain fraction of these atoms can absorb energy from the flame and be raised to an excited electronic state. The excited levels have a short lifetime and drop back to the ground state, emitting photons[2] of characteristic wavelengths, with energy equal to $h\nu$. These can be detected with a conventional monochromator–detector setup.

The intensity of emission is directly proportional to the concentration of the analyte in the solution being aspirated. So a calibration curve of emission intensity as a function of concentration is prepared.

As indicated in the figure, side reactions in the flame may decrease the population of free atoms and hence the emission signal. These will be discussed in Section 17.3.

In the early years of flame photometry, only relatively cool flames were used. We shall see below that only a small fraction of atoms of most elements is excited by flames and that the fraction excited increases as the temperature is increased. Consequently, relatively few elements have been determined routinely by flame emission spectrometry, especially few of those that emit line spectra (several can exist in flames as molecular species, particularly as oxides, which emit molecular band spectra). Only the easily excited alkali metals sodium, potassium, and lithium are routinely determined by flame emission spectrometry in the clinical laboratory. However, with flames such as oxyacetylene and nitrous oxide–acetylene, over 60 elements can now be determined by flame emission spectrometry. This is in spite of the fact that a small fraction of excited atoms is available for emission. Good sensitivity is achieved because, as with fluorescence (Chapter 16), we are, in principle, measuring the difference between zero and a small but finite signal, and so the sensitivity is limited by the response and stability of the detector and the stability (noise level) of the flame aspiration system.

Atoms in the gaseous state exhibit sharp-line spectra since they only undergo electronic transitions.

---

[2] This is in opposition to excited molecules in solution, where there is much greater probability for collisions with solvent and other molecules. In the flame, there is less probability for collision because there are much fewer flame molecules and, therefore, many of the atoms lose their energy of excitation as electromagnetic radiation rather than as heat.

17.2 Distribution between Ground and Excited States— Most Atoms Are in the Ground State

The relative populations of ground-state (N_0) and excited-state (N_e) populations at a given flame temperature can be estimated from the **Maxwell–Boltzmann expression**:

$$\frac{N_e}{N_0} = \frac{g_e}{g_0}\, e^{-(E_e - E_0)/kT} \tag{17.1}$$

Nearly all the gaseous atoms are in the ground state. Atomic emission is still sensitive, for the same reason that fluorescence spectrometry is. We do not have to measure a small decrease in a signal (which has some noise) as in absorption.

where g_e and g_0 are the *statistical weights* of the excited and ground states, respectively; E_e and E_0 are the energies of the two states ($= h\nu$; E_0 is usually zero); k is the Boltzmann constant (1.3805×10^{-16} erg K^{-1}); and T is the absolute temperature. The statistical weights represent the probability that an electron will reside in a given energy level, and they are available from quantum mechanical calculations.[3] See Problem 21 for an example calculation.

Table 17.1 summarizes the relative population ratios for a few elements at 2000, 3000, and 10,000 K. We see that even for a relatively easily excited element such as sodium, the excited-state population is small except at 10,000 K, as obtained in a plasma. Short-wavelength elements (higher energy, $h\nu$) require much more energy for excitation and exhibit poor sensitivity by flame emission spectrometry where temperatures rarely exceed 3000 K. Those with long-wavelength emissions will exhibit better sensitivity. Measurement of ground-state atoms, as in atomic absorption below, will be less dependent on the wavelength or element. We see also from Table 17.1 that the fraction of excited-state atoms is temperature dependent, whereas the fraction in the ground state is virtually constant (since nearly 100% reside there at all temperatures).

In flame emission methods, we measure the excited-state population; and in atomic absorption methods (below), we measure the ground-state population. Because of chemical reactions that occur in the flame, differences in flame emission and atomic absorption sensitivities above 300 nm are, in practice, not as great as one would predict from the Boltzmann distribution. For example, many elements react partially with flame gases to form metal oxide or hydroxide species, and this reaction detracts from the atomic population equally in either method and is equally temperature dependent in either.

---

[3] The statistical weight is given by $2J + 1$, where J is the Russel–Saunders coupling and is equal to $L + S$ or $L - S$; L is the total orbital angular momentum quantum number, represented by the sharp (S), principal (P), diffuse (D), and fundamental (F) series ($L = 0, 1, 2,$ and 3, respectively); and S is spin ($\pm\frac{1}{2}$). The information is supplied in the form of term symbols, $N^M L_J$, where N is the principal quantum number and M is the multiplicity. For example, the transition for the sodium 589.0-nm line, omitting the principal quantum number N, is $^2S_{1/2} - {}^2P_{1/2}$, and $g_e/g_0 = [2(\frac{1}{2}) + 1]/[2(\frac{1}{2}) + 1] = 2/2 = 1$.

Table 17.1
Values of N_e/N_0 for Different Resonance Lines

Line (nm)	N_e/N_0		
	2000 K	3000 K	10,000 K
Na 589.0	9.9×10^{-6}	5.9×10^{-4}	2.6×10^{-1}
Ca 422.7	1.2×10^{-7}	3.7×10^{-5}	1.0×10^{-1}
Zn 213.8	7.3×10^{-15}	5.4×10^{-10}	3.6×10^{-3}

17.3 Atomic Absorption Spectrophotometry

A technique closely related to flame emission spectrometry is atomic absorption spectrophotometry (AAS) because they each use a flame as the atomizer. We discuss here the factors affecting absorption; and because of the close relationship of atomic absorption and flame photometry, we shall make comparisons between the two techniques where appropriate.

PRINCIPLES

The sample solution is aspirated into a flame as in flame emission spectrometry, and the sample element is converted to atomic vapor. The flame then contains atoms of that element. Some are thermally excited by the flame, but most remain in the ground state, as shown dramatically in Table 17.1. These ground-state atoms can absorb radiation of a particular wavelength that is produced by a special source made from that element (see Sources). The wavelengths of radiation given off by the source are the same as those absorbed by the atoms in the flame.

Atomic absorption spectrophotometry is identical in principle to absorption spectrophotometry described in the previous chapter. The absorption follows Beer's law. That is, the *absorbance* is directly proportional to the pathlength in the flame and to the concentration of atomic vapor in the flame. Both of these variables are difficult to determine, but the pathlength can be held constant and the concentration of atomic vapor is directly proportional to the concentration of the analyte in the solution being aspirated. The procedure used is to prepare a calibration curve of concentration in the solution versus absorbance.

> Beer's law is followed in atomic absorption.

The major disadvantage of making measurements by atomic absorption, as we shall see below, is that a different source is required for each element.

INSTRUMENTATION

As in regular absorption spectrophotometry, the requirements for atomic absorption spectrophotometry are a light source, a cell (the flame), a monochromator, and a detector. The flame is placed between the source and the monochromator. A schematic diagram of an atomic absorption spectrophotometer is shown in Figure 17.2. This is for a double-beam instrument that measures the ratio of P_0/P. The source beam is alternately sent through the flame and around the flame by the chopper. The detector measures these alternately and the logarithm of the ratio is displayed. The detector amplifier is tuned to receive only radiation modulated at the frequency of the chopper, and so dc radiation emitted by the flame is discriminated against.

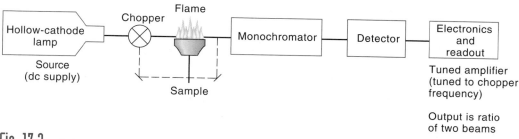

Fig. 17.2. Schematic diagram of atomic absorption instrument. (From G. D. Christian and F. J. Feldman, *Atomic Absorption Spectroscopy. Applications in Agriculture, Biology, and Medicine.* New York: Interscience, 1970. Reproduced by permission of John Wiley & Sons, Inc.)

Double-beam instruments are required for background correction using deuterium continuum lamps (see below). But the beam splitter of a double-beam instrument reduces the radiant energy, causing increased noise levels (decreases signal-to-noise ratios). High-energy source single-beam instruments are available that utilize line-based background corrections (measuring background absorption with a line near the analyte line). These provide good signal-to-noise ratio and are smaller and may even be portable.

The various components of an atomic absorption spectrophotometer are described as follows.

1. Sources. A sharp-line source is required in atomic absorption because the width of the *absorption line* is very narrow, a few thousandths to one-hundredths of a nanometer, at most. Because the absorption line is so narrow, only a small fraction of the radiation from a continuum source passed by the slit and reaching the detector would be absorbed.

> A sharp-line source is used in AAS. The source emits the lines of the element to be measured. These possess the precise energies required for absorption by the analyte atoms.

The source used almost exclusively is a **hollow-cathode lamp** (HCL). This is a sharp-line source that emits specific (essentially monochromatic) wavelengths, and the basic construction is illustrated in Figure 17.3. It consists of a cylindrical hollow cathode made of the element to be determined or an alloy of it, and a tungsten anode. These are enclosed in a glass tube usually with a quartz window since the lines of interest are often in the ultraviolet region. The tube is under reduced pressure and filled with an inert gas such as argon or neon. A high voltage is impressed across the electrodes, causing the gas atoms to be ionized at the anode. These positive ions are accelerated toward the negative cathode. When they bombard the cathode, they cause some of the metal to "sputter" and become vaporized. The vaporized metal is excited to higher electronic levels by continued collision with the high-energy gas ions. When the electrons return to the ground state, the characteristic lines that metallic element are emitted. Also emitted are lines of the filler gas, but these are not usually close enough to the element lines to interfere.

These HCL-emitted lines are passed through the flame and can become absorbed by the test element because they possess just the right energy (the right wavelength) to result in the discrete electronic transitions. The most strongly absorbed line is often, but not always, the one corresponding to the most probable electronic transition, usually from the ground state to the lowest excited state. This is called the **resonance line**. The lines from a hollow-cathode lamp are narrower than the absorption line of the element in the flame because of broadening of the absorption line at the higher temperature and pressure of the flame. So the entire source linewidth is absorbed. Greater specificity also results for the reason that, while with a continuum source an element with an absorption line falling anywhere within the spectral slit width would absorb part of the source radiation, a line source will not be absorbed so long as an element's absorption lines does not overlap with it. There are very few instances where line overlap does occur among lines of different elements.

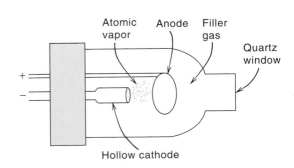

Fig. 17.3. Design of hollow-cathode lamp.

It is sometimes possible to use an alloy of several elements for the hollow cathode, and with such lamps, the lines of all the elements are emitted. These are the so-called multielement hollow-cathode lamps and can be used as a source for usually two or three elements. They may exhibit shorter lifetimes than do single-element lamps due to selective volatilization ("distillation") of one of the elements from the cathode with condensation on the walls of the lamp.

2. Burners. The burner used in most commercial instruments is the **premix chamber burner**, sometimes called the **laminar-flow burner**. This is illustrated in Figure 17.4. The fuel and support gases are mixed in a chamber before they enter the burner head (through a slot) where they combust. The sample solution is aspirated through a capillary by the **Venturi effect** using the support gas, usually air, for the aspiration. The air creates a partial vacuum at the end of the capillary, drawing the sample through the capillary. It is broken into a fine spray at the tip. This is the usual process of **nebulization**.[4] The larger droplets of the resulting aerosol condense and drain out of the chamber. The remaining fine droplets mix with the combustion gases and enter the flame. As much as 90% of the droplets condense out, leaving only 10% to enter the flame.

Premix burners are generally limited to relatively low-burning velocity flames. Although a large portion of the aspirated sample is lost in the chamber, the "atomization efficiency" (efficiency of producing atomic vapor) of that portion of the sample that enters the flame is high because the droplets are finer. Also, the path-length is long. Combustion with premix burners is very quiet. A popular version of the premix burner is the **Boling** burner. This is a three-slot burner head that results in a broader flame and less distortion of the radiation passing through at the

---

[4]In atomic absorption spectrophotometry, we often speak of "atomization" in referring to the process of obtaining *atomic vapor*. This is not to be confused with the above process.

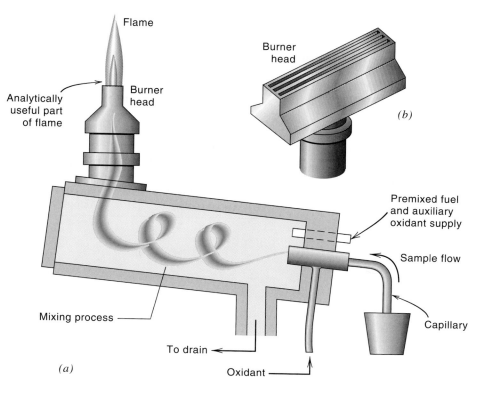

Fig. 17.4. Premix burner. (*a*) Nebulizer, chamber, and burner. (*b*) Burner head. (From G. D. Christian and F. J. Feldman, *Atomic Absorption Spectroscopy. Applications in Agriculture, Biology, and Medicine.* New York: Interscience, 1970. Reproduced by permission of John Wiley & Sons, Inc.)

edges of the flame (see Figure 17.4). This burner warps more easily than others, though, and care must be taken not to overheat it.

The air–acetylene flame is the most popular for AAS. The nitrous oxide–acetylene flame is best for refractory elements.

3. Flames. The chief flames that are used for atomic absorption and emission spectrometry are listed in Table 17.2 together with their maximum burning temperatures. The most widely used flames for atomic absorption are the air–acetylene flame and the nitrous oxide–acetylene flame with premix burners. The latter high-temperature flame is not required and may even be detrimental for many cases in atomic absorption because it will cause ionization of the gaseous atoms (see below). However, it is very useful for those elements that tend to form heat-stable oxides in the air–acetylene flame (the "refractory elements"). The air–acetylene and other hydrocarbon flames absorb a large fraction of the radiation at wavelengths below 200 nm, and an argon–hydrogen-entrained air flame is preferred for this region of the spectrum for maximum detectability. This is a colorless flame, and entrained air is the actual oxidant gas. It is used for elements such as arsenic (193.5 nm) and selenium (197.0 nm), when they are separated from the sample solution by volatilization as their hydrides (AsH_3, H_2Se) and passage of these gases into the flame. This is necessary because this cool flame is more subject to chemical interferences than other flames (see the following paragraphs). A nitrous oxide–acetylene flame offers an advantage in this region of the spectrum when danger of molecular interference exists; the flame absorption is relatively small at short wavelengths.

In flame emission spectrometry, a hot flame is required for the analysis of a large number of elements, and the nitrous oxide–acetylene flame is used. The oxyacetylene flame has a high burning velocity and cannot be used with a conventional premix burner. The nitrous oxide–acetylene flame can, however, be used with a premix burner. Because of its high temperatures, a special, thick, stainless steel burner head must be used to prevent it from melting. A "cool" air–propane or similar flame is preferred for the flame emission spectrometry of the easily excited elements sodium and potassium because of decreased ionization of these elements.

INTERFERENCES

These fall under three classes, spectral, chemical, and physical. We will discuss, these briefly and point out their relative effects in emission and absorption measurements.

1. Spectral Interferences. In emission analyses, when either another emission line or a molecular emission band is close to the emitted line of the test element

Table 17.2

Burning Temperatures and Velocities of Commonly Used Flames

Flame Mixture	Maximum Flame Speed (cm/s)	Maximum Temperature (°C)
Hydrogen–oxygen	—	2677
Hydrogen–air	—	2045
Propane–air	—	1725
Propane–oxygen	—	2900
Acetylene–air	160	2250
Acetylene–oxygen	1130	3060
Acetylene–nitrous oxide	180	2955
Hydrogen–argon-entrained air	—	1577

From G. D. Christian and F. J. Feldman, *Atomic Absorption Spectroscopy. Application in Agriculture, Biology, and Medicine.* New York: Wiley-Interscience, 1970. Reproduced by permission of John Wiley & Sons, Inc.

and is not resolved from it by the monochromator, spectral interference occurs. The most probable danger is from molecular emission, such as from oxides of other elements in the sample. Similar interference would occur in atomic absorption if a dc instrument were used, but it is eliminated if an ac instrument is used. If, on the other hand, an element or molecule is capable of absorbing the source radiation, a positive interference would occur in atomic absorption. With line sources, this danger is minimized but not eliminated.

Light scatter or absorption by solid particles, unvaporized solvent droplets, or molecular species in the flame will cause a positive interference in atomic absorption spectrophotometry. This is especially a problem for wavelengths less than 300 nm, when solutions of high salt content are aspirated because the salt may not be completely desolvated or its molecules dissociated into atoms. Such **background absorption** can be corrected for by measuring the absorbance of a line that is close to the absorption line of the test element but that is not absorbed by the element itself. Since the interfering absorption occurs over a band of wavelengths, the absorbance will be essentially the same at several angstroms removed from the resonance line.

The measurement should be made at least two bandpasses (Chapter 16) away from the absorption line. The line used for correction can be a filler gas line from the hollow-cathode lamp or a "nonresonance" line of the element that is not absorbed, or a nearby line from a second hollow-cathode lamp can be used. A solution of the test element should always be aspirated to check that it does not absorb the background correction line. This technique requires two separate measurements on the sample.

A **background correction** for broadband absorption can also be made in the UV region (where most elements absorb and background absorption is most serious) with a hydrogen or deuterium continuum source. In the visible region, a tungsten continuum source may be used. The monochromator is set at the same wavelength as the resonance line. Sharp-line absorption of the continuum source by the test element is assumed negligible compared to that by the broad background over the bandwidth of the monochromator, so the absorbance of the continuum source can simply be subtracted from the absorbance of the resonance line from the hollow-cathode lamp. This is the basis of commercially available automatic background correctors. A mirror alternately passes the hollow-cathode radiation and the continuum radiation, the absorption of each is measured, and the continuum source absorbance is automatically subtracted from the hollow-cathode absorbance to obtain the net sharp-line absorbance by the test element. So only a single measurement is required.

Light scatter by particles is a common problem in AAS. Since it is broad band in nature, it can be corrected for in a background absorption measurement.

2. Ionization Interference.

An appreciable fraction of alkali and alkaline earth elements and several other elements in very hot flames may be ionized in the flame. Since we are measuring the un-ionized atoms, both emission or absorption signals can be decreased. This in itself is not necessarily serious, except that the sensitivity and linearity may be decreased. However, the presence of other easily ionized elements in the sample will add free electrons to the flame and suppress ionization of the test element. This will result in enhanced emission or absorption and a positive interference. Ionization interference can usually be overcome either by adding the same amount of the interfering element to the standard solutions or more simply by adding large amounts to both the samples and the standards, to make the enhancement constant and ionization minimal. Ionization can usually be detected by noting that the calibration curve has a positive deviation or curvature upward at higher concentrations because a larger fraction of the atoms are ionized at lower concentrations.

Ionization can be suppressed by adding a solution of a more easily ionized element, for example, potassium or cesium.

Refractory compound formation is avoided by chemical competition or by use of a high-temperature flame.

3. Refractory Compound Formation. The sample solution may contain a chemical, usually an anion, that will form a refractory (heat-stable) compound with the test element in the flame. For example, phosphate can react with calcium ions and in the flame to produce calcium pyrophosphate, $Ca_2P_2O_7$. This causes a reduction in the absorbance since the calcium must be in the atomic form to absorb its resonance line. Generally, this type of solution interference can be reduced or eliminated chemically. In the above example, a high concentration (ca. 1%) of strontium chloride or lanthanum nitrate can be added to the solution. Called a releasing agent, the strontium or lanthanum will preferentially combine with the phosphate and prevent its reaction with the calcium. Alternatively, a high concentration of EDTA can be added to the solution to form a chelate with the calcium. This prevents its reaction with phosphate, and the calcium–EDTA chelate is dissociated in the flame to give free calcium vapor. These types of interferences can occur with both atomic absorption and flame emission spectrometry. They may be eliminated also by using a higher-temperature flame such as the nitrous oxide–acetylene flame.

A serious situation occurs when the analyte metal reacts with gases present in the flame. Refractory elements such as aluminum, titanium, molybdenum, and vanadium will react with O and OH species in the flame to produce thermally stable metal oxides and hydroxides. These can be decomposed only by using high-temperature flames. Several of these elements exhibit no appreciable absorption or emission in the conventional air–acetylene flame. A more useful flame for these elements is the nitrous oxide–acetylene flame. It is usually used in the reducing (fuel-rich) condition in which a large red-feather secondary-reaction zone is present. This red zone arises from the presence of CN, NH, and other highly reducing radicals. These (or the lack of oxygen-containing species), combined with the high temperature of the flame, decompose and/or prevent the formation of refractory oxides so that atomic vapor of the metal can be produced.

4. Physical Interferences. Most parameters that affect the rate of sample uptake in the burner and the atomization efficiency can be considered physical interferences. This includes such things as variations in the gas flow rates, variation in sample viscosity due to temperature or solvent variation, high solids content, and changes in the flame temperature. These can generally be accounted for by frequent calibration and use of internal standards. Some instruments offer the capability of using internal standards that can partially compensate for changes in physical parameters. See below.

SAMPLE PREPARATION—SOMETIMES MINIMAL

Sample preparation with flame methods can often be kept to a minimum. As long as chemical or spectral interferences are absent, essentially all that is required is to obtain the sample in the form of a diluted and filtered (for particulates) solution. It often makes no difference what the chemical form of the analyte is because it will be dissociated to the free elemental vapor in the flame. Thus, several elements can be determined in blood, urine, cerebral spinal fluid, and other biological fluids by direct aspiration of the sample. Usually, dilution with water will be required to prevent clogging of the burner.

Note that in the preparation of standards, the matrix of the analyte must always be matched. Thus, if lead in gasoline is to be determined, a simulated solvent matrix must be used for standards, not water.

Chemical interferences can often be overcome by simple addition of (dilution with) a suitable reagent solution. Thus, serum is diluted 1:20 with a solution containing EDTA for the determination of calcium in order to prevent interference from phosphate. Sodium and potassium, in concentrations equal to those in serum, are added to calcium standards to prevent ionization interference.

Reference 9 gives a review of applications of atomic absorption spectrophotometry to biological samples. This technique is widely used for metal analysis in biological fluids and tissues, in environmental samples such as air and water, and in occupational health and safety areas. Routine applications of flame emission spectrometry to biological samples are generally limited to the alkali and alkaline earth metals. Ion-selective electrode measurements (Chapter 13) have largely replaced the flame emission measurements in the clinical chemistry laboratory.

RELATIVE DETECTABILITIES OF ATOMIC ABSORPTION AND FLAME EMISSION SPECTROMETRY

Table 17.3 lists some representative detection limits of various elements by atomic absorption and flame emission spectrometry. We should distinguish here between the sensitivity and detection limits in atomic absorption. The former term is frequently used in the atomic absorption literature. **Sensitivity** is defined as the concentration required to give 1% absorption (0.0044 A). It is a measure of the slope of the analytical calibration curve and says nothing of the signal-to-noise ratio (S/N). **Detection limit** is generally defined as the concentration required to give a signal equal to three times the standard deviation of the baseline (blank)—see Chapter 3.

Generally speaking, atomic absorption shows superior detectability for those elements that emit below 300 nm because of the high thermal energy required to excite the atoms for emission at these wavelengths. But at wavelengths between 300 and 400 nm, either method may exhibit comparable detectability, while flame emission is generally superior in the visible region.

Table 17.3

Representative Detection Limits by Atomic Absorption (AAS) and Flame Emission (FES) Spectrometry

Element	Wavelength (nm)	Detection Limit (ppm)	
		AAS[a]	FES[b]
Ag	328.1	0.001(A)	0.01
Al	309.3	0.1(N)	
	396.2		0.08
Au	242.8	0.03(N)	
	267.7		3
Ca	422.7	0.003(A)	0.0003
Cu	324.8	0.006(A)	0.01
Eu	459.4	0.06(N)	0.0008
Hg	253.6	0.8(A)	15
K	766.5	0.004(A)	0.00008
Mg	285.2	0.004(A)	0.1
Na	589.0	0.001(A)	0.0008
Tl	276.8	0.03(A)	
	535.0		0.03
Zn	213.9	0.001(A)	15

[a]Fuel is acetylene. Letter in parentheses indicates oxidant: A = air, N = nitrous oxide.
[b]Nitrous oxide–acetylene flame.

ELECTROTHERMAL ATOMIZERS—ULTRASENSITIVE

Electrothermal atomization is nearly 100% efficient, compared to 0.1% for flame atomization. Only a few microliters of sample are required.

Although aspiration into a flame is the most convenient and reproducible means of obtaining atomic vapor, it is one of the least efficient in terms of converting all the sample elements to atomic vapor and presenting this to the optical path. The overall efficiency of atomic conversion and measurement of ions present in aspirated solutions has been estimated to be as little as 0.1%. Also, aspiration methods usually require several milliliters of solution for analysis.

Electrothermal atomizers are generally a type of minifurnace in which a drop of the sample is dried and then decomposed at a high temperature to produce an atomic vapor cloud.

Electrothermal atomizers have conversion efficiencies approaching 100%, so absolute detection limits are often 100 to 1000 times improved over those of flame aspiration methods. Our discussion will center on resistively heated atomizers (graphite furnace). Although these are not generally useful for emission measurements, they are well suited for atomic absorption measurements. A schematic of a typical electrothermal atomizer is shown in Figure 17.5.

In most of the electrothermal techniques, a few microliters of sample is placed in a horizontal graphite tube or on a carbon rod or tantalum ribbon. The tube or rod is heated resistively by passing a current through it. The sample is first dried at a low temperature for a few seconds (~100 to 200°C), followed by pyrolysis at 500 to 1400°C to destroy organic matter that produces smoke and scatters the light source during measurement; the smoke from pyrolysis is flushed out by flowing argon gas. Finally, the sample is rapidly thermally atomized at a high temperature, up to 3000°C.

The light path passes over the atomizer (or through the tube). A sharp peak of absorbance versus time is recorded as the atomic cloud passes through the light path (Figure 17.5). Either the height of the observed peak or its area is directly related to the quantity of metal vaporized. The heating is done in an inert atmosphere (e.g., argon gas) to prevent oxidation of the graphite or carbon at the high temperatures involved, and also to prevent formation of refractory metal oxides.

A major difficulty with electrothermal atomization methods is that interelement effects are generally much more pronounced than in flames. The interferences can sometimes be compensated for by using a standard additions method for calibration in which the standard is added to a separate aliquot of the sample and the

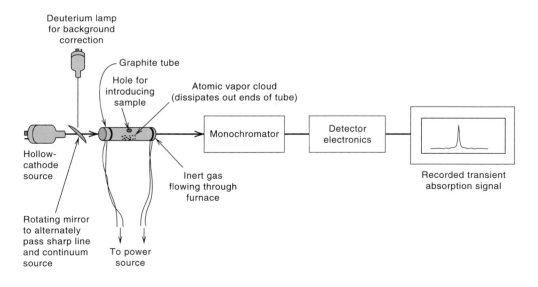

Fig. 17.5. Electrothermal atomization.

increase in the measured signal is proportional to the concentration added. In this manner, the standard is subjected to the same matrix as the sample (see below).

Often when the matrix concentration is changed, there is a change in the peak height and in the *shape* of the analytical peak. In such instances, less influence of the matrix and better accuracy can sometimes be achieved by integrating the signal or measuring its area, rather than measuring its maximum intensity. Of course, this requires more sophisticated instrumentation.

Even with pyrolysis before the atomization step, background absorption in electrothermal methods tends to be more prominent than in flame methods, especially with biological and environmental samples. This is due to residual organic material or vaporized matrix salts. Hence, automatic background correction (see above) is generally required.

Detection limits quoted by the manufacturers of electrothermal atomizers are typically in the range of 10^{-10} to 10^{-12} g or even less! The concentrational detection limit will depend on the sample volume. This will depend on the sample matrix composition and the concentration of the test element. Assume that a 10-μL sample is analyzed for an element with a detection limit of 10^{-11} g. Then the concentration detection limit would be 10^{-11} g/0.01 mL or 10^{-9} g/mL. This is equal to 1 ng/mL or 1 part per billion. The extreme sensitivity of these techniques is, therefore, quite apparent, even when dealing with very small sample volumes.

Electrothermal methods are complementary to flame methods. The latter are better suited when the analyte element is at a sufficiently high concentration to measure and adequate solution volume is available. They provide excellent reproducibility, and interferences are usually readily dealt with. The electrothermal techniques, on the other hand, are required when the concentrations are very small or the sample size is limited. It is possible to analyze solid samples directly without preparing a solution. The calibration and use of electrothermal methods generally requires more care. These are among the most sensitive of all analytical methods.

> Background correction is more critical in electrothermal atomizers.

17.4 Internal Standard and Standard Addition Calibration

In atomic spectrometric methods, signals can frequently vary with time due to factors like fluctuations in gas flow rates and aspiration rates. Precision can be improved by the technique of **internal standards**. As an example, a simple flame emission spectrometer designed for simultaneous measurement of sodium and potassium in serum, using fixed wavelengths and two separate detectors, will usually contain a third fixed-wavelength channel-detector channel for lithium. A fixed concentration of lithium is added to all standards and samples. The instrument records and reads out the *ratios* of the K/Li and Na/Li signals. If the aspiration rate, for example, fluctuates, each signal is affected to the same extent and the ratio, at a given K or Na concentration, remains constant. See the spreadsheet exercise in Section 20.5 for an example of how to perform internal standard calibrations from the signals of the analyte and the internal standard.

The internal standard element should be chemically similar to the analyte element, and their wavelengths should not be too different. See Ref. 8 for a discussion of the selection of the internal standard.

A second difficulty often encountered in flame-spectrometric methods is a suppression (or sometimes enhancement) of the signal by the sample matrix, for example, due to high viscosity or chemical reaction with the analyte. The technique of **standard addition calibration** can be utilized to minimize errors of this type. The sample is measured in the usual way to produce a given signal. A separate

> An internal standard undergoes similar interferences as the analyte. Measurement of the ratio of the analyte to internal standard signals cancels the interferences.

> In standard addition calibration, the standard is added to the sample, and so it experiences the same matrix effects as the analyte.

portion of the sample is taken and spiked with a known amount of standard and followed through the analytical procedure, and a new signal is recorded. The standard, then, is subjected to the same matrix as the unknown analyte. The increase in signal is due to the standard, and the original signal is due to the analyte. It is important to perform blank corrections. A simple proportionality applies, assuming you are in a linear portion of the calibration curve. Two additions of standard are recommended to assure linearity.

 Example 17.1

A serum sample is analyzed for potassium by flame emission spectrometry using the method of standard additions. Two 0.500-mL aliquots are added to 5.00-mL portions of water. To one portion is added 10.0 μL of 0.0500 M KCl solution. The net emission signals in arbitrary units are 32.1 and 58.6. What is the concentration of potassium in the serum?

Solution

The amount of standard added is

$$0.0100 \text{ mL} \times 0.0500 \ M = 5.00 \times 10^{-4} \text{ mmol}$$

This produces a signal of

$$58.6 - 32.1 = 26.5 \text{ arbitrary units}$$

The millimoles potassium in the sample, then, is

$$5.00 \times 10^{-4} \text{ mmol} \times \frac{32.1 \text{ units}}{26.5 \text{ units}} = 6.06 \times 10^{-4} \text{ mmol}$$

This is contained in 0.500 mL serum, so the concentration is

$$\frac{6.06 \times 10^{-4} \text{ mmol}}{0.500 \text{ mL serum}} = 1.21 \times 10^{-3} \text{ mmol/mL serum}$$

Normally in applying the standard addition method, a calibration curve is constructed, similar to Figure 14.6, where the Y axis would be the atomic emission or absorbance signal. If the volumes of added standards are appreciable, the signals are corrected for dilution by multiplying by $(V_i + v)V_i$, or V_t/V_i, where V_i is the initial volume and v is the added volume, and V_t is the total volume.

SPREADSHEET EXERCISE: MULTIPLE STANDARD ADDITIONS

Calcium is determined in a river water sample using electrothermal AAS, using multiple standard additions. Four 25.0-mL aliquots are taken in 50-mL volumetric flasks, and aliquots of a 2.50 ppm standard are added in the amounts of 0, 1.00, 2.00, and 3.00 mL, followed by dilution to volume. Fixed volumes of each solution are measured, giving net absorbance signals of 0.101, 0.183, 0.238, and 0.310, respectively. Use a spreadsheet to prepare the calibration curve and calculate the concentration of calcium in the sample, and the standard deviation of the concentration.

Review Equations 16.18 to 16.21 and the spreadsheet that fcllows those in Chapter 16 for the error analysis. In addition to the statistics described there, we need the standard error in x, S_x:

$$S_x = \frac{S_r}{m} \sqrt{\frac{1}{N} + \frac{(y_c - y_{ave})^2}{m^2 \, S_{xx}}} \qquad (17.2)$$

Note the similarity of this to Equation 16.21. The value of y_c is zero (the volume of the intercept). We use this equation to calculate the standard deviation of the concentration of the unknown:

$$S_c = -\frac{C_x S_x}{b_v} = -\frac{C_x S_x}{-b/m} \qquad (17.3)$$

We place a minus sign in front since the intercept is a negative number, and this gives a positive number for S_c.

The total absorbance, A_t, is the sum of the absorbance by the sample and that due to the added standard:

$$A_t = A_x + A_s \qquad (17.4)$$

$$A_t = kC_x \frac{V_x}{V_t} + kC_s \frac{V_s}{V_t} \qquad (17.5)$$

where V_x is the volume of the unknown (25.0 mL), V_s the volume of added standard, V_t the total volume (50.0 mL), C_x the unknown concentration, C_s the standard concentration (2.50 ppm), and k a porportionality constant. Since V_t is constant,

$$A_t = k'C_x V_x + k'C_s V_s \qquad (17.6)$$

where $k' = k/V_t$. A plot of A_t vs. V_s for a series of solutions should give a straight line with slope $k'C_s$ and intercept of $k'C_x V_x$:

$$m = k'C_s \qquad (17.7)$$
$$b = k'C_x V_x \qquad (17.8)$$

Combining,

$$C_x = \frac{bC_s}{mV_x} \qquad (17.9)$$

The volume intercept is

$$\frac{-b}{m} = -\frac{C_x V_x}{C_s} \qquad (17.10)$$

$$C_x = -\frac{C_s}{V_x} \cdot \frac{-b}{m} \qquad (17.11)$$

We use the spreadsheet to calculate b and m, and from these, C_x.

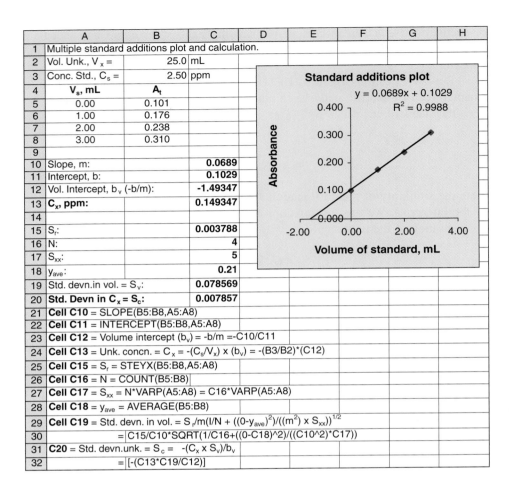

	A	B	C	D	E	F	G	H
1	Multiple standard additions plot and calculation.							
2	Vol. Unk., V_x =	25.0	mL					
3	Conc. Std., C_s =	2.50	ppm					
4	**V_s, mL**	**A_t**						
5	0.00	0.101						
6	1.00	0.176						
7	2.00	0.238						
8	3.00	0.310						
9								
10	Slope, m:		0.0689					
11	Intercept, b:		0.1029					
12	Vol. Intercept, b_v (-b/m):		-1.49347					
13	**C_x, ppm:**		0.149347					
14								
15	S_r:		0.003788					
16	N:		4					
17	S_{xx}:		5					
18	y_{ave}:		0.21					
19	Std. devn.in vol. = S_v:		0.078569					
20	**Std. Devn in C_x = S_c:**		0.007857					
21	**Cell C10** = SLOPE(B5:B8,A5:A8)							
22	**Cell C11** = INTERCEPT(B5:B8,A5:A8)							
23	**Cell C12** = Volume intercept (b_v) = -b/m =-C10/C11							
24	**Cell C13** = Unk. concn. = C_x = -(C_s/V_x) x (b_v) = -(B3/B2)*(C12)							
25	**Cell C15** = S_r = STEYX(B5:B8,A5:A8)							
26	**Cell C16** = N = COUNT(B5:B8)							
27	**Cell C17** = S_{xx} = N*VARP(A5:A8) = C16*VARP(A5:A8)							
28	**Cell C18** = y_{ave} = AVERAGE(B5:B8)							
29	**Cell C19** = Std. devn. in vol. = $S_r/m(I/N + ((0-y_{ave})^2)/((m^2) \times S_{xx}))^{1/2}$							
30	= C15/C10*SQRT(1/C16+((0-C18)^2)/((C10^2)*C17))							
31	**C20** = Std. devn.unk. = S_c = -(C_x x S_v)/b_v							
32	= [-(C13*C19/C12)]							

The standard additions plot shows: Standard additions plot, y = 0.0689x + 0.1029, R^2 = 0.9988, with Absorbance on the y-axis (0.000 to 0.400) and Volume of standard, mL on the x-axis (-2.00 to 4.00).

The concentration is 0.149 ± 0.008 ppm. Note that the volume intercept corresponding to the unknown is -1.49 mL. This corresponds to the addition of 1.49 mL of the 2.50 ppm standard to 50 mL, a 1:33.6 dilution that gives a concentration of 0.0745 ppm in the sample flask. The 25.0-mL sample was diluted twofold, so the sample concentration is 2×0.0745, or 0.149 ppm, as calculated in cell C13:

$$-(C_s/V_x)(-b/m) = -(2.5/25)(-1.49)$$

Or, we calculate that 1.49 mL $\times$ 2.50 μg/mL = 3.72 μg/50 mL = 0.0745 μg/mL in the sample flask (0.0745 ppm), corresponding to 0.149 ppm in the original sample.

Suppose we didn't dilute to 50 mL after adding the standard aliquots. Then V_t would not be constant but must be treated as a variable. For n spike additions, the total spike volume, V_{S_n}, is

$$V_{S_n} = \sum_{i=1}^{n} V_{S_i} \qquad (17.12)$$

So after two spike additions, for example, $V_{S_n} = V_{S_1} + V_{S_2}$, which for the above example is $1.00 + 2.00 = 3.00$ mL. The total volume, V_{T_n}, after n spike additions, is

$$V_{T_n} = V_x + V_{S_n} = V_x + \sum_{i=1}^{n} V_{S_i} \qquad (17.13)$$

In the above example, after two spike additions, V_{T_n} is 25.0 + 3.00 = 28.0 mL.

We can write an equation similar to Equation 17.5 for the total absorbance, A_{T_n}, for n spike additions:

$$A_{T_n} = kC_x \frac{V_x}{V_{T_n}} + kC_s \frac{V_{S_n}}{V_{T_n}} \tag{17.14}$$

where C_x is the unknown concentration in the original sample volume and C_s is the spike concentration in its stock volume. V_{T_n} is a variable, known for each spike. Multiplying by V_{T_n},

$$A_{T_n} V_{T_n} = kC_x V_x + kC_s V_{S_n} \tag{17.15}$$

Compare with Equation 17.6. A plot of $A_{T_n} V_{T_n}$ vs. V_{S_n}, with n being the independent variable, gives a straight line with slope kC_s and intercept $kC_x V_s$:

$$m = kC_s \tag{17.16}$$
$$b = kC_x V_x \tag{17.17}$$

Combining, we arrive at Equation 17.9, and subsequently Equations 17.10 and 17.11, for calculating the unknown concentration. We set up a spreadsheet similar to the one above, but plotting $A_{T_n} V_{T_n}$ vs. V_{S_n}, instead of A_t vs. V_s. The plotting of variable total volume is a bit more complicated than just plotting V_s, but it avoids an experimental step of diluting to volume. The spreadsheet can be set up to automatically calculate V_{T_n} from V_x and V_{S_n}, and from this, $A_{T_n} V_{T_n}$.

Learning Objectives

WHAT ARE SOME OF THE KEY THINGS WE LEARNED FROM THIS CHAPTER?

- Flame emission spectrometry, p. 522
- Distribution of atoms in a flame (key equation: 17.1), p. 524
- Atomic absorption spectrometry—flame and electrothermal, p. 525
- Internal standard and standards addition calibration—using spreadsheets (key equations 17.5, 17.11), pp. 533, 534

Questions

PRINCIPLES

1. What fraction of atoms in a flame are typically in the excited state?
2. Describe the principles of flame emission spectrometry and of atomic absorption spectrophotometry.
3. Compare flame emission and atomic absorption spectrophotometry with respect to instrumentation, sensitivity, and interferences.
4. Why is a sharp-line source desirable for atomic absorption spectroscopy?
5. Explain why flame emission spectrometry is often as sensitive as atomic absorption spectrophotometry, even though only a small fraction of the atoms may be thermally excited in the flame.

6. The Maxwell–Boltzmann expression predicts that the fraction of excited-state atoms in a flame is both highly temperature dependent and wavelength dependent, while the fraction of ground-state atoms remains large in all cases. Yet flame emission and atomic absorption spectrometry in practice do not exhibit large differences in dependence for many elements if the wavelength is greater than about 300 nm. Why is this?

7. Explain why absorption spectra for atomic species consist of discrete lines at specific wavelengths rather than broad bands for molecular species.

8. What causes the red feather in a reducing nitrous oxide–acetylene flame?

9. Explain why electrothermal atomizers result in greatly enhanced sensitivity in atomic absorption spectrophotometry.

10. Explain why an internal-standard element can improve the precision of atomic spectrometry measurements.

INSTRUMENTATION

11. Explain the mechanism of operation of a hollow-cathode lamp.

12. Describe the premix chamber burner. What flames can be used with it?

13. Explain why the radiation source in atomic absorption instruments is usually modulated.

INTERFERENCES

14. Lead in seawater was determined by atomic absorption spectrophotometry. The APCD (ammonium pyrrolidinecarbodithioate) chelate was extracted into methylisobutyl ketone and the organic solvent was aspirated. A standard and reagent blank were treated in a similar manner. The blank reading was essentially zero. Measurements were made at the 283.3-nm line. An independent determination using anodic-stripping voltammetry revealed the atomic absorption results to be high by nearly 100%. Assuming the anodic-stripping voltammetry results are correct, suggest a reason for the erroneous results and how they might be avoided in future analyses.

15. Why is a high-temperature nitrous oxide–acetylene flame sometimes required in atomic absorption spectrophotometry?

16. Why is a high concentration of a potassium salt sometimes added to standards and samples in flame absorption or emission methods?

17. Chemical interferences are more prevalent in "cool" flames such as air–propane, but this flame is preferred for the determination of the alkali metals. Suggest why.

18. An analyst notes that a 1-ppm solution of sodium gives a flame emission signal of 110, while the same solution containing also 20 ppm potassium gives a reading of 125. It was determined that a 20-ppm solution of potassium exhibited no blank reading. Explain the results.

Problems

SENSITIVITY

19. A 12-ppm solution of lead gives an atomic absorption signal of 8.0% absorption. What is the atomic absorption sensitivity?

20. Silver exhibits an atomic absorption sensitivity of 0.050 ppm under a given set of conditions. What would be the expected absorption for a 0.70-ppm solution?

BOLTZMANN DISTRIBUTION

21. The transition for the cadmium 228.8-nm line is a $^1S_0-^1S_1$ transition. Calculate the ratio of N_e/N_0 in an air–acetylene flame. What percent of the atoms is in the excited state? The velocity of light is 3.00×10^{10} cm/s, Planck's constant is 6.62×10^{-27} erg-s, and the Boltzmann constant is 1.380×10^{-16} erg K^{-1}.

QUANTITATIVE CALCULATIONS

22. Calcium in a sample solution is determined by atomic absorption spectrophotometry. A stock solution of calcium is prepared by dissolving 1.834 g $CaCl_2 \cdot 2H_2O$ in water and diluting to 1 L. This is diluted 1:10. Working standards are prepared by diluting the second solution, respectively, 1:20, 1:10, and 1:5. The sample is diluted 1:25. Strontium chloride is added to all solutions before dilution, sufficient to give 1% (wt/vol) to avoid phosphate interference. A blank is prepared, to give 1% $SrCl_2$. Absorbance signals on the strip-chart recorder, when the solutions are aspirated into an air–acetylene flame, are as follows: blank, 1.5 cm; standards, 10.6, 20.1, and 38.5 cm; sample, 29.6 cm. What is the concentration of calcium in the sample in parts per million?

23. Lithium in the blood serum of a manic-depressive patient treated with lithium carbonate is determined by flame emission spectrophotometry, using a standard additions calibration. One hundred microliters of serum diluted to 1 mL gives an emission signal of 6.7 cm on the recorder chart. A similar solution to which 10 μL of a 0.010 M solution of $LiNO_3$ has been added gives a signal of 14.6 cm. Assuming linearity between the emission signal and the lithium concentration, what is the concentration of lithium in the serum, in parts per million?

24. Chloride in a water sample is determined indirectly by atomic absorption spectrophotometry by precipitating it as AgCl with a measured amount of $AgNO_3$ in excess, filtering, and measuring the concentration of silver remaining in the filtrate. Ten-milliliter aliquots each of the sample and a 100-ppm chloride standard are added to separate dry 100-mL Erlenmeyer flasks. Twenty-five milliliters of a silver nitrate solution is added to each with a pipet. After allowing time for the precipitate to form, the mixtures are transferred partially to dry centrifuge tubes and are centrifuged. Each filtrate is aspirated for atomic absorption measurement of silver concentration. A blank is similarly treated in which 10 mL deionized distilled water is substituted for the sample. If the following absorbance signals are recorded for each solution, what is the concentration of chloride in the water sample?

Blank:	12.8 cm
Standard:	5.7 cm
Sample:	6.8 cm

Recommended References

GENERAL

1. L. Ebdon and E. H. Evans, eds., *An Introduction to Analytical Spectrometry*, 2nd. ed. Chichester: Wiley, 1998.

2. J. W. Robinson, *Atomic Spectroscopy*, 2nd. ed. New York: Marcel Dekker, 1996.

3. Web: G. M. Hieftje, "Atomic Spectroscopy—A Primer," www.spectroscopynow. com, link to Education: Tutorials, or News & Features: Archives.

FLAME EMISSION AND ATOMIC ABSORPTION SPECTROMETRY

4. J. A. Dean, *Flame Photometry*. New York: McGraw-Hill, 1960. A classic on the fundamentals.

5. P. M. Hald, "Sodium and Potassium by Flame Photometry," in D. Seligson, ed., *Standard Methods of Clinical Chemistry*, Vol. II. New York: Academic, 1958, pp. 165–185. Describes determination in serum.

6. G. D. Christian and F. J. Feldman, *Atomic Absorption Spectroscopy. Applications in Agriculture, Biology, and Medicine.* New York: Wiley-Interscience, 1970. Describes sample preparation procedures.

7. K. W. Jackson, ed., *Electrothermal and Atomization for Analytical Atomic Spectrometry.* New York: Wiley, 1999.

8. F. J. Feldman, "Internal Standardization in Atomic Emission and Absorption Spectrometry," *Anal. Chem.*, **42** (1970) 719.

9. G. D. Christian, "Medicine, Trace Elements, and Atomic Absorption Spectroscopy," *Anal. Chem.*, **41**(1) (1969) 24A. Lists levels of trace elements in biological samples.

Chapter Eighteen

SAMPLE PREPARATION: SOLVENT AND SOLID-PHASE EXTRACTION

The next chapter introduces chromatographic techniques for analyzing complex samples, whereby multiple analytes are separated on a column and detected as they emerge from the column. But very often, samples need to be "cleaned up" prior to introduction into the chromatographic column. The techniques of solvent extraction and solid-phase extraction and related techniques are very useful for isolating analytes from complex sample matrices prior to chromatographic analysis. Solvent extraction is also useful for spectrophotometric determination.

Solvent extraction involves the distribution of a solute between two immiscible liquid phases. This technique is extremely useful for very rapid and "clean" separations of both organic and inorganic substances. In this chapter, we discuss the distribution of substances between two phases and how this can be used to form analytical separations. The solvent extraction of metal ions into organic solvents is described.

Solid-phase extraction is a technique in which hydrophobic functional groups are bonded to solid particle surfaces and act as the extracting phase. They reduce the need for large volumes of organic solvents.

18.1 Distribution Coefficient

A solute S will distribute itself between two phases (after shaking and allowing the phases to separate) and, within limits, the ratio of the concentrations of the solute in the two phases will be a constant:

$$K_D = \frac{[S]_1}{[S]_2} \qquad (18.1)$$

where K_D is the **distribution coefficient** and the subscripts represent solvent 1 (e.g., an organic solvent) and solvent 2 (e.g., water). If the distribution coefficient is large, the solute will tend toward quantitative distribution in solvent 1.

Neutral organics distribute from water into organic solvents. "Like dissolves like."

Fig. 18.1.   Separatory funnel.

The apparatus used for solvent extraction is the **separatory funnel**, illustrated in Figure 18.1. Most often, a solute is extracted from an aqueous solution into an immiscible organic solvent. After the mixture is shaken for about a minute, the phases are allowed to separate and the bottom layer (the denser solvent) is drawn off in a completion of the separation.

Many substances are partially ionized in the aqueous layer as weak acids. This introduces a pH effect on the extraction. Consider, for example, the extraction of benzoic acid from an aqueous solution. Benzoic acid (HBz) is a weak acid in water with a particular ionization constant K_a (given by Equation 18.4). The distribution coefficient is given by

$$K_D = \frac{[\text{HBz}]_e}{[\text{HBz}]_a} \qquad (18.2)$$

where e represents the ether solvent and a represents the aqueous solvent. However, part of the benzoic acid in the aqueous layer will exist as Bz^-, depending on the magnitude of K_a and on the pH of the aqueous layer; hence, quantitative separation may not be achieved.

18.2 Distribution Ratio

It is more meaningful to describe a different term, the **distribution ratio**, which is the ratio of the concentrations of *all* the species of the solute in each phase. In this example, it is given by

$$\boxed{D = \frac{[\text{HBz}]_e}{[\text{HBz}]_a + [\text{Bz}^-]_a}} \qquad (18.3)$$

We can readily derive the relationship between D and K_D from the equilibria involved. The acidity constant K_a for the ionization of the acid in the aqueous phase is given by

$$K_a = \frac{[\text{H}^+]_a[\text{Bz}^-]_a}{[\text{HBz}]_a} \qquad (18.4)$$

Hence, $$[\text{Bz}^-]_a = \frac{K_a[\text{HBz}]_a}{[\text{H}^+]_a} \qquad (18.5)$$

From Equation 18.2,

$$[\text{HBz}]_e = K_D[\text{HBz}]_a \qquad (18.6)$$

Substitution of Equations 18.5 and 18.6 into Equation 18.3 gives

$$D = \frac{K_D[\text{HBz}]_a}{[\text{HBz}]_a + K_a[\text{HBz}]_a/[\text{H}^+]_a} \qquad (18.7)$$

$$\boxed{D = \frac{K_D}{1 + K_a/[\text{H}^+]_a}} \qquad (18.8)$$

This equation predicts that when $[H^+]_a \gg K_a$, D is nearly equal to K_D, and if K_D is large, the benzoic acid will be extracted into the ether layer; D is maximum under these conditions. If, on the other hand, $[H^+] \ll K_a$, then D reduces to $K_D[H^+]_a/K_a$, which will be small, and the benzoic acid will remain in the aqueous layer. That is, in alkaline solution, the benzoic acid is ionized and cannot be extracted, while in acid solution, it is largely undissociated. These conclusions are what we would intuitively expect from inspection of the chemical equilibria.

Equation 18.8, like Equation 18.1, predicts that the *extraction efficiency will be independent of the original concentration of the solute*. This is one of the attractive features of solvent extraction; it is applicable to tracer (e.g., radioactive) levels and to macrolevels alike, a condition that applies only so long as the solubility of the solute in one of the phrases is not exceeded and there are no side reactions such as dimerization of the extracted solute.

> In solvent extraction, the separation efficiency is usually independent of the concentration.

Of course, if the hydrogen ion concentration changes, the extraction efficiency (D) will change. In this example, the hydrogen ion concentration will increase with increasing benzoic acid concentration, unless an acid–base buffer is added to maintain the hydrogen ion concentration constant (see Chapter 7 for a discussion of buffers).

In deriving Equation 18.8, we actually neglected to include in the numerator of Equation 18.3 a term for a portion of the benzoic acid that exists as the dimer in the organic phase. The extent of dimerization tends to increase with increased concentration, and by Le Châtelier's principle, this will cause the equilibrium to shift in favor of the organic phase with increased concentration. So, in cases such as this, the efficiency of extraction will actually increase at higher concentrations. As an exercise, derivation of the more complete equation is presented in Problem 12.

18.3 Percent Extracted

The distribution ratio D is a constant independent of the volume ratio. However, the fraction of the solute extracted will depend on the volume ratio of the two solvents. If a larger volume of organic solvent is used, more solute must dissolve in this layer to keep the concentration ratio constant and to satisfy the distribution ratio.

The fraction of solute extracted is equal to the millimoles of solute in the organic layer divided by the total number of millimoles of solute. The millimoles are given by the molarity times the milliliters. Thus, the percent extracted it given by

$$\% E = \frac{[S]_o V_o}{[S]_o V_o + [S]_a V_a} \times 100\% \qquad (18.9)$$

where V_o and V_a are the volumes of the organic and aqueous phases, respectively. It can be shown from this equation (see Problem 11) that the percent extracted is related to the distribution ratio by

$$\boxed{\% E = \frac{100D}{D + (V_a/V_o)}} \qquad (18.10)$$

If $V_a = V_o$, then

> Extraction will be quantitative (99.9%) for D values of 1000.

$$\boxed{\% E = \frac{100D}{D + 1}} \qquad (18.11)$$

In the case of equal volumes, the solute can be considered quantitatively retained if D is less than 0.001. It is essentially quantitatively extracted if D is greater than 1000. The percent extracted changes only from 99.5 to 99.9% when D is increased from 200 to 1000.

Example 18.1

Twenty milliliters of an aqueous solution of 0.10 M butyric acid is shaken with 10 mL ether. After the layers are separated, it is determined by titration that 0.5 mol butyric acid remains in the aqueous layer. What is the distribution ratio, and what is the percent extracted?

Solution

We started with 2.0 mmol butyric acid, and so 1.5 mmol was extracted. The concentration in the ether layer is 1.5 mmol/10 mL = 0.15 M. The concentration in the aqueous layer is 0.5 mmol/20 mL = 0.025 M. Therefore,

$$D = \frac{0.15}{0.025} = 6.0$$

Since 1.5 mmol was extracted, the percent extracted is $(1.5/2.0) \times 100\% = 75\%$. Or

$$\% E = \frac{100 \times 6.0}{6.0 + (20/10)} = 75\%$$

Equation 18.10 shows that the fraction extracted can be increased by decreasing the ratio of V_a/V_o, for example, by increasing the organic phase volume. However, a more efficient way of increasing the amount extracted using the same volume of organic solvent is to perform successive extractions with smaller individual volumes of organic solvent. For example, with a D of 10 and $V_a/V_o = 1$, the percent extracted is about 91%. Decreasing V_a/V_o to 0.5 (doubling V_o) would result in an increase of $\% E$ to 95%. But performing two successive extractions with $V_a/V_o = 1$ would give an overall extraction of 99%.

18.4 Solvent Extraction of Metals

To extract a metal ion into an organic solvent, its charge must be neutralized, and it must be associated with an organic agent.

Solvent extraction has one of its most important applications in the separation of metal cations. In this technique, the metal ion, through appropriate chemistry, distributes from an aqueous phase into a water-immiscible organic phase. Solvent extraction of metal ions is useful for removing them from an interfering matrix, or for selectively (with the right chemistry) separating one or a group of metals from others. The technique is widely used for the spectrophotometric determination of metal ions since the reagents used to accomplish the extraction often form colored complexes with the metal ion. It is also used in flame atomic absorption spectrophotometry for introducing the sample in a nonaqueous solvent into the flame for enhanced sensitivity, and removal of matrix effects.

The separation can be accomplished in several ways. You have noted above that the uncharged organic molecules tend to dissolve in the organic layer while the charged anion from the ionized molecules remains in the polar aqueous layer. This is an example of "like dissolves like." Metal ions do not tend to dissolve appreciably in the organic layer. For them to become soluble, their charge must be neutralized and something must be added to make them organiclike. There are two principal ways of doing this.

EXTRACTION OF ION-ASSOCIATION COMPLEXES

In one method, the metal ion is incorporated into a bulky molecule and then associates with another ion of the opposite charge to form an **ion pair**, or the metal ion associates with another ion of great size (organiclike). For example, it is well known that iron(III) can be quantitatively extracted from hydrochloric acid medium into diethyl ether. The mechanism is not completely understood, but evidence exists that the chloro complex of the iron is coordinated with the oxygen atom of the solvent (the solvent displaces the coordinated water), and this ion associates with a solvent molecule that is coordinated with a proton:

$$\{(C_2H_5)_2O : H^+, FeCl_4[(C_2H_5)_2O]_2^-\}$$

Similarly, the uranyl ion UO_2^{2+} is extracted from aqueous nitrate solution into isobutanol by associating with two nitrate ions ($UO_2^{2+}, 2NO_3^-$), with the uranium probably being solvated by the solvent to make it solventlike. Permanganate forms an ion pair with tetraphenylarsonium ion [$(C_6H_5)_4As^+, MnO_4^-$], which makes it organiclike, and it is extracted into methylene chloride. There are numerous other examples of ion-association extractions.

EXTRACTION OF METAL CHELATES

The most widely used method of extracting metal ions is formation of a chelate molecule with an organic chelating agent.

As mentioned in Chapter 9, a chelating agent contains two or more complexing groups. Many of these reagents form colored chelates with metal ions and form the basis of spectrophotometric methods for determining the metals. The chelates are often insoluble in water and will precipitate. They are, however, usually soluble in organic solvents such as methylene chloride. Many of the organic precipitating agents listed in Chapter 10 are used as extracting agents.

EXTRACTION PROCESS FOR METAL CHELATES

Most chelating agents are weak acids that ionize in water; the ionizable proton is displaced by the metal ion when the chelate is formed, and the charge on the organic compound neutralizes the charge on the metal ion. An example is *diphenyl-thiocarbazone (dithizone)*, which forms a chelate with lead ion:

The usual practice is to add the chelating agent, HR, to the organic phase. It distributes between the two phases, and in the aqueous phase it dissociates as a weak acid. The metal ion, M^{n+}, reacts with nR^- to form the chelate MR_n, which then distributes into the organic phase. The distribution ratio is given by the ratio of the metal chelate concentration in the organic phase to the metal ion concentration in the aqueous phase. The following equation can be derived:

$$D = \frac{[MR_n]_o}{[M^{n+}]_a} = K\frac{[HR]_o^{\,n}}{[H^+]_a^{\,n}} \qquad (18.12)$$

where K is a constant that includes K_a of HR, K_f of MR_n, and K_D of HR and MR_n. Note that the distribution ratio is independent of the concentration of the metal ion, provided the solubility of the metal chelate in the organic phase is not exceeded. HR is often in large excess and is considered constant. The extraction efficiency can be affected only by changing the pH or the reagent concentration. A 10-fold increase in the reagent concentration will increase the extraction efficiency the same as an increase in the pH of one unit (10-fold decrease in $[H^+]$). Each effect is greater as n becomes greater. By using a high concentration of reagent, extraction can be performed in more acid solution.

Chelates of different metals extract at different pH values, some in acid to basic solution, some only in alkaline solution. By appropriate adjustment of pH, selectivity can be achieved in the extraction. Also, judicious use of masking agents, complexing agents that prevent one metal ion from reacting with the chelating agent, can enhance the selectivity.

18.5 Accelerated and Microwave-Assisted Extraction

Accelerated solvent extraction is a technique for the efficient extraction of analytes from a solid sample matrix into a solvent. The sample and solvent are placed in a closed vessel and heated to 50 to 200°C. The high pressure allows heating above the boiling point, and the high temperature accelerates the dissolution of analytes in the solvent. Both time of extraction and the volume of solvent needed are greatly reduced over atmospheric extraction.

In **microwave-assisted extraction** (MAE), the solvent is heated by microwave energy. The analyte compounds are again partitioned from the sample matrix into the solvent. This approach is an extension of closed-vessel acid digestion described in Chapter 2. A closed vessel containing the sample and solvent is placed in a microwave oven similar to the one described in Figure 2.27. The kinetics of extraction is affected by the temperature and the choice of solvent or solvent mixture. Atmospheric heating for extraction is limited to the boiling point of the solvent. Closed-vessel temperatures at 175 psig typically reach on the order of 150°C, compared with boiling points of 50 to 80°C for commonly used solvents. Solvent mixtures may be used so long as one of them absorbs microwave energy. Some solvents are microwave transparent, for example, hexane, and do not heat, but a mixture of hexane and acetone heats rapidly.

The closed vessels must be inert to solvents and be microwave transparent. The body is made of polyetherimide (PEI), with a perfluoroalkoxy (PFA) liner. Several sample vessels may be placed in the oven at the same time for multiple extractions.

Microwave extractions may also be performed at atmospheric pressure, without the need for pressurized vessels (see Ref. 6). Heating and cooling cycles are employed to prevent boiling of the solvent. This technique also reduces extraction times substantially. For information on commercial MAE systems, see www.cem.com.

18.6 Solid-Phase Extraction

Liquid–liquid extractions are very useful but have certain limitations. The extracting solvents are limited to those that are water immiscible (for aqueous samples). Emulsions tend to form when the solvents are shaken, and relatively large volumes of solvents are used that generate a substantial waste disposal problem. The operations are often manually performed and may require a back extraction.

Many of these difficulties are avoided by the use of **solid-phase extraction** (SPE), which has become a widely used technique for sample cleanup and concentration prior to chromatographic analysis (next chapter) in particular. In this technique, hydrophobic organic functional groups are chemically bonded to a solid surface, for example, powdered silica. A common example is the bonding of C_{18} chains on silica, with particle sizes on the order of 40 μm. These groups will interact with hydrophobic organic compounds by van der Waals forces and extract them from an aqueous sample in contact with the solid surface. The same solid phases used in high-performance liquid chromatography (Chapter 21) are used for solid-phase extraction.

The powdered phase is generally placed in a small cartridge, similar to a plastic syringe. Sample is placed in the cartridge and forced through by means of a plunger (positive pressure) or a vacuum (negative pressure), or by centrifugation (see Figure 18.2). Trace organic molecules are extracted, preconcentrated on the column, and separated away from the sample matrix. Then they can be eluted with a solvent such as methanol and then analyzed, for example, by chromatography (Chapters 19–21). They may be further concentrated prior to analysis by evaporating the solvent.

The nature of the extracting phase can be varied to allow extraction of different classes of compounds. Figure 18.3 illustrates bonded phases based on van der Waals forces, hydrogen bonding (dipolar attraction), and electrostatic attraction.

When silica particles are bonded with a hydrophobic phase, they become "waterproof" and must be conditioned in order to interact with aqueous samples. This is accomplished by passing methanol or a similar solvent through the sorbent bed. This penetrates into the bonded layer and permits water molecules and analyte to diffuse into the bonded phase. After conditioning, water is passed to remove the excess solvent prior to adding the sample.

In solid-phase extraction, the bonded C_{18} chains take the place of the organic solvent.

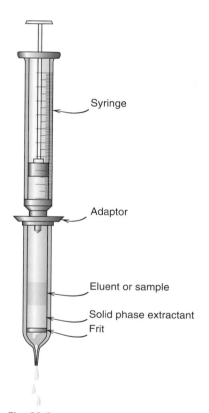

Fig. 18.2. Solid-phase cartridge and syringe for positive pressure elution.

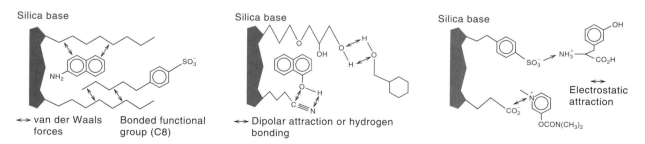

Fig. 18.3. Solid-phase extractants utilizing nonpolar, polar, and electrostatic interactions. (Adapted from N. Simpson, *Am. Lab.*, August, 1992, p. 37. Reproduced by permission of American Laboratory, Inc.)

Figure 18.4 illustrates a typical sequence in a solid-phase extraction. Following conditioning, the analyte and other sample constituents are adsorbed on the sorbent extraction bed. A rinsing step removes some of the undesired constituents, while elution removes the desired analyte, perhaps leaving other constituents behind, depending on the relative strengths of interaction with the solid phase or solubility in the eluting solvent. Such a procedure is used for the determination of organic compounds in drinking water in the official Environmental Protection Agency (EPA) method (Ref. 9).

SPE CARTRIDGES

The SPE sorbent is prepackaged in polypropylene syringe barrels, with typically 500 mg of packing in 3- or 5-mL syringe barrels. Smaller 1-mL syringes packed with 100 mg are becoming more popular because of reduced sample and solvent requirement and faster cleanup times, and even smaller packed beds down to 10 mg are available. These smaller packings, of course, have smaller capacity. Larger ones may be required for preparing large volumes of environmental samples such as polluted water that has large amounts of contaminants to be removed.

The SPE cartridges are used for the isolation and concentration of drugs from biological samples and are typically processed in batches of 12 to 24 using vacuum manifolds (Figure 18.5). There are automated liquid bundling systems to improve the efficiency.

SPE PIPET TIPS

Solid-phase extraction has been automated. The first application systems utilized robotic systems, and then automated *xyz* liquid-handling systems. The automated liquid-handling systems are designed to handle pipet tips to dispense liquids. So

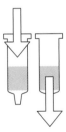

CONDITIONING
Conditioning the sorbent prior to sample application ensures reproducible retention of the compound of interest (the isolate).

RETENTION
■ Adsorbed isolate
● Undesired matrix constituents
▲ Other undesired matrix components

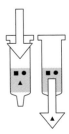

RINSE
▲ Rinse the columns to remove undesired matrix components

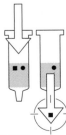

ELUTION
● Undesired components remain
■ Purified and concentrated isolate ready for analysis

Fig. 18.4. Principles of solid-phase extraction. (From N. Simpson, *Am. Lab.*, August, 1992, p. 37. Reproduced by permission of American Laboratory, Inc.)

Fig. 18.5. 16-Port vacuum manifold for use with solid-phase extraction tubes. (Courtesy of Alltech.)

sorbent-filled pipet tips for SPE were introduced (Figure 18.6) for use with the automated systems. These pipet tips can be used with multichannel pipettors (Chapter 2). The flow can be bi-directional, with liquid samples pulled from the bottom and eluent dispensed from the top. Commercially prepared tips are available for specific applications. For example, the Millipore ZipTipC$_4$ can be used to desalt 1 μL of 100 femtomole (10^{-15} mol) amounts of peptides prior to being analyzed by liquid chromatography–mass spectrometry. Pipet tips are used only once and discarded, eliminating any cross-contamination problem. See the EST Analytical Web page for example separations using SPE pipet tips; gas chromatograms of cleaned-up complex samples are illustrated (www.estanalytical.com).

SPE DISKS

The small cross-sectional area SPE pipet tips are prone to plugging by protein samples. So solid-phase extractants are also available in filter form (extraction disks) in which 8-μm silica particles are enmeshed into a web of PTFE [poly(tetrafluoroethylene)] fibrils. Fiberglass-based disks, which are more rigid, are also available. The greater cross-sectional area disks with shorter bed depths allow higher flow rates for large-volume samples with low concentrations of analyte, typically encountered in environmental analysis. The disks are less prone to channeling found with packed cartridges. They tend to plug if samples contain particulate matter, and so a prefilter may have to be used. Disk cartridges are also available that operate like a regular cartridge.

96-WELL SPE PLATES

Liquid chromatography combined with mass spectrometry (Chapter 21) is widely used for rapid and selective drug analysis, and samples can be run in 1 to 3 min. So, faster ways of sample cleanup are needed for processing large numbers of samples. 96-Well plates with small wells (so-called microtiter plates) are widely used for processing large numbers of samples in automated instruments.

Solid-phase extraction systems have been designed in a 96-well microtiter plate format, so they can be processed automatically. Single-block plates with 96 wells contain either packed beds or disks of sorbent particles, in an 8-row ×

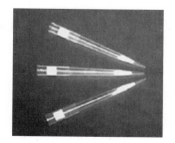

Fig. 18.6. Disposable solid-phase extraction pipet tips. [Courtesy of EST Analytical (www.estanalytical).]

12-column rectangular format (Figure 18.7). The plates sit on top of a 96-well plate collection system. The chemistry is the same as in the above formats. Samples are processed using a vacuum manifold or centrifuge using a microplate carrier. The SPE columns are 1 to 2 mL, with 10 to 100 mg packing of sorbent particles. The bed mass loading determines the solvent and elution volumes, as well as the capacity for analyte and sample matrix constituents. The smallest bed that provides adequate capacity should be used. This minimizes extraction times and the smaller elution volumes require less time for evaporation prior to reconstitution and analysis.

The optimum use of SPE procedures requires investigation of different stationary phases, their masses, the volumes of conditioning, sample load, wash, elution solvents, and the sample size. These variables are readily studied in column format. But it is costly or inconvenient to use only a fraction of the 96 wells to perform all the studies. Hence, modular well plates have been developed that have small removable plastic SPE cartridges that fit tightly in the 96-hole base plate, and only a portion needs to be used to develop a method.

OTHER SORBENTS FOR SOLID-PHASE EXTRACTION

Sorbents are available in long chain lengths (C_{20} and C_{30}) for isolation of hydrophobic molecules. "Universal sorbents" have been developed that will sorb a group of structurally similar compounds. An example in Figure 18.8a is a synthetic polymer of N_2–vinylpyrrolidone (top half of molecule) and divinylbenzene (bottom half). It provides hydrophilicity for wetting and hydrophobicity for analyte retention. A sulfonated version (Figure 18.8b) is a mixed-mode sorbent that has both ion exchange and solvent extraction properties and will retain a range of acidic, neutral, and basic drugs. These wettable sorbents do not require conditioning.

POLYMERIC PHASES

Besides the common silica-based SPE particles, polymer-based supports are also available. These have advantages of being stable over a wide pH range, and they do not possess residual silica groups that can interact with, for example, metal ions

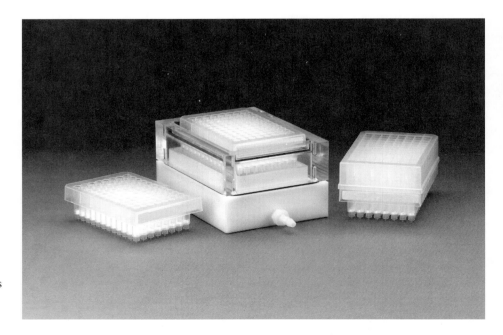

Fig. 18.7. 96-Well extraction plates and vacuum manifold with collection plate.

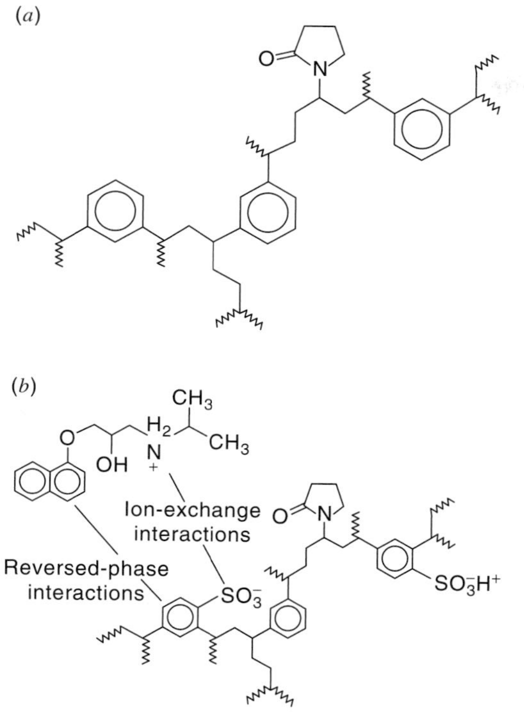

(a)

(b)

Ion-exchange interactions

Reversed-phase interactions

Fig. 18.8. "Universal sorbents": Chemical structures of Waters' Oasis (a) HLB and (b) MXC polymer sorbents. The top structure in (b) is the basic drug propanolol demonstrating drug–sorbent interaction. [From D. A. Wells, *LC.GC*, **17**(7) (1999) 600. Reproduced by permission of LC.GC.]

or other cationic species. The particles are spherical, while silica-based SPE particles are irregular in shape, and the polymeric particles have been designed to be wettable. They typically have higher capacity than silica-based particles.

DUAL PHASES

The use of two different phases can extend the range of compounds extracted. Three types are used, mixed mode, layered, and stacked phases. In the mixed mode, two different types of chemically bonded phases are mixed together in the cartridge. An example is a mixture of C8 and cation exchange particles. In the layered mode, the two different phases are packed one on top of the other. Stacked phases use two cartridges in series to provide enhanced separations. The first two modes are more readily adapted to automation since only a single cartridge is used.

SOLID-PHASE MICROEXTRACTION (SPME)

Solid-phase microextraction is a solvent-less extraction technique, usually used for analyte collection for determination by gas chromatography (Chapter 20) and is based on adsorption. A fused silica fiber is coated with a solid adsorbent or an immobilized polymer, or a combination of the two. Figure 18.9 illustrates an SPME fiber. Typical fiber dimensions are 1 cm × 110 μm. The fiber is inserted in a syringe needle device. Solid, liquid, or gaseous matrices can be sampled by SPME. The fiber is exposed to a gaseous or liquid sample, or the headspace above a solid or liquid sample for a fixed time and temperature; samples are often agitated to increase efficiency of analyte adsorption. Following adsorption, the analyte is thermally desorbed, usually directly in the injection port of a gas chromatograph for introduction into the GC column.

There are limited adsorbents. A widely used one is poly(dimethylsiloxane), which is useful for screening for volatile flavor components of beverages, foods,

Fig. 18.9. Schematic of a solid-phase microextraction device. [From C. L. Arthur, D. W. Potter, K. D. Buchholz, S. Motlagh, and J. Pawliszn, *LC.GC*, **10**(9) (1992) 656. Reproduced by permission of LC.GC.]

and the like. A 100-μm layer coating is used for nonpolar volatile compounds. Another example is an 85-μm layer of polyacrylate. It is relatively nonpolar, due to the presence of methyl groups. It is more polar due to the presence of carbonyl groups, and so extracts polar semivolatile compounds.

Learning Objectives

WHAT ARE SOME OF THE KEY THINGS WE LEARNED FROM THIS CHAPTER?

- Distribution coefficient, distribution ratio (key equations: 18.1, 18.3, 18.8), pp. 541, 542
- Percent extracted (key equation: 18.10), p. 543
- Solvent extraction of metal ions–complexes, chelates, p. 544
- Accelerated and microwave-assisted solvent extraction, p. 546
- Solid-phase extraction, p. 547
- Solid-phase microextraction, p. 551

Questions

1. What is the distribution coefficient? The distribution ratio?
2. Suggest a method for the separation of aniline, $C_6H_5NH_2$, an organic base, from nitrobenzene, $C_6H_5NO_2$ (extremely toxic!).
3. Describe two principal solvent extraction systems for metal ions. Give examples of each.
4. Describe the equilibrium processes involved in the solvent extraction of metal chelates.
5. What is the largest concentration of a metal chelate that can be extracted into an organic solvent? The smallest concentration?
6. Discuss the effect of the pH and of the reagent concentration on the solvent extraction of metal chelates.
7. What is the basis of accelerated solvent extraction?
8. What is the basis of microwave-assisted extraction?
9. How does solid-phase extraction differ from solvent extraction?
10. What is solid-phase microextraction?

Problems

EXTRACTION EFFICIENCIES

11. Derive Equation 18.10 from Equation 18.9.
12. In deriving Equation 18.8, we neglected the fact that benzoic acid partially forms a dimer in the organic phase ($2HBz \rightleftharpoons (HBz)_2$; $K_p = [(HBz)_2]/[HBz]^2$, where K_p is the dimerization constant). Derive an expression for the distribution ratio taking this into account.
13. Ninety-six percent of a solute is removed from 100 mL of an aqueous solution by extraction with two 50-mL portions of an organic solvent. What is the distribution ratio of the solute?

14. The distribution ratio between 3 M HCl and tri-n-butylphosphate for $PdCl_2$ is 2.3. What percent $PdCl_2$ will be extracted from 25.0 mL of a 7.0×10^{-4} M solution into 10.0 mL tri-n-butylphosphate?

15. Ninety percent of a metal chelate is extracted when equal volumes of aqueous and organic phases are used. What will be the percent extracted if the volume of the organic phase is doubled?

MULTIPLE EXTRACTIONS

16. For a solute with a distribution ratio of 25.0, show by calculation which is more effective, extraction of 10 mL of an aqueous solution with 10 mL organic solvent or extraction with two separate 5.0-mL portions of organic solvent.

17. Arsenic(III) is 70% extracted from 7 M HCl into an equal volume of toluene. What percentage will remain unextracted after three individual extractions with toluene?

Recommended References

SOLVENT EXTRACTION

1. G. H. Morrison and H. Freiser, *Solvent Extraction in Analytical Chemistry.* New York: Wiley, 1957. A classic. Detailed coverage of extraction of metals.

2. J. Stary, *The Solvent Extraction of Metal Chelates.* New York: Macmillan, 1964.

ACCELERATED AND MICROWAVE-ASSISTED SOLVENT EXTRACTION

3. B. E. Richter, B. A. Jones, J. L. Ezzell, N. L. Porter, N. Avdalovic, and C. Pohl, Jr., "Accelerated Solvent Extraction: A Technique for Sample Preparation," *Anal. Chem.,* **68** (1996) 1033.

4. A. Zlotorzynski, "The Application of Microwave Radiation to Analytical and Environmental Chemistry," *Crit. Rev. Anal. Chem.,* **25** (1997) 43.

5. Zu B. W. Renoe, "Microwave Assisted Extraction," *Am. Lab.,* August (1994) 34.

6. K. Ganzler, A. Salgo, and K. Valco, "Microwave Extraction. A Novel Sample Preparation Method for Chromatography," *J. Chromatogr.,* **371** (1986) 371.

SOLID-PHASE EXTRACTION

7. N. J. K. Simpson, ed., *Solid-Phase Extraction, Principles, Techniques, and Applications.* New York: Marcel Dekker, 2000.

8. J. S. Fritz, *Analytical Solid-Phase Extraction.* New York: Wiley-VCH, 1999.

9. *Methods for the Determination of Organic Compounds in Drinking Water* (Supplement 1). Cincinnati Environment Monitoring Systems Laboratory, Office of R&D, U.S. Environmental Protection Agency, 1990.

SOLID-PHASE MICROEXTRACTION

10. SPME Applications Guide, Supelco (www.sigma-aldrich.com). Over 600 references, categorized according to application, analyte/matrix, and extraction condition.

11. S. B. Hawthorn, D. J. Miller, J. Pawliszn, and C. L. Arthur, "Solventless Determination of Caffeine in Beverages Using Solid Phase Microextraction with Fused Silica Fibers," *J. Chromatogr.*, **603** (1991) 185.

12. C. Arthur, L. Killiam, K. Buchholz, and J. Pawliszn, "Automation and Optimization of Solid Phase Microextraction," *Anal. Chem.*, **64** (1992) 1960.

13. Z. Zhang and J. Pawilszyn, "Headspace Solid Phase Microextraction," *Anal. Chem.*, **65** (1993) 1843.

14. J. Pawiliszyn and R. M. Smith, eds., *Applications of Solid Phase Microextraction.* Berlin: Springer, 1999.

15. S. A. S. Wercinski, ed., *Solid Phase Microextraction. A Practical Guide.* New York: Marcel Dekker, 1999.

Chapter Nineteen

CHROMATOGRAPHY: PRINCIPLES AND THEORY

In 1906, the Russian scientist Tswett reported separating different colored constituents of leaves by passing an extract of the leaves through a column of calcium carbonate, alumina, and sucrose. He coined the term **chromatography**, from the Greek words meaning "color" and "to write." Tswett's original experiments went virtually unnoticed in the literature for decades, but eventually other methods were developed and today there are several different types of chromatography. Chromatography is taken now to refer generally to the separation of components in a sample by distribution of the components between two phases—one that is stationary and one that moves, usually but not necessarily in a column.

The International Union of Pure and Applied Chemistry (IUPAC) has drafted a recommended definition of chromatography: "Chromatography is a physical method of separation in which the components to be separated are distributed between two phases, one of which is stationary (stationary phase), while the other (the mobile phase) moves in a definite direction" [L. S. Ettre, "Nomenclature for Chromatography," *Pure & Appl. Chem.*, **65**(4) (1993) 819–872]. The stationary phase is usually in a column, but may take other forms, such as a planar phase (flat sheet). Chromatographic techniques have been more valuable in the separation and analysis of highly complex mixtures than any other and revolutionized the capabilities of analytical chemistry. In this chapter, we introduce the concepts and principles of chromatography, including the different types, and describe the theory of the chromatographic process in columns.

The two principal types of chromatography are gas chromatography (GC) and liquid chromatography (LC). Gas chromatography separates gaseous substances based on adsorption on or partitioning in a stationary phase from a gas phase and is described in Chapter 20. Liquid chromatography includes techniques such as size exclusion (separation based on molecular size), ion exchange (separation based on charge), and high-performance liquid chromatography (HPLC—separation based on adsorption or partitioning from a liquid phase). These are presented in Chapter 21, along with thin-layer chromatography (TLC), a planar form of LC, and electrophoresis where separation in an electrical gradient is based on the sign and magnitude of solute charge.

GC and HPLC are the most widely used forms of chromatography.

19.1 Principles of Chromatographic Separations

A solute equilibrates between a mobile and a stationary phase. The more it interacts with the stationary phase, the slower it is moved along a column.

While the mechanisms of retention for various types of chromatography differ, they are all based on establishment of an equilibrium between a stationary phase and a mobile phase. Figure 19.1 illustrates the separation of these components on a chromatographic column. A small volume of sample is placed at the top of the column, which is filled with the chromatographic particles (stationary phase) and solvent.

Courtesy of Merck KGaA. Reproduced by permission.

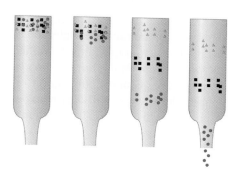

Fig. 19.1. Principle of chromatographic separations.

Mobile-phase solvent is added to the column and is allowed to slowly emerge from the bottom of the column. The individual components interact with the stationary phase to different degrees,

$$X_m \rightleftharpoons X_s \tag{19.1}$$

The distribution equilibrium is described by the distribution constant

$$K_c = \frac{[X]_s}{[X]_m} \tag{19.2}$$

where $[X]_s$ is the concentration of component X on or in the stationary phase at equilibrium and $[X]_m$ its concentration in the mobile phase. This equilibrium constant is governed by the temperature, the type of compound, and the stationary and mobile phases. It is also called the distribution coefficient or the partition coefficient in partition chromatography. Solutes with a large K_c value will be retained more strongly by the stationary phase than those with a small value. The result is that the latter will move along the column (be eluted) more rapidly. Because true equilibrium between the two phases is not achieved, there will be some lag of the analyte molecules between the two phases, which depends on the flow rate of the mobile phase and on the degree of interaction with the stationary phase, and results in band broadening. Figure 19.2 illustrates the distribution of two species A and B along a column as they move down the column. If we measure the concentration of eluted molecules as they emerge from the column and plot this as a function of time or of the volume of mobile phase passed through the column, a chromatogram results. Note that as the substances move down the column, each band becomes more spread out. The areas under the peaks remain the same. Band-broadening effects are treated below. In modern chromatography, a flow cell and

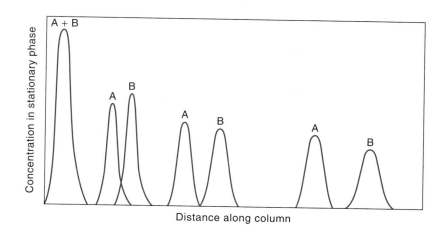

Fig. 19.2. Distribution of two substances, A and B, along a chromatographic column in a typical chromatographic separation.

detector are placed at the end of the column to automatically measure the eluted compounds and print out a chromatogram of the peaks for the separated substances.

Although there are several different forms of chromatography, this simplified model typifies the mechanism of each. That is, *there is nominally an equilibrium between two phases, one mobile and one stationary.* (True equilibrium is never really achieved.) By continually adding mobile phase, the analytes will distribute between the two phases and eventually be eluted, and if the distribution is sufficiently different for the different substances, they will be separated.

19.2 Classification of Chromatographic Techniques

Chromatographic processes can be classified according to the type of equilibration process involved, which is governed by the type of stationary phase. Various bases of equilibration are: (1) adsorption, (2) partition, (3) ion exchange, and (4) pore penetration.

ADSORPTION CHROMATOGRAPHY

The stationary phase is a solid on which the sample components are adsorbed. The mobile phase may be a liquid (*liquid–solid chromatography*) or a gas (*gas–solid chromatography*); the components distribute between the two phases through a combination of sorption and desorption processes. Thin-layer chromatography (TLC) is a special example of adsorption chromatography in which the stationary phase is a *plane*, in the form of a solid supported on an inert plate, and the mobile phase is a liquid.

PARTITION CHROMATOGRAPHY

The stationary phase of partition chromatography is a liquid supported on an inert solid. Again, the mobile phase may be a liquid (*liquid–liquid partition chromatography*) or a gas (*gas–liquid chromatography*, GLC).

In normal-phase chromatography, polar compounds are separated on a polar stationary phase. In reversed-phase chromatography, nonpolar compounds are separated on a nonpolar stationary phase. The latter is more common!

In the normal mode of operations of liquid–liquid partition, a polar stationary phase (e.g., methanol on silica) is used with a nonpolar mobile phase (e.g., hexane). This favors retention of polar compounds and elution of nonpolar compounds and is called **normal-phase chromatography**. If a nonpolar stationary phase is used, with a polar mobile phase, then nonpolar solutes are retained more and polar solutes more readily eluted. This is called **reversed-phase chromatography** and is actually the most widely used.

ION EXCHANGE AND SIZE EXCLUSION CHROMATOGRAPHY

Ion exchange chromatography uses an ion exchange resin as the stationary phase. The mechanism of separation is based on ion exchange equilibria. In size exclusion chromatography, solvated molecules are separated according to their size by their ability to penetrate a sievelike structure (the stationary phase).

These are arbitrary classifications of chromatographic techniques, and some types of chromatography are considered together as a separate technique, such as *gas chromatography* for gas–solid and gas–liquid chromatography. In every case, successive equilibria are at work that determine to what extent the analyte stays behind or moves along with the eluent (mobile phase). In column chromatography, the column may be packed with small particles that act as the stationary phase (adsorption chromatography) or are coated with a thin layer of liquid phase (partition chromatography). In gas chromatography, the more common form today is a capillary column in which microparticles or a liquid are coated on the wall of the capillary tube. We will see in Chapter 20 that this results in greatly increased separation efficiency.

Chromatography Nomenclature and Terms

In the fundamental discussions that follow, we use the IUPAC recommended symbols and terms, published in 1993 (Ref. 5). The listing is very extensive, filling 54 pages. L. S. Ettre, who chaired the IUPAC committee, has published an abbreviated list of symbols and the most significant changes from traditional use [L. S. Ettre, "The New IUPAC Nomenclature for Chromatography," *LC.GC*, **11**(7) (July) (1993) 502]. Majors and Carr published a very useful updated "Glossary of Liquid-Phase Separation Terms," R. E. Majors and P. W. Carr, *LC.GC*, **19**(2) (February) (2001) 124, www. chromatographyonline.com. Full text is also available at www.zirchrom. com/pdf/glossary.pdf. They incorporate the recommended IUPAC terms.

   Some of the older terms and the corresponding recommended terms are given in the following table:

Old		New	
Symbol	Term	Symbol	Term
α	Selectivity factor	α	Separation factor
HETP	Height equivalent to a theoretical plate	H	Plate height
k'	Capacity factor	k	Retention factor
n	Number of theoretical plates	N	Efficiency, number of plates
n_{eff}	Effective number of theoretical plates	N_{eff}	Effective theoretical plates; effective plate number
t_m	Mobile-phase holdup time	t_M	Mobile-phase holdup time
t_r	Retention time	t_R	Retention time
t'_r	Adjusted retention time	t'_R	Adjusted retention time
w	Base peak width	w_b	Bandwidth of peak

   In addition to these terms, we will use a number of others throughout the chapter in describing the properties of gas and liquid chromatography. These are summarized here for easy reference.

A = eddy diffusion term = $2\lambda d_p$
   λ = packing factor
   d_p = average particle diameter

B = longitudinal diffusion term = $2\gamma D_M$

γ = obstruction factor

D_M = diffusion coefficient

C = interphase mass transfer term = $\dfrac{1}{6}\dfrac{d_p^2}{D_M}$

C_m = mobile-phase mass transfer term

C_s = stationary-phase mass transfer term

L = column length

u = mobile-phase linear velocity, cm/s

$\bar{u}$ = average mobile-phase linear velocity, cm/s

v = reduced velocity

h = reduced plate height

R_s = resolution

19.3 Theory of Column Efficiency in Chromatography

The band broadening that occurs in column chromatography is the result of several factors, which influence the efficiency of separations. We can quantitatively describe the efficiency of a column and evaluate the factors that contributed to it.

THEORETICAL PLATES

A theoretical plate represents a single equilibrium step. The more theoretical plates, the greater the resolving power (the greater the number of equilibrium steps).

The separation efficiency of a column can be expressed in terms of the number of theoretical plates in the column. A theoretical plate is a defined concept derived from distillation theory, whereby each theoretical plate in chromatography can be thought of as representing a single equilibrium step. In reality, they are a measure of the efficiency of a column. For high efficiency, a large number of plates is necessary. The **plate height**, H, is the length of a column divided by the number of theoretical plates. To avoid a long column, then, H should be as short (thin or small) as possible. These concepts apply to all forms of column chromatography, but the parameters are easier to determine in gas chromatography.

Experimentally, the plate height is a function of the variance, σ^2, of the chromatographic band and the distance, x, it has traveled through the column, and is given by σ^2/x; σ is the standard deviation of the Gaussian chromatographic peak, and is equal to the width of the peak at the steepest potion of the curve (the inflection point). The width at half-height, w_h, corresponds to 2.35σ, and the peak base width, w_b is equal to 4σ (Figure 19.3). The number of plates, N, for a solute eluting from a column of length, L, is $L/H = Lx/\sigma^2 = L^2/\sigma^2$ (for the full column length: $x = L$) = $16\,L^2/w_b^2$.

The **number of plates** or **efficiency** can be obtained from a chromatogram from the expression

The narrower the peak, the greater the number of plates.

$$N = 16\left(\frac{t_R}{w_b}\right)^2 \tag{19.3}$$

where N is the number of plates of a column *toward a particular compound*, t_R is the *retention time*, and w_b is the peak width measured at the base in the same units

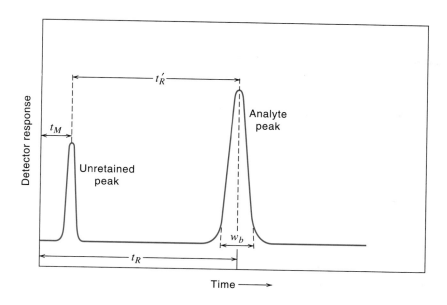

Fig. 19.3. Determination of number of plates. $w_b = 4\sigma$.

as t_R. These are illustrated in Figure 19.3. *Retention volume V_R* may be used in place of t_R. It should be noted that w is not the base width of the peak but the width obtained from the intersection of the baseline with tangents drawn through the inflection points at each side of the peak.

An alternative way to estimate the number of plates is from the width of the peak measured at a height of one-half of the peak height, w_h:

$$N = \frac{5.545 t_R^2}{w_h^2} \tag{19.4}$$

Example 19.1

Calculate the number of plates in the column resulting in the chromatographic peak in Figure 19.3.

Solution

Measuring with a ruler, $t_R = 52.3$ mm and $w_b = 9.0$ mm

$$N = 16\left(\frac{52.3}{9.0}\right)^2 = 54_0$$

This is not a very efficient column, as we will see below.

The *effective plate number* corrects theoretical plates for dead (void) volume and hence is a measure of the true number useful of plates in a column:

$$N_{\text{eff}} = 16\left(\frac{t_R'}{w_b}\right)^2 \tag{19.5}$$

where t'_R is the *adjusted retention time*.

$$t'_R = t_R - t_M \tag{19.6}$$

and t_M is the time required for the mobile phase to traverse the column and is the time it would take an unretained solute to appear. In gas chromatography, an air peak often appears from unretained air injected with the sample, and the time for this to appear is taken as t_M.

The above equations assume a Gaussian-shaped peak, as in Figure 19.3, and the position of the peak maximum is taken for calculations. For asymmetric (tailing) peaks, the efficiency is better determined by the peak centroid and variance by mathematical analysis as described by the *Foley–Dorsey equation* [J. P. Foley and J. G. Dorsey, "Equations for Calculation of Figures of Merit for Ideal and Skewed Peaks," *Anal. Chem.*, **55**(1983) 730–737]. They derived empirical equations based solely on the graphically measurable retention time, t_R, peak width at 10% of peak height, $w_{0.1}$, and the empirical *asymmetry factor*, A/B. $A + B = w_{0.1}$, and are the widths from t_R to the left and right sides, respectively, of the asymmetric peak. (When the peak is symmetrical, $A = B = ½$ the peak width at 10% height).

Foley and Dorsey derived the number of theoretical plates as:

$$N_{sys} = \frac{41.7(t_R/w_{0.1})^2}{A/B + 1.25} \tag{19.7}$$

This equation corrects the retention time and plate count for peak tailing and extra-column sources of broadening.

For a symmetric peak ($A/B = 1$), this becomes $N_{sys} = 18.53(t_R/w_{0.1})^2$, which is close to the theoretical equation of $N_{0.1} = 18.42(t_R/w_{0.1})^2$, that is, the equation holds for ideal as well as asymmetric peaks.

Once the number of plates is known, H can be obtained by dividing the length of the column, L, by N ($H = L/N$). The width of the peak, then, is related to H, being narrower with smaller H. H may be expressed in centimeters/plate, millimeters/plate, and so forth. The *effective* plate height, H_{eff}, is L/N_{eff}.

The term H is usually determined for the last eluting compound. For a well-packed high-performance liquid chromatography (HPLC) column of 5-μm particles H should be about 2 to 3 times the particle diameter. Values of 0.01 to 0.03 mm are typical.

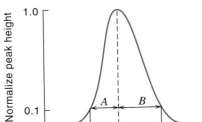

Asymmetric peak and Foley–Dorsey equation terms.

GAS CHROMATOGRAPHY EFFICIENCY— THE VAN DEEMTER EQUATION

We want H to be minimum. But velocities greater than $\bar{u}_{opt}$ are usually used to shorten separation times.

van Deemter showed for a packed gas chromatography column that the broadening of a peak is the summation of somewhat interdependent effects from several sources. The **van Deemter equation** expresses these in terms of the plate height, H:

$$\boxed{H = A + \frac{B}{\bar{u}} + C\bar{u}} \tag{19.8}$$

where A, B, and C are constants for a given system and are related to the three major factors affecting H, and $\bar{u}$ is the average linear velocity of the carrier gas in cm/s. While the van Deemter equation was developed for gas chromatography, it in principle holds for liquid chromatography as well, although the diffusion term becomes less important while the equilibration term becomes more critical (see below). For liquid chromatography, $\overline{u}$ represents the liquid mobile-phase velocity.

The value of $\overline{u}$ is equal to the column length, L, divided by the time for an unretained substance to elute, t_M (Figure 19.3):

$$\overline{u} = \frac{L}{t_M} \tag{19.9}$$

The general flow term for chromatography is the *mobile-phase velocity, u*. But in gas chromatography, the linear velocity will be different at different positions along the column due to compressibility of gases. So we use the *average linear velocity, $\overline{u}$*. In liquid chromatography, compressibility is negligible, and $\overline{u} = u$. Because of this, we will generally use the term u.

The significance of the three terms A, B, and C in packed-column gas chromatography is illustrated in Figure 19.4, which is a plot of H determined as a function of carrier gas velocity. Here, A represents *eddy diffusion* and is due to the variety of tortuous (variable-length) pathways available between the particles in the column and is independent of the gas- or mobile-phase velocity. The heterogeneity in axial velocities (eddy diffusion) is related to particle size and geometry of packing by:

$$A = 2\lambda d_p \tag{19.10}$$

where λ is an empirical column constant (packing factor), with typical values of 0.8 to 1.0 for a well-packed column, and d_p is the average particle diameter. It is minimized by using small and uniform particles, and tighter packing (which creates backpressure, though). But an extremely fine solid support is difficult to pack uniformly, which affects eddy diffusion.

The term B represents **longitudinal** (axial) or **molecular diffusion** of the sample components in the carrier gas or mobile phase, due to concentration gradients within the column. That is, there is a gradient at the interface of the sample zone and the mobile phase, and molecules tend to diffuse to where the concentration is smaller. The diffusion in the mobile phase is represented by:

$$B = 2\gamma D_M \tag{19.11}$$

Peaks are broadened by eddy diffusion, molecular diffusion, and slow mass transfer rates. Small, uniform particles minimize eddy diffusion. Faster flow decreases molecular diffusion but increases mass transfer effects. There will be an optimum flow.

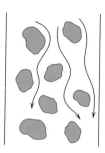

Eddy diffusion.

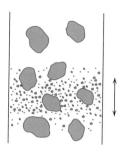

Molecular diffusion.

Molecular diffusion in usually negligible in LC but important in GC.

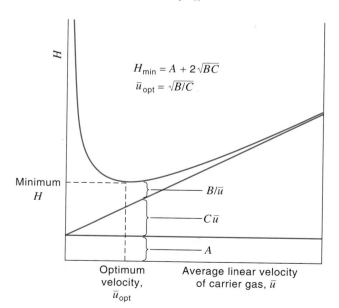

$H_{\min} = A + 2\sqrt{BC}$

$\overline{u}_{\text{opt}} = \sqrt{B/C}$

Minimum H

$B/\overline{u}$

$C\overline{u}$

A

Optimum velocity, $\overline{u}_{\text{opt}}$

Average linear velocity of carrier gas, $\overline{u}$

Fig. 19.4. Illustration of the van Deemter equation.

where γ is an obstruction factor, typically equal to 0.6 to 0.8, and D_M is the diffusion coefficient. Molecular diffusion is a function of both the sample and the carrier gas (in GC where it is important). Since the sample components are fixed in a given analysis, the only way to change B or $B/\overline{u}$ is by varying the type, pressure, and flow rate of the carrier gas. High flow rates reduce molecular diffusion, as do denser gases, such as nitrogen or carbon dioxide versus helium or hydrogen. In liquid chromatography, molecular diffusion in the stationary phase is very small compared to that in gases. In GC, it dominates only at flow rates less than $\overline{u}_{opt}$, and for LC is generally negligible under normal operating conditions. We usually operate at flow rates greater than $\overline{u}_{opt}$ since H does not increase very much and separations are faster.

Mass transfer dominates in LC.

The constant C is the **interphase mass transfer** term and is due to the finite time required for equilibrium of the solute to be established between the two phases as it moves between the mobile and stationary phases. It is dependent on the diffusion coefficient and the particle size (since this influences the distances between particles through which the solute must diffuse, approximated by:

$$C = \frac{1}{6}\frac{d_p^2}{D_M} \qquad (19.12)$$

It is influenced by the partition coefficient and, therefore, by the relative solubility of the solute in the stationary liquid phase (i.e., by the type and amount of liquid phase as well as the temperature). Or, in the case of adsorption chromatography, it is influenced by the adsorbability of the solute on the solid phase. Increasing the solubility of the vapor components of the sample (for gas chromatography) in the stationary liquid phase by decreasing the temperature may decrease C, provided the viscosity of the liquid phase is not increased so much that the exchange equilibrium becomes slower. The term $C\overline{u}$ is also decreased by decreasing the flow rate, allowing more time for equilibrium. In addition, it is minimized by keeping the stationary liquid-phase film as thin as possible to minimize diffusion within this phase. In liquid chromatography, this term dominates due to the slow diffusion in the liquid mobile phase. It is minimized by using small particles, thin stationary phase films, low-viscosity mobile phases, and high temperatures.

From Equations 19.10, 19.11, and 19.12, the rule of thumb for a normal packed column is

$$\boxed{H = 1.5d_p + \frac{D_M}{\overline{u}} + \frac{1}{6}\frac{d_p^2}{D_M}\,\overline{u}} \qquad (19.13)$$

See Problem 11 and your CD for a spreadsheet calculation of a van Deempter equation and plot of the change of A, $B/\overline{u}$, and $C\overline{u}$ as a function of $\overline{u}$.

Since the flow rate of the carrier gas (e.g., L/min) is proportional to the linear velocity, a qualitatively similar curve to Figure 19.4 will be obtained by plotting H as a function of flow rate. The constants A, B, and C would have different numerical values (using flow rate in place of $\overline{u}$ in Equation 19.8).

The conditions (e.g., flow rate) must be adjusted to obtain a balance between molecular diffusion and mass transfer. The three terms, A, $B/\overline{u}$, and $C\overline{u}$ are kept as small as possible to provide the minimum H for the sample solute that is the most difficult to elute (last to be eluted from the column). A van Deemter plot can aid in optimizing conditions. A, B, and C can be determined from three points and a solution of the three simultaneous van Deemter equations. Theoretically, a plot of Equation 19.8 results in a minimum, H_{min}, of $A + 2\sqrt{BC}$ at $\overline{u}_{opt} = \sqrt{B/C}$. Note the importance of the slope beyond $\bar{u}_{opt}$. The smaller the slope, the better since the efficiency will then suffer little at velocities in excess of $\bar{u}_{opt}$.

An efficient packed gas chromatography column will have several thousand theoretical plates, and capillary columns will have in excess of 10,000 theoretical plates. The H value for a 1-m column with 10,000 theoretical plates would be 100 cm/10,000 plates = 0.01 cm/plate. In a high-performance liquid chromatography (below), efficiency on the order of 400 theoretical plates per centimeter is typically achieved, and columns are 10 to 50 cm in length.

REDUCED PLATE HEIGHT IN GAS CHROMATOGRAPHY

For comparing the performance of different columns, a dimensionless plate height term is used, called the *reduced plate height*, h, obtained by dividing by the particle diameter:

$$h = \frac{H}{d_p} \tag{19.14}$$

A well-packed column should have an h value at the optimum flow of 2 or less. For open tubular columns

$$h = \frac{H}{d_c} \tag{19.15}$$

where d_c is the inner diameter of the column.

The reduced plate height is used with the *reduced velocity*, v, for comparing different packed columns over a broad range of conditions; v relates the diffusion coefficient in the mobile phase and the particle size of the column packing:

$$v = \bar{u} \frac{d_p}{D_M} \tag{19.16}$$

(For open-tubular columns, d_p is replaced by d_c.) The *reduced form of the van Deempter equation* is

$$h = A + \frac{B}{v} + Cv \tag{19.17}$$

The reduced form of Equation 19.13 is

$$h = 1.5 + \frac{1}{v} + \frac{v}{6} \tag{19.18}$$

OPEN TUBULAR COLUMNS IN GAS CHROMATOGRAPHY

As we will see in Chapter 20, capillary columns are the most widely used in gas chromatography because of their high efficiency due to large numbers of plates. These columns have no packing, and so the eddy diffusion term in the van Deempter equation disappears. For open tubular columns, the modification of the van Deempter equation, called the *Golay equation*, applies:

There is no eddy diffusion in open tubular columns.

$$H = \frac{B}{\bar{u}} + C\bar{u} \tag{19.19}$$

Golay was a pioneer in the development of capillary columns for GC and recognized the difference from packed columns, both in performance and theory.

HIGH-PERFORMANCE LIQUID CHROMATOGRAPHY: THE HUBER AND KNOX EQUATIONS

LC must contain a correction for mass transfer in both the mobile and stationary phases.

When a van Deempter plot is applied to HPLC, there is a curvature away from the theoretical curve at high velocities. Huber pointed out that in relating plate height to mass transfer kinetic factors, we need to add an additional term to account for mass transfer in both the stationary phase and the mobile phase:

$$H = A + B/\overline{u} + C_m\overline{u} + C_s\overline{u} \tag{19.20}$$

This is known as the *Huber equation*. Here, $\overline{u}$ is the mobile-phase linear velocity. The constant C_m is the mobile-phase mass transfer term and C_s the stationary-phase term. The B (longitudinal diffusion) term, except at very low mobile-phase velocities, is nearly zero and can be neglected. It is a function of the mobile-phase viscosity and the analyte molecule. The A (eddy diffusion) term turns out to be small compared with diffusion in the liquid phases and almost a constant value and is, therefore, usually neglected. So, H is estimated as:

$$H = C_m\overline{u} + C_s\overline{u} \tag{19.21}$$

The term C_s is relatively constant; C_m, in this case, includes stagnant mobile-phase transfer (in the pores of the particles). Representative H versus u plots for HPLC are shown in Figure 19.5 for different size particles. (See below for particle size and efficiency.)

At very slow velocities for small particles, molecular diffusion does become appreciable and H increases slightly. Note the lesser dependence on flow velocity compared to gas chromatography for the smaller particles. For well-packed columns of typical 5-μm particles, H values are usually in the range of 0.01 to 0.03 mm (10 to 30 μm). Note the scale in Figure 19.5 is in that range.

Knox developed an empirical equation for liquid chromatography that contains a term useful for correcting the deviation from the van Deempter equation,

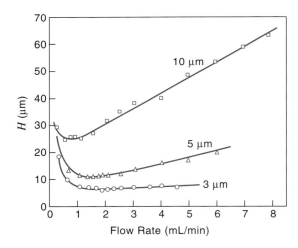

Fig. 19.5. van Deempter plots for different particle sizes in HPLC. The smaller particle sizes are more efficient, especially at higher flow rates. Column i.d.: 4.6 mm; mobile phase: 65% acetonitrile/35% water; sample: *t*-butylbenzene. [From M. W. Dong and M. R. Gant, *LC.GC*, **2** (1984) 294. Reprinted with permission.]

containing the third root of the velocity. The *Knox equation* is usually expressed in the dimensionless reduced form since the physical meaning of the term is not clear:

$$h = Av^{1/3} + \frac{B}{v} + Cv \qquad (19.22)$$

The term A is typically 1 to 2, with a large value representing a poorly packed bed; B is about 1.5 and C about 0.1. So a typical good column follows:

$$h = v^{1/3} + \frac{1.5}{v} + 0.1v \qquad (19.23)$$

EFFICIENCY AND PARTICLE SIZE IN HPLC

Column efficiency is related to particle size. It turns out that for well-packed HPLC columns, H is about two to three times the mean particle diameter, that is,

$$H = (2 \text{ to } 3) \times d_p \qquad (19.24)$$

Figure 19.6 shows the variation of N as a function of linear velocity for different particle sizes. Note N is on a logarithmic scale. Particles of different size distribution packed in columns of equal diameter will exhibit essentially the same longitudinal diffusion. Larger particles exhibit larger stagnant mobile phase transfer (C) since the solute molecule has a longer path to traverse in the pores, which increases band broadening.

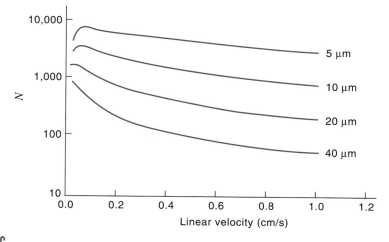

Fig. 19.6. Plate number as a function of linear velocity for different size particles in a 10-cm column. [From J. Maclennan and B. Murphy, *Today's Chemist at Work*, February (1994) 29. Copyright 1994 by the American Chemical Society. Reproduced by permission of Waters Corporation.]

RETENTION FACTOR IN CHROMATOGRAPHY

The **retention factor** k for a sample peak is defined by

Retention factor is a measure of retention time and, therefore, resolution capacity.

$$k = \frac{t_R - t_M}{t_M} = \frac{t'_R}{t_M} \qquad (19.25)$$

where t_R is the **retention time** (time required for the analyte peak to appear) and t_M is the time it would take for an unretained solute to appear (see Equation 19.6). A large retention factor favors good separation. However, large retention factors mean increased elution time, so there is a compromise between separation efficiency and separation time. The retention factor can be increased by increasing the stationary phase volume. A change in the retention factor is an indication of degradation of the stationary phase.

The effective plate number is related to the retention factor and effective plate number via:

$$N_{\text{eff}} = N \left(\frac{k}{k + 1} \right)^2 \qquad (19.26)$$

The volume of a chromatographic column consists of the stationary-phase volume and the **void volume**, the volume occupied by the mobile phase. The latter can be determined from t_M and the flow rate. One void volume of the mobile phase is required to flush the column once.

 Example 19.2

Calculate the retention factor for the chromatographic peak in Figure 19.3.

Solution
Measuring with a ruler, $t_R = 52.3$ mm and $t_M = 8.0$ mm.

$$k = \frac{52.3 - 8.0}{8.0} = 5.5_4$$

The preferred retention factor values are 1 to 5. If too low, the compounds pass rapidly through the column and the degree of separation may be sufficient. And large k values mean long retention time and long analysis times.

RESOLUTION IN CHROMATOGRAPHY

The resolution of two chromatographic peaks is defined by:

You should strive for a resolution of at least 1.0.

$$R_s = \frac{t_{R2} - t_{R1}}{(w_{b1} + w_{b2})/2} \qquad (19.27)$$

where t_{R1} and t_{R2} are the retention times of the two peaks (peak 1 elutes first), and w_b is the baseline width of the peaks. This is a measure of the ability of a column to separate two peaks. A resolution of 0.6 is needed to discern a valley between two peaks of equal heights. A value of 1.0 results in 2.3% overlap of two peaks of equal width and is considered the minimum for a separation to allow good quantitation. A resolution of 1.5 is an overlap of only 0.1% for equal width peaks and is considered sufficient for baseline resolution of equal height peaks.

We can describe resolution in thermodynamic terms, without regard to peak width. The **separation factor**, α, is a thermodynamic quantity that is a measure of the relative retention of analytes, and is given by:

$$\alpha = \frac{t'_{R2}}{t'_{R1}} = \frac{k_2}{k_1} \tag{19.28}$$

where t'_{R1} and t'_{R2} are the adjusted retention times (Equation 19.6) and k_1 and k_2 are the corresponding retention factors (Equation 19.25). This describes how well the peaks are separated without taking peak width into consideration. The resolution can, then, be written as:

$$R_s = \frac{1}{4}\sqrt{N}\left(\frac{\alpha - 1}{\alpha}\right)\left(\frac{k_2}{k_{ave} + 1}\right) \tag{19.29}$$

where k_{ave} is the mean of the two peak capacity factors. This form relates resolution to efficiency, that is, band broadening and retention time (Equation 19.3), and is known as the *resolution equation* or the *Purnell equation*. Note that since N is proportional to L, the resolution is proportional to the square root of the column length, $\sqrt{L}$; this holds strictly only for packed columns. So doubling the length of the column increases the resolution by $\sqrt{2}$ or 1.4. A fourfold increase would double the resolution. Retention times, of course, would be increased in direct proportion to the length of the column. For asymmetric peaks, the centroids of the peaks should be used in calculating retention times for calculating α values.

The number of plates required for a given degree of resolution is given by:

$$N_{req} = 16R^2\left(\frac{\alpha}{\alpha - 1}\right)^2\left(\frac{k_{ave} + 1}{k_2}\right)^2 \tag{19.30}$$

Substituting from Equation 19.26, the number of effective plates required is

$$N_{eff(req)} = 16R^2\left(\frac{\alpha}{\alpha - 1}\right)^2 \tag{19.31}$$

The figure in the margin illustrates how resolution increases differently with increasing values of N, k, or α. Note that increasing k increases the retention time for both peaks and broadens them. In uniformly packed columns, the widths of bands increase with the square root of the distance migrated, while the distance between centers of peaks increases in direct proportion to the distance traveled. Since the bands or peaks move faster than the broadening, separation occurs.

While it is desirable to increase the number of plates, the resolution in a packed column, as noted above, increases only with the square root of N (e.g., by increasing L), and the pressure drop increases. It is more effective to try to increase the selectivity (α) or retention factor (k) by varying the stationary and mobile phases. Increasing the retention time, of course, lengthens the analysis time, and a compromise is generally chosen between speed and resolution.

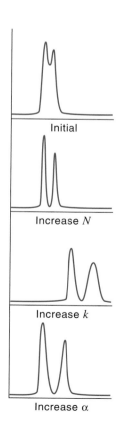

Initial

Increase N

Increase k

Increase α

Example 19.3

Ethanol and methanol are separated in a capillary GC column with retention times of 370 and 385 s, respectively, and base widths (w_b) of 16.0 and 17.0 s. An unretained air peak occurs at 10.0 s. Calculate the separation factor and the resolution.

Solution

Use the longest eluting peak to calculate N (Equation 19.3):

$$N = 16\left(\frac{385}{17.0}\right)^2 = 8.21 \times 10^3 \text{ plates}$$

From Equation 19.28,

$$\alpha = \frac{385 - 10}{330 - 10} = 1.04_2$$

From Equation 19.25,

$$k_1 = \frac{370 - 10}{10.0} = 36.0$$

$$k_2 = \frac{385 - 10}{10.0} = 37.5$$

$$k_{ave} = (36.0 + 37.5)/2 = 36.8$$

From Equation 19.29,

$$R_s = \frac{1}{4} \sqrt{8.21 \times 10^3} \left(\frac{1.042 - 1}{1.04_2}\right)\left(\frac{37.5}{36.8 + 1}\right) = 0.91$$

Or, from Equation 19.27, we obtain

$$R_s = \frac{385 - 370}{(17.0 + 16.0)/2} = 0.91$$

19.4 Chromatography Simulation Software

You are in charge of developing a new chromatographic separation. This involves selecting the proper column (stationary phase) and dimensions, mobile phase, and optimizing variables such as percent organic solvent, solvent or temperature gradient, and so forth. Optimization normally will require many repetitive chromatographic runs. But help is here! There are commercial software packages that assist the analyst in method development and optimization. Some of these are posted on the text website, with detailed descriptions of their capabilities. They are listed here: *DryLab* (LC Resources): www.lcresources.com; *ACD/GC Simulator, ACD/LC Simulator,* and *ACD/ChromManager* (ACD/Labs): www.acdlabs.com; and *ChromSword® AUTO* (Merck KGaA): www.hii.hitachi.com/LC%20ChromSword.htm.

19.5 Freebies: Company Searchable Chromatogram Databases

Agilent Technologies (www.chem.agilent.com) provides a database that contains an extensive library of GC and LC chromatograms based on applications conducted by Agilent chemists. See www.chem.agilent.com/scripts/chromatograms.asp. See also www.chem.agilent.com/scripts/chromatograms.asp for a tutorial site about HPLC columns. A description of these sites is given on the text website, as well as of Supelco (www.sigmaaldrich.com) and Hamilton (www.hamiltoncompany.com) sites for chromatogram databases.

Learning Objectives

WHAT ARE SOME OF THE KEY THINGS WE LEARNED FROM THIS CHAPTER?

- How chemicals are separated on a column, p. 556
- Types of chromatography: adsorption, partition, ion exchange, size exclusion, p. 558
- Chromatographic nomenclature (see tables of terms), p. 559
- Theory of column efficiency, p. 560
    - Plate number (key equation: 19.3), p. 560
    - van Deempter equation for packed GC columns (key equations: 19.8, 19.13), p. 562
    - Golay equation for open tubular GC columns (key equation 19.19), p. 565
    - Huber and Knox equations for HPLC (key equations: 19.20, 19.22), p. 566
- Retention factor (key equation: 19.25), p. 568
- Chromatographic resolution (key equations: 19.27, 19.29), p. 568
- Separation factor (key equation: 19.28), p. 569
- Chromatography simulation software and databases, pp. 570, 571

Questions

1. What is the description of chromatography?
2. Describe the principles underlying all chromatographic processes.
3. Classify the different chromatographic techniques, and give examples of principal types of applications.
4. What is the van Deemter equation? Define terms.
5. How does the Golay equation differ from the van Deemter equation?
6. How do the Huber and Knox equations differ from the van Deempter equation?

Problems

CHROMATOGRAPHY RESOLUTION

7. A gas-chromatographic peak had a retention time of 65 s. The base width obtained from intersection of the baseline with the extrapolated sides of the peak was 5.5 s. If the column was 3 ft in length, what was H in cm/plate?

8. It is desired to just resolve two gas-chromatographic peaks with retention times of 85 and 100 s, respectively, using a column that has an H value of 1.5 cm/plate under the operating conditions. What length columns is required? Assume the two peaks have the same base width.

9. The following gas-chromatographic data were obtained for individual 2-μL injections of n-hexane in a gas chromatograph with a 3-m column. Calculate the number of plates and H at each flow rate, and plot H versus the flow rate to determine the optimum flow rate. Use the *adjusted* retention time t'_R.

Flow rate (mL/min)	t_M (Air Peak) (min)	t'_R (min)	Peak Width (min)
120.2	1.18	5.49	0.35
90.3	1.49	6.37	0.39
71.8	1.74	7.17	0.43
62.7	1.89	7.62	0.47
50.2	2.24	8.62	0.54
39.9	2.58	9.83	0.68
31.7	3.10	11.31	0.81
26.4	3.54	12.69	0.95

10. Three compounds, A, B, and C, exhibit retention factors on a column having only 500 plates of $k_A = 1.40$, $k_B = 1.85$, and $k_C = 2.65$. Can they be separated with a minimum resolution of 1.05?

SPREADSHEET PROBLEM

11. Prepare a spreadsheet for a van Deemter plot for the following hypothetical, A, B, and C terms: $A = 0.5$ mm, $B = 30$ mm · mL/min, and $C = 0.05$ mm · min/mL. Plot H vs. $\bar{u}$ at linear velocities of 4, 8, 12, 20, 28, 40, 80 and 120 mL/min. Also, on the same chart, plot A vs. $\bar{u}$, $B/\bar{u}$ vs. $\bar{u}$, and $C\bar{u}$ vs. $\bar{u}$, and note how they change with the linear velocity, that is, how their contributions to H change. Calculate the hypothetical H_{min} and $\bar{u}_{opt}$ and compare with the H_{min} on the chart. Also calculate $B/\bar{u}_{opt}$ and $C\bar{u}_{opt}$. Look on the chart and see where the $B/\bar{u}$ and $C\bar{u}$ lines cross. Check your results with those in your CD, Chapter 19.

Recommended References

GENERAL

1. J. C. Giddings, *Unified Separation Science*, New York: Wiley-Interscience, 1991.

2. C. E. Meloan, *Chemical Separations: Principles, Techniques, and Experiments*, New York, Wiley, 1999.

3. J. Cazes, ed., *Encyclopedia of Chromatography*, New York: Marcel Dekker, 2001.

4. R. E. Majors and P. W. Carr, "Glossary of Liquid-Phase Separation Terms," *LC.GC*, **19**(2) (2001) 124.

5. L. S. Ettre, "Nomenclature for Chromatography," *Pure Appl. Chem.*, **65** (1993) 819.

HISTORICAL ACCOUNTS

These references give a good review of how separation techniques have evolved, particularly the stationary phases used and why we use the ones we do today.

6. H. J. Isaaq, ed., *A Century of Separation Science*, New York: Marcel Dekker, 2002.

7. L. S. Ettre, "The Birth of Partition Chromatography," *LC.GC*, **19**(5) (2001) 506.

8. M. S. Lesney, "Creating a Central Science. A Brief History of 'Color Writing'," *Today's Chemist at Work*, September, 1998, 67.

9. "Chromatography. Creating a Central Science," *Today's Chemist at Work* (Supplement), September, 2001. An 88-page series of articles on the evolution of the major forms of chromatography.

COMPUTER SIMULATION

10. J. W. Nolan and L. R. Snyder, "Gradient Elution Chromatography," in *Encyclopedia of Analytical Chemistry*, Vol. 13, *Theory and Instrumentation*, 11342 Chichester: Wiley, 2000.

11. R. G. Wolcott, J.W. Dolan, and L. R. Snyder, "Computer Simulation for the Convenient Optimization of Isocratic Reversed-Phase Liquid Chromatography Separations by Varying Temperature and Mobile Phase Strength," *J. Chromatogr. A*, **869** (2000) 3.

12. J. W. Dolan, L. R. Snyder, L. C. Sander, R. C. Wolcott, P. Habner, T. Baczek, and R. Kaliszan, *J. Chromatogr. A*, **857** (1999) 41.

Chapter Twenty

GAS CHROMATOGRAPHY

Gas chromatography (GC) is one of the most versatile and ubiquitous analytical techniques in the laboratory. It is widely used for the determination of organic compounds. The separation of benzene and cyclohexane (bp 80.1 and 80.8°C) is extremely simple by gas chromatography, but it is virtually impossible by conventional distillation. Although Martin and Synge invented liquid–liquid chromatography in 1941, the introduction of gas–liquid partition chromatography by James and Martin a decade later had a more immediate and larger impact for two reasons. First, as opposed to manually operated liquid–liquid column chromatography, GC required instrumentation for application, which was developed by collaboration among chemists, engineers, and physicists; and analyses were much more rapid and done on a small scale. Second, at the time of its development, the petroleum industry was required to have improved analytical monitoring and immediately adopted GC. Within a few short years, GC was used for the analysis of almost every type of organic compound.

Very complex mixtures can be separated by this technique. When coupled with mass spectrometry as a detection system, virtually positive identification of the eluted compounds is possible at very high sensitivity, creating a very powerful analytical system.

There are two types of GC: **gas–solid (adsorption) chromatography** and **gas–liquid (partition) chromatography**. The more important of the two is gas–liquid chromatography (GLC), used in the form of a capillary column. In this chapter, we describe the principles of operation of gas chromatography, the types of GC columns, and GC detectors. The principles of mass spectrometry (MS) are described, along with coupling of the gas chromatograph with a mass spectrometer (GC-MS).

20.1 Performing GC Separations

Analyte in the vapor state distributes between the stationary phase and the carrier gas. Gas-phase equilibria are rapid, so resolution (and the number of plates) can be high.

In gas chromatography, the sample is converted to the vapor state (if it is not already a gas) by injection into a heated port, and the eluent is a gas (the **carrier gas**). The stationary phase is generally a nonvolatile liquid supported on a capillary wall or inert solid particles such as diatomaceous earth (kieselguhr—derived from skeletal remains of microscopic marine single-celled algae, consisting mainly of silica); the kieselguhr is usually calcined to increase particle size, creating what is known as firebrick, sold as Chromosorb P or W, for example. There are a large number of

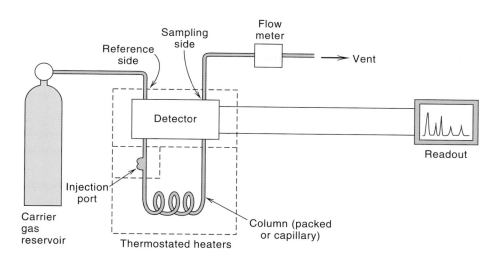

Fig. 20.1. Schematic diagram of gas chromatograph.

liquid phases available, and it is by changing the liquid phase, rather than the mobile phase, that different separations are accomplished. The most important factor in gas chromatography is the selection of the proper column (stationary phase) for the particular separation to be attempted. The nature of the liquid or solid phase will determine the exchange equilibrium with the sample components; and this will depend on the solubility or adsorbability of the sample, the polarity of the stationary phase and sample molecules, the degree of hydrogen bonding, and specific chemical interactions. Most of the useful separations have been determined empirically, although more quantitative information is now available.

A schematic diagram of a gas chromatograph is given in Figure 20.1, and a picture of a modern GC system is shown in Figure 20.2. The sample is rapidly injected by means of a hypodermic microsyringe (see Figure 2.11) through a silicone rubber septum into the column. The sample injection port, column, and detector are heated to temperatures at which the sample has a vapor pressure of at least 10 torr, usually about 50°C above the boiling point of the highest boiling solute. The injection port and detector are usually kept somewhat warmer than the

Fig. 20.2. Modern automated gas chromatography system. (Courtesy of Shimadzu North America.)

column to promote rapid vaporization of the injected sample and prevent sample condensation in the detector. For packed columns, liquid samples of 0.1 to 10 μL are injected, while gas samples of 1 to 10 mL are injected. Gases may be injected by means of a gas-tight syringe or through a special gas inlet chamber of constant volume (gas sampling valve). For capillary columns, volumes of only about 1/100 these sizes must be injected because of the lower capacity (albeit greater resolution) of the columns. Sample splitters are included on chromatographs designed for use with capillary columns that deliver a small fixed fraction of the sample to the column, with the remainder going to waste. They usually also allow splitless injection when packed columns are used (split/splitless injectors).

Separation occurs as the vapor constituents equilibrate between carrier gas and the stationary phase. The carrier gas is a chemically inert gas available in pure form such as argon, helium, or nitrogen. A highly dense gas gives best efficiency since diffusivity is lower, but a low-density gas gives faster speed. The choice of gas is often dictated by the type of detector.

The sample is automatically detected as it emerges from the column (at a constant flow rate), using a variety of detectors whose response is dependent upon the composition of the vapor (see below). Usually, the detector contains a **reference side** and a **sampling side**. The carrier gas is passed through the reference side before entering the column and emerges from the column through the sampling side. The response of the sampling side relative to the reference side signal is fed to a recording device where the chromatographic peaks are recorded as a function of time. By measuring the **retention time** (the minutes between the time the sample is injected and the time the chromatographic peak is recorded) and comparing

Automatic detection of the analytes as they emerge from the column makes measurements rapid and convenient. Retention times are used for qualitative identification. Peak areas are used for quantitative measurements.

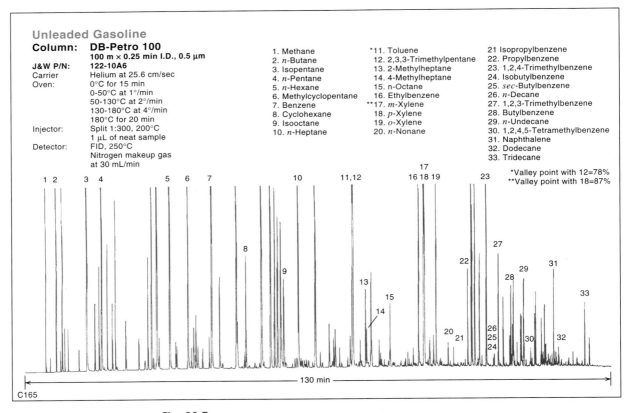

Unleaded Gasoline

Column:	**DB-Petro 100**
	100 m × 0.25 min I.D., 0.5 μm
J&W P/N:	**122-10A6**
Carrier	Helium at 25.6 cm/sec
Oven:	0°C for 15 min
	0-50°C at 1°/min
	50-130°C at 2°/min
	130-180°C at 4°/min
	180°C for 20 min
Injector:	Split 1:300, 200°C
	1 μL of neat sample
Detector:	FID, 250°C
	Nitrogen makeup gas
	at 30 mL/min

1. Methane
2. n-Butane
3. Isopentane
4. n-Pentane
5. n-Hexane
6. Methylcyclopentane
7. Benzene
8. Cyclohexane
9. Isooctane
10. n-Heptane

*11. Toluene
12. 2,3,3-Trimethylpentane
13. 2-Methylheptane
14. 4-Methylheptane
15. n-Octane
16. Ethylbenzene
**17. m-Xylene
18. p-Xylene
19. o-Xylene
20. n-Nonane

21 Isopropylbenzene
22. Propylbenzene
23. 1,2,4-Trimethylbenzene
24. Isobutylbenzene
25. sec-Butylbenzene
26. n-Decane
27. 1,2,3-Trimethylbenzene
28. Butylbenzene
29. n-Undecane
30. 1,2,4,5-Tetramethylbenzene
31. Naphthalene
32. Dodecane
33. Tridecane

*Valley point with 12=78%
**Valley point with 18=87%

C165

130 min

Fig. 20.3. Typical gas chromatogram of complex mixture using a capillary column. (Courtesy of Agilent Technologies.)

this time with that of a standard of the pure substance, it may be possible to identify the peak (agreement of retention times of two compounds does not guarantee the compounds are identical). The area under the peak is proportional to the concentration, and so the amount of substance can be quantitatively determined. The peaks are often very sharp and, if so, the peak height can be compared with a calibration curve prepared in the same manner. Chromatography detection systems usually have automatic readout of the peak area, as well as the retention time.

The separation ability of this technique is illustrated in a chromatogram in Figure 20.3. Since the peaks are automatically recorded, the entire analysis time is amazingly short for complex samples. This, coupled with the very small sample required, explains the popularity of the technique. This is not to exclude the more important reason that many of the analyses performed simply cannot be done by other methods.

With complex mixtures, it is not a simple task to identify the many peaks. Instruments are commercially available in which the gas effluent is automatically fed into a mass spectrometer where they are positively identified according to mass (formula weight and fragmentation pattern). This important analytical technique is called **gas chromatography–mass spectrometry** (GC–MS). The mass spectrometer is a sensitive and selective detector, and when a capillary GC column (very high resolution—see Section 20.2) is used (capillary GC–MS), this technique is capable of identifying and quantifying unbelievably complex mixtures of trace substances. For example, hundreds of compounds may be identified in sewage effluents, and traces of complex drugs in urine or blood or pollutants in water can be determined. For best sensitivity, though, some of the element or compound-type specific detectors listed later offer extraordinary detection limits.

See more on GC–MS below.

What Compounds Can Be Determined by GC?

Many, many compounds may be determined by gas chromatography, but there are limitations. They must be volatile and stable at the temperature employed, typically from 50 to 300°C. GC is useful for:

- All gases
- Most nonionized organic molecules, solid or liquid, containing up to about 25 carbons
- Many organometallic compounds (volatile derivatives of metal ions may be prepared)

If compounds are not volatile or stable, often they can be derivatized to make them amenable to analysis by GC. GC cannot be used for salts nor macromolecules, but these can be determined by HPLC, one of its major uses.

20.2 Gas Chromatography Columns

The two types of columns used in GC are **packed columns** and **capillary columns**. Packed columns were the first type and were used for many years. Capillary columns are more commonly used today, but packed columns are still used for applications that do not require high resolution or when increased capacity is needed.

PACKED COLUMNS

Packed columns can be used with large sample sizes and are convenient to use.

Columns can be in any shape that will fill the heating oven. Column forms include coiled tubes, U-shaped tubes, and W-shaped tubes, but coils are most commonly used. Typical packed columns are 1 to 10 m long and 0.2 to 0.6 cm in diameter. Well-packed columns may have 1000 plates/m, and so a representative 3-m column would have 3000 plates. Short columns can be made of glass, but longer columns may be made of stainless steel so they can be straightened for filling and packing. Columns are also made of Teflon. For inertness, glass is still preferred for longer columns. The resolution for packed columns increases only with the square root of the length of the column. Long columns require high pressure and longer analysis times and are used only when necessary (e.g., for long eluting solutes when high capacity is used). Separations are generally attempted by selecting columns in lengths of multiples of 3, such as 1 or 3 m. If a separation isn't complete in the shorter column, then the next longer one is tried.

The column is packed with small particles that may themselves serve as the stationary phase (adsorption chromatography) or more commonly are coated with a nonvolatile liquid phase of varying polarity (partition chromatography). Gas–solid chromatography (GSC) is useful for the separation of small gaseous species such as H_2, N_2, CO_2, CO, O_2, NH_2, and CH_4 and volatile hydrocarbons, using high surface area inorganic packings such as alumina (Al_2O_3) or porous polymers (e.g., Chromosorb—a polyaromatic crosslinked resin with a rigid structure and a distinct pore size). The gases are separated by their size due to retention by adsorption on the particles. Gas–solid chromatography is preferred for aqueous samples.

The solid support for a liquid phase should have a high specific surface area that is chemically inert but wettable by the liquid phase. It must be thermally stable and available in uniform sizes. The most commonly used supports are prepared from diatomaceous earth, a spongy siliceous material. They are sold under many different trade names. Chromosorb W is diatomaceous earth that has been heated with an alkaline flux to decrease its acidity; it is light in color. Chromosorb P is crushed firebrick that is much more acidic than Chromosorb W, and it tends to react with polar solutes, especially those with basic functional groups.

The polarity of Chromosorb P can be greatly decreased by silanizing the surface with hexamethyldisilazane, $[(CH_3)_3Si]_2NH$. Ottenstein (Ref. 7) has reviewed the selection of solid supports, both diatomaceous earth and porous polymer types.

Column-packing support material is coated by mixing with the correct amount of liquid phase dissolved in a low-boiling solvent such as acetone or pentane. About a 5 to 10% coating (wt/wt) will give a thin layer. After coating, the solvent is evaporated by heating and stirring; the last traces may be removed in a vacuum. A newly prepared column should be conditioned at elevated temperature by passing carrier gas through it for several hours. The selection of liquid phases is discussed below.

Particles should be uniform in size for good packing and have diameters in the range of 60 to 80 mesh (0.25 to 0.18 mm), 80 to 100 mesh (0.18 to 0.15 mm), or 100 to 120 mesh (0.15 to 0.12 mm). Smaller particles are impractical due to high pressure drops generated.

CAPILLARY COLUMNS—THE MOST WIDELY USED

Capillary columns can provide very high resolution, compared with packed columns.

In 1957 Marcel Golay published a paper entitled "Vapor Phase Chromatography and the Telegrapher's Equation" [*Anal. Chem.*, **29** (1957) 928]. His equation predicted increased number of plates in a narrow open-tubular column with the stationary phase supported on the inner wall. Band broadening due to multiple paths (eddy diffusion) would be eliminated. And in narrow columns, the rate of mass transfer is increased since molecules have small distances to diffuse. Higher flow rates can be used due to decreased pressure drop, which decreases molecular diffusion. Golay's work led to the development of various **open-tubular columns** that today provide extremely high resolution and have become the mainstay for gas-chromatographic analyses. These columns are made of thin fused silica (SiO_2) coated on the outside with a polyimide polymer for support and protection of the fragile silica capillary, allowing them to be coiled. The polyimide layer is what imparts a brownish color to the columns, and it often darkens on use. The inner surface of the capillary is chemically treated to minimize interaction of the sample with the silanol groups (Si–OH) on the tubing surface, by reacting the Si–OH group with a silane-type reagent (e.g., dimethyl dichlorosilane).

Capillaries are also made of stainless steel. Stainless steel interacts with many compounds and so is deactivated by treatment with dimethyl dichlorosilane (DMCS), producing a thin lining of fused silica to which stationary phases can be bonded. Stainless steel columns are more robust than fused silica columns and are used for applications requiring very high temperatures.

The capillaries are 0.10 to 0.53 mm i.d., with lengths of 15 to 100 m and can have several hundred thousand plates, even a million. They are sold as coils of about 0.2 m diameter (Figure 20.4). Capillary columns offer advantages of high resolution with narrow peaks, short analysis time, and high sensitivity (with modern detectors) but are more easily overloaded by too much sample. Split injectors by and large alleviate the overload problem.

Figure 20.5 illustrates the improvements in separation power in going from a packed column (6.4 × 1.8 m) to a very long but fairly wide stainless steel capillary column (0.76 mm × 150 m), to a narrow but shorter glass capillary column (0.25 mm × 50 m). Note that the resolution increases as the column becomes narrower, even when the capillary column is shortened.

There are three types of open-tubular columns. **Wall-coated open-tubular** (WCOT) columns have a thin liquid film coated on and supported by the walls of the capillary. The walls are coated by slowly passing a dilute solution of the liquid

Increasing the film thickness increases capacity but increases plate height and retention time.

Fig. 20.4. Capillary GC columns. (Courtesy of Quadrex Corp., Woodbridge, CT.)

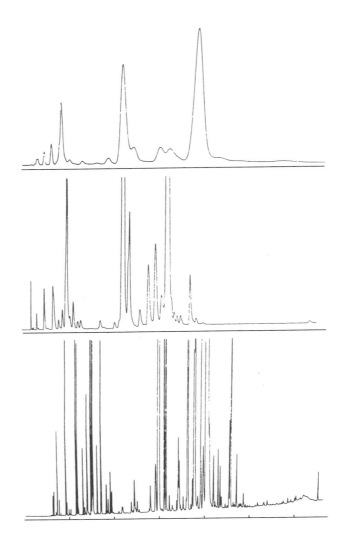

Fig. 20.5. Three generations in gas chromatography. Peppermint oil separation on (top) $\frac{1}{4}$-in. $\times$ 6-ft packed column; (center) 0.03-in. $\times$ 500-ft stainless steel capillary column; (bottom) 0.25-mm $\times$ 50-m glass capillary column. [From W. Jennings, *J. Chromatogr. Sci.*, **17** (1979) 363. Reproduced from the *Journal of Chromatographic Science* by permission of Preston Publications, A Division of Preston Industries, Inc.]

The resolution for open-tubular columns is WCOT > SCOT > PLOT. SCOT columns have capacities approaching those of packed columns.

phase through the columns. The solvent is evaporated by passing carrier gas through the columns. Following coating, the liquid phase is crosslinked to the wall. The resultant stationary liquid phase is 0.1 to 5 μm thick. Wall-coated open-tubular columns typically have 5000 plates/m. So a 50-m column will have 250,000 plates.

In **support coated open-tubular** (SCOT) columns, solid microparticles coated with the stationary phase (much like in packed columns) are attached to the walls of the capillary. These have higher surface area and have greater capacity than WCOT columns. The tubing diameter of these columns is 0.5 to 1.5 mm, larger than WCOT columns. The advantages of low pressure drop and long columns is maintained, but capacity of the columns approaches that of packed columns. Flow rates are faster and dead volume connections at the inlet and detector are less critical. Sample splitting is not required in many cases, so long as the sample volume is 0.5 μL or less. If a separation requires more than 10,000 plates, then a SCOT column should be considered instead of a packed column.

The third type, **porous layer open-tubular** (PLOT) columns, have solid-phase particles attached to the column wall, for adsorption chromatography. Particles of alumina or porous polymers (molecular sieves) are typically used. These columns, like packed GSC columns, are useful for separating permanent gases, as well as volatile hydrocarbons. The resolution efficiency of open-tubular columns is generally in the order: WCOT > SCOT > PLOT. Wide-bore (0.5-mm) open-tubular columns have been developed with thicker stationary liquid phases, up to 0.5 μm, that approach the capacity of SCOT and packed columns, but their resolution is decreased.

Columns can tolerate a limited amount of analyte before becoming overloaded, causing peak distortion and broadening, and shifts in retention time. Sample capacity ranges are from approximately 100 ng for a 0.25-mm-i.d. column with 0.25-μm-thick film, up to 5 μg for a 0.53-mm-i.d. column with a 5-μm-thick stationary phase.

Open-tubular columns, being the mainstay for gas-chromatographic separation, are manufactured by numerous companies. Some of the major ones include Alltech Associates, Inc. (www.alltechweb.com), Agilent Technologies (www.chem.agilent.com/Scripts/PHome.asp), Perkin-Elmer Instruments (http://instruments.perkinelmer.com), Quadrex Corp. (www.quadrexcorp.com), Restek Corp. (www.restekcorp.com), SGE, Inc. (www.sge.com), and Sigma-Aldrich (Supelco) (www.sigmaaldrich.com).

STATIONARY PHASES—THE KEY TO DIFFERENT SEPARATIONS

Over a thousand stationary phases have been proposed for gas chromatography, and numerous phases are commercially available. Hundreds of phases have been used for packed columns, necessitated by their low overall efficiency, and stationary-phase selection is critical for achieving selectivity. Several attempts have been made to predict the proper selection of liquid immobile phase without resorting exclusively to trial-and-error techniques (see below).

Phases are selected based on their polarity, keeping in mind that "like dissolve like." That is, a polar stationary phase will interact more with polar compounds, and vice versa. A phase should be selected in which the solute has some solubility. Nonpolar liquid phases are generally nonselective because there are few forces between the solute and the solvent, and so separations tend to follow the order of the boiling points of the solutes, with the low-boiling ones eluting first. Polar liquid phases exhibit several interactions with solutes such as dipole interactions, hydrogen bonds, and induction forces, and there is not necessarily the same elution correlation with volatility.

Liquid stationary phases are selected based on polarity, determined by the relative polarities of the solutes.

For fused silica columns, the majority of separations can be done with fewer than 10 bonded liquid stationary phases of varying polarity. This is because with their very high resolving power, and selectivity of the stationary phase is less critical. The stationary phases are high-molecular-weight, thermally stable polymers that are liquids or gums. The most common phases are polysiloxanes and polyethylene glycols (Carbowax), with the former the most widely used. The polysiloxanes have the backbone:

Polysiloxanes are the most common stationary phases for capillary GC.

$$\left[O - \underset{\underset{R_2}{|}}{\overset{\overset{R_1}{|}}{Si}} \right]_n$$

The R functional groups determine the polarity, and include methyl (CH_3), phenyl (⬡), cyanopropyl (CH_2CH_2CN), and trifluoropropyl ($CH_2CH_2F_3$). Table 20.1 lists several commonly used stationary phases. Those with cyano functions are susceptible to attack by water and by oxygen. The carbowaxes must be liquid at operating temperatures. Incorporating either phenyl or carborane groups in the

Table 20.1
Capillary Fused Silica Stationary Phases

Phase	Polarity	Use	Max. Temp. (°C)
100% Dimethyl polysiloxane $\left[O-\underset{CH_3}{\overset{CH_3}{Si}} \right]_n$	Nonpolar	Basic general-purpose phase for routine use. Hydrocarbons, polynuclear aromatics, PCBs.	320
Diphenyl, dimethyl polysiloxane $\left[O-Si \right]_{x\%} \left[O-\underset{CH_3}{\overset{CH_3}{Si}} \right]_{100-x\%}$	5% Low 35%, 65% Intermediate 65%, 35% Intermediate	General-purpose, good high-temperature characteristics. Pesticides.	320 300 370
14% Cyanopropylphenyl–86% dimethylsiloxane $\left[O-\underset{}{\overset{CN}{Si}} \right]_{14\%} \left[O-\underset{CH_3}{\overset{CH_3}{Si}} \right]_{86\%}$	Intermediate	Separation of organochlorine pesticides listed in EPA 608 and 8081 methods. Susceptible to damage by moisture and oxygen.	280
80% Biscyanopropyl–20% cyanopropylphenyl polysiloxane $\left[O-\underset{CN}{\overset{CN}{Si}} \right]_{80\%} \left[O-Si \right]_{20\%}$	Very polar	Free acids, polysaturated fatty acids, alcohols. Avoid polar solvents such as water and methanol.	275
Arylenes $\left[O-\underset{R_2}{\overset{R_1}{Si}}-\underset{R_4}{\overset{R_3}{Si}} \right]_n$	Vary R as above to vary polarity	High temperature, low bleed	300–350
Carboranes $CH_3-Si-CH_3$ ... $-Si-O$ (CH_3) open circles = boron filled circles = carbon	Vary R as above to vary polarity	High temperature, low bleed	430
Poly(ethyleneglycol) (Carbowax) $\left[O-CH_2CH_2 \right]_n$	Very polar	Alcohols, aldehydes, ketones, and separation of aromatic isomers, e.g., xylenes	250

siloxane polymer backbone strengthens and stiffens the polymer backbone, which inhibits stationary-phase degradation at higher temperatures, and results in lower column bleed (loss of stationary phase). These columns are important when coupling to a highly sensitive mass spectrometer for detection (see below), where bleeding must be minimized.

RETENTION INDICES FOR LIQUID STATIONARY PHASES

We mentioned above the challenge of selecting the proper packed-column stationary phase from the myriad of possible phases. Methods have been developed that group phases according to their retention properties, for example, according to polarity. The **Kovats indices** and **Rohrschneider constant** are two approaches used to group different materials. Supina and Rose (Ref. 8) have tabulated the Rohrschneider constants for 80 common liquid phases, which enables one to decide, almost by inspection, if it is worth trying a particular liquid phase. Equally important, it is easy to identify phases that are very similar and differ only in trade name. McReynolds described a similar approach, defining phases by their **McReynolds constants** (Ref. 6). McReynolds used a standard set of test compounds for measuring retention times at 120°C on columns with 20% loading to classify stationary phases.

Another useful literature reference for the selection of stationary phases is a booklet entitled *Guide to Stationary Phases for Gas Chromatography*, compiled by Analabs, Inc., North Haven, CT, 1977.

The **Kovats retention index** is useful also for identifying a compound from its retention time relative to those of similar compounds in a homologous series (those that differ in the number of carbon atoms in a similar structure, as in alkane chains). The index I is defined as

$$I = 100\left[n_s + \frac{\log t'_{R(\text{unk})} - \log t'_{R(n_s)}}{\log t'_{R(n_l)} - \log t'_{R(n_s)}}\right] \qquad (20.1)$$

where n_s is the number of carbon atoms in the smaller alkane, and n_l refers to the larger alkane; t'_R is the adjusted retention time (Equation 19.6). The Kovats index for an unknown compound can be compared with cataloged indices on various columns to aid in its identification. The logarithm of the retention time, $\log t'_R$, is generally a linear function of the number of carbon atoms in a homologous series of compounds.

WHAT ABOUT ANALYTE VOLATILITY?

In the above discussions, we have emphasized the role of the polarity of the stationary phase (and of the analyte) in providing effective separations. The other important factor is the relative volatility of the analyte species. The more volatile species will tend to migrate down the column more rapidly. Gaseous species, especially small molecules such as CO, will migrate rapidly. The retention factor, k (see Equation 19.25), is related to volatility by

$$\ln k = \Delta H_v / RT - \ln \gamma + C$$

where ΔH_v is the analyte heat of vaporization, so a higher value (higher boiling point) results in lower volatility and a larger k. Increasing the temperature T decreases this contribution to retention. The $\ln \gamma$ term is a function of the stationary-phase interaction (polarity, etc.), and is an activity term that decreases from unity for the pure state as interaction increases, causing k to increase; C is a constant (and R is the gas constant). Quite a bit of boiling point selectivity and separation tuning capability is provided by the T-dependent term in the equation. This is why people do temperature programming (see below).

So the selection of chromatographic conditions (column, temperature, carrier flow rate) will be influenced by the compound volatility, molecular weight, and polarity.

20.3　Gas Chromatography Detectors

Since the initial experiments with gas chromatography were begun, over 40 detectors have been developed. Some are designed to respond to most compounds in general, while others are designed to be selective for particular types of substances. We describe some of the more widely used detectors. Table 20.2 lists and compares some commonly used detectors with respect to application, sensitivity, and linearity.

Thermal conductivity detectors are very general detectors, but not very sensitive.

The original GC detector was the **thermal conductivity**, or **hot wire**, **detector** (TCD). As a gas is passed over a heated filament wire, the temperature and thus the resistance of the wire will vary according to the thermal conductivity of the gas. The pure carrier is passed over one filament, and the effluent gas containing the sample constituents is passed over another. These filaments are in opposite arms of a Wheatstone bridge circuit that measures the *difference* in their resistance. So long as there is no sample gas in the effluent, the resistance of the wires will be the same. But whenever a sample component is eluted with the carrier gas, a small resistance change will occur in the effluent arm. The change in the resistance, which is proportional to the concentration of the sample component in the carrier gas, is registered on the recorder. The TCD is particularly useful for the analysis of gaseous mixtures, and of permanent gases such as CO_2.

Hydrogen and helium carrier gases are preferred with thermal conductivity detectors because they have a very high thermal conductivity compared with most other gases, and so the largest change in the resistance occurs in the presence of sample component gases (helium is preferred for safety reasons). The thermal conductivity of hydrogen is 53.4×10^{-5} and that of helium is 41.6×10^{-5} cal/°C-mol at 100°C, while those of argon, nitrogen, carbon dioxide, and most organic vapors are typically one-tenth of these values. The advantages of thermal conductivity detectors are their simplicity and approximately equal response for most substances. Also, their response is very reproducible. They are not the most sensitive detectors, however.

The flame ionization detector is both general and sensitive. It is the most commonly used detector.

Most organic compounds form ions in a flame, generally cations such as CHO^+. This forms the basis of an extremely sensitive detector, the **flame ionization detector** (FID). The ions are measured (collected) by a pair of oppositely charged electrodes. The response (number of ions collected) depends on the number of carbon atoms in the sample and on the oxidation state of the carbon. Those atoms that are completely oxidized do not ionize, and the compounds with the greatest number of low oxidation state carbons produce the largest signals. This detector gives excellent sensitivity, permitting measurement of components in the ppb concentration range. This is about 1000 times more sensitive than the thermal conductivity detector. However, the dynamic range is more limited, and samples of pure liquids are generally restricted to 0.1 μL or less. The carrier gas is relatively unimportant. Helium, nitrogen, and argon are most frequently employed. The flame ionization detector is insensitive to most inorganic compounds, including

Table 20.2
Comparison of Gas-Chromatographic Detectors

Detector	Application	Sensitivity Range	Linearity	Remarks
Thermal conductivity	General, responds to all substances	Fair, 5–100 ng, 10 ppm–100%	Good, except thermistors at higher temperatures	Sensitive to temperature and flow changes; concentration sensitive
Flame ionization	All organic substances; some oxygenated products respond poorly. Good for hydrocarbons	Very good, 10–100 pg, 10 ppb–99%	Excellent, up to 10^6	Requires very stable gas flow; response for water is 10^4–10^6 times weaker than for hydrocarbons; mass-sensitive
Flame photometric	Sulfur compounds (393 nm), phosphorus compounds (526 nm)	Very good, 10 pg S, 1 pg P	Excellent	
Flame thermionic	All nitrogen- and phosphorus-containing substances	Excellent, 0.1–10 pg, 100 ppt–0.1%	Excellent	Needs recoating of sodium salts on screen; mass sensitive
Rubidium silicate bead	Specific for nitrogen- and phosphorus-containing substances	Excellent		Mass sensitive
Argon ionization (β-ray)	All organic substances; with ultrapure He carrier gas, also for inorganic and permanent gases	Very good; 0.1–100 ng, 0.1–100 ppm	Good	Very sensitive to impurities and water; needs very pure carrier gas; concentration sensitive
Electron capture	All substances that have affinity to capture electrons; no resonse for aliphatic and naphthenic hydrocarbons	Excellent for halogen-containing substancs, 0.05–1 pg, 50 ppt–1 ppm	Poor	Very sensitive to impurities and temperature changes; quantitative analysis complicated; concentration sensitive
Mass spectrometry	Nearly all substances. Depends on ionization method	Excellent	Excellent	Can provide structural and molecular weight information

water, and so aqueous solutions can be injected. If oxygen is used as the flame support gas in place of air, then many inorganic compounds can be detected because a hotter flame is produced that can ionize them. Since the flame ionization detector is so sensitive, a portion of the sample can be diverted by an appropriate stream splitter so that it can be collected and analyzed further if necessary.

When sulfur and phosphorus compounds are burned in an FID-type flame, chemiluminescent species are produced that produce light at 393 nm (sulfur) and 526 nm (phosphorous). An optical interference filter passes the appropriate light to a photomultiplier tube, a sensitive photon detector. These detectors are known as flame photometric detectors (FPD).

The **flame thermionic detector** is essentially a two-stage flame ionization detector designed to give an increased specific response for nitrogen- and phosphorus-containing substances. A second flame ionization detector is mounted above the first, with the flame gases from the first passing into the second flame. The two stages are divided by a wire mesh screen coated with an alkali salt or base such as sodium hydroxide.

The column effluent enters the lower flame, which acts as a conventional FID whose response may be recorded. A small current normally flows in the second flame due to evaporation and ionization of sodium from the screen. However, if a substance containing nitrogen or phosphorus is burned in the lower flame, the ions resulting from these greatly increase the volatilization of the alkali metal from the screen. This results in a response that is much greater (at least 100 times) than the response of the lower flame to the nitrogen or phosphorus. By recording the signals from both flames, one can obtain the usual chromatogram of a FID; a second chromatogram is obtained where the peaks corresponding to the nitrogen- and phosphorus-containing compounds are amplified over the others, which will be practically missing. This detector is also known as a nitrogen–phosphorous detector (NPD).

In the **β-ray**, or **argon ionization, detector**, the sample is ionized by bombardment with β rays from a radioactive source (e.g., strontium-90). The carrier gas is argon, and the argon is excited to a metastable state by the β particles. Argon has an excitation energy of 11.5 eV, which is greater than the ionization potential of most organic compounds, and the sample molecules are ionized when they collide with the excited argon atoms. The ions are detected in the same manner as in the flame ionization detector. This detector is very sensitive but less accurate than others, and the β-ray source is a potential hazard, although with proper shielding, no danger exists. The sensitivity is about 300 times greater than that of the conventional thermal conductivity cell.

The **electron capture detector** (ECD) is extremely sensitive for compounds that contain electronegative atoms and is selective for these. It is similar in design to the β-ray detector, except that nitrogen or methane doped with argon is used as the carrier gas. These gases have low excitation energies compared to argon and only compounds that have high electron affinity are ionized, by capturing electrons.

The detector cathode consists of a metal foil impregnated with a β-emitting element, usually tritium or nickel-63. The former isotope gives greater sensitivity than the latter, but it has an upper temperature limit of 220°C because of losses of tritium at high temperatures; nickel-63 can be used routinely at temperatures up to 350°C. Also, nickel is easier to clean than the tritium source; these radioactive sources inevitably acquire a surface film that decreases the β-emission intensity and hence the sensitivity. A 30% KOH solution is usually used to clean the sources.

The cell is normally polarized with an applied potential, and electrons (β rays) emitted from the source at the cathode strike gas molecules, causing electrons to be released. The resulting cascade of thermal electrons is attracted to the anode, and establishes a standing current. When a compound possessing electron

The ECD is very sensitive for halogen-containing compounds, for example, pesticides.

affinity is introduced into the cell, it captures electrons to create a large negative ion. The negative ion has a mobility in an electric field about 100,000 less than electrons, and so a decrease in current results.

Relatively few compounds show significant electronegativities, and so electron capture is quite selective, allowing the determination of trace constituents in the presence of noncapturing substances. High-electron-affinity atoms or groups include halogens, carbonyls, nitro groups, certain condensed ring aromatics, and certain metals. The ECD is widely used for pesticides and polychlorinated biphenyls (PCBs). Electron capture has very low sensitivity for hydrocarbons other than aromatics.

Compounds with low electron affinities may be determined by preparing appropriate derivatives. Most important biological compounds, for example, possesses low electron affinities. Steroids such as cholesterol can be determined by preparing their chloroacetate derivatives. Trace elements have been determined at nanogram and picogram levels by preparing volatile trifluoroacetylacetone chelates. Examples are chromium, aluminum, copper, and beryllium. Methylmercuric chloride, present in contaminated fish, can be determined at the nanogram level.

The gas chromatograph may be interfaced with atomic spectroscopic instruments for specific element detection. This powerful combination is useful for speciation of different forms of toxic elements in the environment. For example, a helium microwave induced plasma atomic emission detector (AED) has been used to detect volatile methyl and ethyl derivatives of mercury in fish, separated by GC. Also, gas chromatographs are interfaced to inductively coupled plasma–mass spectrometers (ICP–MS) in which atomic isotopic species from the plasma are introduced into a mass spectrometer (see Section 20.10 for a description of mass spectrometry), for very sensitive simultaneous detection of species of several elements.

Detectors are either concentration sensitive or mass flow sensitive. The signal from a concentration-sensitive detector is related to the concentration of the solute in the detector and is decreased by dilution with a makeup gas. The sample is usually not destroyed. Thermal conductivity, argon-ionization, and electron capture detectors are concentration sensitive. In mass-flow-sensitive detectors, the signal is related to the rate at which solute molecules enter the detector and is not affected by the makeup gas. These detectors usually destroy the sample, such as flame ionization and flame thermionic detectors. Sometimes two-column GC is used to increase resolution, by taking cuts of eluents from an initial column and directing them to a second column for secondary separation. The first detector must be nondestructive or else the eluent split prior to detection, with a portion going to the second column.

20.4 Temperature Selection

The proper temperature selection in gas chromatography is a compromise between several factors. The **injection temperature** should be relatively high, consistent with thermal stability of the sample, to give the fastest rate of vaporization to get the sample into the column in a small volume; decreased spreading and increased resolution result. Too high an injection temperature, though, will tend to degrade the rubber septum and cause dirtying of the injection port. The **column temperature** is a compromise between *speed*, *sensitivity*, and *resolution*. At high column temperatures, the sample components spend most of their time in the gas phase and so they are eluted quickly, but resolution is poor. At low temperatures, they spend more time in the stationary phase and elute slowly; resolution is increased but sensitivity is decreased due to increased spreading of the peaks. The **detector**

Chromatographic conditions represent a compromise between speed, resolution, and sensitivity.

Temperature programming from low to higher temperatures speeds up separations. The more difficult to elute solutes are made to elute faster at the higher temperatures. The more easily eluted ones are better resolved at the lower temperatures.

temperature must be high enough to prevent condensation of the sample components. The sensitivity of the thermal conductivity detector decreases as the temperature is increased and so its temperature is kept at the minimum required.

Separations can be facilitated by **temperature programming**, and most gas chromatographs have temperature programming capabilities. The temperature is automatically increased at a preselected rate during the running of the chromatogram; this may be linear, exponential, steplike, and so on. In this way, the compounds eluted with more difficulty can be eluted in a reasonable time without forcing the others from the column too quickly.

Figure 20.6 shows a temperature programmed separation of a complex hydrocarbon mixture with stepwise linear temperature programming. The first 12 gaseous or light compounds are readily eluted and resolved at a low fixed (100°C) temperature for 5.5 min, while the others require higher temperatures. After 5.5 min, the temperature is linearly increased at 5°C/min for 20 min to 200°C, and then the temperature is held at that value until the last two compounds are eluted.

If the constituent to be determined is not volatile at the accessible temperatures, it may be converted to a **volatile derivative**. For example, nonvolatile fatty acids are converted to their volatile methyl esters. Some inorganic halides are sufficiently volatile that at high temperatures they can be determined by gas chromatography. Metals may be made volatile by complexation, for example, with trifluoroacetylacetone.

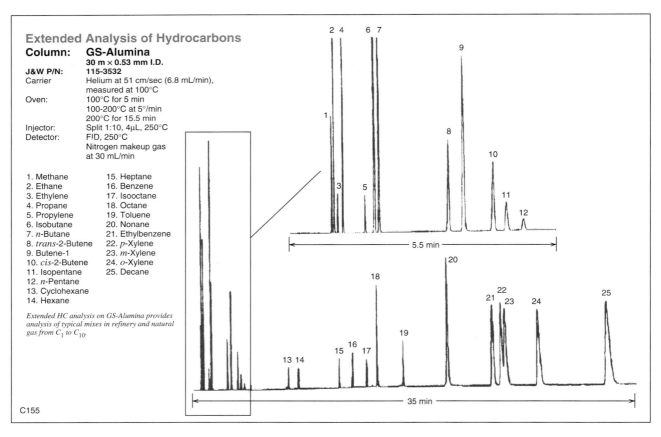

Fig. 20.6. Temperature-programmed analysis. (Courtesy of Agilent Technologies.)

20.5 Quantitative Measurements

The concentrations of eluted solutes are proportional to the areas under the recorded peaks. Electronic integrations in GC instruments print out the areas of peaks, and the retention times of peaks are also generally printed. It is also possible to measure peak height to construct a calibration curve. The linearity of a calibration curve should always be established.

The method of **standard additions** is a useful technique for calibrating, especially for occasional samples. One or more aliquots of the sample are spiked with a known concentration of standard, and the increase in peak area is proportional to the added standard. This method has the advantage of verifying that the retention time of the unknown analyte is the same as that of the standard.

A more important method of quantitative analysis is the use of **internal standards**. Here, the sample and standards are spiked with an equal amount of a solute whose retention time is near that of the analyte. The ratio of the area of the standard or analyte to that of the internal standard is used to prepare the calibration curve and determine the unknown concentration. This method compensates for variations in physical parameters, especially inaccuracies in pipetting and injecting microliter volumes of samples. Also, the *relative* retention should remain constant, even if the flow rate should vary somewhat.

An internal standard is usually added to standard and sample solutions. The ratio of the analyte peak area to internal standard peak area is measured and will remain unaffected by slight variations in injected volume and chromatographic conditions.

SPREADSHEET EXERCISE: INTERNAL STANDARD CALIBRATION

When a driver is arrested for suspicion of driving under the influence of alcohol, the blood alcohol content is determined to see if it exceeds the legal limit. The measurement of breath alcohol is usually done for routine driving arrests because it is noninvasive, and a factor is applied to convert to blood alcohol concentration (BAC). But this is subject to biological variations in individuals. In cases where there is an accident, injury, or death, the blood alcohol is usually determined directly, by analyzing a blood sample by gas chromatography.

A 5.00-mL blood sample from a suspect is spiked with 0.500 mL of aqueous 1% propanol internal standard. A 10-μL portion of the mixture is injected into the GC, and the peak areas are recorded. Standards are treated in the same way. The following results were obtained:

% EtOH (wt/vol)	Peak Area EtOH	Peak Area PrOH
0.020	114	457
0.050	278	449
0.100	561	471
0.150	845	453
0.200	1070	447
Unknown	782	455

Prepare a spreadsheet to construct a calibration curve of the ratio of the EtOH/PrOH areas vs. EtOH concentration, and to calculate the unknown concentration and its standard deviation. Refer to Equations 16.18 to 16.21 and the spreadsheet that follows in Chapter 16 for a refresher on the statistical calculations.

	A	B	C	D	E	F	G	H
1	GC Determination of blood ethanol using propanol internal standard.							
2	%EtOH	Peak area	Peak area	Ratio peak area				
3		EtOH	PrOH	EtOH/PrOH				
4	0.020	114	457	**0.249453**				
5	0.050	278	449	0.619154				
6	0.100	561	471	1.191083				
7	0.150	845	453	1.865342				
8	0.200	1070	447	2.393736				
9	Unknown (y_c)	782	455	1.718681				
10								
11	Slope, m:		**12.0286923**					
12	Intercept, b:		0.01276953					
13	**Unknown % BAC, C:**		0.14182022					
14								
15	S_r:	**0.0345743**						
16	N:	**5**						
17	S_{xx}:	**0.02132**						
18	y_{ave}:	**1.2637535**						
19	M:	**1**						
20	**S_c:**	**0.0033365**						
21								
22	**Cell D4** = ratio =B4/C4 (copy down)							
23	**Cell C11** = m = SLOPE(D4:D8,A4:A8)							
24	**Cell C12** = b = INTERCEPT(D4:D8,A4:A8)							
25	**Cell C13** = C = (y_c-b)/m = (Ratio$_c$ - b)/m = (D9-C12)/C11							
26	**Cell B15** = S_r = STEYX(D4:D8,A4:A8)							
27	**Cell B16** = N = COUNT(A4:A8)							
28	**Cell B17** = S_{xx} = N*VARP(A4:A8) = B16*VARP(A4:A8)							
29	**Cell B18** = y_{ave} = AVERAGE(D4:D8)							
30	**Cell B19** = sample replicates = M =1							
31	**Cell B20** = std. devn. S_c = $S_r/m((1/M + 1/N + (y_c - y_{ave})/m^2 S_{xx}))^{1/2}$							
32		= B15/C11*((1/B19+1/B16+(D9-B18)/C11^2*B17))^0.5							

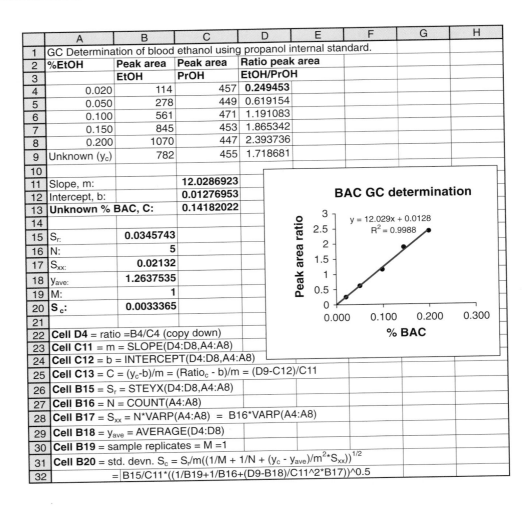

The blood alcohol concentration is 0.142 ± 0.003%, a high level. The legal limit for driving in most states is 0.08% blood alcohol (wt/vol). The influence on the ability to drive is exponential with alcohol concentration and it roughly doubles for every 0.05% BAC, so someone with 0.20% blood alcohol is 8 times as drunk as one with 0.05%!

20.6 Headspace Analysis

Headspace analysis avoids the need for solvent extraction for volatile analytes.

In Chapter 18, we described solvent extraction and solid-phase extraction sample preparation methods, which are applicable to GC analyses as well as others. A convenient way of sampling volatile samples for GC analysis is the technique of **headspace analysis**. A sample in a sealed vial is equilibrated at a fixed temperature, for example, for 10 min, and the vapor in equilibrium above the sample is sampled and injected into the gas chromatograph. A typical 20-mL glass vial is capped with a silicone rubber septum lined with polytetrafluoroethylene (PTFE). A syringe needle can be inserted to withdraw a 1-mL portion. Or the pressurized vapor is allowed to expand into a 1-mL sample loop at atmospheric pressure, and then an auxiliary carrier gas carries the loop contents to the GC loop injector. Volatile compounds in solid or liquid samples can be determined at parts per million or less. Pharmaceutical tablets can be dissolved in a water–sodium sulfate solution for headspace analysis. Figure 1 in your CD, Chapter 20, shows a headspace chromatogram of volatile compounds in a blood sample.

20.7 Thermal Desorption

Thermal desorption (TD) is a technique in which solid or semisolid samples are heated under a flow of inert gas. Volatile and semivolatile organic compounds are extracted from the sample matrix into the gas stream and introduced into a gas chromatograph. Samples are typically weighed into a replaceable PTFE tube liner, which is inserted into a stainless steel tube for heating.

In thermal desorption, the volatile analyte is desorbed from the sample by heating and introduced directly into the GC.

The thermal desorption must take place at a temperature below the decomposition point of other materials in the sample matrix. Solid materials should have a high surface area (e.g., powders, granules, fibers). Bulk materials are ground with a coolant such as solid carbon dioxide prior to weighing. This technique simplifies sample preparation and avoids the necessity of dissolving samples or solvent extraction. Thermal desorption is well suited for dry or homogeneous samples such as polymers, waxes, powders, pharmaceutical preparations, solid foods, cosmetics, ointments, and creams. There is essentially no sample preparation required.

An example of the use of TD is for the analysis of water-based paints for organic volatiles. The TD tube is used in combination with a second tube containing a sorbent that removes the water, which cannot be introduced into the capillary GC column. A small aliquot of paint (e.g., 5 µL) is placed on glass wool in the TD tube. Solids from the paint, which would harm a GC column, remain behind.

See www.markes.com for literature on sorbent selection for thermal desorption.

20.8 Purging and Trapping

The **pure-and-trap** technique is a variation of thermal desorption analysis in which volatiles are purged from a liquid sample placed in a vessel by bubbling a gas (e.g., air) through the sample and collecting the volatiles in a sorbent tube containing a suitable sorbent. The trapped volatiles are then analyzed by thermally desorbing them from the sorbent. This is a form of "headspace" analysis in which analytes are concentrated prior to introduction into the GC. A typical sorbent is a hydrophobic one that can collect organics ranging in volatility from hexane up to C16 or greater. Examples are Tenax TA or graphitized carbon. These sorbents allow bulk polar solvents such as water or ethanol to pass through unretained. Whiskey is analyzed for C4 to C6 ethyl esters, which are markers of maturity, and the alcohol does not interfere in the chromatogram.

Purge-and-trap is a form of headspace analysis in which the volatile analyte is trapped on a sorbent and then thermally desorbed.

Purge-and-trap is suitable for nonhomogeneous samples, since fairly large samples can be taken, and high-humidity samples. Examples include foods such as pizza or fruits. The measurement of malodorous organic volatiles in the headspace vapor above a sample of aged food is used to determine whether it still meets the "freshness" requirements. The food sample, placed in a large purge vessel, is heated under a flow of air, and the effluent air is collected on the sorbent.

Sorbents more selective than Tenax-type sorbents may be used, and two or more may be used in series for measurement of different classes of compounds. Other sorbents include many materials used for chromatography such as alumina, silica gel, Florisil (for PCBs), coconut charcoal, Poropak, and Chromosorb. Some may be coated for specific applications, for example, silica gel coated with sulfuric acid or sodium hydroxide for collecting bases or acids.

Another important use of trapping is for the direct analysis of gaseous samples such as air. The sample is passed directly through the sorbent tube, and the trapped volatiles are subsequently desorbed or removed by extraction. This technique is finding widespread use for indoor and outdoor air monitoring.

For more information about the sorbents and thermal desorption, see SKC (www.skcinc.com) and Markes International (www.markes.com). SKC lists sorbents for specific applications (analytes).

20.9 Small and Fast

Thin-film stationary phases, small diameter and short columns, and a light carrier gas gives fast separations.

The advent of automated, computerized data handling has made possible the design of chromatographic columns with thin liquid-phase films that allow very rapid analyses, for example, of a few seconds. Small diameter and short capillary columns are used with hydrogen as the carrier gas (which allows more rapid mass transfer) and fast temperature program rates. Faster flow rates and higher pressures are used. Fast elution of analytes requires fast detector response time and data acquisition rates because peak widths are on the order of only 0.5 s, compared to 0.5 to 2 s or more for conventional capillary GC, improving analysis times 5- to 10-fold. High-speed chromatography with short columns makes column selection (selectivity) more critical than in conventional capillary GC. Figure 20.7 shows a fast GC chromatographic separation of hydrocarbons in less than 10 s using a 0.32-mm-diameter column of 5 m length, and 0.25-μm-thick stationary phase, compared to 10 min using conventional GC.

Shorter columns and faster temperature programming causes some loss in resolution, but this is partially offset by the smaller internal diameter and thinner liquid film.

There are small portable GCs available from a number of manufacturers. There are limited applications, with isothermal ovens, but they perform well for specific applications. [See C. Henry, "Taking the Show on the Road. Portable GC and GC/MS" (Product Review), *Anal. Chem.*, **69**(5) (1997) 195A].

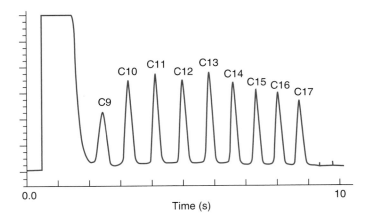

Fig. 20.7. Separation of C9 to C17 hydrocarbons by fast chromatography, with rapid temperature programming. Conditions: 5 m × 0.32 mm, 0.25 μm d_f, 60°C, 19.2°C/s. [From G. L. Reed, K. Clark-Baker, and H. M. McNair. *J. Chromatogr. Sci.*, **37** (1999) 300. Reproduced from the *Journal of Chromatographic Science* by permission of Preston Publications, A Division of Preston Industries, Inc.]

20.10 Gas Chromatography–Mass Spectrometry

The appearance of a chromatographic peak at a particular retention time suggests but does not guarantee the presence of a particular compound. The probability of positive identification will depend on factors such as the type and complexity of the sample and sample preparation procedures employed. A gas chromatogram of an injected blood sample diluted with a solution of an internal standard (to verify retention time and relative peak area) that gives a large peak expected for alcohol strongly suggests the presence of blood alcohol since there are few nontoxic compounds that would likely interfere. Usually, there is indication of alcohol ingestion, and the key legal question is what is the concentration? However, the appearance of a GC peak for cocaine may not be so straightforward in confirming the presence of this drug. Hence, confirmatory evidence is usually sought. Spectral information, such as infrared or ultraviolet spectrometry, may be sought. A very powerful tool is the combination of gas chromatography with mass spectrometry, a technique known as **gas chromatography–mass spectrometry** (GC–MS).

GC–MS is very powerful for positive identification.

GC–MS systems used to fill a room and cost several hundred thousand dollars. Today, relatively inexpensive compact benchtop systems are available and widely used in laboratories. A modern GC–MS instrument is shown in Figure 20.8. We describe first the principles of mass spectrometers and types of instruments, and then discuss how the two techniques of gas chromatography and mass spectrometry are used together.

PRINCIPLES OF MASS SPECTROMETRY

Mass spectrometry is a sophisticated instrumental technique that produces, separates, and detects ions in the gas phase. The basic components of a mass spectrometer are shown in Figure 20.9. A sample with a moderately high vapor pressure is introduced in an inlet system, operated under vacuum (10^{-4} to 10^{-7} torr) and at high temperature (up to 300°C). It vaporizes and is carried to the ionization source. Nonvolatile compounds may be vaporized by means of a spark or other

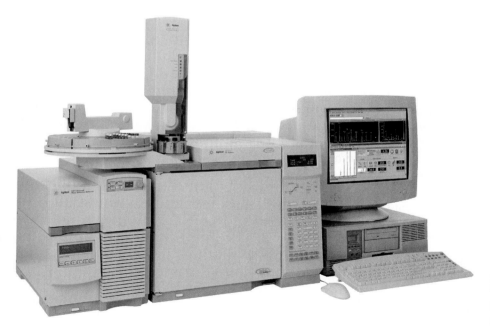

Fig. 20.8. Gas chromatography–mass spectrometry benchtop system. (Courtesy of Agilent Technologies.)

Fig. 20.9. Block diagram of mass spectrometer.

source. Analyte molecules are typically neutral and must be ionized. This is accomplished by various means but typically is done by bombarding the sample with high-energy electrons in an electron-impact source. We describe common ionization sources below.

The ions are separated in the spectrometer by being accelerated through a mass separator described below. Separation actually accomplished based on the mass-to-charge (m/e) ratios of the ions. We will not go into the many useful rules for deducing structural information from mass spectra. But one rule in particular is useful to know for identifying the molecular ion, the **nitrogen rule**. In applying this, M^+ is taken as the highest mass, ignoring isotope contributions. Such a molecular mass will be an even number if it contains an even number (0, 2, 4 . . .) of nitrogen atoms, and will otherwise be an odd number, that is, when it contains an odd number of nitrogens. So if there is no nitrogen, it will be even. There should also be no illogical losses on fragmentation. For example, organic molecules seldom lose more than four H atoms, to give M-4 fragments. Other losses we would expect to see are methyl groups (M-15), NH_2 or O (M-16), OH or NH_3 (M-17), H_2O (M-18), F (M-19), HF (M-20), and C_2H_2 (M-26). So there should be no losses of 4 to 14 or 21 to 25 mass units. If so, the assignment is incorrect or we have a mixture spectrum.

> The nitrogen rule tells you whether you have an even or an odd formula weight compound.

RESOLUTION

> Resolution tells you how accurately you can differentiate two masses. Unit resolution tells you that you can differentiate one mass difference.

In mass spectrometry, the resolving power, that is, the ability to differentiate two masses, is given by the resolution R, defined as the nominal mass divided by the difference between two masses that can be separated:

$$R = \frac{m}{\Delta m} \tag{20.2}$$

where Δm = mass difference between two resolved peaks
m = nominal mass at which peak occurs

The mass difference is usually measured at the mean of some fixed fraction of the peak heights, for example, at 10%. A resolution of 1000 means that a molecule of $m/z = 1000$ would be resolved from $m/z = 1001$ (or $m/z = 10.00$ from 10.01); or $m/z = 500.0$ is resolved from $m/z = 500.5$.

The term **unit resolution** is sometimes used to indicate the ability to differentiate next integer masses. It allows you to distinguish m/z 50 from m/z 51, or m/z 100 from m/z 101, or m/z 500 from m/z 501. Obviously, the higher the molecular weight the better R must be to achieve unit resolution. Resolutions of 500 are suitable for most GC applications.

IONIZATION SOURCES

> Electron-impact ionization produces many fragments.

Table 20.3 lists the common ionization sources used for GC–MS, as well as LC–MS (Chapter 21). The most commonly used ionization source is the **electron-impact** (EI) source. The gaseous molecules are bombarded by a high-energy beam of

Table 20.3

Comparison of Ionization Methods[a]

Ionization Method	Typical Analytes	Sample Introduction	Mass Range	Method Highlights
Electron impact (EI)	Relatively small volatile	GC or liquid–solid probe	To 1000 daltons	Hard method. Versatile, provides structure information
Chemical ionization (CI)	Relatively small, volatile	GC or liquid–solid probe	To 1000 daltons	Soft method. Molecular ion peak [M + H]$^+$
Electrospray (ESI)	Peptides, proteins, nonvolatile	Liquid chromatography or syringe	To 200,000 daltons	Soft method. Ions often multiply charged
Matrix-assisted laser desorption (MALDI)	Peptides, proteins, nucleotides	Sample mixed in solid matrix	To 500,000 daltons	Soft method, very high mass

[a]From Web page of Professor Vicki Wysocki, University of Arizona. Reproduced by permission.

electrons, usually 70 eV, generated from a tungsten filament. An electron that collides with a neutral molecule may impart sufficient energy to remove an electron from the molecule, resulting in a singly charged ion:

$$M + e^- \rightarrow M^+ + 2e^-$$

where M is the analyte molecule and M^+ is the **molecular ion** or **parent ion**. The M^+ ions are produced in different energy states and the internal energy (rotational, vibrational, and electronic) is dissipated by fragmentation reactions, producing fragments of lower mass, which are themselves ionized or converted to ions by further electron bombardment. The fragmentation pattern is fairly consistent for given conditions (electron beam energy). Only a small amount or none of the molecular ion may remain. If it does appear, it will be the highest mass in an EI spectrum, if there are not multiple isotopes. Compounds with aromatic rings, cyclic structures, or double bonds are more likely to give molecular ion peaks because of delocalization effects that decrease fragmentation. Figure 20.10 shows a simple EI spectrum of a small molecule, methanol. Usually, peaks are normalized to the one with the greatest abundance (relative abundance 100%); the largest peak is called the **base peak**. The molecular peak is at $m/z = 32$, the formula weight of CH_3OH. Note the small peak at $m/z = 33$. This is the molecular ion for the ^{13}C isotope of methanol, which has a relative abundance of 1.11% compared to ^{12}C at 100%. The base peak at $m/z = 31$ is from the CH_2OH^+ fragment. Table 20.4 lists the relative abundances of isotopes of some common elements.

CHEMICAL IONIZATION SOURCE

The EI source is what is called a "hard source" and may produce too much fragmentation to allow positive identification of the analyte, and no molecular ion may be present. Consecutive ion fragmentation may result in low-mass ions carrying the large share of the total ion intensity, and the more analytically important primary fragments are of low abundance or missing. **Chemical ionization** (CI) is a "softer" technique that does not cause much fragmentation, and the molecular ion is the dominant one in CI mass spectra. In CI, a reagent gas such as methane, isobutane, or ammonia is introduced into the EI ionization chamber at a high pressure (large excess, 1 to 10 torr) to react with analyte molecules to form ions by

Chemical ionization produces the molecular ion.

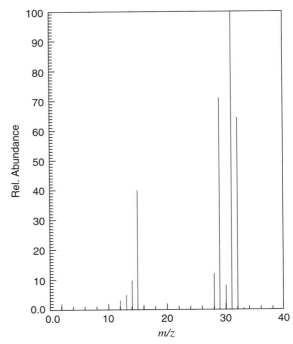

Fig. 20.10. Electron ionization mass spectrum of methanol. From NIST MassSpec Data Center. Reproduced by permission.

either a proton or hydride transfer. The chemical ionization process begins by ionization of the reagent gas. With methane, electron collisions produce CH_4^+ and CH_3^+, which further react with CH_4 to form CH_5^+ and $C_2H_5^+$:

$$CH_4^+ + CH_4 \rightarrow CH_5^+ + CH_3$$
$$CH_3^+ + CH_4 \rightarrow C_2H_5^+ + H_2$$

Table 20.4
Relative Abundances and Exact Masses of Some Common Elements.[a]

Element	Isotope	Mass	Relative Abundance (%)
Hydrogen	^1H	1.007825	100.0
	^2H	2.014102	0.0115
Carbon	^{12}C	12.000000	100.0
	^{13}C	13.003355	1.07
Nitrogen	^{14}N	14.003074	100.0
	^{15}N	15.000109	0.369
Oxygen	^{16}O	15.994915	100.0
	^{17}O	16.999132	0.038
	^{18}O	17.999160	0.205
Sulfur	^{32}S	31.972071	100.0
	^{33}S	32.971450	0.803
	^{34}S	33.967867	4.522
Chlorine	^{35}Cl	34.968852	100.0
	^{37}Cl	36.965903	31.96
Bromine	^{79}Br	78.918338	100.0
	^{81}Br	80.916291	97.28

[a] The most abundant isotope is assigned an abundance of 100%, and the others are listed relative to it.

These react with the sample by transferring a proton (H^+) or by extracting a hydride (H^-) or electron, which imparts a $+1$ charge on the sample molecule:

$$CH_5^+ + MH \rightarrow MH_2^+ + CH_4$$
$$C_2H_5^+ + MH \rightarrow M^+ + C_2H_6$$

MH_2^+ and M^+ may fragment to give the mass spectrum. No M^+ ion may be observed, but the molecular weight is readily obtained from the $M + H$ or $M - H$ ions formed. Weaker acid gas-phase ions further simplify spectra. $C_4H_9^+$ from isobutane and NH_4^+ from ammonia also ionize by proton transfer, but with less energy, and fragmentation of MH_2^+ is minimal. Table 20.5 lists the CI characteristics of different reagents. CI is almost universally available on modern GC–MS instruments for determining molecular weights of eluting compounds otherwise difficult to obtain due to extreme fragmentation with "harder" sources.

Other ion sources are described in Chapters 21 and 25.

MASS ANALYZERS

Mass spectrometers based on magnetic sectors are widely used by organic chemists to determine molecular structure. They deflect ions down a curved tube in a magnetic field based on their kinetic energy determined by the mass, charge, and velocity. The magnetic field is scanned to measure different ions. This mass separator is very powerful and capable of very high resolution, but the instruments are quite large and expensive and not very suitable for use with gas chromatographs. Most GC–MS instruments today are benchtop systems that use more compact, inexpensive mass analyzers with lower resolution. The availability of these is the reason GC–MS is so widely used.

1. Quadrupole Mass Filter. The quadrupole mass analyzer is a "mass filter" that allows only specific ions to pass. Figure 20.11 shows the basic design of the quadrupole analyzer. It consists of four parallel metal rods to which both a dc voltage (U) and an oscillating radiofrequency voltage ($V \cos \omega t$, where ω is the frequency and t the time) is applied. Two opposite poles are positively charged and the other two negatively charged, and their polarities change throughout the experiment. The applied voltages are $U + V \cos \omega t$ and $-(U + V \cos \omega t)$. The applied voltages determine the trajectory of the ions down the flight path between the four poles. As ions from the ionization source enter the RF field along the z axis of the electrodes, they oscillate along the z axis. Only those with a specific mass-to-charge ratio will resonate along the field and have a stable path through to the detector. Others (nonresonating) will be deflected (unstable path) and collide

The quadrupole analyzer is the most commonly used for GC–MS.

Table 20.5
Chemical Ionization Characteristics of Different CI Reagents

Reagent	Adducts Produced	Uses/Limitations
Methane	$M–H^+$, $M–CH_3^+$	Most organic compounds. Adducts not always abundant. Extreme fragmentation.
Isobutane	$M–H^+$, $M–C_4H_9^+$	Less universal. Adducts more abundant. Some fragmentation.
Ammonia	$M–H^+$ (basic compounds) $M–NH_4^+$ (polar compounds)	Polar and basic compounds. Others not ionized. Virtually no fragmentation.

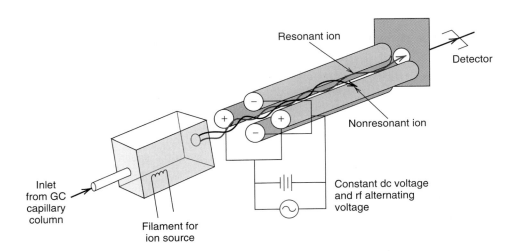

Fig. 20.11. Quadrupole mass spectrometer.

with the electrodes and be lost (they are filtered out). By rapidly varying the voltages, ions of one mass after another will take the stable path and be collected by the detector. Either ω is varied while holding U and V constant or U and V are varied, keeping U/V constant.

The quadrupole analyzer has a number of advantages that make it ideally suited for GC–MS. The path does not depend on the kinetic energy (e.g., velocity) or angular deflection of the incoming ions. So the transmission rate is high. Since only a change in voltage is required, a complete scan can be very fast. As many as eight spectra per second can be recorded over a range of about 800 mass units. Rapid scanning is needed to monitor GC peaks that may be a fraction of a second wide. A resolution of about 1500 can be achieved, and GC systems usually provide unit resolution. Finally, quadrupole instruments are relatively compact and inexpensive.

The TOF analyzer is good for large molecules.

2. Time-of-Flight Analyzer. The time-of-flight (TOF) analyzer is the second most popular mass analyzer for GC–MS. Figure 20.12 shows the basic construction of a TOF mass spectrometer. The ions formed in the ionization chamber are accelerated through the accelerator plates, which have a voltage of 3000 V, pulsating at 3000 to 20,000 times per second, and they enter the drift or flight tube

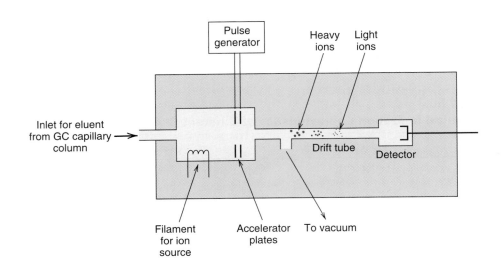

Fig. 20.12. Time-of-flight mass spectrometer.

with constant kinetic energy. Ions of different m/z travel at different velocities. The kinetic energy of ions leaving the source is given by:

$$\frac{mv^2}{2} = Vq \tag{20.3}$$

where m = mass of ion
        v = velocity of ion
        V = accelerating voltage
        q = ion charge

Rearranging,

$$v = \sqrt{\frac{2Vq}{m}} \tag{20.4}$$

So the ion velocity is inversely proportional to the square root of the mass. The time t to reach the detector is $t = L/v$, where L is the length of the flight tube. The difference, Δt, in arrival time that separates two ions is

$$\Delta t = L\,\frac{(m_1)^{1/2} - (m_2)^{1/2}}{(2Vq)^{1/2}} \tag{20.5}$$

depending, then, on the square root of the masses.

Pulsed-mode acceleration is needed since continuous ionization and acceleration would lead to a continuous stream of all ions with overlapping masses. The sequence of events for pulsed operation is to turn on the electron source for 10^{-9} s to form a packet of ions, and then the accelerating voltage for 10^{-4} s to draw the ions into the drift tube. Then the power is turned off for the remainder of the pulse interval as the ion packets drift down the tube to the detector.

TOF analyzers, like quadrupoles, scan the mass spectrum rapidly. Resolutions of 500 can be obtained. These analyzers are popular for high mass ion detection since they have no real upper mass limit.

MASS SPECTROMETRY WITH GAS CHROMATOGRAPHY

One of the biggest problems in the early development of GC–MS was interfacing the column outlet to the mass spectrometer. Packed columns were used, and the high volumes of both sample and carrier gas overwhelmed the MS system, which operates under low pressure, and special interfaces had to be built. The advent of fused silica capillary columns meant that the GC–MS interface could be dispensed with, and the column eluent is introduced directly into the ion source. It is essential that column bleeding be minimized since the mass spectrometer will detect the stationary-phase materials. Bleeding is prevented by chemically bonding alkylsiloxanes to the column wall. Other low bleeding stationary phases are mentioned above.

Of course, monitoring GC peaks (several times per second) generates enormous amounts of data. The evolution of fast and high-capacity computers is the other technology advancement that made GC–MS routine.

The separated ions are detected by means of an electron multiplier, which is similar in design to photomultiplier tubes described in Chapter 16. Detection sensitivities at the nanogram level are common.

The mass spectrometer may be operated in various modes. In the **total ion current** (TIC) monitoring mode, it sums the currents from all fragment ions as

Nanogram quantities are detected by MS.

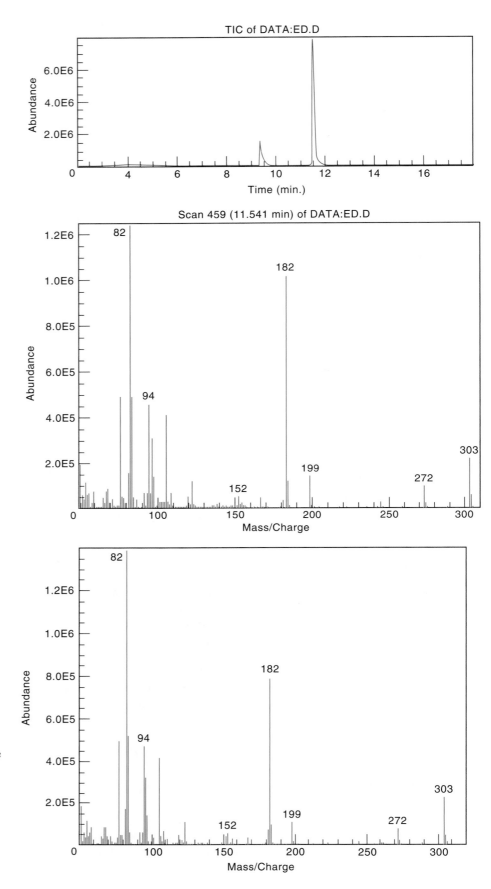

Fig. 20.13. Confirmation of cocaine by GC–MS. *Top:* Total ion current gas chromatogram of cocaine in urine sample. *Middle:* Mass spectrum taken from peak at 11.5 min. *Bottom:* Mass spectrum taken from GC peak of cocaine standard at same elution time.

a molecule (or molecules) in a GC peak passes through the detector, to provide a conventional looking gas chromatogram of several GC peaks. In the **selective ion** mode, a specific m/z ratio is monitored, and so only molecules that give a molecular or fragment ion at that ratio will be sensed. The **mass spectrum** of each molecule detected is stored in the system's computer, and so the mass spectrum corresponding to a given GC peak can be read out. The mass spectrum is generally characteristic for a given compound (if only one compound is present under the GC peak), giving a certain "fingerprint" of peaks at various m/z ratios. Certain peaks will dominate in intensity.

Figure 20.13 illustrates the application of GC–MS for positive identification of cocaine in a suspected powder sample, dissolved in methanol. Shown at the top is the gas chromatogram of the sample obtained from the TIC. The peak at 11.5 minutes corresponds to the retention time expected for cocaine. The middle figure is the mass spectrum corresponding to the compound at that peak, and the bottom one is the mass spectrum of a cocaine standard. The mass spectrum of the sample compound is essentially the same as for the cocaine standard. Furthermore, the parent ion peak is present at the m/z corresponding to M^+ for cocaine (formula weight 303.35). (There is a small peak at m/z 304 corresponding to MH^+, which is often formed in the ion chamber.)

> The highest m/z peak often corresponds to MH^+. It may be very small.

The fragmentation pattern often exhibits peaks corresponding to loss of specific groups in the molecule, for example, $-CO_2$ or $-NH$, which lends further credence to the presence of a given molecule or which can be used to gain structural information about a molecule. Manufacturers of mass spectrometers provide computer libraries of mass spectra of thousands of compounds, and spectral computer searches can be made to match an unknown spectrum.

The NIST/EPA/NIH Mass Spectral Library 1998 database (www.nist.gov/srd/analy.htm) is the product of a multiyear, comprehensive evaluation and expansion of the world's most widely used mass spectral reference library, and is sold in ASCII or Windows versions. It contains 108,000 compounds with electron ionization spectra, chemical structures, and molecular weights. It is available with the NIST MS Search Program for GC/MS deconvolution, MS interpretation, and chemical substructure analysis. The NIST chemistry WebBook (http://webbook.nist.gov) is a *free* online system that contains the mass spectra of over 12,000 compounds (this Standard Reference Data Program also has IR and UV–Vis spectra).

The Wiley/NIST Registry of Mass Spectral Data, 7th edition, 1999, contains over 390,000 reference spectra. The registry has 32-bit search software to identify unknown mass spectra. You can view reference spectra by mass, molecular weight, or peaks.

The marriage of capillary gas chromatography with mass spectrometry provides an extremely powerful analytical tool. Capillary GC, with thousands of theoretical plates, can resolve hundreds of molecules into separate peaks, and mass spectrometry can provide identification. Even if a peak contains two or more compounds, identifying peaks can still provide positive identification, especially when combined with retention data.

Learning Objectives

WHAT ARE SOME OF THE KEY THINGS WE LEARNED FROM THIS CHAPTER?

- Gas chromatograph, p. 575
- GC columns—packed, capillary, p. 578
- Stationary phases—polar to nonpolar, p. 581

- GC detectors (see Table 20.2), p. 584
- Temperature programming, p. 588
- Quantitative measurements—internal standards, spreadsheets for calculation, p. 589
- Headspace analysis, thermal desorption, purging, and trapping, pp. 590, 591
- Small columns for fast separations, p. 592
- GC–MS, mass analyzers, pp. 593, 597, 599

Questions

1. Describe the principles of gas chromatography.
2. What compounds can be determined by gas chromatography?
3. What are the main types of gas chromatography?
4. Compare packed and capillary columns in number of plates.
5. Compare WCOT, SCOT, and PLOT columns.
6. Describe the principles of the following gas chromatography detectors: (a) thermal conductivity, (b) flame ionization, (c) electron capture.
7. Compare the detectors in Question 6 with respect to sensitivity and types of compounds that can be detected.
8. How does temperature programming improve separations?
9. What is required for fast GC analysis?
10. Describe the principles of gas chromatography–mass spectrometry. What are its benefits?
11. What is the molecular ion?
12. What is the nitrogen rule?
13. What ion sources are commonly used for GC–MS?
14. What mass analyzers are commonly used for GC–MS?

Problems

15. What mass spectrometry resolution is required for unit resolution of molecular mass 600 and 601?
16. A mass spectrometer has a resolution of 5000. How closely are two peaks at nominal mass 600 resolved?
17. Gas reduction valves used on gas tanks in gas chromatography usually give the pressure in psig (pounds per square inch above atmospheric pressure). Given that atmospheric pressure (760 torr) is 14.7 psi, calculate the inlet pressure to the gas chromatograph in torr, for 40.0 psig, if the ambient pressure is 745 torr.

SPREADSHEET PROBLEM

18. A water sample is analyzed for traces of benzene using headspace analysis. Samples and standards are spiked with a fixed amount of toluene as internal standard. The following data are obtained:

ppb Benzene	Peak Area Benzene	Peak Area Toluene
10.0	252	376
15.0	373	370
25.0	636	371
Sample	533	368

What is the concentration of benzene in the sample? Prepare a spreadsheet similar to the one described in the chapter, and print the calibration curve.

LITERATURE SEARCH

19. Using Chemical Abstracts or SciFinder Scholar (the online access to Chemical Abstracts—see Appendix A) if your library subscribes to it, find at least one article on the gas chromatography determination of ethanol in blood. Read the journal article and write a summary of the principle of the method and prepare a synopsis of the procedure employed, including any sample preparation. Is there information on the accuracy and precision of the method?

Recommended References

GENERAL

1. H. M. McNair and J. M. Miller, *Basic Gas Chromatography*. New York: Wiley-Interscience, 1997.

2. W. Jennings, E. Mittlefehldt, and P. Stremple, eds., *Analytical Gas Chromatography*, 2nd ed. San Diego: Academic, 1997.

3. D. Rood, *A Practical Guide to Care, Maintenance, and Troubleshooting of Capillary Gas Chromatography*. New York: Wiley, 1999.

4. A. Van Es, *High-Speed Narrow-Bore Capillary Gas Chromatography*, Heidelberg: Huthig, 1992.

5. B. Kolb and L. S. Ettre, *Static Headspace-Gas Chromatography*. New York: Wiley, 1997.

6. W. O. McReynolds, "Characterization of Some Liquid Phases," *J. Chromatogr. Sci.*, **8** (1970) 685.

7. D. M. Ottenstein, "Column Support Materials for Use in Gas Chromatography," *J. Gas Chromatog.*, **1**(4) (1963) 11.

8. W. R. Supina and L. P. Rose, "The Use of Rohrschneider Constants for Classification of GLC Columns," *J. Chromatogr. Sci.*, **8** (1970) 214.

GAS CHROMATOGRAPHY/MASS SPECTROMETRY

9. M. C. McMaster and C. McMaster, *GC/MS: A Practical User's Guide*. New York: Wiley, 1998.

10. H.-J. Hubschmann, *Handbook of GC/MS*. New York: Wiley, 2001.

11. M. Oehme, *Practical Introduction to GC-MS Analysis with Quadrupoles*. New York: Wiley, 1999.

WEB PAGES

12. Scientific Instrument Services, www.sisweb.com. Has exact mass calculator.

13. JEOL USA, Inc., www.jeol.com. Has tutorials on basic mass spectrometry and on mass analyzers, and links to other MS websites.

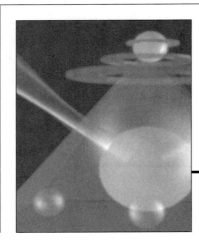

Chapter Twenty-One

LIQUID CHROMATOGRAPHY

HPLC is the liquid chromatography analog of GC. The secret to its success is small uniform particles to give small eddy diffusion and rapid mass transfer.

Gas chromatography (GC), because of its speed and sensitivity and quite broad applicability, has been more widely used since its development than the various modes of liquid chromatography. But the latter techniques have potentially broader use because approximately 85% of known compounds are not sufficiently volatile or stable to be separated by gas chromatography. The wealth of chromatographic theory accumulated, primarily from gas chromatography, has led to the development of techniques of **high-performance liquid chromatography (HPLC)** that rival gas chromatography in performance and allows separations and measurements to be made in a matter of minutes. The driving force for the rapid acceptance of gas chromatography in the 1950s was its immediate application to the petrochemical industry. Conversely, the applicability of HPLC to the pharmaceutical industry made it the mainstay of pharmaceutical laboratories in the 1970s, and today the HPLC market for new instruments is larger than the more mature GC market.

In this chapter, we describe the principles of HPLC and developments that have led to its success. The techniques of normal and reverse-phase liquid chromatography (separations based on polarity), size exclusion chromatography (separations based on molecular size), and ion exchange chromatography (separations based on charge) are described. Thin-layer chromatography, a planar form, is discussed. In addition, we describe electrophoresis, in which separations are accomplished in an electric field gradient, based on the sign and magnitude of charge on the analyte.

21.1 High-Performance Liquid Chromatography

In 1964, J. Calvin Giddings from the University of Utah published a paper entitled, "Comparison of the Theoretical Limit of Separation Ability in Gas and Liquid Chromatography" [*Anal. Chem.*, **36** (1964) 1890]. Conventional liquid chromatography up to that time was primarily accomplished in large columns with large particles under gravity feed, with manual collection of fractions of eluents for measurement in a spectrophotometer. Giddings predicted improved liquid chromatography performance if one could use small particles under increased flow pressure and that theoretically very high plate numbers could be achieved. A couple of years later, Csaba Horvath and colleague S. R. Lipsky at Yale University built the first HPLC, and called it "high pressure liquid chromatography." But it was not until the early

1970s that the technology of producing small silanized silica particles allowed the use of small-volume longer columns necessary to give the high-resolution performance. Today, HPLC has become known as "high performance liquid chromatography."

PRINCIPLES

Classical liquid chromatography has largely been supplanted by the much more powerful and analytically useful form of high-performance liquid chromatography. Figure 21.1 shows the basic components of an HPLC system, and Figure 21.2 illustrates a modern HPLC instrument. These instruments tend to be assembled in modular form, unlike most GC instruments, allowing the user to change different components.

The rate of distribution of solutes between the stationary and the mobile phase in traditional liquid chromatography is largely diffusion-controlled. Diffusion in liquids is extremely slow compared to that in gases. To minimize diffusion and the time required for the movement of sample components to and from the interaction sites in the column, two criteria should be met. First, the packing should be finely divided and have high spherical regularity to allow for optimum homogeneity and packing density; and second, the stationary phase should be in the form of a thin uniform film with no stagnant pools. The former results in a smaller A value in the van Deemter equation (smaller eddy diffusion) and the latter results in a small C value (more rapid mass transport between the phases—necessary for high flow rates). Because molecular diffusion in liquids is small, the B term in Equation 19.8 is small. Hence, the detrimental increase in H at slow flow rates does not occur as in Figure 19.4. This is illustrated in Figure 21.3 and is expressed by the Huber equation (Equation 19.21) and the Knox equation (Equation 19.23) in Chapter 19.

Molecular or longitudinal diffusion in liquids is slow and can be neglected.

Mass transfer is the primary determinant of H in HPLC.

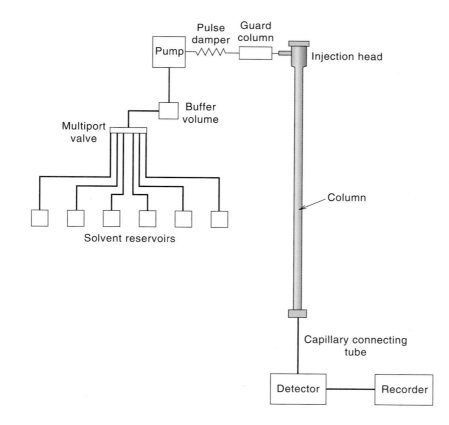

Fig. 21.1. Basic components of high-performance liquid chromatograph. (Adapted from Analabs, Inc. *Research Notes.* Copyright © 1971. Reproduced by permission.)

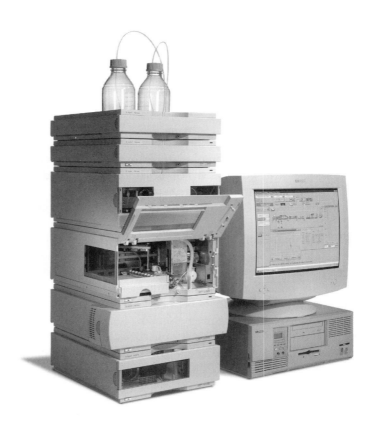

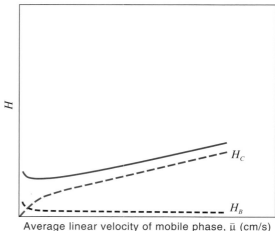

Fig. 21.2. High-performance liquid chromatograph. (Courtesy of Agilent Technologies.)

STATIONARY PHASES

Original HPLC microparticles were irregularly shaped porous silica gel or alumina particles of 10 μm or less. Since then, spherical particles have been developed that can be packed more homogeneously and provide improved efficiency (Figure 21.4). The particles are **high-purity silica**, low in trace metal content, and are typically 5 to 10 μm in diameter, although 3-μm particles are finding more use for high-speed chromatography (discussed below). Pore sizes are in the 60- to 100-Å range, although pore sizes of 300 Å or larger are used for larger biomolecules to allow them to penetrate the pores.

Most HPLC is performed in the liquid–liquid (partition chromatography) mode, but adsorption chromatography is useful for many applications. Liquid stationary phases are either coated on the particles or are chemically bonded.

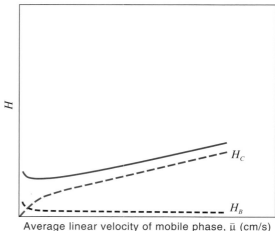

Fig. 21.3. van Deempter plot for HPLC.

The most commonly used particles are **microporous** or **diffusive particles**, permeable to solvent (Figure 21.5a). The majority of the surface area is within the pores. Mobile phase moves around the particles and solute diffuses into the stagnant mobile phase within the pores to interact with the stationary phase, and then diffuses out into the moving mobile phase. The use of small particles minimizes the pathlength of the diffusion pathlength, and hence band broadening. Those in Figure 21.4 are microporous. Figure 21.6 shows an aggregated spherical particle with 50% porosity, with 100-Å pore size. Also shown is a widely used one with a spongelike structure with about 70% porosity and the same nominal pore size with greater surface area, but with greater distribution in pore size and less resistance to solubility in alkaline solution.

Silica tends to dissolve above pH 8, and crosslinked polymeric particles, for example, polystyrene or polymethacrylates, are used for separation of bases. These can withstand strongly basic mobile phases but exhibit somewhat lower efficiency.

Silica particles have surface silanol groups, —SiOH. These are used for chemical bonding of stationary phases by silination reactions with chlorosilanes:

$$-\underset{\underset{\displaystyle CH_3}{|}}{\overset{\overset{\displaystyle CH_3}{|}}{Si}}-OH + Cl-\underset{\underset{\displaystyle CH_3}{|}}{\overset{\overset{\displaystyle CH_3}{|}}{Si}}-R \rightarrow -\underset{\underset{\displaystyle CH_3}{|}}{\overset{\overset{\displaystyle CH_3}{|}}{Si}}-O-\underset{\underset{\displaystyle CH_2}{|}}{\overset{\overset{\displaystyle CH_3}{|}}{Si}}-R + HCl$$

About half the silanol groups are chemically bonded, and the remainder are "endcapped" with trimethyl silyl groups to render them inert. Zorbax also makes particles in which the siloxane bonds are protected by steric hindrance provided by larger isopropyl groups, $-\underset{\underset{\displaystyle CH_3}{|}}{\overset{\overset{\displaystyle CH_3}{|}}{C}}-CH_3$, in place of the two methyl groups on the organochlorosilane derivatizing agent.

The most common nonpolar bonded phases (for reversed-phase chromatography) are C_{18} and C_8 (shown above), with C_{18} the most popular (known as ODS for octadecylsilane); C_8 is intermediate in hydrophobicity, and C_{18} is very nonpolar. Phenyl groups are also useful [R = $(CH_2)_3C_6H_5$]. C_5 particles are used for

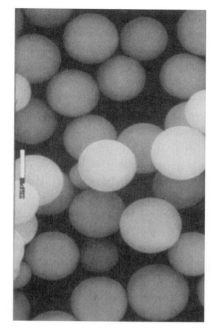

Fig. 21.4. Spherical porous silica particles, 10 μm, 800× magnification. Particles are fully porous with 100-Å pores. They are available as base silica for adsorption chromatography, or with bonded phases. (Astrosil® from Stellar Phases, Inc. Courtesy of Stellar Phases, Inc.)

Endcapping. The middle free SiOH group has been endcapped.

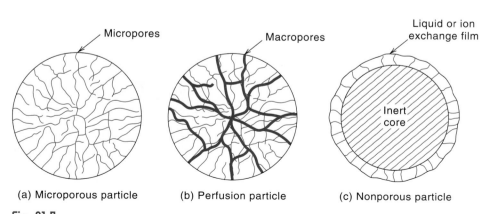

(a) Microporous particle   (b) Perfusion particle   (c) Nonporous particle

Fig. 21.5. Structural types of particles used in high-performance liquid chromatography. (From D. C. Scott in *Modern Practice of Liquid Chromatography*, J. J. Kirkland, ed. New York: Wiley-Interscience, 1971; with permission.)

Chemically bonded phases (functional groups) are more stable.

(a)

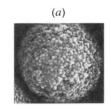

ZORBAX Rx-SIL (Silica Sol)

(b)

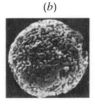

Xerogel Silica

Fig. 21.6. (*a*) Zorbax porous silica microsphere particle, 50% porosity, 100-Å pores. (*b*) Xerogel silica particle, 70% porosity, 100-Å pores. (Courtesy of Agilent Technologies.)

HPLC–MS because of their low bleed properties. Polar phases for normal-phase chromatography in increasing order of polarity include cyano [R = $(CH_2)_3CN$], diol [R = $(CH_2)_2OCH_2CH(OH)CH_2OH$], amino [R = $(CH_2)_3NH_2$], and dimethylamino [R = $(CH_2)_3N(CH_3)_2$].

When particles are coated with liquid phases, rather than bonded, they are first treated with trimethylcholosilane to deactivate the silanol groups.

Perfusion packings, developed by Fred Regnier and co-workers at Purdue University, are made of a mixture of large and small pores (larger through pores and small diffusive pores, Figure 21.5*b*). The diffusive pores, as in microporous particles, provide the sorption capacity. The through pores allow the mobile phase to pass directly through the packing, hence increasing the rate of mass transfer in the mobile phase. Since the solute spends less time undergoing mass transfer, peaks are narrower; the process is actually a combination of diffusion plus convection. These packings are larger than the microporous ones, being about 12 μm in diameter. They can be used at higher flow rates and give better efficiency for large molecules such as proteins. They are also useful for preparative chromatography.

Nonporous packings (Figure 21.5*c*), either silica or resin, have much smaller particle size, 1.5 to 2.5 μm diameter, with a thin porous layer. They eliminate the occurrence of a stagnant mobile phase, which allows for much faster rate of mass transfer. Molecules, large or small, can be separated in a few minutes. But the thin layer is limited to very small loading capacities, and the backpressure is much greater than in columns with 3- or 5-μm porous packings. The backpressure is inversely proportional to the square of the particle diameter, so cutting the size in half causes the pressure to increase fourfold. For these reasons, small porous particles are preferred for most applications, and a 3.5-μm porous packing can give comparable separation times as a column containing 1.5-μm nonporous silica bonded phase, with less pressure, when operated at high flow rates. Nonporous columns are useful for separating complex peptide mixtures in seconds to minutes, and are used in ion chromatography (discussed below).

Monolithic columns are another approach to provide lower pressure drops and higher rates of mass transfer. These are continuous solid columns of porous silica stationary phase instead of packed particles. Like perfusion packings, they have a bimodal pore structure (Figure 21.7). Macropores, which act as flowthrough pores, are about 2 μm in diameter. The silica skeleton contains mesopores with diameters of about 13 nm (130 Å). It can be surface modified with stationary phases like C_{18}. The rod is shrink-wrapped in a polyetheretherketone (PEEK) plastic holder to prevent "wall effects" of solution flowing along the walls. The surface area of the mesopores is about 300 m^2/g, and the total porosity is 80%, compared with 65% for packed particles. The column exhibits a van Deemter curve approximating

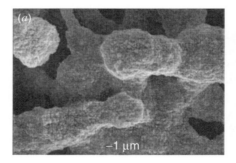

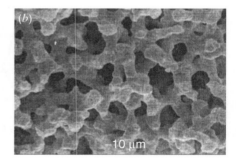

Fig. 21.7. Electron micrograph of (*a*) mesoporous and (*b*) macroporous structures in monolithic rod. [From D. Lubda, K. Cabrera, W. Krass, C. Schaefer, and D. Cunningham, *LC·GC* **19** (12) (2001) 1186. Reproduced by permission.]

that for 3.5-μm packed particles, giving significantly better efficiency than conventional 5-μm particles. Because of low pressure drops, flow rates as high as 9 mL/min can be used, resulting in fast separations. Mass transport is facilitated by convection in addition to diffusion, making these columns well suited for efficient separations of proteins, peptides, and polynucleotides, as well as small molecules. Polymeric monolithic columns are rather short, and the plate counts are too low to perform isocratic separation of complex mixtures, although several columns can be linked in series. These columns are manufactured by Merck KGaA, Darmstadt, Germany (www.merck.de), and are called Chromolith.

Various approaches have been taken to develop HPLC particles, besides polymer-based ones, that can be used at high pH, as well as in acid solution. High pH is required for the separation of basic compounds. Silica particles are generally limited to about pH 8 because the silane bonds hydrolyze about this pH, and they have a lower limit of about pH 2. Waters Corporation (www.waters.com) developed a **hybrid silica/polymer** particle they call Xterra. The particles are derived from a mixture of two high-purity monomers, one forming SiO_2 units and the other $RSiO_{1.5}$ units, where R is a methyl group. The organasilane groups are incorporated throughout the structure and are then surface-bonded to attach a variety of different reverse-phase (polar) groups (e.g., C_{18}). The low pH stability is in part due to the fact that the particles are formed from trifunctional silanes rather than monofunctional silanes. Because the base particle already contains methylsiloxane units, it gives bonded phases with 30 to 50% lower concentrations of residual silanol groups, which gives reduced peak tailing. The particles exhibit good stability at both high (greater than pH 11) and low pH.

Peter Carr at the University of Minnesota has developed **porous zirconia** (ZrO_2) particles that are chemically stable over the pH range 1 to 14 and are also thermally stable up to 200°C. These are manufactured by ZirChrom Separations, Inc. (www.zirchrom.com). They are produced through the controlled polymerization-induced aggregation of 1000-Å colloidal zirconia, resulting in monodisperse 3-μm porous zirconia spheres. These are then sintered at temperatures up to 900°C to produce a monoclinic crystallographic form of zirconia. The particles are bonded with a variety of reverse-phase stationary phases. Thin layers of polybutadiene or polystyrene are coated on the zirconia. Particles are also produced with a very thin layer of elemental carbon that can be used as a stationary phase. A proprietary covalent bonding technology is used to graft C_{18} groups to the carbon surface. Visit the ZirChrom website for a list of representative uses for each type of particle. (Its site also has a free automated buffer preparation calculator.) See also *Anal. Chem.*, **73** (2001) 598A for an excellent review of the properties of these phases.

EQUIPMENT FOR HPLC

The price that must be paid for faster, more efficient separations using finer column packings is pressure and the special hardware to handle this. Pressures of 1000 to 3000 psi are required to provide flow rates of 1 to 2 mL/min in columns of 3- to 5-mm diameter and 10 to 30 cm long, although in certain instances pressures up to 6000 psi may be required. Probably 80 to 90% of HPLC separations are performed with pressures of less than 1200 psi, and even some polyurethane column materials require very low pressures near atmospheric pressures.

High-performance liquid chromatography apparatus consists of four principal parts:

1. Mobile-Phase Supply System. This system contains a pump to provide the high pressures required and usually contains some means of providing gradient elution (i.e., changing concentrations of the eluent, such as solvent, salt, or H^+).

The solvent reservoirs can be filled with a range of solvents of different polarities, provided they are miscible, or they can be filled with solutions of different pH and are mixed in the buffer volume. The solvents must be pure and be degassed to avoid formation of gas bubbles when they prevent proper check valve function or enter the piston chamber. And they generate spurious peaks when they pass through the detector. The problem is most serious when mixing solvents (usually acetonitrile or methanol with water) because the solubility of air in mixtures is less than in the same proportion of pure solvents; and, when solvents saturated with air are mixed, bubbles are released. Degassing systems only need to remove enough air to reduce it to below saturation levels. The most common ways of degassing are sparging with helium, which removes about 80% of the air from solution, or vacuum degassing, which removes around 60%. When sparging with helium, some of the more volatile solvent is evaporated, and commercial systems incorporate pressure reservoirs to minimize this. The air is replaced with another gas, helium, but its solubility is such that outgassing is not a problem. Many manufacturers incorporate an online vacuum degassing system in which the solvent passes through a thin-walled porous polymer tube, for example, PTFE or Teflon AF, an amorphous fluoropolymer with two- to threefold higher gas diffusion rates, in a vacuum chamber. Some workers prefer to first briefly sparge with helium, and then rely on the vacuum degasser to maintain low gas content.

Typical flow rates are 1 to 2 mL/min for conventional 4.6-mm i.d. columns. Solvents used for HPLC should be of "HPLC" grade, that is, solvents that have been filtered through 0.2-μm filters. This extends the pump life by preventing scoring and reduces contamination or plugging of the column.

The most commonly used pump for HPLC is the reciprocating pump. This has a small cylindrical piston chamber that is alternately filled with mobile phase and emptied via back-and-forth movement of the piston. This produces a pulsed flow that must be damped. Reciprocating pumps have a number of advantages. They have a small internal volume, are capable of high output pressures, and they can readily be used for gradient elution. They provide constant flow rates, independent of solvent viscosity or column backpressure. Other pumps used are motor-driven syringe pumps and pneumatic (constant-pressure) pumps.

2. Sample Injection System. A typical injection system is shown in Figure 21.8. This consists of a stainless steel ring with six different ports, one to the column. A movable Teflon cone within the ring has three open segments, each of which connects a pair of external ports. Two of the ports are connected by an external sample loop of known or fixed volume. In one configuration, the cone permits direct flow of effluent into the column, and the loop can be filled with the sample. The cone can then be rotated 30° to make the sample loop part of the moving stream, which sweeps the sample into the column. Samples of a few microliters can be injected at pressures up to 6000 psi.

Samples can be introduced manually into the valve with a syringe to fill the sample loop. Automated sampling valves are routinely used today in which samples are taken from an autosampler for unattended operation. The major limitation of valve injectors is that the sample size is fixed, and the loop must be changed in order to vary the injected sample size. There are automated motor-driven adjustable syringes that provide enough pressure to inject the sample past a check valve that prevents backflow.

3. Column. Straight lengths of stainless steel tubing make excellent columns (Figure 21.9). These come in various diameters and lengths, depending on the particular application. Usually the internal diameter is 3.9 or 4.6 mm. A well-packed

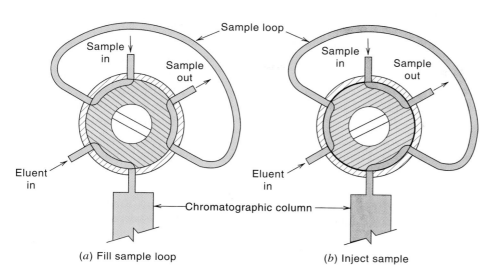

Fig. 21.8. Sample loop injector. (From D. C. Scott in *Modern Practice of Liquid Chromatography*, J. J. Kirkland, ed. New York: Wiley-Interscience, 1971. With permission.)

4.6-mm column of 5-μm-diameter (d_p) particles should give a plate count on the order of 60,000 to 90,000 plates/meter at flow rates of 1 mL/min. A typical 15-cm-long column with 4.6-mm i.d. will give 15,000 plates with 3-μm particles, 9000 plates with 5-μm particles, and 5000 plates with 10-μm particles. The smaller particles are limited to 25 cm length or shorter columns because of high-pressure drops. At 25 cm, 50,000 plates are possible at a pressure of 5000 psi with water–methanol mixtures. They can resolve 50 to 100 peaks, about the upper limit for conventional HPLC. We will see below that fast chromatography can be performed with wide-bore, short columns, but narrow columns give better resolution.

Some rules for selecting column stationary and mobile phases are given below. Temperature control of the column is usually not necessary for liquid–solid chromatography, unless it has to be operated at elevated temperatures, but is generally required for other forms of liquid chromatography (liquid–liquid, size exclusion, ion exchange). Some detectors, especially refractometers, are very sensitive to temperature changes; and so, if the column is operated at greater than ambient

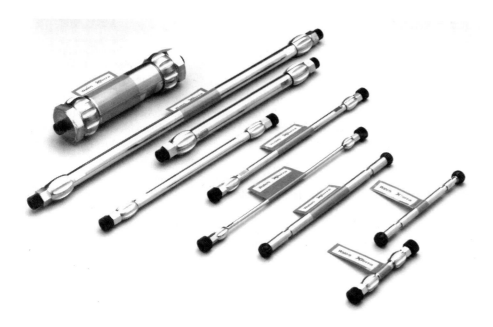

Fig. 21.9. Typical HPLC columns. (Courtesy of Waters Corporation.)

A guard column extends the life of the analytical column and improves separations by retaining strongly sorbed compounds and debris.

temperature, a cooling jacket should be placed between the end of the column and the detector to bring the mobile phase back to ambient temperature.

A small, 3- to 10-cm **guard column** or **precolumn** is placed between the injector and the analytical column. This generally contains the same packing as the analytical column. It is placed there for two reasons. First, it will retain debris (e.g., pump-seal fragments) and sample particulate matter that would otherwise get on the analytical column and foul it, changing column efficiency and selectivity. Second, it retains highly sorbed compounds that would be caught on and not be eluted from the analytical column. The guard column extends the life of the analytical column. It must be replaced or regenerated periodically. Some analytical columns have an integrated guard column, designed to eliminate almost all dead volume, for example, by butting it up next to the analytical column, in order to minimize extra column broadening. An in-line filter may be placed between the pump and the injector.

4. Detector. Detectors with high sensitivity are required in high-performance liquid chromatography, usually with sensitivities in the microgram to nanogram range. Widely used detectors are refractometer detectors and ultraviolet (UV) detectors. The **differential refractometer detector** is often called a "universal detector." This measures changes in refractive index of the eluent that result from the presence of solutes as they emerge from the column. It cannot be used effectively with gradient elution due to a change in the baseline (a change in the solvent index of refraction as the gradient is changed) nor when the solvent has an index of refraction close to that of the solutes. As mentioned, it is very sensitive to temperature changes. This detector is rugged and will detect concentrations of about 10^{-5} to 10^{-6} g/mL (10 to 1 ppm). The **ultraviolet detector** has much better sensitivity, about 10^{-8} g/mL (0.01 ppm). It is not temperature sensitive, is relatively inexpensive, and can be used with gradient elution. It is sensitive to a large number of organic compounds. Because of its advantages, the UV detector represents about 80% of the measurements made. Of course, it cannot be used with solvents that have significant absorption in the UV or with sample components that do not absorb in the UV.

Many UV detectors are simple interference filter devices that can measure the absorbance at only a few selected wavelengths. The more expensive detectors have a monochromator that allows selection of a particular wavelength. Scanning of the spectrum can even be achieved for qualitative identification by momentarily stopping the flow of the mobile phase. The most popular HPLC detector is the variable wavelength UV–Vis detector. It can measure nanogram amounts of UV absorbing analytes or those with suitable chromophores that absorb in the visible region. Analytical columns can handle 100-μL samples, so these detectors can measure 10 ppb concentrations in favorable cases.

A diode array detector provides additional resolving power and fingerprinting.

A common feature of modern HPLC instruments is a **diode array** detector, as described in Chapter 16. The instantaneous recording of absorption spectra provides a powerful qualitative tool. The focused radiation source passes through the detector flow cell and is dispersed by a grating to a photodiode array for detection. The ability to mathematically resolve overlapping spectra can provide additional separating ability when a chromatographic peak may consist of two or more analytes.

Fluorescence detectors can give improved selectivity over ultraviolet absorption detectors because fewer compounds fluoresce than absorb (Chapter 16). Sensitivities at least as good as and perhaps better than the UV detector are achieved, depending on the geometry of the excitation source–detector arrangement, the intensity of the source, and the quantum efficiency of the fluorophore. The **amperometric detector** (see Chapter 15) is useful for detecting electroactive substances and has found considerable use in biological applications, for example, in the HPLC separation and detection of trace quantities of catecholamines from the brain.

In arranging the apparatus, there must be a minimum of "dead volume" between the injection port and column and between the column and detector, to minimize spreading of the peaks and to obtain maximum efficiency. A 20-cm length of stainless steel capillary tubing can generally be used to connect the column to the detector without significantly affecting column performance. The detector volume must also be small, and typical volumes are on the order of 1 μL or so, with high-performance detectors used with capillary LC systems having submicroliter volumes. A spectrophotometric flow cell typically has a "Z-cell" configuration with quartz windows, in which the light path is along the axis of a horizontal length of the effluent flow to increase the absorbance pathlength and the sensitivity.

HPLC METHOD DEVELOPMENT: WHAT COLUMN? WHAT SOLVENT?

HPLC is used either in the liquid–solid **adsorption chromatography** mode or the liquid–liquid **partition chromatography** mode. The most common is partition chromatography, either normal or reversed-phase. Both partition and adsorption chromatography operate on differences in solute polarity since polarity is important in determining both adsorption and solubility. Liquid–liquid partition processes are quite sensitive to small molecular weight differences and so are preferred for the separation of members of a homologous series. Adsorption processes, on the other hand, are sensitive to steric effects and are preferred for the separation of isomers having different steric configurations. The most common adsorption phases are alumina or silica particle.

As a general rule, highly polar materials are best separated using partition chromatography, while very nonpolar materials are separated using adsorption chromatography. Between extremes, either process might be applicable. Compound polarity follows the approximate order of: hydrocarbons and derivatives < oxygenated hydrocarbons < proton donors < ionic compounds; that is, RH < RX < RNO_2 < ROR (ethers) = RCO_2R (esters) = RCOR (ketones) = RCHO (aldehydes) = RCONHR (amides) < RNH_2, R_2NH, R_3N (amines) < ROH (alcohols) < H_2O < ArOH (phenols) < RCO_2H (acids) < nucleotides < $^+NH_3RCO_2^-$ (amino acids). In adsorption chromatography the adsorbent is usually kept constant, and the eluting solvent polarity is increased until elution is achieved. Some commonly used solvents in order of increasing polarity are: light petroleum solvents (hexane, heptane, petroleum ether) < cyclohexane < trichloroethane < toluene < dichloromethane < chloroform < ethyl ether < ethyl acetate < acetone < n-propanol < ethanol < water. In some circumstances, such a highly polar solvent may be required for elution that many solutes are eluted together rather than being separated. A less polar adsorbent should then be used.

In **normal-phase chromatography** (NPC), the stationary phase is polar. A nonpolar mobile phase is used, such as n-hexane, methylene chloride, or chloroform. The stationary phase is a bonded siloxane with a polar functional group (polarity order: cyano < diol < amino < dimethylamino). These phases retain polar compounds in preference to nonpolar compounds.

In **reversed-phase chromatography** (RPC), a relatively nonpolar stationary phase is used, with a polar mobile phase such as methanol, acetonitrile, tetrahydrofuran, water, or usually a mixture of water with one of the organic solvents. The organic solvent is called the **modifier**, and acetonitrile is the most common one. The water content is varied for adjusting the polarity. Methanol is used for acidic compounds and acetonitrile for basic compounds. Tetrahydrofuran is used for those with large dipoles. These solvents are UV transparent and have low viscosity. The most common bonded phases are n-octyldecyl (C_{18}) or n-decyl (C_8) chains, or phenyl groups. Polar reversed-phase columns such as polyethylene glycol (PEG)

Reversed-phase HPLC is used to separate organic compounds.

contain ether groups that interact with polar analytes. They are useful for pheno-lic compounds and multiaromatic ring, hydroxyl-containing compounds. While many of these may by separated on C$_{18}$ columns, there may be overlaps. Different orders of elution can be expected.

A wide range of organic compounds can dissolve in the mixed water–organic solvent phases and be separated, so RPC is by far the most popular form of HPLC. For those that can be separated either by NPC or RPC, the elution order is gener-ally reversed, although not always exactly. Of course, if a sample is very nonpo-lar and insoluble in water mixtures, normal–phase chromatography is used. Otherwise, reversed-phase chromatography is used.

1. Solvent Selection. In contrast to gas–liquid chromatography, where the mo-bile gas phase is fixed and separation conditions are changed by varying the polar stationary phase, in liquid–liquid partition chromatography we can also vary the mo-bile phase, that is, the "solvent strength." This is easier to do than changing columns, and adjusting the mobile phase is part of developing an LC separation method.

In most applications, a pure solvent will not provide efficient separation of a range of compounds, and a blend of two or more solvents is used. The weak sol-vent is designated the A solvent and the strong solvent the B solvent. Trial-and-error can be used to obtain an optimum % B solvent. We are interested in two fac-tors to achieve efficient separations with baseline resolution: the retention factor, *k*, a *polarity* term, and the separation factor, *α*, a *selectivity* term (see Equation 19.28). Generally, we try to adjust the solvent strength to position all solute bands within a *k* range of about 0.5 to 20. Adjusting for acceptable retention will provide adequate resolution for many compounds. If we have incomplete separation, a change in % B from say 35 to 40 or 45% will often result in significant change in band spacing and give better resolution. But while some bands may now be better separated, others may now overlap that did not before. A mixture of two strong solvents, for example, acetonitrile and methanol, with water can provide interme-diate band spreading and acceptable resolution, as illustrated in Figure 21.10. The two experiments with 80 and 40% acetonitrile (ACN) gives an idea of *k*-value range

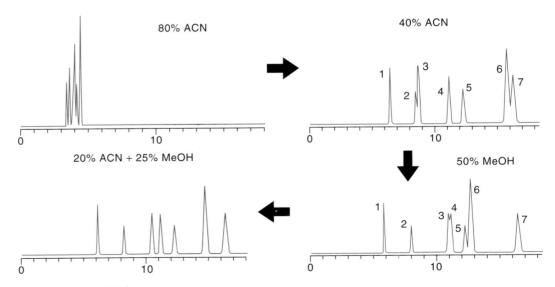

Fig. 21.10. Hypothetical series of method development experiments, beginning with a strong mobile phase of 80% acetonitrile–water. (From L. R. Snyder, J. J. Kirkland, and J. L. Glajch, *Practical HPLC Method Development*, 2nd ed. New York: Wiley-Interscience, 1998. With permission.)

with solvent strength; 40% ACN gives increased but still incomplete resolution. Changing from ACN to 50% methanol (MeOH) changes the spacing but results in new overlaps. Finally, mixing the two solvents (equal volumes of 40% ACN and 50% MeOH used before) gives a separation intermediate between the two, with acceptable resolution. This multi-solvent selection method works for many cases.

2. Column Differences. There are over 100 different brands of reversed-phase columns. Column manufacturers provide information such as plate count, unique selectivity, reproducibility, pressure drop, and so forth. While these are important to know, they are of little help to you in selecting the most appropriate column for most of your applications.

While indexes similar to those available for gas chromatography columns have not been developed, there have been attempts to try to classify columns. One way is to use a single compound such as aspirin to characterize a series of columns for a fixed polarity analyte, to determine the relative time it elutes. Others use a chemometrics (statistical) approach to group columns according to efficiency, peak symmetry, tailing, and free silane. These methods at present are of limited value in discerning relatively small differences. The bottom line is that empirical evaluation, starting with manufacturers' recommended applications, is the best way to settle on the appropriate column.

See www.mac-mod.com/ comparison_guide.html for a very detailed comparison of 60 commonly used C_{18} phases. They are categorized by relative hydrophobicity and polarity, and column efficiencies for neutral and basic compounds are given. See also http://ois.nist.gov/srmcatalog/ certificates/870.pdf complete NIST report.

3. Gradient Elution. In gas chromatography, we described how separations can be improved and speeded up by using temperature programming, but in HPLC this is a limited option (but see the discussion below on fast chromatography for exceptions). A more viable approach, not available in gas chromatography, is to employ mobile-phase gradient elution. Gradient elution LC is one of the most effective ways of separating several analytes with differing relative retention times. In isocratic elution, there is generally poor resolution of peaks early in the chromatogram and broadening of peaks at the longer retention times. In RPC, the gradient is accomplished by increasing the fraction of organic modifier solvent (e.g., methanol). This allows weakly retained compounds to be eluted later and be better resolved than they would in isocratic elution, and longer retained ones to be eluted more quickly, giving a chromatogram with compact, well-spaced peaks, with improved peak shape and lower detection limits since band broadening is lessened.

As in GC temperature programming, the gradient can be stepwise or continually ramped. The pumps are programmed to keep the total flow rate constant. The starting solvent should have a polarity that rapidly elutes and resolves the first components, and the solvent strength (polarity) is increased to a value that resolves the last peaks in a reasonable time. The multi-solvent method described above can help determine the starting and finishing polarities.

One of the problems encountered with gradient elution in reversed-phase chromatography is that the column has to be reequilibrated with the beginning solvent between runs. This requires flushing with 15 to 20 column volumes of the initial mobile phase. John Dorsey at Florida State University developed a procedure to dramatically reduce the reequilibration time by controlling the solvation of the bonded alkyl chains throughout the chromatographic run [L. A. Cole and J. G. Dorsey, *Anal. Chem.*, **62** (1990) 16]. A 3% solution of 1-propanol provides nearly monolayer coverage of a C_{18} surface, and adding 3% 1-propanol to each mobile-phase component reduces the reequilibration time about 75%. Also, the wetting effect provided by the 1-propanol may give increased column efficiencies at the beginning of the chromatogram with highly aqueous initial mobile phase. See also D. L. Warner and J. G. Dorsey, *LC·GC*, **15**(3) (1997) 254, for additional details on practical applications.

Gradient elution works well with bonded-phase partition chromatography and adsorption chromatography but is difficult to use with liquid–liquid chromatography.

FAST LIQUID CHROMATOGRAPHY: SMALL IS BETTER

Faster separations result from using smaller particles and short columns.

By making columns shorter and packing them with smaller particles, separations can be performed in one-tenth the time of conventional HPLC (typically on the order of 20 min), and some separations require only a few seconds. Fast liquid chromatography (LC) columns have the same diameter as conventional columns (4.6 mm) but are only 1 to 3 cm in length, instead of the typical 25 cm, and are packed with 3-μm particles instead of 5 μm, operating in the 1- to 3-mL/min range. The H vs. $\bar{u}$ plot is fairly flat at high flow rates (see Figure 19.5) because the smaller particles are less resistant to mass transfer (C term in the van Deemter equation). The columns are short because of the increased backpressure with the smaller particles, but a typical column can generate 4000 plates, enough for many separations. As many as 20 peaks have been separated in 30 s. While not high resolution, such separations solve many of the analytical problems in the pharmaceutical industry, which accounts for about half the HPLC market. In pharmaceutical quality control, analysts generally need to look for a few active ingredients in a tablet or formulation.

Why use fast LC? Besides the obvious advantage of faster analyses, there is 50 to 80% less solvent use (and less waste generated). Sensitivity is increased 3 to 5 times because there is less dilution of the analyte peaks with the smaller column volumes as a result of tighter packing. These densely packed columns are more prone to clogging. A scavenger column or filter is placed before the injector, in addition to the guard column placed before the column.

If resolution needs to be increased when using these short columns, the mobile phase can be adjusted so that all analytes are eluted in a k range of 1 to 10, with slower elution, for example, 10 min instead of 30 s. A modifier of 50% acetonitrile may be used in high-speed LC, with only 20% for high-resolution LC.

As in high-speed gas chromatography, high-speed LC requires fast-responding detectors and electronics to measure the narrow (0.5 s or 3 μL), fast eluting peaks. The detector should have a time constant capable of measuring rates of change of signal on the order of 50 to 100 ms or less. The sampling rate of the data system should be >10 points/second, and diode array spectrometers should read >5 scans/second. These measurements are more subject to extra column broadening effects, and dead volumes must be minimized, meaning a small detector flow cell (<3 μL), a low dispersion injector (these are commercially available), and shorter small-diameter connection tubing. Smaller flow cells will have shorter pathlength, and so there is some compromise in sensitivity.

NARROW-BORE COLUMNS—HIGHER SENSITIVITY, LOWER SOLVENT USE

Narrow columns improve sensitivity.

If we make columns narrower, lower detection results from narrower and taller peaks. If a 4-μL sample is injected into a 2.1-mm-diameter column, it will be diluted about fivefold less, and the peak will be five times higher than when injected into a 4.6-mm column of the same length. The flow rate is proportional to the square of the column diameter, and the optimum flow rate for the same resolution is about five times less for the smaller column, resulting in less volumetric dilution of the sample. Of course, the same sensitivity can be achieved with conventional columns by injecting a fivefold larger sample, 20 μL. So only if we are sample limited do we really gain any sensitivity advantage. But solvent use is cut fivefold. Injected sample volumes for the 2.1-mm columns should be no more than 5 μL to minimize the injection contribution to the peak volumes. And the flow cell volume should be 3 to 5 μL. The main use of these columns is when interfacing to a mass spectrometer (LC–MS—see below) because of the decreased peak volumes and low volumetric flow rates required for introduction into the mass spectrometer.

Example 21.1

Calculate the volume, in microliters, of a sample zone corresponding to 1 mm in (a) a 2.1-mm i.d. column and (b) a 4.6-mm i.d. column, assuming that the mobile phase occupies 65% of the column volume.

Solution

(a) Calculate the volumes in cm³, and so use centimeter dimensions:

$$V = l\pi r^2 = 0.1 \text{ cm} \times 3.14 \ (0.105 \text{ cm})^2 = 0.0035 \text{ cm}^3 = 3.5 \ \mu\text{L}$$

Since the mobile phase occupies 65% of the volume, multiply by 0.65 to obtain the actual volume:

$$3.5 \ \mu\text{L} \times 0.65 = 2.3 \ \mu\text{L}$$

(b) $0.1 \text{ cm} \times 3.14 \ (0.23 \text{ cm})^2 \times 0.65 = 0.0108 \text{ cm}^3 = 10.8 \ \mu\text{L}$

This means that the peak height will be the same with a 2.1-mm i.d. column for a 2.3-μL sample as for a 10.8-μL sample with a 4.6-mm i.d. column, about one-fifth the sample size. It also means a small column is not needed for high sensitivity if you are not sample limited.

Example 21.2

If the optimum volumetric flow rate for a 4.6-mm i.d. column is 1.5 mL/min with 5-μm particles, what would it be for a 2.1-mm i.d. column with the same particles? How much solvent would be consumed for a 10-min separation for each column?

Solution

The flow rate is proportional to the square root of the column i.d.:

$$\text{Flow rate (2.1 mm i.d.)} = 1.5 \text{ mL/min} \times \frac{(2.1 \text{ mm})^2}{(4.6 \text{ mm})^2}$$

$$= 0.31 \text{ mL/min}$$

For the 4.6-mm i.d. column, we would use 15 mL solvent, while for the 2.1-mm i.d., column, we would use 3.1 mL, one-fifth as much.

Figure 21.11 shows the difference in throughput for a narrow-bore (2.1-mm $\times$ 20-cm long) column packed with 2.5 μm C_{18} beads, compared with a 4.6-mm $\times$ 5-cm, 5-μm d_p column. A threefold increase in achieved. The narrow-bore column is heated to reduce the solvent viscosity for flow through the more densely packed column.

Fused silica capillaries of 0.25 or 0.32 mm i.d. packed with 3- or 5-μm particles are widely used for LC–MS. These columns require small samples, on the order of a microliter or less. The volmetric flow rate is several microliters per minute. Resolution is comparable to that of standard analytical columns.

WHAT ABOUT TEMPERATURE?

The zirconia particles described above are stable at high temperatures. Polystyrene-coated porous zirconia particles have been used at 200°C. Peter Carr at the University of Minnesota, in studies on high-temperature fast LC, demonstrated that column efficiency at high velocity improves at higher temperatures, especially for solutes that are strongly retained. Also, the van Deemter plot flattens out significantly as the temperature is increased. Carr performed separations at 150°C at flow rates of 15 mL/min in a 4.6-mm $\times$ 5-cm column with 3-μm particles.

LIQUID CHROMATOGRAPHY–MASS SPECTROMETRY

Mass spectrometry detection in liquid chromatography, like with gas chromatography, has become a powerful analysis tool for sensitive and selective mass detection in characterizing complex samples. Review the principles of mass spectrometry and the types of instruments used for chromatography detection in Chapter 20.

It is more difficult to interface a liquid chromatograph to a mass spectrometer because of the necessity to remove the solvent. Also, the analytes are nonvolatile and may be thermally labile but must be presented in gaseous form. Hence, the combination of LC and MS was termed an "unlikely marriage," and it took several years to develop interfaces to the stage of reliable and easy use. Today, there are several types of interfaces that make LC–MS a routine technique. Compact and benchtop systems are commercially available (Figure 21.12).

Electrospray ionization is the most common HPLC–MC interface.

The commonly used interfaces are the electrospray ionization (ESI) source, thermospray ionization (TSI), atmospheric pressure chemical ionization (APCI), and particle beam ionization. The choice depends largely on the polarity and thermal stability of the analyte. Electrospray ionization is preferred for polar and ionic and very large molecules such as proteins and peptides. Atmospheric pressure

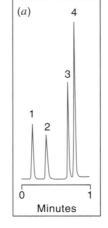

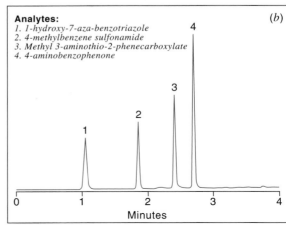

Analytes:
1. *1-hydroxy-7-aza-benzotriazole*
2. *4-methylbenzene sulfonamide*
3. *Methyl 3-aminothio-2-phenecarboxylate*
4. *4-aminobenzophenone*

Fig. 21.11. Increased throughput with narrow-bore column. (*a*) Column: 2.5 μm C_{18} particles, 2.1 mm i.d., 2.0 cm long, 60°C. Injection volume 1 μL. (*b*) 5.0 μm C_{18} particles, 4.6 mm i.d., 5.0 cm long, 30°C. Injection volumes 3 μL. (Courtesy of Waters Corporation.)

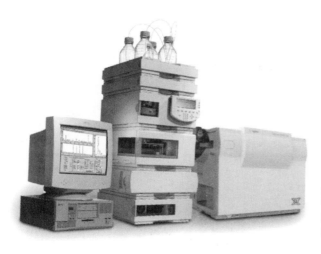

Fig. 21.12. Liquid chromatography–mass spectrometry benchtop system. (Courtesy of Agilent Technology.)

chemical ionization is also suited for large molecules and for nonpolar compounds. Thermospray, the first really successful technique, used for polar as well as non-polar compounds, has largely been replaced by the atmospheric techniques. Particle beam ionization is useful for relatively volatile, small polar and nonpolar molecules (<1000 daltons).

A schematic of a **particle beam** interface is shown in Figure 21.13. The eluent from the HPLC column is nebulized using helium gas to form an aerosol in a reduced pressure chamber heated at 70°C. A cone with a small orifice is at the end of the chamber, which leads into a lower pressure area. The difference in pressure causes a supersonic expansion of the aerosol. The helium and the solvent molecules are lighter than the analyte molecules and tend to diffuse out of the stream and are pumped away. The remaining stream passes through a second cone into a yet lower pressure area, and then the analyte vapor passes into the ion source. The particle beam interface produces electron ionization (EI) spectra similar to those of GC–MS, so the vast knowledge of EI spectra can be used for analyte identification.

In the **thermospray** interface, the sample from the column passes through a heated tube and rapidly expands as a jet spray into a heated vacuum chamber. The solvent mist is electrostatically charged. The solvent is rapidly pumped away from the droplets, and a static charge is imparted to the particle, which enters through a skimmer into the mass spectrometer. Thermospray produces "soft ionization" with little or no fragmentation, allowing gentle ionization of nonvolatile, thermally labile organic compounds. The spectrum generally displays a protonated molecular ion.

The **electrospray ionization** interface is the most popular. It is also a soft ionization technique. The sample solution is sprayed across a high potential difference of a few thousand volts from a needle into an orifice in the interface (Figure 21.14). Heat and gas flow desolvate the charged droplets, giving charged analyte

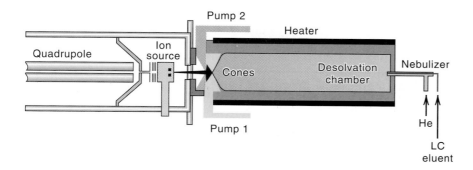

Fig. 21.13. Particle beam interface. [From A. N. Eaton, *Today's Chemist at Work*, October (1994) 34. Copyright 1994 by the American Chemical Society. Reproduced by permission of Micromass Ltd.]

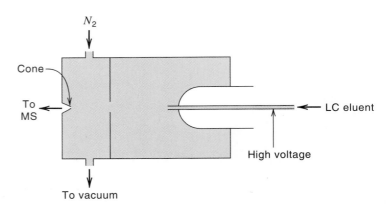

Fig. 21.14. Electrospray ionization interface for LC–MS.

molecules that enter the mass spectrometer. Electrospray ionization can produce multiply charged ions, with the number of charges increasing with molecular weight. For example, multiple charge sites are produced on proteins and peptides. This allows determination of relatively large molecular weights, well beyond the normal 2000- to 3000-dalton m/z ratio limit of the quadrupole mass spectrometer since the m/z ratio decreases by a factor of, for example, 5. Proteins of 50,000 daltons or larger can be measured. Of course, the number of charges needs to be evaluated. Techniques include: comparing two charge states that differ by a single charge and solving simultaneous equations; identifying species with the same charge but different adduct masses; and looking at m/z ratios for resolved isotopic clusters. Electrospray is good for charged, polar, or basic compounds.

Atmospheric pressure chemical ionization uses an atmospheric pressure ionization interface. The interface is similar to that used for ESI, but a corona discharge is used to ionize the analyte in an atmospheric pressure region. (Note: This is different than the chemical ionization source described in Chapter 20 for GC–MS.) The gas-phase ionization of less polar analytes is more efficient than with ESI, but the mass range is limited to about 2000 daltons. ESI and APCI are complementary. APCI has largely replaced thermal ionization as an interface in commercial instruments.

The quadrupole mass filter is the most popular mass analyzer for LC–MS because of its low cost, compactness, and ruggedness. In recent years, time-of-flight analyzers have been developed for interfacing with electrospray ionizers.

CHECK OUT SOME HPLC APPLICATIONS

High-performance liquid chromatography is widely used for many clinical, forensic, environmental, and industrial applications. It has replaced many gas chromatography procedures for forensic drug-screening analysis. You can find examples of specific types of analyses by perusing the chromatogram databases listed in the websites at the end of Chapter 19. For example, go to the Supelco database and look for determination of explosives extracted from water using solid-phase microextraction (SPME—Chapter 18). Or look at rapid drug profiling (of an over-the-counter drug). You will find many other examples.

21.2 Size Exclusion Chromatography

Molecules that can penetrate the gel particles are separated based on size and shape. Others pass straight through the column.

Size exclusion chromatography (also called molecular or gel chromatography) is a type of chromatography in which the stationary phase is a **molecular sieve**. It is used both in open column, gravity-fed form for preparative separations, and in high-performance separations. These sieves are polymeric carbohydrates and acrylamides that have an open network formed by the crosslinking of the polymeric chains. They are hydrophilic and, therefore, capable of absorbing water, whereupon swelling causes an opening of this structure. The degree of crosslinking will determine the size of the "holes."

Solvated molecules larger than the largest pores of the swollen gel cannot penetrate the gel particles and, therefore, will pass straight through the column through the spaces between the individual particles. Smaller molecules, however, will penetrate the open network in the particles to a varying degree, depending on their size and shape. They are, therefore, retarded to varying degrees and will be eluted in order of decreasing molecular size. Gels with a high degree of swelling are used to fractionate large molecules (generally high-molecular-weight substances), whereas the denser (lower swelling) gels are used for separation of low-molecular-weight compounds.

Names such as **gel filtration chromatography** (mobile phase is water), used by biochemists, and **gel permeation chromatography** (mobile phase is an organic solvent), used by polymer chemists, describe this technique. **Size exclusion chromatography**, however, is the recommended term. Molecular weight distribution of polymers can be obtained by this technique, and proteins, enzymes, peptides, nucleic acids, hormones, polysaccharides, and so on can be separated.

The **exclusion limit** is the molecular weight of that molecule that will just permeate the gel and be retarded. This can range from a molecular weight of 1000 to several million, depending on the gel. It should be emphasized that separations are based on a molecule's size and configuration rather than just its molecular weight, but there is generally a correlation. Generally, *molecules smaller than the exclusion limit can be fractionated down to a certain limiting size* (see Table 21.1).

The gels must be equilibrated for a few hours to a day or more with the solvent to be used, depending on the solvent uptake. Those with loose crosslinking designed for high-molecular-weight substances require the longer periods of soaking.

Sephadex is a popular molecular-sieve material for the separation of proteins. It is a polymeric carbohydrate material that, because of hydroxyl groups along the polymer chain, is fairly polar and so will adsorb water. The amount of crosslinking

Proteins can be separated by molecular exclusion chromatography.

Table 21.1
Sephadex Gels[a] and Bio-Gels[b]

Sephadex Type	Fractionation Range[c] for Peptides and Globular Proteins (MW)	Bio-Gel Type	Fractionation Range (MW)
G-10	Up to 700	P-2	100–1,800
G-15	Up to 1,500	P-4	800–4,000
G-25	1,000–5,000	P-6	1,000–6,000
G-50	1,500–30,000	P-10	1,500–20,000
G-75	3,000–70,000	P-30	2,500–40,000
G-100	4,000–150,000	P-60	3,000–60,000
G-150	5,000–400,000	P-100	5,000–100,000
G-200	5,000–800,000	P-150	15,000–150,000
		P-200	30,000–200,000
		P-300	60,000–400,000

[a] Courtesy of Pharmacia Fine Chemicals Inc.
[b] Courtesy of Bio-Rad Laboratories.
[c] Upper limit is the exclusion limit.

in the preparation can be carefully controlled to give different pore sizes and exclusion limits. Gels are characterized with respect to their swelling ability by their "water regain." This represents the amount of water imbibed by the gels on swelling. The type numbers of the Sephadex gels refer to the water-regaining values of the gels. Sephadex G-10, thus, has a water-regaining value of about 1 mL/g dry gel, and Sephadex G-200 has a value of about 20 mL/g. Several types of Sephadex gels and the fractionation range of molecules are listed in Table 21.1. These gels are insoluble in water and are stable to mild redox agents as well as to bases and weak acids.

Bio-Gel is a more chemically inert series of molecular-sieve gels, consisting of polyacrylamides. These are insoluble in water and common organic solvents and can be used in the pH range of 2 to 11. The inert gel decreases the possibility of adsorption of polar substances; adsorption can be a variable with Sephadex, causing changes in the chromatographic behavior of these substances. Table 21.1 lists the different Bio-Gel preparations and their separation properties.

Styragel is a polystyrene gel that is useful for purely nonaqueous separations in methylene chloride, toluene, trichlorobenzene, tetrahydrofuran, cresol, dimethylsulfoxide, and so on. It cannot be used with water, acetone, or alcohols. Gels of this can be prepared with exclusion limits for molecular weights of from 1600 to 40 million.

Molecular sieves are useful for the desalting of proteins that have been partially fractionated by salting out with a high concentration of some salt. A gel with a low exclusion limit, such as Sephadex 25, will allow the proteins to pass right through the column while the salts are retained. The protein dilution is limited to the elution volume of the column (the volume external of the swelled gel to fill the column).

For high-performance analytical applications, small polystyrene or microporous silica particles of 5- to 10-μm diameter are used, with pore sizes of a few nanometers to several hundred nanometers. The controlled pore silica particles are coated with a hydrophilic phase to reduce solute adsorption. The polymeric particles can be used over a wider pH range (2 to 12) since silica is limited to pH 2 to 7.

21.3 Ion Exchange Chromatography

Cations or anions are separated by ion exchange chromatography.

While most other types of chromatography are used principally for separations of complex organic substances, ion exchange chromatography is particularly well suited for the separation of inorganic ions, both cations and anions, because the separation is based on exchange of ions in the stationary phase. It has also proved to be extremely useful for the separation of amino acids.

The stationary phase in ion exchange chromatography consists of beads made of a polystyrene polymer crosslinked with divinylbenzene. The crosslinked polymer (resin) has free phenyl groups attached to the chain, which can easily be treated to add ionic functional groups. There are basically four types of ion exchange reins used in analytical chemistry, and these are summarized in Table 21.2. Like size exclusion chromatography, ion exchange is used either in packed open tube, gravity-fed form, or in a high-performance mode known as *ion chromatography* (discussed below).

CATION EXCHANGE RESINS

These resins contain acidic functional groups added to the aromatic ring of the resin. The strong-acid cation exchangers have sulfonic acid groups, —SO_3H, which

Table 21.2

Types of Ion Exchange Resins

Type of Exchanger	Functional Exchanger Group	Trade Name
Cation		
Strong acid	Sulfonic acid	Dowex[a] 50; Amberlite[b] IR120; Ionac[c] CGC-240; Rexyn[d] 101; Permutit[e] Q
Weak acid	Carboxylic acid	Amberlite IRC 50; Ionac CGC-270; Rexyn 102; Permutit H-70
Anion		
Strong base	Quaternary ammonium ion	Dowex 1; Amberlite IRA 400; Ionac AGA-542; Rexyn 201; Permutit S-1
Weak base	Amine group	Dowex 3; Amberlite IR 45; Ionac AGA-316; Rexyn 203; Permutit W

[a] Dow Chemical Company.
[b] Mallinckrodt Chemical Works.
[c] J. T. Baker Chemical Company.
[d] Fisher Scientific Company.
[e] Matheson Coleman & Bell.

are strong acids much like sulfuric acid. The weak-acid cation exchangers have carboxylic acid groups, $-CO_2H$, which are only partially ionized. The protons on these groups can exchange with other cations:

$$n\mathrm{RzSO_3^-H^+} + M^{n+} \rightleftharpoons (\mathrm{RzSO_3})_n M + n\mathrm{H^+} \qquad (21.1)$$

and

$$n\mathrm{RzCO_2^-H^+} + M^{n+} \rightleftharpoons (\mathrm{RzCO_2})_n M + n\mathrm{H^+} \qquad (21.2)$$

where Rz represents the resin. The equilibrium can be shifted to the left or right by increasing $[\mathrm{H^+}]$ or $[M^{n+}]$ or decreasing one with respect to the amount of resin present.

Cation exchange resins are usually supplied in the hydrogen ion form, but they can easily be converted to the sodium ion form by treating with a sodium salt. The sodium ions then undergo exchange with other cations. The **exchange capacity** of a resin is the total number of equivalents of replaceable hydrogen per unit volume or per unit weight of resin, and it is determined by the number and strength of fixed ionic groups on the resin. The ion exchange capacity affects solute retention, and exchangers of high capacity are most often used for separating complex mixtures, where increased retention improves resolution.

Weak-acid cation exchange resins are more restricted in the pH range in which they can be used, from 5 to 14, while the strong-acid resins can be used from pH 1 to 14. At low pH values, the weak-acid exchangers will "hold on" to the protons too strongly for exchange to occur. Also, the weak-acid cation exchangers will not completely remove the cations of very weak bases, while strong-acid resins will. This is analogous to the incompleteness of a weak acid–weak base reaction. Weak-acid resins are generally used for separating strongly basic or multifunctional ionic substances such as proteins or peptides that are often firmly retained on strong-acid exchangers, while strong-acid resins are more generally preferred, especially for complex mixtures.

Strong-acid resins are used for most separations. Weak-acid resins are preferred for proteins and peptides that are retained too strongly by the strong acids.

ANION EXCHANGE RESINS

Basic groups on the resin in which the hydroxyl anion can be exchanged with other anions make up the anion exchange resins. There are strong-base groups (quaternary ammonium groups) and weak-base groups (amine groups). The exchange reactions can be represented by

$$n\text{RzNR}_3{}^+\text{OH}^- + \text{A}^{n-} \rightleftharpoons (\text{RzNR}_3)_n\text{A} + n\text{OH}^- \tag{21.3}$$

and

$$n\text{RzNH}_3{}^+\text{OH}^- + \text{A}^{n-} \rightleftharpoons (\text{RzNH}_3)_n\text{A} + n\text{OH}^- \tag{21.4}$$

where R represents organic groups, usually methyl.

The strong-base exchangers can be used over the pH range 0 to 12, but the weak-base exchangers only over the range of 0 to 9. The latter exchangers will not remove very weak acids, but they are preferred for strong acids that may be retained by strong-base resins, such as sulfonates.

Strong-base resins are generally applicable. Weak-base resins are used for separating strong acids.

CROSSLINKAGE

The greater the crosslinkage of the resin, the greater the difference in selectivities. Increasing the crosslinkage is expressed by manufacturers as percent of divinylbenzene. Generally, crosslinkage also increases the rigidity of the resin, reduces swelling, reduces porosity, and reduces the solubility of the resin. In general, medium-porosity materials are used for low-molecular-weight ionic species and high-porosity materials are used for high-molecular-weight ionic species. The degrees of crosslinkage is expressed by the manufacturers as percent of divinylbenzene. Generally, crosslinkage of 8 to 10% is used.

EFFECT OF pH—SEPARATION OF AMINO ACIDS

The ionic forms of many substances will be affected by the pH of the effluent solution. Hydrolysis of metal ions and of salt of weak acids and bases is controlled by adjusting the pH. Weak acids will not dissociate in high acid concentrations and will not exchange, and the same is true for weak bases in high alkaline concentrations. Control of pH is especially important in the separation of amino acids, which are **amphoteric** (can act as acids or bases). There are three possible forms:

Amino acids may be positively or negatively charged or neutral (isoelectric point). These three forms may be separated by a combination of cation and anion exchange resins.

$$\underset{A}{\text{R}-\underset{\underset{\text{NH}_3{}^+}{|}}{\text{CH}}-\text{CO}_2\text{H}} \overset{+\text{H}^+}{\rightleftharpoons} \underset{B}{\text{R}-\underset{\underset{\text{NH}_3{}^+}{|}}{\text{CH}}-\text{CO}_2{}^-} \overset{-\text{H}^+}{\rightleftharpoons} \underset{C}{\text{R}-\underset{\underset{\text{NH}_2}{|}}{\text{CH}}-\text{CO}_2{}^-}$$

Form B, called a **zwitterion**, is the dominant form at the pH corresponding to the **isoelectric point** of the amino acid. The isoelectric point is the pH at which the net charge on the molecule is zero. In more acid solutions than this, the $-\text{CO}_2{}^-$ group is protonated to form a cation (form A), while in more alkaline solutions, the $-\text{NH}_3{}^+$ group loses a proton to form an anion (form C). The isoelectric point will vary from one amino acid to another, depending on the relative acidity and basicity of the carboxylic acid and amino groups. Thus, group separations based on the isoelectric points are possible by pH control.

At a given pH, the amino acids can be separated into three groups by being passed successively through an anion and a cation exchange column. The uncharged zwitterions (isoelectric point) will pass through both columns, while the positively and negatively charged amino acids will each be retained by one of the columns. The groups can be further subdivided by changing the pH.

Moore and Stein [*J. Biol. Chem.*, **192** (1951) 663] successfully separated up to 50 amino acids and related compounds on a single Dowex-50 cation exchange column by a combination of pH and temperature control. (The temperature affects the equilibria involved.)

Automatic amino acid analyzers based on ion exchange separation are commercially available. The elution of each amino acid is automatically recorded by measuring the color formed between the amino acid and ninhydrin as it is eluted. By operating at pressures of several hundred psi, these perform separations in about 200 min. They have proved valuable to the biochemist as an aid in the elucidation of protein structure. The protein fragments are degraded to amino acids, which must be determined.

EFFECT OF COMPLEXING AGENTS—SEPARATION OF METAL IONS ON ANION EXCHANGE COLUMNS

Many metals can be separated on anion exchange columns by being converted to anions by complexation. The complexing agent is an anion such as chloride, bromide, or fluoride. Uncharged complexing agents also affect the equilibrium if they form a complex that must dissociate before the metal is exchanged or if they change its size. Many complexing agents are either weak acids or bases or are salts of these, and so a complicated interdependence of pH and complexation often results.

Some of the most successful separations of metals have been on anion exchange columns. A complexing acid is added in high concentration to form anionic complexes of the metals. Concentrated hydrochloric acid forms anionic chlorocomplexes with all the common metals, with the exception of the alkali and alkaline earth metals and Al(III), Ni(II), and Cr(III), and so all of these can be adsorbed on a quaternary ammonium anion exchange column.

Negatively charged chloro complexes of metal ions in HCl solutions are retained by anion exchange resins. They are dissociated and eluted by dilution of the acid.

21.4 Ion Chromatography

The application of HPLC techniques to ion exchange chromatography has become known as **ion chromatography**. This technique combines the separating power of ion exchange with the universality of the conductivity detector. In ordinary ion exchange chromatography, a conductivity detector is limited in use because of the high background conductance (millimhos) of the eluting agent. In 1970, William Bauman at Dow Chemical Company suggested a way to remove the background eluent using a second ion exchange column, and thus permit detection of analyte ions with a highly sensitive conductivity detector (micromhos). This second column is called the **suppressor column**. For anion analysis, this is a cation exchange column in the acid form; and for cation analysis, it is an anion exchange column in the base form. For ion chromatography, weak exchange resins are usually used, although strong ones are also used. The principles are illustrated in Figure 21.15.

Suppose the salt MA of an anion A^- is separated on an anion exchange resin and eluted with NaOH. Eluting from the column will be a mixture of MA and NaOH. Upon passing through the cation exchange suppressor column, the following reactions take place:

Ion chromatography is the high-performance form of ion exchange chromatography.

The suppressor column removes the eluent ions and exchanges the analyte ion for H^+ (cations) or OH^- (anions), so that a high-sensitivity conductivity detector can be used.

$$RzSO_3^-H^+ + M^+ + A^- \rightarrow RzSO_3^-M^+ + H^+ + A^-$$
$$RzSO_3^-H^+ + Na^+ + OH^- \rightarrow RzSO_3^-Na^+ + H_2O$$

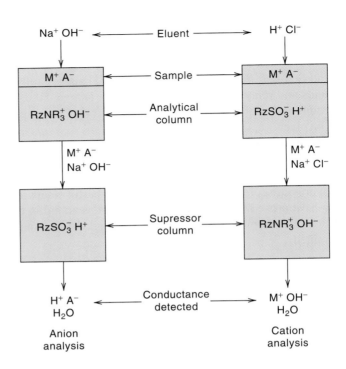

Fig. 21.15. Principles of ion chromatography.

So the NaOH is converted to H_2O, and the analyte ion is converted to the corresponding acid HA. The acid HA is detected by the conductivity detector. For cation analytes, a mixture of MA and HCl emerges from the first (analytical) column. These react in the suppressor anion exchange column as follows:

$$RzNR_3^+OH^- + M^+ + A^- \rightarrow RzNR_3^+A^- + M^+ + OH^-$$

$$RzNR_3^+OH^- + H^+ + Cl^- \rightarrow RzNR_3^+Cl^- + H_2O$$

The suppressor column must be regenerated periodically.

The HCl is converted to H_2O and the analyte cation is converted to the corresponding base MOH, which is detected by the conductivity detector.

The suppressor column obviously will eventually become depleted and will have to be regenerated (with HCl for the cation exchanger and with NaOH for the anion exchanger). The suppressor column is usually a small-volume bed of a high-capacity resin to minimize band spreading in the column. Since microgram or smaller quantities of analytes are usually separated, a low-capacity analytical column is used coupled with relatively low eluent concentration (1 to 10 mM).

Ion chromatography is particularly useful for the determination of anions. A typical eluting agent consists of a mixture of $NaHCO_3$ and Na_2CO_3, and these are converted to low-conductivity carbonic acid. Anions such as F^-, Cl^-, Br^-, I^-, NO_2^-, NO_3^-, SO_4^{2-}, PO_4^{3-}, SCN^-, IO_3^-, and ClO_4^-, as well as organic acids or their salts, can be readily determined in a matter of minutes, down to parts per million or lower levels.

If solutions are too dilute for direct analysis, analytes can be concentrated first on an ion exchange concentrator column. Low parts-per-billion concentrations can be measured in this way.

Electronic suppression of background conductance avoids the use of a suppressor column.

The advent of high-performance conductivity detectors with a wide dynamic range and electronic suppression of background conductance has allowed the development of ion chromatography without a suppressor column. A low-capacity analytical column is combined with low-concentration eluent, typically phthalate buffers, for anion measurements. The advantages of avoiding a suppressor column

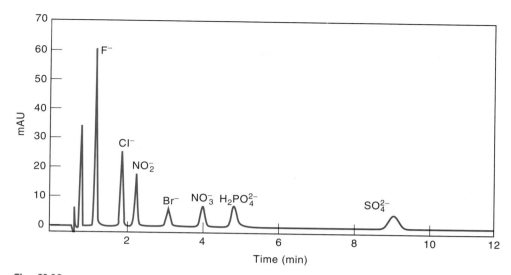

Fig. 21.16. Ion chromatography separation of anions. (Courtesy of Hewlett-Packard Company.)

are that band broadening from the column is eliminated, and anions of weak acids, such as cyanide and borate, are more readily detected because they are only weakly ionized in neutral or weakly acidic solution.

Ion exchange particles for ion chromatography are either resin particles with functionalized surfaces or nonporous silica particles coated with an ion exchange film or small ion exchange particles (Figure 21.4). Resins include functionalized polystyrene–divinylbenzene copolymers and polymethacrylate.

Figure 21.16 shows a typical anion analysis by ion chromatography. Such analyses would be difficult to perform by other methods. Ion chromatography forms the basis of automatic amino acid analysis.

Samples are often pretreated by passing through a small ion exchange pre-column. This concentrates the ions of interest and allows the sample matrix material to pass through unretained. Then the ions are eluted onto the analytical column. The same eluent used for the separation can be used for stripping the concentrator column, provided the capacity of the concentrator column is less than 40% of that of the analytical column. By concentrating ions from a 20-mL sample, low parts-per-billion can be measured. See C. A. Lucy, *LC·GC*, **14**(5) (1996) 406, for a good discussion of the practical aspects of ion chromatographic separation.

21.5 Thin-Layer Chromatography

Thin-layer chromatography (TLC) is a planar form of chromatography useful for wide-scale qualitative analysis screening and can also be used for quantitative analysis. The stationary phase is a thin layer of finely divided adsorbent supported on a glass or aluminum plate, or plastic strip. Any of the solids used in column liquid chromatography can be used, provided a suitable binder can be found for good adherence to the plate.

A sample is spotted onto the plate with a micropipet, and the chromatogram is "developed" by placing the bottom of the plate or strip (but not the sample spot) in a suitable solvent (see Figure 21.17). The solvent is drawn up the plate by capillary action, and the sample components move up the plate at different rates,

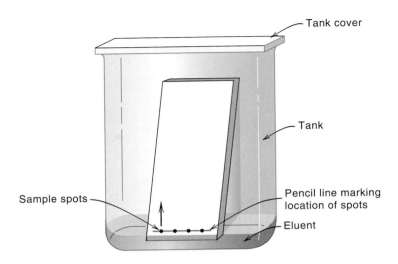

Fig. 21.17. Thin-layer chromatography setup.

depending on their solubility and their degree of retention by the stationary phase. Following development, the individual solute spots are noted or are made visible by treatment with a reagent that forms a colored derivative. The spots will generally move at a certain fraction of the rate at which the solvent moves, and they are characterized by the **R_f value**:

$$R_f = \frac{\text{distance solute moves}}{\text{distance solvent front moves}} \qquad (21.5)$$

where the distances are measured from the center of where the sample was spotted at the bottom of the plate. The solvent front will be a line across the plate. The distance the solute moves is measured at the center of the solute spot or at its maximum density, if tailing occurs. The R_f value, then, is characteristic for a given stationary phase and solvent combination. Because of slight variation in plates, it is always a good idea to determine the R_f value on each set of plates.

DEVELOPING THE CHROMATOGRAM

A typical setup for thin-layer chromatography is shown in Figure 21.17. A thin pencil line is drawn across the plate a few centimeters from the bottom, and the sample is spotted on this for future reference in R_f measurements. The spot must be made as small as possible for maximum separation and minimum tailing. It is best done dropwise with a warm air blower (e.g., a hair dryer) to evaporate the solvent after each drop. The plate is placed in a chamber with its end dipping in the developing solvent. A closed (presaturated) chamber must be used to saturate the atmosphere with the solvent and prevent it from evaporating from the surface of the plate as it moves up. The developing may take 10 min to 1 h, but it requires no operator time. The amount of development will depend on the complexity of the mixture of solutes being separated. If a wide plate is used, several samples and standards can be spotted along the bottom and developed simultaneously.

Development times of 5 min can be accomplished by using small microscope slides for TLC plates, and preliminary separations with these can be conveniently used to determine the optimum developing conditions. Typically, sample sizes range from 10 to 100 μg per spot (e.g., 1 to 10 μL of a 1% solution). Sample spots should be 2 to 5 mm in diameter.

A principle advantage of this technique is that greater separating power can be achieved by using **two-dimensional thin-layer chromatography**; a large square TLC plate is used, and the sample is spotted at a bottom corner of the plate. After development with a given solvent system, the plate is turned 90° and further development is obtained with a second solvent system. Thus, if two or more solutes are not completely resolved with the first solvent, it may be possible to resolve them with a second solvent. Proper control of the pH is often important in achieving efficient separations.

DETECTION OF THE SPOTS

If the solutes fluoresce (aromatic compounds), they can be detected by shining an ultraviolet light on the plate. A pencil line is drawn around the spots for permanent identification. Color-developing reagents are often used. For example, amino acids and amines are detected by spraying the plate with a solution of ninhydrin, which is converted to a blue or purple color. After the spots are identified, they may be scraped off and the solutes washed off (eluted) and determined quantitatively by a micromethod.

Frequently, colorless or nonfluorescent spots can be visualized by exposing the developed plate to iodine vapor. The iodine vapor interacts with the sample components, either chemically or by solubility, to produce a color. Thin-layer plates and sheets are commercially available that incorporate a fluorescent dye in the powdered adsorbent. When held under ultraviolet light, dark spots appear where sample spots occur due to quenching of the plate fluorescence.

A common technique for organic compounds is spraying the plate with a sulfuric acid solution and then heating it to char the compounds and develop black spots. This precludes quantitative analysis by scraping the spots off the plate and eluting for measurement.

STATIONARY PHASES FOR TLC

The stationary phase consists of a finely divided powder (particle size 10 to 50 μm). It can be an adsorbent, an ion exchanger, or a molecular sieve, or it can serve as the support for a liquid film. An aqueous slurry of the powder is prepared, usually with a binder such as plaster of paris, gypsum, or poly(vinyl alcohol) to help it adhere to the backing material. The slurry is spread on the plate in a thin film, typically 0.1 to 0.3 mm, using a spreading adapter to assure uniform thickness. Adapters are commercially available. The solvent is evaporated off and adsorbents are activated by placing in an oven at 110°C for several hours. Commercially prepared plates and strips on plastic are available.

The same stationary phases that are used in column chromatography can be used in TLC.

The most commonly used stationary phases are **adsorbents**. Silica gel, alumina, and powdered cellulose are the most popular. Silica gel particles contain hydroxyl groups on their surface which will hydrogen bond with polar molecules. Adsorbed water prevents other polar molecules from reaching the surface, so the gel is activated by heating to remove the adsorbed water. Alumina also contains hydroxyl groups or oxygen atoms. Alumina is preferred for the separation of weakly polar compounds, but silica gel is preferred for polar compounds such as amino acids and sugars. Magnesium silicate, calcium silicate, and activated charcoal may also be used as adsorbents. Adsorbents are sometimes not activated by heating, in which case the residual water acts as the stationary phase.

Adsorbents are used most frequently.

Thin-film **liquid stationary phases** can be prepared for separation by liquid–liquid partition chromatography. The film, commonly water, is supported on materials such as silica gel or diatomaceous earth, as in column chromatography. Either silica gel or diatomaceous earth may be silanized to convert the surfaces to non-polar methyl groups for reversed-phase thin-layer chromatography.

Ion exchange resins are available in particle sizes of 40 to 80 μm, suitable for preparing thin-layer plates. Examples are Dowex 50W strong-acid cation exchange and Dowex 1 strong-base anion exchange resins, usually in the sodium or hydrogen or the chloride forms, respectively. An aqueous slurry of six parts resin to one part cellulose powder is suitable for spreading into a 0.2- to 0.3-mm layer.

Size exclusion thin layers can be prepared from Sephadex Superfine. The gel is soaked in water for about 3 days to complete the swelling, and then spread on the plate. The plates are not dried, but stored wet. The capillary action through these molecular sieves is much slower than with most other thin layers, typically only 1 to 2 cm/h, and so development takes 8 to 10 h, compared to about 30 min for other stationary phases.

MOBILE PHASES FOR TLC

Use the same guidelines as for column chromatography.

In adsorption chromatography, the eluting power of solvents increases in the order of their polarities (e.g., from hexane to acetone to alcohol to water). A single solvent, or at most two or three solvents, should be used whenever possible, because mixed solvents tend to chromatograph as they move up the thin layer, causing a continual change in the solvent composition with distance on the plate. This may result in varying R_f values depending on how far the spots are allowed to move up the plate.

The developing solvent must be of high purity. The presence of small amounts of water or other impurities can produce irreproducible chromatograms.

QUANTITATIVE MEASUREMENTS

The powerful resolving power of two-dimensional thin-layer chromatography has been combined with quantitative measurements by optically measuring the density of chromatographic spots. This can be done by measuring the transmittance of light through the chromatographic plate or the reflectance of light, which is attenuated by the analyte color. Or, fluorescence intensity may be measured upon illumination with ultraviolet radiation. Full spectrum recording and multiple wavelength scanning (with diode arrays) capabilities are commercially available.

HIGH-PERFORMANCE THIN-LAYER CHROMATOGRAPHY (HPTLC)

High-performance TLC uses finer particles for fast and efficient separations using smaller samples.

The power of thin-layer chromatography has been enhanced by consideration of chromatography principles to improve the speed and efficiency of separation and by the development of instrumentation to automate sample application, development of the chromatogram, and detection, including accurate and precise *in situ* quantitation as mentioned above. The use of a very fine particle layer results in faster and more efficient separations. The particle size has a narrower distribution range, with an average size of 5 μm, instead of the average 20 μm for conventional TLC. Mechanical applicators permit reproducible application and reduction in the diameter of the starting spots. Smaller volume samples are used compared with conventional TLC, about one-tenth, and separation times are reduced by a factor of 10. In addition to precoated silica gel layers, a range of chemically bonded phases, similar to those used in normal- and reversed-phase high-performance liquid chromatography, are available.

The very fine particles used in HPTLC slows the movement of the mobile phase after a relatively short distance. To overcome this limitation, a "forced-flow" technique has been employed, using a pressurized chamber. The mobile phase is delivered with the aid of a pump at a constant velocity through a slit in a plastic sheet covering the stationary phase. For details, see the studies by Kalász and co-workers [*J. Chromatogr. Sci.*, **18** (1980) 324; *Chromatographia*, **18** (1984) 628].

Modern thin-layer chromatography can be complementary to HPLC. It allows the processing of many samples in parallel, providing low-cost analysis of simple mixtures for which the sample workload is high. The TLC plates acts as "storage detectors" of the analyte if they are saved.

21.6 Electrophoresis

Electrophoretic methods are used to separate substances based on their charge-to-mass ratios, using the effect of an electric field on the charges of these substances. These techniques are widely used for charged colloidal particles or macromolecular ions such as those of proteins, nucleic acids, and polysaccharides. There are several types of electrophoresis, **zone electrophoresis** being one of the most common.

In zone electrophoresis, proteins are supported on a solid so that, in addition to the electric migration forces, conventional chromatographic forces may enter into the separation efficiency. There are several types of zone electrophoresis according to the different supports. The common supports include starch gels, polyacrylamide gels, polyurethane foam, and paper. Starch gel electrophoresis has been a popular technique, although it is now somewhat superseded by the use of polyacrylamide gels, which minimize convection and diffusion effects. A block or "plate" of starch gel is prepared, and the sample is applied in a narrow band (line) across the block about midway between the ends, which are contacted with electrodes through a connecting bridge. When current is passed through the cell, the different components of a mixture move with velocities that depend on their electric charges, their sizes, and their shapes. As electrophoresis proceeds, the negatively charged components migrate toward the anode and the positively charged components migrate toward the cathode. The result is a series of separated bands or lines of sample constituents, such as visualized by a stain.

Very complex mixtures can be resolved with zone electrophoresis. For example, the starch gel-electrophoretic separation of plasma proteins reveals 18 components. A densitometer can be used to measure the intensity of the colored zones and thereby obtain quantitative information. Capillary gel electrophoresis is a powerful variant of this (discussed below).

The migration rate of each substance depends on the applied voltage and on the pH of the buffer employed. The applied voltage is expressed in volts per centimeter. It is up to 500 V in low-voltage electrophoresis and can be several thousand volts in high-voltage electrophoresis. The latter is used for high-speed separation of low-molecular-weight substances. Macromolecules have lower ionic mobilities and are less amenable to high-voltage separations.

Zone electrophoresis is used largely in clinical chemistry and biochemistry for separating amino acids and proteins. These contain amino and carboxylic acid groups that can ionize or protonate, depending on the pH. At a certain pH, the net charge of an amino acid is zero and it exists as a **zwitterion** (see Chapter 8) that exhibits no electrophoretic mobility. This pH is the **isoelectric point** of the amino acid.

In Chapter 25, we describe the use of gel electrophoresis for separation of nucleic acids in DNA sequencing.

Large molecules, such as proteins, migrate in an electric field based on their charge-to-mass ratios, but also interact chromatographically with the support.

Mobility is affected by the applied voltage and the pH (which influences the charge on the analyte).

Proteins do not migrate at the pH of their isoelectric point.

21.7 Capillary Electrophoresis

A relatively new separation technique that is capable of separating minute quantities of substances in relatively short time with high resolution is **capillary electrophoresis** (CE). It offers the ability to analyze a nanoliter (10^{-9} L) of sample, with over 1 million theoretical plates and a detection sensitivity of injected components at the attomole (10^{-18} mol) level or less!

The basic setup for the technique is illustrated in Figure 21.18. The instrumentation requirements (except for perhaps the detector) are actually very simple, and the system is easy to use. The separation medium is a fused silica capillary tube (e.g., 25 to 75 μm i.d., 25 to 100 cm long) containing an appropriate electrolyte. A small volume of sample is introduced into one end of the capillary (see below) and then each end of the capillary is inserted in an electrolyte buffer solution (usually the same as in the capillary tube). Platinum electrodes immersed in each solution are connected to a direct current (dc) high-voltage source, capable of delivering currents up to ca. 250 μA at voltages ranging from 1000 to 30,000 V.

A detector, for example, a UV absorbance detector, through which the solution flows, is placed near or at one end of the capillary. A focused beam is passed through the capillary and may be collected by an optical fiber coupled to a photomultiplier tube. The short pathlengths (10 to 100 μm) involved make sensitive detection a challenge. But the small peak volumes, often less than 1 nL, lead to very low detection limits, even with moderately sensitive detectors (i.e., the solute is concentrated in a very small volume). The use of laser sources, especially for fluorescence detection, has pushed detection limits to zeptomoles (10^{-21} mol)! A capillary electrophoresis instrument is shown in Figure 21.19.

HOW DOES CE WORK? THE POWER OF ELECTROOSMOTIC FLOW

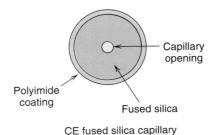

CE fused silica capillary

Capillary opening

Polyimide coating

Fused silica

Electroosmosis is the bulk flow of solvent (solution) through an electric field. All analytes flow in the same direction, with the positive ones migrating the most rapidly and the negative ones least rapidly.

The capillary is made of fused silica. The surface of the internal walls contains ionizable silanol groups (SiOH). The capillary is filled with a buffer. Above about pH 2, the silanol groups ionize to produce a negative charge on the capillary surface, call the **zeta potential**. This attracts cations from the buffer solution to create an electrical double layer along the walls (Figure 21.20). When a high dc voltage is applied, the mobile-phase positive charges in the diffuse outer double layer migrate in the direction of the cathode. Because the ions are solvated, the buffer fluid is dragged along by the migrating charge, creating a solution flow of the bulk solvent up to several hundred nanoliters per minute (depending on the pH, buffer concentration, and other factors that affect the zeta potential). This is called **electroosmotic**

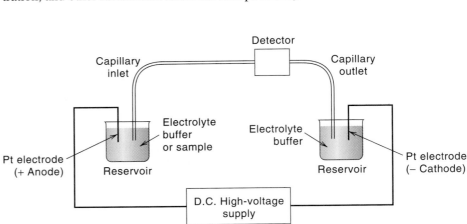

Fig. 21.18. Capillary electrophoresis system.

Fig. 21.19. Capillary electrophoresis instrument. (Courtesy of Agilent Technology.)

flow (EOF). Meanwhile, the analyte molecules are subjected to **electrophoretic mobility**, that is, the cations are attracted toward the anode and the anions toward the cathode. But the solution flow toward the cathode results in unidirectional flow of all analytes, regardless of the charge. The smaller, more positively charged ions migrate most rapidly and will be detected first; and the larger, more negatively charged ones will migrate the most slowly. Neutral molecules migrate at the electro-osmosis flow rate, since they are not accelerated or retarded by the electric field, and are unresolved.

WHY DOES CE HAVE SUCH HIGH RESOLVING POWER? AGAIN, THE MAGIC OF ELECTROOSMOTIC FLOW

We can get an understanding of the difference between pressure-driven chromatography systems and electroosmotically driven CE systems by comparing the flow profiles (Figure 21.21). In pressure-driven flow, the flow profile is parabolic, with the flow at the center moving at twice the average velocity (**laminar flow**). This results in band broadening, which is a reason GC or LC peaks become broader the

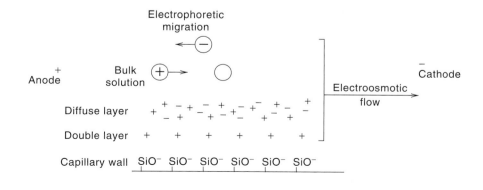

Fig. 21.20. Charge distribution and electroosmotic flow in fused silica capillary.

Fig. 21.21. Pressure-driven versus electroosmotically driven flow profiles.

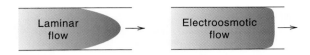

further they migrate. But in CE, the electroosmotic flow is generated along the entire length of the capillary, producing a constant flow all along the capillary (except right at the wall where the double layer is fixed). As a result, the flow profile is pluglike, and analyte molecules are swept along at the same rate across the capillary, which minimizes sample dispersion and generates very sharp peaks.

In CE, there are no eddy diffusion or mass transfer effects, only molecular diffusion broadening. The separation efficiency of CE is 10–100 times that of HPLC.

There is no packing material and no stationary phase. So there is no eddy diffusion (A term) and no equilibrium mass transfer (C term), only molecular diffusion (B term). A key to the high separation efficiency of this technique is the large surface area-to-volume ratio of the capillary, which allows efficient cooling by heat dissipation through the capillary walls. This minimizes band broadening by thermal effects caused by resistive heating. In fact, the Joule heating generated by the application of a high voltage is what limits the applicable voltage in other electrophoresis techniques, and hence the speed of separations. The CE capillary has a thick wall to help dissipate the heat.

CE is not really a chromatographic method.

You have probably noticed that the CE mechanism actually does not include a chromatographic distribution mechanism. Consequently, it is as readily applicable to macromolecules as to smaller ones. Hence, it is valuable for the separation of large biomolecules. Chemical modification of the silica wall or addition of detergents to the background electrolyte is often required to eliminate wall adsorption of proteins.

SAMPLE INTRODUCTION IN CE

The sample, typically a few nanoliters, can be introduced into the capillary by **hydrostatic injection** (gravity, pressure, or vacuum) or by **electromigration**. The sample volume should generally be less than 2% of the total capillary length. For gravity introduction, the capillary sample end is dipped into the sample (which may be as small as 5 μL) and raised for a short predetermined time to allow sample to flow into the capillary. Or, it is inserted into a pressurized vial to force sample into the capillary. Or it is drawn in by suction from the other end of the capillary. After injection, the sample vial is replaced with a buffer reservoir. Alternatively, the sample end is immersed in the sample solution and a relatively low voltage is applied for a few seconds, for example, 2000 V for 10 s. This injects the small volume of sample by electroosmosis.

Reproducibility with hydrostatic injection is on the order of 1 to 2%. The advantage of electroosmosis injection is that more sample can be introduced, improving detection limits. The disadvantage is that because of differences in ion mobilities, the sample plug that enters the capillary is not representative of the sample.

One way to sharpen the sample zone in electroosmotic injection and improve resolution, sensitivity, and representation is by dissolving the sample in a lower ionic strength solution than that of the separation electrolyte when injecting electroosmotically. In this case, the field strength is higher in sample zone. The sample ions migrate rapidly toward the capillary buffer until they encounter the electrolyte boundary and a lower electric field. This is a technique called "stacking" and focuses the sample zone into a narrow band about 10-fold smaller. The sample should be dissolved in a 1:10 dilution of the capillary electrolyte or, ideally, water.

DETECTORS IN CE

The detector is placed near or at the cathode end of the capillary, where solution flows. The short pathlength across the capillary makes sensitive detection a challenge. But the small peak volumes, often less than 1 nL, lead to very low detection limits, even with moderately sensitive detectors (i.e., the solute is in a very small volume).

The most common detector employed is a UV absorbance detector. A focused beam may be passed through the capillary (where a portion of the protective sheath is removed), and may be collected by an optical fiber coupled to a photomultiplier tube. To increase the pathlength, the light beam may enter the capillary at an angle through a hole cut in the sheath on one side of the capillary and be internally reflected down the capillary and exit at a hole downstream on the other side. A reflective coating may be placed on the capillary wall.

A flow cell may be inserted in the capillary. Figure 21.22 shows the cross section of a commercial micro-Z-cell in which the light path is along the horizontal z axis for increased pathlength. Fluorescence detection is used for analytes that fluoresce. The use of laser sources (laser-induced fluorescence) has pushed detection limits to zeptomoles (10^{-21})!

Other detectors include electrochemical, either conductometric or amperometric. Also, mass spectrometry detection has become quite popular, using an electrospray-type interface for introduction into a quadrupole mass spectrometer, similar to that used for HPLC. The strong electric field at the end of the capillary creates an aerosol of charged microdroplets, and the solvent evaporates to give gaseous ions.

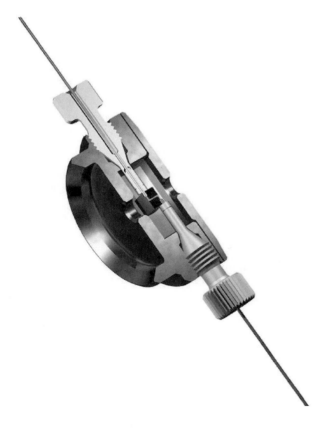

Fig. 21.22. Cross section of capillary electrophoresis Z-flow cell. (Courtesy of Agilent Technology.)

THE BASICS: PROPERTIES OF CE SEPARATIONS

We can characterize the separation efficiency in CE as we did for chromatography. The major dispersive effect in CE is molecular diffusion of the solute. The plate height, absent other external broadening (and with Joule heating minimized) is given by

$$H = \frac{B}{\bar{u}} \tag{21.6}$$

Here, $\bar{u}$ is virtually equal to u since the flow is constant throughout the column. The net or apparent mobility of a solute, μ_{net}, is the sum of the electrophoretic mobility, μ_{ep}, and the electroosmotic flow, μ_{es}:

$$\mu_{net} = \mu_{ep} + \mu_{es} \tag{21.7}$$

The electrophoretic mobility is given by

$$\mu_{ep} = \frac{z}{6\pi\eta r} \tag{21.8}$$

where z = number of ionic charges on solute
           η = solution viscosity
           r = ionic radius

It will be positive for a cation and negative for an anion. The number of plates is determined by three factors:

$$N = \frac{\mu_{net} V}{2D} \tag{21.9}$$

where V is the applied voltage and D is the solute diffusion coefficient (cm²/s). So plate count is increased by increasing the voltage. Large molecules such as proteins and nucleotides have very small diffusion coefficients, and so it is possible to obtain several million plates for these molecules. Note that, in contrast to chromatography, the number of plates is independent of the column length. But we can use a higher voltage with longer columns, which increases the plate count.

 Example 21.3

For a representative 50-cm capillary column operated at 30,000 V, the typical net mobility for a 10-min migration time is about 2×10^{-8} m² V⁻¹ S⁻¹. Calculate the number of plates for a small ion, Li⁺ ($D = 1.0 \times 10^{-9}$ m² s⁻¹) and for a 100,000 MW protein with $D = 3 \times 10^{-11}$ m² s⁻¹.

Solution
For Li⁺:

$$N = \frac{(2 \times 10^{-8} \text{ m}^2 \text{ V}^{-1})(30{,}000 \text{ V})}{2(1.0 \times 10^{-9} \text{ m}^2 \text{ s}^{-1})} = 300{,}000 \text{ plates}$$

For the protein:

$$N = \frac{(2 \times 10^{-8} \text{ m}^2 \text{ V}^{-1})(30{,}000 \text{ V})}{2(3 \times 10^{-11} \text{ m}^2 \text{ s}^{-1})} = 10{,}000{,}000 \text{ plates}$$

The number of plates for the protein is over 30-fold greater! Practical considerations usually prevent us from achieving such high plate count because of other zone-broadening sources. The injected sample has a finite width, there is some heating in the capillary, the detector has some dead volume, and so forth. We can routinely obtain 100,000 plates or more, but sometimes on the order of 1 million.

• •

The migration rate, v, of an ion in an electric field is a function of the column length L:

$$v = \mu_{net}E = \mu_{net}\frac{V}{L} \tag{21.10}$$

where E is the electric field strength (V/cm). The time to reach the detector, t_D, is the distance, l_D, to the detector divided by the velocity:

$$t_D = \frac{l_D}{v} = \frac{l_D L}{\mu_{net}V} \tag{21.11}$$

If the detector is at the end of the capillary, then $t_D = L^2/(\mu_{net} V)$. So, for rapid separations, we should apply a high voltage across a short column. The migration rate is influenced by temperature via the viscosity term in Equation 21.8, and the buffer is less viscous at higher temperatures.

We can determine the electroosmotic flow using a neutral marker molecule such as methanol, acetone, or benzene, from the time it takes to reach the detector and the electric field:

$$\mu_{eo} = \frac{1}{Et_D} \tag{21.12}$$

We can calculate the electrophoretic mobility of an ion, then, from Equation 21.7. The electrosmotic flow is influenced by the zeta potential, which is determined primarily by the pH.

The separation power and sensitivity of CE is illustrated in Figure 21.23. A mixture of 18 amino acids is separated in 30 min, at quantities ranging from 2 to 7 attomoles. The amino acids were derivatized with fluorescein isothiocyanate (FITC) to form fluorescent derivatives, and a fluorometric detector system was employed. The detection limits ranged down to 10^{-20} mol in 1 nL, corresponding to 10^{-11} M and 6000 molecules!

SEPARATION OF SMALL IONS BY CE

Capillary electrophoresis has become an attractive alternative to ion chromatography and atomic absorption spectrometry for measuring inorganic ions because of its multi-ion capability, high speed, high resolution and sensitivity, and relatively low cost. Because of their relatively high electrophoretic mobility, analysis times can be shortened for small ions by having them move electrophoretically in the

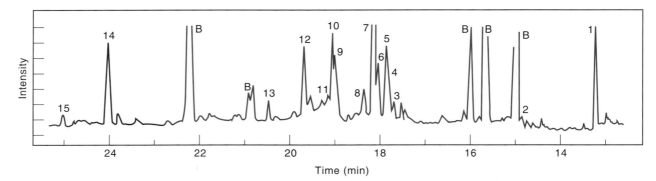

Fig. 21.23. Capillary zone electrophoresis separation of between 2 and 7 attomoles of 18 amino acids. The separation is driven by a 25-kV potential and a pH 10 buffer is used for separation. Injection was for 10 s at 2 kV. Amino acids: peak 1, Arg; 2, Lys; B are peaks associated with the reagent blank; 3, Leu; 4, Ile; 5, Trp; 6, Met; 7, Phe, Val, His, and Pro; 8, Thr; 9, Ser; 10, Cys; 11, Ala; 12, Gly; 13, Tyr; 14, Glu; and 15, Asp. (From N. J. Dovichi and Y. F. Cheng, *Am. Biotech. Lab.*, February (1989) (Reproduced by permission.)

same direction as the electroosmotic flow. For cations, no change is needed since they move in the same direction and have total migration times that are very fast. But, on the other end of the scale, a small -1 ion would have a slower total migration than a large -1 ion. The electrophoretic mobility of the small -1 ion is high, and this causes the total velocity to be slow since its electrophoretic mobility carries it away from the detector. For anions, the wall of the capillary may be treated with an alkylammonium salt, R_4N^+, for example, cetyltrimethylammonium bromide. The positively charged R_4N^+ ions attach to the negatively charged surface silanol groups and in turn create a double layer of anions, which reverses the electroosmotic flow (toward the anode). The detector, of course, is then placed at the anode end. Another approach for anions is to operate at a low pH where the silanol groups are not ionized, and electrophoretic migration of anions dominates toward the anode.

SEPARATION OF NEUTRAL MOLECULES: MECC

We mentioned that neutral molecules migrate electroosmotically together. So they are not separated in the type of CE we have discussed, which is called capillary-zone electrophoresis (CZE). If an anionic surfactant that forms micelles, such as sodium dodecylsulfate, is incorporated in the mobile phase, then we can separate neutrals. Above a certain concentration, called the **critical micelle concentration** (CMC), the surfactant molecules will self-aggregate, forming micelles in which the hydrophilic tails are turned inward, to give a nonpolar core in which neutral solutes can partition. The negatively charged hydrophobic head groups form an outer shell.

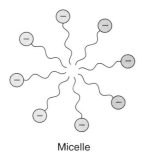

Micelle

The neutral solutes will partition between the micelles and the buffer, similar to what happens in HPLC, and different molecules interact differently. Since the micelles are negatively charged, their electrophoretic migration is toward the anode, but the net flow is toward the cathode, due to electroosmotic flow, but at a slowed rate. Remember, neutral molecules move at the electroosmotic rate. When they equilibrate with the micelle, their movement is slowed. The more time they spend inside the micelles, the longer the time to the detector. This technique is called **micellar electrokinetic capillary chromatography** (MECC); it is a form of chromatography in that the micelles act as a pseudo-stationary phase. The movement of cations is also affected because of electrostatic interactions with the charged micelles. MECC is useful for water-insoluble neutral compounds such as steroids.

CAPILLARY ELECTROCHROMATOGRAPHY

We can combine features of capillary electrophoresis and HPLC to form a hybrid technique that has some of the best features of each, called **capillary electrochromatography** (CEC), for the separation of neutral molecules as done in HPLC. The capillary is filled with reversed-phase chromatographic stationary-phase particles, for example, 1.5- to 3-μm C_{18} silica particles. There is no backpressure developed in electroosmotic pumping, so very fine particles can be used. Also, long columns can be used. A polar solvent mobile phase is driven by electroosmotic flow rather than pressure. The carrier buffer generally has 40 to 80% of organic solvent such as acetonitrile or methanol. Separation is based on the distribution of the solutes between the mobile phase and the stationary phase, as in HPLC. But the electroosmotic pumping results in a flat plug profile instead of a parabolic one, leading to narrow bands and high separation efficiencies. Very small volumes of sample can be injected. Typically, CEC exhibits about twice the number of plates as HPLC for the same size particles and same length. The combination of smaller particles and longer columns in CEC makes it possible to use columns with 100,000 to 500,000 plates, compared with 25,000 in HPLC.

CEC resolution is about double that in HPLC due to the plug electro-osmotic flow.

This is a true chromatographic technique. It is useful for separation of biomolecules. While in CZE, very high numbers of plates can be achieved for these molecules when the molecules are large, the differences in the charge-to-mass become small, making resolution by CZE difficult, and CEC offers a separation advantage.

CAPILLARY GEL ELECTROPHORESIS

This is a CE analog of conventional zone gel electrophoresis for the separation of macromolecules based on size. The capillary is filled with a porous polymer gel, and molecular sieving occurs as the molecules move through the gel, that is, separation is based on both electrophoretic mobility and molecular size. Very high resolution is achieved. The trend is to fill the capillary with a liquid gel matrix (pumpable gel solutions, such as derivatized celluloses dissolved in the run buffer). This allows replacement of the gel in the capillary to eliminate contamination problems from the sample matrix that occurs with fixed gels. This technique is widely used for separation of nucleotides in deoxyribonucleic acid (DNA) sequencing (Chapter 25).

Learning Objectives

WHAT ARE SOME OF THE KEY THINGS WE LEARNED FROM THIS CHAPTER?

- HPLC, p. 604
  - Stationary phases, particles, p. 606
  - Instrument components, p. 609
  - Method development—column, solvent selection, p. 613
  - Fast LC, p. 616
  - Narrow-bore columns, p. 616

Questions

1. Describe how high-performance liquid chromatography differs from conventional liquid chromatography.

2. Describe some commonly used detectors in liquid chromatography and their bases of operation.

3. In what order would the following compounds be eluted from an alumina column using n-hexane as the eluting solvent? CH_2CH_2OH; CH_3CHO; CH_3CO_2H.

4. What solvent would you choose to separate a group of hydrocarbons, $CH_3(CH_2)_xCH_3$, on an alumina column?

5. What is normal-phase chromatography? Reversed-phase chromatography?

6. What is the most common form of HPLC? Why?

7. Why are silica particles endcapped in bonded reversed-phase particles?

8. What are some commonly used nonpolar bonded phases for reversed-phase HPLC? What are bonded polar phases for normal-phase chromatography?

9. Briefly describe the differences between microporous particles, perfusion particles, and nonporous particles. What are their unique features/uses?

10. What is a guard column and why is it used?

11. What is the multi-solvent selection method? What two factors are optimized in solvent selection?

12. Why is mobile-phase gradient elution used in HPLC?

13. How does fast LC differ from conventional HPLC?

14. What advantage do narrow-bore columns have in HPLC?

15. How does temperature affect HPLC separations?

16. What are common interfaces/ionizers for LC–MS?

17. Describe the principle of size exclusion chromatography. What is the exclusion limit?

18. What is a molecular sieve?

19. Explain the difference between a cation exchange resin and an anion exchange resin.

20. Describe the factors that affect the selectivity of ion exchange resins.

21. Describe the principles of ion chromatography.

22. What is the retention time? The R_f value?

23. What are the largest and smallest R_f values possible?

24. Describe the basis of separation in electrophoresis.

25. Describe the principles of capillary electrophoresis. What are its advantages?

Problems

ION EXCHANGE CHROMATOGRAPHY

26. Alkali metal ions can be determined volumetrically by passing a solution of them through a cation exchange column in the hydrogen form. They displace an equivalent amount of hydrogen ions that appear in the effluent and can be titrated. How many millimoles of potassium ion are contained in a liter of solution if the effluent obtained from a 5.00-mL aliquot run through a cation exchange column requires 26.7 mL of 0.0506 M NaOH for titration?

27. The sodium ion in 200 mL of a solution containing 10 g/L NaCl is to be removed by passing through a cation exchange column in the hydrogen form. If the exchange capacity of the resin is 5.1 meq/g of dry resin, what is the minimum weight of dry resin required?

28. What will be the composition of the effluent when a dilute solution of each of the following is passed through a cation exchange column in the hydrogen form? (a) NaCl; (b) Na_2SO_4; (c) $HClO_4$; (d) $FeSO_4$; $(NH_4)_2SO_4$.

Recommended References

HIGH-PERFORMANCE LIQUID CHROMATOGRAPHY

1. V. R. Meyer, *Practical High-Performance Chromatography*, 3rd ed. New York: Wiley, 1999.

2. L. R. Snyder, J. J. Kirkland, and J. L. Glajch, *Practical HPLC Method Development*, 2nd ed. New York: Wiley, 1997.

3. A. Weston and P. R. Brown, *HPLC and CE: Principles and Practice*. San Diego: Academic, 1997.

4. U. D. Neue and M. Zoubair, *HPLC Columns: Theory, Technology, and Practice*. New York: Wiley, 1997.

5. R. P. W. Scott, *Liquid Chromatography for the Analyst*. New York: Dekker, 1994.

6. P. C. Sadek, *The HPLC Solvent Guide*. New York: Wiley, 1996.

7. L. Huber and S. A. George, eds., *Diode-Array Detection in HPLC*. New York: Dekker, 1993.

8. R. L. Cunico, K. M. Gooding, and T. Wehr, *Basic HPLC and CE of Biomolecules*. Hercules, CA: Bay Analytical Laboratory, 1998.

9. W. M. A. Niessen, *Liquid Chromatography-Mass Spectrometry*, 2nd ed. New York: Marcel Dekker, 1999.

SIZE EXCLUSION CHROMATOGRAPHY

10. *Sephadex-Gel Filtration in Theory and Practice*, Pharmacia Fine Chemicals, Upsalla, Sweden. This company also supplies yearly booklets of literature references and gel permeation chromatography.

ION EXCHANGE AND ION CHROMATOGRAPHY

11. J. Inczedy, *Analytical Applications of Ion Exchangers*. Oxford: Pergamon, 1996.

12. O. Samuelson, *Ion Exchange Separations in Analytical Chemistry*. New York: Wiley, 1963.

13. J. S. Fritz and D. T. Gjerde, *Ion Chromatography*, 3rd ed. New York: Wiley, 2000.

14. H. Small, *Ion Chromatography*. New York: Plenum, 1989.

15. H. Small and B. Bowman, "Ion Chromatography: A Historical Perspective," *Am. Lab.*, October 1998, 56C. By a pioneer in ion chromatography.

16. P. K. Dasgupta, "Ion Chromatography. The State of the Art," *Anal. Chem.*, **64** (1992) 775A.

THIN-LAYER CHROMATOGRAPHY

17. B. Fried and J. Sherma, *Thin-Layer Chromatography*, 4th ed. New York: Marcel Dekker, 1999.

18. J. Sherma and B. Fried, eds., *Handbook of Thin-Layer Chromatography*, 2nd ed. New York: Marcel Dekker, 1996.

CAPILLARY ELECTROPHORESIS

19. G. Lunn, *Capillary Electrophoresis Methods for Pharmaceutical Analysis*. New York: Wiley, 1999.

20. S. M. Palfrey, ed., *Clinical Applications of Capillary Electrophoresis*. Totowa, NJ: Humana, 1999.

CAPILLARY ELECTROCHROMATOGRAPHY

21. I. S. Krull, R. L. Stevenson, K. Mistry, and M. E. Schwartz, *Capillary Electrochromatography and Pressurized Flow Capillary Electrochromatography: An Introduction*. New York: HNB, 2000.

22. A. L. Crego, A. Gonzalez, and M. L. Marina, "Electrochromatography," *CRC Crit. Rev. Anal. Chem.*, **26** (1996) 261.

23. K. D. Altria, N. W. Smith, and C. H. Turnbull, "A Review of the Current Status of Capillary Electrochromatography Technology and Applications," *Chromatographia*, **46** (1997) 664.

24. L. A. Colon, K. J. Reynolds, R. Aliceamaldonado, and A. M. Fermier, "Advances in Capillary Electrochromatography," *Electrophoresis*, **18** (1997) 2162.

25. **Website:** www.ceandcec.com. Very basic coverage of both capillary electrophoresis and capillary electrochromatography, but has links to many other useful websites.

Chapter Twenty-Two

KINETIC METHODS OF ANALYSIS

"There is nothing permanent but change."
—Heraclitus

We mentioned in Chapters 6 and 14 the use of catalysts to alter the reaction rate in certain redox reactions; the titration reaction of As(III) with Ce(IV) is catalyzed by OsO_4. The catalyst is added in sufficiently high concentration to make the reaction occur immediately. If the catalyst concentration is low and the reaction slow, then we can measure the rate of the reaction and relate it to the catalyst concentration. In this chapter, we describe the basic kinetics of rate-limited reactions. Then we discuss reactions catalyzed by specific catalysts called enzymes and the measurement of the reaction rate to either determine the enzyme activity (concentration) or the concentration of the catalyst's substrate by adding a fixed amount of enzyme to the solution.

22.1 Kinetics—The Basics

Kinetics is the description of **reaction rates**. The **order** of a reaction defines the dependence of reaction rates on the concentrations of reacting species. Order is determined empirically and is not necessarily related to the stoichiometry of the reaction. Rather, it is governed by the **mechanism** of the reaction, that is, by the number of species that must collide for the reaction to occur.

The order of a reaction defines the number of species that must react, not the ratio in which they react (stoichiometry).

FIRST-ORDER REACTIONS

Reactions in which the rate of the reaction is directly proportional to the concentration of a single substance are known as first-order reactions. Consider the reaction

$$A \rightarrow P \qquad (22.1)$$

Substance A might be a compound that is decomposing to one or more products. The rate of the reaction is equal to the rate of disappearance of A, and it is proportional to the concentration of A:

$$-\frac{dA}{dt} = k[A] \qquad (22.2)$$

This is a **rate expression**, or **rate law**. The minus sign is placed in front of the term on the left side of the equation to indicate that A is disappearing as a function of time. The constant k is the **specific rate constant** at the specified temperature and has the dimensions of reciprocal time, for example, s^{-1}. The **order of a reaction** is the sum of the exponents to which the concentration terms in its rate expression are raised. Thus, this is a first-order reaction and its rate depends only on the concentration of A.

Equation 22.2 is known as the **differential form** of the first-order rate law. The **integrated form** of the equation is

$$\log[A] = \log[A]_0 - \frac{kt}{2.303} \qquad (22.3)$$

where $[A]_0$ is the initial concentration of A ($t = 0$) and $[A]$ is its concentration at time t after the reaction is started. This equation gives the amount of A that has reacted after a given time interval. It is a straight-line equation, and if t is plotted versus $\log[A]$ (which can be measured at different times), a straight line with slope $-k/2.303$ and intercept $\log[A]_0$ is obtained. Thus, the rate constant can be determined.

The rate of reaction slows down with time.

Note that, from Equation 22.2, the rate of the reaction (*not* the rate constant) will decrease as the reaction proceeds because the concentration of A decreases. Since $[A]$ decreases logarithmically with time (see Equation 22.3), it follows that the rate of the reaction will decrease exponentially with time. The time for one-half of a substance to react is called the **half-life** of the reaction, $t_{1/2}$. The ratio of $[A]/[A]_0$ at this time is $\frac{1}{2}$. By inserting this in Equation 22.3 and solving for $t_{1/2}$, we see that for a first-order reaction

$$t_{1/2} = \frac{0.693}{k} \qquad (22.4)$$

Consider a reaction complete after 10 half-lives. It really takes an infinite time for completion.

After the reaction is half complete, then one-half of the remaining reacting substances will react in the same time $t_{1/2}$, and so on. This is the exponential decrease we mentioned. Theoretically, it would take an infinitely long time for the reaction to go to completion, but for all practical purposes, it is complete (99.9%) after 10 half-lives. It is important to note that the half-life, and hence the time for the reaction to go to completion, is independent of the concentration for first-order reactions.

Radioactive decay is an important example of a first-order reaction.

SECOND-ORDER REACTIONS

Suppose we have the following reaction:

$$A + B \rightarrow P \qquad (22.5)$$

The rate of the reaction is equal to the rate of disappearance of either A or B. If it is empirically found to be

$$-\frac{dA}{dt} = -\frac{dB}{dt} = k[A][B]$$ (22.6)

then the reaction is first order with respect to [A] and to [B] and second order overall (the sum of the exponents of the concentration terms is 2). The specific reaction rate constant has the dimensions of reciprocal time and molarity, for example, $s^{-1} M^{-1}$.

The integrated form of Equation 22.6 depends on whether the initial concentrations of A and B ($[A]_0$ and $[B]_0$) are equal. *If they are equal*, the equation is

$$kt = \frac{[A]_0 - [A]}{[A]_0[A]}$$ (22.7)

If $[A]_0$ and $[B]_0$ *are not equal*, then

$$kt = \frac{2.303}{[B]_0 - [A]_0} \log \frac{[A]_0[B]}{[B]_0[A]}$$ (22.8)

If the concentration of one species, say B, is very large compared with the other and its concentration remains essentially constant during the reaction, then Equation 22.6 reduces to that of a first-order rate law:

By making the concentration of one species large compared to the other, a second-order reaction behaves as a (pseudo) first-order reaction.

$$-\frac{dA}{dt} = k'[A]$$ (22.9)

where k' is equal to $k[B]$; the integrated form becomes

$$kt = \frac{2.303}{[B]_0} \log \frac{[A]_0}{[A]}$$ (22.10)

Since $[B]_0$ is constant, Equation 22.10 is identical in form to Equation 22.3. This is a **pseudo first-order reaction**.

The half-life of a second-order reaction in which $[A]_0 = [B]_0$ is given by

The half-life of a second-order reaction depends on the concentrations.

$$t_{1/2} = \frac{1}{k[A]_0}$$ (22.11)

Thus, unlike the half-life of a first-order reaction, the half-life here depends on the initial concentration.

A reaction between A and B need not necessarily be second order. Reactions of a fraction rate order are common. A reaction such as $2A + B \rightarrow P$ may be third order (rate $\propto [A]^2[B]$), or it may be second order (rate $\propto [A][B]$), or a more complicated order (even a fractional order).

REACTION TIME

The time for a reaction to go to completion will depend on the rate constant k and, in the case of second-order reactions, the initial concentrations. A first-order reaction is essentially instantaneous if k is greater than 10 s^{-1} (99.9% complete in less than 1 s). When k is less than 10^{-3} s^{-1}, the time for 99.9% reaction exceeds 100 min. Although it is more difficult to predict the time for second-order reactions, they can generally be considered to be instantaneous if k is greater than about 10^3 or 10^4 s^{-1} M^{-1}. If k is less than 10^{-1} s^{-1} M^{-1}, the reaction requires hours for completion.

22.2 Enzyme Catalysis

Enzymes are proteins that are nature's catalysts in the body.

Enzymes are remarkable naturally occurring proteins that catalyze *specific* reactions with a high degree of efficiency. Enzymes range in formula weight from 10,000 to 2,000,000. They are, of course, intimately involved in biochemical reactions in the body, that is, the life process itself. The determination of certain enzymes in the body is, therefore, important in the diagnosis of diseases. But aside from this, enzymes have proved extremely useful for the determination of **substrates**, the substances whose reaction the enzymes catalyze.

ENZYME KINETICS

We can describe the rate equation for enzyme reactions from a simple reaction model. The typical enzyme-catalyzed reaction can be represented as follows:

$$E + S \underset{k_2}{\overset{k_1}{\rightleftharpoons}} ES \overset{k_3}{\rightarrow} P + E \tag{22.12}$$

The reaction rate is first order with respect to substrate and enzyme. If [S] is large, the reaction becomes zero order with respect to S.

where E is the enzyme, S is the substrate, ES is the **activated complex** that imparts a lower energy barrier to the reaction, P is the product(s), and the k's are the rate constants for each step. That is, the enzyme forms a complex with the substrate, which then dissociates to form product (Figure 22.1). The rate of reaction, R, is proportional to the complex concentration and, therefore, to the substrate and enzyme concentrations:

$$R = k_3[\text{ES}] = k[\text{S}][\text{E}] \tag{22.13}$$

Assuming that k_1 and k_2 are much larger than k_3, we find the rate of the reaction to be limited by the rate of dissociation of the activated complex.

Fig. 22.1. Mechanism of enzyme activity. [D. Leja, National Human Genome Research Institute (www.nhgri.nih.gov). Reproduced by permission.]

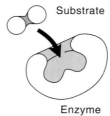
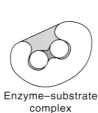
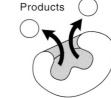

Substrate

Products

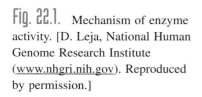

Enzyme   Enzyme–substrate complex

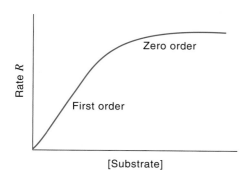

[Substrate]

Fig. 22.2. Dependence of enzyme-catalyzed reaction rate on substrate concentration. At high concentrations, the enzyme becomes saturated with substrate and the reaction rate becomes maximum and constant since [ES] becomes constant (Equation 22.13).

The dependence of an enzyme-catalyzed reaction rate on substrate concentration is illustrated in Figure 22.2. An enzyme is characterized by the number of molecules of substrate it can complex per unit time and convert to product, that is, the **turnover number**. As long as the substrate concentration is small enough with respect to the enzyme concentration that the turnover number is not exceeded, the reaction rate is directly proportional to substrate concentration, that is, it is first order with respect to substrate (Equation 22.13). If the enzyme concentration is held constant, then the overall reaction is first order and directly proportional to substrate concentration ($k[E] = $ constant in Equation 22.13). This serves as the basis for substrate determination.[1] However, if the amount of substrate exceeds the turnover number for the amount of enzyme present, the enzyme becomes *saturated* with respect to the number of molecules it can complex (saturated with respect to substrate), and the reaction rate reaches a maximum value. At this point, the reaction becomes independent of further substrate concentration increases, that is, becomes **pseudo zero order** if the enzyme concentration is constant (Figure 22.2); in Equation 22.13, [ES] becomes constant and $R = $ constant.

When the enzyme is saturated with respect to substrate, then the overall reaction is first order with respect to enzyme concentration ($k[S] = $ constant in Equation 22.13). This becomes the basis for enzyme determination since a linear relationship between reaction rate and enzyme concentration will exist. Since substrate is consumed in the reaction, however, it must be kept at a high enough concentration that the reaction remains zero order with respect to substrate during the time of the reaction (i.e., the enzyme remains saturated). Eventually, at high enzyme concentrations, insufficient substrate will be available for saturation, and a plot similar to Figure 22.2 will result.

When the enzyme is saturated with substrate, the reaction rate is proportional to the enzyme concentration.

PROPERTIES OF ENZYMES

The rate of an enzymatic reaction depends on a number of factors, including the temperature, pH, ionic strength, and so forth. The rate of the reaction will increase as the temperature is increased, up to a point. Above a certain temperature, the activity of the enzyme is decreased because, being a protein, it becomes *denatured*, that is, the tertiary structure of the enzyme is destroyed as hydrogen bonds are broken. The steric nature of an enzyme is critical in its catalytic mechanism. Most animal enzymes become denatured at temperatures above about 40°C.

As with other catalytic reactions, temperature changes as small as 1 or 2°C may result in changes as high as 10 to 20% in the reaction rate under analytical

Above the optimum temperature, the enzyme becomes denatured. When you cook an egg, the protein is denatured.

---

[1] Substrates need not be determined from the reaction rate. Instead, the reaction may be allowed to proceed until the substrate is completely converted to product. The concentration of the product is measured before (blank) and after the reaction. Each of these techniques is discussed in more detail below under the determination of enzymes and of enzyme substrates.

conditions. So it is important that the temperature be controlled during the measurements of enzyme reactions.

Enzymes should be stored at 5°C or less since they are eventually deactivated over a period of time at even moderate temperatures. Some enzymes lose activity when frozen.

There is also an optimum pH for enzyme reactions.

The reaction rate will be at a maximum at a certain pH, owing to complex acid–base equilibria such as acid dissociation between the substrate, the activated complex, and the products. Also, the maximum rate may depend on the ionic strength and on the type of buffer used. For example, the rate of aerobic oxidation of glucose in the presence of the enzyme glucose oxidase is maximum in an acetate buffer at pH 5.1, but in a phosphate buffer of the same pH, it is decreased.

The **activity** of an enzyme preparation will vary from one source to another because the enzymes are usually not purified to 100% enzyme. That is, the percent enzyme will vary from one preparation to another. The activity of a given preparation is expressed in **international units** (I.U.). An international unit has been defined by the International Union of Biochemistry as "the amount that will catalyze the transformation of one micromole of substrate per minute under defined conditions." The defined conditions will include temperature and pH. For example, a certain commercial preparation of glucose oxidase may have an activity of 30 units per milligram. Thus, for the determination of a substrate, a certain number of units of enzyme is taken. The **specific activity** is the units of enzyme per milligram of *protein*. **Molecular activity** is defined as units per molecule of enzyme, that is, it is the number of molecules of substrate transformed per minute per molecule of enzyme. The **concentration** of an enzyme in solution should be expressed as international units per milliliter or liter.

Enzyme concentrations are usually expressed in activity and not molar units.

ENZYME INHIBITORS AND ACTIVATORS

Although enzymes catalyze only certain reactions or certain types of reaction, they are still subject to interference. When the activated complex is formed, the substrate is adsorbed at an *active site* on the enzyme. Other substances of similar size and shape may be adsorbed at the active site. Although adsorbed, they will not undergo any transformation. However, they do compete with the substrate for the active sites and slow down the rate of the catalyzed reaction. This is called **competitive inhibition**. For example, the enzyme succinic dehydrogenase will specifically catalyze the dehydrogenation of succinic acid to form fumaric acid. But other compounds similar to succinic acid will competitively inhibit the reaction. Examples are other diprotic acids such as malonic and oxalic acids. Competitive inhibition can be reduced by increasing the concentration of the substrate relative to that of the interferent so that the majority of enzyme molecules combine with the substrate.

Noncompetitive inhibition occurs when the inhibition depends only on the concentration of the inhibitor. This is usually caused by adsorption of the inhibitor at a site other than the active site but one which is necessary for activation. In other words, an inactive derivative of the enzyme is formed. Examples are the reaction of the heavy metals mercury, silver, and lead with sulfhydryl groups (—SH) on the enzyme. The sulfhydryl group is tied up by the heavy metal ($ESH + Ag^+ \rightarrow ESAg + H^+$), and this reaction is irreversible. This is why heavy metals are poisons; they inactivate enzymes in the body.

Some enzymes require the presence of a certain metal for activation, perhaps to form a complex of the proper stereochemistry. Any substance that will complex with the metal ion may then become an inhibitor. For example, magnesium ion is required as an activator for a number of enzymes. Oxalate and fluoride will complex the magnesium, and they are inhibitors. Activators of enzymes are sometimes called **coenzymes**.

Substrate inhibition sometimes occurs when excessive amounts of substrate are present. In cases such as this, the reaction rate actually decreases after the maximum velocity has been reached. This is believed to be due to the fact that there are so many substrate molecules competing for the active sites on the enzyme surfaces that they block the sites, preventing other substrate molecules from occupying them.

THE MICHAELIS CONSTANT

As explained previously, an enzyme at a given concentration eventually becomes saturated with respect to substrate as the substrate concentration is increased, and the reaction rate becomes maximum, R_{max}. The **Lineweaver–Burk equation** describes the relationship between the enzyme effectiveness as a catalyst and the maximum rate:

$$\frac{1}{R} = \frac{1}{R_{max}} + \frac{K_m}{R_{max}[S]} \qquad (22.14)$$

where K_m is the **Michaelis constant**. The Michaelis constant is a measure of the enzyme activity and can be shown to be equal to $(k_2 + k_3)/k_1$ in Equation 22.12. It is also equal to the substrate concentration at one-half the maximum rate, $R_{max}/2$, as derived from Equation 22.14 by setting $R = R_{max}/2$. A plot of $1/[S]$ versus $1/R$ gives a straight line whose **intercept** is $1/R_{max}$ and whose **slope** is K_m/R_{max}. Thus, the Michaelis constant, which is characteristic for an enzyme with a substrate, can be determined.

Example 22.1

An enzyme reaction gave the following reaction rate absorbance data as a function of substrate concentration:

[S] (M)	R (ΔA/min)
0.0400	0.093
0.0100	0.231
0.0400	0.569
0.0800	0.758
0.120	0.923
0.160	0.995
0.240	1.032

Prepare a spreadsheet for a Lineweaver–Burk plot and calculate the maximum rate and the Michaelis constant.

Solution

	A	B	C	D	E
1	**Lineweaver-Burk Plot**				
2	**[S], M**	**R, ΔA/min**	**1/[S], M^{-1}**	**1/R, ΔA^{-1}min**	
3	0.00400	0.093	**250**	**10.753**	
4	0.0100	0.231	100	4.329	
5	0.0400	0.569	25	1.757	
6	0.0800	0.758	12.5	1.319	
7	0.120	0.923	8.333	1.083	
8	0.160	0.995	6.25	1.005	
9	0.240	1.032	4.167	0.969	
10					
11	**Cell C3 = 1/A3.** Copy down.				
12	**Cell D3 = 1/B3.** Copy down.				
13	Slope (ΔA^{-1}min/M^{-1}) =			0.0395	
14	**Intercept = 1/R$_{max}$ =**			0.737	
15	**R$_{max}$ (ΔAmin^{-1}) = 1/D14 =**			1.357	
16	**K$_m$ (M) = slope x R$_{max}$ = D13*D15 =**			0.0536	
17					
18					
19					

Lineweaver-Burk Plot

$y = 0.0395x + 0.7369$
$R^2 = 0.9979$

The significance of K_m is illustrated in Figure 22.3 (similar to Figure 22.2). When the reaction rate increases rapidly with substrate concentration, K_m is small (curve 1). The substrate that gives the lowest K_m for a given enzyme is often (but not necessarily) the enzyme's natural substrate, hence the reason for the rapid increase in rate with increased substrate concentration. A small K_m indicates that the enzyme becomes saturated at small concentrations of substrate. Conversely, a large K_m indicates that high concentrations of substrate are required to achieve maximum reaction velocity. In such a case, it would be difficult to achieve zero-order rate with respect to substrate, and the substrate would not be a good one for determining the enzyme.

A small K_m means a fast reaction rate and easy substrate saturation.

ENZYME SPECIFICITY

In general, there are four types of enzyme specificity: (1) **absolute specificity**, in which the enzyme will catalyze only one reaction; (2) **group specificity**, in which the enzyme will act on molecules with certain functional groups, such as amino, phosphate, or methyl groups; (3) **linkage specificity**, in which the enzyme will act

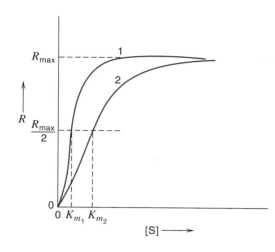

Fig. 22.3. Illustration of relationship of Michaelis constant, $K_m = ([S]$ at $R_{max}/2$), to reaction rate. Curve 1. Small K_m. Curve 2. Large K_m.

on a particular type of chemical bond; and (4) **stereochemical specificity**, in which the enzyme will act on a particular steric or optical isomer.

In addition to the substrate being acted on, many enzymes require a second cosubstrate. Such a cosubstrate may activate many enzymes and is an example of a **cofactor** or **coenzyme** (described previously). An example is nicotinamide adenine dinucleotide (NAD$^+$), which is a cofactor for many dehydrogenase reactions by acting as a hydrogen acceptor:

$$SH_2 + NAD^+ \overset{\text{enzyme}}{\rightleftharpoons} S + NADH + H^+ \qquad (22.15)$$

NAD$^+$ is a common cofactor in clinical chemistry measurements. The reactions are monitored by measuring the NADH concentration.

where SH$_2$ is the reduced form of the substrate, S is its oxidized (dehydrogenated) form, and NADH is the reduced form of NAD$^+$.

ENZYME NOMENCLATURE

Enzymes are classified according to the type of reaction and the substrate, that is, according to their reaction specificity and their substrate specificity. Most enzyme names end in "ase." Enzymes can be divided into four groups based on the kind of chemical reaction catalyzed: (1) Those that catalyze addition (*hydrolases*) or removal (*hydrases*) of water. Hydrolases include esterases, carbohydrases, nucleases, and deaminases, while hydrases include enzymes such as carbonic anhydrase and fumarase. (2) Those that catalyze the transfer of electrons: *oxidases* and *dehydrogenases*. (3) Those that catalyze transfer of a radical, such as *transaminases* (amino groups), *transmethylases* (methyl groups), or *transphosphorylases* (phosphate groups). (4) Those that catalyze splitting or forming of a C—C bond: *desmolases*. For example, α-glucosidase acts on any α-glucoside. The rate of reaction may be different for different glucosides. More generally, however, enzymes show absolute specificity to one particular substrate. Thus, glucose oxidase catalyzes the aerobic (oxygen) oxidation of glucose to gluconic acid plus hydrogen peroxide:

$$C_6H_{12}O_6 + O_2 + H_2O \xrightarrow{\text{glucose oxidase}} C_6H_{12}O_7 + H_2O_2 \qquad (22.16)$$

Actually, this enzyme shows almost complete specificity for β-D-glucose, α-D-Glucose reacts at a rate of 0.64 relative to 100 for the β form. In the latter form, all the hydrogens are axial and the hydroxyl groups are equatorial, allowing

Glucose oxidase is used for determining glucose.

the molecule to lie down flat on the enzyme active site and form the enzyme–substrate complex. The α form does not have the same arrangement of hydrogens and hydroxyls and cannot lie flat on the enzyme. Thus, the aerobic conversion of glucose (usually 36% α and 64% β) depends on the mutarotation of the α form to the β form. The mutarotation (equilibrium) is shifted as the β form is removed. Another enzyme, mutarotase, will affect the mutarotation, but this is usually not necessary. There is one other substance, 2-deoxy-D-glucose, that is affected by glucose oxidase. The relative rate of its reaction is about 10% of that of β-D-glucose, and it is usually not present in blood samples being analyzed for glucose.

There are thousands of enzymes in nature, and most of these exhibit absolute specificity.

DETERMINATION OF ENZYMES

Enzyme activities are measured by determining the rate of substrate conversion, under pseudo zero-order substrate conditions.

Enzymes themselves can be analyzed by measuring the amount of substrate transformed in a given time or the product that is produced in a given time. The substrate should be in excess so that the reaction rate depends only on the enzyme concentration. The results are expressed as international units of enzyme. For example, the activity of a glucose oxidase preparation can be determined by measuring manometrically or amperometrically the number of micromoles of oxygen consumed per minute. On the other hand, the use of enzymes to develop specific procedures for the determination of substrates, particularly in clinical chemistry, has proved to be extremely useful. In this case, the enzyme concentration is in excess so the reaction rate is dependent on the substrate concentration.

DETERMINATION OF ENZYME SUBSTRATES

Two general techniques may be used for measuring enzyme substrates. First, **complete conversion** of the substrate may be utilized. Before and following completion of the enzymatically catalyzed reaction, a product is analyzed or the depletion of a reactant that was originally in excess is measured. The analyzed substance (net change) is then related to the original substrate concentration. These reactions are often not stoichiometric with respect to the substrate concentration because of possible side reactions or instability of products or reactants. Also, the reaction may require extraordinarily long times for completion. For these reasons, the analytical procedure is usually standardized by preparing a calibration curve of some type in which the measured quantity is related to known concentrations or quantities of the substrate.

The enzyme is not consumed, so its concentration just needs to be held constant for rate measurements.

The second technique employed for substrate determination is the measurement of the **rate** of an enzymatically catalyzed reaction, as is used to determine enzyme activity. This may take one of three forms. First, the time required for the reaction to produce a preset amount of product or to consume a preset amount of substrate may be measured. Second, the amount of product formed or substrate consumed in a given time may be measured (see Experiment 35 for glucose determination). These are single-point measurements (called *end-point measurements*) and require well-defined reaction conditions. They are easy to automate or may be performed manually. A third procedure is continuous measurement of a product or substrate concentration as a function of time to give the slope of the reaction rate curve, $\Delta c/\Delta t$. These are the so-called true rate measurements. The measurements must generally be made during the early portion of the reaction where the rate is pseudo first order.

Rate methods are generally more rapid than end-point methods or complete conversion reactions. Complete conversion reactions, on the other hand, are less subject to interference from enzyme inhibitors or activators as long as sufficient time is allowed for complete conversion. See also the discussion of enzyme electrodes in Chapter 13 for a different approach to measuring substrates.

Example 22.2

The blood alcohol content of an individual is determined enzymatically by reacting ethanol with NAD^+ in the presence of the enzyme alcohol dehydrogenase to produce NADH (Table 22.1). The rate of formation of NADH is measured at 340 nm (Figure 22.4). The following absorbances are recorded for a 0.100% (wt/vol) alcohol standard and the unknown, treated in the same way. Use a spreadsheet to calculate the rates of absorbance changes and from these, the unknown concentration.

T (s)	A_{std}	A_{unk}
0	0.004	0.003
20	0.052	0.036
40	0.099	0.070
60	0.147	0.098
80	0.201	0.132
100	0.245	0.165

Solution

	A	B	C	D	E	F	G
1	T, sec	A std	Net	ΔA/ΔT	A unk	Net	ΔA/ΔT
2	0	0.004			0.003		
3	20	0.052	**0.048**	**0.00240**	0.036	**0.033**	**0.00165**
4	40	0.099	0.095	0.00238	0.070	0.067	0.00168
5	60	0.147	0.143	0.00238	0.098	0.095	0.00158
6	80	0.201	0.197	0.00246	0.132	0.129	0.00161
7	100	0.245	0.241	0.00241	0.165	0.162	0.00162
8	Ave.:			**0.00241**			**0.00163**
9	Std. Devn:			**0.00003**			**0.00004**
10	Sample concentration:		**0.068**	%(wt/vol)			
11							
12	**Cell C3** =	B3-$B$2	Copy down				
13	**Cell D3** =	C3/A3	Copy down				
14	**Cell F3** =	E3-$E$2	Copy down				
15	**Cell G3** =	F3/A3	Copy down				
16	**Cell D8** =	AVERAGE(D3:D7)					
17	**Cell G8** =	AVERAGE(G3:G7)					
18	**Cell D9** =	STDEV(D3:D7)					
19	**Cell G9** =	STDEV(G3:G7)					
20	**Cell C10** =	0.1*(G8/D8)					

EXAMPLE ENZYMATIC ANALYSES

Spectrophotometric methods are widely used to measure enzyme reactions. The reaction product may have an absorption spectrum quite different from the substrate, allowing simple measurement of the product or substrate. In other cases, a dye-forming reagent is employed that will react with the product or the substrate

Dehydrogenase reactions are monitored by measuring the UV absorbance of NADH.

and the increase or decrease in color is measured. Frequently, the chromogen is enzymatically coupled with the product using a second enzyme.

1. Dehydrogenase Reactions. The reduced (NADH) and oxidized (NAD$^+$) forms of nicotinamide adenine dinucleotide exhibit marked differences in their ultraviolet absorption spectra and are, therefore, widely used for following the course of dehydrogenase reactions. The ultraviolet absorption spectra for NAD$^+$ and NADH are given in Figure 22.4. NAD has negligible absorption at 340 nm while NADH has an absorption maximum, and so it is a simple matter to monitor the increase or decrease in NADH concentration.

An example using NADH for measurement is the determination of the enzyme **lactic acid dehydrogenase** (LDH), which is important in confirming myocardial infarction (heart attack). NAD$^+$ is required in the LDH-catalyzed oxidation of lactic acid to pyruvic acid:

$$
\begin{array}{ccc}
\text{CH}_3 & & \text{CH}_3 \\
| & & | \\
\text{CHOH} + \text{NAD}^+ \underset{}{\overset{\text{LDH}}{\rightleftharpoons}} & & \text{C}{=}\text{O} + \text{NADH} + \text{H}^+ \\
| & & | \\
\text{CO}_2\text{H} & & \text{CO}_2\text{H} \\
\text{lactic acid} & & \text{pyruvic acid}
\end{array}
\qquad (22.17)
$$

The reaction is reversible and can be employed in either direction. In the forward reaction, serum containing an unknown amount of LDH would be added to a solution containing enzyme saturating concentrations of lactic acid and NAD, and the increase in absorbance at 340 nm would be measured as a function of time.

NADH can sometimes be used to follow enzyme reactions in which it is not directly involved by using it as a coupling agent in a secondary reaction with the product. For example, serum glutamic–oxaloacetic transaminase (GOT) catalyzes the reaction of α-ketoglutarate and asparate, and the product is reduced by NADH in the presence of another enzyme, malic acid dehydrogenase (MDH):

$$
\alpha\text{-ketoglutarate} + \text{asparate} \overset{\text{GOT}}{\rightleftharpoons} \text{glutamate} + \text{oxaloacetate} \qquad (22.18a)
$$

$$
\text{oxaloacetate} + \text{NADH} \overset{\text{MDH}}{\rightleftharpoons} \text{malate} + \text{NAD}^+ \qquad (22.18b)
$$

The second reaction is fast compared to the first in the presence of an excess of MDH, and so the rate of decrease of NADH concentration is directly proportional to the GOT activity.

2. Commonly Determined Substrates. A list of some substrates determined in blood or urine is given in Table 22.1. They are discussed in order in the paragraphs following.

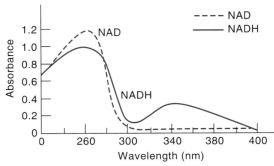

Fig. 22.4. Ultraviolet absorption spectra of NAD and NADH. (Courtesy of Worthington Biochemical Corporation.)

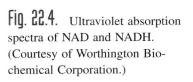

Table 22.1

Examples of Commonly Used Enzyme Reactions for Determining Substrates in Clinical Chemistry

Substrate Determined	Enzyme	Reaction
Urea	Urease	$NH_2-\overset{\overset{\displaystyle O}{\|\|}}{C}-NH_2 + H_2O \rightarrow 2NH_3 + CO_2$
Glucose	Glucose oxidase	$\underset{\text{(glucose)}}{C_6H_{12}O_6} + H_2O + O_2 \rightarrow \underset{\text{(gluconic acid)}}{C_6H_{12}O_7} + H_2O_2$
Uric acid	Uricase	$\underset{\text{(uric acid)}}{C_5H_4O_3} + 2H_2O + O_2 \rightarrow \underset{\text{(allantoin)}}{C_4H_6O_3N_4} + CO_2 + H_2O_2$
Galactose	Galactose oxidase	D-Galactose $+ O_2 \rightarrow$ D-galactohexodialdose $+ H_2O_2$
Blood alcohol	Alcohol dehydrogenase	Ethanol $+ NAD^+ \rightarrow$ acetaldehyde $+ NADH + H^+$

Urease was the first enzyme to be isolated and crystallized. It quantitatively converts urea to ammonia and carbon dioxide. The amount of urea is calculated from a determination of either the ammonia or the carbon dioxide produced, usually the former. This can be done spectrophotometrically by reacting the ammonia with a color reagent.

Glucose is usually determined by measuring the hydrogen peroxide produced upon addition of glucose oxidase. This is done spectrophotometrically by coupling the hydrogen peroxide as it is produced with a reagent such as *o*-toluidine. This coupling occurs in the presence of a second enzyme, horseradish peroxidase, and a colored product results. Commercial preparations of glucose oxidase usually contain impurities that react with and consume part of the hydrogen peroxide and so the conversion is not stoichiometric. Catalase is an enzyme impurity, for example, that is specific for the decomposition of hydrogen peroxide. Nevertheless, the fraction of hydrogen peroxide converted to the dye product is constant, and a calibration curve can be prepared using different concentrations of glucose.

Oxidase reactions are monitored by measuring O_2 depletion or H_2O_2 production.

There are a number of possible inhibitors in the glucose determination. Most of them, however, occur in the second enzymatic reaction. The glucose oxidase method would be more specific, then, if the hydrogen peroxide were measured directly without the need for a second enzyme. For example, added iodide ion, in the presence of a molybdenum(VI) catalyst, is rapidly oxidized to iodine. The iodine concentration can be followed amperometrically (Chapter 15). An alternative is to measure the depletion of oxygen amperometrically.

Uric acid is usually determined by measuring its ultraviolet absorption at 292 nm. However, the amount of uric acid in blood is small and the absorption is not specific. So, after the measurement, the uric acid is destroyed by adding the enzyme uricase. The absorbance is measured again. The *difference* in absorbance is due to the uric acid present. Since only uric acid will be decomposed by uricase, the method becomes specific. A similar procedure for the colorimetric determination of uric acid involves the oxidation of uric acid with molybdate to form molybdenum blue, a molybdenum(V) compound. In principle, uric acid could be determined as glucose was, but impurities in uricase preparations usually rapidly destroy the very small amount of hydrogen peroxide produced.

Galactose is determined in the same manner as glucose, by oxidizing the chromogen by H_2O_2 in the presence of peroxidase. Blood alcohol can be determined by UV measurement of the NADH produced.

Coupled enzyme reactions are often used for detection reactions.

3. Commonly Determined Enzymes. Table 22.2 summarizes reactions used to determine the activity of some enzymes frequently determined in the clinical laboratory. The pyruvate formed in the GPT reaction is coupled with NADH in the presence of added LDH for measurement. CK catalyzes the transfer of a phosphate group from creatinine phosphate to the nucleotide adenosine diphosphate (ADP) to produce adenosine triphosphate (ATP). The ATP is reacted with glucose in the presence of the enzyme hexokinase to form glucose-6-phosphate, which can then be reacted with NAD in the presence of glucose-6-phosphodehydrogenase. CK is also determined now by high-performance liquid chromatography, size exclusion chromatography, ion exchange chromatography, or electrophoresis.

Natural LDH consists of five components called **isozymes**, or **isoenzymes**, the ratio of which varies with the tissue source. The two electrophoretically fastest components occur in high percentage in LDH from heart muscle, and the level of these is preferentially increased in blood after heart muscle damage. The LDH method measures total LDH isozymes, which will usually indicate the heart damage. The two heart muscle isozymes mentioned, however, more readily catalyze the reduction of α-ketobutyrate than the slower moving components of hepatic origin and are referred to as α-hydroxybutyrate dehydrogenase (HBD) active. Elevated HBD can be determined by the reaction given and is more specific for myocardial infarction than LDH; it also remains elevated for longer periods after infarct.

The **phosphatases** are determined by measuring the blue color of thymolphthalein in highly alkaline solution after a specific time; the high alkalinity stops the enzyme reactions.

Enzyme reactions can, of course, be measured by other techniques than spectrophotometry. Techniques that have been used include amperometry, conductivity, coulometry, and ion-selective electrodes. Certain enzyme reactions have been

Table 22.2
Examples of Commonly Determined Enzymes in Clinical Chemistry

Enzyme	Abbreviation	Reaction
Glutamic–pyruvic transaminase	GPT	α-Ketoglutarate + L-alanine $\overset{\text{GOT}}{\rightleftharpoons}$ glutamate + pyruvate
		Pyruvate + NADH + H$^+$ $\overset{\text{LDH}}{\rightleftharpoons}$ lactate + NAD$^+$
Glutamic–oxaloacetic transaminase	GOT	α-Ketoglutarate + aspartate $\overset{\text{GOT}}{\rightleftharpoons}$ glutamate + oxaloacetate
		Oxaloacetate + NADH + H$^+$ $\overset{\text{MDH}}{\rightleftharpoons}$ malate + NAD$^+$
Creatinine phosphokinase	CK	Creatinine phosphate + ADP $\overset{\text{CK}}{\rightleftharpoons}$ creatine + ATP
		ATP + glucose $\overset{\text{hexokinase}}{\rightleftharpoons}$ ADP + glucose-6 phosphate
		Glucose 6-phosphate + NAD$^+$ $\overset{\text{G-6PDH}}{\longrightarrow}$ 6-phosphogluconate + NADH + H$^+$
Lactate dehydrogenase	LDH	L-Lactate + NAD$^+$ $\overset{\text{LDH}}{\rightleftharpoons}$ pyruvate + NADH + H$^+$
α-Hydroxybutyrate dehydrogenase	HBD	α-Ketobutyrate + NADH + H$^+$ $\overset{\text{HBD}}{\rightleftharpoons}$ α-hydroxybutyrate + NAD$^+$
Alkaline phosphatase		Na thymolphthalein monophosphate $\overset{\text{pH 10.1}}{\longrightarrow}$ Na thymolphthalein + phosphate
Acid phosphatase		Same as alkaline phosphatase, except pH 6.0

measured using enzyme electrodes whose response is specific for the particular reaction. These are described in Chapters 13 and 15.

Enzyme inhibitors and **activators** may be determined by employing enzyme reactions. The easiest technique is to measure the decrease or increase in the rate of the enzymatic reaction. Or the enzyme may be "titrated" with an inhibitor (or vice versa), and the amount of inhibitor required to completely inhibit the reaction measured. Trace elements, for example, have been determined by their inhibition or activation of enzyme reactions.

Many enzymes ideally represent the analytical chemist's dream of an absolutely specific reactant, but because of inhibitor effects, as well as problems associated with pH and ionic strength control, they must be used with some caution.

Most of the procedures discussed above can be adapted to automatic rate monitoring systems for enzymatic analysis. See, for example, Ref. 7.

Learning Objectives

WHAT ARE SOME OF THE KEY THINGS WE LEARNED FROM THIS CHAPTER?

- First-order reactions, half-life (key equations: 22.3, 22.4), p. 643
- Second-order reactions, half-life (key equations: 22.7, 22.8), p. 645
- Enzyme catalysis—Michaelis constant (key equation: 22.14); using spreadsheets to calculate, pp. 646, 649
- Substrate determinations (Table 22.1), enzyme determinations (Table 22.2), p. 653

Questions

GENERAL

1. Distinguish between a first-order and a second-order reaction.
2. What is the half-life of a reaction? How many half-lives does it take for a reaction to go to completion?
3. What is a pseudo-first-order reaction?
4. Suggest a way to determine whether a particular reaction between two substances A and B is first order or second order.

ENZYMES

5. What is an international unit?
6. What is the difference between competitive inhibition and noncompetitive inhibition of an enzyme?
7. Why are heavy metals often poisons in the body?
8. What are coenzymes?
9. Suggest a way to test whether an enzyme inhibitor is competitive or noncompetitive.
10. Suggest how a Lineweaver–Burk plot can be used to determine whether an inhibitor is competitive or noncompetitive.

Problems

KINETICS

11. A first-order reaction requires 10.0 min for 50% conversion to products. How much time is require for 90% conversion? For 99% conversion?

12. A first-order reaction required 25.0 s for 30% conversion to products. What is the half-life of the reaction?

13. A solution is 0.100 M in substances A and B, which react by a second-order reaction. If the reaction is 15.0% complete in 6.75 min, what is its half-life under these conditions? What would be the half-life if A and B were each at 0.200 M, and how long would it take for 15.0% completion of the reaction?

14. Sucrose is hydrolyzed to glucose and fructose:

$$C_{12}H_{22}O_{11} + H_2O \rightarrow C_6H_{12}O_6 + C_6H_{12}O_6$$

In dilute aqueous solution, the water concentration remains essentially constant, and so the reaction is pseudo first order and follows first-order kinetics. If 25.0% of a 0.500 M sucrose solution is hydrolyzed in 9.00 h, in how much time will the glucose and fructose concentration be equal to one-half the concentration of the remaining sucrose?

15. Hydrogen peroxide decomposes by a second-order reaction,

$$2H_2O_2 \xrightarrow{\text{catalyst}} 2H_2O + O_2$$

If 35.0% of a 0.1000 M solution decomposes in 8.60 min, how much time is required for the evolution of 100 mL O_2 from 100.0 mL of a 0.1000 M solution of H_2O_2 at standard temperature and pressure?

ENZYME

16. The activity of a glucose oxidase preparation is determined by measuring the volume of oxygen gas consumed as a function of time. A 10.0-mg fraction of the preparation is added to a solution containing 0.01 M glucose and saturated in oxygen. After 20.0 min, it is determined that 10.5 mL oxygen is consumed at standard temperature and pressure (STP). What is the activity of the enzyme preparation expressed in enzyme units per milligram? If the purified enzyme has an activity of 61.3 units/mg, what is the percent purity of this enzyme preparation?

SPREADSHEET PROBLEM

17. When an apple is sliced, it turns brown on exposure to air due to catalysis of the oxidation of phenols in the apple by o-diphenyl oxidase enzyme. An experiment is performed to determine the Michaelis constant of o-diphenyl oxidase in which fresh pieces of apple are ground up and then centrifuged to produce a supernatant that will serve as the enzyme source (see www.ultranet. com/~jkimball/BiologyPages/E/EnzymeKinetics.html). Catechol is used as the substrate. A fixed amount of the enzyme preparation is added to a tube containing 0.300 mM catechol, and the change in absorbance is measured at 540 nm at 1-min intervals for several minutes. The experiment is repeated with three other tubes containing 0.600, 1.20, and 4.80 mM catechol. The following results are obtained (mM catechol/ΔA/min): 0.30/0.020; 0.60/0.035; 1.20/0.048; 4.80/0.081. Prepare a spreadsheet to determine K_m.

KINETICS

1. H. H. Bauer, G. D. Christian, and J. E. O'Reilly, eds., *Instrumental Analysis.* Boston: Allyn and Bacon, 1978, Chapter 18, "Kinetic Methods," by H. B. Mark, Jr.
2. R. A. Greinke and H. B. Mark, Jr., "Kinetic Aspects of Analytical Chemistry," *Anal. Chem.,* **46** (1974) 413R.
3. H. L. Pardue, "A Comprehensive Classification of Kinetic Methods of Analysis Used in Clinical Chemistry," *Clin. Chem.,* **23** (1977) 2189.
4. D. Perez-Bendito and M. Silva, *Kinetic Methods in Analytical Chemistry.* New York: Wiley, 1988.
5. H. A. Mottola, *Kinetic Aspects of Analytical Chemistry.* New York: Wiley-Interscience, 1988.

ENZYMES

6. H. U. Bergmeyer, *Methods of Enzymatic Analysis*, 3rd ed. New York: Wiley. A series of 12 volumes plus index volume, 1983–1987.
7. N. Gochman and L. J. Bowie, "Automated Rate-Monitoring Systems for Enzyme Analyses," *Anal. Chem.,* **48** (1976) 1189a.

WEBSITES

8. www.worthington-biochem.com. Link to the Worthington enzyme manual that gives the properties of enzymes, their assay, and references. Plus a beginning biochemistry introduction to enzymes excerpted from their 1972 *Manual of Enzyme Measurements.*
9. www.chem.qmul.ac.uk/iubmb/enzyme or www.chem.qmw.ac.uk/iubmb/kinetics. Enzyme nomenclature from the International Union of Biochemistry (IUB) and the International Union of Pure and Applied Chemistry (IUPAC).

Chapter Twenty-Three

AUTOMATION IN MEASUREMENTS

"If you don't know how to do something,
you don't know how to do it with a computer."
—Anonymous (From J. F. Ryan, *Today's Chemist at Work*)

The services of the analytical chemist are constantly increasing as more and better analytical tests are developed, particularly in the environmental and clinical laboratories. The analyst often must handle a large number of samples and/or process vast amounts of data. Instruments are available that will automatically perform many or all of the steps of an analysis, greatly increasing the load capacity of the laboratory. The data generated can often be processed best by computer techniques; computers may even be interfaced to the analytical instruments. An important type of automation is in process control whereby the progress of an industrial plant process is monitored in real time (i.e., online), and continuous analytical information is fed to control systems that maintain the process at preset conditions.

In this chapter, we briefly consider the types of automated instruments and devices commonly used and the principles behind their operation. Their application to process control is discussed.

23.1 Principles of Automation

There are two basic types of automation equipment. **Automatic devices** perform specific operations at given points in an analysis, frequently the *measurement step*. Thus, an *automatic titrator* will stop a titration at the end point, either mechanically or electrically, upon sensing a change in the property of the solution. **Automated devices**, on the other hand, control and regulate a *process* without human intervention. They do so through mechanical and electronic devices that are regulated by means of *feedback information* from a sensor or sensors. Hence, an *automated titrator* may maintain a sample pH at a preset level by addition of acid or base as drift from the set pH is sensed by a pH electrode. Such an instrument is called a *pH-stat* and may be used, for example, in maintaining the pH during an enzyme reaction which releases or consumes protons. The rate of the reaction can actually be determined from a recording of the rate of addition of acid or base to keep the solution pH constant.

Automated devices are widely used in process control systems, whereas automatic instruments of various sophistication are used in the analytical laboratory for performing analyses. The latter may perform all steps of an analysis, from sample pickup and measurement through data reduction and display.

Automatic instruments improve the analyst's efficiency by performing some of the operations done manually. Automated instruments control a system based on the analysis results.

23.2 Automated Instruments: Process Control

In process analysis, analytical measurements are performed on chemical processes to provide information about the progress of the process or the quality of product. There are various ways in which process analysis may be performed, as illustrated in Figure 23.1. Samples may be taken intermittently and transported to the laboratory for measurement. This allows access to the usual laboratory instruments and the ability to perform a variety of measurements. But this procedure is relatively slow and the chemical process is usually complete before the analytical information is available. Hence, laboratory analysis is more suited for *quality control*, to ascertain the quality of a product. More efficient measurements can be made if the

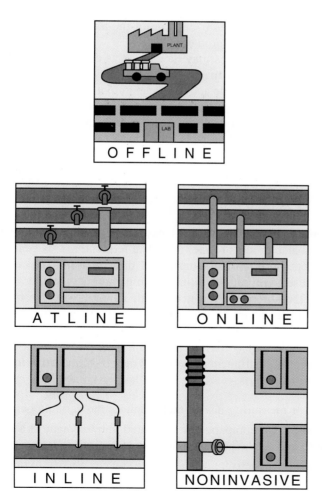

Fig. 23.1. Methods for process analysis. [From J. B. Callis, D. Illman, and B. R. Kowalski, *Anal. Chem.*, **59** (1987) 624A. Copyright 1987 by the American Chemical Society. Reprinted by permission of the copyright owner.]

instrument is transferred to the chemical plant. But for true real-time analysis, instruments should be interfaced directly to the chemical process, with automatic sampling and analysis. A more idealized approach would be to place a sensor directly inline so that measurements are continuous and no chemical treatment of the sample is needed. Such sensors are more limited in scope and availability since they must be selective for a given analyte and must withstand the chemical environment, not be poisoned, and remain in calibration. An even more idealized approach is a noninvasive measurement. For example, an analyte may exhibit an absorbance spectrum that allows selective measurement by the passage of light through the chemical system. Again, these types of process measurements will be limited.

The use of analytical data for automated process control can save millions of dollars in improved production efficiencies and product quality.

An important aspect of real-time process analysis is the use of the analytical data to control the chemical process via feedback of the information to a controller that can alter the addition of chemical reactants to maintain an intermediate, for example, at a preset level. The application of online measurements with feedback control of the chemical process can save a chemical company millions or even hundreds of millions of dollars in a year by providing optimization of the process for achieving maximum reaction efficiency and product formation and avoiding failed reactions, detecting contaminants, and the like. This application of analytical chemistry has become a critical part of industrial production. The University of Washington Center for Process Analytical Chemistry (CPAC) is supported by industrial sponsors for conducting research on state-of-the-art process measurement technology (see www.cpac.washington.edu).

The measurement devices may be classed as continuous or discrete (batch) instruments. The **continuous instrument** constantly measures some physical or chemical property of the sample and yields an output that is a continuous (smooth) function of time. A **discrete**, or **batch**, **instrument** analyzes a discrete or batch-loaded sample, and information is supplied only in discrete steps. In either case, information on the measured variable is fed back to monitoring or control equipment. Each technique utilizes conventional analytical measurement procedures and must be capable of continuous unattended operation.

CONTINUOUS ANALYZERS

Continuous-process control instruments may make measurements directly in a flowing stream or a batch process reactor such as a fermentor. This generally precludes any analytical operation on the sample, and direct sensing devices such as electrodes must be used. If a sample dilution, temperature control, or reagent addition is required, or measurements are made with nonprobe-type instruments, then a small fraction of the stream is diverted into a test stream where reagents may be mixed continuously and automatically with the sample, and the test measurement is made. The sample may be passed through a filter prior to measurement.

Process control instruments operate by means of a **control loop**, which consists of three primary parts:

1. A **sensor** or measuring device that monitors the variable being controlled.
2. A **controller** that compares the measured variable against a reference value (set point) and feeds the information to an operator.
3. An **operator** that activates some device such as a valve to bring the variable back to the set point.

The control loop operates by means of a **feedback mechanism**, illustrated in Figure 23.2. The process may be any industrial process that produces some desired product. It has one or more inputs that can be controlled to provide the desired

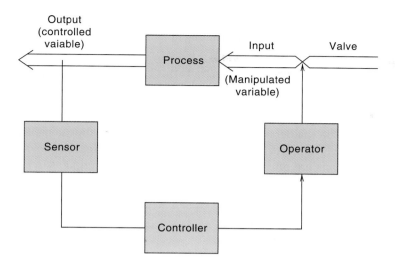

Fig. 23.2. Feedback control loop.

product (output). The sensor measures the variable to be controlled (e.g., pH, temperature, reactant). The information is fed to the controller, which compares the measured variable against a reference set point. The difference is fed to the operator that actuates the valve (opens or closes it) or some other appropriate device to adjust the variable back to the set point.

These devices are characterized by their **dead time**. This is equal to the time interval, after alteration of the variable at the input, during which no change in the variable is sensed by the detector at the output. It includes the analytical dead time. It can be minimized by keeping the detector as near the input as possible and by high flow rate. The sensor may actually be placed at the input, before the process. In this manner, corrective action on the manipulated variable (if it changes erroneously) is taken before the process occurs, and an error need not occur in the output prior to corrective action. However, the result is not detected at the output. Such systems may be called **feedforward systems**, and are **open-loop controls**, as opposed to **closed-loop controls** in feedback systems. Sophisticated algorithms, using computers, are used for control, based on chemometric methods and multivariate analysis. These are beyond the scope of this text.

DISCRETE ANALYZERS

In discrete analyzers, a batch sample is taken at selected intervals and then analyzed, with the information being fed to the controller and operator in the usual fashion. Obviously, the sampling and analytical dead times are increased over continuous analyzers, and the manipulated variable is held at a fixed value between measurements. If a transient error occurs between measurements, it may not be detected and corrected for. On the other hand, a short transient error may be detected during the measurement interval and a correction for this applied during the entire interval between measurements.

Discrete measurements must be made when the sensing instrument requires discrete samples, as in a chromatograph, or a flow injection analyzer (see below).

INSTRUMENTS USED IN AUTOMATED PROCESS CONTROL

In principle, any conventional measuring device or technique can be used in process control. But laboratory instruments normally are not suited for online measurement. The instruments must be more robust and are designed for unattended operation, with minimal skill required for operation, since plant personnel are often not trained analytical chemists. (But you, as the resident analytical chemist will be in charge of selecting the measurement technique, the instrument, and making sure reliable data are obtained!)

The choice is dictated by cost and applicability to the problem. The most widely used methods include spectrophotometry to measure color, ultraviolet or infrared absorption, turbidity, film thickness, and the like; electrochemistry, primarily potentiometry, for the measurement of pH and cation or anion activity; and gas and liquid chromatography, especially in the petrochemical industry where complex mixtures from distillation towers are monitored. Spectrophotometric and other measurements are often rapidly made using flow injection analysis.

23.3 Automatic Instruments

Automatic instruments relieve the analyst from several operations. The precise nature of automatic operations improves the precision.

Automatic instruments, as mentioned before, are not feedback control devices but rather are designed to automate one or more steps in an analysis. They are generally intended to analyze multiple samples, either for a single analyte or for several analytes.

Automatic instruments will perform one or more of the following operations:

1. Sample pickup (e.g., from a small cup on a turntable or assembly line)
2. Sample dispensing
3. Dilution and reagent addition
4. Incubation
5. Placing of the reacted sample in the detection system
6. Reading and recording the data
7. Processing of the data (correct for blanks, correct for nonlinear calibration curves, calculate averages or precisions, correlate with the sample number, etc.).

Clinical instruments may also contain a feature for deproteinizing the sample. Instruments that provide only a few of these steps, primarily electronic data processing, are called **semiautomatic instruments**.

While all automatic instruments are discrete in the terminology of automated process analyzers, in that they analyze individual discrete samples, they may be classified as follows:

1. **Discrete sampling instruments.** In discrete sampling, each sample undergoes reaction (and usually measurement) in a separate cuvet or chamber. These samples may be analyzed sequentially or in parallel (see below).
2. **Continuous-flow sampling instruments.** In continuous-flow sampling, the samples flow sequentially and continuously in a tube, perhaps being separated by air bubbles. They are each sequentially mixed with reagents in the same tube at the same point downstream and then flow sequentially into a detector.

Discrete samplers have the advantage of minimizing or avoiding cross contamination between samples. But continuous-flow instruments require fewer mechanical manipulations and can provide very precise measurements.

Discrete instruments may be designed to analyze samples for one analyte at a time. These are the so-called **batch instruments**, or **single-channel analyzers**. See Figure 23.3. However, those that analyze the samples in parallel, that is,

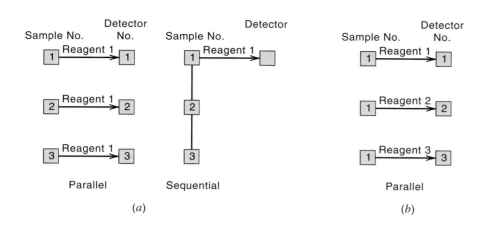

Fig. 23.3. Discrete analyzers. (*a*) Single channel (batch). (*b*) Multichannel.

simultaneously rather than sequentially one at a time, can analyze a large number very quickly; and they can readily be changed to perform different analyses. Discrete analyzers may also analyze separate aliquots of the same sample (one in each cup) in parallel for several different analytes. These are **multichannel analyzers**.

Continuous-flow instruments may also be single-channel (batch) instruments that analyze a continuous series of samples sequentially for a single analyte (Figure 23.4). Or they may be multichannel instruments in which the samples are split at one or more points downstream into separate streams for different analyte analyses, or separate aliquots of samples may be taken with separate streams in parallel.

Modern-day instruments are so sophisticated that they actually possess *automated* features whether they perform an analysis automatically or not. For example, they may monitor sample chamber temperature and by feedback to a regulator maintain it constant (important in enzyme reactions).

23.4 Flow Injection Analysis

PRINCIPLES

Flow injection analysis (FIA) is based on the injection of a liquid sample into a moving, nonsegmented continuous carrier stream of a suitable liquid. The injected sample forms a zone, which is then transported toward a detector. Mixing with the reagent in the flowing stream occurs mainly by diffusion-controlled processes, and a chemical reaction occurs. The detector continuously records the absorbance, electrode potential, or other physical parameter as it changes as a result of the passage of the sample material through the flow cell.

FIA is like HPLC without a column. It is low pressure, and there is no separation. The injected sample mixes and reacts with the flowing stream. A transient signal (peak) is recorded.

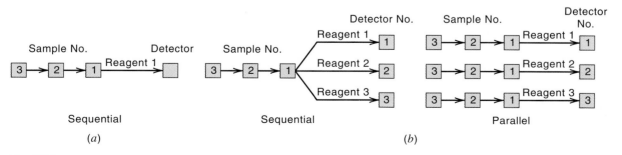

Fig. 23.4. Continuous-flow analyzers. (*a*) Single channel (batch.) (*b*) Multichannel.

An example of one of the simplest FIA methods, the spectrophotometric determination of chloride in a single-channel system, is shown in Figure 23.5. This is based on the release of thiocyanate ions from mercury(II) thiocyanate and its subsequent reaction with iron(III) and measurement of the resulting red color (for details, see Experiment 37). The samples, with chloride contents in the range 5 to 75 ppm chloride, are injected (S) through a 30-μL valve into the carrier solution containing the mixed reagent, pumped at a rate of 0.8 mL/min. The iron(III) thiocyanate is formed on the way to the detector (D) in a mixing coil (50 cm long, 0.5 mm i.d.), as the injected sample zone disperses in the carrier stream of reagent. The mixing coil minimizes band broadening (of the sample zone) due to centrifugal forces, resulting in sharper recorded peaks. The absorbance A of the carrier stream is continuously monitored at 480 nm in a micro-flow-through cell (volume of 10 μL) and recorded (Figure 23.5b). To demonstrate the reproducibility of the analytical readout, each sample in this experiment was injected in quadruplicate, so that 28 samples were analyzed at seven different concentrations of chloride. As this took 14 min, the average sampling rate was 120 samples per hour. The fast scan of the 75- and 30-ppm sample peaks (shown on the right in Figure 23.5b) confirms that there was less than 1% of the solution left in the flow cell at the time when the next sample (injected at S2) would reach it, and that there was no carryover when the samples were injected at 30-sec intervals.

A key feature of FIA is that since all conditions are reproduced, dispersion is very controlled and reproducible. That is, all samples are sequentially processed in exactly the same way during passage through the analytical channel, or, in other words, what happens to one sample happens in exactly the same way to any other sample.

A peristaltic pump is typically used to propel the stream. For process analysis, these are not suitable because the pump tubing must be frequently changed, and more rugged pumps are used, such as syringe pumps, or pumping is by means of air displacement in a reservoir. The injector may be a loop injector valve as used in high-performance liquid chromatography. A bypass loop allows passage of

FIA measurements are very rapid.

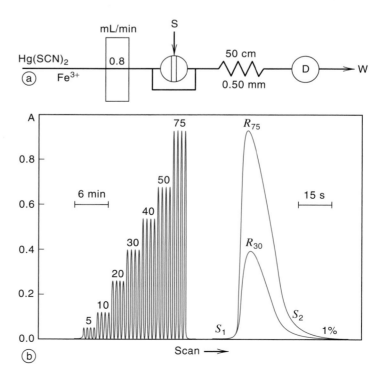

Fig. 23.5. (a) Flow injection diagram for the spectrophotometric determination of chloride: S is the point of sample injection, D is the detector, and W is the waste. (b) Analog output showing chloride analysis in the range of 5 to 75 ppm Cl with the system depicted in (a).

carrier when the injection valve is in the load position. The injected sample volumes may be between 1 and 200 μL (typically 25 μL), which in turn requires no more than 0.5 mL of reagent per sampling cycle. This makes FIA a simple, microchemical technique that is capable of having a high sampling rate and minimum sample and reagent consumption. The pump, valve, and detector may be computer controlled for automated operation.

FIA is a general solution-handling technique, applicable to a variety of tasks ranging from pH or conductivity measurement to colorimetry and enzymatic assays. To design any FIA system properly, one must consider the desired function to be performed. For pH measurement, or in conductimetry, or for simple atomic absorption, when the original sample composition is to be measured, the sample must be transported through the FIA channel and into the flow cell in an undiluted form in a highly reproducible manner. For other types of determinations, such as spectrophotometry, the analyte has to be converted to a compound measurable by a given detector. The prerequisite for performing such an assay is that during the transport through the FIA channel, the sample zone is mixed with reagents and sufficient time is allowed for production of a desired compound in a detectable amount.

Besides the single-line system, described in Figure 23.5, a variety of manifold configurations may be used to allow application to nearly any chemical system. Several are shown in Figure 23.6. The two-line system (B) is the most commonly used, in which the sample is injected into an inert carrier and then merges with the reagent. In this manner, the reagent is diluted by a constant amount throughout, even when the sample is injected, in contrast to the single-line system; reagent dilution by the sample in a single-line system is all right as long as there is excess reagent and the reagent does not exhibit a background response that would shift upon dilution. If two reagents are unstable when mixed, then they may be mixed online (C or D), or they may merge with the sample following injection (E). Mixing coils may be interspersed between confluence points to allow dispersion before merging.

Only a few microliters of sample are required.

A two-line system is most often used.

DISPERSION COEFFICIENT

The degree of dispersion or dilution in an FIA system is characterized by the dispersion coefficient, D. Let us consider a simple dispersion experiment. A sample solution, contained within the valve cavity prior to injection, is homogeneous and has the original concentration C^o that, if it could be scanned by a detector, would yield a square signal the height of which would be proportional to the sample concentration (Figure 23.7). When the sample zone is injected, it follows the movement of the carrier stream, forming a dispersed zone whose form depends on the

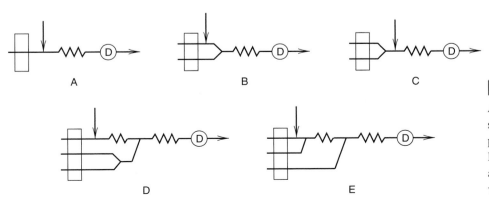

Fig. 23.6. Types of FIA manifolds. A, single line; B, two-line with a single confluence point; C, reagent premixed into a single line; D, two-line with a single confluence point and reagent premix; E, three-line with two confluence points.

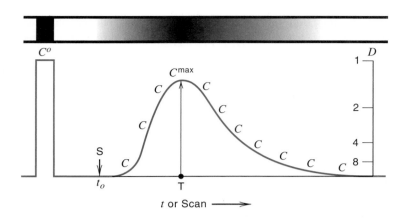

Fig. 23.7. Originally homogeneous sample zone (top left) disperses during its movement through tubular reactor (top center), thus changing from an original square profile (bottom left) of original concentration C^o to a continuous concentration gradient with maximum concentration C^{max} at the apex of the peak.

geometry of the channel and the flow velocity. Therefore, the response curve has the shape of a peak, reflecting a continuum of concentrations (Figure 23.7), forming a concentration gradient, within which no single element of fluid has the same concentration of sample material as the neighboring one. It is useful, however, to view this continuum of concentrations as being composed of individual elements of fluid, each of them having a certain concentration of sample material C, since each of these elements is a potential source of a readout.

To design an FIA system rationally, it is important to know (a) how much the original sample solution is diluted on its way toward the detector and (b) how much time has elapsed between the sample injection and the readout. For this purpose the **dispersion coefficient** D has been defined as the ratio of concentrations of sample material before and after the dispersion process has taken place in that element of fluid that yields the analytical readout:

The dispersion coefficient is a measure of the extent of dilution of the injected sample at readout.

$$D = \frac{C^o}{C} \qquad (23.1)$$

If the analytical readout is based on maximum peak height measurement, the concentration within that (imaginary) element of fluid, which corresponds to the maximum of the recorded curve (C^{max}), has to be considered. Thus, by relating C^{max} to the original concentration of injected sample solution C_s^o (Figure 23.7)

$$D_s^{max} = \frac{C_s^o}{C_s^{max}} \qquad (23.2)$$

and sample (and reagent) concentrations may be estimated. Note at the peak that the definition of the dispersion coefficient considers only the physical process of dispersion and not the ensuing chemical reactions, since D refers to the concentrations of sample material prior to and after the dispersion process alone has taken place.

Typical dispersions for reactions are 3 to 10.

The sample solution, when $D = 2$, for example, has been diluted 1:1 with the carrier stream. For convenience, sample dispersion has been defined as *limited* ($D = 1$ to 3), *medium* ($D = 3$ to 10), and *large* ($D > 10$), and the FIA systems designed accordingly have been used for a variety of analytical tasks. Limited dispersion is preferred when the injected sample is simply being carried to a detector (e.g., ion-selective electrode, atomic absorption spectrophotometer). Medium dispersion is employed when the analyte must mix with and react with the carrier reagent to form a product to be detected. Large dispersion is used only when the sample must be diluted to bring it into measurement range.

The simplest way of measuring the dispersion coefficient is to inject a well-defined volume of a dye solution into a colorless carrier stream and to monitor the absorbance of the dispersed dye zone continuously by a colorimeter. To obtain the D^{max} value, the height (i.e., absorbance) of the recorded peak is measured and then compared with the distance between the baseline and the original signal obtained when the cell has been filled with the undiluted dye. Provided that the Beer–Lambert law is obeyed, the ratio of respective absorbances yields a D^{max} value that describes the FIA manifold, detector, and method of detection.

The FIA peak is a result of two kinetic processes, which occur simultaneously; the *physical* process of zone dispersion and the *chemical* processes resulting from reactions between sample and reagent species. The underlying physical process is well reproduced for each individual injection cycle; yet it is not a homogeneous mixing, but a dispersion, the result of which is a concentration gradient of sample within the carrier stream.

FACTORS AFFECTING PEAK HEIGHT

The degree of dispersion, and therefore the recorded peak height, is determined by a number of factors, including injected sample volume, channel geometry and length, and flow rate.

1. Sample Volume. Consider a one-line FIA system in which the pumping rate Q is 1.5 mL/min, the tube length L is 20 cm, and the inner diameter of the tube is 0.5 mm. When increasing volumes of a dye solution are injected, a series of curves will be recorded (Figure 23.8a), all starting from the same point of injection S, where the height of the individual peaks will increase until an upper limit "steady

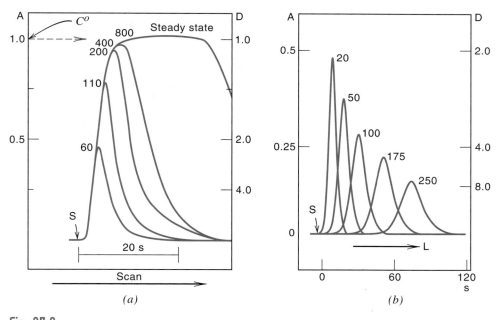

(a)   *(b)*

Fig. 23.8. (*a*) Response curves as function of injected sample volume. The peak height increases with volume of sample injected into FIA system until steady-state signal is reached. All curves recorded from same starting point S, with sample volumes of 60, 110, 200, 400, and 800 μL. Note that $D = 1$ for steady state, and that peak width increases with injected volume. (*b*) Dispersion of injected sample zone in FIA system as function of tube length traversed. The sample volume is 60 μL; L is given in centimeters; the tube inside diameter is 0.5 mm; and Q in all experiments is 1.5 mL/min.

state" has been reached. At this final level the recorded absorbance will correspond to the concentration of undiluted dye C^o, and $D = 1$. The rising edges of all curves coincide and have the same shape regardless of the injected volumes, and thus:

$$\frac{C^{max}}{C^o} = 1 - \exp(-kS_v) = 1 - \exp(-0.693n) = 1 - 2n = \frac{1}{D^{max}} \quad (23.3)$$

where $n = S_v/S_{1/2}$ and $S_{1/2}$ is the volume of sample solution necessary to reach 50% of the steady-state value, corresponding to $D = 2$. By injecting two $S_{1/2}$ volumes, 75% of C^o is reached, corresponding to $D = 1.33$, and so on. Therefore, $D = 1$ can never truly be reached; yet, injection of five $S_{1/2}$ volumes results in $D = 1.03$, and injection of seven $S_{1/2}$ volumes results in $D = 1.008$, corresponding to 99.2% of C^o. Since the concept of steady state is not used in FIA, the maximum sample requirements will not exceed two $S_{1/2}$ for limited dispersion, and less than one $S_{1/2}$ in all other applications. Since the first portion of the rising curve might be considered nearly linear up to approximately 50% C^o (i.e., $D = 2$), it follows that for FIA readouts with medium and large dispersion coefficients, *the peak height is directly proportional to the injected volume.*

The peak height is proportional to the injected sample volume for medium dispersion ($S_v < S_{1/2}$).

It has been shown that $S_{1/2}$ is a function of the geometry and of the volume of the flow channel.

2. Channel Length and Flow Rate.

The microreactor between the injection port and the detector may have different lengths, diameters, and geometries. The influence of coil length L and inner radius R of the tubing on the dispersion has been studied in detail. The use of tubing of a small diameter will result in lower $S_{1/2}$ values because the same sample volume will occupy a longer length of tube (θ). That is, the sample will be less easily mixed and dispersed. The sample volume S_v injected is equal to $\pi R^2 \theta$. If the tube radius R is halved, the sample will occupy a fourfold-longer portion of the tube (θ), and hence the $S_{1/2}$ value will be four times smaller. Therefore, if a **limited dispersion** is desired, a sample volume of minimum one $S_{1/2}$ should be injected into a manifold consisting of the shortest possible piece of a narrow tube connecting the injection port and the detector.

Doubling the tube diameter increases the sample and reagent volume requirements fourfold.

Even if a **medium dispersion** FIA system is required, it is economical to use narrow channels. The sample and reagent economy is improved when narrow channels are used because, for the same linear flow velocity, the pumping rate Q in a tube of radius R is only one-fourth of that required for a tube of radius $2R$. The optimum internal diameters of tubes connecting the injection port and the detector are 0.5 to 0.8 mm. The narrower channels are more subject to clogging.

When designing systems with medium dispersion, where the sample has to be mixed and made to react with the components of the carrier stream, one would first tend to increase the tube length L to increase the mean residence time T. One can expect, however, that dispersion of the sample zone will increase with the distance traveled, and this band broadening will eventually result in a loss of sensitivity and lower sampling rate (see Figure 23.8). Thus, one obtains, upon increasing the tube length, a series of curves, the height of which decreases with the increase of tube length. It has been shown that dispersion in an FIA system caused by the flow in an open narrow tube increases with the square root of the mean residence time T (or the distance traveled L).

Thus, although the zone broadening becomes progressively smaller relative to the distance traveled, the increase in T obtained through an increase of length L (for increased dispersion) is not worthwhile above a certain limit. It is desirable to keep L short for narrow peaks and for rapid throughput. Because of the physical distances between the individual components of the FIA system (injection port, reaction coils, flow-through cell), a compromise must be made, and therefore, in practice, the overall length of a well-designed FIA manifold is between 10 and 100 cm of 0.5-m tubing. Residence times up to 20 s can be obtained readily by selection of the flow rate.

To summarize, the dispersion of the sample zone increases with the square root of the distance traveled through the tubular conduit and the square of the radius of the tube. It generally increases somewhat with increasing flow rate but may behave in the opposite manner at very slow flow rates where diffusion effects predominate.

$D \propto L^{1/2} R^2$

3. Channel Geometry. Rather than a straight tube, some sort of coiled or curved tube is usually used for the FIA microreactor. This is to increase the degree of radial mixing, by inducing secondary flow due to centrifugal forces as the solution goes around a curve. This increased radial dispersion reduces the parabolic profile in the axial direction, formed when the sample zone is injected into a laminar flow of carrier stream. (In laminar flow, the velocity at the center is twice the mean velocity, while that at the tube walls is zero, resulting in a parabolic profile in the axial direction.) Thus, by inducing secondary flow, the reagent becomes more readily mixed with the sample and the axial dispersion of the sample zone is reduced.

Relaxation of the laminar profile in the radial direction is best achieved by creating a local turbulence whereby the direction of flow is suddenly changed. In this way the elements of fluid that are lagging because they are close to the walls of the channel are moved into the rapidly advancing central stream line, while those elements of fluid that have advanced are reshuffled closer to the tube wall. The more frequently this process is repeated, the more symmetrical the concentration gradient within the dispersed sample zone will be, and the peak shape will change from an asymmetric to a symmetric (Gaussian) one. A higher and narrower peak results. This is sometimes accomplished by preparing a knotted reactor, that is, by tying a plastic tube in several closely knit knots.

Inducing secondary flow by going around a curve sharpens peaks and increases their height.

Example 23.1

A dye solution is continuously flowed through the flow cell of a spectrophotometric detector, and the recorded absorbance is 0.986. The same dye solution is injected into the carrier stream using a 25.0-μL injection loop, and the maximum absorbance of the resulting FIA peak is 0.327. What is (a) the dispersion coefficient and (b) the $S_{1/2}$ value?

Solution

$$D^{max} = \frac{A^o}{A^{max}} = \frac{0.986}{0.327} = 3.01$$

From Equation 23.3,

$$\frac{1}{D^{max}} = 1 - 2\frac{S_v}{S_{1/2}}$$

$$\frac{1}{3.01} = 1 - 2\frac{25.0\ \mu L}{S_{1/2}}$$

$$S_{1/2} = 74.9\ \mu L$$

Example 23.2

If the sample volume in Example 23.1 is 25.0 μL, the flow rate is 1.00 mL/min and the reactor coil length is 50.0 cm of 0.800-mm i.d. tubing, what would be the dispersion coefficient (a) if the sample volume is changed to 50.0 μL and (b) if the coil length is changed to 100 cm, the tubing i.d. to 0.500 mm, and the flow rate to 0.750 mL/min?

Solution

(a) From Equation 23.3

$$\frac{1}{D^{\max}} = 1 - 10^{-kS_v}$$

$$\frac{1}{3.01} = 1 - 10^{-k(25.0\,\mu\text{L})}$$

$$\log 0.668 = -k(25.0\,\mu\text{L})$$

$$k = 7.01 \times 10^{-3}\,\mu\text{L}^{-1}$$

For $S_v = 50.0\,\mu$L,

$$\frac{1}{D^{\max}} = 1 - 10^{-(7.01 \times 10^{-3}\,\mu\text{L}^{-1})(50.0\,\mu\text{L})}$$

$$\frac{1}{D^{\max}} = 1 - 10^{-0.351} = 0.554$$

$$D^{\max} = 1.80$$

(b) $$D \propto L^{1/2}R^2 = kL^{1/2}R^2$$

Therefore

$$3.01 = k(50.0\,\text{cm})^{1/2}\,(0.0400\,\text{cm})^2$$

$$k = 266\,\text{cm}^{-2.5}$$

For $L = 100$ cm and $R = 0.0250$ cm

$$D = 266\,\text{cm}^{-2.5}\,(100\,\text{cm})^{1/2}\,(0.0250\,\text{cm})^2$$

$$D = 1.66$$

STOPPED-FLOW MEASUREMENTS

Kinetic measurements can be made in FIA.

FIA is capable of very high precision, better than 1%, as a result of very controlled sample dispersion in a continuous flow stream. Kinetic FIA measurements can be made by precisely stopping (under computer control) the flow of the analyte product in the detector cell at a position corresponding to the peak or some point past it, depending on concentrations. Then the change of signal, for example, absorbance, is recorded for a period of time to obtain the rate. The slope of this recorded signal is plotted for different concentrations to prepare a calibration curve.

Such a technique is useful for enzymatic analyses (Chapters 22 and 24). An advantage of such kinetic measurements is that the rate curve is selective for the analyte, and the rate signal is not part of a background signal (although a blank should still always be run to establish either the flat base signal to measure from at the selected part of the peak or to correct for any background enzymatic reaction).

You are referred to the references at the end of the chapter for complete descriptions of the many variations and applications of FIA.

SEQUENTIAL INJECTION ANALYSIS

Sequential injection analysis (SIA) is a computer-controlled, single-line, injection technique that simplifies the manifold and is more robust for unattended operation. The flow is intermittent, and only a few microliters of reagent are used. Instead of an injection valve for sample introduction, SIA uses a multiport selection valve as shown in Figure 23.9. The common port of the valve (in the center) can access any of the other ports by electrical rotation of the valve. It is connected to a reversible piston pump via a two-position valve; carrier is drawn into the pump in one position and delivered toward the valve in the other. A holding coil is placed between the pump and the valve to prevent aspirated (injected) solution from entering the pump. The different ports of the valve are connected to sample, reagent, standards, waste reservoir, and the detector. In operation, the pump is first filled with carrier. The valve is switched to the detector position, and the piston pushes carrier through the system until it exits at the waste end. The valve is then switched to the sample position, and a few microliters of sample are drawn in by reversing the flow of the pump using precise timing. (Before beginning the analysis, the sample and reagent tubes are filled by aspirating some excess into the holding coil and expelling it to waste via the auxiliary waste port.) The pump is stopped during rotation of the valve to avoid pressure surges. After the sample is introduced, the reagent is aspirated, next to the sample. Thus, the sample and reagent solutions are sequentially injected into the holding coil, and hence the name sequential injection analysis. Finally, the valve is switched to the detector port, the pump is changed to forward flow, and the injected solutions are propelled through the reaction coil to the detector flow cell. The solutions merge via diffusion and secondary forces and react to form a product that is detected, resulting in a transient signal as in conventional FIA.

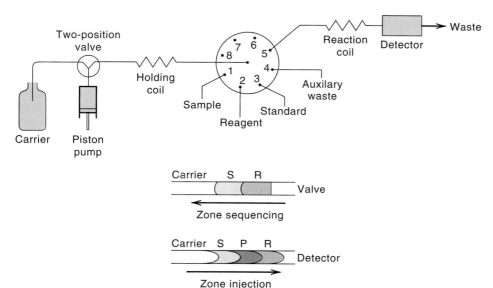

Fig. 23.9. SIA manifold and sequential injection. S, P, and R are sample, product, and reagent, respectively.

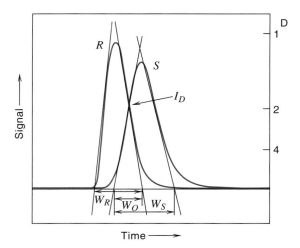

Fig. 23.10. Zone penetration in SIA. [Reprinted with permission from T. Gübeli, G. D. Christian, and J. Ruzicka, *Anal. Chem.*, **63** (1991) 2407. Copyright 1991 the American Chemical Society.]

Only one pump and valve are required in SIA. The entire operation is computer controlled for the precise timing required, and commercial systems have programmable software to set sequences and aspiration times. The software also collects peak height data, prepares calibration curves, and calculates concentrations, with precision data. Different reagents can be introduced to perform a different analysis, simply by programming from the keyboard to aspirate from a different reagent port, rather than changing the plumbing as required in FIA. Two-reagent chemistries may be used by sandwiching a small volume of sample between them so that all three zones overlap before reaching the detector.

The parameter of prime importance in SIA is the degree of penetration or overlap of adjacent zones. This is dependent on the relative volumes, in addition to the usual parameters of tubing size and length, reaction coil geometry, and so forth. The degree of penetration and the dispersion determine the relative signal that is recorded. We may define the zone penetration

$$p = \frac{2w_0}{w_s + w_r} \tag{23.4}$$

where w_0 is the baseline width of the zone overlap and w_s and w_r are the baseline widths of the sample and reagent, respectively. See Figure 23.10. When $w_s + w_r = 2w_0$, there is complete overlap (rare). Normally, p is between zero and unity, that is, there is partial overlap. In Figure 23.10 I_D is the isodispersion point. The goal is to have enough overlap to have a dispersion coefficient of at least 2, so there is excess reagent. Note in Figure 23.10 that the reagent zone is first (shorter time of flow) since it was injected second, closest to the valve. The sample zone also travels a longer distance to the detector, increasing its dispersion. We can reduce the sample dispersion if necessary by reversing the order of injections. For more details, see Ref. 11. See www.flowinjection.com for examples of commercial SIA and FIA systems.

23.5 Microprocessors and Computers

You can purchase analytical instruments of varying sophistication, depending on need and budget. But nearly all have modern digital electronic features that make operation and data collection and processing easy. Many are interfaced to a personal computer, and powerful software can be used for controlling the instrument and data processing. Some software will be specific for the instrument, while some, especially for data collection and processing, may be more generic commercial packages. An instrument may have a dedicated microcomputer built in, so-called smart instruments. The computer is a **microprocessor** and peripherals constructed from microelectronic circuits. Many instruments are preprogrammed in "machine language," but others have basic programs that can be modified by the operator to perform specific dedicated functions. The design and engineering of a microprocessor may require tens or hundreds of thousands of dollars, but the electronic circuits are etched in microchips and can be mass-produced inexpensively. An example is the popular digital watch that can be purchased inexpensively and has a variety of stored information such as dates, and has programs in which a specific time to buzz an alarm can be entered or calculations made.

Microprocessors control instruments (e.g., wavelength, scan rate). They may collect data and process it.

Communication between the analyst and the instrument is established through a keyboard array that accepts command inputs (e.g., a repetitive scan button) and numerical data inputs (e.g., to enter the number of scans desired). The instrument provides digital display and printout in addition to the recorded spectrum.

Microprocessor-controlled instruments can provide a number of other functions, depending on the type of instrument and the desired capabilities. These include such parameters as automatic background correction, first- or second-derivative spectra, integration or averaging of signals for a preselected interval for improved signal-to-noise ratio, and statistical evaluation of data, for example, the standard deviation. Calibration to read out in any desired concentration units is simply a matter of putting in a standard, keying in its concentration, and pressing an appropriate calibrate button. If nonlinear analytical curves are obtained, the microprocessor may determine the algorithm of the curve and automatically correct the nonlinearity. This is particularly valuable in atomic absorption instruments. Periodic rereading of one or more standards may be used to correct automatically for calibration drift. Least-squares fitting of calibration curves may be utilized.

Instruments designed for enzyme and other kinetic assays will be programmed to measure the signal continuously over a given time interval (e.g., absorbance) and calculate the rate of the reaction: if properly calibrated, the activity of the enzyme or the concentration of the substrate will be presented.

The instrument may contain closed-loop feedback control (automated) features. The temperature of the reaction chamber may be controlled for kinetic measurements. An automatic titrator may sense the approach of an end point and slow down the addition of titrant near the end point for improved precision in locating the end of the titration, just as a human operator would, but probably in a more reproducible fashion. The instrument may monitor its various functions and shut itself down if it malfunctions.

Learning Objectives

WHAT ARE SOME OF THE KEY THINGS WE LEARNED FROM THIS CHAPTER?

Questions

1. Distinguish between an automatic instrument and an automated instrument.
2. Distinguish between discrete and continuous automated devices.
3. Distinguish between discrete and continuous-flow sampling automatic instruments.
4. What is a feedback loop?
5. Describe the principles of flow injection analysis.
6. Describe the principles of sequential injection analysis and how it differs from conventional flow injection analysis.
7. Summarize the principal uses of computers in the analytical laboratory.
8. What are some of the functions a microprocessor can perform?

Problems

FLOW INJECTION ANALYSIS

9. The dispersion coefficient for an FIA system is 4.00. What would it be if (a) the injected sample volume were doubled, (b) the tube diameter were doubled, or (c) the tube length were doubled? Compare the result for (a) with Example 23.2.
10. Calculate the $S_{1/2}$ volume for an FIA system with a dispersion coefficient of 4.00 and an injection volume of 50.0 μL.

Recommended References

GENERAL

1. G. D. Christian and J. E. O'Reilley, eds., *Instrumental Analysis*, 2nd ed. Boston: Allyn and Bacon, 1986. Chapter 25, "Automation in Analytical Chemistry," by K. S. Fletcher and N. C. Alpert. Provides an excellent detailed but brief description of various types and operations of automated and automatic instruments.

FLOW INJECTION ANALYSIS

2. J. Ruzicka and E. H. Hansen, *Flow Injection Analysis*, 2nd ed. New York: Wiley, 1988.
3. M. Valcarcel and M. D. Luque de Castro, *Flow Injection Analysis, Principles and Applications*. Chichester: Ellis Horwood, 1987.
4. Z. Fang, *Flow Injection Separation and Preconcentration*. Weinheim: VCH, 1993.
5. Z. Fang, *Flow Injection Atomic Absorption Spectrometry*. New York: Wiley, 1995.
6. J. L. Burguera, ed., *Flow Injection Atomic Spectroscopy*. New York: Marcel Dekker, 1989.
7. M. M. Calatayud, *Flow Injection Analysis of Pharmaceuticals: Automation in the Laboratory*. London: Taylor and Francis, 1997.

SEQUENTIAL INJECTION ANALYSIS

8. M. Trojanowicz, *Flow Injection Analysis: Instrumentation and Applications.* River Edge, NJ: World Scientific, 2000. The only book that discusses sequential injection analysis.

9. G. D. Christian, "Sequential Injection Analysis for Electrochemical Measurements and Process Analysis," *Analyst,* **119** (1994) 2309.

10. P. J. Baxter and G. D. Christian, "Sequential Injection Analysis: A Versatile Technique for Bioprocess Monitoring," *Accounts Chem. Res.,* **29** (1996) 515.

11. T. Gübeli, G. D. Christian, and J. Ruzicka, "Fundamentals of Sinusoidal Flow Sequential Injection Spectrophotometry," *Anal. Chem.,* **63** (1991) 2407.

WEB DATABASES

12. Hansen's FI Bibliography, www.flowinjection.com/search.html. Over 8000 references on flow injection analysis.

13. Chalk's Flow Analysis Database, www.fia.unf.edu/fad.lasso. Over 11,000 references on flow methods.

PROCESS ANALYSIS

14. G. D. Christian and E. D. Yalvac, "Process Analytical Chemistry," in R. Kellner, J.-M. Mermot, M. Otto, and H. M. Widner, eds. *Analytical Chemistry,* Weinheim: Wiley-VCH, 1998.

15. J. B. Callis, D. L. Illman, and B. R. Kowalski, "Process Analytical Chemistry," *Anal. Chem.,* **59** (1987) 624A.

16. M. T. Riebe and D. J. Eustace, "Process Analytical Chemistry. An Industrial Perspective," *Anal. Chem.,* **62** (1990) 65A.

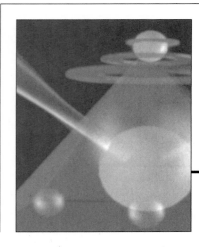

Chapter Twenty-Four

CLINICAL CHEMISTRY

The previous chapters have dealt with the methodology and tools that are basic to all types of analyses. In this chapter and the next two, we shall discuss specifically the practical aspects of analyzing some different types of materials. We will draw on your knowledge gained from the previous chapters, and these subsequent chapters will be more of a reference nature for specified analyses, suggesting approaches that may be taken for solving real analytical problems.

In this chapter, we discuss specifically the practical aspects of clinical analyses. The clinically significant constituents of blood and urine are described, including major electrolytes, proteins, and organic substances. Some of the commonly used analytical procedures for important clinical determinations, that is, the normal physiological ranges of the constituents and the conditions under which they may fall outside this range, are given. Also, the sensitive technique of immunoassay is described.

Spectrophotometric techniques are widely used for many clinically important analytes. When you have a physical examination, a blood chemistry analysis will be performed to screen for common disease indicators such as glucose, cholesterol, lipids, and urea nitrogen. The multiple analyses are generally performed on an automatic analyzer based on spectrophotometric detection; different chemistries will be employed for different analytes, such as enzymatic or immunochemical. Electrolytes such as sodium or potassium will probably be measured using an automatic instrument incorporating ion-selective electrodes. Many more specific analyses may require separation techniques such as gas or liquid chromatography. Illicit drug screening will likely be done using immunoassays, with confirmation using chromatography (e.g., GC–MS). The sections below provide information on how to collect and prepare samples and list some representative techniques for several common analytes.

24.1 Composition of Blood

Most clinical chemistry analyses are performed on blood serum.

Whole blood can be broken down into its general components as follows: **plasma**, which contains the *serum* and *fibrinogen*; and the **cellular elements**, which contain the *erythrocytes*, *leukocytes*, and *platelets*. These are described in Chapter 1. Plasma is the liquid portion of circulating blood. The cells are separated from the plasma by centrifuging whole blood. If blood is allowed to clot, the fibrinogen is removed with the cells, leaving **serum**. The majority of clinical analyses are performed on whole blood, plasma, or serum, and most of these use serum. Urine is also frequently analyzed.

Table 24.1 summarizes the normal range of concentrations of some clinically important constituents in human blood. We should emphasize that these normal ranges are approximations. Table 5.3 summarizes the major electrolyte composition (cations and anions) in blood. The analysis of some of the more commonly determined constituents will be discussed below. The physiological significance of the results is also discussed.

See Table 5.3 for the electrolyte composition of blood.

Table 24.1
Information Pertinent to Blood Samples for Chemical Examinations[a]

Determination	Sample[b]	Normal Range
Albumin	S	4–5 g/dL
Amino acid nitrogen	B	4–6 mg/dL
Ammonia	B	40–125 μg/dL
Amylase	S	Up to 150 mg/dL
Bilirubin	S	Direct up to 0.4 mg/dL
		Total up to 1.0 mg/dL
Calcium	S	4.5–5.5 meq/L
Carbon dioxide content	S	25–32 meq/L
Chloride, serum	S	100–108 meq/L
Cholesterol, total	S	140–250 mg/dL[c]
Cholesterol, esters	S	50–65% of total
Creatinine	S	0.7–1.7 mg/dL
Creatinine clearance		100–180 mL/min
Lipase	S	Up to 1.5 units
Lipids, total	S	350–800 mg/dL
Fatty acids	S	200–400 mg/dL
Globulins, total	S	2.5–3.5 g/dL
alpha-1		0.1–0.4 g/dL
alpha-2		0.3–0.7 g/dL
beta		0.4–0.9 g/dL
gamma		0.6–1.3 g/dL
Iron, serum	S	50–180 μg/dL
Magnesium	S	1.5–2.5 meq/L
Nonprotein nitrogen (NPN)	B	25–40 mg/dL
Phosphatase, acid	S	Up to 4 Gutman units
Phosphatase, alkaline	S	Up to 4 Bodansky units
Phospholipids	S	100–250 mg/dL
as phosphorus		4–10 mg/dL
Phosphorus, inorganic	S	3–4.5 mg/dL
Potassium	S	3.8–5.6 meq/L
Protein, total serum	S	6.5–8 g/dL
Protein-bound iodine (PBI)	S	3.5–8 μg/dL
Sodium	S	138–146 meq/L
Sugar, blood (glucose)	B	65–90 mg/dL
Transaminase (SGO)	S	Up to 40 units
Uric acid	S	3–6 mg/dL
Urea nitrogen (BUN)	B	Up to 20 mg/dL

[a] Adapted from J. S. Annino, *Clinical Chemistry*. 3rd ed. Boston: Little, Brown and Company, 1960.
[b] S, serum; B, blood.
[c] This range varies with age.

24.2   Collection and Preservation of Samples

Blood and urine samples are often collected after the patient has fasted for a period of time (e.g., overnight), particularly for cholesterol or glucose analysis. One study indicates that an average breakfast has no significant effect on the concentration of the blood content of carbon dioxide, chloride, sodium, potassium, calcium, urea nitrogen, uric acid, creatinine, total protein, and albumin. Serum phosphorus is slightly depressed at 45 min after the meal but not after 2 h.

CAUTION: The collection and handling of blood and other biological specimens requires great care because of the possible presence of transferable disease constituents, particularly from HIV or AIDS. Rubber gloves must be worn and often masks, as well. You will not draw blood or take other biological samples, nor handle samples from individuals. All experiments are designed to use commercially available or synthetic materials that are safe (e.g., freeze-dried serum). Animal serum, such as horse serum or bovine serum, can be purchased (www.sigma-aldrich.com). One experiment uses urine, and that can be either your own, from an animal, or synthetic.

When serum is required for the analysis, the blood is collected in a clean and dry tube to prevent contamination and **hemolysis**. Hemolysis is the destruction of red cells, with the liberation of hemoglobin and other cell constituents into the surrounding fluid (serum or plasma). When hemolysis occurs, the serum is noticeably red instead of its normal straw color. A substance may occur at a much higher concentration in the cellular portion of the blood than in the serum or plasma. If it is the concentration in the serum or plasma that is clinically significant, then hemolysis would give erroneously high serum or plasma results, as in the case of potassium, iron, magnesium, zinc, urea, protein (from hemoglobin), and others. For this reason, blood samples should be centrifuged as soon as possible to separate serum or plasma from the cells.

> Hemolysis contaminates the plasma or serum with cellular constituents.

If plasma or whole blood is required for analysis, then the blood is collected in a tube containing an **anticoagulant**. *Heparin* (sodium salt) is frequently used. However, its effect is temporary and heparin is expensive. Therefore, a more widely used anticoagulant is *potassium oxalate*, about 1 mg per mL blood. Oxalate precipitates blood calcium, and the calcium is required in the clotting process. Obviously, plasma prepared in this way cannot be analyzed for calcium or potassium; many other metals are precipitated by oxalate, and so serum is usually analyzed for these. The usual practice in preparing whole blood or plasma samples is to add the required amount of oxalate in solution form to the collection tube and then to dry the tube in an oven at 110°C. By this procedure, the collected blood is not diluted. For example, 0.5 mL of a 2% potassium oxalate solution would be taken and dried for a 10-mL blood sample. Potassium oxalate causes red cells to shrivel, with the result that the intracellular water diffuses into the plasma. Thus, the plasma should be separated as soon as possible.

> An anticoagulant must be added if plasma is to be separated.

If a sample must be kept anaerobically, as in the case of CO_2 analysis, mineral oil is added to the collection tube. This is lighter than blood and will cover it. A cork stopper should be used in these cases because the oil causes rubber stoppers to swell.

Sometimes a preservative is added to the sample, usually along with an anticoagulant. Sodium fluoride is widely used as a preservative for samples to be analyzed for glucose. This is an enzyme inhibitor that prevents the enzymatic breakdown of glucose, or **glycolysis**. One milligram sodium fluoride per milliliter blood is adequate. Since it also inhibits other enzymes, including urease, sodium fluoride should not be added to samples to be analyzed for enzymes or for urea based on urease catalysis.

Samples can generally be stored for one or two days by refrigerating them. This slows down enzymatic and bacteriological processes but does not eliminate them, and so it is best to analyze fresh samples when possible. Samples are always brought to room temperature before analyzing. Freezing samples will preserve them for long periods of time, at the same time suspending the activity of enzymes in the blood. Whole blood should not be frozen because the red cells will become ruptured thereby. Serum and plasma samples fractionate into layers of different composition when frozen and so should be shaken gently after thawing.

Samples are more stable if a **protein-free filtrate** (PFF) is prepared. There are a number of procedures for preparing them. In the *Folin-Wu* or *tungstic acid method*, 1 volume of blood, serum, or plasma is mixed with 7 volumes of water and 1 volume of 0.33 M sulfuric acid and allowed to turn brown. Then, 1 volume of 10% sodium tungstate (wt/vol, $Na_2WO_4 \cdot 2H_2O$) is added, and after 2 min, the precipitated protein is filtered or centrifuged. In the *trichloroacetic acid* (TCA) *method*, 1 volume of blood, serum, or plasma is mixed with 9 volumes of 5% TCA (wt/vol), and after proteins are precipitated, the mixture is filtered or centrifuged. Both of these yield acidic filtrates but are useful for the analysis of many substances.

> A protein-free filtrate is sometimes needed to prevent matrix interferences from the proteins.

When glucose is analyzed by being oxidized (e.g., with alkaline copper tartrate; see below), high results are obtained if the above filtrates are used. A PFF prepared with *barium hydroxide* and *zinc sulfate* removes most interfering nonglucose-reducing substances (see Experiment 35). This provides essentially a neutral PFF because the $Zn(OH)_2$ product is precipitated:

$$Ba(OH)_2 + ZnSO_4 \rightarrow \underline{BaSO_4} + \underline{Zn(OH)_2}$$

Glucose is stable in a protein-free filtrate because the glycolytic enzymes are removed.

24.3 Clinical Analysis—Common Determinations

Table 24.2 summarizes some of the more frequently determined constituents in blood and the principles of some of the procedures employed. These are only representative procedures, and in many cases, there are numerous variations or different procedures offering varying degrees of convenience, speed, sensitivity, accuracy, precision, and so on.

The major **serum electrolytes**—sodium, potassium, calcium, magnesium, chloride, and bicarbonate (CO_2)—are fairly easy to determine. The metals are most readily determined by the use of flame-spectrophotometric or atomic absorption methods, although colorimetric methods exist for calcium and magnesium. Calcium and, less frequently, magnesium are also titrated with EDTA. Ion-selective electrodes are used for the routine analysis of sodium, potassium, and calcium. Bicarbonate is analyzed also by titration against standard acid (see Experiment 8) in addition to a manometric method. Chloride is widely determined by automatic coulometric titration with electrogenerated silver ion.

Blood glucose and **blood urea nitrogen** (BUN) are probably the two most frequently performed clinical tests. In the procedures described in Table 24.2, the total of all reducing sugars is measured, and so results tend to be high. But these methods have been adopted as standard ones for many years. The enzymatic determination of glucose (Chapter 22) is an established method, and dedicated enzymatic glucose analyzers are now widely used.

> There are probably more blood glucose analyses than any other analytical measurement in the world.

Uric acid is more specifically determined enzymatically than by the described method. The uric acid in a separate aliquot of the sample is destroyed with the

Table 24.2
Some Procedures Used in Clinical Analysis

Substance Determined	Sample Analyzed[a]	Measurement Method	Principle
Barbiturates	S	Ultraviolet spectrophotometry	Extract barbiturate into $CHCl_3$, back-extract into NaOH solution, and measure UV absorption at alkaline pH at 252 nm
Calcium	S	Atomic absorption spectrophotometry	Dilute 1:20 with 10,000 ppm Na_2EDTA and measure absorbance at 422.7 nm
Carbon dioxide	S	Manometric; electrode	Van Slyke manometer method; CO_2 electrode
Chloride	S	Volumetric; coulometric	Titrate with Ag^+ or Hg(II)
Creatinine	S	Spectrophotometric	PFF (TCA) reacted with alkaline picrate and absorbance measured at 490 nm
Iron	S	Spectrophotometric	Iron in PFF is reduced to Fe(II) and complexed with bathophenanthroline; absorbance measured at 535 nm
Magnesium	S	Atomic absorption spectrophotometry	Dilute 1:20 with 1% Na_2EDTA and measure absorbance at 285.2 nm
Nonprotein nitrogen (NPN)	B	Spectrophotometry	Digest PFF to form NH_4HSO_4, add alkaline mercuric iodide (Nessler's reagent) to complex ammonia, and measure absorbance at 480 nm
Phosphatase, acid	S	Spectrophotometry	Incubate with sodium glycerophosphate for 1 h at pH 4.9 to liberate phosphate; determine phosphate as with inorganic phosphorus
Phosphatase, alkaline	S	Spectrophotometry	Same, but incubate at pH 9.6
Phosphorus, inorganic	S	Spectrophotometry	PFF (TCA); react with Mo(VI) to form phospho-molybdic acid and reduce this to molybdenum blue; measure absorbance at 660 nm
Potassium	S	Flame photometry; electrode	Dilute 1:50 with 0.15 M NaCl and measure at 766 nm; K^+ i.s.e.
Protein, total	S	Spectrophotometry	Biuret reagent (alkaline copper tartrate); forms complex with proteins; measure absorbance at 550 nm after 30 min
Protein-bound iodine (PBI)	S	Spectrophometry (catalytic)	Precipitate proteins with TCA, ash or digest the protein, and measure the catalytic effect of I^- on the Ce(IV)–As(III) reaction rate by measuring absorbance of Ce(IV) at 420 nm after 20 min
Salicylate	S	Spectrophotometry	Form complex with ferric nitrate in acid solution and measure absorbance at 535 nm
Sodium	S	Flame photometry; electrode	Dilute 1:50 and measure at 589 nm; Na^+ i.s.e.
Sugar (glucose)	B or S	Spectrophotometry	$Ba(OH)_2$–$ZnSO_4$ PFF; oxidize sugar with alkaline Cu(II) to form Cu_2O; the Cu_2O is determined by allowing it to reduce phosphomolybdic acid to form molybdenum blue, which is measured at 420 nm or react sugar with o-toluidine and measure at 635 nm
Uric acid	S	Spectrophotometry	Tungstic acid PFF; oxidize with alkaline phospho-tungstate and measure the blue reduction product of phosphotungstate at 680 nm
Urea nitrogen (BUN)	B	Spectrophotometry (enzymatic)	Tungstic acid PFF; incubate with urease enzyme at pH 6.8 to produce NH_3; determine the NH_3 with Nessler's reagent

[a] S, serum; B, blood.

enzyme uricase and the decrease in the blue phosphotungstate reduction product is measured. This second aliquot effectively serves as a blank. The uric acid may also be measured directly by its absorbance in the ultraviolet region at 290 nm, before and after incubation with uricase.

Albumin and **globulins** can be analyzed by the same procedure as for total proteins after fractionation by salting out with either sodium sulfate or sodium sulfite. More detailed information may be required about the protein fractions (α, β, γ globulins), in which case starch gel electrophoresis can be used to separate the proteins. Micro-Kjeldahl analysis of proteins is used when highly accurate data are required; the biuret method is accurate to about 4%.

Acid and alkaline phosphatases are enzymes contained in the blood and are active in acid and alkaline solution, respectively. The substrate used for their analysis is sodium glycerophosphate, at the appropriate pH. These enzymes hydrolyze the substrate to release phosphate. Other phosphate substrates can be used.

Barbiturates, which are drugs not normally present in the blood, are extractable (un-ionized) from blood into methylene chloride. They can then be back-extracted into 0.45 M sodium hydroxide as the ionized form. The ionized form absorbs in the ultraviolet region, whereas the nonionized form does not. Plotting an absorption spectrum can qualitatively confirm the presence of barbiturates. The ionized form at pH 13 to 14 exhibits an absorption maximum between 252 and 255 nm, with a minimum between 234 and 237 nm; and at pH 9.8 to 10.5, a different ionized form exhibits a maximum at 240 nm.

A number of the above tests are also performed with urine samples, and frequently the same procedures can be used. Because of different concentrations and interferences, modifications in the procedures may be introduced. Analyses are usually performed with a 24-h sample, that is, urine collected over a 24-h period, since the total daily excretion of most substances has much more significance than the concentration in a random sample. The composition of urine compared with that of serum is illustrated in Figure 24.1.

There is a defined need for **point-of-care analysis** for certain tests, those for which results are needed quickly. For example, the length of time between symptom onset and initiation of treatment is critical for minimizing cardiac injury during a heart attack. The emergency care unit (ECU) is very dependent on quick blood test results to confirm a heart attack and render the appropriate care. Transporting samples to a laboratory to obtain results can be very inefficient. So ECUs may have a bedside enzyme analyzer requiring minimal operation. A blood sample may be inserted, a button pushed, and results obtained in less than one minute.

One of the most successful biosensor analyzers is the blood glucose analyzer that is used by diabetics as personal monitors. The patient simple places a small drop of blood (from a sterile lancet-prick) on a test strip, and glucose oxidase on the strip catalyzes the oxidation of glucose to produce hydrogen peroxide, which is then measured either electrochemically or photometrically. See, for example, http://abbott.com/diabetes/diabetes.html and www.diabeticsupply.com.

24.4 Immunoassay

Immunoassay techniques are important for the specific determination of hormones, drugs, vitamins, and other compounds at nanogram and smaller levels. In these techniques, an antigen and a (specific) antibody react to form a complex or precipitate. The first analytical application was in the form of **radioimmunoassay (RIA)** in which Berson and Yalow demonstrated the ability to selectively measure

Immunoassays can be very selective and sensitive.

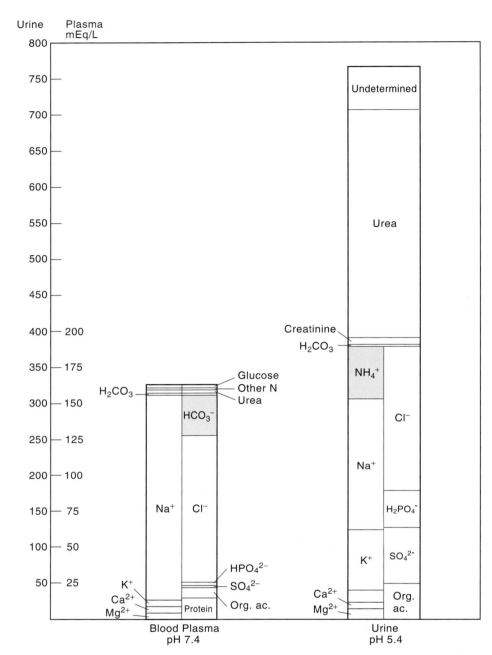

Fig. 24.1. Composition of urine compared with that of blood plasma. Nonelectrolytes are expressed as millimoles on the milliequivalent scale. Values on the scale inclusive for all constituents (the sum of all of them). (From J. A. Gamble, *Chemical Anatomy, Physiology, and Pathology of Extracellular Fluid.* Cambridge, MA: Harvard University Press, 1950. Reproduced by permission of Harvard University Press.)

small quantities of insulin (Ref. 10). It was not until the late 1960s and early 1970s that RIA became widely available for routine analyses. At this time, the methods moved from the research laboratory to the clinical laboratory in what must be record time, demonstrating the real need that existed. Immunoassays and related competitive binding assays are widely used now in the clinical chemistry laboratory. The importance attached to the technique is further evidenced by the fact that the Nobel prize in physiology was awarded to Rosalyn Yalow in 1977, following Berson's death, for its development (www.almaz.com/nobel/nobel.html).

Immunoassay techniques generally involve a competitive reaction between an analyte antigen and a standard antigen that has been tagged, for limited binding sites on the antibody to that antigen. The tag may be a radioactive tracer, an enzyme, or a fluorophore. We describe below the immunoassay technique as well

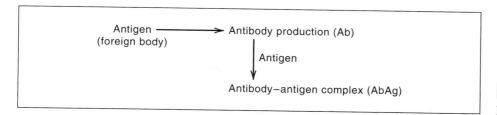

Fig. 24.2. Principles of immunological reactions.

as the principles that are common to all the techniques. Fluorescence and enzyme immunoassay are also described.

PRINCIPLES OF IMMUNOASSAY

Immunoassay combines the sensitivity of radiochemistry, fluorescence, or enzymatic tags with the specificity of immunology. Immunology is the study of antigens and their reactions with antibodies, that is, an organism's defense mechanism to foreign bodies through antibodies (Figure 24.2). An **antigen** (e.g., a hormone) is a (foreign) substance capable of producing antibody formation in the body and is able to react with (bind to) that antibody. An antigen is always a large molecule such as a protein. An **antibody** is a protein endowed with the capacity to recognize, by stereospecific association, a substance foreign to the organism it has invaded, for example, bacteria and viruses.

An antibody is a high-molecular-weight globulin protein of about 150,000 (see Figure 24.3 for a model). When the protein exhibits antibody activity, it is referred to as an immunoglobulin (Ig). There are actually five major immunoglobulins in human blood (IgA, IgD, IgE, IgG, and IgM), but IgG is the most abundant. The Ig consists of two light polypeptide chains of about 214 amino acid residues and one heavy chain of about 430 residues. These are linked via disulfide bridges into a flexible Y-shaped structure (Figure 24.4). When treated enzymatically with papain, three fragments of molecular weight about 50,000 each are formed. Two are identical and retain the ability to bind antigens, hence are referred to as Fab (Fragment, antigen binding). The third fragment does not bind antigen by itself but can be crystallized from solution, hence it is called Fc (Fragment, crystallizable). The Fc fragment is of fairly consistent composition, and the Fab fragments have portions of variable composition and will specifically bind to a given antibody. The key domains lie at the terminal ends of the Fab regions (shaded regions in Figure 24.4), which form the binding sites for the antigen (the Fc fragment can also bend

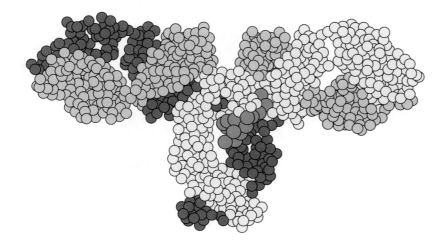

Fig. 24.3. Space-filling model of antibody structure.

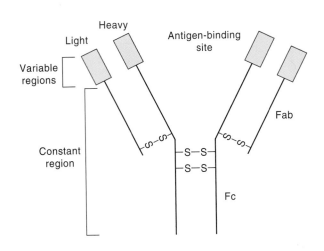

Fig. 24.4. Antibody.

around so its end binds the antigen, but the Fab is the main binder). All antibodies are similar in structure except for the variable, antigen-binding portions of the Fab fragments.

The antibody will specifically react with an antigen to form an **antigen–antibody complex**. It is produced in the organism (where it will remain present for some time) only after the organism has had at least one exposure to the intruder (through vaccination, either spontaneous or artificial). An antibody is produced for use in immunoassay by injecting the antigen into an animal species to which it is foreign and recovering the serum that contains the resultant antibody (**anti-serum**).

The strength of the antigen–antibody complex is called the **affinity** or **avidity**. Affinity refers to the intrinsic association constant between an antibody and a univalent antigen, while avidity refers to the overall binding energy between antibodies and a multivalent antigen. We can write the overall binding reaction as

$$Ab + Ag = AbAg \qquad (24.1)$$

And the formation constant is

$$K = \frac{[AbAg]}{[Ab][Ag]} \qquad (24.2)$$

The formation constants are quite large, typically 10^8 to 10^{10}. The binding forces are quite weak, van der Waals, electrostatic, and hydrophobic, but there are many binding groups. The bonds are broken (the complex dissociated) by addition of salts, or by increasing pH, temperature, or solvent polarity.

All immunoassay procedures are based on the original discovery by Berson and Yalow that low concentrations of antibodies to the antigen hormone insulin could be detected radiochemically by their ability to bind radiolabeled (^{131}I) insulin. The determination of unknown concentrations of antigen, then, is based on the fact that radiolabeled antigen and unlabeled antigen (from the sample or a standard) compete physiochemically for the binding sites on the antibodies (radioimmunoassay, RIA). This is illustrated in Figure 24.5.

The initial reaction vessel contains antibody solution (antiserum), labeled antigen, and the serum sample that may contain unlabeled (natural) **antigen** (*the substance to be determined*). Upon incubation, the antibody (Ab) will form an antigen–antibody immunocomplex (Ag–Ab). In the absence of unlabeled antigen, a certain fraction of the labeled antigen (Ag*) is bound (as Ag*–Ab). But when

The analyte antigen competes with the tagged antigen for sites on the antibody. The displaced tagged antigen is measured, either directly or indirectly.

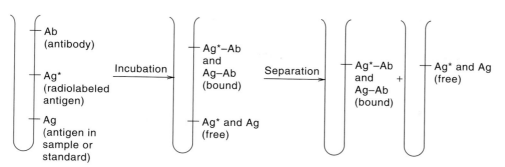

Fig. 24.5. Principles of radio-immunoassay.

increasing amounts of unlabeled antigen (Ag) are added, the limited binding sites of the antibody are progressively saturated and the antibody can bind less of the radiolabeled antigen. Following incubation, the bound antigens are separated from the unbound (free) antigens, and the labeled portion (radioactivity, fluorescence, etc.) of either or both phases is measured to determine the percent bound of the labeled antigen.

The antibody solution is initially diluted so that, in the absence of unlabeled standard or unknown antigen, about 50% of the tracer dose of Ag* is bound. A diminished binding of labeled antigen when sample is added indicates the presence of unlabeled antigen. A calibration curve is prepared using antigen standards of known concentration by plotting either the percent bound labeled antigen or the ratio of percent bound to free (B/F) as a function of the unlabeled antigen concentration. From this, the unknown antigen concentration can be ascertained.

A typical calibration curve is shown in Figure 24.6. Note the wide range of concentrations (10^{-2} to 10^{3} ng/mL in the most linear region) as well as the high sensitivity. The slope of the calibration curve, and hence the range and sensitivity, is dependent on the initial dilution of the antibody. Sensitivity is greatest with high dilutions of the antibody, but a wider range of antigen concentrations is covered when a more concentrated antibody is used. Also, as mentioned above, to obtain a very high sensitivity, it is advisable to use a dilution of antiserum that will bind about 50% of the labeled antigen (using a very small amount) in the absence of unlabeled antigen. This is because a decrease of 50% of the bound/free (B/F) ratio from an initial value of 1.0 (i.e., 50%, initially bound) can be more accurately determined than can a similar percentage decrease from an initial value of, say, 10 (91% bound). In the former case, there would be a decrease from 50% bound (B/F = 1.0) to 33% bound (B/F = 0.5), while in the latter case there would be a decrease from 91% (B/F = 10) to 83% bound (B/F = 5). The sensitivity of the initial part of the curve may be increased by using minimal amounts of tracer antigen.

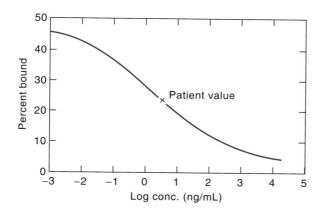

Fig. 24.6. Typical immunoassay calibration curve.

The competitive principle of immunoassays can be applied to nonimmunologic systems. Any substance that can bind specifically to a macromolecule can be measured using the principle of **competitive protein binding**. The specific binding macromolecule can be a serum protein such as thyroxine binding globulin, a specific receptor, or an antibody.

SPECIFICITY OF IMMUNOASSAYS

No antiserum used in immunoassays is completely specific for a particular antigen. The specificity is influenced by (a) heterogeneity of the antibody, (b) cross reaction with other antigens, and (c) possible interferences of the antigen–antibody reaction from low-molecular-weight substances that may alter the environment of the reaction. A given antigen induces the formation of multiple antibodies. It will combine with the multiple antibodies to various degrees, depending on the respective equilibrium constants. In addition, a single type of antibody may have a variable number and location of binding sites. The problem of heterogeneity has been diminished with the development of improved techniques for antisera purification. Also, the synthetic production of **monoclonal antibodies**, that is, single antibodies, provides high specificity.

The problem of cross reaction with other antigens is a complex one and must be considered separately for each type of antigen being measured. Special purification techniques may be required.

Nonspecific factors that may modify the rate of the antigen–antibody reaction include pH, ionic strength, high temperatures, the composition of the incubating medium buffer, heparin, urea, and high bilirubin concentrations. Antigen standards and unknowns should be prepared (diluted) in antigen-free plasma to swamp out differences in composition. The same buffer or other diluent should be used. If antigen-free plasma of the same species is not available, then plasma from a non-cross-reacting species can be used.

PREPARATION OF THE ANTIBODY

This is not the same as the titer described in Chapter 5.

A suitable antiserum is, of course, required for the immunoassay. The concentration of the antibody (called its **titer**) is important, but the main criterion for establishing a suitable antiserum is its specificity and affinity for the antigen being assayed.

The general method of inducing antibody formation is to inject 0.2 to 2 mg of the pure antigen mixed with "Freund's adjuvant" (a mixture of mineral oil, waxes, and killed bacilli, which enhances and prolongs the antigenic response). Animals used include rabbits, sheep, guinea pigs, goats, chickens, or monkeys—depending on the volume of antiserum desired and the degree of foreign activity of the antigen.

Hapten analytes must be linked to larger molecules to induce antibody formation.

Molecules that are too small to induce antibody formation (formula weight 1000 to 5000, called **haptens** as opposed to antigens, e.g., smaller polypeptides, steroids) are linked to carriers such as proteins or synthetic compounds before they are injected. Common carriers are human gamma globulin and albumin, synthetic peptides, and polymers.

Once an animal has been immunized, it can be injected several times to obtain different lots of antisera. Antisera are commercially available for most assays for which labeled antigens are available. Diluted antisera can be stored for long periods of time when frozen. Once thawed, they should be stored at 4°C.

INCUBATION PERIOD FOR THE ASSAY

The time required to reach equilibrium in the antigen–antibody reaction during the assay varies from a few hours to several days, depending on the specific antigen being measured. Lengthy incubations are generally not desirable (the antigen may be damaged by prolonged exposure to the high concentrations of plasma proteins, oxidants, etc.), and so some procedures may not be carried to the point of equilibrium.

SEPARATION OF THE BOUND AND FREE ANTIGEN

Various techniques are used to separate the bound antigen from the free antigen after incubation so that the percent bound can be determined from the label response. The immune complex is a protein-containing substance that can be **precipitated** (denatured) with high concentrations of various salts [e.g., $(NH_4)_2SO_4$, Na_2SO_4] or organic solvents (e.g., acetone, ethanol). Precipitation is one of the most widely used separation techniques. The precipitated complex is generally separated by centrifugation (although filtration can be used), and then the label response in either of the phases can be measured.

> These are heterogeneous assays. Homogeneous assays (below) do not require separations.

The **double-antibody technique** is also widely used. Here, a second antibody is employed to precipitate the primary antigen–antibody complex. This second antibody is produced by injecting a second animal with the gamma globulin produced in the first animal used to prepare the first antibody. Although the use of a second antibody introduces another source of error, the double-antibody technique has the advantage of being applicable for almost any immunoassay, and the separation is complete.

Other separation techniques used include electrophoresis (Chapter 21) and bonding of the antigen or antibody to a solid phase for use as reagent.

FLUORESCENCE IMMUNOASSAY

Antibodies or antigens can be labeled with fluorescent dyes and the fluorescence used for measurements (see Ref. 12). This technique is popular because of the greater ease of handling the chemicals than with radiochemicals and the problem of radioactivity decrease with time. Fluorescein isothiocyanate (FITC) and lissamine rhodamine B (RB 200) are commonly used dyes. The protein to be conjugated (reacted) with the dye is usually a mixture of immunoglobulins and should be free of other serum protein fractions because albumin and other globulins label more readily than gamma globulin and would render the method nonspecific. After labeling, unreacted dye is removed and the labeled protein purified using size exclusion chromatography.

> Here, the tag is a fluorophore. Its fluorescence is quenched by the immunochemical reaction.

The fluorescence of a tagged antigen is often quenched upon immunochemical reaction, and the decrease in fluorescence can be measured without going through a physical separation. This serves as the basis for **homogeneous immunoassays**, in contrast to **heterogeneous immunoassays**, which require a separation.

ENZYME IMMUNOASSAY

Here, the tag employed is an enzyme, for example, peroxidase, and the activity of the unreacted enzyme-tagged antigen is measured using an appropriate enzyme reaction. Again, in these techniques, the activity of the enzyme is often inhibited upon immunochemical reaction and the decrease in activity can be measured in a homogeneous immunoassay system. Homogeneous enzyme immunoassays are commonly used for the measurement of low-molecular-weight molecules such as digoxin, amphetamines, and prescription drugs.

> The activity of the enzyme tag is inhibited by the immunochemical reaction.

Enzyme immunoassays are broadly known as **enzyme-linked immunosorbent assays** (ELISA). See Ref. 13 for the original description. The different types are illustrated in Figure 24.7. These assays are conveniently performed by immobilizing

> In ELISA, the antigen or antibody is adsorbed on a plastic surface.

$$Ab-E + Ag \rightarrow Ag-Ab-E$$

Noncompetitive binding enzyme immunoassay. E is inhibited upon binding.

$$\begin{array}{ccc} \xi-Ab + Ag-E & & \xi-Ab-Ag \\ + & \rightarrow & + \\ Ag & & \xi-Ab-AgE \end{array}$$

Competitive ELISA. The amount of bound Ag—E decreases as Ag increases.

$$\xi-Ab + Ag \rightarrow \xi-Ab-Ag$$

$$\xi-Ab-Ag + Ab-E \rightarrow \xi-Ab-Ag-Ab-E$$

Noncompetitive sandwich ELISA. The bound AbE increases in proportion to the amount of Ag.

Fig. 24.7. ELISA assays. Ag is the antigen analyte.

In competitive binding, the analyte antigen and tagged antigen compete for sites on the adsorbed antibody.

In sandwich assays, the analyte antigen binds to the adsorbed antibody, and then tagged antibody binds to the bound antigen.

Indirect ELISA uses a universal-labeled antibody.

the antigen or antibody onto a solid surface, such as glass or plastic particles. Centrifugation is avoided by adsorbing onto tubes or disks made of polypropylene or polystyrene. These are readily washed, and they adsorb a reproducible amount of antibody or antigen. Small plastic wells impressed in a sheet (e.g., 96 wells) are commonly used so that many samples and standards can be processed simultaneously. There are automatic pipetors and measurement devices that are used with these.

If an enzyme tag is placed on the antibody, it is inhibited upon reacting with the analyte antigen; in this case, we have a **noncompetitive** protein binding assay (see, e.g., Ref. 11). In **competitive binding** ELISA, the antibody to the antigen analyte is adsorbed onto the solid phase, which occurs via hydrophobic interactions. Then a known amount of enzyme-labeled antigen is added along with the sample containing unlabeled antigen. Following incubation, the wells are washed and enzyme substrate is added to produce usually a colored product via the enzyme-catalyzed reaction. Maximum coloration (maximum enzyme reaction) occurs if there is no antigen in the sample to compete with binding of the labeled antigen, and this diminishes in proportion to the amount of antigen in the sample. This technique requires relatively large amounts of purified antigen for uniform labeling.

Sandwich assays are noncompetitive, whereby the sample is added to the wells containing the adsorbed antibody, followed (after incubation) by addition of enzyme-labeled antibody, which will bind in proportion to the amount of bound antigen. The unbound-labeled antibody is washed away, and the solid phase then contains the antigen sandwiched between unlabeled and labeled antibody. The color produced upon enzymatic reaction is then directly proportional to the amount of antigen.

In **indirect** ELISA, the sample antigen is first adsorbed on the solid phase, either directly or via an antibody as above. The unlabeled primary antibody is added, incubated, and washed. Finally, a secondary labeled antibody (raised against the immunoglobulin class of the species from which the primary antibody originated) is added, incubated, and washed. Again, the enzyme activity (color intensity produced) is proportional to the antigen. The advantage of this technique is that a "universal" antibody conjugate may be used as the secondary antibody against all primary antibodies used from the same immunoglobulin class of the appropriate species. So individually labeled primary antibodies are not needed for each antigen analyte. Indirect ELISA methods are also usually more sensitive than direct, probably due to the binding of more than one labeled secondary antibody to each primary antibody. And noncompetitive systems are usually more sensitive than competitive assays.

Learning Objectives

WHAT ARE SOME OF THE KEY THINGS WE LEARNED FROM THIS CHAPTER?

- The makeup of blood, what is measured? (Figure 24.2, Table 24.1, Table 24.2), p. 678
- Immunoassay—antigen, antibody; fluorescence IA; enzyme IA (ELISA), p. 683

Questions

1. What are the principal components of blood?
2. What is hemolysis and why is it important?
3. Why is sodium fluoride often added to blood samples collected for glucose analysis?
4. Why should whole blood samples not be frozen?
5. What are two of the most frequently performed clinical analyses?
6. What is an antibody?
7. Explain the principles of immunoassays. What labels are used? What are some variations in applications?

Recommended References

GENERAL

1. C. A. Burtis and E. R. Ashwood, eds., *Tietz Fundamentals of Clinical Chemistry*, 4th ed., Wasington, DC: American Association of Clinical Chemistry, 1995.
2. C. A. Burtis and E. R. Ashwood, eds. *Tietz Textbook of Clinical Chemistry*, 3rd ed., Washington, DC: American Association of Clinical Chemistry, 1998.
3. M. Reiner and D. Seligson, eds., *Standard Methods in Clinical Chemistry*. San Diego: Academic. Multivolume series started in 1953.
4. D. Glick, ed., *Methods of Biochemical Analysis*. New York: Wiley-Interscience. A series of annual volumes, started in 1954.
5. D. S. Young and R. B. Friedman, *Effect of Diseases on Clinical Laboratory Tests*, 4th ed. Washington, DC: American Association of Clinical Chemistry, 2001.
6. D. S. Young and R. B. Friedman, *Effect of Drugs on Clinical Laboratory Tests*, 5th ed. Washington, DC: American Association of Clinical Chemistry, 2000.
7. J. Wang, *Electroanalytical Techniques in Clinical Chemistry and Laboratory Medicine*. New York: VCH, 1988.

IMMUNOASSAYS

8. J. T. Wu, *Quantitative Immunoassay: A Practical Guide for Assay Establishment, Troubleshooting, and Clinical Applications.* Washington, DC: American Association of Clinical Chemistry, 2000.

9. C. P. Price and D. J. Newman, *Principles and Practice of Immunoassay*, 2nd ed., Basingstoke Hampshire, UK: Macmillan, 1997.

10. S. A. Berson and R. S. Yallow, "Immunoassay of Plasma Insulin," *Immunoassay of Hormones, Ciba Found. Colloq. Endocronol.*, **14** (1962) 182.

11. T. A. Kelly and G. D. Christian, "Homogeneous Enzymatic Fluorescence Immunoassay of Serum IgG by Continuous-Flow Injection Analysis," *Talanta*, **29** (1982) 1109.

12. C. M. O'Donnell and S. C. Suffin, "Fluorescence Immunoassays," *Anal. Chem.*, **51** (1979) 33A.

13. E. Engvall and P. Perlman, "Enzyme-Linked Immunosorbent Assay (ELISA). Quantitative Assay of Immunoglobulin G," *Immunochemistry*, **8** (1971) 871.

Chapter Twenty-Five

CENTURY OF THE GENE— GENOMICS AND PROTEOMICS: DNA SEQUENCING AND PROTEIN PROFILING

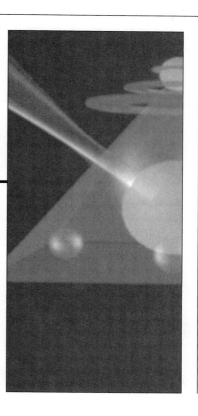

The century of the gene began when James Watson and Francis Crick unraveled the secret of the building block of genes—DNA exists as a double helix. For this pioneering work, they were awarded the Nobel Prize in 1962. It took the next 40 years to actually determine the entire human gene sequence (the genome). Analytical chemistry techniques and tools have played key roles. In this chapter, we will describe the basis for DNA sequencing and how this is used to construct the genome sequence. And we will discuss the tools being used for the more challenging goal of determining protein function. First, a little biology.

As a good introduction to this chapter, go to www.ornl.gov/hgmis/ project/info.html and read *From the Genome to the Proteome.*

25.1 Of What Are We Made?

The real building blocks of our body are proteins. Nearly all biological activity is carried out by proteins. Our muscles, teeth, liver, skin, blood—just about all of our bodies—use proteins to create the chemistry of life. Where do the proteins come from? Their unique syntheses are directed by our genes (DNA)—the process of "gene expression" is how genes actually do their work of controlling biological functions. Genes code for specific amino acids, the building blocks of proteins. Everything cells do is coded in their DNA, which directs which cells will make hair and what color it will be, among many other things.

Genes make up our chromosomes. The nucleus of nearly every one of our 100 trillion cells contains 23 pairs of chromosomes, that is, 46 individual chromosomes (Figure 25.1). Exceptions are sperm and egg cells—each contains 23 chromosomes. Each parent contributes 23 chromosomes to a child. Each child gets only a portion (a subset) of each parent's DNA, and the pieces received are

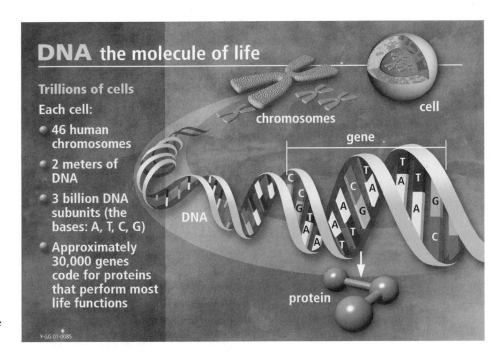

Fig. 25.1. From chromosomes to proteins. (Courtesy of the U.S. Department of Energy Human Genome Program; www.ornl.gov/hgmis.)

random, so each child gets a different subset of each parent's genes, except for identical twins, who have identical genes.

Chromosomes, when observed under a microscope, are fatty, tubular-looking substances. They are, in fact, made up of a series of genes, which are tightly twisted double strands of DNA molecules (Figure 25.2). The two strands are held together by weak hydrogen bonding between pairs of nucleotide bases on each strand, which may be viewed as "steps" in the DNA "ladder."

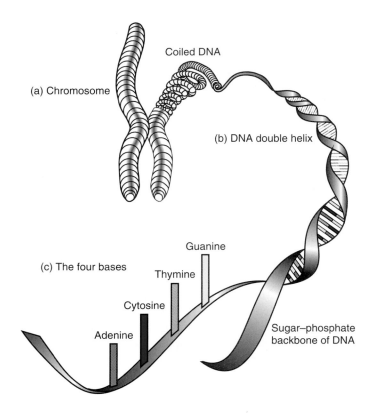

Fig. 25.2. Chromosome structure. Each chromosome consists of tightly coiled double strand of DNA, which when uncoiled reveals double-helix structure. (Courtesy of the Department of Energy Joint Genome Institute. Credit is given to the University of California, Lawrence Livermore National Library, and the Department of Energy under whose auspices the work was performed.)

25.2 What Is DNA?

DNA stands for **deoxyribonucleic acid**. It is a long, polymer-type molecule in which the replicating groups are **nucleotides**. Nucleotides consist of a sugar ring, a phosphate group, and a nucleic acid ring (which contains a basic nitrogen group and hence is the **base** in the nucleotide). The sugar–phosphate backbone does not change, but the base group can be different throughout the chain. The basic structure of a nucleotide is shown in Figure 25.3. These nucleotides form bonds between the 3′ group and the 5′ group (Figure 25.4). There are only four nucleic acid bases in the DNA nucleotides, to give four different nucleotides. They are adenine (A), cytosine (C), guanine (G), and thymine (T). Their structures are shown in Figure 25.5.

The nucleotide bases in DNA bond (pair) in *only two combinations*: A with T and G with C (Figure 25.6). The various combinations of these nucleotides in the very long DNA double strand are all that is necessary to write a code for every part of the body. Does that seem possible? Well, consider Morse code. It consists of only dots (·) and dashes (–). These, in combination, are used to create the alphabet (a = · –, b = – · · ·, c = – · – ·, etc.) Then the 26 letters are used to create words (with spaces between letters); words become sentences; sentences, paragraphs; paragraphs, chapters; chapters, books; books, encyclopedias, and so on. As we shall see later, in DNA, different combinations of three nucleotides in a row are used for coding the DNA expression. Figure 25.4 shows three nucleotides bonded together to form a DNA **oligonucleotide**. Note that the sugar rings at the 3′ OH group link with the phosphate group at the 5′ position of the sugar. So one end of the DNA oligo always ends with a 5′ group, and the other ends with a 3′ group. Now, when a single strand of DNA such as this links with another identical (complementary) single strand, they do so in opposite directions, and so the structure is said to be antiparallel. In other words, the sugars are "right side up" in one and "upside down" in the other, and the 3′ and 5′ ends of each strand are at opposite ends. The two chains are positioned so their bases can interact with each other in the middle and pair up. An A nucleotide is always opposite a T nucleotide, and G opposite a C. While the base interactions are weak, given enough of them, they are sufficient to hold the double strand together.

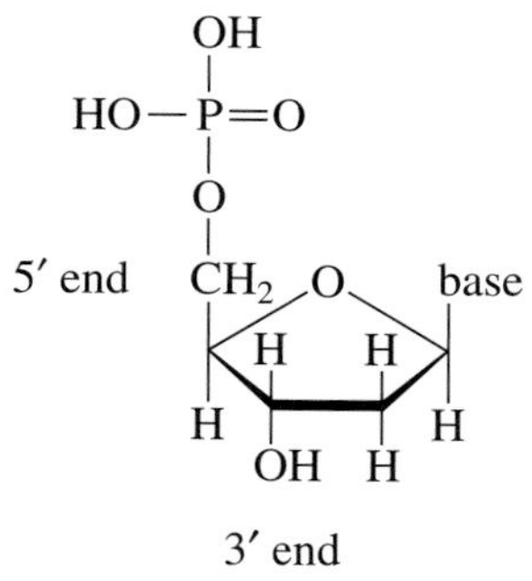

25.3 Human Genome Project

The Human Genome Project is an International Human Genome Sequencing Consortium (www.nhgri.nih.gov/NEWS/profiles.html) that was formed in 1990, with the goal of identifying all 3 billion bits (base pairs) of DNA that comprises the entire human genetic code in the genome and put them in proper sequence. The consortium consists of some 20 groups from the United States, United Kingdom, France, Germany, Japan, and China, funded mainly by the U.S. National Institutes of Health (NIH) and the Wellcome Trust of London, to the tune of about $2.5 billion. The U.S. scientific effort is managed by the National Human Genome Research Institute and the U.S. Department of Energy Human Genome Program.

The goal was originally to finish assembling the human genome by 2003. The consortium laboratories sequence the genome one bit at a time, very methodically, "clone-by-clone" sequencing (more on this later). Then along came a private group, called Celera Genomes Group, founded by Dr. Craig Venter, a former NIH researcher, who announced in 1998 it would launch a competing project and finish 3 years sooner. They and co-workers at 13 other academic, nonprofit, and industrial sites initiated the sequencing in September 1999, and finished a draft by

Before or after reading this section, go to www.pbs.org/wgbh/nova/genome/dna.html, and read *Journey into DNA.*

Only four bases code for all biological functions.

Fig. 25.3. Basic structure of nucleotide.

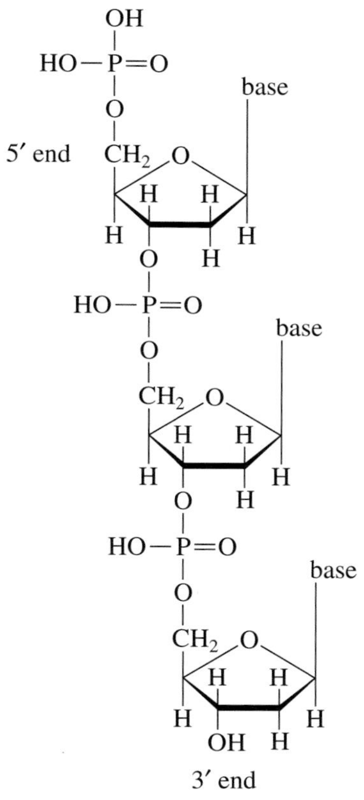

Fig. 25.4. Oligonucleotide. These are anionic at physiological pH.

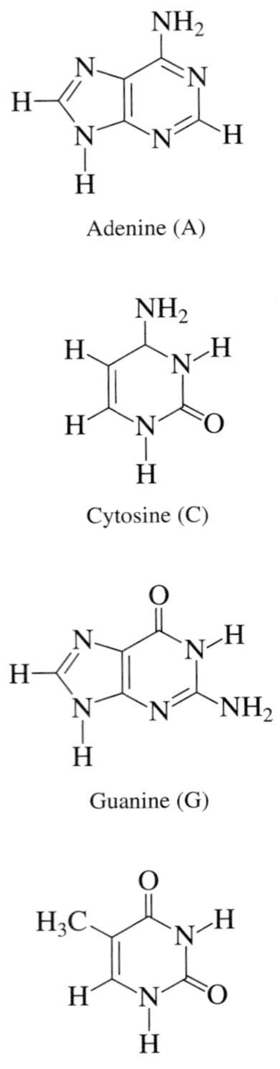

Adenine (A)

Cytosine (C)

Guanine (G)

H₃C

Thymine (T)

Fig. 25.5. Nucleic acids in DNA. A hydrogen bonds with T, and C hydrogen bonds with G in oligonucleotides.

For the role of analytical chemistry in the Human Genome Project, see E. Zubritsky, "How Analytical Chemists Saved the Human Genome Project," *Anal. Chem.*, **74** (2002) 23A (Feature article).

Check out these historical announcements.

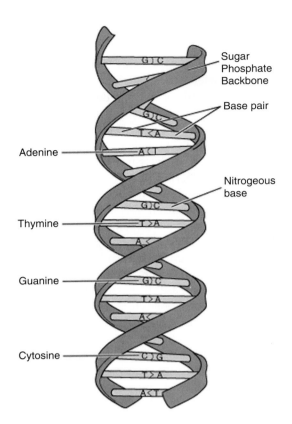

Fig. 25.6. DNA double helix, made up of sugar-phosphate backbone, with pairing between A–T and C–G bases to create coiled shape. (D. Leja, National Human Genome Research Institute. Reproduced by permission.)

October 2000, a remarkable feat. How could they do this? Two ways. First, they built upon the vast array of data already gathered by the consortium (which is entirely open and free to the public). Second, they developed a radically new way of sequencing, known as "whole genome shotgun cloning" (more on that later). And they began using new faster automated analytical technology based on capillary zone electrophoresis, rather than gel electrophoresis, built by their sister company, Applied Biosystems, the ABI Model 3700 (more about methodologies later). Much of the consortium switched to the same technology to keep pace, and in June 2000, both sides agreed to announce they had largely completed the gene sequence. In February 2001, each published the drafts of their results, the consortium's in *Nature*, **409** (2001) 860, and Celera's in *Science*, **291** (2001) 1304. You may see the publication announcements at www.nature.com/genomics/human and www.sciencemagazine.org/content/vo129/issue5501. The consortium data are publicly available. Celera arranged an agreement with *Science* that permitted the Celera team to post its data in a proprietary database. Anyone has free access to it, but not for commercial purposes.

Actually, the published "drafts" of the human genome were only about 89% complete, and finishing the job and filling in details requires another 3 years.

What are some of the findings of sequencing the human genome? One of the most surprising is that there are only about 30,000 to 35,000 genes, those that produce proteins, representing only about 1% of the genome. (As more information is being gathered, indications are this may be an underestimate for primates.) This is less than half the 80,000 to 100,000 genes that had been predicted. Most of the genome consists of long stretches of largely repetitive noncoding regions of DNA. And about one-fourth of the genome is desertlike in that it contains long gene-free segments. This complexity has been one of the challenges in sorting out the genes.

25.4 How Are Genes Sequenced?

Gene sequencing is a complicated, but now fairly routine, process. Figures 25.7 and 25.8 summarize the steps involved, which will be described below. Overlapping fragments of genes are sequenced, and from the overlapping portions, the entire gene can be deduced. The process involves first isolating and multiplying (cloning) large segments of the genes to provide enough to work with. These cloned segments constitute a genomic library and are provided to sequencing laboratories. The long overlapping segments are broken into shorter splices, using a random shearing technique in which the sample is forced through a needle and the shear forces break them up. Originally, they were spliced using a *nuclease* enzyme, but the shearing technique gives more randomness. (Nucleases degrade nucleic acids. They may act on only single strands or on double strands, or on both. A nuclease may degrade from one end—an exonuclease—or begin in the middle—an endonuclease.) The splices are replicated by pairing with the appropriate complementary nucleotides. Beginning at one end, one nucleotide at a time is added. Ordinarily, this would go on until the entire fragment is replicated. But we have a way of randomly stopping the replication at different points along the DNA, for each of the four nucleotides. We also have a way to measure what the terminating nucleotide is, whether it is A, C, G, or T. If, among the millions or billions of fragment clones, we obtain some of every possible length along the nucleotide sequence, and if we can separate them by length and line them up, we will have the complete ordered set of nucleotides in the original DNA template. Suppose we have isolated a series of strings, in the order of length, each terminated with a nucleotide:

_____ T

_____ A

_____ A

_____ C

_____ G

Then the original template was TAACG. Now we will describe how this is done.

Before or after reading the following sections on sequencing, go to www.pbs.org/wgbh/nova/genome/sequener.html and read *Sequence for Yourself.*

25.5 Replicating DNA: The Polymerase Chain Reaction

DNA has a unique property. It is denatured by heating (much as proteins are), but it does so by separation of the two strands. This occurs if we heat DNA in solution

Fig. 25.7. Creating genomic library.

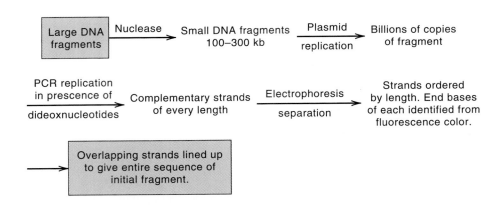

Fig. 25.8. Sequencing genomic library.

to the **critical temperature** (T_m). The separated strands are still complementary, and when the solution is cooled, they will, by random molecular motion line up and stick together again, reforming the double-stranded DNA. This process is called **annealing**, or **hybridization**.

We wish to multiply the DNA in order to have enough to work with for sequencing and detection. We can actually rebuild each of the separated complementary strands, using an enzyme called a **DNA polymerase**, in the presence of the four nucleotides and a **primer**. A primer is a short piece of DNA [oligonucleotide, generally about 20 bases (20 b) or nucleotides (20 nt) long]. The primer is complementary to and sticks to one end of the DNA template that we wish to replicate. Here is a typical primer, called T3:

$$5'\text{-ATT-AAC-CCT-CAC-TAA-AGG-GA-}3'$$

It is 20 nt long.

The DNA polymerase extends the primer to create the complement of the DNA strand, hence the primer is used to *initiate* the reaction. The DNA polymerase extends the 3' end of the primer. It "reads" each base in the single-strand template and attaches the complementary base (i.e., A matching a T, or T with A; and C with G or G with C). As it moves down the strand, it creates the exact complement (i.e., for a 5' to 3' strand, we now have the complementary 3' to 5' strand). The same thing happens with the other (complementary) separated strand. So we have now doubled the number of DNA molecules. We can repeat this process by cycling the temperature, until we have sufficient DNA to work with, each time doubling the number of molecules. A single copy of a DNA strand can be amplified to obtain billions of replicates. This process is called a **polymerase chain reaction** or **PCR**.

There are different classes of polymerases. The one commonly used in PCR is Taq polymerase, which is stable at high temperatures and is isolated from the bacterium *Thermophilis aquaticus*.

The polymerase chain reaction can be used to detect and measure extremely small quantities of DNA, even in mixtures, and is useful in forensics and in clinical diagnosis. This was used, for example, in the O.J. Simpson case to amplify the DNA from small blood samples at the crime scene. The replicated DNA is sequenced in the usual fashion, as described below, and the nucleotide pattern is matched against a suspect's pattern.

DNA polymerase adds complementary nucleotides to a template DNA strand.

25.6 Plasmids and Bacterial Artificial Chromosomes (BACs)

For the genome project, "libraries" of large pieces of genomic DNA that have been cloned were assembled. It is these that researchers sequence. The cloning for the libraries takes advantage of nature's ability to do the replicating for us. Different types of DNA are used for replicating the genomic DNA fragments, and these are called **vectors**. There are two common types, plasmids and BACs.

PLASMIDS

A plasmid is a small circular piece of DNA found in bacteria. Plasmids can be isolated in large quantities. The isolated plasmids are spliced and the DNA we want to sequence is added to (*inserted in*) the splices. The added DNA is called the **insert**. The plasmid containing the vector and the insert is put back into the bacteria, and the bacteria will replicate it along with their own DNA, producing billions of copies, which we then isolate. We usually know the sequence of the vector, but not of the insert. Typical plasmids are pGEM, pBR322, and PUC18. Plasmid DNAs are limited in size, about 2 to 20 kilobases (2 to 20,000 nt).

BACTERIAL ARTIFICIAL CHROMOSOME

BACs are used to replicate larger DNA strands, much like with plasmids. They are cloning vectors, capable of inserting nucleotides of 100 to 400 kilobases (100,000 to 400,000 nt)—very large. Like plasmids, they are placed in host bacteria that grow and replicate the BAC; they really produce minichromosomes. Major pieces of the human genome DNA have been cloned this way, to create a **genomic library** of BACs for distribution to consortium laboratories for sequencing. Each genomic library contains a large piece of the genomic DNA, around 100,000 to 300,000 nt long. A library will contain enough of these random BACs that a given gene will probably be on several overlapping BACs, and there is a 99% probability that any given piece of DNA from the splice of genome DNA is present. Entire genomes are cloned into BAC libraries, and the BAC clones are sequenced fairly quickly using the technique of **shotgun sequencing**.

Beginning of Gene Sequencing

The basis for the tremendous human genome sequencing effort goes back to the 1970s, when Frederick Sanger in Cambridge, England, developed a method for sequencing DNA, using electrophoretic slab gels. These were, of course, run manually and were slow by today's standards, but this was groundbreaking. Sanger and his co-workers were the first to sequence genomes; the first was a simple virus, followed by a mitochondrial genome. They later began sequencing bacterial genomes. Sanger received two Nobel Prizes for his work on developing the DNA sequencing method (1980, www.almaz.com/nobel/chemistry/1980c.html) and work on the structure of proteins (1958, www.almaz.com/nobel/chemistry/1958a.html).

Dr. Maynard Olson (now director of the University of Washington Genome Center) pioneered a yeast artificial cloning system and introduced certain mapping markers in the 1980s that were critical for initiating the sequencing of the human gene. Another critical piece was the development of powerful software to sort out the billions of bits of information. Dr. Phillip Green, a colleague of Olson's, created two software programs, called Phred and Phrap that have played critical roles in analyzing raw DNA data.

25.7 DNA Sequencing

So now we have different segments of the genome, ready to sequence. These segments are of different length and will overlap. To complete the puzzle, we will sequence each one of them; then line them up to see where the overlaps are, and *voilà*, we know the genome for all 23 chromosome pairs. (This is a bit simplified because of the varied structure of the genome.)

The actual sequencing process is very similar to the PCR process. We will cycle through denaturation and annealing to replicate the DNA fractions. The replication or hybridization is done in the presence of the template DNA, a Taq polymerase, and a primer to initiate the replication. We should note that accurate sequencing often begins about 50 bases from the primer. (The selection of a primer is important, to place it at an appropriate position on the DNA templates, but we will not go into the specifics. A primer should be placed a maximum distance from the 3′ end, since it extends in that direction.)

There is one important difference here from conventional PCR. Besides adding all four nucleotides for constructing the complement strand, we also add a small percentage, about 5%, of isomers of each nucleotide in which the 3′ hydroxy group on the sugar is replaced with a hydrogen; these are called **dideoxynucleotides**. Consider just the A nucleotide. Most of the time when the polymerase sees a T, it will lay down the complementary A, and the hybridization process continues. But about 5% of the time, the dideoxy-A will be laid down. The polymerase is incapable of growing the strand from a dideoxynucleotide, and the elongation is terminated. The same happens with the other nucleotides. When we have billions of copies of the DNA template, chances are certain that we will terminate a fraction of these at all possible lengths, each terminating with either A, C, G, or T dideoxy molecule; there will be millions of each. They will be separated together by electrophoresis and measured (below). Now there is one more important feature of the dideoxynucleotides. They are chemically tagged, usually with a fluorophore that allows us to see them as they are eluted from an electrophoresis apparatus. Each has a different tag, which fluoresces at different wavelengths. So from the four different fluorescent colors, we know which nucleotide the DNA fragment ends in.

If we look at only the A tag, the separated pieces might look like this on the gel:

Chain growth terminates with the addition of a dideoxynucleotide.

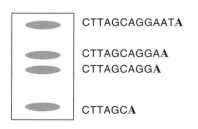

CTTAGCAGGAAT**A**

CTTAGCAGGA**A**
CTTAGCAGG**A**

CTTAGC**A**

The shorter fragments migrate toward the bottom faster (and elute first). Here, from the bottom, the first and second fragments are separated by two fragments (there were no A terminations corresponding to those two lengths), and the top two are separated by one.

Now, of course, we really have a mixture of fragments that end in all four nucleotides, and there will be some at every possible length. We separate and detect each one of these, knowing which of the four colors it is (the terminal group). In the above chromosome, it would be filled in as:

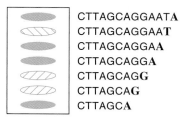

CTTAGCAGGAAT**A**
CTTAGCAGGAA**T**
CTTAGCAGGA**A**
CTTAGCAGG**A**
CTTAGCAG**G**
CTTAGCA**G**
CTTAGC**A**

So the sequence is CTTAGC**AGGAATA**. The bold letters are the ones we have determined. We look at overlap with other analyzed fragments to determine the entire sequence. So we might have two fragments that look like:

CTTAGC**AGGAATA**

CTTAGC**TAGGCCT**

The **TA** bases overlap, and the sequence is then **AGGAATAGGCCT**. The actual lengths we deal with are much larger, but this illustrates the principle. By convention, sequences are always written from the 5′ to 3′.

The actual separation and measurements are done with an automated DNA sequencing instrument. The separations are done by electrophoresis, either gel or capillary electrophoresis. There may be 96 gel lanes of samples run at a time. Figure 25.9 shows the reconstructed order of bands for a typical run. An ultraviolet

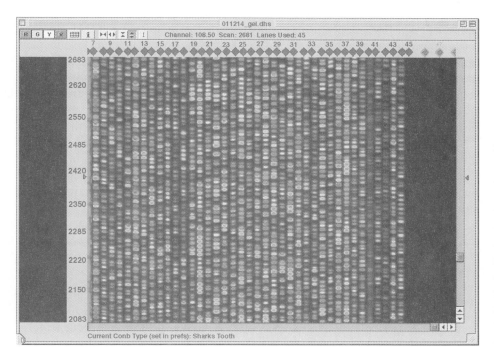

Fig. 25.9. Gel electrophoresis separation of nucleotides. Each vertical lane represents a different sample. In each lane, bases are eluted in order of size, the smaller ones traveling the fastest. Each band has one of four base colors, which identifies the end base on the oligonucleotide corresponding to that band. (Courtesy of University of Washington Genome Center.) (See your CD, auxiliary data, for a colored picture of the lanes, showing the four different nucleotide colors.)

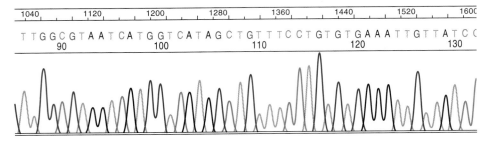

Fig. 25.10. Chromatogram of separated bands. Each peak corresponds to one of four colors, and terminating base is printed above peak, in the same color. (Courtesy of University of Michigan DNA Sequencing Core.) (See your CD, auxiliary data, for a colored picture of the separated peaks, showing the four nucleotide colors matched to the terminating bases.)

laser in the instrument scans each lane of the gel near the bottom, and the instrument detects the fluorescence from each nucleotide band (at the emission wavelengths of each of the four nucleotide tags).

As each nucleotide is eluted and sensed (the smaller ones coming off first), the fluorescence signals from the bands are plotted to give a chromatogram. Figure 25.10 shows a partial chromatogram from a single lane, scanned from the smallest to the largest fragment. The instrument's computer, knowing the color of each band that gives rise to a peak, automatically prints the nucleotide symbol above each peak. The figure shows only a small segment of the entire chromatogram. Typically, up to 700 nucleotides are measured in a sample. Beyond that, the band spreading becomes too great to achieve accurate and sensitive detection.

What we have just described is the approach used by the publicly funded Human Genome Project consortium. The National Science Foundation and the National Institutes of Health have funded the preparation of libraries of BAC clones. Participating laboratories replicate the libraries and perform the sequencing using the shotgun sequencing technique.

To summarize the shotgun genome sequencing procedure, sequencing laboratories obtain BAC clones from the libraries of BAC clones. Each BAC contains a large piece of the genomic DNA, up to 300 kb. They occur randomly, so that any one gene is likely to be present on overlapping BACs. The BACs are replicated by the laboratory, using a DNA polymerase. Only a few hundred nucleotides can be sequenced at a time, so the multiplied BACs are usually fragmented using shear forces, and the fragments are inserted into a vector (usually a plasmid) to replicate them. The multiplied fragments are then sequenced using a mixture of the four nucleotides, their dideoxy isomers, and a primer to form complementary strands of different lengths. The strands are separated by electrophoresis, in order of size, and the end nucleotide of each is identified from its color. Then the sequence of the nucleotides is written in the direction of 5′ to 3′. The sequence of overlapping strands gives the complete BAC piece sequence.

Phred and Phrap

DNA sequencing generates huge amounts of data. The individual chromatographic peaks may overlap, be shifted somewhat from expected position, be distorted in shape, and so forth. There are two problems with which sequencing laboratories must deal. The first is to reliably identify the nucleotide peaks, and the second is to take the enormous amounts of data and assemble them into the proper overlapping sequences to construct the genome sequence. Professor Phillip Green at the University of Washington, and a member of the U.W. Genome Center, has written two powerful software programs to accomplish both of these tasks that are widely used by sequencing laboratories. They are called Phred (for "phragment read") and Phrap ("phragment assembly program" or "Phil's revised assembly program").

Phred reads and identifies DNA sequence trace data (bases), assigns quality values to the bases, and writes the base identities and quality values to output files. It uses Fourier methods to examine the base traces (chromatographic peaks or profiles) in the region where a base is anticipated, that is, it determines where the peak would be centered if there were no distortions or other factors shifting the peaks from their "true" locations.

Phrap is a program for shotgun sequencing assembly. It delineates the likely accurate base identifications in each "read" and eliminates repeats. It constructs a contiguous sequence as a mosaic of the highest quality parts of the reads, providing sequences with less than 1 error per 10 kilobases. A probability error is assigned for each consensus sequence position, which allows scientists to focus on manual editing of particular regions, as needed.

25.8 Whole Genome Shotgun Sequencing

The approach taken by the private Celera group is somewhat different. They bypass the use of BAC libraries of long DNA pieces. Instead they break up the entire genome into fragments only 2000 to 10,000 nt long and sequence each of these, in chunks of up to about 700 bases (the resolution limit of the instruments). If we have 3 billion bases to worry about, this means a lot of pieces to sort out and assemble in the proper order (actually, only a portion of those are sorted, but it is still a lot). Powerful computers do the sorting to put the puzzle pieces together.

To more rapidly process these millions of pieces, the whole operation is done automatically, performing separations by capillary electrophoresis, which has higher resolving power and is faster than gel electrophoresis. Detection is done by passing a laser beam through the eluting drops at the ends of multiple capillaries. The public consortium has largely switched to this instrument as well (the ABI Model 3700).

There is controversy about which approach provides more accurate results. Both groups are continuing on filling the remaining gaps (remember, only 89% of the genes were initially determined).

25.9 Single-Nucleotide Polymorphisms

The decoding of the human genome revealed that human beings are remarkably similar in genetic makeup (in fact, our genetic code does not differ much from that of animals). About 99.9% of the genes in humans are identical, and only 0.1% of our genes result in the differing features that make each of us different. Also, within this 0.1% are the genes that contribute to disease and disfunction. These genes

SNPs are what differentiate all of us. They constitute 0.1% of the genes.

typically differ by a single base. Of course, 0.1% of 3 billion is still 3 million genes. But there are genetic markers that narrow the search for genomic regions of concern, called **single-nucleotide polymorphisms** (SNPs, pronounced "snips"). It is estimated that 80 to 90% of human variation is due to SNPs. Detection of SNPs is a very active area of research for clinical and pharmocogenic studies, to identify genetic defects and to discover drugs to treat them.

SNPs usually occur in the same genomic locations in all people. About 3 million have been identified, and there are perhaps 10 million SNPs in the entire human population. Many have little significance, but the presence of SNPs in specific regions have high correlations with specific diseases. Most diseases have hundreds of different SNPs. There are only about 200,000 SNPs that have reasonable frequencies, and nearly all common genetic illnesses are found in those 200,000.

SNPs are identified by sequencing. New technologies based on DNA microarrays are being developed for rapid SNP assays.

25.10 DNA Chips

DNA microarrays contain known DNA strands. Hybridized complements of specific strands from added samples are detected by fluorescence.

To rapidly and inexpensively obtain information about an individual's genetic characteristics, particularly for identifying genetic markers for specific diseases, scientist have developed new technologies. These are based on **DNA microarrays**, or what has become known as **DNA chips**. They are used to assay a genetic sample to see if it contains particular DNA sequences, for example, certain SNPs.

DNA chips are based on the fact that complementary DNA strands stick together. Single-stranded DNA segments of known sequence are prepared (by denaturization) and spotted (tethered) on microarray plates. Each spot has a different tethered nucleotide sequence. The unknown sample is placed on each spot. When an unknown sequence finds its complement, it sticks to it. Fluorescent markers are added that bind to the hybridized strands. Fluorescence occurs only when the complements hybridize or bond. Hence, one can immediately identify the unknown sequences from the known nucleotide tethers that they bind to, from the fluorescing spots.

Hundreds of thousands of different DNA sequences can be placed on a microscopic grid just a few centimeters across. The chips are mass produced by commercial manufacturers, labeled, and sold to users. There are also commercial hand-held DNA chip analyzers with on the order of 100 test sites (see www.nanogen.com).

SNP ANALYSIS

DNA chips are used to determine mutated SNPs that cause diseases.

Identification of an individual's SNPs that might be related to a specific disease will be an important feature in the future for medical diagnosis, prevention, and treatment. DNA chip technology will be key to such determinations. Chips are made that contain many possible variations, due to specific diseases, of one gene. A DNA sample is taken from the patient to be tested, and the DNA is amplified using PCR. The amplified sample is placed on the chip, and luminescence identifies the particular sequence variant the patient might have. Different drugs can be administered and more tests performed to see which ones might eliminate the variant.

An example of using DNA chips is to determine whether a woman possesses mutations of the BRCA1 gene. Mutation of this gene, and of BRCA2, are believed to be responsible for up to 60% of hereditary breast and ovary cancers. The DNA chips contain fragments of normal BRCA1 DNA. The sample of DNA from the patient's blood, and a control sample that does not contain a mutation, is denatured to give single strands that are cut into smaller manageable fragments, and labeled with fluorescent dyes. The patient's DNA is labeled with a green dye, and the control is labeled with a red dye. Both sets of labeled DNA samples are placed on the chip and allowed to hybridize (bind) to the BRCA1 DNA fragments on the chip. If the patient's DNA does not exhibit mutations, it will hybridize with the same BRCA1 fragments that the normal sample does, and the patient's spots will be identified by the green fluorescence and the controls by the red. If, however, there are mutations in the patient's genes, then her DNA will not hybridize in the region where the mutation is located.

EXPRESSION PROFILING

A gene expresses itself by transcribing its DNA sequence into a complementary copy of **messenger ribonucleic acid** (mRNA). The RNA sequence is translated, by a ribosome enzyme, into the string of amino acids that make up a particular protein. A cell's response to a stimulus, such as a toxin, can be evaluated by determining the effect on a gene's expression, that is, whether the gene is turned on or off.

Gene expression is determined by measuring mRNA on DNA chips.

Expression analysis can also be used to classify types of cancers or other diseases. This is done by obtaining patterns of gene expression of patients with known variants of the disease, and then comparing the mRNA sequences with a new patient's mRNA gene expression. In this case, the known RNA complements for the disease variants are placed on the chips as templates.

Expression analysis will be very important in the future for classifying different types of cancer and other diseases, based on patterns of gene activity in the tumor cells. This will enable more specific treatment targeted to each specific type of cancer and also allow scientists to determine which therapies are most effective for a given type.

25.11 Draft Genome

The construction of the working draft of the human genome was accomplished by the team of sequencing centers by producing multiple overlapping data. Over 22 billion bases of raw sequence data were produced, which provided sevenfold coverage of the entire genome (sequenced seven times). There were 3.9 billion overlapping bases in the fragments. More than 30% of this draft was high quality, finished sequence, with 8- to 10-fold coverage and 99.9% accuracy. There were only a few gaps in the 89% of the finished genome portion. The continued work fills in the genes and sequence of the remainder of the genome (2003 completion date).

25.12 Genomes and Proteomics: The Rest of the Story

The fundamental unit of all living systems is the cell. The DNA in the nucleus provides the instructions to direct the cell activities, primarily through the production of proteins. But it is the proteins that perform most life functions, often as protein–protein complexes. To understand protein function, its relation to disease, and to develop drugs for specific diseases, we need to study the expression of genes in producing proteins, and we need to determine protein structure and function. Analytical tools and techniques are used for these studies.

We define **genomics** as the study of DNA and the encoding process leading to protein formation. **Proteomics** is the study of proteins in a cell—identification of the entire protein complement of a cell, tissue, or organisms (the proteome). It includes the study of protein interaction and involves protein sequencing to determine the primary protein structure.

Proteins are large molecules made up of 20 different kinds of **amino acids** (see Chapters 10 and 21). A series of several linked amino acids is called a **peptide**. Amino acids link between carbonyl and amino groups by splitting out a water

$$\overset{\overset{\textstyle O}{\|}}{-C}-\overset{\overset{\textstyle H}{|}}{N}-$$

in a condensation reaction, to form —C—N— bonds. Genes encode the synthesis of proteins from amino acids. Within the genes, a sequence of three specific bases (called **codons**) direct the protein-synthesizing chemistry in the cell to add a specific amino acid, to grow a peptide chain. The nucleotide base sequence of ATG, for example, encodes for the addition of a methionine molecule. The average size gene contains about 3000 bases, so there are 1000 three-base codons. Hence, a protein coded by the gene will contain about 1000 amino acids. This is the **genetic code**—a series of codons that direct the formation of specific proteins. The particular amino acid content of a protein causes it to fold into a specific three-dimensional structure that determines how it will function in the cell (see Figure 25.11).

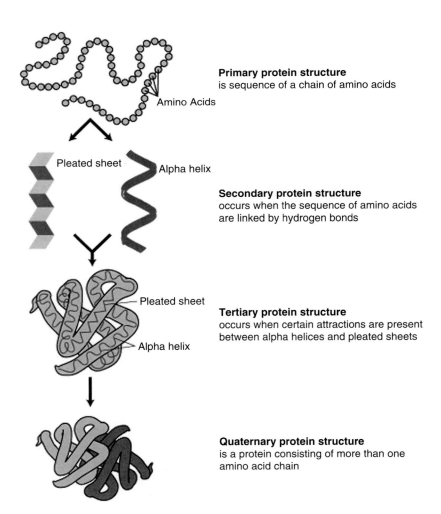

Primary protein structure
is sequence of a chain of amino acids

Amino Acids

Pleated sheet   Alpha helix

Secondary protein structure
occurs when the sequence of amino acids
are linked by hydrogen bonds

Pleated sheet

Alpha helix

Tertiary protein structure
occurs when certain attractions are present
between alpha helices and pleated sheets

Quaternary protein structure
is a protein consisting of more than one
amino acid chain

Fig. 25.11. Protein structure.
(D. Leja, National Human Genome Research Institute. Reproduced by permission.)

The coding instructions from the gene are transmitted by formation of a complementary single-strand intermediate messenger ribonucleic acid (mRNA), by a process called **transcription**. The mRNA migrates out of the nucleus into the cytoplasm, where it becomes the template for protein synthesis.

We described above the approach of studying gene expression using DNA chips or microarrays, and how to use the information to identify mutations of genes that cause specific diseases. The field of proteomics goes a step further, to obtain structural information about proteins. There are powerful tools for separating and studying protein mixtures.

TWO-DIMENSIONAL PAGE

Two-dimensional polyacrylamide gel electrophoresis (2-D PAGE) is the primary technique used to separate complex protein mixtures. The proteins are resolved, first based on charge and then based on molecular size. The sample is placed at the top of a tube of gel for initial resolution using **isolectric focusing**, in which the proteins move in an electric field based on their charge (their isoelectric points—see Chapter 21). There is an immobilized pH gradient in the gel that enhances the separation efficiency, by changing the charge as the proteins move down the gel, much like temperature programming does in gas chromatography. After this first-dimension separation, the gel is placed at the top of a sodium dodecyl sulfate (SDS) polyacrylamide gel, and the proteins at each position in the isolectric focusing gel are further separated by electrophoresis, based on size (SDS-PAGE). This 2-D separation based on orthogonol separation principles provides very high resolution. Several thousand proteins, or protein clumps, can be resolved. See Figure 25.12 for a typical resolved mixture.

MALDI–TOF

Once the proteins have been separated, there is the problem of identifying them or those identified as being of interest. (Often, in protein expression studies, the protein spots, e.g.—from a cancer sample—are compared with those of a control, and only those spots that appear different are analyzed in more detail. The spots are stained with a fluorescent dye to illuminate the differences.)

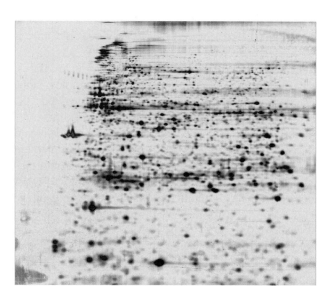

Fig. 25.12. 2-D gel separation of proteins. (From Harefield Hospital, Middlesex, UK. Courtesy of Dr. Mike Dunn. Reproduced by permission.)

Protein identification is achieved by a sequence analysis protocol in which the proteins in individual spots are excised from the gel and are degraded into a mixture of peptides. The protein is cut using a proteolytic enzyme, trypsin, which is very selective in the peptide bonds that are cleaved (the C-terminal side of either argenine or lysine is cleaved in the protein). The peptide mixture is then analyzed by mass spectrometry to identify the individual peptides.

Analysis by mass spectrometry requires an ionization source and a mass analyzer (see Chapter 20). The ionization technique of choice is **matrix-assisted laser desorption ionization** (MALDI), while the mass analyzer is a **time-of-flight** (TOF) analyzer. The combination is MALDI–TOF. MALDI is a "soft ionization" technique that induces very little degradation of high-molecular-weight samples. Often, the intact (protonated) peptide is detected, making identification easy. The peptide sample is mixed with a large excess of a matrix material, a small UV-absorbing organic molecule such as dihydrobenzoic acid, and the mixture is dried on a plate to provide a crystalline matrix. The matrix selected happens to absorb energy at the wavelength of the nitrogen laser used (377 nm), converting it to heat. This causes the sample to vaporize and ionize, by protonation of the peptide to a +1 charge. This vapor is introduced into the TOF analyzer via electrostatic lenses. MALDI, besides providing a single mass spectrum peak corresponding to one ion species, is very sensitive. Only femtomole (10^{-15} mol) quantities of sample are required to provide a good mass spectrum.

Time-of-flight is the simplest of mass analyzers. It determines the m/z ratio of ions by measuring the time for the ions to fly through a fixed length path to an ion detector. Each ion has the same kinetic energy when introduced into the analyzer, but the speed varies with the mass, causing each ion to arrive at the detector at different times.

IDENTIFICATION OF THE PROTEIN

The trypsin digestion splits the protein into a mixture of numerous polypeptides of, say, 4 to 20 amino acids in length. From the MALDI–TOF experiment, we can identify several of these. There are databases of proteins that contain amino acid sequence data for over 100,000 proteins (see *Protein Information Resource*: http://pir.georgetown.edu). Computer database searching programs compare the experimental fingerprint with the peptide sequences in every protein of a given organism, and only one protein can produce the observed series of peptides. Accurate protein identification is sometimes possible with 5 or 6 peptide matches. It is possible to compare observed peptide fingerprint data with every possible human protein.

There are other techniques and procedures for analyzing protein mixtures, many in the experimental stage, in order to speed up processing of samples, since progress in the field requires investigating large numbers of samples, but the methodologies and techniques we have described here continue as workhorses.

This chapter has presented the basics of DNA sequencing and analysis and complex protein analyses. There are many technical details involved to implement these technologies that are beyond the scope of this text. You should refer to biochemistry and molecular biology texts for more information on the biology of DNA and proteins. The Web references given throughout the chapter and those at the end of the chapter provide helpful tutorial information on these subjects.

Learning Objectives

WHAT ARE SOME OF THE KEY THINGS WE LEARNED FROM THIS CHAPTER?

- The makeup of DNA, p. 695
- The human genome project—start and finish, p. 695
- The sequencing of genes—the polymerase chain reaction (PCR); plasmids and BACs, pp. 697, 699
- DNA sequencing—separating and identifying the nucleotides; gel and capillary electrophoresis, p. 700
- Whole genome shotgun sequencing, p. 703
- Single-nucleotide polymorphisms (SNPs) as genetic markers, p. 703
- DNA chips, p. 704
- Proteomics—separating and identifying protons: 2-D PAGE to separate, MALDI–TOF mass spectrometry to identify, pp. 705, 707

Questions

1. What are genes and chromosomes?
2. What is DNA made of?
3. What is the polymerase chain reaction?
4. What is a plasmid? A BAC?
5. What is a genomic library?
6. Explain the shotgun sequencing of DNA.
7. What is a SNP?
8. What is a DNA chip?
9. What is expression profiling?
10. What is genomics? Proteomics?
11. What are the basic chemical constituents of a protein?
12. How does a gene encode for a protein?
13. What is 2-D PAGE?
14. What is MALDI–TOF?
15. How are proteins identified from amino acid fingerprints?

Problem

16. You perform electrophoretic separation of the nucleotide fragments from a piece of DNA. The primer is GATCCA. The following terminal tags are identified from the four fluorescent colors, in increasing order of length: ATTGCAT. This overlaps with another DNA fragment using the same primer, which gives terminal tags, in increasing order of length, of CATTCCGTA. What is the sequence of the overlapping fragments?

Recommended References

WEBSITES

1. www.ornl.gov/hgmis. Site containing U.S. Department of Energy HGR information. The *Education* link has an excellent primer on molecular genetics, which includes mapping and sequencing the human genome and a glossary of terms. The link S*cience Behind the Project* gives succinct tutorial on basic cell functions, from the genome to the proteome.

2. www.nhgri.nih.gov. Site of the National Human Genome Research Institute of the National Institutes of Health (NIH). Contains an alphabetized list of genetic terms and genetic illustrations, as well as the HGP.

3. www.genome.washington.edu/UWGC. University of Washington Genome Center. Has excellent tutorial, including shotgun sequencing. Gives detailed technical information and listings of genes.

4. http://science-education.nih.gov. The link *Snapshots of Science & Medicine* has feature articles, for example, on DNA chips.

5. http://seqcore.brcf.med.umich.edu. University of Michigan DNA Sequencing Core. Excellent tutorial on sequencing.

6. www.ensembl.org. Excellent site for chromosome details. Has pictures of each chromosome. By clicking on a chromosome, you can get a detailed picture of the genes and SNPs at different positions in the chromosome.

7. www.pbs.org/wgbh/nova/genome. Good visual tutorials on DNA and sequencing.

8. www.ncgr.org. The National Center for Genetic Resources. Concentrates on software technology.

THE GENOME

9. K. Davies, *Cracking the Genome: Inside the Race to Unlock Human DNA.* New York: Free Press, 2001.

PROTEOMES

10. S. P. Gygi, G. L. Corthals, Y. Zhang, Y. Rochan, and R. Aebersold, "Evaluation of Two-Dimensional Gel Electrophoresis-Based Proteome Analysis Technology," *Proc. National Acad. Sci.*, **97** (2000) 9390.

11. M. Mann, R. C. Hendrickson, and A. Pandey, "Analysis of Proteins and Proteomes by Mass Spectrometry," *Ann. Rev. Biochem.*, **70** (2001) 437.

12. S. Pennington and M. J. Dunn, eds., *Proteomics: From Protein Sequence to Function.* Berlin: Springer, 2001.

BIOINFORMATICS

13. G. R. Grant and W. J. Ewens, *Statistical Methods in Bioinformatics: An Introduction.* Berlin: Springer, 2001.

14. S. Misener and S. A. Krawetz, eds., *Bioinformatics: Methods and Protocols.* Totowa, NJ: Humana, 2000.

15. D. W. Mount, *Bioinformatics: Sequence and Genome Analysis.* Cold Spring Harbor, NY: Cold Spring Harbor Laboratory, 2001.

16. A. D. Baxevanis and B. F. Francis Ouellete, eds., *Bioinformatics: A Practical Guide to the Analysis of Genes and Proteins*, 2nd ed. New York: Wiley-Liss, 2001.

MICROARRAYS

17. M. Schena, ed., *Microarray Biochip Technology*. Westborough, MA: Eaton, 2000.

18. M. Shena, ed., *DNA Microarrays: A Practical Approach*. Oxford: Oxford University Press, 1999.

19. J. B. Rampal, ed., *DNA Arrays: Methods and Protocols*. Totowa, NJ: Humana, 2001.

Chapter Twenty-Six

ENVIRONMENTAL SAMPLING AND ANALYSIS

Analytical measurements are key to understanding the environment.

Environmental problems (air, water, solid waste, and occupational health and safety) and their control receive a great deal of interest and publicity. We have come to realize that this is a very complex area; and while many advances have been made, much is yet to be learned concerning the environment. Analytical chemistry plays a very important role in both defining and controlling environmental pollution. In this chapter, we briefly describe some of the analytical techniques used to collect and analyze environmental samples.

26.1 Getting a Meaningful Sample

Obtaining a representative sample is probably the most important step in environmental analysis. Environmental sampling errors often greatly exceed analytical procedural errors. It was found in a survey, for example, that for geochemical samples, the analytical variance was only 0.1% of the total variance, while the sampling variance was 43% of the total, with 53% attributable to the actual geochemical variance [M. H. Ramsey, "Appropriate Precision: Matching Analytical Precision Specifications the Particular Application," *Anal. Proceed.*, **30**(1993) 112]. For reliable interpretation of chemical analyses, the combined sampling and analytical variance should not exceed 20% of the total variance. Sample inhomogeneity can be a problem. You should review sampling considerations in Chapter 3 for estimating the number of samples needed to achieve an acceptable sampling error.

We will address sampling concerns and selected sampling procedures and preparation for air, water, and solids below. But first, let's look at some generic sampling techniques.

The analysis of liquid and solid samples very often requires some form of solvent extraction to isolate organic constituents. Conventional solvent extraction can be used, and the pH may be adjusted to achieve some selectivity, for example, extracting from acid solution to prevent basic compounds from extracting. More efficient means of extraction are commonly employed today, such as microwave-assisted or accelerated solvent extraction for solid samples (Chapter 19). See

Field analysis

Courtesy of Merck KGaA. Reproduced by permission.

www.sampleprep.duq.edu for an overview of microwave-assisted solvent extraction. This site lists EPA methods that use microwave-assisted extraction. Methods for acid digestion for inorganics are given, using a periodic table to select a particular element and the recommended acid or acids.

Solid-phase extraction (SPE) is widely used for isolating and concentrating analytes prior to chromatography measurement (Chapter 19).

26.2 Air Sample Collection and Analysis

It was once thought that when a particular pollutant was admitted into the atmosphere, its chemical composition did not change and that an analysis of that pollutant would give an indication of the degree of contamination. However, it is now recognized that many chemicals undergo photochemical decomposition and reaction in the atmosphere, forming different pollutants that may be even more toxic than their precursors. The familiar smog, for example, is generally considered to be related to the interaction of nitrogen oxide, hydrocarbons, and sunlight. These factors point to the importance of proper sampling. Once the sample has been collected, any number of standard measurement techniques may be employed.

Air is made up primarily of N_2, O_2, and Ar, which comprise 99.9% of dry air. There is a variable amount of water vapor, and many minor and trace gaseous components, as well as aerosol and particulate species. Table 26.1 lists some atmospheric gaseous components of environmental interest, along with representative concentrations in the troposphere. Typically, gaseous concentrations are expressed as mixing ratios, that is, volume/volume concentrations. A 1-ppm concentration represents 1 volume in 10^6 volumes of air. Such mixing ratios are independent of temperature and pressure. Environmental effects, though, may be quantitatively related to mass concentrations, and concentrations may be reported as mass per unit volume, usually mg/m^3 of air, under specific conditions of temperature and pressure. Aerosols and particulates are reported in this way.

Table 26.1
Some Gases of Environmental Importance

Compound	Approximate Concentrations (when present) (v/v)	Representative Measurement
CO	100 ppb to 20 ppm	Electrochemical; GC
CO_2	345 ppm	
CH_4	2 ppm	
$CFCl_3$ (Freon 11)	200 ppt	GC, electron capture
CF_2Cl_2 (Freon 12)	350 ppt	GC, electron capture
Hydrocarbons	1 ppt to 1 ppb	IR
NO	5 ppt to 1 ppb	UV; chemiluminescence
NO_2	1 to 150 ppb	Spectrophotometry; chemiluminescence
N_2O	300 ppb	IR; GC, electron capture
O_3	1 to 100 ppb	UV; chemiluminescence
SO_2	1 to 100 ppb	Flame photometric; spectrophotometric

GENERAL CONSIDERATIONS

Working with gases is different from liquids and solids. We deal with large volumes. The weight of analyte in the sample will generally be determined.

In environmental analysis, samples are generally collected for one of several important reasons: to establish hazardous levels in the environment, to understand the chemistry of the environment, to evaluate the efficiency of environmental control measures, or to determine the source of a pollutant. Hence, the mode of sampling can be important. Measurements will include monitoring ambient air, emissions from industrial or other processes, and indoor air for specific toxic substances. Ambient air components will exhibit both short- and long-term changes. Urban ozone will undergo daily cycles, but stratospheric ozone varies seasonally and has been undergoing a long-term decline. Gases such as carbon dioxide, nitrous oxide, and methane are exhibiting gradual increases, on the order of 0.8% per year.

Industrial emissions may be monitored as the gas leaves a stack or exhaust, where concentrations are much higher than when dispersed. Remote monitoring may be made to obtain an integrated measurement of emissions.

1. Size of Sample. The volume of air sampled is governed by the minimum chemical concentration that must be measured, the sensitivity of the measurement, and the information desired. The concentration range of a chemical may be unknown, and the sample size will have to be determined by trial and error. Trial samples of more than 10 m³ may be required to determine ambient concentrations.

2. Rate of Sampling. The useful sampling rate will vary with the sampling device and should be determined experimentally. Most sampling devices for gaseous constituents have permissible flow rates of 0.003 to 0.03 m³/min. The collection efficiency need not be 100% as long as it is reproducible and can be calibrated with known standards. The efficiency should be at least 75%, however. All gaseous samplers have a threshold level below which their efficiency drops to near zero. This varies with the sampling device and must be determined under the conditions used. Some sampling devices are described below.

3. Duration of Sampling. The time of day and duration of sampling will be determined by the information that is desired. Remember that the sampling period will give an indication of only the *average* concentration during that period. It would be more meaningful to sample city air samples for carbon monoxide content during the rush hour and between rush hours to obtain a realistic indication of the overall CO exposure of an individual in the city. Only a series of relatively short sampling times may reveal concentrations that are known to be deleterious. A sampling device capable of efficient operation at high flow rates will be required for short sampling times. It may be desirable in cases requiring many short-interval samplings to employ instead automatic continuous monitoring using highly sensitive detection devices.

4. Sample Storage. Storage of air samples should be kept to a minimum. They should be protected from heat and light. Care should be taken that the desired test component does not react with other constituents or with the container. Gaseous samples are sometimes collected by adsorption onto a solid and the gases must not be lost by desorption prior to analysis.

THE SAMPLING TRAIN

The requirements for intermittent air sampling are a vacuum source, a means of measuring the amount of air sampled, and a collector or combination of collectors. An interval timer may be used to control the time and duration of sampling. The **sampling device** (collector) should be the first unit in the sampling train, followed by the **metering device** and then the **vacuum pump**. Some of the more commonly used devices are described below.

1. Vacuum Sources. A vacuum is used to draw the sample through the collection device. Motor- or hand-driven vacuum pumps, aspirators, and automobile vacuums are commonly used. When vacuum devices are being used to draw samples through a filter, in which pressure loss may build up during sampling (e.g., as the filter becomes partially clogged), it is recommended that some constant-flow device be attached.

2. Metering Devices. Flow measurement devices are of two general types: those that measure rate and those that measure volume. The former are small and inexpensive but have the disadvantage of measuring only instantaneous rates of flow; so they must be checked frequently during the sampling period. The latter record the total flow passing through them and are therefore more useful. They are, however, bulkier and generally more expensive.

A **rotameter** is a rate-measuring device that consists of a spherical float within a tube that has a self-contained scale (see Figure 26.1). The bore of the tube is tapered, being larger at the top. The gas enters the bottom of the tube, and the stainless steel float rises and falls in direct proportion to the rate of gas flow. In effect, a variable orifice is supplied. The scale is calibrated for the particular gas (e.g., air) whose flow is to be measured. Graphs are frequently made available with the meters for applying temperature or pressure corrections, and the scale can be calibrated for measuring other gases.

The **dry-test meter** and the wet-test gas meter are volume-measuring devices. A set of plastic bellows is alternately filled and emptied, thereby driving the dial points via a system of bell cranks; very little pressure is required. A thermometer and manometer are provided with the meter, for temperature and pressure corrections. Dry-test gas meters are useful for large-volume measurements. The **wet-test meter** is generally more accurate that the dry-test meter for smaller volumes. The gas drives a rotor, which in turn drives the meter. The meter housing is partially filled with water through which the rotor turns. It is calibrated by the manufacturer at a given level of water.

We can measure either the rate of flow of a gas and the time, or its total volume directly.

Fig. 26.1. Rotameter. (Courtesy of Fisher Scientific Co.)

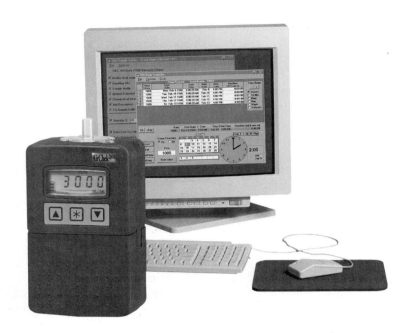

Fig. 26.2. Air sampling pump.
[Courtesy of SKC (www.skcinc.com).]

Air sampling pumps may be self-metering, that is, they have a metering device built in. The pump may be operated by a programmable computer to set start/stop times and for intermittent, repeated, or continuum sampling (Figure 26.2). The one shown has an isothermal flow sensor. Fluctuations in temperature or pressure are automatically corrected for. The sampling data can be downloaded to the computer.

All gas-measuring devices should be calibrated before and after use. A usual method employed is to measure the volume of a liquid (such as water) displaced by the gas flowing through the meter. Figure 26.3 illustrates a simple calibration method. A saturator (to saturate the air with water) is placed before the meter to prevent evaporation of part of the water from the carboy. The displaced water from the carboy is weighed or its volume is measured. By measuring the pressure with a manometer in series and the temperature, accurate calculation of the gas volume at standard conditions is obtained.

3. Sampling Devices. The third component of the sampling train is the collector, which may be of a variety of types, depending on the particular application. Included are filters, fritted-glass scrubbers, and impingers. These and others are described below for the sampling of aerosol constituents and of gaseous constituents.

We may collect solid or liquid aerosols, particles, or gases and vapors.

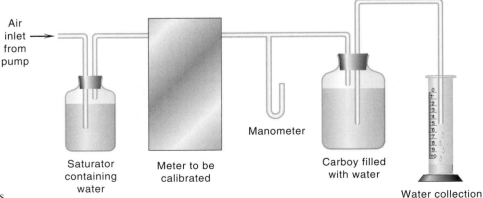

Air inlet from pump

Saturator containing water

Meter to be calibrated

Manometer

Carboy filled with water

Water collection

Fig. 26.3. Calibration of gas meters.

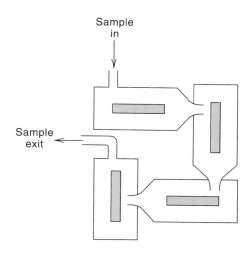

Fig. 26.4. Cascade impactor.

(a) Aerosol Constituents. One of the most commonly used means of collecting aerosol constituents is **filtration**. After collection on the filter, the aerosol content may be determined by weighing, by chemical analysis, or by particle sizing. Fiber filters (wood fiber paper, glass fiber), granular filters (fritted glass or metal, porous ceramic, sand), and membrane filters (cellulose ester) are used. The last is best for particle sizing. Most filters cannot be used at high temperatures or under moist conditions, but some glass filters can be used at temperatures up to 800°C.

A second type of collection device for aerosol constituents is the **impinger**; this collects both solid and liquid aerosols. In dry impingers, also called **impactors**, the aerosols impinge on a surface exposed to the airstream. The cascade impactor is shown in Figure 26.4. This consists of a series of progressively smaller size jets impinging at right angles on conventional microscope slides. Impactors are efficient for the collection of particles down to 2 μm in size. These impingers are well suited for collection of aerosols for microscopic examination.

In **wet impingers**, the aerosols impinge on a surface submerged in a liquid (Figure 26.5). A glass tube with an orifice at the end is directed toward a flat surface, which can be the bottom of the collection vessel or a glass platform connected to the tube. The particles are retained by the liquid in the collection tube. The liquid should not be a solvent for the particles to be collected. These impingers can collect particles as small as 0.1 μm.

More sophisticated collectors include **electrostatic precipitators** and **thermal precipitators**, which will collect particles down to 0.001 to 0.01 μm.

(b) Gaseous Constituents. Gases and vapors may be collected by absorption in a liquid, adsorption on a solid surface, freezing or condensation, or filling an evacuated container. The distinction between gases and vapors is that the latter exist as liquids at ordinary temperatures and can be easily condensed.

Gas adsorption tubes are the same as the thermal desorption tubes described in Chapter 20, and are used to collect volatile organic compounds (VOCs). A single sorbent or multiple sorbents may be used in the tube. Once the analytes are collected, they can be extracted from the sorbent, or thermally desorbed. Solid-phase microextraction (SPME—Chapter 18) can be used for spot sampling of air by letting the fiber equilibrate with the air. If the fiber is contained within a needle, sample can be drawn over it for integral sampling. Headspace and purge-and-trap sampling can also be used, as described in Chapter 20.

Air sample bags are convenient for sampling gases and vapors in which the analyte concentration is well above the detection limits of the measuring system

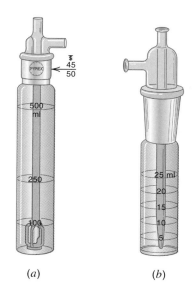

Fig. 26.5. Wet impingers. (*a*) Graduated impinger. (*b*) Midget impinger. (Courtesy of Arthur H. Thomas Co.)

(Figure 26.6). These are made of Teflon or Tedlar. The bag generally should not be filled more than 80% of capacity. Collected samples should not be shipped by air unless in a pressurized cabin; the reduced barometric pressure could cause the bag to burst.

See www.skcinc.com for examples of sampling pumps and collectors for air analysis. The site provides information about EPA air sampling standards, as well as OSHA, NIOSH, and ASTM standards.

AIR SAMPLE ANALYSIS

Most gases of environmental interest are chemically active and can be selectively absorbed from a known volume of air in a suitable reagent solution, followed by a chemical or physical measurement. Gases with a covalent bond, except for nonpolar diatomic gases such as O_2, N_2, and Cl_2, generally exhibit characteristic absorption spectra in the infrared region and may absorb in the UV region. A gas analyzer can be made selective for a particular gas by appropriate wavelength selection; there are nondispersive-type IR and UV analyzers available based on filters for wavelength isolation.

Representative procedures are described below for some important atmospheric constituents as indications of the general approaches taken.

A highly sensitive method for the determination of **nitrogen dioxide** in the atmosphere involves absorption of the nitrogen dioxide in a solution of sulfanilic acid that also contains an azo-dye-forming reagent. A stable pink color is produced within 15 min. Concentrations of 0.005 to 5 ppm nitrogen dioxide in the atmosphere may be measured. **Total oxides of nitrogen**, excluding nitrous oxide, may be determined in gaseous effluents from combustion by collecting the gas sample in an evacuated flask containing an oxidizing absorbent of hydrogen peroxide in dilute sulfuric acid. The oxides of nitrogen are converted to nitric acid by the

There are many ways of measuring chemicals in the atmosphere. Gas chromatography is common for trace constituents.

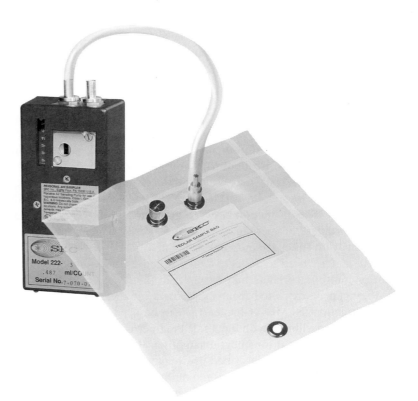

Fig. 26.6. Air sample bag. [Courtesy of SKC (www.skcinc.com).]

absorbing solution, and the resulting nitrate ion is reacted with phenol disulfonic acid to produce a yellow product that can be measured colorimetrically. From five to several thousand parts per million of oxides of nitrogen measured as nitrogen dioxide can be determined. **NO_2** may be reacted with luminol in alkaline solution, to produce chemiluminescence with photon emission at 425 nm; 30 ppt NO_2 may be detected.

NO can be measured by UV absorption. A highly sensitive and selective method is based on reaction of NO with O_3 to produce an excited NO_2^* species that emits radiation in the 590- to 2600-nm range upon returning to the ground state. **Nitrous oxide**, N_2O, is the most abundant oxide of nitrogen in the atmosphere and can be measured by IR absorption or gas chromatography with electron capture detection.

Ozone can be measured by UV absorption. If reacted with ethylene, C_2H_2, it produces an excited CH_2O^* species that emits a photon at 430 nm when returning to the ground state; 0.1 to 1000 ppb O_3 can be measured.

For the determination of **sulfur dioxide** in the atmosphere, the measured air sample is drawn through a solution of sodium tetrachloromercurate. The sulfur dioxide is absorbed by formation of the dichlorosulfitomercurate(II) complex ion, $HgCl_2SO_3^{2-}$. This complex resists oxidation by oxygen from the air and it reacts with formaldehyde and pararosaniline in acid solution to form a pararosaniline-methylsulfonic acid, which is highly colored and absorbs at 560 nm. This method is sensitive and relatively free from interferences. Metal traces such as iron and manganese interfere and, if they are not removed as particulates by prefiltration, their interference is eliminated by adding a chelating agent such as EDTA to the collection solution. As little as 0.003 ppm sulfur dioxide can be measured in the atmosphere. A monitor for SO_2 is based on flame photometric measurement, using a fuel-rich air–hydrogen flame; S_2 is formed that exhibits emission lines at 384.0 and 394.1 nm, the latter being detected with a photomultiplier tube.

Total hydrocarbons in the air may be determined using infrared spectrophotometry. The hydrocarbons are collected in a condensation trap immersed in liquid oxygen. The hydrocarbons absorb in the 3- to 4-μm region of the infrared spectrum using a 20-m pathlength cell. They are expressed as parts per million hexane, and the instrument is calibrated using a hexane standard.

Chlorofluorocarbons are chemically stable and are separated and measured at parts per trillion levels using gas chromatography with electron capture detection.

Gases from **emissions** may be measured by placing sensors in the gas stream. Or they may be measured remotely spectrophotometrically via a light beam that traverses the exhaust gas and is reflected back to the detector. Extractive sampling may be done by inserting a sampling probe tube in the duct and pumping gas to the analyzer or sample collection device.

Needless to say, we have only touched on a few of the many analytical procedures involving air samples. Other analyses performed include the determination of acetylene, total aldehydes, ammonia, formaldehyde, formic acid, and total organic acids. Various aerosol fractions in the air are commonly analyzed.

Example 26.1

Benzene in air at 1.00 atm pressure and 75.0°F is sampled by drawing air through an adsorption tube for 10.0 min at a flow rate of 1.00 L/min. The benzene is thermally desorbed and measured by gas chromatography. The sample contained 88 ng benzene. What is the concentration of benzene in air in parts per billion (vol/vol)?

Solution

$$\frac{88 \times 10^{-9} \text{ g}}{10.0 \text{ L}_{air}} = 8.8 \times 10^{-9} \text{ g/L}_{air}$$

The temperature is 23.9°C + 273.2 = 297.1 K. From Bolye's law, 1 mol pure benzene vapor has a volume of:

$$V = \frac{nRT}{P} = \frac{(1 \text{ mol})(0.08206 \text{ L atm/K mol})(297.1 \text{ K})}{1.00 \text{ atm}} = 24.4 \text{ L}$$

$$\frac{(8.8 \times 10^{-9} \text{ g/L}_{air})(24.4 \text{ L/mol})}{78.11 \text{ g/mol}} = 2.7_5 \times 10^{-9} \text{ L/L}_{air} = 2.7_5 \text{ ppb (vol/vol)}$$

• •

HOW DO WE CALIBRATE GAS ANALYZERS?

Proper calibration of gas measuring devices is critical. We must prepare gas mixtures of known composition (if not available commercially). This can be done statically by successive addition of the gaseous components of interest to a cylinder and measuring the weight. Or standards are prepared dynamically by dilution of a concentrated standard of a gas or mixture of gases into a stream of a base gas, for example, air or inert nitrogen. A stream of mixed gases results, containing known concentrations of the test gases. The dilution is accomplished using either a permeation or diffusion tube. The test compound, for example, NO or SO_2, from a container permeates through a membrane or diffuses along a capillary. The permeation or diffusion rate depends on the vapor pressure of the compound, the temperature, the membrane, and geometry. The rate is determined by measuring the weight loss from the container over a period of time.

26.3 Water Sample Collection and Analysis

Some of the many potential sources of water contaminants include industries such as the petroleum industry, the iron and steel industry, the pulp and paper industry, the coal industry, the chemistry industry, and the food industry, in addition to the private sources in the home and public and private sewage disposal. Land runoff erosion and mine wastes can be serious problems. This enumeration merely serves to point out the complexity involved in monitoring water purity. It is again apparent that water analysis involves a multitude of substances with a correspondingly larger multitude of possible analytical methods. Government regulations require many industries to regularly monitor effluents or discharges.

SAMPLING OF WATER

There are several types of water matrices that are sampled and analyzed, including surface waters (rivers, lakes, and runoff water), groundwater and springwater, potable (drinking) water, estuarine waters, saline water, water from the atmosphere (rainwater, snow, fog, dew), steam, and process waters. Some of these can be quite heterogeneous, either spatially or temporally. Sampling considerations will depend on the type of water sample.

In general, samples may be obtained from faucet outlets, at different points in pipe systems, from the surface of rivers and lake waters, and at different depths. The most important consideration is that the frequency and duration of sampling be sufficient to obtain a representative and reproducible sample. In some cases, composite samples may be used, in which individual samples taken at frequent intervals are combined.

Depth samplers are used to collect samples from large bodies of water at a specific depth. These contain some mechanism for removing a stopper after the bottle has been lowered to the desired depth. Commercial samplers are available from laboratory supply houses. Samples to be shipped should have an air space of 10 to 25 mL to allow room for expansion.

GROUNDWATER SAMPLES

Groundwater is monitored by drilling a small monitoring well to access the water. Before sampling, the stagnant water in the well is purged by drawing it out and allowing the well to refill; several well volume purges may be needed to assure a fresh sample. Common devices for obtaining samples include electric submersible pumps, peristaltic pumps, positive displacement bladder pumps, bailers (which can be used also to purge the well), and dialysis membranes. Suction or vacuum pumps may cause volatile constituents to be outgassed, but they have the advantage of keeping heavy solids in suspension. Peristaltic pumps operate by a multiroller rotating head that squeezes a rubber tube (much like if you rotated your fingers in a fist over a tube), drawing fluid through the tube. The pump tubing, connected to the sampling tubing, should be medical-grade silicone rubber to avoid organic peroxides found in other grades. Teflon connecting tubing should be used if organics are to be analyzed.

Bailers are useful for sampling small-diameter shallow wells. The disadvantage is that the bail tends to mix the sample being probed, particulates are collected from the well bottom, the sample may get aerated, and volatiles may get degassed in the open vessel. The bailer may be modified with a bottom draw valve, and it should be lowered slowly into the water.

Samples should be analyzed as soon after collection as possible for maximum accuracy. For certain constituents and physical values, immediate analysis in the field is necessary to obtain reliable results because the composition of the sample may change before it arrives at the laboratory. Included are pH and temperature and dissolved gases such as oxygen, hydrogen sulfide, and carbon dioxide. In certain cases, the gases may be fixed by reaction with a reagent and the analysis completed at the laboratory. Care must be taken to avoid contamination from oxygen or carbon dioxide in the air. Containers for collection of water samples must be clean and must not contaminate or adsorb analytes. Teflon containers, while expensive, are preferred for storing samples with trace analytes. Glass containers, particularly, should be washed with acid and perhaps stored filled with EDTA solution to minimize trace metal leaching by the sample.

Obviously, on-site measurement or monitoring devices are desirable or necessary for many analyses. For example, amperometric oxygen sensors are used to obtain immediate oxygen values and may be immersed at different points and depths to gather data of interest. Automatic samplers and instruments are used for regular or continuous monitoring. For example, a flow injection apparatus (Chapter 23) can be controlled by a computer or timer, to automatically take a sample at set intervals—perhaps through a filter sampler—and to inject and measure it. The detection may be by means of a simple colorimeter consisting of a small flow cell, a light-emitting diode (LED) source, and a diode detector. The monitoring of phosphate or nitrate, for example, can be accomplished using appropriate reagents.

On-site or automatic measurements are required for some analyses.

ANALYSIS OF WATER SAMPLES

The problem of analyzing water samples is essentially no different from that of analyzing aqueous solutions in general. You are referred to Ref. 4 for a compilation of some of the many commonly analyzed substances in water and the procedures employed for their analysis. Measurements made include acidity or alkalinity, biochemical oxygen demand, carbon dioxide, chlorine, dissolved oxygen, electrical conductivity, fluoride, particulate and dissolved matter, ammonia, phosphate, nitrate, silica, sulfate, sulfite, sulfides, turbidity, various metal ions, bacteria, microorganisms, and so forth.

26.4 Soil and Sediment Sampling

Sediments in contact with surface waters vary seasonally by climate as flow changes and by depth. Many organic and inorganic constituents absorb on sediment, and the concentrations are much higher than in the water. The analysis of sediments is important for evaluating transport of trace constituents from the sediment to the overlying water and vice versa.

Samples are usually obtained by dredging the bottom to obtain a grab sample, or dill core samples are taken. Grab samples are easy to obtain and can be large. The disadvantage is that only disturbed samples are collected, and fine particles may be carried away by out-flowing water. Core samples retain fine particles, but they cover only a small area, so larger numbers of samples may have to be taken. The vertical integrity of layers in the sediment is maintained.

Composite sediment samples are often used for analysis. Cored sediment samples are subsectioned and frozen for preservation prior to analysis; the core may be frozen first and then subsectioned. Sediment samples may be separated by particle sizes by wet or dry sieving (using polyethylene or nylon sieves for inorganics, or stainless steel for organics). They may be air dried, or heated to dry, depending on the volatility of the analytes, followed by grinding to aid in extraction of analytes. Volatile analytes are extracted from wet samples (stored by freezing). For trace metal analysis, samples are usually dry ashed or wet digested (like biological materials are).

Surface soils are sampled for recent spills. If the constituents are volatile or extensive time has passed, then deeper samples may be required. Shallow samples, 15 to 30 cm deep, can be taken with scoops, trowels, or shovels. Stainless steel tools should be used when sampling for organics and high-density polyethylene for inorganics. A soil punch (thin-walled steel tube device) is better for taking reproducible samples, by pushing into the soil to the desired depth.

For samples deeper than about 30 cm, trenching may be done to obtain a profile, but this is usually costly. Augers, either hand operated or powered, are often used, or corers. Samples are taken from auger cuttings. A better approach is to auger to the desired depth, and then drive a soil probe into the soil. This retains the sample as it is withdrawn.

26.5 Sample Preparation for Trace Organics

The determination of trace organic compounds in environmental samples requires getting them separated from the sample matrix prior to measurement, and usually involves extraction of some sort. Common trace organics include aliphatic and aromatic hydrocarbons, aldehydes, phenols, chlorinated solvents, polychlorinated biphenyls (PCBs), pesticides, and phthalates and adipates (from the polymer industry).

Water samples are conveniently extracted using water-immiscible solvents in a separatory funnel. Polar solvents such as ethylacetate or dimethylether are used to extract organic acids. Nonpolar solvents such as hexane, n-heptane, cyclohexane, or dichloromethane are used to extract neutral lipids, for example, triglycerides, and other nonpolar constituents such as aliphatic hydrocarbons and organopesticides. Solid-phase extraction is increasingly used to separate and preconcentrate trace organics from water samples.

Moist samples such as soils are usually dried in a vacuum oven and then ground. Before extraction, the samples are rehydrated by adding an aqueous buffer, which helps the transfer of analytes into a water-miscible extracting solvent such as acetone.

The solid–solvent mixture is stirred or shaken and then filtered or centrifuged to separate the matrix. Typically three extractions, using small portions of solvent, are needed to assure quantitative extraction. Heating may help in some cases. Or ultrasonic energy (ultrasonic batch or ultrasonic probe) can be used to increase efficiency. Continuous Soxhlet extraction is often used. A Soxhlet extractor has a round-bottom distillation pot, with a porous thimble above it in a siphon chamber, into which the sample is placed. Solvent is continuously cycled through the sample by distilling the solvent to a condenser centered over the thimble. The condensed solvent permeates through the matrix and the thimble, and is siphoned back into the pot, where it is recycled. The extracted material is concentrated in the pot.

Analytes extracted from environmental samples are concentrated prior to measurement by evaporating the solvent, using low temperature and reduced pressure for rapid solvent removal. Volatiles in soils and sediments may be sampled directly using headspace analysis for gas chromatography.

Once you have the extracted and preconcentrated organic analytes, you may have to perform additional cleanup, for example, by running them through a column of adsorbent packing material such as silica or alumina. Then chromatography is most often used for measurement. Pesticides are commonly determined by gas chromatography with electron capture detection or GC–MS. A nonpolar capillary GC column is used. Trace PCB and polycyclic aromatic hydrocarbon (PAH) determinations can be done using HPLC with UV detection.

26.6 Contaminated Land Sites—What Needs to Be Analyzed?

Assessing the extent of pollution in contaminated land sites is one of the most commonly faced environmental problems. Sampling is critical to make proper assessment. Obviously, samples will be taken to represent parts of the site where contamination is expected, or where contaminants are expected to be transported. You should examine surface soil samples, core samples at various depths, runoff in nearby streams, rivers or lakes, sediments, and vegetation from surface aquatic systems and at the land site.

26.7 EPA Methods and Performance-Based Analyses

The Environmental Protection Agency has published hundreds of official methods for measurement of organic and inorganic constituents in air, drinking water, wastewater, solid waste, and the like, and for pesticides and toxic substances. See Recommended References, 15 and 16 for Web lists of EPA test methods and sources for many of them (some are available online).

For many years, analysts were required to use only EPA methods and techniques when performing regulatory analyses. This was true, even if newer techniques or methodologies exhibited superior performance in sensitivity, selectivity, speed, or cost. That rule has been relaxed and laboratories may develop new methods, provided they meet certain criteria. The EPA Environmental Monitoring Management Council (EMMC) has accepted use of a performance-based measurement system (PBMS), which conveys "what" needs to be accomplished, but not prescriptively "how" to do it. EPA defines PBMS as "A set of processes wherein the data needs, mandates or limitations of a program or project are specified, and serve as criteria for selecting appropriate methods to meet those needs in a cost-effective manner." (See www.epa.gov/sw-864/pbms.htm.)

The focus is on the outcome, rather than on the method used to achieve that outcome. An acceptable level of performance is defined for a particular analysis using criteria such as accuracy, precision, detection limits, specificity, sensitivity, and robustness. If a laboratory can demonstrate that it meets the criteria, the method becomes an accepted one. See Chapter 4 for more details on getting a method validated.

Learning Objectives

WHAT ARE SOME OF THE KEY THINGS WE LEARNED FROM THIS CHAPTER?

- Collecting air samples—the sample train, sampling devices, p. 713
- Some example air analyses, p. 718
- Collecting water samples—surface, groundwater, p. 720
- Collecting soil and sediment samples, p. 722
- Extracting trace organics, p. 722
- EPA performance-based analyses, p. 723

Questions

1. What are the main components in a sampling train for air samples?
2. What are the principal uses of impingers? Scrubbers?
3. What precautions should be taken to protect air samples before analysis?
4. What are some of the commonly analyzed substances or parameters in water samples?
5. What are some parameters of water samples that are best measured in the field?

Problems

6. A 25-L air sample is determined to have 2.8 ppm (vol/vol) of carbon monoxide. How many grams are contained in 1 L at 20°C and 1 atm pressure?
7. Toluene in drinking water is determined using EPA Method 502.2 for volatile organic carbon (VOC), a purge-and-trap method, with measurement by capil-

lary gas chromatography. Helium is bubbled through a 5.00-mL sample in the purging chamber, and the toluene is trapped in a tube of Tenax sorbent. The tube is heated and backflushed with helium to desorb the VOCs into a capillary GC column. A standard of 0.50 μg/L of toluene in water is treated the same way, giving a peak area of 128, compared to 97 for the sample. What is the concentration of toluene in the drinking water in parts per trillion?

Recommended References

GENERAL

1. D. Pérez-Bendito and S. Rubio, *Environmental Analytical Chemistry*. Amsterdam: Elsevier, 1998.

2. F. W. Fifield and P. J. Haines, eds, *Environmental Analytical Chemistry*, 2nd ed. Oxford: Blackwell Science, 2000.

3. S. E. Manahan, *Environmental Chemistry*, 7th ed. Washington, DC: American Association of Clinical Chemistry, 2000.

4. L. H. Keith, ed., *Principles of Environmental Sampling*, 2nd ed. Washington, DC: American Chemical Society/Oxford, 1996.

5. J. M. Van Emon, C. L. Gerlach, and J. C. Johnson, eds., *Environmental Immunochemical Methods*. Washington, DC: American Chemical Society, 1996.

6. W. L. Budde, *Analytical Mass Spectrometry. Strategies for Environmental and Related Applications*. Washington, DC: American Chemical Society/Oxford, 2001.

7. R-K. Smith, *Handbook of Environmental Analysis*, 4th ed. 1999; R. E. Wagner, *Guide to Environmental Analytical Methods*, 4th ed. 1998 Amsterdam, NY: Genium. Available on CD-ROM along with the hardcopies.

AIR ANALYSIS

8. G. D. Wright, *Fundamentals of Air Sampling*. Boca Raton, FL: CRC Press, 1994.

9. K. R. Spurny, *Analytical Chemistry of Aerosols*. Boca Raton, FL: Lewis, 1999.

10. E. R. Kennedy, T. J. Fischbach, R. Song, P. M. Eller, and S. A. Shulman, *Guidelines for Air Sampling and Analytical Methods Development and Evaluation*. Cincinnati: NIOSH, USDHH, publication no. 95-117, 1995.

WATER/SOLIDS ANALYSIS

11. *Standard Methods for the Examination of Water and Wastewater*, 20th ed. New York: American Public Health Association, 1998.

12. L. H. Keith, *Compilation of EPA's Sampling and Analysis Methods*, 2nd ed. Boca Raton, FL: CRC Press, 1996.

13. H. Giergielewicz-Mozajska, L. Dabrowski, and J. Namiesnik, "Accelerated Solvent Extraction (ASE) in the Analysis of Environmental Solid Samples—Some Aspects of Theory and Practice," *CRC Critical Rev. Anal. Chem.*, **31** (2001) 149.

14. G. LeBlanc, "A Review of EPA Sample Preparation Techniques for Organic Compound Analysis of Liquid and Solid Samples," *LC · GC*, **19**(11) (2001) 1120.

WEBSITES

15. www.epa.gov/Standards.html *Environmental Test Methods and Guidelines.*
16. www.epa.gov/epahome/index *Index to EPA Test Methods.*

EXPERIMENTS

"Theory guides, experiment decides."
—I. M. Kolthoff

"I hear and I forget.
I see and I remember.
I do and I understand."
—Confucius

The following experiments follow the order of coverage in the text, after introductory experiments on the use of the analytical balance and volumetric glassware. Before beginning experiments, you should review the material in Chapter 2 on the basic tools of analytical chemistry and their use and the material in Appendix D on safety in the laboratory. It will also be helpful to review data handling in Chapter 3, particularly significant figures and propagation of errors, so you will know how accurately to make each required measurement. You will also use the spreadsheets described in that chapter and Chapter 16 for calibration plots and unknown calculations, including precisions.

Use of Apparatus

EXPERIMENT 1   USE OF THE ANALYTICAL BALANCE

Before beginning this experiment, familiarize yourself with the principles of the balance and the general rules for weighing discussed in Chapter 2.

Principle

You will use either an electronic balance or a mechanical single-pan balance (if you have a double-pan balance, your instructor will provide instructions on its use). The electronic balance is set to zero using the zero (tare) control bar or button before weighing an unknown object. For mechanical balances, the zero point is determined before weighing.

Solutions and Chemicals Required
None.

Procedure

1. *Zeroing or determination of the zero point.* Dust the pan slightly with a small brush. Color the doors of the balance.

    (a) **Electronic balance.** Turn the balance on and touch the zero (tare) bar or button. The reading should adjust to 0.0000 g.

    (b) **Mechanical balance.** The balance has a knob that will first partially release the beam and the pan and then completely release them (two positions). Release the beam and pan and adjust the zero reading. Arrest the pan and the beam and then repeat the operation. Throughout the remainder of this experiment, do all zero-point and rest-point determinations in *duplicate*; thereafter, single determinations should suffice.

       The zero or zero point of the balance will not remain constant. It may vary by several tenths of a milligram from day to day, or even from hour to hour. Hence, a new determination of the zero point must be made every time a series of weighings is to be made or, if the zero has changed, rezero the balance.

2. *Weighing of objects.* Determine the weights of three test objects supplied by the instructor. One of these may be the crucible cover to be weighed by difference in part 3 below.

    (a) **Electronic balance.** Make sure the balance is still zeroed. Place the test object on the pan using paper or tongs. Close the doors and after the reading stabilizes, record the weight to the nearest 0.1 mg.

    (b) **Mechanical balance.** If more than an hour has elapsed, or if the balance has been used by others since the zero point of the empty balance was adjusted, it should be rechecked. Place the test object on the pan using paper or tongs (never place or remove an object without the arrest engaged). Close the doors. Move the beam arrest knob to the first position. This partially releases the beam but arrests it if it is so much off balance that it swings beyond a certain position. This prevents damage to the knife edges when heavy objects are on the pan and it is far off balance. Dial in weights from the weight knobs to the nearest 0.1 g until the beam is in motion (near balance). Start with the larger weights to find the nearest 10 g, then the unit weights (nearest 1 g), and finally the 0.1-unit weights (nearest 0.1 g). Then, completely release the beam and read the weight to the nearest 0.1 mg from the vernier or digital counter after it has come to rest. Record the total weight.

3. *Weighing by difference.* Obtain a crucible and cover (or any two objects of about this weight) from your laboratory drawer. Weigh the cover of the crucible as in part 2 and record this weight. Weigh the two objects combined and record this weight. Remove the crucible cover and obtain the weight of the crucible. The difference in the last two weighings represents the weight of the crucible cover. This should agree within 0.5 mg with the weight obtained by direct weighing.

4. *Electronic balance taring.* Place the cucible cover on the pan, and push the tare bar or button. The reading should come to zero. Then place the crucible on the pan along with the cover, and record the weight. How close does this agree with the weight for the crucible obtained in part 3?

EXPERIMENT 2   USE OF THE PIPET AND BURET AND STATISTICAL ANALYSIS

Principle

Practice in the use of pipets is checked by weighing the amount of water delivered in successive pipettings. Precision of delivery of different volumes is determined. Similar experiments may be done with a buret. The experiments may also be used for calculation (calibration) of the volumes of the glassware if the temperature of the water is measured, as described in Table 2.4.

Solutions and Chemicals Required

Cleaning solution; distilled water.

Procedure

1. *Cleaning the glassware.*   Check your buret and pipets for cleanliness by rinsing with distilled water and allowing to drain. Clean glassware will retain a continuous, unbroken film of water when emptied. If necessary, clean the buret with a detergent and a buret brush. For pipets, try warm water, then detergent plus warm water. If these do not work, use a commercial cleaning solution. Rinse the buret or pipet several times with tap water and finally with distilled water. Leave the buret filled with distilled water when not in use. Always check your volumetric glassware for cleanliness before use, and clean whenever necessary. Burets if thoroughly rinsed and filled with distilled water immediately after use should remain clean for weeks; pipets, however, become contaminated easily and must be cleaned frequently.

2. *Pipetting.*   Practice filling a 25-mL pipet and adjusting the meniscus to the calibration mark until you are proficient. The forefinger used to adjust the level should be only slightly moistened. If it is too wet, it will be impossible to obtain an even flow.

   Weigh a clean, dry 50-mL Erlenmeyer flask and a rubber stopper to the nearest milligram (0.001 g).

   Transfer 25 mL water from the pipet to the flask; allow the water to run out with the tip of the pipet touching the side of the flask, being careful that no water splashes or is otherwise deposited on the neck of the flask. It is important that the neck of the flask and the stopper remain dry throughout the experiment. The pipet must be held vertically. Allow to drain for 10 s before removing the pipet. Do not blow out the last drop. Replace the stopper and weigh.

   Perform the procedure at least two times. Only the outside and the neck of the flask need be carefully dried for subsequent weighings. Calculate the standard deviation and the range of delivery in parts per thousand as described in Chapter 3.

   Repeat, using 1-, 5-, 10-, and 50-mL pipets. Compare the precision of delivery for the different pipets.

3. *Use of the buret.*   Check the stopcock (if ground glass) of your buret for proper lubrication, and be sure the bore and tip are clear. Fill with distilled water and place in the buret clamp; the water should be at room temperature. Displace any air bubbles from the tip of the buret, and adjust to near the zero mark. Allow to stand for a few minutes to check for leakage and read the initial volume to the nearest 0.01 mL. Use a meniscus reader, and be careful to avoid parallax.

   Using a procedure similar to that for the pipets, draw off about 5 mL into the weighed flask. Touch the tip of the buret to the wall of the flask to remove the adhering drop, and read the volume to the nearest 0.01 mL. Insert the stopper and weigh to the nearest milligram. Repeat the operation, adding to the flask the next 5 mL water, and weigh again. Continue in this way until the entire 50 mL have been weighed.[1]

Subtract the weight of the empty flask from each of the succeeding weights to get the weight of 5, 10, 15, and so forth, milliliters of water at the temperature observed.

Repeat the entire procedure. At each approximate volume, the weight, compared to the exact measured volume, should be within 0.03 g of that predicted from the first measurement. Recheck any points that are not.

4. *Calibration of glassware.*   If your instructor wishes you to use the data from these experiments to calibrate the volumes of the glassware, use Table 2.4 to convert the weight in air to the volume. The temperature of the water delivered must be determined; the Table 2.4 spreadsheet is on your CD, and you can enter data in that by copying it to your computer desktop.

5. *Statistics: The normal error curve.*   In this experiment, you will make 20 measurements of the weight of water delivered by a 1-mL pipet, as described above, by adding successive pipet volumes to a small Erlenmeyer flask and calculating the weight increase each time. Be sure to rapidly stopper the flask each time. Plot the frequencies of the weights on graph paper to prepare a normal error curve. Calculate the standard deviation and relate this to the plotted curve. Do about 68% of the readings fall within one standard deviation?

Gravimetry

EXPERIMENT 3   GRAVIMETRIC DETERMINATION OF CHLORIDE

Principle
The chloride content of a soluble sample is determined by precipitating silver chloride with added silver nitrate, filtering, drying, and weighing the precipitate. The Cl content is calculated from the weight of AgCl.

Equation

$$Ag^+ + Cl^- \rightarrow \underline{AgCl}$$

Solutions Required
1. *Provided.*   Conc. HNO_3, conc. NH_3, dil. (3 M) HCl.
2. *To prepare*

   (a) **0.1 M AgNO$_3$.** Dissolve about 3 g solid $AgNO_3$ (need not be dried) in about 180 mL distilled water. Store in a brown bottle. Do not get $AgNO_3$ on your hands. If you do, wash immediately with a sodium thiosulfate solution. If left on the skin, brown-black spots (metallic silver) will appear within several hours and will take 2 to 3 weeks to wear off.

   (b) **Wash solution.** Add approximately 2 mL chloride-free concentrated nitric acid to about 600 mL distilled water in a wash bottle.

---

[1]To minimize evaporation errors, your instructor may direct you to refill the buret for each volume delivery (0–5, 0–10, . . . 0–50 mL) and deliver it into an empty (dry) flask.

Things to Do before the Experiment

1. *Obtain your unknown from your instructor.*

    (a) Solution. Give your instructor a clean, labeled, 250-mL volumetric flask with stopper. The flask should be cleaned and rinsed with three or four portions of distilled water. It need not be dry. After receiving the unknown solution in the flask, dilute to the mark with distilled water. Mix well (see Chapter 2 for proper procedure).

    (b) Solid. Obtain a soluble chloride sample from your instructor and place it in a clean weighing bottle. The weighing bottle and cap are placed in a beaker marked with your name and covered with a watch glass during the drying. The weighing bottle should not be stoppered with the cap. Dry in the oven for 1 to 2 h at 120 to 160°C (about 1 h in a forced-air oven, or 2 h in a gravity convection oven). Store in a desiccator after drying until ready for weighing (wait until cooled before placing the cap on the bottle, to avoid creation of a vacuum in the capped bottle).

2. *Prepare three filter crucibles as described below.*

3. *Prepare the 0.1 M AgNO₃ solution.*[1]

Preparation of Crucibles

1. *Sintered-glass filter crucible.* Prepare three crucibles. Use crucibles of fine or medium porosity and *not* coarse porosity. Clean each crucible from surface contamination using soap and water, rinse, then place it in a crucible holder in a suction flask. If chemical cleaning is required, follow the procedure at the end of the experiment. With gentle suction, draw several small portions of distilled water through the filter. Place three crucibles thus prepared in a beaker (marked with your initials), cover with a watch glass, and put the beaker and crucibles in the oven at 120° to 130°C for 1 to 2 h. With clean crucible tongs, transfer the hot crucibles to the desiccator, cool for 30 to 40 min, and weigh accurately. Repeat the heating for 0.5 h, cool, and weigh. Continue until weights constant within ±0.3 to 0.4 mg are obtained.

2. *Porcelain crucibles with glass fiber filter paper.* Place two glass fiber filter papers in the crucible to just cover the bottom of the porcelain crucible (Gooch crucible). Wash with water as above and heat to constant weight (±0.3 to 0.4 mg is obtained).

Procedure

1. *Preparation of the sample*

    (a) Solution. Pipet three 50-ml, aliquots into three separate 500-mL beakers. Dilute each sample with 100 to 150 mL distilled water. Add about 0.5 mL concentrated nitric acid. Cover the beakers with clean watch glasses.

    (b) Solid. After the dried sample has cooled in a desiccator for at least 30 to 40 min, weigh accurately triplicate samples of about 0.2 to 0.3 g into numbered 400-mL beakers. (Samples may contain carbonate, which is hygroscopic. If so, your instructor will advise you to weigh the samples by difference.) Dissolve the sample in distilled

---

[1]Your laboratory will very likely have a collection bottle for leftover silver solutions and precipitates. Both economy and pollution abatement make it important not to dump these materials down the drain.

water and dilute to about 150 mL. Add about 0.5 mL chloride-free concentrated nitric acid. (If carbonate-containing samples are used, add slowly until the foaming action stops.)

2. *Precipitation.*   Assume the solid sample to be pure sodium chloride, and calculate the millimoles of silver nitrate required to precipitate the chloride. The slow addition of silver nitrate to the solution is best accomplished by use of a buret, with stirring. Wash the buret well with tap water, rinse with three or four portions of distilled water, and finally fill it with approximately 0.1 *M* silver nitrate solution. To each sample, add silver nitrate solution slowly with good stirring until an excess of 10% over the calculated amount is added. Heat the suspensions nearly to boiling, with frequent or constant stirring, to coagulate the silver chloride; the stirring helps prevent bumping of the solution during heating and the danger of loss of precipitate. Let the precipitate settle, and test for complete precipitation by carefully adding a few drops of silver nitrate to the clear supernatant liquid; if more precipitate or a cloudiness appears, add a few more milliliters silver nitrate solution, stir well, heat, let the precipitate settle, and test again. Continue in this way until precipitation is complete. Let the covered beakers with their contents stand in the desk, *protected from light,*[2] or at least 2 h before filtration; standing overnight, or from one laboratory period to the next, may be required, and this is all right as long as they are kept in the dark.

3. *Filtration and washing of the precipitate.*   Decant the solution from the first sample through the first weighed crucible, pouring the solution down a stirring rod, and using gentle suction. The precipitate should be disturbed as little as possible. To the precipitate in the beaker add about 25 mL of the wash solution, stir well, let the precipitate settle, and decant the solution through the filter crucible. The nitric acid wash solution replaces silver nitrate adsorbed on the surface of the precipitate. An electrolyte is required to prevent peptization of the sample. The nitric acid is volatilized during the drying step. Repeat the washing by decantation four times, and finally bring the precipitate onto the filter; use small portions of the wash solution for transfer. Remove with a rubber policeman any solid particles adhering to the beaker. Continue washing the precipitate in the crucible with the wash solution until the last portions of washings give a negative test for silver ion. Silver ion is tested for by adding a drop of dilute hydrochloric acid. Ten or more washings may be required. Filter and wash the second and the third samples in the same way.

4. *Drying and weighing of the precipitate.*   Place the crucibles containing the precipitates in a covered beaker in the oven for 2 h at 120 to 130°C. Cool the crucibles in the desiccator, and weigh accurately. Reheat for 1-h periods as necessary to obtain weights constant within ±0.3 to 0.4 mg.

Cleaning the Crucibles

After completing the analysis, clean the crucibles as follows. Remove the cake of silver chloride, then place the crucibles in a beaker, put about 5 mL concentrated ammonia solution in each crucible, and cover the beaker with a watch glass. After 10 to 15 min, transfer the crucible to the crucible holder in a suction flask and wash with several portions of distilled water. If the filter plate is dark, empty and rinse

---

[2]When exposed to light, AgCl decomposes as follows: $AgCl \rightarrow Ag + \frac{1}{2}Cl_2 \uparrow$, resulting in a purple coloration on the surface of the precipitate due to metallic silver, and a loss in weight. Some coloration cannot be avoided and is not serious as long as strong light (especially sunlight) is avoided as much as possible.

the ammonia from the suction flask and treat the crucible in the same way with a few milliliters concentrated nitric acid. Finally, wash well with distilled water.

Calculations

1. *Solution.*   Calculate and report the grams of Cl contained in your unknown. Since one-fifth of the sample was taken for each determination (50 mL out of 250 mL), the weight determined in each aliquot must be multiplied by 5. Report each individual result, the mean, and the relative standard deviation in ppm.

2. *Solid.*   Calculate and report the percent Cl in your unknown for each portion analyzed. Report each individual result, the mean, and the relative standard deviation.

EXPERIMENT 4   GRAVIMETRIC DETERMINATION OF SO₃ IN A SOLUBLE SULFATE

Principle
Sulfate is precipitated as barium sulfate with barium chloride. After the filtering with filter paper, the paper is charred off, and the precipitate is ignited to constant weight. The SO_3 content is calculated from the weight of $BaSO_4$.

Equation
$$SO_4^{2-} + Ba^{2+} \rightarrow \underline{BaSO_4}$$

Solutions and Chemicals Required

1. *Provided.*   Dil. (0.1 M) AgNO₃, conc. HNO₃, conc. HCl.

2. *To prepare.*   0.25 M BaCl₂. Dissolve about 5.2 g BaCl₂ (need not be dried) in 100 mL distilled water.

Things to Do before the Experiment

1. *Obtain your unknown and dry.*   Check out a sample of a soluble sulfate from the instructor and dry it in the oven at 110 to 120°C for at least 2 h. Allow it to cool in a desiccator for at least 0.5 h.

2. *Prepare crucibles.*   Clean three porcelain crucibles and covers. Place them over Tirrill burners and heat at the maximum temperature of the burner for 10 to 15 min; then place in the desiccator to cool for at least 1 h. Weigh accurately each crucible with its cover. Between heating and weighing, the crucibles and covers should be handled only with a pair of tongs.

3. Prepare the 0.25 M BaCl₂ solution.

Procedure

1. *Preparation of the sample.*   Weigh accurately to four significant figures, using the direct method, three samples of about 0.5 to 0.7 g each. Transfer to 400-mL beakers, dissolve in 200 to 250 mL distilled water, and add 0.5 mL concentrated hydrochloric acid to each.

2. *Precipitation.*   Assume the sample to be pure sodium sulfate and calculate the millimoles barium chloride required to precipitate the sulfate in the largest sample. Heat the solutions nearly to boiling on a wire gauze over a Tirrill burner and adjust the burner to keep the solution just below the boiling point. Add slowly from a buret, drop by drop, 0.25 M barium chloride solution until 10% more than the above calculated amount is added; stir vigorously

throughout the addition. Let the precipitate settle, then test for complete precipitation by adding a few drops of barium chloride without stirring. If additional precipitate forms, add slowly, with stirring, 5 mL more barium chloride; let settle, and test again. Repeat this operation until precipitation is complete. Leave the stirring rods in the beakers, cover with watch glasses, and digest on the steam bath until the supernatant liquid is clear. (The initial precipitate is fine particles. During digestion, the particles grow to filterable size.) This will require 30 to 60 min or longer. Add more distilled water if the volume falls below 200 mL.

3. *Filtration and washing of the precipitate.* Prepare three 11-cm No. 42 Whatman filter papers or equivalent for filtration; the paper should be well fitted to the funnel so that the stem of the funnel remains filled with liquid, or the filtration will be very slow. See the discussion in Chapter 2 for the proper preparation of filters. Filter the solutions while hot; be careful not to fill the paper too full, as the barium sulfate has a tendency to "creep" above the edge of the paper. Wash the precipitate into the filter with hot distilled water, clean the adhering precipitate from the stirring rod and beaker with the rubber policeman, and again rinse the contents of the beaker into the filter. Examine the beaker very carefully for particles of precipitate that may have escaped transfer. Wash the precipitate and the filter paper with hot distilled water until no turbidity appears when a few milliliters of the washings acidified with a few drops of nitric acid are tested for chloride with silver nitrate solution. During the washing, rinse the precipitate down into the cone of the filter as much as possible. Examine the filtrate for any precipitate that may have run through the filter.

4. *Ignition and weighing of the precipitate.* Loosen the filter paper in the funnels and allow to drain for a few minutes. Fold each filter into a package enclosing the precipitate, with the triple thickness of paper on top. Place in the weighed porcelain crucibles and gently press down into the bottom. Inspect the funnels for traces of precipitate; if any precipitate is found, wipe it off with a small piece of moist ashless filter paper and add to the proper crucible. Place each crucible on a triangle on a tripod or the ring of a ring stand, in an inclined position with the cover displaced slightly. Heat gently with a small flame (Tirrill burner) until all the moisture has been driven off and the paper begins to smoke and char. Adjust the burner so that the paper continues to char without catching fire. If the paper inflames, cover the crucible to smother the fire, and lower the burner flame. When the paper has completely carbonized and no smoke is given off, gradually raise the temperature enough to burn off the carbon completely. A red glowing of the carbon as it burns is normal, but there should be no flame. The precipitate should finally be white with no black particles. Allow to cool. Place the crucible in a vertical position in the triangle, and moisten the precipitate with three or four drops of dilute (1:4) sulfuric acid. Heat *very gently* until the acid has fumed off. (This treatment converts any precipitate that has been reduced to barium sulfide by the hot carbon black to barium sulfate.) Then, cover the crucibles and heat to dull redness in the full flame of the Tirrill burner for 15 min.

Allow the covered crucibles to cool in a desiccator for at least 1 h and then weigh them. Heat again to redness for 10 to 15 min, cool in the desiccator, and weigh again. Repeat until two successive weighings agree within 0.3 to 0.4 mg.

Calculation

Calculate and report the percent SO_3 in your unknown for each portion analyzed. Report also the mean and the relative standard deviation.

EXPERIMENT 5 GRAVIMETRIC DETERMINATION OF NICKEL IN A NICHROME ALLOY

Principle

Nickel forms a red chelate with dimethylglyoxime (DMG), which is quite suitable for gravimetric analysis. Precipitation of the chelate is complete in an acetic acid–acetate buffer or in an ammoniacal solution. Acetate buffer is generally used when Zn, Fe, or Mn is present in the alloy. The sample given to you is a nichrome alloy that has Ni (approximately 60%), Cr, and Fe as the major constituents. Interference from Cr and Fe is removed by complexation with tartrate or citrate ions. Precipitation is then carried out in an ammoniacal solution. The Ni content is calculated from the weight of the precipitate (see Table 10.2 for the formula).

Equation

$$Ni^{2+} + 2HDMG \rightarrow Ni(DMG)_2 + 2H^+$$

Solutions and Chemicals Provided

HCl (1:1), HNO_3 (1:1), NH_3 solution (conc. and 1:1), 1% DMG in ethanol, 20% citric acid solution, 30% ethanol.

Things to Do before the Experiment

1. *Obtain the alloy sample from your instructor.*

2. *Prepare three sintered-glass filter crucibles.* Dry to a constant weight of ±0.3 to 0.4 mg. (Repeated heatings, see Experiment 3.)

Procedure

1. *Dissolving the sample.* Weigh accurately triplicate 0.10- to 0.12-g samples of the alloy. Place in separate numbered 400-mL beakers and add 25 mL each of HCl (1:1) and HNO_3 (1:1). Heat moderately (under the hood, please) until the brown fumes are driven off—40 to 50 min. Keep the beakers covered partially with watch glasses. (You may notice some undissolved carbon particles in the residue. If this occurs, then, using quantitative technique, filter the contents of each beaker through Whatman filter paper No. 40 into separate 600-mL beakers. Wash the 400-mL beaker several times with small amounts of water and transfer the washings to the 600-mL beaker through the filter. Wash the filter several times with small amounts of water. Remove the filter paper and wash the funnel and stem with small amounts of water before removing from the receiving beaker.)

2. *Precipitation.* Add 4 to 6 mL citric acid solution and 100 to 150 mL water to each beaker. Add conc. NH_3 dropwise with stirring until the solution is slightly basic (smell of ammonia in the solution). This should take 2 to 3 mL. Test to see that the pH is about 8 by touching the wet stirring rod to pH paper. Be careful that no drops adhere to the rod to keep solution loss negligible. If a precipitate is formed, insufficient citric acid has been added. Dissolve the precipitate by adding dropwise HCl (1:1, about 3 mL) until the pH is about 3. Add 1 to 2 mL more citric acid solution. Check again by adding conc. NH_3. If no precipitate is formed add HCl (1:1) until the solution is slightly acidic (no ammonia odor in solution). Heat the beakers on the hot plate to about 70 to 80°C. Do not boil. Remove from the hot plate and add about 50 mL 1% DMG solution dropwise with stirring. If a red precipitate forms, add HCl (1:1) dropwise until it dissolves. The DMG should be homogeneous throughout the solution before the precipitation is begun. Add NH_3 (1:1) dropwise with stirring until the solution *strongly* smells of ammonia; this may take 3 to 4 mL. A red precipitate is formed. It is important

that the solution be distinctly alkaline at this point. Since the nose may retain the odor of the ammonia reagent, take care that the odor comes from the beaker. Test to be sure the pH is greater than 8.5 by touching the wet stirring rod to pH paper to test the pH. This is best done *after* the bulk of the precipitate is formed and allowed to settle. Cover the beakers with watch glasses and set aside for at least 2 h (preferably overnight).

3. *Filtering and washing the precipitate.*   Filter the precipitate through previously weighed glass crucibles using gentle suction. After transferring the bulk of the precipitate, wash the precipitate with lukewarm distilled water and transfer quantitatively into the crucible. If traces of the precipitate (which are difficult to transfer with a rubber policeman) stick to the sides of the beaker, dissolve in a few drops of hot HCl (1:1) and reprecipitate by adding a few drops of 1% DMG and NH_3 (1:1). The precipitate now obtained will not stick and can be transferred to the crucibles. Wash the precipitate in the crucible at least three to four times with warm water. Finally, wash the precipitate *once* with 30% alcohol, which dissolves any excess DMG from the precipitate.

4. *Drying and weighing the precipitate.*   Place the crucibles in beakers (labeled with your name and covered with watch glasses). Dry in the oven for 1–2 h at 110 to 130°C. Cool the crucibles in a desiccator (for 30 to 40 min) and weigh accurately. Reheat for 1-h periods necessary to obtain weights constant to within ±0.3 to 0.4 mg.

Calculation

Calculate and report the percent nickel in your unknown for each portion analyzed. Report each individual result, the mean, and the relative standard deviation.

Acid–Base Titrations

EXPERIMENT 6   DETERMINATION OF REPLACEABLE HYDROGEN IN ACID BY TITRATION WITH SODIUM HYDROXIDE

Principle

One-tenth molar sodium hydroxide is prepared and standardized against primary standard potassium acid phthalate (KHP). The unknown is titrated with the standardized sodium hydroxide.

Equations

$$HA^- + OH^- \rightarrow A^{2-} + H_2O$$

Solutions and Chemicals Required

1. *Provided.*   KHP, 0.2% phenolphthalein solution in 90% ethanol, 50% NaOH solution.

2. *To prepare.*[1]   0.1 *M* NaOH solution. This solution must be carbonate free. The Na_2CO_3 in NaOH pellets is insoluble in 50% NaOH solution. The precipitated

---

[1] If you are to use this solution in Experiment 7 to standardize HCl, prepare 1 L instead of 500 mL.

Na_2CO_3 will settle to the bottom of the container after several days to a week, and the clear, syrupy supernatant (approximately 19 *M* in NaOH) can be carefully withdrawn or decanted from the container to obtain carbonate-free NaOH. The solution is prepared in a borosilicate (Kimax or Pyrex) beaker (it gets very warm) and after cooling is stored in a polyethylene bottle. Your instructor will provide the carbonate-free 50% solution to you.

Distilled water free from carbon dioxide will be needed in this experiment. To prepare this, fill a 1000-mL Florence flask nearly to the shoulder with distilled water, insert a boiling rod with the cupped end down (the cup must be empty), heat to boiling on a wire gauze over a Meker burner, and boil for 5 min. Cover the flask with an inverted beaker and allow to cool overnight, or cool under the cold water tap. Prepare an additional 800 mL the same way.

Fill a 500-mL bottle, with rubber stopper, half full of CO_2-free water, add 2.5 mL clear 50% NaOH solution, stopper, and swirl to mix.[2] Avoid exposing the solutions to the atmosphere as much as possible. Finally, add more CO_2-free water to nearly fill the bottle, stopper, and shake thoroughly.

The sodium hydroxide thus prepared is approximately 0.1 *M*. It will be standardized against primary standard potassium acid phthalate.

Things to Do before the Experiment

1. *Obtain KHP and dry.* Dry about 4 g primary standard potassium acid phthalate in a weighing bottle at 110 to 120°C for 1 to 2 h. Cool in a desiccator for at least 30 to 40 min before weighing.

2. *Obtain your unknown and dry.* The unknown may be either a liquid or a solid. If it is a solid, obtain it in a weighing bottle before the day of the experiment and dry for 2 h at 110 to 120°C. Store in the desiccator. If the unknown is a liquid, obtain it in a clean 250-mL volumetric flask and dilute to volume with CO_2-free water. Mix well.

Procedure

1. *Standardization of the 0.1 M NaOH solution against potassium acid phthalate.* Weigh accurately three portions of the dried potassium acid phthalate of about 0.7 to 0.9 g each, and transfer to clean 250-mL wide-mouth Erlenmeyer flasks. (These quantities are designed for titrations using 50-mL burets. If you are supplied a 25-mL buret, your instructor will direct you to adjust the quantities of KHP and the unknown accordingly.) The direct method of weighing, using a tared weighing dish, may be used with this material. Dissolve each sample in about 50 mL CO_2-free distilled water. Rinse your buret with three small portions of the 0.1 *M* NaOH solution, fill, and adjust to near zero. Record the initial volume reading to the nearest 0.02 mL. Add 2 to 3 drops phenolphthalein indicator to each KHP sample and titrate with the 0.1 *M* NaOH to a faint pink end point. The color should persist at least 30 s. Split drops at the end of the titration. Estimate the buret reading to the nearest 0.02 mL.

   Calculate the molarity of the NaOH to four significant figures from the weight of KHP used (three significant figures if the molarity is slightly less than 0.1 *M*). Use the average of the results.

   Keep the NaOH bottle stoppered with a rubber stopper, and protect the solution from the air as much as possible to minimize absorption of CO_2. Proceed as soon as possible to the determination of the unknown acid; standard

---

[2]If the NaOH were pure, only 2 mL (2 g NaOH) would be required for 500 mL of 0.1 *M* NaOH. Two and one-half milliliters is required to compensate for the water and Na_2CO_3 content of the pellets.

sodium hydroxide should be used within one week of standardization. After that time, changes in molarity may have occurred, and restandardization will be necessary.[3]

2. *Determination of replaceable hydrogen in an unknown acid.* If the unknown is a weak acid (e.g., KHP, acetic acid, or vinegar) you will be instructed to use phenolphthalein indicator (color change same as above). If it is a strong acid, you may use another indicator, such as chlorophenol red (color change from yellow to violet).

If the unknown is a solid, dry it for 2 h at 110°C. Cool for 30 min. Your instructor will inform you of the approximate weight of sample to take so that about 30 to 40 mL NaOH will be used. If it is a KHP unknown, approximately 1 g may be taken. Weigh three samples to the nearest 0.1 mg and transfer them into numbered 250-mL wide-mouth Erlenmeyer flasks. Dissolve in 50 mL CO_2-free water, warming if necessary. Add two drops of indicator and titrate with 0.1 M NaOH until the color change persists 30 s.

If the unknown is a liquid, transfer with a pipet three 50-mL aliquots from the 250-mL volumetric flask and titrate as above.

Calculation

1. *Solids.* Calculate and report the percent of replaceable hydrogen or percent KHP in your unknown for each portion analyzed.

2. *Liquid.* Calculate and report the milligrams replaceable hydrogen in your unknown. Since one-fifth of the sample was taken for each determination (50 mL out of 250 mL), the weight determined in each aliquot must be multiplied by 5.

EXPERIMENT 7 DETERMINATION OF TOTAL ALKALINITY OF SODA ASH

Principle

One-tenth molar hydrochloric acid is standardized against primary standard sodium carbonate. Phenolphthalein is used to approximate the halfway point of the titration, and then either modified methyl orange indicator or bromcresol green indicator is used to detect the final end point; the last indicator is used with boiling of the solution near the end point to remove CO_2 (see discussion of Figure 8.11). Alternatively, the hydrochloric acid is standardized against standardized sodium hydroxide from Experiment 6 using phenophthalein indicator. The soda ash sample is titrated with the hydrochloric acid solution, with the addition of 2 mol hydrogen per mole Na_2CO_3.

Equations

$$CO_3^{2-} + 2H^+ \rightarrow H_2CO_3$$
$$H_2CO_3 \rightleftharpoons H_2O + CO_2 \uparrow$$

Solutions and Chemicals Required

1. *Provided.* 0.2% phenolphthalein in 90% ethanol, either modified methyl orange (mixture of methyl orange and xylene cylanole FF available from Eastman Kodak Co.) or 0.1% bromcresol green in 0.001 M NaOH, and either primary standard Na_2CO_3 or standardized 0.1 M NaOH.

---

[3]If you are to use this solution in Experiment 7 to standardize the HCl, your instructor will advise you to save the remaining solution after completing this experiment. In that event, you should prepare and standardize the HCl within one week of standardizing the NaOH.

2. *To prepare.* 0.1 M HCl. Concentrated hydrochloric acid has a density of 1.18 and contains 37% by weight HCl. Hence, about 4 mL concentrated acid should be diluted to 500 mL to make 0.1 M acid. Measure about 0.5 mL more than this amount in a 10-mL graduated cylinder and pour into water in a 500-mL glass-stoppered bottle that is filled to the shoulder with distilled water.[1] Shake until the solution is homogeneous.

Things to Do before the Experiment

1. *Obtain and dry the Na_2CO_3.* If you are to standardize your HCl against Na_2CO_3, obtain about 2 g primary standard Na_2CO_3 in a weighing bottle. Place the weighing bottle and cover (removed) in a 150-mL beaker marked with your name and dry at 160°C for 2 h or more. Cool at least 30 min in a desiccator before weighing.

2. *Obtain and dry your unknown.* Obtain the unknown in a weighing bottle and dry as above for the Na_2CO_3.

Procedure

Standardize the HCl by either method 1 or method 2 below.

1. *Standardization of HCl against Na_2CO_3.* Weigh accurately (to 0.1 mg) the weighing bottle containing the dried sodium carbonate. Keep the stopper in place while weighing. Transfer quantitatively (for four-figure accuracy) about 0.2 g to a clean 250-mL Erlenmeyer flask. The difference is the weight of sodium carbonate taken. Anything between 0.16 and 0.24 g will be satisfactory. (This procedure is called "weighing by difference" and is necessary because Na_2CO_3 is hygroscopic.) Weigh out a second and third portion in the same way. Add about 50 mL distilled water and one to two drops phenolphthalein indicator solution to each flask.

   The Na_2CO_3 will be titrated first to a phenolphthalein end point (pink to colorless; 1 H^+ added: $CO_3^{2-} + H^+ \rightarrow HCO_3^-$) to approximate where the final end point should be (modified methyl orange or bromcresol green indicator; 1 more H^+ added: $HCO_3^- + H^+ \rightarrow H_2CO_3$). Rinse the buret three times with small portions (about 5 mL each) of the approximately 0.1 M acid prepared above, then fill and adjust to near the zero mark. Record the volume to the nearest 0.02 mL. Titrate the first sample of the sodium carbonate, adding the acid no faster than 0.5 mL per second, swirling the flask constantly until the pink color disappears. At this point, about half the total volume of acid necessary has been added (actually a slight excess). Use this number to estimate where the final end point will occur, keeping in mind that slightly less acid should be required than has been added. Continue the titration by either method (a) or method (b) below.

   (a) Modified methyl orange indicator. Before beginning, check with your instructor which indicator you are to use. Add two to three drops of modified methyl orange indicator and titrate until the indicator color changes from green to gray. The proper color can be more easily discerned by comparing with the color of two drops of the indicator in a solution prepared by adding 0.20 g potassium acid phthalate in 100 mL distilled water. The pH of this solution is 4.0, the same as the end-point pH in the presence of CO_2. *Use the same technique for the standardization and the unknown titrations.*

---

[1]If you are to do Experiment 19, "pH Titration of Unknown Soda Ash," prepare twice this amount and use this standardization for that experiment.

(b) Bromcresol green indicator (alternative method). Add two to four drops bromcresol green. The color change for this indicator is from blue through pale green to yellow. Titrate to a blue-green color (just before the end point); interrupt the titration at this point and boil the solution carefully for 2 to 3 min to drive off the carbon dioxide. The color should revert to blue. Cool the solution to room temperature and continue the titration to the pale-green color. This marks the final end point.

Titrate the other two samples in the same manner as the first and calculate the molarity of the HCl from the weights of Na_2CO_3 taken, remembering that each carbonate has reacted with two protons. Use the mean of the three determinations for calculations involving the unknown.

2. *Standardization of HCl against standardized 0.1 M NaOH (alternate procedure).*[2] Rinse your 25-mL pipet with 0.1 M HCl, and add with the pipet 25 mL to a 250-mL Erlenmeyer flask. Add two to three drops phenolphthalein indicator solution. Rinse your buret with three small portions of the 0.1 M NaOH solution prepared and standardized in Experiment 6. Fill and adjust to near zero. The NaOH is placed in the buret rather than in the flask to protect it more from atmospheric CO_2. Record the volume to 0.02 mL. Titrate the acid until a faint pink color is obtained that lasts for at least 30 s. Repeat the titration two more times, using your 50-mL pipet for the second and third titrations if less than 25 mL NaOH was required in the first. Calculate the molarity of the HCl.

3. *Determination of sodium carbonate in soda ash.* Accurately weigh out by difference three samples of about 0.25 to 0.35 g each, dissolve each in about 60 mL water, and, titrate with 0.1 M HCl, following the same procedure with respect to indicators and end point used in the standardization procedure against Na_2CO_3 or, if not used, the one specified by your instructor.

Calculation

Calculate the percentage of Na_2CO_3 or Na_2O in your unknown for each portion analyzed. Your instructor will specify which one is to be reported.

EXPERIMENT 8   DETERMINATION OF BICARBONATE IN BLOOD USING BACK-TITRATION

Principle

About 95% of the total carbon dioxide in human blood exists as HCO_3^-, the remainder existing as dissolved CO_2. The HCO_3^- concentration, for most clinical work, can be used as a diagnostic aid. It is determined by adding an excess of 0.01 M HCl, to volatilize the HCO_3^- as CO_2, swirling to allow the CO_2 to escape, and then back-titrating the excess HCl with 0.01 M NaOH. The 0.01 M HCl and NaOH solutions are prepared by diluting standardized 0.1 M solutions.

Equations

$$HCO_3^- + H^+ \rightarrow H_2O + CO_2 \uparrow$$
$$\text{excess } H^+ + OH^- \rightarrow H_2O$$

---

[2]The sodium hydroxide solution is a **secondary standard**, and any errors in standardization will be represented in the standardization of the hydrochloric acid. The sodium hydroxide should be used within one week of standardization. If this experiment is done before Experiment 6, this procedure can be used to standardize the sodium hydroxide solution for that experiment.

Solutions and Chemicals Required

1. *Provided.* 0.1% Phenol red (phenolsulfonphthalein) solution in 0.003 M NaOH, 1% saline (NaCl) solution in CO_2-free water (see Experiment 6), Antifoam A (Dow Corning Corp.)

2. *To prepare.* Standard 0.1 M and 0.01 M HCl and 0.1 M and 0.01 M NaOH solutions. The molarity of the standard HCl and standard NaOH needs to be known only to three significant figures. If you have prepared standard HCl or NaOH in Experiment 6 or 7, use these for the present experiment. If you have not, prepare 250 mL standard 0.1 M NaOH as in Experiment 6. Since only three significant figures are required, you may use one-tenth the amount of KHP for titration, in which case the end point will occur at about 4.0 to 4.5 mL. A 10-mL buret should be used in these titrations. Standardize a 0.1 M HCl solution by titrating 5.00 mL of it with the standard 0.1 M NaOH as in Experiment 7. The phenol red indicator may be used. If you have only a standard 0.1 M HCl solution, use this to standardize 0.1 M NaOH solution.

   Prepare 500 mL of 0.01 M HCl and 0.01 M NaOH solutions by diluting 50 mL of the 0.1 M solutions to 500 mL with the saline solution. These should be prepared fresh on the day of use. The saline aids in the volatilization of the CO_2 from the acidified solution by decreasing its solubility.

Things to Do before the Experiment

Prepare and standardize the 0.1 M HCl and 0.1 M NaOH. This will require drying primary standard KHP ahead of time if either standard HCl or NaOH is not available.

Procedure[1]

1. *Preparation of the sample.* Either serum or plasma (oxalated or heparinized) may be used for the determination. This may be freshly drawn blood from an animal. (*Do not* do this yourself; your instructor will supply the sample.) See Chapter 1 for a discussion of the differences between serum, plasma, and whole blood. A 10- to 15-mL sample (20 to 30 mL whole blood) should be adequate for triplicate determinations by a class of 30 students. Fluoride should be added to prevent glycolysis, or breakdown of glucose, which can change the pH. The fluoride inhibits the enzyme catalysis causing glycolysis and stabilizes the pH for about 2 h. The tube used for collecting the sample can be rinsed with a solution of 100 mg sodium heparin plus 4 g sodium fluoride per 100 mL. The sample should be kept anaerobically, that is, stoppered to keep out atmospheric CO_2. Since the analysis should be done on the day the blood is drawn, the solutions should be prepared ahead of time.

2. *Preparation of comparison solution.* Prepare a standard for color comparison at the end point as follows. Place 6 mL of 1% saline solution in a 25-mL Erlenmeyer flask and add 0.10 mL serum or plasma. Add two drops phenol red indicator, insert a stopper, and rotate gently to mix the contents. The transition range of this indicator is pH 8.4 to 6.7 (yellow to red). Because of the buffering capacity of the blood, the end point occurs in this range.

3. *Titration of the sample.* The pooled serum or plasma sample will have been prepared by touching the end of a stirring rod to some Antifoam A and rotating it in the pooled sample. This will prevent excess foaming when the sample is swirled. Place 0.100 mL of serum or plasma in a 25-mL Erlenmeyer

---

[1] This determination may be performed on a macroscale using 5.00 mL acid and 1.00 mL sample, or using 2.00 mL acid and 0.500 mL sample. In the former case, the back-titration will require about 2.4 mL of 0.01 M NaOH, and in the latter case, about 0.7 mL.

flask and add 1.00 mL of 0.01 M HCl and 4 mL of 1% saline. Swirl the flask vigorously for at least 1 min to allow the CO_2 to escape. Add two drops of indicator and then titrate with 0.01 M NaOH dropwise, but rapidly, until a pink color matching the standard persists for at least 15 s. The NaOH may be added carefully with a graduated 1-mL measuring pipet and read to the nearest 0.01 mL.

The normal value of blood bicarbonate is about 26 meq/L (25 to 32 meq/L), or 0.026 meq/mL. Meq HCO_3^- = mmol HCO_3^-. Since 0.1 mL blood was taken for analysis, it should consume about 0.0026 mmol HCl, or 0.26 mL of 0.01 M HCl. Hence, since 1 mL of 0.01 M HCl was taken, about 0.74 mL should remain unreacted, and the back-titration should take about 0.7 mL of 0.01 M NaOH.

Calculation

Calculate the bicarbonate content of the sample in meq/L. (See Chapter 5 for definition of milliequivalents of serum electrolytes.)

Complexometric Titration

EXPERIMENT 9   DETERMINATION OF WATER HARDNESS WITH EDTA

Principle

Water hardness, due to Ca^{2+} and Mg^{2+}, is expressed as mg/L $CaCO_3$ (ppm). The total of Ca^{2+} and Mg^{2+} is titrated with standard EDTA using an Eriochrome Black T indicator.[1] A standard EDTA solution is prepared from dried (do not exceed 80°C) $Na_2H_2Y \cdot 2H_2O$ (purity 100.0 ± 0.5%). If the sample does not contain magnesium, Mg–EDTA is added to the titration flask to provide a sharp end point with the Eriochrome Black T, since calcium does not form a sufficiently strong chelate with the indicator to give a sharp end point. See the discussion in Chapter 9 for a more complete description.

Equations
Titration:

$$Ca^{2+} + H_2Y^{2-} \rightarrow CaY^2 + 2H^+$$
$$Ca^{2+} + MgY^2 \rightarrow CaY^2 + Mg^{2+}$$

End point:

$$Mg^{2+} + HIn^{2-} \rightarrow MgIn^- + H^+$$
$$MgIn^- + H_2Y^2 \rightarrow MgY^2 + HIn^{2-} + H^+$$
$$\text{(red)} \qquad \text{(colorless) (colorless)} \quad \text{(blue)}$$

The free acid parent of the indicator is H_3In, and that of the titrant EDTA is H_4Y.

Solutions and Chemicals Required

1. *Provided.*   0.5% (wt/vol) Eriochrome Black T indicator solution in ethanol, 0.005 M Mg–EDTA (prepared by adding stoichiometric amounts of 0.01 M

---

[1]If it is desired to titrate only Ca^{2+}, this can be done at pH 12 (use NaOH), where $Mg(OH)_2$ is precipitated and does not titrate. Hydroxynaphthol blue indicator is used.

EDTA and 0.01 M $MgCl_2$). The indicator solution should be prepared fresh every few days, as it is unstable. A portion of the Mg–EDTA solution, when treated with pH 10 buffer and Eriochrome Black T, should turn a dull violet color; one drop 0.01 M EDTA should change this to blue and one drop 0.01 M $MgCl_2$ should change it to red.

2. *To prepare*

   (a) **NH_3–NH_4Cl buffer solution, pH 10.** Dissolve 3.2 g NH_4Cl in water, add 29 mL conc. NH_3, and dilute to about 50 mL. The buffer solution is best stored for long periods of time in a polyethylene bottle to prevent leaching of metal ions from glass.

   (b) **Standard 0.01 M EDTA solution.** Dry about 1.5 g reagent-grade $Na_2H_2Y \cdot 2H_2O$ in a weighing bottle at 80°C for 2 h. Cool in a desiccator for 30 min and accurately weigh (to the nearest milligram) approximately 1.0 g. Transfer to a 250-mL volumetric flask. (This is the disodium salt of EDTA; the free acid is insoluble). Add about 2.00 mL distilled, *deionized* water and shake or swirl periodically until the EDTA has dissolved. EDTA dissolves slowly and may take 0.5 h or longer. If possible, the solution should be allowed to stand overnight before using. If any undissolved particles remain, addition of three pellets of NaOH may aid dissolution, but there is danger of adding metallic impurities. After the EDTA is dissolved, dilute to 250.0 mL and shake thoroughly to prepare a homogeneous solution. Then, rinse a clean polyethylene bottle with three small portions of the EDTA solution and transfer the remainder of the solution to the bottle for storage. (Polyethylene is preferable to glass for storage because EDTA solutions gradually leach metal ions from glass containers, resulting in a change in the concentration of free EDTA.) Calculate the molarity of the EDTA solution.

Things to Do before the Experiment

Dry the $Na_2H_2Y \cdot 2H_2O$ and prepare the standard EDTA solution.

Procedure

Obtain a water sample from your instructor. Add with a pipet or a buret a 50-mL aliquot of the sample to a 250-mL wide-mouth Erlenmeyer flask, add 2 mL of the buffer solution, 0.5 mL of the Mg–EDTA solution, and five drops of the indicator solution. (If the unknown contains magnesium, addition of Mg–EDTA is not necessary—consult your instructor.) Avoid adding too much indicator with dilute solutions or the end-point change may be too gradual. The indicator will not become wine red until magnesium is added. *The buffer should be added before the indicator*, so that small amounts of iron present in the water will not react with the indicator. [An end-point color change from wine red to violet indicates a high level of iron in the water. This interference can be eliminated by adding a few crystals of potassium cyanide. *Caution must be used with this. Add it only if instructed to and after the alkaline buffer is added, since HCN is volatile and very toxic.* After the titration, add 1 g $FeSO_4 \cdot 7H_2O$ to convert CN^- to harmless $Fe(CN)_6^{4-}$.] If the water sample is likely to contain copper, add a few crystals of hydroxylamine hydrochloride. This reduces copper(II) to copper(I), which does not interfere.

Titrate with 0.01 M EDTA until the color changes from wine red through purple to a pure blue. The reaction (color change) is slow at the end point, and titrant must be added slowly and the solution stirred thoroughly in the vicinity of the end point. A comparison solution for the proper color at the end point may be prepared by adding 2 mL of pH 10 buffer to 50 mL distilled water, five drops indicator, a few drops Mg–EDTA, and a few drops EDTA.

If the end point for the first titration is less than 10 mL, double the volume of sample for the remaining two titrations.

Calculation

Calculate and report the hardness of the water as ppm $CaCO_3$ for each portion analyzed.

Precipitation Titrations

EXPERIMENT 10   DETERMINATION OF SILVER IN AN ALLOY: VOLHARD'S METHOD

Principle

Primary standard $AgNO_3$ is used to standardize a 0.1 M KSCN solution, using a ferric alum indicator. The unknown silver alloy is analyzed by titrating with the standardized KSCN solution.

Equations

$$SCN^- + Ag^+ \rightarrow \underline{AgSCN}$$
$$Fe^{3+} + SCN^- \rightarrow \underline{Fe(SCN)^{2+}} \qquad \text{(indicator reaction)}$$
$$\text{red}$$

Solutions and Chemicals Required

1. *Provided.*   Primary standard $AgNO_3$, ferric alum indicator [$KFe(SO_4)_2 \cdot 12$ H_2O, saturated solution], 6 M HNO_3 (free of oxides of nitrogen). Nitric acid free from lower oxides of nitrogen should be colorless and can be prepared, if necessary, by boiling 1:1 HNO_3 until NO_2 is expelled.

2. *To prepare.*   0.1 M KSCN. Weigh approximately 5.0 g KSCN, dissolve in water in a 500-mL or 1-L bottle and add 500 mL distilled water. Shake well to ensure a homogeneous solution. This solution is approximately 0.1 M and will be standardized against primary standard $AgNO_3$.

Things to Do before the Experiment

Prepare and standardize (below) the 0.1 M KCSN solution. This will require drying and cooling $AgNO_3$.

Procedure

1. *Standardization of KSCN solution.*   Obtain about 3 g primary standard $AgNO_3$ in a weighing bottle from your instructor. Dry in the oven at 110°C for 1 to 2 h, *but no longer.* Cool in a desiccator for 30 to 40 min. This will be weighed by the direct method.

   Place a clean, dry weighing dish on the balance and determine its weight to the nearest 0.1 mg. Add 0.70 to 0.75 g $AgNO_3$ to the weighing dish and determine the increase in weight to the nearest 0.1 mg. Transfer quantitatively to a clean 250-mL wide-mouth Erlenmeyer flask. Keep away from strong light as much as possible until ready to titrate. Weigh two other 0.70- to 0.75-g portions of $AgNO_3$ and transfer to separate (numbered) 250-mL Erlenmeyer flasks.

   Add 50 mL distilled water to the flask ready for titration, 50 mL of 6 M nitric acid free from lower oxides of nitrogen, and 2 mL ferric alum indicator.

(Lower oxides of nitrogen form nitroso complexes with Fe^{3+} that are red in color and will interfere with the end point.) Fill your 50-mL buret with the KSCN solution, record the initial volume to the nearest 0.01 mL, and titrate with constant vigorous agitation until a faint reddish-brown color appears in the solution; this is more easily seen if the precipitate is allowed to settle after each addition near the end point. It will be helpful to compare the color with a solution made by adding 5 mL of 6 *M* nitric acid and 2 mL ferric alum to 75 mL water. The color must be permanent after strong shaking. Titrate the other two $AgNO_3$ samples in the same manner and calculate the molarity of the KSCN solution from each titration. Use the mean of the three determinations.

2. *Determination of silver in an alloy.*[1]   Obtain a sample of a silver alloy from the instructor.[2] Place in a 250-mL beaker, add 20 mL dilute (1:1) HNO_3, cover with a watch glass, and warm on the steam bath in the hood until the alloy has completely dissolved. When solution is complete, remove the watch glass, rinse it off with a jet of water from your wash bottle, catching the rinsings in the beaker, and continue heating with the beaker uncovered until all brown fumes have disappeared and the solution is colorless. Cool to room temperature and transfer the solution quantitatively to a clean 250-mL volumetric flask with the aid of a funnel and stirring rod to pour the solution down. Rinse the beaker several times with distilled water; adding the rinsings to the flask. Dilute to the mark and shake thoroughly to ensure a homogeneous solution.

   Transfer with a pipet two samples, 50 mL each, into 250-mL Erlenmeyer flasks, add 2 mL ferric alum indicator, and titrate with standard 0.1 *M* KSCN solution to the appearance of a faint reddish-brown color, which is permanent even after strong agitation. The two titrations should agree within 0.05 mL. If they do not, pipet and titrate two more. When finished, put all silver-containing solutions in a jar for this purpose.

Calculations
Calculate and report the number of grams silver in your alloy sample.

EXPERIMENT 11   DETERMINATION OF CHLORIDE IN A SOLUBLE CHLORIDE: FAJANS' METHOD.

Principle
The sample is titrated with a standard $AgNO_3$ solution, using a dichlorofluorescein adsorption indicator end point.

Equations
$$Cl^- + Ag^+ \rightarrow AgCl$$

Solutions and Chemicals Required
1. *Provided.*   0.1% dichlorofluorescein indicator solution, dextrin.
2. *To prepare.*   Standard 0.1 *M* $AgNO_3$. Obtain from the instructor about 4.5 g primary-standard-grade silver nitrate in a clean, dry weighing bottle; dry in the oven at 110°C for 1 to 2 h, *but no longer*. This material will be used to

---

[1]The unknown may instead be impure silver nitrate, in which case dissolve it in 50 mL water and proceed as in the standardization of the KSCN above.
[2]You may be instructed to weigh out 0.3 to 0.4 g of the alloy and then report the percentage of silver in the sample.

prepare a standard solution by the direct method. Cool for 30 to 40 min in a desiccator.

Using a weighing dish, weigh accurately (nearest 0.1 mg) about 4.3 g $AgNO_3$. Transfer to a 250-mL beaker and dissolve in about 100 mL distilled water. Carefully pour this solution into a 250-mL volumetric flask and rinse the beaker several times with distilled water, adding the rinsings to the flask. Dilute to the mark, shake well, and pour into a clean, dry 500-mL glass-stoppered amber bottle. Shake again to ensure a homogeneous solution. Keep away from strong light as much as possible. Store away from light. Calculate the molarity of the solution.

Things to Do before the Experiment

1. Prepare the 0.1 M $AgNO_3$ solution.

2. Obtain a sample of unknown chloride from your instructor and dry it in the oven at 120°C for 1 h or longer. It is okay to dry it overnight. Cool in a desiccator at least 30 to 40 min before weighing.

Procedure

Weigh out in the weighing dish three samples of about 0.25 to 0.30 g each (vary the sample weights, but weigh each exactly), transfer to 250-mL Erlenmeyer flasks, and dissolve in about 50 mL distilled water. Add 10 mL 1% dextrin suspension (shake well before using) and 10 drops dichlorofluorescein indicator solution. The dextrin prevents excessive coagulation of the precipitate at the end point. This keeps a larger surface area for adsorption of the indicator, which enhances the sharpness of the end point. Instead of a suspension of dextrin, 0.1 g of the solid may be added.

The pH of the solution should be between 4 and 10. If it is too acid (e.g., due to hydrolysis of other chloride salts in the unknown), it may be neutralized by adding solid $CaCO_3$ until excess remains in suspension. The suspension does not interfere with the end point. Results are low at pH values above 4 because the indicator tends to displace chloride from the precipitate. (More accurate results are obtained by standardizing the $AgNO_3$ solution against dried reagent-grade NaCl, using the same conditions as for the unknown titration.)

Titrate with 0.1 M $AgNO_3$. Thorough mixing by vigorous swirling of the flask throughout the titration is essential to achieve a good end point. Do not let direct sunlight strike the flask; if you are working next to a window on a very bright day, it would be well to draw the shade. The end point is marked by a change from a pale yellow to a distinct pink color. If the titration is overrun, add a few grains of NaCl or KCl and practice the determination of the end point a few times before titrating the second and third samples. When finished, put all silver-containing solutions in a jar provided for this purpose.

Calculations

Calculate and report the percent Cl in your sample for each portion titrated.

Potentiometric Measurements

EXPERIMENT 12   DETERMINATION OF THE pH OF HAIR SHAMPOOS

Principle

A calibrated glass pH electrode–SCE pair is used to measure the pH of commercially available shampoos, hair conditioners and rinses, and depilatories, both concentrated and diluted 1:10. An estimate is made whether the pH is controlled primarily by a buffered solution or a strong electrolyte. An evaluation is made of the relative importance of what is probably the best pH for a shampoo and why—does the pH affect the cleaning power of the shampoo or does it affect the hair itself?

Equations
Electrode Response

$$E = k - \frac{2.303RT}{F} \text{pH}$$

Solutions and Chemicals Required

Provided. Standard pH 3, 5, 7, 9, and 11 buffers (approximate—other values near these may be used). Each member of the class should bring one or more samples of commercially available hair shampoos, conditioners, rinses, and depilatories. These may include dandruff shampoos and hair removers such as Neet or Nair. All the samples will be shared by the class. There should be enough of each to provide about 15 to 20 mL to each student.

Procedure

1. *Calibration of the electrodes.* The experiment will be performed by first calibrating the electrodes at pH 7 and making approximate pH measurements of the various samples. The samples are then separated into five groups: those with pH < 4, pH 4 to 5.9, pH 6 to 8, pH 8.1 to 10, and pH > 10. The glass electrode should have been soaked in 0.1 M KCl solution at least one day prior to its use; store the electrode in dilute KCl solution when not in use. Calibrate the pH meter for one group of samples at a time (pH 3 for the first group, pH 11 for the last group) using the procedure described by your instructor. (Your instructor may decide to use only three buffers for calibration, e.g., pH 4, 7, and 9.) This will consist essentially of adjusting the meter to read the pH of the standard buffer at room temperature with the electrodes immersed in the buffer solution. Rinse the electrodes with distilled water after calibration and blot with tissue paper; do not wipe them as this may impart a static charge to the glass membrane. Be careful to turn the pH meter to "standby" when removing electrodes from solution. If only a small quantity of buffer is used, it would be better to discard it rather than chance contamination of the entire supply.

2. *pH Measurements of samples.* Prepare a 1:10 dilution of each sample in a 100-mL beaker by pipetting 5.00 mL into 45.0 mL distilled water measured from a 50-mL buret. Measure the pH of first the diluted sample and then the concentrated sample. For the latter, if a small beaker is used (e.g., 25 mL) it may be possible to make the measurement with 5 mL of sample; just enough to cover the bulb of the glass electrode. Alternatively, a combination pH–reference electrode may be used, in which case the sample can be placed in a test tube and the single probe dipped in this. (A combination electrode contains both a reference electrode and a glass electrode in one probe. A salt bridge wick for the reference electrode is above the pH glass bulb, and so the probe will have to be immersed a bit deeper to make contact with both electrodes. There will be two wire leads from the probe, one for each electrode.)

Immerse the electrode in the test solution and swish or agitate a few seconds. Then, allow the pH reading to equilibrate and record to the nearest 0.01 pH. Rinse the electrode well between measurements and blot off the water. It is best to make the measurements from low pH to high, or vice versa. If the pH reading of any of the samples falls outside the range of 4 to 10, the electrodes should be recalibrated with a buffer closer to the sample pH (if depilatories are measured, their pH will exceed 10 and a buffer at pH 11 should be used.)

Report

Arrange the samples into either shampoos or conditioners and rinses (depilatories are separate). Arrange the samples in each group in order of increasing pH readings of the concentrated samples. Your instructor may compare your pH readings with the mean of the class. From the changes in readings on dilution, list whether you think the pH of each solution was determined by a buffer system or by a strong acid or base. Consult the reference J. J. Griffin, R. F. Corcoran, and K. K. Akana, *J. Chem. Ed.*, **54** (1977) 553 and include in your report a discussion of the relevance of pH in hair cleansing or conditioning and in hair damage. Propose mechanisms for the pH action. How do depilatories work? Within what pH range should children's shampoos be?

EXPERIMENT 13   POTENTIOMETRIC DETERMINATION OF FLUORIDE IN DRINKING WATER USING A FLUORIDE ION-SELECTIVE ELECTRODE[1]

Principle

Fluoride in a water sample is determined by measurement with a fluoride ion-selective combination electrode (contains reference electrode built in). First, you will determine whether the electrode response is Nernstian over a wide range of concentrations. Then, you will determine fluoride in the unknown by comparing potential measurements with standards over a narrower range, bracketing the unknown; a calibration curve will be prepared.

Equations

$$E = k - \frac{2.303RT}{F} \log a_{F^-}$$

$$= k' - \frac{2.303RT}{F} \log[F^-] \text{ (if ionic strength is held constant)}$$

Solutions and Chemicals Required

1. *Provided.*   TISAB (total ionic-strength adjustment buffer) solution, which is prepared with 57 mL glacial acetic acid, 58 g sodium chloride, and 4 g CDTA (cyclohexylenedinitrilotetraacetic acid) in about 500 mL water, adjusted to pH 5.0 to 5.5 with 5 *M* NaOH and diluted to a total volume of 1 L. A 1:1 dilution of *all* samples with this solution serves the following:

   (a) It provides a high total ionic strength background, swamping out variations in ionic strength between samples.

   (b) It buffers the samples at pH 5 to 6. (In acid media, HF forms; while in alkaline media, OH^- ion interferes in the electrode response.)

   (c) The CDTA preferentially complexes with polyvalent cations present in water (e.g., Si^{4+}, Al^{3+}, Fe^{3+}), which otherwise would complex with F^- and change its concentration.

---

[1]This experiment can also be used to determine % NaF or % SnF_2 in toothpaste by preparing a suspension of the paste in water and pipetting the supernatant, or to determine the fluoride content of children's fluoride tablets or drops. See T. Light and C. Cappucino, *J. Chem. Ed.*, **52** (1975) 247.

2. *To prepare*

    (a) Stock standard 0.1 *M* fluoride solution. Dry about 1 g NaF at 110°C for 1 h, cool in a desiccator for 30 min. Weigh out 0.45 to 0.50 g of the dried NaF (to the nearest milligram), transfer to a 100-mL volumetric flask, dissolve, and dilute to volume with distilled deionized water. Shake thoroughly and transfer to a polyethylene bottle (rinse with a few small portions first). Fluoride tends to adsorb on glass and should be stored in plastic containers. **Caution:** Fluoride is poisonous. Handle with care. Commercially prepared fluoride solutions may be used.

    (b) Linearity standards. By serial dilution of the stock solution with distilled deionized water, prepare a series of solutions of about 10^{-2}, 10^{-3}, 10^{-4}, 10^{-5}, and 10^{-6} *M* fluoride (calculate accurate concentrations). (Do *not* pipet by mouth!) For example, dilute the stock solution 10:100, 1:100, and 1:1000 mL to prepare the first three solutions. Then, dilute the 10^{-4} *M* solution 10:100 and 1:100 mL to prepare the last two. Transfer to polyethylene bottles. These solutions should be prepared on the day of use.

    (c) Calibration standards. The unknown concentration will be within 1×10^{-3} and 1×10^{-2} M fluoride, and a calibration curve will be prepared using concentrations of fluoride to bracket the unknown. Using the above procedure, prepare additional standards of 2×10^{-3} and 4×10^{-3} *M* fluoride (calculate the accurate concentrations). Prepare on the day of use.

Things to Do before the Experiment

Prepare the stock NaF solution. This will require drying of NaF.

Procedure

    1. *Determination of range of response and range of linearity.* Connect the electrode leads to an expanded-scale pH meter. Add with a pipet 10 mL of the 10^{-6} *M* standard solution and 10 mL TISAB solution to the small plastic beaker provided. Place the electrode in the beaker. Stir the solution with a magnetic stirrer and small stirring bar during measurement. You may make readings in pH units. (1 pH = 59.2 mV at 25°C; arbitrarily take pH 0 as 0 mV). The meter should also allow reading in millivolts. When a steady reading is obtained, record the value. Rinse and blot the electrode and repeat, going from dilute to concentrated standard solutions. Prepare a spreadsheet of the data, and chart mV vs. log *C*. Report the slope in mV/decade, the intercept, and the r^2 value. Report also the range of linearity.

    2. *Standardization for unknown.* Record the readings of the standard solution from 1×10^{-3} *M* to 1×10^{-2} *M*, and plot a calibration curve as above over this range. Enter the formula in one cell for calculating an unknown concentration from the millivolt reading and the slope and intercept of the calibration curve.

    3. *Analysis of unknown.* After preparing the calibration curve, obtain an unknown fluoride sample. This may be a synthetic solution, in which case obtain the unknown in a 250-mL volumetric flask. Immediately dilute to volume with distilled deionized water and transfer to a polyethylene bottle. Add 10 mL of the unknown with a pipet to a small plastic beaker followed by 10 mL TISAB. Record the mV reading as above. Make at least three separate runs (separate additions and potential readings). **Note:** The unknown should be measured at the same time the calibration curve is prepared. The mV scale should not be adjusted between calibration and sample measurements. If it is, take a new reading of one of the standards, and readjust to the original reading.

Calculations

From the spreadsheet calibration curve, determine the concentration of fluoride in the unknown solution. Report the results in parts per million fluoride, along with the standard deviation for the three measured samples.

Reduction–Oxidation Titrations

EXPERIMENT 14   ANALYSIS OF AN IRON ALLOY OR ORE BY TITRATION WITH POTASSIUM DICHROMATE

Principle

An iron alloy or ore is dissolved in HCl and the iron is then reduced from Fe(III) to Fe(II) with stannous chloride ($SnCl_2$). The excess $SnCl_2$ is oxidized by addition of $HgCl_2$. The calomel formed (insoluble Hg_2Cl_2) does not react at an appreciable rate with the titrant. The Fe(II) is then titrated with a standard $K_2Cr_2O_7$ solution to a diphenylamine sulfonate end point.

Equations

Reduction:

$$2Fe^{3+} + Sn^{2+} \text{ (slight excess)} \rightarrow 2Fe^{2+} + Sn^{4+} + Sn^{2+} \quad \text{(due to excess)}$$

$$Sn^{2+} + 2Hg^{2+} + 2Cl^{-} \rightarrow Sn^{4+} + \underline{Hg_2Cl_2} \quad \text{(white precipitate)}$$

If too much Sn^{2+} is added, then

$$Sn^{2+} + \underline{Hg_2Cl_2} \rightarrow Sn^{4+} + 2Hg^{0} + 2Cl^{-}$$

(Black Hg^{0} precipitate makes end-point determination impossible. The sample must be discarded because Hg^{0} reacts with $Cr_2O_7^{2-}$.)

Titration:

$$6Fe^{2+} + Cr_2O_7^{2-} + 14H^{+} \rightarrow 6Fe^{3+} + 2Cr^{3+} + 7H_2O$$

Solutions and Chemicals Required

1. *Provided.*   0.28% (wt/vol) of the sodium salt of *p*-diphenylamine sulfonate indicator, 0.5 *M* $SnCl_2$ in 3.5 *M* HCl (with mossy tin added to stabilize against air oxidation: $Sn^{4+} + Sn^{0} \rightarrow 2Sn^{2+}$), saturated $HgCl_2$ solution, conc. HCl, 6 *M* HCl, conc. H_3PO_4-conc. H_2SO_4 mixture (15 mL each added to 600 mL water and cooled to room temperature), 0.1 *M* $FeCl_3$ in 6 *M* HCl.

2. *To prepare.*[1]   Standard 1/60 *M* $K_2Cr_2O_7$ solution (approx. 0.017 *M*). Dry about 3 g primary standard $K_2Cr_2O_7$ in a weighing bottle at 120°C for at least 2 h. Drying in the oven for longer periods of time (i.e., until the next lab

---

[1]If you are to use this solution also for Experiment 15 and/or 25, prepare 500 mL, taking accurately about 2.5 g $K_2Cr_2O_7$, dissolving in 200 mL water, and transferring to a 500-mL volumetric flask. You need to save at least 150 mL for Experiment 15 and 100 mL for Experiment 25. The 250 mL you prepare for this experiment will be enough if you are careful not to waste any.

period) will do no harm. Place the $K_2Cr_2O_7$ in the desiccator to cool for 30 to 40 min. Weigh accurately to the nearest milligram, about 1.3 g in a weighing dish, and transfer quantitatively to a 200-mL beaker. Dissolve in about 200 mL water and transfer quantitatively to a 250-mL volumetric flask. Dilute to volume and mix thoroughly. Rinse a clean 1-L bottle with three small portions of the solution and transfer the remainder of the solution to the bottle for storage. Calculate the molarity of the solution. $K_2Cr_2O_7$ may also be prepared approximately and standardized against electrolytic iron wire (primary standard) using the same procedure as given below for an alloy sample.

Things to Do before the Experiment

1. *Dry the necessary amount of $K_2Cr_2O_7$.*

2. *Obtain and dry or dissolve your unknown as required.*

(a) **Alloy sample.** This may be in the form of a wire. Your instructor will provide you with three separate (weighed) pieces of the unknown, each to be placed in separate, labeled 500-mL Erlenmeyer flasks containing 10 mL conc. HCl. Cover with inverted 100-mL beakers and store these dissolving samples in your desk overnight or longer. Alternatively, the samples can be dissolved on the day of the experiment by heating on a hot plate or steam bath in 400-mL beakers covered with ribbed watch glasses in the hood to hasten dissolution. (After dissolution, rinse the cover glass and the sides of the beaker, catching all the rinsings in the beaker. Use as little water as possible. The final volume should not be more than 50 mL.)

(b) **Ore sample.** Check out a sample of an iron ore from the instructor. Dry in the oven at 110 to 120°C for at least 2 h; longer drying will do no harm.

Procedure

1. *Reduction of the iron and trial titration.* Before titrating your unknown, it is advisable to perform one or two trial titrations. This can be done while the (ore) samples are dissolving. Add approximately 10 mL 0.1 M $FeCl_3$ solution in 6 M HCl to a 600-mL beaker and add about 50 mL water. (This dilutes the sample sufficiently that when all the Fe^{3+} is reduced, the solution will be nearly colorless. If the volume is less, then the pale green of the Fe^{2+} makes detection of complete reduction more difficult). Place a ribbed watch glass on the beaker and heat nearly to boiling on a hot plate in the hood; the solutions must be very close to the boiling point, perhaps simmering gently, but not boiling violently since $FeCl_3$ can be lost due to volatilization. Add 0.5 M stannous chloride solution with a dropper through the lip of the beaker until the color begins to fade; then, continue the addition drop by drop, swirling the beaker and allowing each drop to react before adding the next, until the solution is colorless. It will first become pale yellow and then will gradually turn more clear. It may never get completely colorless but may instead go to a pale green due to the ferrous ion (this will depend on the amount of iron). Whichever you get (colorless or pale green) stop addition and allow the solution to heat for two more minutes. If the yellow color returns, add a few more drops of $SnCl_2$ until it becomes colorless or pale green again. Repeat dropwise addition of $SnCl_2$ until the solution does not return to the yellow color. At this point, add two drops excess, no more. (If more than two drops are added, the stannous chloride can be oxidized with a few drops of potassium permanganate solution and the above reduction process repeated.)

Remove from the hot plate, rinse down the cover glass and sides of the beaker, and cool quickly to room temperature by immersing the bottom of the beaker in cold water. Two to three samples may be taken this far together; *the remainder of the procedure must be carried out with each sample individually without interruption through the titration.* If any sample turns yellow again while awaiting its turn, it must be reheated and sufficient stannous chloride added to discharge the color, with two drops excess. Fill your 50-mL buret with the standard $K_2Cr_2O_7$ and have it ready for titration.

To one sample, which must be at room temperature, add 100 mL water and then add rapidly 15 mL saturated mercuric chloride solution, previously measured out, while stirring and immediately mix thoroughly. A slight, white precipitate should form. If either a heavy gray or black precipitate (elemental mercury) or no precipitate forms, too much or not enough (to reduce all the Fe^{3+}) stannous chloride has been added; in either case, the sample must be discarded. Mix for 2 min, then add 100 mL of the H_3PO_4–H_2SO_4 mixture and six to eight drops diphenylamine sulfonate indicator. Titrate immediately with the $K_2Cr_2O_7$ solution, stirring constantly, until the green color changes to a purple or violet blue that remains for at least 1 min. (The acid mixture provides the protons consumed in the titration reaction and forms a nearly colorless phosphate complex with the Fe^{3+} titration product, which sharpens the end point.)

2. *Alloy sample.* The sample should by now be dissolved in the Erlenmeyer flasks (or in the heated 600-mL beakers). Adjust the volume to 40 to 60 mL with distilled water. Heat nearly to boiling and follow the same procedure as used in the trial titration, starting with addition of stannous chloride. All three samples can be taken up to the point to just before the addition of mercuric chloride and then must be treated one at a time up through the titration.

3. *Ore sample.* After cooling the dried sample in a desiccator for 30 to 40 min, weigh out by difference (the iron ore may be hygroscopic) three samples; consult the instructor as to the size of the samples. Transfer to 600-mL beakers. Add 10 mL water and swirl until the sample is completely moistened and in suspension; then cover with ribbed watch glasses and add 10 mL concentrated HCl, pouring it through the lip of the beaker. Heat on a hot-plate in the hood until the iron ore has dissolved to give a clear, red-brown solution; with some samples there may be an insoluble sandy residue, which may be disregarded. [Silica or insoluble sulfides (black) or silicates may remain.] The hot plate should be adjusted to keep the solutions just barely at the boiling point; vigorous boiling should be avoided since it may cause loss of material and excessive evaporation of acid. If necessary, add 6 *M* HCl to keep the volume about 20 mL. When all the iron has been dissolved, the insoluble residue (if any) will be gray or white, with no black or reddish particles, after adding stannous chloride to reduce the iron. When the solution appears clear, add distilled water to bring the volume to about 50 mL and follow the same procedure used in the trial titration starting with addition of stannous chloride to the hot solution. All three samples can be carried up to the point just before addition of mercuric chloride and then must be treated one at a time up through the titration.

Calculations

1. *Alloy sample.* Calculate and report the milligrams iron in each portion of the unknown analyzed, along with the mean and the precision.

2. *Ore sample.* Calculate and report the percent iron in each portion of the unknown analyzed, along with the mean and the precision.

EXPERIMENT 15 ANALYSIS OF COMMERCIAL HYPOCHLORITE OR PEROXIDE SOLUTION BY IODOMETRIC TITRATION

Principle

The oxidizing power (percent $NaOCl$ or H_2O_2) of the solution is determined iodometrically by reacting it with an excess of iodide in acetic acid solution and titrating the iodine produced (I_3^- in the presence of excess iodide) with standard sodium thiosulfate solution. The sodium thiosulfate is standardized against primary standard potassium iodate, and a starch indicator is used.

Equations

Standardization of $Na_2S_2O_3$:

$$IO_3^- + 8I^- + 6H^+ \rightarrow 3I_3^- + 3H_2O$$
$$I_3^- + 2S_2O_3^{2-} \rightarrow 3I^- + S_4O_6^{2-}$$

Sample titration:

$$ClO^- + 3I^- + 2H^+ \rightarrow Cl^- + I_3^- + H_2O$$

or

$$H_2O_2 + 3I^- + 2H^+ \xrightarrow[\text{catalyst}]{Mo(VI)} 2H_2O + I_3^-$$

I_3^- is titrated with $S_2O_3^{2-}$ as above.

Solution and Chemicals Required

1. *Provided.* $6\,M$ HCl, KI, primary standard KIO_3, glacial acetic acid, 3% ammonium molybdate solution (for peroxide samples), Na_2CO_3, dil. (1:4) H_2SO_4.

2. *To prepare*

   (a) Starch solution. Prepare a 1% solution by mixing 0.5 g soluble starch with 2 to 3 mL of distilled water; add a pinch of HgI_2. Pour this mixture into 50 mL boiling distilled water with stirring and continue heating for 2 to 3 min until the solution is clear or only faintly opalescent. Cool to room temperature. The HgI_2 stabilizes the starch indefinitely; otherwise, it should be prepared fresh on the day of use. **Note:** Approximately 0.4 g of the commercial indicator Thiodene may be used in place of the prepared starch solution.

   (b) Standard 0.01 *M* KIO₃ solution. This will be used to standardize the $Na_2S_2O_3$ solution. (If you prepared a standard 1/60 *M* $K_2Cr_2O_7$ solution in Experiment 14, you may titrate 50-mL aliquots of this iodometrically for the standardization of the $Na_2S_2O_3$. If so, consult your instructor for directions.) A standard solution of KIO_3 is prepared and aliquots of this are titrated with the $Na_2S_2O_3$ solution. This procedure is used instead of titrating individually weighed portions of KIO_3. The reason is that KIO_3 has a low equivalent weight and only about 0.1-g portions can be titrated. Hence, it is more accurate to prepare a standard solution. This requires special care in the accurate preparation of the solution since only one solution is prepared.

   Dry about 1.5 g primary standard KIO_3 at 120°C for 1 to 2 h and cool in a desiccator for 30 to 40 min. Accurately weigh (to the nearest 0.1 mg) 1.0 to 1.4 g and dissolve in a small amount of distilled water in a 200-mL beaker. Quantitatively transfer, with rinsing, to a

500-mL volumetric flask, using a glass funnel and stirring rod to direct the solution into the flask. Dilute to the calibration mark. Calculate the molarity of the solution.

(c) **0.1 M Na$_2$S$_2$O$_3$ solution.** Boil about 1200 mL distilled water for 5 to 10 min to ensure sterility and to expel carbon dioxide. Cool to room temperature. (Sodium thiosulfate solutions are subject to bacterial attack, which may change the molarity after a time. Thus, all water and glassware used to prepare and store the solution should be sterilized. If any turbidity or bacteria or mold growth appears, the solution should be discarded. Removal of carbon dioxide is also beneficial, because thiosulfate is more stable in neutral solution.) Sterilize a 0.5-L bottle with hot distilled water, near boiling, and then rinse with the cool boiled distilled water. Weigh out on a watch glass, using a rough balance, 12.5 g sodium thiosulfate crystals, Na$_2$S$_2$O$_3$ · 5H$_2$O. Transfer to the sterilized bottle, add 500 mL of the freshly boiled and cooled distilled water, add 0.05 g sodium carbonate, and shake thoroughly until the solution is homogeneous. (A small amount of sodium carbonate is added to keep the solution neutral or slightly alkaline and thereby stabilize it against decomposition to elemental sulfur.) Store in a refrigerator if possible, but let warm up to room temperature before using.

Things to Do before the Experiment

1. *Dry the required amount of primary standard KIO$_3$.*

2. *Prepare the 0.1 M Na$_2$S$_2$O$_3$ solution.* Although this can be prepared on the day of the experiment, it is preferable to prepare it at least a day before it is to be standardized. The solution tends to lose some of its titer right after preparing.

Procedure

1. *Standardization of the Na$_2$S$_2$O$_3$ solution.* The solution should be standardized on the day of the experiment. Consult your instructor if you are to use the standard K$_2$Cr$_2$O$_7$ solution from Experiment 14 for standardization. Otherwise, proceed as follows.

Rinse the 50-mL buret several times with small portions of the thiosulfate solution and fill it with thiosulfate solution. Adjust to near the zero mark and record the volume reading to the nearest 0.02 mL. Add with a pipet a 50.00-mL aliquot of the potassium iodate solution to a clean 250-mL wide-mouth Erlenmeyer flask. Add about 2 g solid potassium iodide and swirl to dissolve. Add, with rapid mixing, 5 mL dilute H$_2$SO$_4$. Mix thoroughly.

Titrate *immediately* with thiosulfate solution. (In strongly acid solution, the excess iodide is rapidly air-oxidized to I$_3^-$, and so the titration must be performed quickly.) Thorough, continuous mixing throughout the titration is essential; the thiosulfate must not be allowed to accumulate in local excess in the acid solution or else some decomposition into H$_2$SO$_3$ and S may occur. Titrate until the yellow color (due to I$_3^-$) *almost* disappears. It will become a pale yellow. Then, add 2 to 3 mL starch solution and titrate until the blue color just disappears (properly done, this should occur within 0.5 mL after adding the starch solution).

The standardization should be repeated until you are sure of the titration volume to within one part per thousand (e.g., ±0.03 mL at a titration volume of 30 mL). Calculate the molarity of the Na$_2$S$_2$O$_3$ solution.

2. *Determination of hypochlorite or H$_2$O$_2$ in unknown*[1]

(a) **Weighing and diluting the unknown.** Roughly calibrate a weighing bottle by pouring into it about 12 mL water and noting the level to which it fills the bottle. Empty and thoroughly dry the weighing bottle and weigh it to the nearest milligram. The hypochlorite or peroxide solutions to be analyzed will be supplied in the commercial bottles, fitted with siphon delivery tubes. Clean off any solid crust on the tip and discard a few drops to flush out the tip. Withdraw about 12 mL into the calibrated and weighed weighing bottle; it is essential that the upper portion of the bottle, particularly the ground-glass part, remain dry. Replace the stopper and weigh to the nearest milligram. Empty the weighing bottle into a 250-mL volumetric flask containing about 100 mL water, using a funnel. Wash out the weighing bottle and the funnel with a jet of water from your wash bottle, catching the rinsings in the volumetric flask. Dilute to the mark and mix thoroughly. Transfer with a pipet three 50-mL aliquots of the solution into 250-mL Erlenmeyer flasks containing about 50 mL water; rinse down the walls of the flasks in such a way as to form a layer of water above the sample. From this point on, handle each sample individually through the remainder of the procedure.

(b) **Titration.** Fill your buret with standard 0.1 *M* sodium thiosulfate solution. Measure out and have ready 10 mL glacial acetic acid, 2 g potassium iodide, and, if the sample is a peroxide, a dropping bottle containing 3% ammonium molybdate catalyst solution. Hypochlorite requires no catalyst. When ready to titrate, add glacial acetic acid, potassium iodide, and if the sample is a peroxide, add three drops of the catalyst solution. Titrate immediately, swirling the flask constantly. When the color has faded to a pale yellow, add about 2 mL starch solution and continue the titration drop by drop until the solution just becomes colorless. Complete the other samples in the same way.

Calculation

Calculate the percentage by weight of NaClO or H$_2$O$_2$ in the solution and the relative standard deviation. (**Note:** Commercial bleach should contain at least 5.25% NaClO. If less than this is present, it cannot be called "bleach.")

EXPERIMENT 16 IODOMETRIC DETERMINATION OF COPPER

Principle

A copper metal sample is dissolved in nitric acid to produce Cu(II), and the oxides of nitrogen are removed by adding H$_2$SO$_4$ and boiling to SO$_3$ fumes. The solution is neutralized with NH$_3$ and then slightly acidified with H$_3$PO$_4$. [The H$_3$PO$_4$ also complexes any iron(III) that might be present and prevents its reaction with I$^-$.] Finally, the solution is treated with excess KI to produce Cu̲I̲ and an equivalent amount of I$_3^-$, which is titrated with standard Na$_2$S$_2$O$_3$ solution, using a starch indicator. KSCN is added near the end point to displace absorbed I$_2$ on the CuI by forming a layer of CuSCN. For best accuracy, the Na$_2$S$_2$O$_3$ is standardized against

---

[1] This experiment should be completed in a single laboratory period by all students, including calculation and reporting of results. The hypochlorite and peroxide solutions are subject to decomposition, with a resultant change in concentration. The instructor may take the average of the class results as the correct value, or he or she may perform an analysis alone for comparison.

high-purity copper wire since some error occurs from reduction of copper(II) by thiocyanate.

Equations

$$2Cu^{2+} + 5I^- \rightarrow 2CuI + I_3^-$$
$$I_3^- + 2S_2O_3^{2-} \rightarrow 3I^- + S_4O_6^{2-}$$

Solutions and Chemicals Required

1. *Provided.* 6 *M* HNO$_3$, conc. H$_2$SO$_4$, 3 *M* H$_2$SO$_4$, conc. H$_3$PO$_4$, 6 *M* NH$_3$, KSCN.

2. *To prepare.*

   (a) Starch solution. Prepare the day of the experiment as described in Experiment 15, or use Thiodene indicator.

   (b) 0.1 *M* Na$_2$S$_2$O$_3$ solution. Prepare as described in Experiment 15.

Things to Do before the Experiment

Prepare the 0.1 *M* Na$_2$S$_2$O$_3$ solution. Although this can be prepared on the day of the experiment, it is preferable to prepare it at least a day before it is standardized. The solution tends to lose some of its titer right after preparing.

Procedure

1. *Standardization of the 0.1 M Na$_2$S$_2$O$_3$ solution.* The same procedure is used as will be used for analyzing the sample. Weigh out three 0.20- to 0.25-g samples of pure electrolytic copper foil and add to 250-mL Erlenmeyer flasks. In a hood, dissolve in 10 mL 6 *M* HNO$_3$, heating on a steam bath if necessary. Do in a fume hood. Add 10 mL conc. H$_2$SO$_4$ and evaporate to copious white SO$_3$ fumes. Cool and add carefully 20 mL water. Boil 1 to 2 min and cool. Add 6 *M* NH$_3$ dropwise with swirling of the sample solution until the first dark blue of the Cu(NH$_3$)$_4^{2+}$ complex appears. Then, add 3 *M* H$_2$SO$_4$ until the blue color just disappears, followed by 2.0 mL conc. H$_3$PO$_4$. Cool to room temperature.

   From this point, each sample must be treated separately. Dissolve about 2 g KI in 10 mL water and add to one of the flasks. Titrate *immediately* with the Na$_2$S$_2$O$_3$ solution until the yellow color of I$_3^-$ *almost* disappears. Add 2 to 3 mL of the starch solution or approximately 0.4 g Thiodene indicator, and titrate until the blue color begins to fade (should be less than 0.5 mL). Finally, add about 2 g KSCN and continue the titration until the blue color just disappears.

   Repeat with the other two samples. Calculate the molarity of the Na$_2$S$_2$O$_3$.

2. *Determination of copper in an unknown.* Add 10 mL of 6 *M* HNO$_3$ to each of three clean 250-mL Erlenmeyer flasks and take these to your instructor, who will add an unknown sample to each. (This unknown may be copper foil as used in the standardization.) Dissolve each and titrate as described for the standardization of Na$_2$S$_2$O$_3$. **Note:** In place of H$_3$PO$_4$, approximately 2 g ammonium bifluoride, NH$_4$HF$_2$ or NH$_4$F · HF, may be added to complex any iron and at the same time adjust the solution to the proper acidity. This experiment is also suitable for determining copper in about 0.3 g brass.

Calculation

Calculate the grams copper in each unknown sample and report the values of each and the relative standard deviation. (If a weighed brass sample is analyzed, report the percent copper.)

EXPERIMENT 17   DETERMINATION OF ANTIMONY BY TITRATION WITH IODINE

Principle

Antimony(III) is titrated to antimony(V) in neutral or slightly alkaline solution with iodine to a blue starch end point. The iodine is standardized against primary standard arsenic(III) oxide. Tartaric acid is added to complex the antimony and prevent its hydrolysis to form insoluble basic salts such as SbOCl and SbO_2Cl (which form in slightly acid and neutral solution).

Equations

Standardization:

$$H_2AsO_3^- + I_3^- + H_2O \rightarrow HAsO_4^{2-} + 3I^- + 3H^+ \text{ (pH 8)}$$

Sample titration:

$$SbOC_4H_4O_6^- + I_3^- + H_2O \rightarrow SbO_2H_4O_6^- + 3I^- + 2H^+$$

Solutions and Chemicals Required

1. *Provided.*   Primary standard As_2O_3, 1 *M* NaOH, Na_2CO_3, $NaHCO_3$, KI, tartaric acid, 1 *M* HCl. For stibnite ore: KCl, 0.1% (wt/vol) methyl red indicator in 60% ethanol, conc. HCl, 6 *M* HCl, 6 *M* NaOH.

2. *To prepare*

    (a) **Starch solution.** Prepare as described in Experiment 15, or use Thiodene indicator.

    (b) **0.05 *M* iodine solution.** Weigh 6.5 g iodine crystals and 20 g potassium iodide. Grind the iodine in a mortar with repeated small portions of the weighed KI crystals and water, pouring off the solution frequently into a glass-stoppered bottle until the solids are completely dissolved. [I_2 is only slightly soluble in water, but forms soluble KI_3 (I_3^- complex) in the presence of excess KI.]

    Avoid pouring undissolved iodine into the bottle. Dilute the solution to about 500 mL and mix thoroughly. Check for any undissolved iodine. Preferably let stand overnight before standardizing to ensure complete dissolution of the iodine. Alternatively, before diluting the solution, add more KI until all iodine is dissolved.

Things to Do before the Experiment

1. *Obtain and dry your unknown.*   Obtain a sample in a weighing bottle from your instructor and dry at 110 to 120°C for at least 1 to 2 h. Cool for at least 30 to 40 min before weighing.

2. *Dry the As_2O_3.*   Obtain and dry about 1 g primary standard As_2O_3 at 110 to 120°C for 1 to 2 h. Cool in a desiccator at least 30 to 40 min before weighing.

3. *Prepare the 0.05 M I_2 solution.*   If your unknown is a stibnite ore that will require some time to dissolve (as opposed to a water-soluble synthetic sample), it is advisable to also standardize the iodine solution before the day of the experiment to allow sufficient time to complete the experiment (procedure below).

Procedure

1. *Standardization of iodine solution.*   Weigh directly and accurately three 0.15- to 0.20-g portions of dried primary standard As_2O_3. Transfer to

250-mL Erlenmeyer flasks and dissolve in 10 to 20 mL 1 M NaOH, heating if necessary to aid dissolution. No undissolved particles should remain. Cool and add 1 M HCl until the solution is just acidic to litmus paper. Add 3 to 4 g solid $NaHCO_3$. No further CO_2 evolution should occur on addition of the last portion of $NaHCO_3$. If it does, add more $NaHCO_3$. The pH of the solution should be 7 to 8. Wash the walls of the flask down and add 50 mL water and 3 mL starch solution or about 0.4 g Thiodene indicator. Titrate with the iodine solution to the appearance of the first tinge of blue that persists for at least 30 s. From the three titrations, calculate the molarity of the I_2 solution. Use the average of the results.

2. *Determination of antimony in unknown*

   (a) **Water-soluble synthetic sample.** Consult your instructor for the proper size sample to weigh so that it will contain about 2 mmol antimony. Weigh three portions of the dried unknown and dissolve in 50 mL water in 500-mL Erlenmeyer flasks. Dissolve 4 g $NaHCO_3$ and 2 g tartaric acid in 100 mL water and add this solution to the antimony solution. The solution should be clear at this point with no hydrolyzed antimony chloride. Add 3 mL starch solution and titrate to a blue color that persists at least 30 s.

   (b) **Insoluble stibnite ore.** Consult your instructor for the proper size sample to weigh so that it will contain about 2 mmol antimony. Into the dry 250-mL beakers weigh accurately triplicate samples of the dried ore. Add about 0.3 g finely powdered potassium chloride, nearly cover the beaker with a watch glass, and carefully add 10 mL conc. hydrochloric acid by pouring it down the side of the beaker. (A high concentration of chloride is necessary to prevent hydrolysis during the dissolution before tartaric acid is added. $SbCl_3$ is formed.) Warm (do not boil) in the hood until the ore is decomposed; the mixture should no longer give an odor of hydrogen sulfide, and any residue (silica) should be white or only slightly gray. A stibnite ore consists of antimony sulfide, Sb_2S_3, silica, and small amounts of other substances. When all the antimony is dissolved, no more hydrogen sulfide should be evolved. Do not allow the solution to evaporate to dryness, which might result in loss of antimony trichloride; add more HCl as necessary. When decomposition is complete, add 3 g finely powdered tartaric acid and continue the heating for 10 to 15 minutes. Add water in portions of about 5 mL with good stirring until the solution is diluted to about 100 mL. (The solution must be diluted slowly, since some of the antimony may be hydrolyzed by local excesses of water.) If, during dilution, a red-orange precipitate (Sb_2S_3) appears, heat gently until the precipitate has dissolved before continuing the dilution. If a white precipitate of basic salts forms, the determination should be discarded. When the dilution is complete, boil the solution for 1 min.

   Rinse off the watch glass into the solution, and carefully neutralize the solution with 6 M sodium hydroxide by using a few drops of methyl red indicator. Then, add 6 M HCl dropwise until the solution is just acidic, carefully avoiding an excess.

   In 600-mL beakers or 500-mL Erlenmeyer flasks, prepare solutions containing 4 g sodium bicarbonate in 200 mL water. Pour the sample into the sodium bicarbonate solution, avoid loss by effervescence, and rinse several times with a stream of water from the wash bottle to obtain a complete transfer of the solution. Add 3 mL starch indicator or about 0.4 g Thiodene indicator and titrate with standard iodine solution to the appearance of the first permanent blue color. A fading or indistinct end point is due to insufficient sodium bicarbonate in the solution to consume the acid produced in the titration. Add 1 g additional $NaHCO_3$ and complete the titration to a permanent blue color.

Calculation

Calculate and report the percent Sb_2O_3 for each portion of your sample analyzed. Report also the mean of your values and the precision.

EXPERIMENT 18 MICROSCALE QUANTITATIVE ANALYSIS OF HARD-WATER SAMPLES USING AN INDIRECT POTASSIUM PERMANGANATE REDOX TITRATION[1]

Principle

The calcium in an unknown hard-water sample is precipitated as calcium oxalate in ammonia solution, and the precipitate is quantitatively filtered and washed, and is then dissolved in dilute sulfuric acid. The oxalic acid is titrated with standardized potassium permanganate solution.

Note: This experiment may serve as a template for microscale versions of some other experiments in the text. Statistical comparison of the microscale experiment with a similar conventional macroscale experiment using a 50-mL buret shows comparable precision.[1] But there is a slight negative determinate error in the microscale experiment, averaging about 50 ppm for ca. 500-ppm samples, probably due to physical loss of some of the precipitate, using a large funnel. Richardson suggests using a microfunnel to minimize this problem. For additional information on microscale titrations, see M. M. Singh, C. B. McGown, Z. Szafran, and R. M. Pike, *J. Chem. Educ.*, **77** (2000) 625.

Equations

$$Ca^{2+} + C_2O_4^{2-} \rightarrow CaC_2O_4$$

$$CaC_2O_4 + 2H^+ \rightarrow Ca^{2+} + H_2C_2O_4$$

$$5H_2C_2O_4 + 2MnO_4^- + 6H^+ \rightarrow 10CO_2 + 2Mn^{2+} + 8H_2O$$

Solutions and Chemicals Required

1. *Provided.* Conc. H_2SO_4, conc. HCl, conc. HNO_3, primary standard $Na_2C_2O_4$ (dried at 120°C for 1 h), 0.35 *M* $NH_4C_2O_4$, 0.10 *M* $AgNO_3$, dil. (1:10) NH_3 solution, methyl red (0.02% in 60% ethanol—dissolve first in the ethanol portion). Provide concentrated acids and $AgNO_3$ in dropper bottles.

2. *Approx. 0.2 M KMnO₄.* One liter is enough for 30 students. Prepare as follows (see Section 14.6). The preparation may be scaled up for more students. Calculate the weight of potassium permanganate required to make 1 L of 0.02 *M* solution. Weigh out in a weighing dish about 0.05 g more than this amount. Transfer to two 600-mL beakers, placing about half of the permanganate in each, add 500 mL of distilled water to each, cover with watch glasses, heat to boiling, and boil *gently* for 1 to 2 min, *not longer*. (Longer boiling, 0.5 to 1 h, is desirable but will evaporate the solution and change its

[1]Courtesy of Professor J. N. Richardson, Shippensburg University. Details of statistical analysis of the experiment in *J. Chem. Educ.*, in press (2002).

concentration. The solution may be heated for a longer period at a temperature just below its boiling point.) Allow to stand for at least 24 h before proceeding with the next step. Keep covered with watch glasses at all times to exclude dust and vapors, and to retard evaporation.

Mount a sintered-glass filter in a filter flask, and filter the permanganate solution through the crucible. Do not stir or swirl the solution; a sediment will have settled to the bottom that would clog the filter and make filtration very slow. For this reason, discard the last few milliliters of the solution from the first beaker. Use a filter trap; if any tap water backs up from the suction pump, the permanganate will be contaminated. Pour the filtered solution into a clean brown glass bottle with glass stopper and shake until homogeneous. The solution must never come into contact with organic material, including corks and rubber stoppers.

3. *Unknown hard-water solution.* Prepared by dissolving ca. 20 g of dried $CaCO_3$ in a minimum volume of 1 *M* HCl. To this solution is added a few drops of methyl red indicator, followed by dropwise addition of 1 *M* NaOH until the methyl red end point is noted (red to yellow). The resulting solution is then quantitatively transferred to a 2.000-L volumetric flask and diluted to the mark followed by thorough mixing. Student unknowns may then be dispensed from a buret into individual 100.00-mL volumetric flasks. Typical aliquots range from 10.00 to 20.00 mL, resulting in diluted unknowns with concentrations ranging from ca. 400 to 800 ppm Ca^{2+}.

Microburet

Constructed using a 2.000-mL graduated pipet, 2-cm length of latex tubing, 10-mL plastic syringe barrel, and an automatic delivery pipet tip. A detailed illustration of the microburet, as well as documentation regarding its construction, are provided in M. M. Singh, C. B. McGown, Z. Szafran, and R. M. Pike, *J. Chem. Educ.,* **75** (1998) 371.

Procedure

1. *Standardization of KMnO₄.* Each bench prepares 200 mL of dilute (1:20 v:v) sulfuric acid using concentrated H_2SO_4. (**Caution:** Acid should be added *slowly* to water with stirring.) Remove dissolved gases from this solution by boiling for 5 to 10 min with a glass stirring rod in the beaker to keep solution from bumping. Cool the solution to room temperature using an ice bath and store in a tightly capped, appropriately labeled polypropylene bottle. All students at a bench will share this solution (with typically three or four students occupying a lab bench). *Do not proceed past this point unless the standardization titration will be attempted the same day.*

To a 100.00-mL volumetric flask, each bench adds 0.5000 g of primary standard $Na_2C_2O_4$, dissolves it in the dilute H_2SO_4 prepared previously, and dilutes to the mark, mixing thoroughly. Again, this solution is shared among all members of a lab bench. From this point on, each student works individually. Rinse a 2-mL microburet with the primary standard solution and then fill it with the solution. Transfer 1.500 to 2.000 mL of the solution into a clean 30-mL beaker, recording the volume delivered exactly. Dilute to a total volume of about 5 mL using the dilute H_2SO_4. Prepare three more samples this way, using a different volume of $Na_2C_2O_4$ solution each time.

Clean a 2-mL microburet, rinse with and then fill with the 0.02 *M* KMnO₄ solution. Calculate the approximate volume of this titrant needed to reach an end point for the primary standard sample with the smallest volume. Add rapidly all but about 0.2 mL of this amount from the buret, with constant but gentle stirring with a stirring rod. Let this solution stand until the color disappears, which may take a minute or two. Heat the solution to 55 to 60°C, and complete the titration at this temperature (the temperature can be monitored by a thermometer, or one can watch for initial formation of steam). The remaining titrant is added dropwise, allowing each drop to react before adding the next. The end point is the first perceptible pink that persists for at least 30 s. Repeat this procedure for each of the remaining samples of $Na_2C_2O_4$. Calculate the molarity of the $KMnO_4$ for each titration, along with the average molarity and standard deviation.

2. *Analysis of an Unknown Hard-Water Sample.* Each student is assigned a labeled 100.00-mL volumetric flask containing an unknown hard-water sample. Add distilled water to the mark and invert the flask to mix thoroughly. Use a microburet to transfer an aliquot of hard water of about 1 mL into a clean 30 mL beaker. The volume transferred should be recorded to the precision of the buret. To this sample add 10 mL of distilled water, 7 drops of concentrated HCl, and a drop or two of methyl red indicating solution. Heat the resulting solution to nearly boiling, and add 1 mL of 0.35 M $(NH_4)_2C_2O_4$. (**Note:** students typically obtain this solution from a community macroburet or from a small graduated cylinder because they are only concerned with having an excess of oxalate ion in the solution.)

Prepare and fill a microburet with dilute (1 : 10 v : v) NH_3 solution. Add this solution dropwise to the unknown sample in the beaker, noting that the solution should become cloudy as the precipitate (CaC_2O_4) starts to form. Continue the addition of NH_4OH until the solution becomes alkaline to methyl red as indicated by a change from pink to pale yellow. Addition of excess NH_3 solution should be avoided. Cover the beaker with Parafilm, and allow the precipitate to digest overnight. If the pink returns, add more NH_3 solution to obtain the methyl red end point. Repeat this procedure for two more hard-water samples, using a slightly different aliquot size each time.

Gravity filter the precipitated CaC_2O_4 in each sample through 42.5-mm diameter No. 1 Whatman filter paper seated in a long-stem filter funnel.[2] Note that transfer of the precipitate must be quantitative. Wash the filtered precipitate with cold distilled water until the filtrate is clear upon addition of $HNO_3/AgNO_3$. This is done by collecting a few drops of filtrate in a test tube and adding a drop or two of the acid, followed by a drop or two of $AgNO_3$ solution.

Remove the filter paper from the funnel and place in a clean 30-mL beaker. Add about 3 mL of 1 : 10 (v : v) sulfuric acid (previously boiled as in step 1). Then stir the mixture until the solid CaC_2O_4 is dissolved and the filter paper is torn apart. Add about 10 mL of distilled water to the mixture, and heat to just below the boiling point. Titrate the mixture with the standardized $KMnO_4$ solution as follows: (1) add 0.10 mL of titrant and allow the mixture to stand until the color fades, and (2) continue the titration as normal until an end point is reached. The temperature of the solution should be maintained above 55°C for the duration of the titration. Repeat this procedure for each of the other unknown samples.

---

[2]Improved recovery may be achieved by using a smaller funnel to fit the paper.

Calculation

Using the data obtained, along with a balanced chemical reaction equation, calculate the concentration of calcium in parts per million for each trial. Report the mean and standard deviation.

Potentiometric Titrations

EXPERIMENT 19   pH TITRATION OF UNKNOWN SODA ASH

Principle

The unknown soda ash is titrated with standard HCl using a potentiometric (pH) end point measured with a pH meter using a pH glass electrode–saturated calomel reference electrode combination. The end-point breaks are compared with indicator color changes.

Equations

$$CO_3^{2-} + H^+ \rightarrow HCO_3^-   \text{(phenolphthalein end point)}$$

$$HCO_3^- + H^+ \rightarrow H_2O + CO_2   \text{(methyl purple end point)}$$

Note that between the first and second end points, a gradual decrease in pH due to the HCO_3^-/CO_2 buffer system will occur. This will give a poor visual end point, unless the buffer couple is destroyed. In practice, the visual titration used for standardization is continued until the methyl purple end point is reached, at which time the solution is gently boiled to remove the CO_2, leaving only the remaining HCO_3^-, which is then titrated to completion (see Chapter 8 for a more detailed discussion).

Solutions and Chemicals Required

   1. *Provided.*   0.2% phenolphthalein in 95% ethanol, 0.1% methyl purple in water, primary standard Na_2CO_3, standard pH 7 buffer.

   2. *To prepare.*   Standard 0.1 M HCl solution. Use the solution prepared in Experiment 7. If this solution is not available, prepare and standardize 500 mL as described in Experiment 7. Alternatively, the acid may be standardized against the primary standard Na_2CO_3 by pH titration as described below for the unknown soda ash.

Things to Do before the Experiment

Prepare and standardize the HCl solution. This will require prior drying of primary standard Na_2CO_3.

Obtain the unknown soda ash from your instructor and dry for at least 2 h at 160°C. Cool at least 30 min in a desiccator before weighing.

Procedure

The glass electrode to be used for pH measurements should have been soaked and stored in 0.1 M KCl for at least one day prior to its use. Always store the electrode in KCl solution when not in use. Calibrate the pH meter as described by your instructor, using the pH 7 standard buffer. This will consist essentially of adjusting the meter to read pH 7.00 with the electrodes immersed in the buffer solution. If only small quantities of buffer are used, it would be better to discard it rather than to chance contamination of the entire supply.

1. *Trial titration.* The purpose of this titration is to locate quickly and approximately the two end points. Weigh accurately by difference a dried sample of unknown soda ash (0.2 to 0.3 g) and add it to a 400-mL beaker containing a magnetic stirring bar. Add approximately 50 mL water and a few drops phenolphthalein indicator. The indicators are for the purpose of making a comparison between the potentiometric end points and the indicator color changes. Place the beaker on a magnetic stirrer, immerse the electrodes, and start the stirrer, being careful not to touch the electrodes to the stirring bar. Titrate with standard HCl, taking readings about every 2 mL. After the phenolphthalein color disappears, add a few drops methyl purple indicator and titrate at 2-mL increments until the second end point is reached. Add a few increments beyond the end point. The correct color for the second end point can be determined by comparison with the color of a few drops of the indicator in a solution of 0.20 g potassium acid phthalate in 100 mL water. Prepare a spreadsheet to plot your titration curve of pH (ordinate) versus volume of HCl (abscissa), and also the first derivative plot, and locate the approximate end points. See Chapter 14 and your CD for the preparation of a spreadsheet for derivative titrations.

2. *Final titration.* Weigh accurately another sample of the unknown and titrate as before, but make pH readings every 5 mL to within 3 mL of each end point (both sides of end point). Then, make readings of 0.50- to 1-mL intervals within 1 mL of the end point. Near the end point, *take readings as quickly as possible because the pH will tend to drift as CO_2 escapes from the solution.* Note and record the points at which the indicators change color. Use the spreadsheet you prepared above to plot your titration curve and its first derivative. Print out the titration curve and indicate on this curve the range in which the indicators change color. Determine the end point from the second inflection point of the curve. Repeat the titration on two more portions of the unknown. Be sure to rinse the electrodes between titrations.

Calculations

Calculate and report the percent Na_2CO_3 and Na_2O in your unknown for each portion analyzed. Hand in the plots of the titration curves with your report. Report also the mean percent value and the precision.

EXPERIMENT 20 POTENTIOMETRIC TITRATION OF A MIXTURE OF CHLORIDE AND IODIDE

Principle

The mixture is titrated with a standard solution of silver nitrate, and the potentiometric end points are indicated with a silver wire electrode–glass electrode pair using a pH meter for potential measurements. Because the pH during the titration remains essentially constant, the glass electrode's potential remains constant, and this electrode serves as the reference electrode; this eliminates the necessity of preparing a chloride-free salt bridge for the reference electrode. AgI ($K_{sp} = 1 \times 10^{-16}$) precipitates first since it is less soluble than AgCl ($K_{sp} = 1 \times 10^{-10}$). The AgCl starts precipitating near the equivalence point of the iodide titration (when $[Ag^+][Cl^-] = 1 \times 10^{-10}$; $[Ag^+]$ at the iodide equivalence point is $\sqrt{1 \times 10^{-16}} = 1 \times 10^{-8} M$). The potential (i.e., pX) rise of the iodide titration curve will level off at the point when the chloride starts precipitating, that is, near the iodide equivalence point inflection. This will be followed by the typical S-shaped chloride potentiometric end point. The error in determining the iodide end point is small if it is taken at the point at which the potential levels off. (It should be noted that while mixtures of chloride and iodide can be titrated, mixtures of bromide with either chloride or iodide cannot generally be titrated because of mixed crystal formation—see isomorphous replacement in Chapter 10.)

Equations

$$I^- + Ag^+ \rightarrow \underline{AgI}$$
$$Cl^- + Ag^+ \rightarrow \underline{AgCl}$$

Solutions and Chemicals Required

0.1 M standard $AgNO_3$. Prepare as described in Experiment 11.

Things to Do before the Experiment

Dry the primary standard $AgNO_3$ for 1 to 2 h at 110 to 120°C (*no longer*). Store in a desiccator until ready for weighing.

Obtain and dry your unknown at 120°C for 1 to 2 h. Store in desiccator until ready for weighing.

Procedure

Weigh directly three 0.5- to 0.6-g samples of the dried unknown into 400-mL beakers. Dissolve in 150 mL distilled water, add a magnetic stirring bar, and place the beaker on a magnetic stirrer. (Dissolve and titrate only one portion at a time to minimize air oxidation of the iodide.) Immerse the electrodes in the solution, taking care that they do not hit the magnetic stirrer. Connect the silver wire electrode to the reference terminal of the pH meter and the glass electrode to its usual terminal.[1] Stir the solution and titrate the sample with the standard $AgNO_3$. Take "pH" readings (actually pX), or millivolt readings, at 2-mL increments until the change is greater than 0.4 pH unit or 25 mV. Then add 0.2-mL increments. After the first end point is reached, add 2-mL increments until the second end point is approached and then 0.2-mL increments. Using a spreadsheet, plot the potential versus volume of $AgNO_3$ and determine the end point for the iodide and the chloride (inflection point of second potential break). Use these values to estimate the end point for the other two samples and repeat the above procedure for these samples. Titrant may be added rapidly up to within 2 or 3 mL of the end point. Be sure to rinse the electrodes between titrations.

Calculations

Calculate and report the percent iodide (from the volume required to reach the first end point) and chloride (from the volume required to go from the first end point to the second end point) in your unknown for each portion analyzed. Hand in the plots of the titration curves with your report.

Report also the mean values and the precision.

---

[1]The silver electrode is actually the indicating electrode and the glass electrode is the reference electrode. But most glass pH electrodes require a special plug and will not fit the reference terminal. The above arrangement is satisfactory and simply means the potential will be of the opposite sign from usual and change in the opposite direction.

Spectrochemical Measurements

EXPERIMENT 21   SPECTROPHOTOMETRIC DETERMINATION OF IRON

Principle

A complex of iron(II) is formed with 1,10-phenanthroline, $Fe(C_{12}H_8N_2)_3^{2+}$, and the absorbance of this colored solution is measured with a spectrophotometer. The spectrum is plotted to determine the absorption maximum. Hydroxylamine (as the hydrochloride salt to increase solubility) is added to reduce any Fe^{3+} to Fe^{2+} and to maintain it in that state.

Equations

$$4Fe^{3+} + 2NH_2OH \rightarrow 4Fe^{2+} + N_2O + 4H^+ + H_2O$$

| 1,10-phenanthroline | tris(1,10-phenanthroline) iron(II) |

Solutions and Chemicals Required

1. *Standard iron(II) solution.*   Prepare a standard iron solution by weighing 0.0176 g ferrous ammonium sulfate, $Fe(NH_4)_2(SO_4)_2 \cdot 6H_2O$. Quantitatively transfer the weighed sample to a 250-mL volumetric flask and add sufficient water to dissolve the salt. Add 0.7 mL conc. sulfuric acid, dilute exactly to the mark with distilled water, and mix thoroughly. This solution contains 10.0 mg iron per liter (10 ppm); if the amount weighed is other than specified above, calculate the concentration.

2. *1,10-Phenanthroline solution.*   Dissolve 25 mg 1,10-phenanthroline monohydrate in 25 mL water. Store in a plastic bottle.

3. *Hydroxylammonium chloride solution.*   Dissolve 10 g hydroxylammonium chloride in 100 mL water.

4. *Sodium acetate solution.*   Dissolve 10 g sodium acetate in 100 mL water.

Procedure

Into a series of 100-mL volumetric flasks, add with pipets 1.00, 2.00, 5.00, 10.00, and 25.00 mL of the standard iron solution. Into another 100-mL volumetric flask, place 50 mL distilled water for a blank. The unknown sample will be furnished in another 100-mL volumetric flask. To each of the flasks (including the unknown) add 1.0 mL of the hydroxylammonium chloride solution and 5.0 mL of the 1,10-phenanthroline solution. Buffer each solution by the addition of 8.0 mL of the sodium acetate solution to produce the red color of ferrous 1,10-phenanthroline. [The iron(II)–phenanthroline complex forms at pH 2 to 9. The sodium acetate neutralizes the acid present and adjusts the pH to a value at which the complex forms.] Allow at least 15 min after adding the reagents before making absorbance measurements so that the color of the complex can fully develop. Once developed, the color is stable for hours. Dilute each solution to exactly 100 mL. The standards will correspond to 0.1, 0.2, 0.5, 1, and 2.5 ppm iron, respectively.

Obtain the absorption spectrum of the 2.5-ppm solution by measuring the absorbance from about 400 to 700 nm (or the range of your instrument). Take readings at 25-nm intervals except near the vicinity of the absorption maximum, where you should take readings at 5- or 10-nm intervals. Follow your instructor's directions for the operation of your spectrophotometer. The blank solution should be used as the reference solution. Plot the absorbance against the wavelength and select the wavelength of the absorption maximum. This may be done with a spreadsheet. From the molar concentration of the iron solution and the cell pathlength, calculate the molar absorptivity of the iron(II)–phenanthroline complex at the absorption maximum.

Prepare a calibration curve by measuring the absorbance of each of the standard solutions of the wavelength of maximum absorbance. Measure the unknown in the same way. Using a spreadsheet, prepare a calibration curve by plotting the absorbance of the standards against concentration in ppm. From this plot and the unknown's absorbance, determine the final concentration of iron in your unknown solution. Perform the calculations of the measured concentration by entering the formula in a cell of the spreadsheet using the slope and intercept values and the measured absorbance. (See Chapters 3 and 16 for preparing the spreadsheet.) Report the number of micrograms of iron in your unknown along with the molar absorptivity and the spectrum of the iron(II)–phenanthroline complex.

EXPERIMENT 22   DETERMINATION OF NITRATE NITROGEN IN WATER[1]

Principle

Nitrate, NO_3^-, is reacted with phenoldisulfonic acid to give a yellow color with an absorption maximum at 410 nm. Chloride interference is removed by precipitating the chloride. Nitrite, NO_2^-, levels in excess of 0.2 mg/L cause positive interference, but these concentrations rarely occur in surface waters.

Solutions and Chemicals Required

1. *Provided.*   1 *M* NaOH solution; conc. NH_3, phenoldisulfonic acid reagent prepared by dissolving 25 g phenol in 150 mL conc. H_2SO_4, adding 75 mL fuming H_2SO_4 (15% free SO_3), stirring well, and heating for 2 h on a hot-water bath.

2. *To prepare*

   (a) **Silver sulfate solution.** (This will not be required if synthetic chloride-free unknowns are provided.) Dissolve 0.44 g Ag_2SO_4, free from nitrate, in 100 mL distilled water. One milliliter is equivalent to 1 mg chloride.

   (b) **EDTA solution.** Prepare a paste of 50 g $Na_2EDTA \cdot 2H_2O$ in 20 mL water, add 60 mL conc. NH_3, and mix well to dissolve the paste.

   (c) **Stock nitrate solution, 100 mg/L N.** Dissolve 0.722 g anhydrous KNO_3 and dilute to 1 L.

   (d) **Standard nitrate solution, 10 μg/mL N (44 μg/mL NO_3^-).** Dilute 50 mL of the stock solution to 500 mL with distilled water.

---

[1]Nitrate may also be determined fluorometrically by reacting with fluorescein in conc. H_2SO_4 and measuring the fluorescence quenching of the fluorescein. See *J. Chem. Ed.*, **51** (1974) 682.

Procedure

1. *Removal of chloride interference.* (This step may be eliminated if synthetic chloride-free nitrate unknowns are prepared in distilled water.) Small amounts of chloride cause negative interferences. If the chloride content is above 10 mg/L, the chloride should be removed. (Your instructor will provide an estimate of the chloride concentration.) Treat a 100-mL sample with an equivalent amount of silver sulfate solution and remove the precipitated silver chloride by centrifugation or filtration. If necessary, coagulate the precipitate by heating the solution or by allowing to stand overnight away from strong light (only if sample is free from nitrifying organisms—see below).

2. *Determination of nitrate.* The sample should not exhibit appreciable color. To prevent any change in the nitrogen balance through biological activity, natural waters should be analyzed promptly after sampling. They may, however, be stored near freezing by adding 0.8 mL H_2SO_4/L as preservative. If the sample is acidified, it should be neutralized just before the analysis is started.

   Neutralize the chloride-free prepared sample from above or a 100-mL fresh sample if already chloride free to about pH 7 with dilute NaOH. Transfer to a casserole and evaporate to dryness. Mix the residue with 2.0 mL phenoldisulfonic acid reagent, using a glass rod to help dissolve the solids; heat on a hot-water bath if necessary to aid dissolution. Dilute with 20 mL distilled water and then add 6 to 7 mL ammonia until maximum color is developed. If a flocculent hydroxide forms, dissolve it by adding the EDTA reagent dropwise with stirring. (Alternatively, the sample may be filtered.) Transfer the clear solution to a 50-mL volumetric flask and dilute to volume with distilled water.

   Prepare standards in the same manner, using the same volumes of reagents, by evaporating 10, 25, and 50 mL of the standard nitrate solution, respectively; these represent 0.10, 0.25, and 0.50 mg N, respectively. Omit the chloride precipitation step. Prepare a blank using the same volumes of reagents.

   Read the absorbance of the solution at 410 nm, correct for the blank, prepare a calibration curve, and calculate the concentration of nitrate nitrogen in your sample in mg/L.

   As little as 1 μg nitrate nitrogen can be detected, representing 0.01 mg/L in a 100-mL sample. The nitrate concentration in drinking water usually falls below 10 mg/L. If concentrations are high, measurements can be extended sixfold by measuring at 480 nm, or twofold by diluting prepared samples to 100 mL instead of 50 mL.

EXPERIMENT 23 SPECTROPHOTOMETRIC DETERMINATION OF LEAD ON LEAVES USING SOLVENT EXTRACTION[1]

Principle

Lead on the surfaces of leaves is dissolved by shaking with nitric acid solution. The lead is extracted as the dithizone complex into methylene chloride at pH above 9. The intensity of the color of the complex is measured spectrophotometrically and compared to a calibration curve prepared similarly from lead standards to calculate the amount of lead. Cyanide and sulfite may be added as masking agents to eliminate most interference from other metals.

---

[1]The instructions for this experiment include the use of a cyanide-containing solution, which serves as a masking agent for certain metal ions. *You will not use this solution unless your instructor directs you to.* For safety and illustrative purposes you can assume the measurements are due only to lead.

Equation

$$\tfrac{1}{2}Pb^{2+} + \begin{array}{c} \phi-NHNH \\ \\ \phi-N{=}N \end{array}\!\!\!\!>\!C{=}S \rightleftharpoons \begin{array}{c} \phi-NHN \\ \\ \phi-N{=}N \end{array}\!\!\!\!>\!\!\begin{array}{c} Pb^{2+}/2 \\ | \\ C{-}S \end{array} + H^+$$

green   red

Solutions and Chemicals Required

1. *Provided.* 1 M HNO$_3$, 0.1 M HNO$_3$, thymol blue indicator solution (0.1% in water), 2 M NH$_3$ solution, ammonia–cyanide–sulfite solution (350 mL conc. NH$_3$ solution, 30 mL 10% NaCN, and 1.5 g Na$_2$SO$_3$ diluted to 1 L; the pH is about 11).

2. *To prepare*

   (a) **Stock 1000-ppm standard lead solution.** Dissolve 0.160 g Pb(NO$_3$)$_2$ and dilute to 100 mL in a volumetric flask.

   (b) **Standard 10-ppm working solution.** On the day of the experiment, dilute 1 mL of the stock solution to 100 mL in a volumetric flask.

   (c) **Dithizone solution.** Dissolve 7.5 mg dithizone in 300 mL methylene chloride. This should be prepared fresh on the day of the experiment.

Caution: *Do not discard* any of the solutions used in this experiment down the drain. The aqueous solutions may contain cyanide and should be added to a reservoir provided for collection that contains FeSO$_4$ [to convert CN$^-$ to Fe(CN)$_6^{4-}$]. *No acid must ever be added to this reservoir!* All glassware should be rinsed with an alkaline solution and the washings added to the reservoir. The methylene chloride solutions should also be added to a reservoir provided.

Things to Do before the Experiment

Collect leaf samples. These can be from trees near a road or highway and from some that are more isolated for comparison. Collect at least two large leaves from each tree and place each in a clean plastic bag and seal. The leaves selected should be reasonably free from dirt or other visible contamination. Record the location of the leaves. Your instructor will advise you of the number of trees you should sample.

Procedure

1. *Preparation of calibration curve.* This should be done at the time the samples are to be analyzed. The chelate formation and solvent extraction are to be performed in clean 6-oz vials with caps. To each of six labeled vials add with pipets 0 (blank), 2, 4, 6, or 8 mL of the 10-ppm lead standard and sufficient water to bring the volume to about 20 mL. Add about 60 mL of the ammonia–cyanide–sulfite solution (*only if instructed to*) using a graduated cylinder and 25 mL of the CH$_2$Cl$_2$–dithizone solution using a pipet (*not by mouth*). Stopper the vial and shake for about a minute. Using a pipet, withdraw most of the heavier methylene chloride layer and filter through dry filter paper (Whatman No. 40) into a dry Bausch and Lomb Spectronic 20 measuring tube or the equivalent, or centrifuge before transferring. (The samples should be prepared for measurement now so they can be measured when the standards are.)

   Using one of the standards, measure the absorbance from 400 to 600 nm in 20-nm increments to determine the wavelength of maximum absorption. Using this wavelength, measure the absorbance of each standard, using the blank to zero the instrument. Plot absorbance against micrograms of lead taken to prepare a calibration curve, using a spreadsheet (Chapters 3 and 16).

2. *Determination of lead on leaves.* For each plastic bag containing a leaf sample, heat 20 mL of 0.1 M HNO$_3$ to about 70°C. Add 20 mL to each bag, close, and shake for about 2 min. Pour into clean 100-ml- beakers. Add one drop thymol blue indicator solution to each, followed by dropwise addition of 2 M NH$_3$ until the indicator color change is complete (to blue) and add a couple of extra drops. *The solution should smell of ammonia.* Then, add (*only if instructed to*) 60 mL of the ammonia–cyanide–sulfite solution—and then add with a pipet 25 mL of the CH$_2$Cl$_2$–dithizone solution and proceed with the extraction measurement as with the standards.

Calculation

Blot each leaf dry, place on a sheet of paper, and trace the outline of the leaf. Cut out the leaf outline and weigh on an analytical balance to three figures. Cut out a 10-cm × 10-cm (100 cm^2) square from the same paper and weigh. Calculate the area of the leaf in cm^2.

From the measured absorbance of each sample and the calibration curve, calculate the micrograms of lead on the leaf and report the amount of lead in μg Pb/100 cm^2 leaf. Is there any correlation of lead content with proximity of the tree to a roadway?

EXPERIMENT 24 SPECTROPHOTOMETRIC DETERMINATION OF INORGANIC PHOSPHORUS IN SERUM[1]

Principle

The inorganic phosphorus in a protein-free filtrate is reacted with ammonium molybdate [Mo(VI)] to form ammonium phosphomolybdate. This is reduced with a mild reducing agent to produce "molybdenum blue," a heteropoly molybdenum(V) species. Molybdates are not reduced under these conditions. The blue color of the solution is measured spectrophotometrically.

Equations

$$7H_3PO_4^{3-} + 12(NH_4)_6Mo_7O_{24} + 51H^+ \rightarrow 7(NH_4)_3PO_4 \cdot 12MoO_3 + 51NH_4^+ + 36H_2O$$

$$(NH_4)_3PO_4 \cdot 12MoO_4 + \text{mild reducing agent} \rightarrow Mo(V) \text{ species (blue)}$$

[Although normal ammonium molybdate $(NH_4)_2MoO_4$ can be crystallized, the common crystalline form is $(NH_4)_6Mo_7O_{24} \cdot 4H_2O$ or $3(NH_4)_2O \cdot 7MoO_3 \cdot 4H_2O$.]

Solutions and Chemicals Required

1. *Provided.* 5% (wt/vol) Trichloroacetic acid solution; 5 M H$_2$SO$_4$; aminonaphtholsulfonic acid reducing solution prepared as follows: Add 0.50 g 1,2,4-aminonaphtholsulfonic acid and 5.0 mL sodium sulfite solution (20 g anhydrous Na$_2$SO$_3$/100 mL) to 195 mL sodium bisulfite solution (15 g NaHSO$_3$/100 mL) in a brown glass-stoppered bottle. Stopper and shake until the powder is dissolved. If solution is not complete, add 1-mL increments of sodium sulfite solution with continued shaking until solution is complete. Avoid excess sodium sulfite. Store in refrigerator. The solution is stable for about 1 month.

---

[1]A synthetic serum sample may be prepared as described in Experiment 30, footnote 1, and adding 60 g of albumin as a source of protein (e.g., bovine serum albumin, BSA.) Serum contains about 6% (w/w) protein. The solution will contain about 0.41 mg P/dL, which can be varied from unknown to unknown.

2. *To prepare*

(a) **Stock phosphorus standard solution (100 mg/dL P).** Dissolve 0.439 g KH_2PO_4 in water and dilute to 100 mL in a volumetric flask.

(b) **Working phosphorus standards.** Add with a pipet 1 mL of the stock solution to a 100-mL volumetric flask and dilute to volume with 5% trichloroacetic acid (TCA). (**CAUTION: TCA is very corrosive. Avoid contact with the skin. It should never be pipetted by mouth.**) This contains 1 mg/dL phosphorus and will be used to prepare the serial standards. Transfer with a pipet 2 and 5 mL of this solution into 10-mL volumetric flasks and dilute to volume with 10% TCA. You now have standards of 0.2, 0.5, and 1 mg/dL P. These correspond to serum concentrations of 2, 5, and 10 mg/dL, respectively, in the procedure below, since the sample is diluted 1:10.

(c) **Ammonium molybdate solution.** Dissolve 0.62 g ammonium molybdate, $(NH_4)_6Mo_7O_{24} \cdot 4H_2O$, in 2.0 mL water and add 8 mL of 5 M H_2SO_4. The solution should be stable indefinitely. Discard if blanks show a blue color.

Procedure

1. *Serum.* Perform the analysis in duplicate. Place 9.50 mL of 5% TCA in a 12-mL centrifuge tube. Add 0.500 mL serum, mix well, and let stand for 5 min. Centrifuge at 1500 rpm until the supernatant is clear (ca. 5 min). If a centrifuge is not available, the sample should be filtered through dry Whatman No. 42 filter paper into a dry beaker. Since the filter paper may contain reducing impurities, the blank should be prepared using filtered 5% TCA.

Transfer 5.00 mL of the clear supernatant to a 15 × 150 mm test tube. Prepare a blank and standards by pipetting 5.00 mL of 5% TCA and of the 0.2, 0.5, and 1.0 mg/dL P standard solutions into four separate test tubes. To all the test tubes, add 1.00 mL of the molybdate reagent and mix well. Finally, add 0.40 mL of the aminonaphtholsulfonic acid reagent and mix well. Allow to stand for 5 to 10 min or longer and measure the absorbance for each solution in a cuvet at 690 nm, setting the zero absorbance with distilled water. Subtract any blank reading from all standard and sample readings.

Plot the net absorbance of the standards against concentration using a spreadsheet (Chapters 3 and 16). From this plot and the net absorbance of the sample, determine the concentration of phosphorus in the protein-free filtrate. Multiply by 20 to obtain the concentration in the original serum sample. The normal range of phosphorus in serum is about 3.0 to 4.5 mg/dL for adults and 4.5 to 6.5 mg/dL for children.

EXPERIMENT 25 SPECTROPHOTOMETRIC DETERMINATION OF MANGANESE AND CHROMIUM IN MIXTURE

Principle

Manganese and chromium concentrations may be determined simultaneously by measurement of the absorbance of light at two wavelengths, after the metals have been oxidized to $Cr_2O_7^{2-}$ and MnO_4^-. Beer's law has been shown to apply closely if the solutions are at least 0.5 M in H_2SO_4. $Cr_2O_7^{2-}$ has an absorption maximum at 440 nm and MnO_4^- has one at 545 nm. (A somewhat more intense maximum is at 525 nm, but there is less interference from $Cr_2O_7^{2-}$ at 545 nm.) Equations similar to Equations 16.16 and 16.17 are solved for the unknown concentrations from the measured absorbances at the two wavelengths. The four constants

($\epsilon b = k$) are determined by measurements of absorbance at the two wavelengths using pure solutions of known concentration; a calibration curve is prepared at each wavelength for both $Cr_2O_7^{2-}$ and MnO_4^- and the slopes of the curves (A versus C) are used to obtain an average k value.

Equations

The unknown contains Cr^{3+} and Mn^{2+}. The former is oxidized to $Cr_2O_7^{2-}$ by heating with peroxydisulfate (persulfate) in the presence of a silver catalyst:

$$2Cr^{3+} + 3S_2O_8^{2-} + 7H_2O \xrightarrow[\Delta]{Ag^+} Cr_2O_7^{2-} + 6SO_4^{2-} + 14H^+$$

Mn^{2+} is oxidized in part by peroxydisulfate, but also by periodate:

$$2Mn^{2+} + 5S_2O_8^{2-} + 8H_2O \xrightarrow[\Delta]{Ag^+} 2MnO_4^- + 10SO_4^{2-} + 16H^+$$

$$2Mn^{2+} + 5IO_4^- + 3H_2O \longrightarrow 2MnO_4^- + 5IO_3^- + 6H^+$$

For the mixture,

$$A_{440} = k_{Cr,440}C_{Cr} + k_{Mn,440}C_{Mn}$$
$$A_{545} = k_{Cr,545}C_{Cr} + k_{Mn,545}C_{Mn}$$

The k values are determined from the slopes of the calibration curves of the pure solutions:

$$k_{Cr,440} = A_{440}/C_{Cr} \qquad k_{Cr,545} = A_{545}/C_{Cr}$$
$$k_{Mn,440} = A_{440}/C_{Mn} \qquad k_{Mn,545} = A_{545}/C_{Mn}$$

Solutions and Chemicals Required

1. *Provided.* 18 M H_2SO_4, $K_2S_2O_8$, KIO_4, $AgNO_3$.

2. *To prepare*

    (a) **Standard 0.002 M $MnSO_4$ solution.** Dry about 1 g $MnSO_4$ at 110°C for 1 h, cool for 30 min, and weigh out about 0.08 g (to the nearest tenth milligram). Transfer to a 250-mL volumetric flask, dissolve, and dilute to volume. Calculate the molarity of the solution and the concentration of Mn in mg/L (ppm).

    (b) **Standard 0.0178 M $K_2Cr_2O_7$ solution.** Use the solution prepared in Experiment 14 or prepare 100 mL as directed there (weigh to the nearest milligram). Calculate the molarity of the solution and the concentration of Cr in mg/L (remember there are two Cr atoms per molecule of $K_2Cr_2O_7$).

    (c) **0.1 M $AgNO_3$.** Dissolve about 0.2 g $AgNO_3$ in about 12 mL water.

Things to Do before the Experiment

Prepare the standard $MnSO_4$ and $K_2Cr_2O_7$ solutions. This will require drying $MnSO_4$ and $K_2Cr_2O_7$.

Procedure

1. *Calibration (determination of k values).* **Note:** The absorbance of the calibration solutions and the unknown should be read at the same time. Therefore, get all solutions prepared before making any readings. They are all

sufficiently stable that they could be allowed to set until another laboratory period but it is best not to.

(a) **Manganese.** Add with pipets aliquots of 10, 15, and 25 mL of the standard $MnSO_4$ solution into three different 250-mL Erlenmeyer flasks. Add distilled water to bring the volume in each flask to about 50 mL. To each flask add 10 mL conc. H_2SO_4 (CAREFULLY, using a graduated cylinder) and 0.5 g solid KIO_4 (potassium periodate or metaperiodate, depending on the manufacturer). Heat each to boiling for about 10 min, cool, transfer quantitatively to 250-mL volumetric flasks, and dilute to the mark with distilled water. Determine the absorbance of each solution at 440 and 545 nm, using 0.5 M H_2SO_4 as a blank solution. Permanganate solutions containing periodate are stable. The absorbance at 440 nm will be less than 0.1 and, hence, the spectrophotometric error (precision) will be large (see Figure 16.27). But this is acceptable, in fact, desirable, because the correction for manganese absorption at this wavelength is small; that is, a relatively large error in determining a small correction results in only a small error.

(b) **Chromium.** Add 10-, 15-, and 25-mL aliquots of the standard $K_2Cr_2O_7$ solution to 250-mL volumetric flasks, add about 100 mL distilled water and 10 mL conc. H_2SO_4, mix thoroughly, and dilute to 250 mL with distilled water. Determine the absorbance of each solution at 440 and 545 nm, using 0.5 M H_2SO_4 as the blank solution. The absorbance in this case will be small (<0.1) at 545 nm.

(c) **Determination of k values.** Using a spreadsheet, plot absorbance versus concentration in units of mg/L for each solution at each wavelength, and plot the least-squares straight line through each set of data points. The lines should intercept at zero absorbance and zero concentration; under Chart Options, you can instruct Excel to plot the intercept at zero. The slopes of these lines are the coefficients ($k = A/C$) to be used in determining the concentrations of chromium and manganese in the unknown. These slopes relate absorbance and concentration for the instrumental parameters used. Therefore, one should use the same instrument, cuvets, cuvet position, and volumes of solutions for all determinations in this experiment.

2. *Analysis of unknown.*   Obtain a mixture of Mn^{2+} and Cr^{3+} or $Cr_2O_7^{2-}$ unknown in a 250-mL volumetric flask and dilute to volume. Transfer with a pipet three 50-mL aliquots into 250-mL Erlenmeyer flasks. The procedure may be stopped up to this point. But once the peroxydisulfate is added, the oxidation should be completed.

To each flask add 5 mL conc. H_2SO_4 (BE CAREFUL!), then mix well. Add about 1 or 2 mL of 0.1 M $AgNO_3$ solution and 1.0 g solid potassium peroxydisulfate ($K_2S_2O_8$). BE CAREFUL! PEROXYDISULFATE IS A STRONG OXIDIZING AGENT THAT CAN REACT VIOLENTLY WITH REDUCING AGENTS! USE ONLY AS DIRECTED. Do not spill. Dissolve peroxydisulfate and heat the solution to boiling and boil gently for about 5 min. Cool the solution and then add 0.5 g KIO_4. Again heat to boiling for 5 min.

Cool each solution to room temperature, quantitatively transfer to 250-mL volumetric flasks, and dilute to volume. The solutions at this point (or before dilution) are stable and can be saved until another laboratory period if necessary. Also save the serial standards to calibrate the instrument at the same time.

From the unknown absorbances at the two wavelengths, calculate the parts per million of Cr and Mn in your unknown using Beer's law for the mixture and the determined constants. Use a spreadsheet similar to the one given in Chapter 16 and in your CD to perform the mixture calculation. The calculated results will have the same units as used in determining the constants. Keep in mind the dilutions made. Report the results for each portion analyzed.

EXPERIMENT 26 ULTRAVIOLET SPECTROPHOTOMETRIC DETERMINATION OF ASPIRIN, PHENACETIN, AND CAFFEINE IN APC TABLETS USING SOLVENT EXTRACTION

Principle

APC tablets are a mixture of aspirin, phenacetin, and caffeine. Each of these substances has characteristic absorption in the ultraviolet region, with the principal maxima lying at 277 nm for aspirin, 275 nm for caffeine, and 250 nm for phenacetin. In the procedure, a powdered tablet is dissolved in methylene chloride, and the aspirin is separated from the phenacetin and caffeine by extracting it into aqueous sodium bicarbonate solution. The separated aspirin is back-extracted into methylene chloride by acidifying the aqueous layer and is then measured spectrophotometrically at 277 nm. The phenacetin and caffeine that remain in the original methylene chloride layer are determined in mixture as described in Chapter 16 (Equations 16.16 and 16.17).

Equations

aspirin (A)
(acetylsalicylic
acid)

phenacetin (P)
(acetophenetidin)

caffeine (C)
(Theine; 1,3,7-
trimethylxanthine)

Solutions and Chemicals Required

1. *Provided.* CH_2Cl_2, 4% (wt/vol) $NaHCO_3$ solution (chilled), conc. HCl, 1 *M* H_2SO_4.

2. *To prepare*[1]
   Standard solutions. Prepare individual standard solutions of about 100 mg/L, 20 mg/L, and 10 mg/L each for aspirin, phenacetin, and caffeine in methylene chloride as follows. Weigh about 25 mg (to the nearest 0.1 mg) of each, transfer to 100-mL volumetric flasks, dissolve, and dilute to volume with methylene chloride. Dilute 2 and 1 mL of this solution to 25 mL in 25-mL volumetric flasks to prepare the 20 and 10 mg/L solutions, respectively.

---

[1]Caffeine and phenacetin are available from Sigma-Aldrich.

Procedure[2]

Weigh accurately and record the weight of one tablet. This should be equivalent to about 220 mg aspirin, 160 mg phenacetin, and 30 mg caffeine. To minimize required dilutions and save on solvents, cut the tablet into quarters and weigh out a one-quarter portion to be analyzed. Crush to a fine powder in a beaker. Add, with stirring, 20 mL methylene chloride; then transfer the mixture quantitatively to a 60-mL separatory funnel, rinsing *all* particles in with a little more methylene chloride. Extract the aspirin from the methylene chloride solution with two 10-mL portions of chilled 4% sodium bicarbonate to which has been added two drops hydrochloric acid, and then with one 5-mL portion of water. Wash the combined aqueous extracts with three 10-mL portions of methylene chloride and add the methylene chloride wash solutions to the original methylene chloride solution. Leave the aqueous extract in the separatory funnel. Filter the methylene chloride solution through paper previously wetted with methylene chloride (to remove traces of water) into a 50-mL volumetric flask and dilute to the mark with methylene chloride. Then dilute further a 1-mL aliquot of this solution to 50 mL with methylene chloride in a volumetric flask.

Acidify the bicarbonate solution (aqueous extract), still in the separatory funnel, with 6 mL of 1 *M* sulfuric acid. This step should be performed without delay, to avoid hydrolysis of the aspirin. The acid must be added slowly in small portions. Mix well only after most of the carbon dioxide evolution has ceased. The pH at this point should be 1 to 2 (pH test paper). Extract the acidified solution with eight separate 10-mL portions of methylene chloride and filter through a methylene chloride-wet paper into a 100-mL volumetric flask. Dilute to volume. Then, dilute further a 5-mL portion of this solution to 25 mL with methylene chloride in a volumetric flask.

Record absorbance versus wavelength curves for the standard solutions and unknown solutions between 200 to 300 nm. (This step may be deleted if you do not have a recording ultraviolet spectrophotometer.) Does the wavelength of 277 nm appear to be the most suitable wavelength for the determination of aspirin? Do the wavelengths of 250 and 275 nm appear to be the best wavelengths for the measurement of the absorbance for the mixture of phenacetin and caffeine? Explain.

Using the absorbances of the standard and the unknown aspirin solution at 277 nm, calculate the percent aspirin in the APC tablets and the number of milligrams of aspirin per tablet keeping in mind the dilutions.

To calculate the concentrations of phenacetin and caffeine, the absorbances of the phenacetin and caffeine standards and of the methylene chloride extract of the sample must all be read at both 250 and 275 nm. Using these absorbances, calculate the percent phenacetin and caffeine in the APC tablets and the milligrams of each per tablet. See Chapter 16 for the spectrophotometric determination of mixtures. Use a spreadsheet similar to the one in Chapter 16 and your CD to do the mixture calculations.

EXPERIMENT 27   INFRARED DETERMINATION OF A MIXTURE OF XYLENE ISOMERS

Principle

Meta- and *para*-xylene are determined in mixtures using *ortho*-xylene as an internal standard, to compensate for variation in cell length between runs. The infrared spectrum of the unknown mixture is recorded and the relative height of peaks of

---

[2]Aspirin tends to decompose in solution, and analyses should be performed as soon as possible after preparing solutions.

the two compounds are compared with those of standard mixtures, using the base-line technique.

Solutions and Chemicals Required

1. *Provided.* Ortho-, meta-, and *para*-xylene.

2. *To prepare.* Meta-xylene/*para*-xylene standards. Prepare a series of standards (use available burets), all containing 30% (vol/vol) o-xylene as internal standard, by mixing the appropriate volumes of the three isomers to give 25, 35, and 45 vol % of m-xylene. The corresponding concentrations of p-xylene will be 45, 35, and 25%, respectively.

Procedure

Consult your instructor on the proper operation of your instrument. Handle the infrared cell carefully, avoiding contact with water and the fingers. Fill the cell with pure m-xylene and obtain a spectrum on this from 2 to 15 μm, being sure to record the last peak just before 15 μm (692 cm^{-1}). Each time you run a sample, be sure to check 0% T by placing a card in the sample beam and adjust the pen to 0% T. Empty the cell, rinse and fill with p-xylene, and run a spectrum on this. Repeat for o-xylene. Run spectra on each of the standard mixtures. From the spectra of the pure substances, choose a peak of each isomer to measure. Using the baseline method (see Figure 16.11), measure P_0/P for the peak for each compound. Prepare a calibration curve of the ratio of $\log(P_0/P)_{meta}/\log(P_0/P)_{ortho}$ and of $\log(P_0/P)_{para}/\log(P_0/P)_{ortho}$ versus concentration for the meta and para isomers, respectively. See Chapter 20 and your CD for spreadsheet preparation using an internal standard.

Obtain an unknown mixture of meta and para isomers from your instructor. Prepare a mixture of this with o-xylene by adding 70 parts of the unknown to 30 parts o-xylene. Run the spectrum on this mixture and, using the baseline method and the same peaks as before, measure P_0/P for the three compounds and calculate $\log(P_0/P)/\log(P_0/P)_{ortho}$ for the two unknown isomers. Compare with the calibration curve to determine the percent concentrations of the meta and para isomers; use the spreadsheet for calculations. Remember to divide by 0.7 to convert to initial concentrations.

EXPERIMENT 28   FLUOROMETRIC DETERMINATION OF RIBOFLAVIN (VITAMIN B$_2$)

Principle

Riboflavin is strongly fluorescent in 5% acetic acid solution. The excitation and fluorescence spectra are obtained to determine the wavelengths of excitation and emission to use, and an unknown is determined by comparison to standards.

Chemicals and Solutions Required

Riboflavin standards. Prepare a 100-ppm riboflavin stock solution by accurately weighing about 50 mg riboflavin, transferring to a 500-mL volumetric flask, and diluting to volume with 5% (vol/vol) acetic acid. This should be stored in a cool, dark place. On the day of the experiment, dilute an aliquot of this 1:10 to obtain a 10-ppm working standard solution. Dilute aliquots of this with 5% acetic acid to prepare standards of 0 (blank), 0.2, 0.4, 0.6, 0.8, and 1.0 ppm riboflavin.

Procedure

Record the excitation and emission spectra of the 0.6-ppm solution to determine the best excitation wavelength and best detection wavelength. If a filter instrument rather than a recording spectrofluorometer is used, take readings with different arrangements of the filters to give the maximum reading. Using these wavelengths, set the instrument gain to give a reading of 100% with the 1-ppm solution. Read the fluorescence of the other standards and prepare a calibration curve. Obtain your unknown in a 50-mL volumetric flask and dilute to volume with 5% acetic acid. Read the fluorescence of this, and, from the calibration curve, calculate the micrograms riboflavin in your unknown. Use a spreadsheet.

Atomic Spectrometry Measurements

EXPERIMENT 29 DETERMINATION OF CALCIUM BY ATOMIC ABSORPTION SPECTROPHOTOMETRY

Principle
The effects of instrumental parameters and of phosphate and aluminum on calcium absorption are studied [see, e.g., W. Hoskins et al., *J. Chem. Ed.*, **54** (1977) 128]. Calcium in an unknown synthetic or serum sample is determined by comparing the absorbance with that of standards.

Solutions and Chemicals Required
500 ppm Ca, 4% $SrCl_2$, 2000 ppm NaCl, 100 ppm phosphate, ethanol.

1. *Provided.* Ethanol, 4%, $SrCl_2$ solution (wt/vol), stock solution of 140 meq/L Na and 4.1 meq/L K (see footnote below).

2. *To prepare*

    (a) **500 ppm Ca stock solution.** Dissolve 1.834 g (accurately weighed) $CaCl_2 \cdot 2H_2O$ in water and dilute to 1 L in a volumetric flask. Dilute this 1 : 10 to prepare a 50-ppm stock solution. Use this to prepare the solutions required below. (Commercial 1000-ppm Ca^{2+} solutions may be used.)

    (b) **2000-ppm Na stock solution.** Dissolve 0.51 g NaCl in 100 mL water.

    (c) **100 ppm phosphate.** Dissolve 0.15 g Na_2HPO_4 in 1 L water.

    (d) **100-ppm Al stock solution.** Dissolve 0.18 g $Al_2(SO_4)_3 \cdot K_2SO_4 \cdot 24H_2O$ in 100 mL water. ($AlCl_3$ may be used, but take care when adding water.)

Study of Instrumental Parameters
Follow your instructor's directions for the operation of the instrument. If you have a single-beam instrument, the hollow-cathode lamp should be allowed to warm up for 30 min before the experiment. A few minutes should be adequate with a double-beam instrument. An air–acetylene flame should be used with a premix burner.

1. *Burner height.* Adjust the fuel and support gas pressures until the flame is near stoichiometric (just a slight yellow color to the flame). Then, turn up the fuel pressure to impart a strong yellow glow to the flame (fuel rich). The yellow glow is due to unburned carbon particles in the rich flame. In a lean flame, an excess of oxidant is present and the flame appears blue. Prepare and aspirate a 5-ppm calcium solution and note its absorbance at 422.67 nm. Adjust the wavelength setting to obtain maximum absorbance. The monochromator is now set exactly at the calcium line. With the burner height adjusting knob, raise the burner so that the light beam just passes over the tip

of it (base of the flame). Use distilled water to zero the instrument, and then measure the absorbance of the 5-ppm calcium solution. Lower the burner in increments (six to eight steps) and record the absorbance at each height.

Plot the absorbance against heights of observation in the flame and select the optimum height.

2. *Fuel/air ratio.* Hold the air pressure constant and adjust the fuel pressure in increments from a very fuel-rich to a lean flame. Record the absorbance of 5 ppm Ca at each increment.

Select the optimum fuel pressure and vary the air pressure in a similar manner. Plot absorbance against gas pressure for both the fuel and the air, noting the pressure setting of the one held constant. Select the optimum fuel, and air settings. Is this a rich, stoichiometric, or lean flame?

Interference Studies

1. *Effect of phosphate.* Prepare a solution containing 5 ppm Ca and 10 ppm phosphate. Record the absorbance of this solution, using the optimum conditions determined above, and compare to that of 5 ppm Ca. Explain the results.

Prepare a solution containing 5 ppm Ca, 10 ppm phosphate, and 1% $SrCl_2$. Prepare also a solution of 5 ppm Ca and 1% $SrCl_2$. Record the absorbance of the solutions. Compare the absorption of the first solution with that of the phosphate-containing solution above and with that of the 1% $SrCl_2$-containing solution. Explain.

2. *Effect of sodium.* Prepare a solution containing 5 ppm Ca and 1000 ppm Na. Record the absorbance and compare with that of 5 ppm Ca. Explain any difference.

3. *Effect of aluminum.* Prepare a solution containing 5 ppm Ca and 10 ppm Al. Record the absorbance and compare with that of 5 ppm Ca by itself. Suggest a possible reaction for the results.

Determination of Calcium in an Unknown

(The method of standard additions below may be used instead of the following procedures.)

1. *Synthetic unknown.* Obtain an unknown from your instructor and dilute to give a concentration of 5 to 15 ppm Ca. Prepare a series of calcium standards of 0, 2.5, 5, 7.5, 10, and 15 ppm from the 50-ppm stock solution. If the unknown contains phosphate, add $SrCl_2$ to standards and the unknown to give a final concentration of 1%. Record the absorbance (or % absorption and convert to absorbance) of these and prepare a calibration curve of absorbance versus concentration. Determine the concentration of the unknown in the usual manner.

2. *Serum.* Calcium in serum or an "artificial serum" as described in the footnote to Experiment 30 is determined by diluting 1:20 with 1% $SrCl_2$ solution. The normal calcium content of serum is about 100 ppm, and so that analyzed solution contains about 5 ppm Ca. Sodium and potassium equal to that in the sample are added to the standards.

Add 0.5 mL of the unknown serum or the "artificial serum" to a 10-mL volumetric flask and dilute to volume with 1% $SrCl_2$. (If the method of standard additions is to be used also for comparison, dilute 2.5 mL of unknown to 50 mL with 1% $SrCl_2$). Prepare standards of 0, 3, 4, 5, 6, and 8 ppm Ca, each also containing 1% $SrCl_2$, 6.9 meq/L Na, and 0.21 meq/L K.[1]

Prepare a calibration curve from the absorbance of the standards and from this determine the concentration of calcium in the unknown. Use a spreadsheet.

Method of Standard Additions

This procedure may be used, instead of the one above, to analyze the unknowns, and it illustrates the usefulness of the method of standard additions for compensating for matrix effects. Dilute your unknown as described above, using distilled water to give a concentration of about 5 ppm Ca (1 : 20 for serum, e.g., 2.5 mL diluted to 50 mL with 1% $SrCl_2$). Transfer with a pipet separate 10.0-mL aliquots of the diluted unknown to three separate clean and dry test tubes or flasks. Add to these 50.0, 100, and 150 μL, respectively, of the 500-ppm standard calcium solution (or 25.0, 50.0, and 75.0 μL of a 1000-ppm solution if available). This results in an increase in the calcium concentration in the diluted unknown of about 2.5, 5.0, and 7.5 ppm, respectively, depending on the exact concentration of the standard, and brackets the unknown. The volume changes can be considered negligible. Use an appropriate syringe microliter pipet if available (e.g., a 50-μL Eppendorf pipet or Finn-pipette) or else a 0.1-mL graduated measuring pipet. For best accuracy, the pipet should be calibrated (see Chapter 2).

Zero the instrument with distilled water and aspirate the diluted unknown and the standard addition samples. The absorbance increases in the latter are due to the added calcium. Using a spreadsheet, prepare a plot of absorbance against added concentration of calcium (starting at zero added, i.e., the sample). From the x-axis intercept of the plot, determine the concentration of calcium in the diluted unknown. See the spreadsheet in Chapter 17 for preparation of a standard additions plot and unknown calculation. Calculate the concentration in the original sample. How does this method account for phosphate interference?

EXPERIMENT 30   FLAME EMISSION SPECTROMETRIC DETERMINATION OF SODIUM

Principle

The intensity of sodium emission in a flame at 589.0 nm is compared with that of standards. If an internal standard instrument is available, the ratio of sodium to lithium emission is measured.

Solutions and Chemicals Required

1. *Stock standard NaCl solution (1000 ppm Na).*   Dry about 1 g NaCl at 120°C for 1 h and cool for 30 min. Weigh and dissolve 0.254 g NaCl in water and dilute to 1 L. Care must be taken to avoid sodium contamination, especially from the water and glassware. A blank must be run to correct for sodium in the water.

2. *LiNO₃ internal standard solution (1000 ppm Li).*   (Lithium is not required if a direct-intensity instrument, rather than an internal standard instrument, is used.) Dissolve 0.99 g $LiNO_3$ in water and dilute to 100 mL.

---

[1]A stock solution of 140 meq/L Na and 4.1 meq/L K (20 times the concentration in the standard solution) can be prepared by dissolving 8.1 g NaCl and 0.21 KCl in 1 L water. This contains the normal levels of Na and K in serum and compensates for ionization interference due to these elements in the serum. Dilute this solution 1 : 20 in the standards.

3. *Working standard solutions*[1]

     (a) Direct-intensity instrument. Prepare standards of 0, 10, 20, 30, and 40 ppm Na by diluting 0, 5, 10, 15, 20, and 25 mL of the stock NaCl solution to 50 mL. (It may be better with some instruments to prepare solutions five times more dilute than this by adding 0, 1, 2, 3, 4, and 5 mL of solution to the flasks. This may result in a more linear calibration curve. Follow your instructor's directions.)

     (b) Internal standard instrument. Prepare the same solutions as above for the direct-reading instrument, but add 5 mL of the stock lithium solution to each flask. (This results in 100 ppm Li in each solution. The recommended concentration may vary from one manufacturer to another, and so your instructor may direct you to add a different amount.)

Things to Do before the Experiment

Dry the NaCl at 120°C for 1 hour and cool in a desiccator.

Procedure

Have your instructor add your unknown to a 100-mL volumetric flask.

1. *Direct-reading instrument.*   Dilute your unknown to volume with water. Follow your instructor's directions for the operation of the instrument. Several atomic absorption instruments can be used for measuring emission. Set the zero reading while aspirating distilled water (the blank). Aspirate each standard and the unknown and record their emission intensity readings. With some instruments, the 100% reading is set with the most concentrated standard.

     Using a spreadsheet, plot the emission readings for the standards against concentration and determine the concentration in the unknown solution from the calibration curve. From this, calculate the micrograms of solution in your unknown if it is water, and ppm or meq/L if it is serum (see footnote 1).

2. *Internal standard instrument.*   Add the same amount of lithium solution to your unknown as was added to your standard, and dilute to volume with water. Follow your instructor's directions for the operation of the instrument. The lithium emission line is 670.8 nm. Prepare a calibration curve as above for the direct-reading instrument, but record the ratio of the Na/Li emission intensities. Determine the concentration of sodium in the unknown solution and report micrograms of sodium in the unknown if it is water, and ppm or meq/L if it is serum (see footnote 1). Use a spreadsheet as described in Chapter 20 for the internal standard plot and calculation.

---

[1]The unknown may be simply a sodium chloride solution, or it may be serum. If serum is analyzed, then the standards should be prepared over a narrower concentration range to better bracket the unknown. Sodium in serum (approximately 140 meq/L, or 3200 ppm Na) may be determined by simple 1:100 dilution (e.g., 0.1 mL diluted to 10 mL) or 1:500 if required by the instrument. An alternative unknown is an "artificial serum" prepared by dissolving the following salts in water and diluting to 1 L:

NaCl	8.072 g
KCl	0.21g
KH_2PO_4	0.18 g
$CaCl_2 \cdot 2H_2O$	0.37 g
$MgSO_4 \cdot 7H_2O$	0.25 g

This contains 138.1 meq/L Na, which can be varied from unknown to unknown. Bovine serum albumin (60 g) may be added as a source of protein. Serum contains about 6% (wt/wt) protein.

Chromatography

EXPERIMENT 31   THIN-LAYER CHROMATOGRAPHY SEPARATION OF AMINO ACIDS

Principle

The amino acids are separated on a TLC sheet, e.g., silica gel, using a choice of two developing solvents. The locating reagent is ninhydrin.

Solutions and Chemicals Required

Provided

(a) **Developing solvent:** No. 1, butyl alcohol–acetic acid–water (80 : 20 : 20 vol/vol); No. 2, propyl alcohol–water (7 : 3 vol/vol).

(b) **Locating reagent:** Ninhydrin solution—0.3% ninhydrin (1,2,3-triketohydrindene, Eastman No. 2495) in butyl alcohol containing 3% glacial acetic acid.

Chromatographic Equipment

Fisher Scientific TLC Kit A or equivalent TLC sheets, sprayer (for application of reagent), developing apparatus (e.g., Fisher Scientific TLC Kit A: www.fisheredu. com or equivalent).

Procedure

Obtain an unknown mixture from your instructor. The mixtures to be separated will contain approximately 1 mg of each amino acid per milliliter 0.5 M alcoholic hydrochloric acid solution (see discussion below). Spot approximately 1 μL of the sample solution on the Chromagram sheet about 2 cm from the lower edge. (Activation of the sheet is not necessary for this separation.) The unknown should contain acids whose R_f values are sufficiently different with the solvent system used so that they will be readily separated. Your instructor will advise you which standard amino acid solutions to run with your unknown so that you can distinguish between two amino acids that might have an R_f value close to one in your mixture. Allow 15 to 20 min drying to ensure complete evaporation of the hydrochloric acid.

Develop the Chromagram sheet in the solvent of choice for a distance of 10 cm or for approximately 90 min. (See the list of R_f values in the table below and follow your instructor's directions for the solvent of choice for your unknown mixture.) Dry the developed Chromagram sheet and spray with the ninhydrin solution. Heat gently for several minutes until separated zones appear clearly visible.

Results

The table following lists the approximate R_f values obtained when the two solvent systems are applied to the separation of 13 different amino acids. From this table and the standard acids you run, determine what amino acids are present in your unknown.

Discussion

Due to the limited solubility of various amino acids, great care must be taken in the preparation of the sample before applying chromatography. The use of alcoholic 0.5 M HCl is suggested in the procedure. However, it may become necessary to add significant amounts of water to achieve solubility. When this is the case, it becomes extremely important to keep the spot area small and to allow enough time for complete evaporation of the spotting solvents before starting development.

When a greater degree of resolution is required, it is possible to improve results by using the two-dimensional technique of development. The spot containing the components to be separated is placed in the lower left-hand corner of a 20 × 20-cm sheet, 2 cm from each edge. The sheet is developed in the normal fashion, removed, and dried. It is then turned counterclockwise 90° and developed again in a different solvent system to separate components that were not resolved in the first migration.

The R_f values of separated amino acids depend on a number of factors, including the concentrations of the amino acids as well as other components in the sample mixture. It is for this reason that standards used should be as nearly like the actual samples as possible.

The visualization with ninhydrin has limits of detection that may vary from 0.01 to 0.5 μg, depending on the particular amino acids as well as the method of separation employed. A degree of ninhydrin color stabilization can be achieved by spraying the visualized TLC sheet with the following solution: 1 mL saturated aqueous copper nitrate solution dissolved by 0.2 mL 10% nitric acid in 100 mL ethanol (95%). The sheet is then exposed to ammonia vapors and a red copper complex is obtained. The color is stable only in the absence of acid.

Amino Acid	Approx. R_f Value Developing Solvent No. 1	Approx. R_f Value Developing Solvent No. 2
Alanine	0.29	0.50
Arginine monohydrochloride	0.15	0.15
Asparagine	0.20	0.43
Cystine	0.12	0.22
Glutamic acid	0.33	0.40
Glycine	0.22	0.39
Leucine	0.57	0.69
Lysine	0.10	0.20
Methionine	0.47	0.63
Serine	0.25	0.45
Tryptophan	0.55	0.71
Tyrosine	0.52	0.69
Valine	0.40	0.60

EXPERIMENT 32   GAS CHROMATOGRAPHIC ANALYSIS OF A TERTIARY MIXTURE

Principle

A mixture of pentane, hexane, and heptane is separated by gas chromatography. A number of different types of columns can be used, and a simple thermal conductivity or hot wire detector is satisfactory. The instrument response for each compound is calibrated by running a standard mixture of the compounds. The order of separation is determined by running the individual compounds.

An alternative experiment is to analyze a two-component mixture of *n*-hexane and *n*-heptane, and use *n*-pentane as an internal standard.

Solutions and Chemicals Required

1. *Provided.*   Acetone.

2. *To prepare.*   Standard mixture. Prepare the following standard mixture: *n*-pentane (5.00 mL), *n*-hexane (10.00 mL), *n*-heptane (15.00 mL). All mixtures

should be fresh the day of use and kept in plastic-stoppered vials. Alternatively, the standards may be weighed and results reported on a weight/weight basis (or use density to calculate weights from volumes).

Caution. Exercise great care not to damage the Hamilton syringes. Syringes require proper cleaning and handling to give consistent results. After each use, remove the plunger, rinse thoroughly with acetone, and allow to dry. Insert the syringe through a septum on one end of a glass tube and insert a clean syringe containing acetone into the other end. Force acetone through the tubing and the dirty syringe until clean. Dry by drawing air through the barrel of the syringe with a water aspirator. The glass will appear frosted when completely dry.

Procedure

Obtain an unknown mixture from the instructor. Check the instrument instructions and your instructor regarding the operation of the chromatograph. Do not make any temperature adjustments. Using the appropriate syringe for the instrument, obtain chromatograms of each separate component of the mixture, three chromatograms of your standard mixture, and three chromatograms of your unknown.

Having obtained the necessary data, leave the instrument in the manner prescribed by the instructor. Especially important is the decrease in gas flow and adjustment of filament conditions to a "stand-by" state. Also, clean, rinse, and dry with acetone all syringes and vials.

Analysis of Data

The peak areas on the chromatographic curves are equal to height times width at half-height.[1] Make measurements with a millimeter ruler. Your instrument may automatically print peak areas for each compound, in which case, use these.

Individual components are recognized by the positions of their peaks with respect to the origin (the retention time).

The standard mixture is used for quantitative calibration. Since the detector does not exhibit equal response to all compounds, then a calibration factor must be determined for each, using the standard mixture. One simple way of calibration is just to make a direct comparison of the absolute peak area for each compound with its percent in the mixture. Thus, if 25% compound A gives a peak area of 40, then a peak area of 80 for that compound would correspond to 50% A. Obviously, greatest accuracy would be obtained by preparing a calibration curve of area versus percent compound over a range of percentages for each compound. This would compensate for the fact that the net volume of a given total volume of the individual compounds may vary with the composition.

From the average of the measured chromatographic areas for each compound of the unknown, calculate the average volume percentage of each compound.

Internal Standard Calibration

Your instructor may give you a two-component unknown of *n*-hexane and *n*-heptane, and direct you to add *n*-pentane to all solutions as an internal standard. In this

---

[1] In place of determining the areas under the peaks by measurement and using these for the computations, a somewhat more accurate method is to photocopy the chromatograms first and then cut out from the original chromatograms carefully with scissors each peak area. These pieces are then weighed on the analytical balance and the weights are used in the calculations. If you use this method, hand in the "cutouts" with your report. The method of choice is electronic integration.

case, prepare the standards as before, but for the unknown, take 25.00 mL and add 5.00 mL of *n*-pentane. Calculate the ratios of the analyte to *n*-pentane areas for the calibration. See Chapter 20 for a description of the use of internal standards.

EXPERIMENT 33 QUALITATIVE AND QUANTITATIVE ANALYSIS OF FRUIT JUICES FOR VITAMIN C USING HIGH-PERFORMANCE LIQUID CHROMATOGRAPHY[1]

Principle

Samples of several juices are chromatographed directly on a strong-base anion exchange column using a UV detector at 254 nm. By comparison of retention times with that of a vitamin C standard (*l*-ascorbic acid), the presence or absence of vitamin C is ascertained. Peak area measurements are used for quantitative determination of the concentration of vitamin C in those juices in which it is found.

Solutions and Reagents Required

1. *Provided.* Mobile-phase solution (1.36 g KH_2PO_4/L distilled deionized water), *l*-ascorbic acid.

2. *To prepare.*   Vitamin C standard. Weigh 50 mg of *l*-ascorbic acid to the nearest 0.1 mg, dissolve in a 50-mL volumetric flask, and dilute to volume with distilled deionized water. Prepare the day of use. This is a 0.1% solution. Prepare serial dilution of this to obtain also 0.05% and 0.02% standards.

Procedure

1. *Calibration.* Consult your instructor for the operation of your instrument, including the proper attenuation. It will contain a column of a strong-base anion exchange resin. At a typical pressure of 1000 psi, the flow rate will be about 0.5 mL/min.

   Using a 10- or 25-μL syringe, inject a 10-μL aliquot of the 0.02% vitamin C standard and record the chromatogram. Repeat for the 0.05% and 0.1% standards. Measure the retention times and the areas. The peak areas may be obtained from the height times the width at half-height for each. Alternatively, the peaks may be cut out and weighed. Or, the instrument may print out the integrated areas. Using a spreadsheet, plot a calibration curve of the peak area against concentration.

2. *Unknown.*   Your instructor will provide you three or more unknown juices. These may be such products as "Hi C" drinks, grape drinks, orange juice, and the like. Inject 10-μL portions of each and record the chromatograms. Several peaks may be obtained. If you have time, run at least two chromatograms on each. From a comparison of retention times, identify the juices that contain vitamin C. Measure the areas of the vitamin C peaks, and from the calibration curve determine and report the concentrations in the unknowns. If two or more chromatograms of the unknowns were run, report the average concentration and the standard deviation if more than two. The vitamin C peak may be partially overlapped by another. In this case, extrapolate to the baseline and measure the area from the baseline.

---

[1]The oxidized form of ascorbic acid, dehydroascorbic acid, does not absorb in the UV region and is not detected by this method. For more quantitative results, the samples should be treated with a reducing agent such as H_2S [see Roe et al., *J. Biol. Chem.*, **174** (1984) 201] to reduce any dehydroascorbic acid.

EXPERIMENT 34 ANALYSIS OF ANALGESICS USING HIGH-PERFORMANCE LIQUID CHROMATOGRAPHY

Principle

The compositions of the analgesics Bufferin, Anacin, Empirin, and several brands of aspirin tablets are qualitatively determined by comparing their chromatograms with those of standard aspirin, phenacitin, and caffeine samples. The aspirin content of each is quantitatively determined by comparison of peak areas with those of standards. Separations are performed on a strong anion exchange column, using a UV detector.

Solutions and Reagents Required

1. *Provided.*[1] Aspirin (acetylsalicyclic acid), phenacetin (acetophenetidin), caffeine (theine; 1,3,7-trimethylxanthine), methanol, mobile-phase solution (0.01 M sodium borate + 0.005 M ammonium nitrate). See Experiment 26 for the structures of aspirin, phenacetin, and caffeine.

2. *To prepare*[2]

   (a) **Stock solutions of aspirin, phenacetin, and caffeine.** Weigh 50 mg each of phenacetin and caffeine to the nearest milligram and transfer to labeled 50-mL volumetric flasks. Dissolve in methanol and dilute to volume. Weigh about 150 mg aspirin to the nearest 0.1 mg, transfer to a labeled 50-mL volumetric flask, dissolve in methanol, and dilute to volume with methanol. Prepare on the day of use.

   (b) **Aspirin working standards.** Prepare serial dilutions of the stock 300 mg/dL solution to obtain standards of about 200, 100, and 50 mg/dL in methanol.

Procedure

1. *Calibration.* Consult your instructor for the operation of your instrument, including the proper attenuation for each sample. It will contain a column of a strong anion exchange resin. At a typical operating pressure of 1000 psi, the flow rate will be about 0.5 mL/min.

   Using a 5- or 10-μL syringe, inject a 5-μL aliquot of each 100 mg/dL standard solution to obtain chromatograms for aspirin, phenacetin, and caffeine. Similarly, obtain chromatograms for the serial aspirin standards (one chromatogram for each, unless you have time for more).

   Measure the retention times. For the aspirin chromatograms, also measure the areas. The peak areas may be obtained from the height times the width at half-height for each. Alternatively, the peaks may be cut out and weighed. Or the instrument may print out the integrated areas. Using a spreadsheet, plot a calibration curve of peak area against concentration.

2. *Unknowns.* Your instructor will provide you with five tablets each of Bufferin, Anacin, Empirin, Bayer aspirin, and one or two other brands of aspirin. Prepare solutions of each in methanol by grinding the five tablets with a mortar and pestle, weighing to the nearest 0.1 mg about 175 mg, and transferring to a 50-mL volumetric flask. Dissolve in methanol and dilute to the mark. Mix well and allow any residue to settle for about 10 min, then obtain chromatograms as above, taking 5-μL aliquots. If time permits, run two chromatograms on each. It will take about 10 to 15 min for each chromatogram.

---

[1]Caffeine and phenacetin are available from Sigma-Aldrich.
[2]Aspirin tends to decompose in solution, and measurements should be made as soon as possible after preparing solutions.

From a comparison of retention times, identify the components of each analgesic. Measure the areas of the aspirin peaks, and from the calibration curve determine and report the concentrations in the various tablets. If there is any overlap of the aspirin peak with another, extrapolate to the baseline and measure the area from the baseline. If two chromatograms were run, report the average concentration.

Kinetic Analysis

EXPERIMENT 35   ENZYMATIC DETERMINATION OF GLUCOSE IN BLOOD[1]

Principles

A protein-free filtrate of a 0.5-mL sample of whole blood, serum, or plasma is prepared by precipitating proteins with zinc hydroxide. The glucose in an aliquot of this is reacted with a mixture of the enzymes glucose oxidase and horseradish peroxidase, and the hydrogen peroxide produced is coupled with the chromogenic hydrogen donor *o*-dianisidine to form a color with absorption maximum at 540 nm.

Equations

$$\text{glucose} + O_2 + H_2O \xrightarrow{\text{glucose oxidase}} \text{gluconic acid} + H_2O_2$$

$$H_2O_2 + o\text{-dianisidine} \xrightarrow[\text{peroxidase}]{\text{horseradish}} \text{chromogen (absorption max 540 nm)}$$

Solutions and Chemicals Required[2]

1. *Provided*

    (a) **Zinc sulfate solution, 2.2% (from ZnSO$_4$ · 7H$_2$O); 3 *M* H$_2$SO$_4$.**

    (b) **Barium hydroxide solution, saturated.** Add about 80 g Ba(OH)$_2$ · 8H$_2$O to 1 L boiling water. Remove heat, insert a stopper with a soda lime trap, mix thoroughly, and *set aside for several days to settle.* At 25°C, the solution will be about 0.22 *M*.

    (c) **Barium hydroxide solution 0.06 *M*.** Dilute 270 mL of the saturated barium hydroxide solution to 1 L with CO$_2$-free (boiled and cooled) water. A volume of 9.00 ± 0.10 mL of this should be required to neutralize 10.00 mL of the zinc sulfate solution. Test by adding 10.00 mL zinc sulfate and 25 mL water to a 100-mL Erlenmeyer flask, adding two drops 0.2% phenolphthalein solution (in alcohol), and titrating with the barium hydroxide solution to a faint but permanent pink color. Dilute one or the other solution to proper equivalence.

2. *To prepare*

    (a) **Glycerol–buffer solution, pH 7.0.** Dissolve 0.35 g Na$_2$HPO$_4$ and 0.21 g KH$_2$PO$_4$ in 60 mL water and add 40 mL reagent-grade glycerol.

    (b) **Enzyme reagent.** Grind 250 mg glucose oxidase, 5 mg horseradish peroxidase, and 1 mL of the glycerol–buffer solution in a clean dry mortar. Wash into a 100-mL graduated cylinder using the glycerol–buffer solution, and bring to a volume of 100 mL with the same solution. Filter into a clean and dry flask. A Buchner funnel with light suction will facilitate the filtration.

---

[1]See footnote 3, concerning aqueous unknowns.
[2]Commercially packaged reagents are available, which eliminates much of the solution preparation.

Dissolve 10 mg *o*-dianisidine in 1.0 mL absolute methanol by intermittent shaking. Pour into an amber bottle and add the filtered enzyme solution. Pour the initial portion rapidly to prevent precipitation of the chromogen. This reagent is stable for 3 weeks when stored in a refrigerator.

(c) **Glucose standards.** Prepare a 1% stock solution by dissolving 1 g (accurately weighed) of oven-dried (110°C, 1 h) reagent-grade glucose in water and diluting with 0.25% benzoic acid solution in a volumetric flask. The benzoic acid acts as a preservative.

Prepare working standards containing 100, 200, 300, and 400 mg glucose per 100 mL by dilution of the stock solution with 0.25% benzoic acid (e.g., 5.0, 10.0, 15.0, and 20.0 mL diluted to 50 mL).

Things to Do before the Experiment
Dry the glucose at 110°C for 1 h and cool in a desiccator.

Procedure
Add with a pipet a 0.5-mL blood, serum, or plasma sample[3] into a 50-mL Erlenmeyer flask and add 4.5 mL 0.06 *M* barium hydroxide solution. Swirl to mix and then slowly add 5.00 mL 2.2% zinc sulfate solution. Mix thoroughly and allow to stand 5 min. Filter into a dry flask, or centrifuge in a dry centrifuge tube.

Prepare a blank filtrate and standard filtrates in the same manner by using 0.5 mL water for the blank and 0.5-mL aliquots of the working standards.

Add with a pipet a 0.2-mL aliquot of each of the filtrates into a clean, dry test tube, and at 30-s intervals add 1.00 mL enzyme reagent. Immediately place in a 37°C water bath. (Reactions may be run for 45 minutes at room temperature with some loss in sensitivity.) Run duplicate aliquots of the unknown filtrate. After exactly 30 minutes, remove the tubes one at a time (30-s intervals) and add 50 mL 3 *M* H₂SO₄.[4] The sulfuric acid effectively stops the enzyme reaction.

After 5 min, *but* before 1 h, determine the absorbance of the solutions at 540 nm, using the blank as reference. Using a spreadsheet, construct a calibration curve and report the mg/dL glucose in your sample.

Flow Injection Analysis

EXPERIMENT 36   CHARACTERIZATION OF PHYSICAL PARAMETERS OF A FLOW INJECTION ANALYSIS SYSTEM

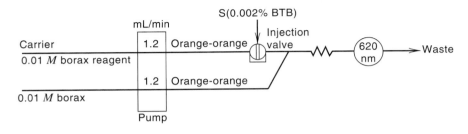

---

[3]Standard commercial control serum preparations (e.g., Versatol) may be used. Preparations with "abnormal" glucose concentrations are available, so the range of normal (ca. 90 mg/dL) to abnormal (300 mg/dL or greater) levels can be covered.

Alternatively, aqueous glucose unknowns may be used, in which case the protein precipitation step is eliminated and 9.5 mL water is added in place of the barium hydroxide and zinc sulfate solution.

[4]The reaction is pseudo first order at low concentrations of glucose, but a slower rate is observed at high concentrations of glucose. Thus, the absorbance may not be a linear function of incubation time at high concentrations.

Principles

The purpose of this experiment is to learn the operation of your flow injection apparatus and to estimate (a) the flow rates of the carrier and reagent streams, (b) the volumes of the sample injection loop and of the flow channels, (c) the dispersion, D^{max}, of the sample plug at the peak maximum (two-line and single line), and (d) the maximum sampling frequency. The two-line system is used for all experiments, except for a single-line dispersion experiment.

Equations

$$D = C^0/C = H^0/H$$
$$D^{max} = C^0/C^{max} = H^0/H^{max}$$

H^0 is the recorded steady-state signal for pure sample, and H^{max} is the recorded peak maximum for the injected sample.

Solutions and Chemicals Required

1. *Provided.* Stock 0.1 M borax solution ($Na_2B_4O_7$), stock 0.4% (wt/vol) bromothymol blue (BTB) solution (0.4 g dissolved in 25 mL 96% ethanol and made to 100 mL with 0.01 M borax). **NOTE:** The bromothymol blue should not be injected in acid solution since the acid form of the indicator adsorbs on the plastic tubing.

2. *To prepare.* Dilute 100 mL of 0.1 M borax solution to 1 L to prepare 0.01 M working solution. Dilute 1 mL of the 0.4% bromothymol blue solution to 200 mL with the 0.01 M borax solution. This 0.002% BMT solution will serve as the working solution. The absorbance of this solution in a 1-cm-long cell is about 1.2.

Assembly of Apparatus

Assemble the peristaltic pump tubes, injector, reactor, and detector as described in the instrument operation manual or by your instructor. There will be three peristaltic pump tubes, one for the carrier (0.89 mm i.d., orange–orange color-coded stops), one for the reagent (0.89 mm i.d., orange–orange), and one for the sample (0.89 mm i.d., orange–orange).

Connect the sampling tube to the inlet of the injection loop and the sample pump tube to the outlet of the loop (the sample will be drawn into the loop by means of the pump).

Procedure

1. *Checking the flow system.* Turn on the detector and recorder and allow to warm up. Fill the carrier bottle and the reagent bottle with the 0.01 M sodium borate solution. Fill a small beaker approximately half full with the 0.002% bromothymol blue solution. Place the appropriate tubes in the bottles. Place the injector in the load position and turn on the peristaltic pump. The channels will fill and solution will exit to the waste bottle. The blue sample solution will fill the injector sample loop. Allow to flow until all air bubbles are gone. **NOTE:** Occasionally an air bubble may get stuck in the detector, as indicated by a deflection on the recorder. This is most conveniently dislodged by introducing a large air bubble in the carrier stream by momentarily pulling the carrier tube out of the carrier solution.

    Set the detector at 620 nm. The injected sample should provide a peak maximum of about 0.15 absorbance unit with a 1-cm path flow cell. Adjust your recording device (strip chart recorder or computer—use a chart speed of 0.5 cm/s if a recorder) to accommodate this range. With the carrier being pumped, turn the injection valve to the inject position. Note the blue sample plug as it passes through the channels. If sample continues to be aspirated (to waste) in the inject position of your valve, in order to conserve sample, the sampling tube may be removed from the sample solution, causing air to be aspirated. When the sample reaches the detector, there will be a deflection followed by return to baseline. If necessary, adjust the recording sensitivity so that the deflection will be about two-thirds full scale. After the sample has passed the detector, turn the valve to the load position to refill the sample loop. Continue the flow until the sample liquid reaches the end of the injector waste tube (this assures proper flushing of the sample loop with the new sample), then inject the sample and record the peak. Continue several injections in this manner until you are satisfied the system is operating properly, the detector is not drifting, and reproducible peaks are being obtained. You are now ready to perform the experiment.

2. *Determination of flow rates.*   Fill a 10-mL graduated cylinder with distilled water, record the volume, and insert the carrier tube in the cylinder. Turn on the pump and simultaneously start a stop watch (or begin timing with a clock). Pump for 5 min, remove the tube, and turn off the pump. Record the volume of water remaining in the cylinder. Calculate the flow rate in milliliters per minute.

   Perform a similar experiment for the reagent flow and then for the sample flow (injector in the load position).

   Finally, measure the flow rate of the waste stream from the detector, by collecting the waste for 5 min.

   If pump tubing of equal internal diameter is used for all channels, the flow rates should be similar. Also, the flow rate of the waste stream should total the combined flow rates of the carrier and reagent streams.

   Note that flow rate is directly proportional to the square of the internal radius of a pump tube, so rates can be appropriately adjusted by changing the pump tubes (i.e., flow is proportional to cross-section area $= \pi R^2$).

3. *Estimation of sample loop volume.*   With the pump turned on, place the valve in the load position and remove the sampling tube from the sample. This will allow the loop to fill with air. Then, insert the tube in the sample solution and, with a stopwatch, measure the time from when the sample just enters the loop to when it leaves the loop. Perform the measurement several times and take the average. From a knowledge of the sample flow rate determined above and the measured time to fill the loop, calculate the sample loop volume in microliters. **NOTE:** This is an estimate and will not include the dead volume of the holes in the injector rotor where the loop is connected.

4. *Estimation of the flow system volume.*   Fill the sample loop with air and inject the air into the carrier. Measure the time to reach the merging point of the carrier and reagent streams. From a knowledge of the carrier flow rate determined above and the measured time, calculate the volume from the injection valve to the merging point, in microliters. Perform a similar experiment, but measure the time for the air to go from the merging point to the detector. From a knowledge of the combined flow of the carrier and reagent streams in the reactor module determined above and the measured time, calculate the volume from the reactor module inlet to the detector, in microliters.

5. *Determination of total dead time.*   With the recording device on, simultaneously inject a sample and start a stopwatch. Measure the time for the sample to reach the detector, that is, the time for the recorder to begin deflection. This represents the dead time from injection to initial measurement.

6. *Time to flush sampling tube and sample loop.*   When injecting different samples, it will be necessary to flush the previous sample completely from the sampling tube and the injection loop. Determine the time required to just

flush the previous sample by introducing air into the tube and then reinserting the tube in the sample solution with the pump running. Measure the time for the sample solution to reach and fill the sample loop. Use at least three times this time for introduction of each sample in all future experiments, to allow adequate washing of the loop by the new sample.

From a knowledge of the sample flow rate, also calculate the volume of solution required to flush the line.

7. *Determination of dispersion: Two-line system.* Adjust the recording sensitivity so the peak deflection for an injected sample is about one-fifth full scale. Make several injections of the sample and determine the average peak height. This represents H^{max}. Turn the pump off and insert both the carrier tube and the reagent tube in the sample solution. Turn the pump on and record the signal until a steady-state signal is obtained. This represents H^0. **NOTE:** Alternatively, H^0 may be determined by inserting the exit waste tube in the sample solution and reversing the pump flow to fill the cell with sample solution. Calculate the dispersion at the peak height, D^{max}. **NOTE:** If transmittance is recorded rather than absorbance, convert the readings to absorbance for calculating D^{max}.

8. *Determination of dispersion: Single-line system.* Also determine the dispersion by clamping the reagent tube to convert to a single-line system; remove the tube from the pump. Determine the dispersion as above and compare with that of the two-line system.

9. *Determination of maximum sampling frequency.* Expand the recording time axis 10-fold. Inject a sample and record the rise and fall of the peak. Measure the distance from baseline to baseline on the peak and convert this to seconds. This will represent the minimum time between injections. Report the maximum sampling frequency in samples per hour.

At the end of the experiment, wash the system thoroughly by pumping distilled water for a few minutes. Include flushing the sampling tube and valve loops. Then release the pump cassettes and pump tubes. If the system is not to be used for an extended time, your instructor will instruct you to empty it.

In your report, list all instrument parameters and solutions employed.

EXPERIMENT 37   SINGLE-LINE FIA: SPECTROPHOTOMETRIC DETERMINATION OF CHLORIDE

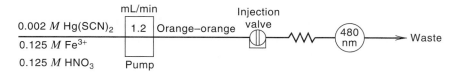

Principle

The analytical procedure is based on the following reactions:

$$Hg(SCN)_2 + 2Cl^- \rightarrow HgCl_2 + 2SCN^-$$

$$2SCN^- + Fe^{3+} \rightarrow Fe(SCN)_2^+$$

The carrier stream contains $Hg(SCN)_2$ and Fe(III). The chloride of the injected sample reacts with $Hg(SCN)_2$, liberating SCN^-, which in turn forms with Fe(III) the red-colored complex ion $Fe(SCN)_2^+$, which is measured spectrophotometrically at 480 nm. The height of the recorded absorbance peak is then proportional to the concentration of chloride in the sample. Besides $Fe(SCN)_2^+$, other (higher) complex ions between Fe(III) and SCN^- might be formed, causing nonlinearity in the calibration curve at high concentrations.

Perform Experiment 36 first to characterize and learn how to use the system, or else read through it.

Solutions and Chemicals Required

1. *Reagent.* The carrier stream is prepared by dissolving 0.157 g of mercury(II) thiocyanate, 7.6 g of iron(III) nitrate, 0.8 mL of concentrated nitric acid, and 40 mL of methanol in water, making the final volume up to 250 mL.

2. *Standard solutions.* Standard solutions in the 5- to 75-ppm Cl range are made by suitable dilution of the stock solution containing 1000 ppm Cl (0.165 g of sodium chloride in 100 mL).

Procedure

Assemble the flow injection apparatus in the single-line mode as described by the manufacturer or your instructor. Use 0.89-mm i.d. pump tubing for the carrier and the sampling tubes (orange–orange color-coded stops on the peristaltic pump tubing). This should provide a flow rate of about 1.15 mL/min when using 25 rpm for the peristaltic pump.

Turn on the detector and allow to warm up for several minutes to stabilize. If a monochromator is used, set to 480 nm. (Since the color produced is specific for the analyte, a simple visible light source–detector system may be used without a monochromator.) Use injections of the highest concentration standard into the pumped reagent carrier to adjust the recording sensitivity to give about 75% deflection. Use a recorder speed of about 0.5 cm/s, if using a strip chart recorder. Each time the injection loop is filled with a new solution, it should be flushed with at least three loop volumes of the solution before filling with the aliquot to be injected. That is, run 100 μL through a 25-μL loop and then stop.

Reagent carrier solution is pumped through the system and the individual chloride standards are injected successively in triplicate, yielding a series of peaks for each. Determine the precision of the procedure by injecting a single standard (in the middle concentration range) 10 times. Report the percent relative standard deviation.

Obtain an unknown from your instructor and inject at least three times to obtain an average peak height.

Prepare a calibration curve from the peaks recorded for the standard solutions and, from this, calculate the concentration of chloride in your unknown solution.

At the end of the experiment wash the system thoroughly by pumping distilled water for a few minutes. Include flushing the sampling tube and valve loop. Then release the pump cassettes and pump tubes. If the system is not to be used for an extended time, your instructor will instruct you to empty it.

In your report, list all instrument parameters and solutions employed.

EXPERIMENT 38 THREE-LINE FIA: SPECTROPHOTOMETRIC DETERMINATION OF PHOSPHATE

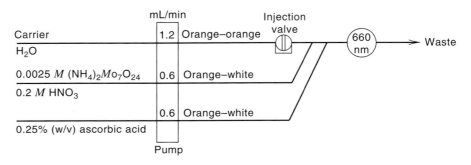

Principle

The analytical procedure is based upon the following reactions:

$$7H_3PO_4 + 12(NH_4)_6Mo_7O_{24} + 51H^+ \rightarrow 7(NH_4)_3PO_4 \cdot 12MoO_3 + 51NH_4^+ + 36H_2O$$

Mo(VI)–YELLOW COMPLEX + ascorbic acid → Mo(V)–BLUE COMPLEX

It is based on the same procedure as Experiment 24. The carrier is distilled water. Reagent 1 is a 0.0025 M solution of heptamolybdate. Reagent 2 is 5% ascorbic acid (wt/vol) in a 10% aqueous solution of glycerine. The glycerine aids in preventing the colored complex from adhering to the walls of the flow-through cell upon injection of the sample into the carrier stream. The sample merges successively with the reagents and passes through the reaction coils. The phosphate in the sample combines with the heptamolybdate, forming a yellow-colored complex. This yellow complex then reacts with the ascorbic acid, which reduces the molybdenum from the +6 state to the +5 state, forming a blue-colored complex, which has an extremely high absorptivity and is measured spectrophotometrically at 660 nm. The height of the recorded peak is proportional to the concentration of phosphate. The calibration curve linearity and slope depend upon the extent of reaction, that is, how much of the blue complex has been formed. This is a function of the kinetic nature of the FIA procedure in that the reaction may not reach its steady-state value, but rather some fraction of the steady-state value, since it is a slow reaction. The extent of reaction is dependent upon the particular reaction system. Addition of antimony(III) catalyzes the reduction by ascorbic acid.

Perform Experiment 36 first to characterize and learn how to use the system, or else read through it.

Solutions and Chemicals Required

Reagent 1. In a 100-mL volumetric flask, place approximately 50 mL of distilled water and carefully add 1.3 mL of concentrated nitric acid. To this add 0.309 g of ammonium heptamolybdate, $(NH_4)_6Mo_7O_{24} \cdot 4H_2O$, and dilute to the mark with distilled water. This results in a solution that is 0.2 M in nitric acid and 0.0025 M in heptamolybdate.

Reagent 2. In a 100-mL volumetric flask place 5 g of ascorbic acid, approximately 50 mL of distilled water, and 10 mL of glycerine. Mix thoroughly prior to diluting to the mark with distilled water. This results in a solution that is 5% (wt/vol) in ascorbic acid.

Standard solutions. Standard solutions of phosphate (as P) in the 10- to 100-ppm P range are made by suitable dilution of a stock solution containing 100 ppm P (0.440 g of anhydrous KH_2PO_4 per liter).

Procedure

Assemble the injector, reactor, and detector as described by the manufacturer or your instructor. Orange–white color-coded peristaltic pump tubing (0.64 mm i.d.) should be used for reagents 1 and 2, while orange–orange (0.89-mm i.d.) pump tubing should be used for the H_2O carrier and sample. The sample injection size should be approximately 25 μL. This corresponds to a sample loop length of approximately 13 cm when using 0.5-mm inside diameter Micro-line tubing. Each time the injection loop is filled with a new solution, it should be flushed with at least three loop volumes of the solution before filling with the aliquot to be injected. That is, run 100 μL through a 25-μL loop and then stop.

Turn on the detector and allow it to warm up for several minutes to stabilize. If a monochromator is used, set to 660 nm. (Since the color produced is specific for the analyte product, a simple visible white light source detector may be used without a monochromator.) Use a chart speed of approximately 0.5 cm/s if a strip chart recorder is used. With the water carrier and the two reagents being pumped, inject the highest concentration standard solution. Adjust the recording sensitivity to give about 75% deflection with this standard.

Inject the individual phosphate standards successively in triplicate, thereby yielding a series of peaks for each standard. The precision of the procedure is determined by injecting the 50-ppm standard 10 times. Report the percent relative standard deviation.

Obtain an unknown from your instructor and inject it at least three times to obtain an average peak height. Using a spreadsheet, prepare a calibration curve from the peaks recorded for the standard solutions. From this calibration curve, calculate the concentration of phosphate in your unknown solution.

Your report should include all instrument operation parameters, as well as reagents and solutions used in performing the experiment.

Catalyzed Reaction

The sensitivity of this reaction can be enhanced through the use of a catalyst or by using stopped-flow techniques. Antimony(III) catalyzes the reaction, and phosphorus concentrations in the 1- to 10-ppm range can be analyzed with this system without the need for elevated temperatures for increasing reaction rate (the uncatalyzed reaction is actually quite slow, requiring up to 10 min for complete reaction to occur). To perform the catalyzed reaction, a stock antimony(III) solution is prepared by placing 1.10 g of potassium antimony(III) tartrate, $KSbOC_4H_4O_6 \cdot \frac{1}{2}H_2O$, in a 100-mL volumetric flask and diluting to the mark with distilled water. This makes a 0.033 M or 4000-ppm solution in antimony(III). A 2.5-mL aliquot of this solution is then added during the preparation of reagent 2, prior to diluting to the mark, yielding a 5% ascorbic acid solution, which is also 100 ppm in antimony(III). Appropriate dilutions of the 100-ppm phosphate standard solution are then made to yield concentrations from 1 to 10 ppm. The experiment is then run in the same manner as before, substituting the new ascorbic acid solution containing antimony and the new phosphate standards. The recorder sensitivity is adjusted using the 10-ppm standard.

Shutdown Procedure

At the end of the experiment, flush the system thoroughly by pumping distilled water through *all pump tubes* for several minutes. After this has been done, release the clamps that hold the pump tubes and release the pump tubes from around the pump rollers. If the system is not to be used for an extended time, your instructor will advise you as to the proper procedure for emptying it.

Compare with Manual Spectrophotometry

You made triplicate measurements of each standard and the unknown in this experiment. About how long did it take to make these measurements (once the reagents were prepared)? You will have probably performed some spectrophotometric experiments. How long do you think the same number of measurements would take using conventional spectrophotometry, considering the cuvet has to be rinsed for each measurement. How much reagent would be consumed for each measurement compared with the amount used in this experiment?

Team Experiments

EXPERIMENT 39   METHOD VALIDATION AND QUALITY CONTROL STUDY

Goal

Your instructor will select one experiment for teams to perform validation studies. An example is a gas chromatography experiment such as Experiment 32, but for one analyte. A flow injection analysis (FIA) experiment, such as Experiment 37, would be a good choice as well, since multiple measurements can be made rapidly. The team will determine *linearity, accuracy, precision, sensitivity, range, limit of detection, limit of quantitation,* and *robustness* (repeatability) of the method. In addition, a *control chart* will be prepared over at least one laboratory period. The instructor will have available a reference standard to use for accuracy studies. Plan for two laboratory periods for the completed study. A report of the method will be prepared and documented. Before beginning the experiment, you should review method validation in Chapter 4.

Equation

$$\text{Response factor (RF)} = (\text{signal} - y\ \text{intercept})/\text{concentration}$$

Things to Do before the Experiment

Prepare the solutions needed for the selected experiment that are not provided. You will need to prepare a variety of concentrations of standards. Some of these may be prepared as you perform the different parts of the experiment.

Procedure

Work in teams of four. Pairs in the team will work together on the different tasks to complete the project. *Team A* will prepare the *calibration curve* and determine *linearity, accuracy,* and *repeatability.* They will also prepare the *control chart.* *Team B* will determine *precision, sensitivity, range, limit of detection,* and *limit of quantitation.* (These, assignments may be adjusted if the teams and the instructor agree on the assignments, planning for timely completion.) The whole team will prepare the report and sign and date it. The team will be graded together on the thoroughness and care of the study, and of the report. This is a good exercise in teamwork.

Team A

Prepare the Calibration Curve. It is preferable to measure each standard three times and plot the average; the standard deviation of each point can be calculated, and the range of one standard deviation or the range of data for each point can be marked on the calibration curve. Using Excel, plot the trendline, with the regression line equation (slope and intercept).

Linearity. Calculate r^2 and the y intercept as a percentage of the midrange response. Calculate the response factor (RF) for each experimental point on the line, and using Excel, plot the RF vs. concentration. From the slope (RF/unit concentration), calculate the RF change over the range of experimental points and calculate this as a percentage of the average RF value.

Accuracy. Obtain the reference standard from your instructor. You will not know its concentration until after you make measurements and provide the results. Measure it seven times and calculate the standard deviation. Report the results to your instructor and obtain the "true" value and the standard deviation if known.

Repeatability. Make small variations in one or more parameters in the experiment that should not proportionally or directly impact the measurement. This could be reagent concentration, the pH, time of reaction, size of measured sample (e.g., increase the injected volume of a standard in a chromatography measurement by 10%, adjusting for the increased quantity injected in calculating the predicted response.)

Control Chart. You will probably prepare this on the second day, after the method has been developed. You will be given an unknown (blind) sample. Make a measurement on it every 20 min throughout one laboratory period, intermittently with other measurements the team is making; you should make at least eight measurements. Plot the determined concentration vs. time of day. After you have done this, and shown it to your instructor, obtain the "known" value from the instructor, and draw a horizontal line on the chart at that concentration. From the precision that Team B has determined for the method (at midrange), draw lines for inner and outer control limits at 2 and 2.5 standard deviations, respectively. Are your values within the control lines? Is there a trend in one direction?

Team B
Precision. Take a standard at the low end, the midrange, and the high end of the calibration curve. Measure each seven times and calculate the standard deviations.

Sensitivity. Take two standards that are about two standard deviations apart, in the midrange of the calibration curve, measuring each three times. Report the mean and standard deviations. Can you distinguish them? How many significant figures should you report?

Range. Select an acceptable precision within the measurement range. This will depend on what the precision is at midrange. Let's take the acceptable precision to be twice that precision. From the precision determined at the low and high ends of the calibration curve, is the calibration curve within the acceptable range? If so, do more experiments at the low end, taking smaller concentrations and determining the standard deviation to establish what the lower range is.

Limit of Detection. Make seven blank readings. Calculate the (absolute) standard deviation. Measure a standard seven times that should give a response 5 to 10 times the standard deviation above the blank signal. Calculate the detection limit as the concentration that gives a net response of 3 standard deviations above the blank.

Limit of Quantitation (LOQ). Calculate the LOQ two ways. First, calculate the concentration that gives a response 10 blank standard deviations above the blank signal. Also, determine the concentration that gives a relative standard deviation of 15%. (You may have this information in the range study above.)

Report
Prepare a report on validation of the method and the quality control (control chart). Document with original data, which may include copies of recordings,

Excel printouts, calculations, literature references, and so forth. Make some conclusions about the validity of the method.

EXPERIMENT 40 PROFICIENCY TESTING: DETERMINATION OF Z VALUES OF CLASS EXPERIMENTS

Goal

To calculate z values for all the class results of one or more experiments, and to compare the z value of your results with those of the remainder of the class. Review proficiency testing in Chapter 4 before beginning the experiment.

Equation

$$z = \frac{\overline{X}_i - \hat{X}}{s}$$

where $\overline{X}_i$ = mean of your results
   $\hat{X}$ = "true value," or the class mean
   s = standard deviation of the true value

or of the class mean (the class standard deviation may be used for the true value if needed. You will calculate the class standard deviation).

Procedure

Your instructor will provide you with coded results for the entire class. You will know the identity for the code of only your experiment. The instructor will either provide the true value (used for grading the experiment) or ask you to use the class average (which you will calculate) to determine the z value.

Calculations

Calculate the z values for each coded experiment. Arrange them from high ($+$) to low ($-$). Plot them, using Excel, in bar form (like Figure 4.4). Identify your value, and give the chart to your instructor. Would you accept your result as accurate?

Appendix A

LITERATURE OF ANALYTICAL CHEMISTRY

"To reinvent the wheel is a waste of time and talent."
—Anonymous

When defining a problem, one of the first things the analyst does is to go to the scientific literature and see if the particular problem has already been solved in a manner that can be employed. There are many reference books in selected areas of analytical chemistry that describe the commonly employed analytical procedures in a particular discipline and also some of the not-so-common ones. These usually give reference to the original chemical journals. For many routine or specific analyses, prescribed standard procedures have been adopted by the various professional societies.

If you do not find a solution to the problem in reference books, then you must resort to the scientific journals. *Chemical Abstracts* is the logical place to begin a literature search. This journal contains abstracts of all papers appearing in the major chemical publications of the world. Yearly and cumulative indexes are available to aid in the literature search. The element or compound to be determined as well as the type of sample to be analyzed can be looked up to obtain a survey of the methods available. Author indexes are also available. Your library may subscribe to *SciFinder Scholar*, the online access to *Chemical Abstracts*. You can search by chemical substance, topic, author, company name, and access abstracted journals. Once you have an article online, you can link to referenced articles in the paper.

You can also locate many relevant references for a specific problem by using Web search engines. Following is a selected list of some references in analytical chemistry. References to the various methods of measurements are included at the ends of the chapters covering these methods throughout the text.

A.1 Journals[1]

1. *American Laboratory*

2. *Analytical Biochemistry*

---

[1]The *Chemical Abstracts* abbreviation of each title is indicated by italics.

3. *Analytical Chimica Acta*
4. *Analytical Abstracts*
5. *Analytical Chemistry*[2]
6. *Analytical Instrumentation*
7. *Analytical Letters*
8. *Analyst*, The
9. *Applied Spectroscopy*
10. *Clinica Chimica Acta*
11. *Clinical Chemistry*
12. *Electroanalysis*
13. Journal of *AOAC International*
14. Journal of *Chromatographic Science*
15. Journal of *Chromatography*
16. Journal of *Electroanalytical Chemistry and Interfacial Electrochemistry*
17. *Microchemical Journal*
18. *Spectrochimica Acta*
19. *Talanta* (G. D. Christian, Joint Editor-in-Chief—take a look!)
20. Zeitschrift für *analytische Chemie*

A.2 General References

Some general references are given in Chapter 1, including encyclopedias, and more specific references and useful websites are given in the chapters throughout the text. Following are several classical references, general and substance specific, that provide much useful information for the analyst. They are available in many libraries.

1. *Annual Book of ASTM Book of Standards, 75 vols.* Philadelphia: American Society for Testing and Materials, 2000.
2. R. Belcher and L. Gordon, eds., *International Series of Monographs on Analytical Chemistry.* New York: Pergamon. A multivolume series.
3. N. H. Furman and F. J. Welcher, eds., *Scott's Standard Methods of Chemical Analysis*, 6th ed., 5 vols. New York: Van Nostrand, 1962–1966.
4. I. M. Kolthoff and P. J. Elving, eds., *Treatise on Analytical Chemistry.* New York: Interscience. A multivolume series.
5. I. M. Kolthoff, E. B. Sandell, E. J. Meehan, and S. Bruckenstein, *Quantitative Chemical Analysis*, 4th ed. London: Macmillan, 1969.
6. L. Meites, ed., *Handbook of Analytical Chemistry.* New York: McGraw-Hill, 1963.
7. C. N. Reilly, ed., *Advances in Analytical Chemistry and Instrumentation.* New York: Interscience. A multivolume series.
8. A. I. Vogel, *A Textbook of Quantitative Inorganic Analysis, Including Elementary Instrumental Analysis*, 3rd ed. London: Longman, 1972.

---

[2]This journal has a biannual volume every April that reviews in alternate years the literature of various analytical techniques and their applications in different areas of analysis.

9. C. L. Wilson and D. W. Wilson, *Comprehensive Analytical Chemistry*, G. Sveha, ed., New York: Elsevier. A multivolume series.

10. *Official Methods of Analysis of AOAC International*, 17th ed., W. Horwitz, ed. Gaithersburg, MD: AOAC International, 2000. Available on CD-ROM.

A.3 Inorganic Substances

1. *ASTM Methods for Chemical Analysis of Metals.* Philadelphia: American Society for Testing and Materials, 1956.

2. F. E. Beamish and J. C. Van Loon, *Analysis of Noble Metals.* New York: Academic, 1977.

3. T. R. Crompton, *Determination of Anions: A Guide for the Analytical Chemist.* Berlin: Springer, 1996.

A.4 Organic Substances

1. J. S. Fritz and G. S. Hammond, *Quantitative Organic Analysis.* New York: Wiley, 1957.

2. T. S. Ma and R. C. Rittner, *Modern Organic Elemental Analysis.* New York: Marcel Dekker, 1979.

3. J. Mitchell, Jr., I. M. Kolthoff, E. S. Proskauer, and A. W. Weissberger, eds., *Organic Analysis*, 4 vols. New York: Interscience, 1953–1960.

4. S. Siggia, Jr., and J. G. Hanna, *Quantitative Organic Analysis via Functional Group Analysis*, 4th ed. New York: Wiley, 1979.

5. A. Steyermarch, *Quantitative Organic Microanalysis*, 2nd ed. New York: Academic, 1961.

A.5 Biological and Clinical Substances

1. J. S. Annino, *Clinical Chemistry. Principles and Procedures*, 2nd ed. Boston: Little, Brown, 1960.

2. G. D. Christian and F. J. Feldman, *Atomic Absorption Spectroscopy. Applications in Agriculture, Biology, and Medicine.* New York: Wiley-Interscience, 1970.

3. D. Glick, ed., *Methods of Biochemical Analysis.* New York: Interscience, A multivolume series.

4. R. J. Henry, D. C. Cannon, and J. W. Winkelman, eds., *Clinical Chemistry. Principles and Techniques*, 2nd ed. Hagerstown, MD: Harper & Row, 1974.

5. M. Reiner and D. Seligson, eds., *Standard Methods of Clinical Chemistry.* New York: Academic. A multivolume series from 1953.

6. N. W. Tietz, ed., *Fundamentals of Clinical Chemistry*, 2nd ed. Philadelphia: W. B. Saunders, 1976.

A.6 Gases

1. C. J. Cooper and A. J. DeRose, *The Analysis of Gases by Gas Chromatography*. New York: Pergamon, 1983.

A.7 Water and Air Pollutants

1. *Quality Assurance Handbook for Air Pollution Measurement Systems*, U.S.E.P.A., Office of Research and Development, Environmental Monitoring and Support Laboratory, Research Triangle, NC 27711. Vol. I, *Principles.* Vol. II, *Ambient Air Specific Methods.*
2. *Standard Methods for the Examination of Water and Wastewater.* New York: American Public Health Association.

A.8 Occupational Health and Safety

1. National Institute of Occupational Health and Safety (NIOSH), P. F. O'Connor, ed., *Manual of Analytical Methods*, 4th ed. Washington, DC: DHHS (NIOSH) Publication No. 94-113 (August 1994).

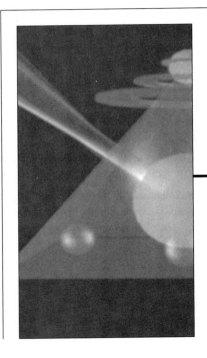

Appendix B

REVIEW OF MATHEMATICAL OPERATIONS: EXPONENTS, LOGARITHMS, AND THE QUADRATIC FORMULA

B.1 Exponents

It is convenient in mathematical operations, even when working with logarithms, to express numbers semiexponentially. Mathematical operations with exponents are summarized as follows:

$$N^a N^b = N^{a+b} \qquad \text{e.g., } 10^2 \times 10^5 = 10^7$$

$$\frac{N^a}{N^b} = N^{a-b} \qquad \text{e.g., } \frac{10^5}{10^2} = 10^3$$

$$(N^a)^b = N^{ab} \qquad \text{e.g., } (10^2)^5 = 10^{10}$$

$$\sqrt[a]{N^b} = N^{b/a} \qquad \text{e.g., } \sqrt{10^6} = 10^{6/2} = 10^3$$

$$\sqrt[3]{10^9} = 10^{9/3} = 10^3$$

The decimal point of a number is conveniently placed using the **semiexponential form**. The number is written with the decimal placed in the units position, and it is multiplied by 10 raised to an integral exponent equal to the number of spaces the decimal was moved to bring it to the units position. The exponent is negative if the decimal is moved to the right (the number is less than 1) and it is positive if the decimal is moved to the left (the number if 10 or greater). Some examples are:

Number	Semiexponential Form
0.00267	2.67×10^{-3}
0.48	4.8×10^{-1}
52	5.2×10^{1}
6027	6.027×10^{3}

Any number raised to the zero power is equal to unity. Thus, $10^0 = 1$, and 2.3 in semiexponential form is 2.3×10^0. Numbers between 1 and 10 do not require the semiexponential form to place the decimal.

B.2 Taking Logarithms of Numbers

It is convenient to place numbers in the semiexponential form for taking logarithms or for finding a number from its logarithm. The following rules apply:

$$N = b^a$$
$$\log_b N = a$$

or

$$N = 10^a$$
$$\log_{10} N = a$$

For example

$$\log 10^2 = 2$$
$$\log 10^{-3} = -3$$

Also,

$$\log ab = \log a + \log b$$

for example

$$\log 2.3 \times 10^{-3} = \log 2.3 + \log 10^{-3}$$
$$= 0.36 - 3$$
$$= -2.64$$
$$\log 5.67 \times 10^7 = \log 5.67 + \log 10^7$$
$$= 0.754 + 7$$
$$= 7.754$$

The exponent is actually the **characteristic** of the logarithm of a number, and the logarithm of the number between 1 and 10 is the **mantissa**. Thus, in the example $\log 2.3 \times 10^{-3}$, -3 is the characteristic and 0.36 is the mantissa.

B.3 Finding Numbers from Their Logarithms

The following relationships hold:

$$\log_{10} N = a$$
$$N = 10^a = \text{antilog } a$$

For example,

$$\log N = 0.371$$
$$N = 10^{0.371} = \text{antilog } 0.371 = 2.35$$

In general, to find a number from its logarithm, write the number in exponent form, and then break the exponent into the mantissa (m) and the characteristic (c). Then, take the antilog of the mantissa and multiply by the exponential form of the characteristic:

$$\log_{10} N = mc$$
$$N = 10^{mc} = 10^m \times 10^c$$
$$N = (\text{antilog } m) \times 10^c$$

For example,

$$\log N = 2.671$$
$$N = 10^{2.671} = 10^{.671} \times 10^2$$
$$= 4.69 \times 10^2 = 469$$
$$\log N = 0.326$$
$$N = 10^{0.326} = 2.12$$
$$\log N = -0.326$$
$$N = 10^{-0.326} = 10^{0.674} \times 10^{-1}$$
$$= 4.72 \times 10^{-1} = 0.472$$

Whenever the logarithm is a negative number, the exponent is broken into a negative integer (the characteristic) and a positive noninteger less than 1 (the mantissa), as in the last example. Note that in the example, the sum of the two exponents is equal to the original exponent (-0.326). Another example is

$$\log N = -4.723$$
$$N = 10^{-4.723} = 10^{0.277} \times 10^{-5}$$
$$= 1.89 \times 10^{-5} = 0.0000189$$

B.4 Finding Roots with Logarithms

It is a simple matter to find a given root of a number using logarithms. Suppose, for example, you wish to find the cube root of 325. Let N represent the cube root:

$$N = 325^{1/3}$$

Taking the logarithm of both sides,

$$\log N = \log 325^{1/3}$$

This $\frac{1}{3}$ can be brought out front:

$$\log N = \tfrac{1}{3} \log 325 = \tfrac{1}{3}(2.512) = 0.837$$
$$N = 10^{0.837} = \text{antilog } 0.837 = 6.87$$

B.5 The Quadratic Formula

A quadratic equation of the general form

$$ax^2 + bx + c = 0$$

can be solved by use of the quadratic formula:

$$x = \frac{-b \pm \sqrt{b^2 - 4ac}}{2a}$$

Quadratic equations are frequently encountered in calculation of equilibrium concentrations of ionized species using equilibrium constant expressions. Hence, an equation of the following type might require solving:

$$\frac{x^2}{1.0 \times 10^{-3} - x} = 8.0 \times 10^{-4}$$

or

$$x^2 = 8.0 \times 10^{-7} - 8.0 \times 10^{-4}x$$

Arranging in the quadratic equation form above, we have

$$x^2 + 8.0 \times 10^{-4}x - 8.0 \times 10^{-7} = 0$$

or

$$a = 1 \qquad b = 8.0 \times 10^{-4} \qquad c = -8.0 \times 10^{-7}$$

Therefore,

$$x = \frac{-8.0 \times 10^{-4} \pm \sqrt{(8.0 \times 10^{-4})^2 - 4(1)(-8.0 \times 10^{-7})}}{2(1)}$$

$$= \frac{-8.0 \times 10^{-4} \pm \sqrt{0.64 \times 10^{-6} + 3.2_0 \times 10^{-6}}}{2}$$

$$= \frac{-8.0 \times 10^{-4} \pm \sqrt{3.8_4 \times 10^{-6}}}{2}$$

$$= \frac{-8.0 \times 10^{-4} \pm 1.9_6 \times 10^{-3}}{2} = \frac{1.1_6 \times 10^{-3}}{2}$$

$$x = 5.8_0 \times 10^{-4}$$

A concentration can only be positive, and so the negative value of x is not a solution. You can use Excel Solver for solving quadratic equations. See Chapter 6.

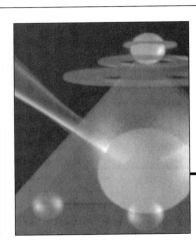

Appendix C

TABLES OF CONSTANTS

Table C.1
Dissociation Constants for Acids

Name	Formula	Dissociation Constant, at 25°C			
		K_{a1}	K_{a2}	K_{a3}	K_{a4}
Acetic	CH_3COOH	1.75×10^{-5}			
Alanine	$CH_3CH(NH_2)COOH^a$	4.5×10^{-3}	1.3×10^{-10}		
Arsenic	H_3AsO_4	6.0×10^{-3}	1.0×10^{-7}	3.0×10^{-12}	
Arsenious	H_3AsO_3	6.0×10^{-10}	3.0×10^{-14}		
Benzoic	C_6H_5COOH	6.3×10^{-5}			
Boric	H_3BO_3	6.4×10^{-10}			
Carbonic	H_2CO_3	4.3×10^{-7}	4.8×10^{-11}		
Chloroacetic	$ClCH_2COOH$	1.51×10^{-3}			
Citric	$HOOC(OH)C(CH_2COOH)_2$	7.4×10^{-4}	1.7×10^{-5}	4.0×10^{-7}	
Ethylenediaminetetraacetic	$(CO_2^-)_2NH^+CH_2CH_2NH^+(CO_2^-)_2{}^a$	1.0×10^{-2}	2.2×10^{-3}	6.9×10^{-7}	5.5×10^{-11}
Formic	$HCOOH$	1.76×10^{-4}			
Glycine	$H_2NCH_2COOH^b$	4.5×10^{-3}	1.7×10^{-10}		
Hydrocyanic	HCN	7.2×10^{-10}			
Hydrofluoric	HF	6.7×10^{-4}			
Hydrogen sulfide	H_2S	9.1×10^{-8}	1.2×10^{-15}		
Hypochlorous	$HOCl$	1.1×10^{-8}			
Iodic	HIO_3	2×10^{-1}			
Lactic	$CH_3CHOHCOOH$	1.4×10^{-4}			
Leucine	$(CH_3)_2CHCH_2CH(NH_2)COOH^b$	4.7×10^{-3}	1.8×10^{-10}		
Maleic	$cis\text{-}HOOCCH : CHCOOH$	1.5×10^{-2}	2.6×10^{-7}		
Malic	$HOOCCHOHCH_2COOH$	4.0×10^{-4}	8.9×10^{-6}		
Nitrous	HNO_2	5.1×10^{-4}			
Oxalic	$HOOCCOOH$	6.5×10^{-2}	6.1×10^{-5}		
Phenol	C_6H_5OH	1.1×10^{-10}			
Phosphoric	H_3PO_4	1.1×10^{-2}	7.5×10^{-8}	4.8×10^{-13}	
Phosphorous	H_3PO_3	5×10^{-2}	2.6×10^{-7}		
o-Phthalic	$C_6H_4(COOH)_2$	1.2×10^{-3}	3.9×10^{-6}		
Picric	$(NO_2)_3C_6H_5OH$	4.2×10^{-1}			
Propanoic	CH_3CH_2COOH	1.3×10^{-5}			

TABLE C.2 DISSOCIATION CONSTANTS FOR BASES 805

Table C.1 (continued)
Dissociation Constants for Acids

Name	Formula	Dissociation Constant, at 25°C			
		K_{a1}	K_{a2}	K_{a3}	K_{a4}
Salicyclic	$C_6H_4(OH)COOH$	1.0×10^{-3}			
Sulfamic	NH_2SO_3H	1.0×10^{-1}			
Sulfuric	H_2SO_4	$\gg 1$	1.2×10^{-2}		
Sulfurous	H_2SO_3	1.3×10^{-2}	5×10^{-6}		
Trichloroacetic	Cl_3COOH	1.29×10^{-1}			

[a]The first two carbonyl protons are most readily dissociated, with K_a values of 1.0 and 0.032, respectively. The protons on the more basic nitrogens are most tightly held (K_{a3} and K_{a4}).

[b]K_{a1} and K_{a2} for stepwise dissociation of the protonated form R—CH—CO$_2$H.

$$\mid$$

$$NH_3^+$$

Table C.2
Dissociation Constants for Bases

Name	Formula	Dissociation Constant, at 25°C	
		K_{b1}	K_{b2}
Ammonia	NH_3	1.75×10^{-5}	
Aniline	$C_6H_5NH_2$	4.0×10^{-10}	
1-Butylamine	$CH_3(CH_2)_2CH_2NH_2$	4.1×10^{-4}	
Diethylamine	$(CH_3CH_2)_2NH$	8.5×10^{-4}	
Dimethylamine	$(CH_3)_2NH$	5.9×10^{-4}	
Ethanolamine	$HOC_2H_4NH_2$	3.2×10^{-5}	
Ethylamine	$CH_3CH_2NH_2$	4.3×10^{-4}	
Ethylenediamine	$NH_2C_2H_4NH_2$	8.5×10^{-5}	7.1×10^{-8}
Glycine	$HOOCCH_2NH_2$	2.3×10^{-12}	
Hydrazine	H_2NNH_2	1.3×10^{-6}	
Hydroxylamine	$HONH_2$	9.1×10^{-9}	
Methylamine	CH_3NH_2	4.8×10^{-4}	
Piperidine	$C_5H_{11}N$	1.3×10^{-3}	
Pyridine	C_5H_5N	1.7×10^{-9}	
Triethylamine	$(CH_3CH_3)_3N$	5.3×10^{-4}	
Trimethylamine	$(CH_3)_3N$	6.3×10^{-5}	
Tris(hydroxymethyl)aminomethane	$(HOCH_2)_3CNH_2$	1.2×10^{-6}	
Zinc hydroxide	$Zn(OH)_2$		4.4×10^{-5}

Table C.3
Solubility Product Constants

Substance	Formula	K_{sp}
Aluminum hydroxide	$Al(OH)_3$	2×10^{-32}
Barium carbonate	$BaCO_3$	8.1×10^{-9}
Barium chromate	$BaCrO_4$	2.4×10^{-10}
Barium fluoride	BaF_2	1.7×10^{-6}
Barium iodate	$Ba(IO_3)_2$	1.5×10^{-9}
Barium manganate	$BaMnO_4$	2.5×10^{-10}
Barium oxalate	BaC_2O_4	2.3×10^{-8}
Barium sulfate	$BaSO_4$	1.0×10^{-10}
Beryllium hydroxide	$Be(OH)_2$	7×10^{-22}
Bismuth oxide chloride	$BiOCl$	7×10^{-9}
Bismuth oxide hydroxide	$BiOOH$	4×10^{-10}
Bismuth sulfide	Bi_2S_3	1×10^{-97}
Cadmium carbonate	$CdCO_3$	2.5×10^{-14}
Cadmium oxalate	CdC_2O_4	1.5×10^{-8}
Cadmium sulfide	CdS	1×10^{-28}
Calcium carbonate	$CaCO_3$	8.7×10^{-9}
Calcium fluoride	CaF_2	4.0×10^{-11}
Calcium hydroxide	$Ca(OH)_2$	5.5×10^{-6}
Calcium oxalate	CaC_2O_4	2.6×10^{-9}
Calcium sulfate	$CaSO_4$	1.9×10^{-4}
Copper(I) bromide	$CuBr$	5.2×10^{-9}
Copper(I) chloride	$CuCl$	1.2×10^{-6}
Copper(I) iodide	CuI	5.1×10^{-12}
Copper(I) thiocyanate	$CuSCN$	4.8×10^{-15}
Copper(II) hydroxide	$Cu(OH)_2$	1.6×10^{-19}
Copper(II) sulfide	CuS	9×10^{-36}
Iron(II) hydroxide	$Fe(OH)_2$	8×10^{-16}
Iron(III) hydroxide	$Fe(OH)_3$	4×10^{-38}
Lanthanum iodate	$La(IO_3)_3$	6×10^{-10}
Lead chloride	$PbCl_2$	1.6×10^{-5}
Lead chromate	$PbCrO_4$	1.8×10^{-14}
Lead iodide	PbI_2	7.1×10^{-9}
Lead oxalate	PbC_2O_4	4.8×10^{-10}
Lead sulfate	$PbSO_4$	1.6×10^{-8}
Lead sulfide	PbS	8×10^{-28}
Magnesium ammonium phosphate	$MgNH_2PO_4$	2.5×10^{-13}
Magnesium carbonate	$MgCO_3$	1×10^{-5}
Magnesium hydroxide	$Mg(OH)_2$	1.2×10^{-11}
Magnesium oxalate	MgC_2O_4	9×10^{-5}
Manganese(II) hydroxide	$Mn(OH)_2$	4×10^{-14}
Manganese(II) sulfide	MnS	1.4×10^{-15}
Mercury(I) bromide	Hg_2Br_2	5.8×10^{-23}
Mercury(I) chloride	Hg_2Cl_2	1.3×10^{-18}
Mercury(I) iodide	Hg_2I_2	4.5×10^{-29}
Mercury(II) sulfide	HgS	4×10^{-53}
Silver arsenate	Ag_3AsO_4	1.0×10^{-22}
Silver bromide	$AgBr$	4×10^{-13}
Silver carbonate	Ag_2CO_3	8.2×10^{-12}
Silver chloride	$AgCl$	1.0×10^{-10}
Silver chromate	Ag_2CrO_4	1.1×10^{-12}
Silver cyanide	$Ag[Ag(CN)_2]$	5.0×10^{-12}

TABLE C.4 FORMATION CONSTANTS FOR SOME EDTA METAL CHELATES **807**

Table C.3 *(continued)*
Solubility Product Constants

Substance	Formula	K_{sp}
Silver iodate	$AgIO_3$	3.1×10^{-8}
Silver iodide	AgI	1×10^{-16}
Silver phosphate	Ag_3PO_4	1.3×10^{-20}
Silver sulfide	Ag_2S	2×10^{-49}
Silver thiocyanate	$AgSCN$	1.0×10^{-12}
Strontium oxalate	SrC_2O_4	1.6×10^{-7}
Strontium sulfate	$SrSO_4$	3.8×10^{-7}
Thallium(I) chloride	$TlCl$	2×10^{-4}
Thallium(I) sulfide	Tl_2S	5×10^{-22}
Zinc ferrocyanide	$Zn_2Fe(CN)_6$	4.1×10^{-16}
Zinc oxalate	ZnC_2O_4	2.8×10^{-8}
Zinc sulfide	ZnS	1×10^{-21}

Table C.4
Formation Constants for Some EDTA Metal Chelates
($M^{n+} + Y^{4-} \rightleftharpoons MY^{n-4}$)

Element	Formula	K_f
Aluminum	AlY^-	1.35×10^{16}
Bismuth	BiY^-	1×10^{23}
Barium	BaY^{2-}	5.75×10^7
Cadmium	CdY^{2-}	2.88×10^{16}
Calcium	CaY^{2-}	5.01×10^{10}
Cobalt (Co^{2+})	CoY^{2-}	2.04×10^{16}
(Co^{3+})	CoY^-	1×10^{36}
Copper	CuY^{2-}	6.30×10^{18}
Gallium	GaY^-	1.86×10^{20}
Indium	InY^-	8.91×10^{24}
Iron (Fe^{2+})	FeY^{2-}	2.14×10^{14}
(Fe^{3+})	FeY^-	1.3×10^{25}
Lead	PbY^{2-}	1.10×10^{18}
Magnesium	MgY^{2-}	4.90×10^8
Manganese	MnY^{2-}	1.10×10^{14}
Mercury	HgY^{2-}	6.30×10^{21}
Nickel	NiY^{2-}	4.16×10^{18}
Scandium	ScY^-	1.3×10^{23}
Silver	AgY^{3-}	2.09×10^7
Strontium	SrY^{2-}	4.26×10^8
Thorium	ThY	1.6×10^{23}
Titanium (Ti^{3+})	TiY^-	2.0×10^{21}
(TiO^{2+})	$TiOY^{2-}$	2.0×10^{17}
Vanadium (V^{2+})	VY^{2-}	5.01×10^{12}
(V^{3+})	VY^-	8.0×10^{25}
(VO^{2+})	VOY^{2-}	1.23×10^{18}
Yttrium	YY^-	1.23×10^{18}
Zinc	ZnY^{2-}	3.16×10^{16}

Table C.5
Some Standard and Formal Reduction Electrode Potentials

Half-Reaction	E^0 (V)	Formal Potential (V)
$F_2 + 2H^+ + 2e^- = 2HF$	3.06	
$O_3 + 2H^+ + 2e^- = O_2 + H_2O$	2.07	
$S_2O_8^{2-} + 2e^- = 2SO_4^{2-}$	2.01	
$Co^{3+} + e^- = Co^{2+}$	1.842	
$H_2O_2 + 2H^+ + 2e^- = 2H_2O$	1.77	
$MnO_4^- + 4H^+ + 3e^- = MnO_2 + 2H_2O$	1.695	
$Ce^{4+} + e^- = Ce^{3+}$		1.70 (1 M HClO$_4$); 1.61 (1 M HNO$_3$); 1.44 (1 M H$_2$SO$_4$)
$HClO + H^+ + e^- = \frac{1}{2}Cl_2 + H_2O$	1.63	
$H_5IO_6 + H^+ + 2e^- = IO_3^- + 3H_2O$	1.6	
$BrO_3^- + 6H^+ + 5e^- = \frac{1}{2}Br_2 + 3H_2O$	1.52	
$MnO_4^- + 8H^+ + 5e^- = Mn^{2+} + 4H_2O$	1.51	
$Mn^{3+} + e^- = Mn^{2+}$		1.51 (8 M H$_2$SO$_4$)
$ClO_3^- + 6H^+ + 5e^- = \frac{1}{2}Cl_2 + 3H_2O$	1.47	
$PbO_2 + 4H^+ + 2e^- = Pb^{2+} = 2H_2O$	1.455	
$Cl_2 + 2e^- = 2Cl^-$	1.359	
$Cr_2O_7^{2-} + 14H^+ + 6e^- = 2Cr^{3+} + 7H_2O$	1.33	
$Tl^{3+} = 2e^- = Tl^+$	1.25	0.77 (1 M HCl)
$IO_3^- + 2Cl^- + 6H^+ + 4e^- = ICl_2^- + 3H_2O$	1.24	
$MnO_2 + 4H^+ + 2e^- = Mn^{2+} + 2H_2O$	1.23	
$O_2 + 4H^+ + 4e^- = 2H_2O$	1.229	
$2IO_3^- + 12H^+ + 10e^- = I_2 + 6H_2O$	1.20	
$SeO_4^{2-} + 4H^+ + 2e^- = H_2SeO_3 + H_2O$	1.15	
$Br_2(aq) + 2e^- = 2Br^-$	1.087[a]	
$Br_2(l) + 2e^- = 2Br^-$	1.065[a]	
$ICl_2^- + e^- = \frac{1}{2}I_2 + 2Cl^-$	1.06	
$VO_2^+ + 2H^+ + e^- = VO^{2+} + H_2O$	1.000	
$HNO_2 + H^+ + e^- = NO + H_2O$	1.00	
$Pd^{2+} + 2e^- = Pd$	0.987	
$NO_3^- + 3H^+ + 2e^- = HNO_2 + H_2O$	0.94	
$2Hg^{2+} + 2e^- = Hg_2^{2+}$	0.920	
$H_2O_2 + 2e^- = 2OH^-$	0.88	
$Cu^{2+} + I^- + e^- = CuI$	0.86	
$Hg^{2+} + 2e^- = Hg$	0.854	
$Ag^+ + e^- = Ag$	0.799	0.228 (1 M HCl); 0.792 (1 M HClO$_4$)
$Hg_2^{2+} + 2e^- = 2Hg$	0.789	0.274 (1 M HCl)
$Fe^{3+} + e^- = Fe^{2+}$	0.771	
$H_2SeO_3 + 4H^+ + 4e^- = Se + 3H_2O$	0.740	
$PtCl_4^{2-} + 2e^- = Pt + 4Cl^-$	0.73	
$C_6H_4O_2 \text{ (quinone)} + 2H^+ + 2e^- = C_6H_4(OH)_2$	0.699	0.696 (1 M HCl, H$_2$SO$_4$, HClO$_4$)
$O_2 + 2H^+ + 2e^- = H_2O_2$	0.682	
$PtCl_6^{2-} + 2e^- = PtCl_4^{2-} + 2Cl^-$	0.68	
$I_2(aq) + 2e^- = 2I^-$	0.6197[b]	
$Hg_2SO_4 + 2e^- = 2Hg + SO_4^{2-}$	0.615	
$Sb_2O_5 + 6H^+ + 4e^- = 2SbO^+ + 3H_2O$	0.581	
$MnO_4^- + e^- = MnO_4^{2-}$	0.564	
$H_3AsO_4 + 2H^+ + 2e^- = H_3AsO_3 + H_2O$	0.559	0.577 (1 M HCl, HClO$_4$)
$I_3^- + 2e^- = 3I^-$	0.5355	
$I_2(s) + 2e^- = 2I^-$	0.5345[b]	

Table C.5 *(continued)*
Some Standard and Formal Reduction Electrode Potentials

Half-Reaction	E^0 (V)	Formal Potential (V)
$Mo^{6+} + e^- = Mo^{5+}$		0.53 (2 M HCl)
$Cu^+ + e^- = Cu$	0.521	
$H_2SO_3 + 4H^+ + 4e^- = S + 3H_2O$	0.45	
$Ag_2CrO_4 + 2e^- = 2Ag + CrO_4^{2-}$	0.446	
$VO^{2+} + 2H^+ + e^- = V^{3+} + H_2O$	0.361	
$Fe(CN)_6^{3-} + e^- = Fe(CN)_6^{4-}$	0.36	0.72 (1 M HClO$_4$, H$_2$SO$_4$)
$Cu^{2+} = 2e^- = Cu$	0.337	
$UO_2^{2+} + 4H^+ + 2e^- = U^{4+} + 2H_2O$	0.334	
$BiO^+ + 2H^+ + 3e^- = Bi + H_2O$	0.32	
$Hg_2Cl_2(s) + 2e^- = 2Hg + 2Cl^-$	0.268	0.242 (sat'd KCl—SCE); 0.282 (1 M KCl)
$AgCl + e^- = Ag + Cl^-$	0.222	0.228 (1 M KCl)
$SO_4^{2-} + 4H^+ + 2e^- = H_2SO_3 + H_2O$	0.17	
$BiCl_4^- + 3e^- = Bi + 4Cl^-$	0.16	
$Sn^{4+} + 2e^- = Sn^{2+}$	0.154	0.14 (1 M HCl)
$Cu^{2+} + e^- = Cu^+$	0.153	
$S + 2H^+ + 2e^- = H_2S$	0.141	
$TiO^{2+} + 2H^+ + e^- = Ti^{3+} + H_2O$	0.1	
$Mo^{4+} + e^- = Mo^{3+}$		0.1 (4 M H$_2$SO$_4$)
$S_4O_6^{2-} + 2e^- = 2S_2O_3^{2-}$	0.08	
$AgBr + e^- = Ag + Br^-$	0.071	
$Ag(S_2O_3)_2^{3-} + e^- = Ag + 2S_2O_3^{2-}$	0.01	
$2H^+ + 2e^- = H_2$	0.000	
$Pb^{2+} + 2e^- = Pb$	−0.126	
$CrO_4^{2-} + 4H_2O + 3e^- = Cr(OH)_3 + 5OH^-$	−0.13	
$Sn^{2+} + 2e^- = Sn$	−0.136	
$AgI + e^- = Ag + I^-$	−0.151	
$CuI + e^- = Cu + I^-$	−0.185	
$N_2 + 5H^+ + 4e^- = N_2H_5^+$	−0.23	
$Ni^{2+} + 2e^- = Ni$	−0.250	
$V^{3+} + e^- = V^{2+}$	−0.255	
$Co^{2+} + 2e^- = Co$	−0.277	
$Ag(CN)_2^- + e^- = Ag + 2CN^-$	−0.31	
$Tl^+ + e^- = Tl$	−0.336	−0.551 (1 M HCl)
$PbSO_4 + 2e^- = Pb + SO_4^{2-}$	−0.356	
$Ti^{3+} + e^- = Ti^{2+}$	−0.37	
$Cd^{2+} + 2e^- = Cd$	−0.403	
$Cr^{3+} + e^- = Cr^{2+}$	−0.41	
$Fe^{2+} + 2e^- = Fe$	−0.440	
$2CO_2(g) + 2H^+ + 2e^- = H_2C_2O_4$	−0.49	
$Cr^{3+} + 3e^- = Cr$	−0.74	
$Zn^{2+} + 2e^- = Zn$	−0.763	
$2H_2O + 2e^- = H_2 + 2OH^-$	−0.828	
$Mn^{2+} + 2e^- = Mn$	−1.18	
$Al^{3+} + 3e^- = Al$	−1.66	
$Mg^{2+} + 2e^- = Mg$	−2.37	
$Na^+ + e^- = Na$	−2.714	
$Ca^{2+} + 2e^- = Ca$	−2.87	
$Ba^- + 2e^- = Ba$	−2.90	
$K^+ + e^- = K$	−2.925	
$Li^+ + e^- = Li$	−3.045	

[a]E^0 for Br$_2$(l) is used for saturated solutions of Br$_2$ while E^0 for Br$_2$(aq) is used for unsaturated solutions.
[b]E^0 for I$_2$(s) is used for saturated solutions of I$_2$, while E^0 for I$_2$(aq) is used for unsaturated solutions.

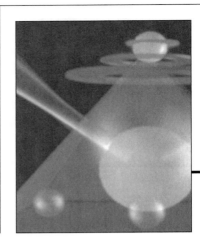

Appendix D

SAFETY IN THE LABORATORY

General Safety Rules

Good housekeeping practices will assure the safest working conditions in the laboratory. Always clean up spilled chemicals; do not leave broken or chipped glassware lying around; and put away all chemical bottles and apparatus when finished with them. Neutralize acid spills with sodium bicarbonate and alkali spills with boric acid. Mercury spills should be vacuumed up with a suction flask or dusted with sulfur powder. Clean up the mercury thoroughly because mercury vapors from fine droplets are highly toxic.

Most states by law require a person working in a chemical laboratory to wear protective eyeglasses. It is good practice in any event, to use them. You should locate fire extinguishers, exits, safety showers, eye fountains, and fire blankets in your laboratory. Any dangerous or potentially dangerous laboratory situation should be brought immediately to the attention of the laboratory supervisor.

Perform only the authorized experiments and never work alone. Never eat or drink in the laboratory. When working with volatile chemicals, as when heating acids or when using organic solvents, use the fume hood. Use a safety shield when working with potentially dangerous reactions. Special care should be taken when working with organic solvents. Many are flammable and many have been identified as acute or chronic toxic substances, frequently carcinogenic. Use rubber gloves when possible and avoid breathing fumes.

Appendix E

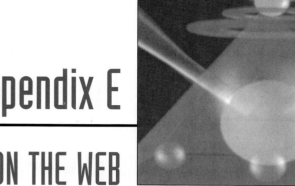

PERIODIC TABLES ON THE WEB

Atomic weights and atomic numbers of the elements, listed alphabetically, are given on the front inside cover of this text. The values are a composite of the best agreements among the tables in the websites listed below. Additional information about each element can be found from various versions of the periodic table. There are many available on the Web. Following are some recommended ones. In most, you can click on an element and obtain atomic weight and number and other properties such as isotopes, physical properties, and history of the element.

1. http://chemlab.pc.maricopa.edu/periodic/periodic.html. A pictorial periodic table. Elements are searchable by a number of parameters such as atomic weight, atomic radius, boiling point, and the like. There are links to other types of periodic tables, such as Mendeleev's table and a spiral table. Also included are the lyrics to Tom Leher's *The Elements.*

2. http://periodic.lanl.gov/default.html. This is from the Los Alamos National Laboratory Chemistry Division. This was named in 2001 as one of the best Web sources in science and technology by ScientificAmerican.com, part of *Scientific American* magazine.

3. www.chem.qmul.ac.uk/iupac/AtWt/table.html. This version of the periodic table is based on that recommended by the Commission on the Nomenclature of Inorganic Chemistry and published in *IUPAC Nomenclature of Inorganic Chemistry, Recommendations 1990.* Atomic weights are given to five significant figures. IUPAC references are given for more precise values.

4. www.chemsoc.org. This chemistry societies network links to a Virtual Element periodic table. You can download it as a screensaver for your computer.

5. www.stanford.edu/~glassman/chem/index.html. Besides an interactive periodic table, this site has a number of other useful pages, including molar conversions, unit conversions, an equation balancer, and links related to chemistry.

Appendix F

ANSWERS TO EVEN-NUMBERED PROBLEMS

CHAPTER 2

16. $V_{25°} = 25.071$ mL; $V_{20°} = 24.041$ mL

18. 0.05138 M

CHAPTER 3

4. (a) 5 (b) 4 (c) 3

6. 68.9466

8. $162._2$

10. To the nearest 0.01 g for three significant figures

12. (a) 128.0 g (b) 128.1 g (c) 1.9 g

14. (a) 0.052% S.D., 0.16% c.v. (b) 0.0021% S.D., 8.8% c.v.

16. (a) 0.052% (b) 0.026% (c) 0.027%

18. (a) 0.014 ± 0.0003 (b) 1.34 ± 0.03 (c) $11,990 \pm 40$

20. 0.5024–0.5030 M

22. ± 3.9 ppm

24. 0.1064–0.1072 M

26. $t = 0.8_7$, $t_{Table} = 2.365$. Therefore high probability both methods give same result.

28. $F = 2.79$, $F_{Table} = 4.88$. No significant difference.

30. $t = 3._6$ ($>T_{Table}$), 95% probability significant difference.

32. Zn: $Q = 0.70$, $Q_{Table} = 0.970$. Therefore all valid. Sn: $Q = 0.75$, $Q_{Table} = 0.970$. Therefore all valid.

34. 0.44 ppm

36. 139.0–140.2; 138.4–140.8

38. 3.05 ppm

40. 0.915

42. $t = 1.8_6$, $t_{Table} = 2.262$ at 95% C.L. No significant difference. $R = 0.999$, $r^2 = 0.998$, so high degree of correlation.

44. 1.6 g

CHAPTER 4

20. 16%

CHAPTER 5

10. (a) 5.23% (b) 55.0% (c) 1.82%

12. (a) 1.98 (b) 4.20 (c) 1.28 (d) 3.65 (e) 2.24 (f) 1.31

14. (a) 5.84×10^4 mg (b) 3.42×10^4 mg (c) 1.71×10^3 mg (d) 284 mg (e) 7.01×10^3 mg (f) 2.25×10^3 mg

16. 0.0333, 0.100, 0.001, 0.0333 mmol/mL of Mn^{2+}, NO_3^-, K^+, SO_4^{2-}

18. (a) 0.408 M (b) 0.300 M (c) 0.147 M

20. (a) 1.40 g (b) 8.08 g (c) 4.00 g

22. (a) 11.6 M (b) 15.4 M (c) 14.6 M (d) 17.4 M (e) 14.8 M

24. 1.06×10^3 mg/L

26. (a) 10.0 ppm (b) 27.8 ppm (c) 15.8 ppm (d) 16.3 ppm (e) 13.7 ppm (f) 29.8 ppm

28. (a) 0.123% (b) 1.23‰ (c) 1.23×10^3 ppm

30. 156 mL

32. 5.00 M

34. 100 mL

36. 0.172%

38. 390 mg

40. 9.25 mg/dL

42. 15.3%

44. 7.53 mL

46. 22.7 mL

48. 84.4%

50. 47.1%

52. 1.52 mg

54. 20.0 mg Fe_2O_3/mL $KMnO_4$

56. (a) 36.46 g/eq (b) 85.67 g/eq (c) 389.91 g/eq (d) 41.04 g/eq (e) 60.05 g/eq

58. (a) 128.1 g/eq (b) 64.05 g/eq

60. (a) 151.91 g/eq (b) 17.04 g/eq (c) 17.00 g/eq (d) 17.00 g/eq

62. 0.608 eq/L

64. 0.180 eq/L

66. 0.474 g $KHC_2O_4 \cdot H_2C_2O_4$/g $Na_2C_2O_4$

68. 84.5 meq/L

70. 8.76 g/L

72. (a) 0.132_8 g (b) 1.31_0 g

CHAPTER 6

2. $[A] = 4.3 \times 10^{-7}$ M, $[B] = 0.30$ M, $[C] = 0.80$ M

4. 0.085%

5. 1.1×10^{-22}

8. (a) $3[Bi^{3+}] + [H^+] = 2[S^{2-}] + [HS^-] + [OH^-]$ (b) $[Na^+] + [H^+] = 2[S^{2-}] + [HS^-] + [OH^-]$

12. $2[Ba^{2+}] = 3([PO_4^{3-}] + [HPO_4^{2-}] + [H_2PO_4^-] + [H_3PO_4])$

14. (a) 0.30 (b) 0.90 (c) 0.90 (d) 3.3

16. 0.96_5

18. 0.0019 M

20. (a) $K_a f_{H^+} f_{CN^-}$ (b) $K_b f_{NH_4^+} f_{OH^-}$

CHAPTER 7

6. (a) pH = 1.70, pOH = 12.30 (b) pH = 3.89, pOH = 10.11 (c) pH = −0.08, pOH = 14.08 (d) pH = pOH = 7.00 (e) pH = 6.55, pOH = 7.45

8. (a) 3.8×10^{-10} M (b) 5.0×10^{-14} M (c) 1.0×10^{-7} M (d) 5.3×10^{-15} M

10. pH = 12.70, pOH = 1.30

12. 2.3×10^{-7} M; 6.64 pH

14. 3.2%

16. 18_2 g/mol

18. 8.80

20. 0.014_5 M

22. 11.54

24. Fourfold

26. 10.57

28. 2.92

30. 8.72

32. 7.00

34. 2.54

36. 4.16

38. 12.96

40. 10.98

42. 4.05

44. 5.12

46. 9.24

48. 7.3 g

50. 2.62

52. 2.0×10^{-3} M HPO_4^{2-}, 9.6×10^{-4} M $H_2PO_4^-$

54. 0.868 M HOAc and OAc$^-$

56. 0.24 mL H_3PO_4, 5.4 g KH_2PO_4

60. $K_b = K_b^\circ / f_{BH^+} f_{OH^-}$

62. pH = 3.88, [OAc$^-$] = $10^{-3.88}$ M

66. [H$_2$A] $\approx C_{H_2A} = 10^{-3}$ M in very acid solution; log [HA$^-$] = −6.40 + pH; slope = +1 for pH < pK_{a1}; [A^{2-}] $\approx C_{A^{2-}} = 10^{-3}$ M in very basic solution; log [HA$^-$] = 2.05 − pH, slope = −1 for pH > pK_{a2}

CHAPTER 8

14. 0.1025 M

16. 0.1174 M

18. 0 mL: 13.00, 10.0 mL: 12.70, 25.0 mL: 7.00, 30.0 mL: 1.90

20. 0 mL: 11.12, 10.0 mL: 9.84, 25.0 mL: 9.24, 50.0 mL: 5.27, 60.0 mL: 2.04

22. 0%: 9.72, 25%: 7.60, 50%: 7.12, 75%: 6.64, 100%: 4.54, 150%: 1.96

24. 35.5 mL

26. 62.8%

28. 0.0200 M H_3PO_4, 0.0300 M HCl

30. 403 mg Na_2CO_3, 110 mg NaOH

32. 3.00 mmol Na_2CO_3, 1.5_0 mmol NaOH

34. $25_.2$%

CHAPTER 9

4. 0.0033 M

6. 2.0×10^{-5} M

8. (a) $2.7_8 \times 10^7$ (b) $3.9_0 \times 10^{17}$

10. 1.2_6. Yes

12. 0.1039_8 M

14. 10.01 (mg $CaCO_3$/L H_2O)/mL EDTA

16. 9.93 mg/dL; 4.95 meq/L

18. 3.04 ppm

CHAPTER 10

10. 16.2 g

12. 0.2138, 1.902, 0.1314, 0.6474

14. 98.68%

16. 1.071%

18. 26 mL

20. 24.74 g

22. 42.5% Ba, 37.5% Ca

24. 0.846 g AgCl, 1.154 g AgBr

26. 8.20×10^{-19}

28. 1.9×10^{-8} M

30. 5.1×10^{-7} M; 1.0×10^{-9} M

32. 1.6×10^{-4} g

34. AB: $s = 2 \times 10^{-9}$ M, $AC_2 : s = 1 \times 10^{-6}$ M

36. 0.33 M excess F^- needed. Is feasible.

38. 2.0×10^{-5} M

CHAPTER 11

4. 6.1×10^{-3} M CaF_2, 1.20×10^{-2} M HF, $8.0_1 \times 10^{-5}$ M F^-

6. 8.4×10^{-3} M AgCl, $1.2_0 \times 10^{-8}$ M Ag^+, $5.9_6 \times 10^{-5}$ M $Ag(en)^+$, 8.4×10^{-3} M $Ag(en)_2^+$

8. $2_.7 \times 10^{-5}$ M

10. 5.434 g/L

CHAPTER 12

10. Ni, H_2S, Sn^{2+}, V^{3+}, I^-, Ag, Cl^-, Co^{2+}, HF

12. (a) Pt/Fe^{2+}, $Fe^{3+}//Cr_2O_7^{2-}$, Cr^{3+}, H^+/Pt (b) Pt/I^-, $I_2//IO_3^-$, I_2, H^+/Pt
    (c) $Zn/Zn^{2+}//Cu^{2+}/Cu$ (d) Pt/H_2SeO_3, SeO_4^{2-}, $H^+//Cl_2$, Cl^-/Pt

14. 1.34 V

16. 0.65 V

18. $PtCl_6^{2-} + 2V^{2+} = PtCl_4^{2-} + 2V^{3+} + 2Cl^-$, 0.94V

20. $2VO_2^+ + U^{4+} = 2VO^{2+} + UO_2^{2+}$, 0.67 V

CHAPTER 13

8. $5_4 \times 10^{-13}$

10. (a) -0.028 V (b) 0.639 V (c) 0.84 V

12. (a) 0.845 V (b) -0.020 V (c) -0.497 V

14. (a) $1 \times 10^{-4}\%$ (b) 0.08_4 V (c) 0.261 V

16. -0.505 V

18. (a) 5.78 (b) 10.14 (c) 12.32 (D) 13.89

20. 11.2

22. 0.015 M

24. 4.8×10^{-4} M

26. 0.020

28. $1.0_7 \times 10^4$

CHAPTER 14

6. (a) $IO_3^- + 5I^- + 6H^+ = 3I_2 + 3H_2O$ (b) $2Se_2Cl_2 + 3H_2O = H_2SeO_3 + 3Se + 4Cl^- + 4H^+$ (c) $H_3PO_3 + 2HgCl_2 + H_2O = H_3PO_4 + Hg_2Cl_2 + 2H^+ + 2Cl^-$

8. 1.12_7 V

10. 10.0 mL: 0.715 V, 50.0 mL: 0.771 V, 100 mL: 1.36 V, 200 mL: 1.46 V

12. $E = [(1)\ E^0_{Fe^{3+}Fe^{2+}} + (2)\ E^0_{Sn^{4+}Sn^{2+}}]/[(1) + (2)]$

14. (a) 0.319 V, -0.780 V, 1.6×10^{12} (b) 0.691 V, -0.154 V, 2.6×10^{22}

16. 4.96 meq/L

18. 1.47 mL

CHAPTER 15

10. $[Fe^{3+}]/[Fe^{2+}] = 5:1$

CHAPTER 16

30. 0.25 μm, 250 nm

32. $20,000 - 150,000$ Å; $5000 - 670$ cm^{-1}

34. 0.70 A, 0.10 A, 0.56 T, 0.10 T

36. 4.25×10^3 cm^{-1} mol^{-1} L

38. 5.15×10^4 cm^{-1} mol^{-1} L

40. 0.528 g

42. (a) 0.193 g/day (b) 4.91 mmol/L (c) 0.25

44. 0.405 ppm

46. 2.61 mg/L

CHAPTER 17

20. 13%

22. 190 ppm

24. 84 ppm

CHAPTER 18

12. $D = K_D(1 + 2K_P K_D[HB_z])_a/(1 + K_a[H^+]_a)$

14. 48%

16. 96.2% extracted with 10 mL; 99.45% extracted with 2×5 mL

CHAPTER 19

8. $1.0_6 \times 10^3$ cm

10. $R_s = 0.96$. Not quite resolved.

CHAPTER 20

16. $\Delta m = 0.12$

18. 20.9 ppm

CHAPTER 21

26. 270 mmol/L

28. (a) HCl (b) H_2SO_4 (c) $HClO_4$ (d) H_2SO_4

CHAPTER 22

12. 48.5 s

14. 12.7 h

16. 3.82%

CHAPTER 23

10. 133 μL

CHAPTER 25

16. ATTCGATTCCGTA

CHAPTER 26

6. 3.3×10^{-6} g/L_{air}

Index

GC, *see* Gas chromatography
GC-MS, 577 (*see also* Gas chromatography-mass
  spectrometry)
Gel filtration chromatography, 621
Gel permeation chromatography, 621
Gene expression, 693
Gene expression profiling, 705
Gene sequencing, 695
Genetic code, 706
Genomic library, 695, 699
Genomics, 706
GF, *see* Gravimetric factor
gfw, 142
Gibbs free energy, 192 (*see also* Free energy)
  and equilibrium constant, 192
Gibbs-Stockholm convention, 357
Glass pH electrode,
  acid error, 389
  alkaline error, 388
  calibration, 385, 747
  high pH (full range), 389
  mechanism, 386
  principle of, 384
Glassware calibration, 39, 730
Glassware cleaning, 729
Globar, 485
Globulin determination, 683
GLP, *see* Good laboratory practice
Glucose,
  enzymatic determination, 785
  spectrophotometric determination, 655
  standard solution, 786
Glucose electrode, 453
Glucose oxidase, 651
Glycolysis, 680
GMP, *see* Good manufacturing practice
gmw, 142
Golay equation, 565
Gooch crucible, 47
Good laboratory practice, 124, 125
Good manufacturing practice, 135
Grab sample, 8
Grades of chemicals, *inside back cover*
Gradient elution, 615
  reequilibration time, 615
Gram-atomic weight, 141
Gram-formula weight, 142
Gram-molecular weight, 142 (*see also* Formula weight)
Gran plot, 437
  advantage, 439
  from first derivative, 439
  volume correction, 439
Graphing,
  using spreadsheets, 78
Grating, *see* Diffraction grating
Gravimetric analysis, 11
  examples, 324
  precipitation conditions, 316
  precipitation process, 315
  steps of, 314
  weight relationships, 180
Gravimetric calculations, 320
  mixtures, 324
Gravimetric factor, 181, 320
  examples, 183

Gross sample, 8
Ground state, 462
Groundwater sampling, 721
Guanine, 695
Guard columns, 612
Gylcolyis,
  prevention, 741

H
Hair shampoo pH determination, 746
Half-life,
  first order reaction, 644
  second-order reaction, 645
Half-reaction, 414
Half-reaction potential, 356, 415
Half-wave potential, 449
Hamilton syringe cleaning , 782
Hapten, 688
Headspace analysis, 590
Hemolysis, 680
Henderson-Hasselbalch equation, 235
  thermodynamic equilibrium constant, 255
Heparin, 680
Heptane, GC determination, 781
Heterogeneous equilibria, 209
Hexane, GC determination, 781
High performance liquid chromatography, 604
  (*see also* HPLC)
  gradient elution, 615
  high speed, 616
  high temperature, 618
  narrow-bore columns, 616
  principles of, 605
  stationary phases, 606 (*see also* HPLC
    stationary phases)
High speed gas chromatography, 592
High speed HPLC, 616
High-performance thin-layer chromatography, 630
Hollow cathode lamp, 526
Holographic grating, 488
Horseradish peroxidase, 655
HPLC, 604 (*see also* High performance liquid
  chromatography)
  backpressure, 608
HPLC equipment, 609
  column designs, 610
  detectors, 612
  mobile-phase system, 609
HPLC injectors, 610
HPLC method development, 613
  column selection, 615
  solvent selection, 614
HPLC-MS, 618
  atmospheric pressure chemical ionization
    interface, 620
  electrospray ionization interface, 619
  low bleed bonded phases for, 608
  narrow-bore columns, 617
  particle beam interface, 619
  quadrupole mass filter, 620
  thermospray interface, 619
HPLC solvents, 610
HPLC stationary phases,
  high-purity silica, 606
  hybrid silica/polymer, 609

microporous particles, 607
  monolithic, 608
  nonpolar bonded, 607
  nonporous packings, 608
  perfusion packings, 608
Huber equation, 566
Human Genome Project, 695
Hybridization, 698
Hydrochloric acid,
  standard solution of, 44
  standardization, 739
Hydrogen electrode, 373 (*see also* NHE)
Hydrolysis, 230
Hydrolysis constant,
  salt of weak acid, 230
  salt of weak base, 233
Hydronium ion, 222
Hydrophilic, 318
Hydrophobic, 318
Hydrostatic injection, 634
8-Hydroxyquinoline, *see* Oxine
Hyperchromic shift, 465
Hyperchromism, 465
Hypochlorite,
  iodometric titration, 753
Hypochromism, 465
Hypsochromic shift, 465

I
Immunoassay,
  heterogeneous, 689
  homogeneous, 689
  principles of, 685
Immunoassay specificity, 688
Immunoglobulin, 685
Immunology, 685
Impactor, 717
Impedance, high input, 380
Impinger, 717
Inclusion in precipitates, 318
Indeterminate error, 67 (*see also* Random error)
Indicator, 159, 270 (*see also* pH indicator)
  adsorption, 349, 350
  chelometric, 305
  pH transition range, 271
  precipitation, 349
Indicator electrode, 370, 374
Infrared radiation,
  absorption of, 469
Infrared spectra, 469
  quantitative measurements, 483
Ingamell's sampling constant, 114
Inner filter effect, 509
Instability constant, 296
Instrumental analysis, 11
Intercept,
  spreadsheet calculation, 109
  standard deviation of, 104
Interferogram, 499
Interferometer, 485
Interferometer operation, 499
Internal conversion, 506
Internal standard,
  in atomic spectrometry, 533
  in gas chromatography, 589

pH INDICATORS[a]

pH Indicators			pH Transition Intervals		
Cresol red	pink	0.2		1.8	yellow
m-Cresol purple	red	1.2		2.8	yellow
Thymol blue	red	1.2		2.8	yellow
p-Xylenol blue	red	1.2		2.8	yellow
2,2′,2″,4,4′-Pentamethoxy-triphenylcarbinol	red	1.2		3.2	colorless
2,4-Dinitrophenol	colorless	2.8		4.7	yellow
4-Dimethylaminoazobenzene	red	2.9		4.0	yellow–orange
Bromochlorophenol blue	yellow	3.0		4.6	purple
Bromophenol blue	yellow	3.0		4.6	purple
Methyl orange	red	3.1		4.4	yellow–orange
Bromocresol green	yellow	3.8		5.4	blue
2,5-Dinitrophenol	colorless	4.0		5.8	yellow
Alizarinsulfonic acid sodium salt	yellow	4.3		6.3	violet
Methyl red	red	4.4		6.2	yellow–orange
Methyl red sodium salt	red	4.4		6.2	yellow–orange
Chlorophenol red	yellow	4.8		6.4	purple
Hematoxylin	yellow	5.0		7.2	violet
Litmus extra pure	red	5.0		8.0	blue
Bromophenol red	orange–yellow	5.2		6.8	purple
Bromocresol purple	yellow	5.2		6.8	purple
4-Nitrophenol	colorless	5.4		7.5	yellow
Bromoxylenol blue	yellow	5.7		7.4	blue
Alizarin	yellow	5.8		7.2	red
Bromothymol blue	yellow	6.0		7.6	blue
Phenol red	yellow	6.4		8.2	red
3-Nitrophenol	colorless	6.6		8.6	yellow–orange
Neutral red	bluish-red	6.8		8.0	orange–yellow
4,5,6,7-Tetrabromophenolphthalein	colorless	7.0		8.0	purple
Cresol red	orange	7.0		8.8	purple
1-Naphtholphthalein	brownish	7.1		8.3	blue–green
m-Cresol purple	yellow	7.4		9.0	purple
Thymol blue	yellow	8.0		9.6	blue
p-Xylenol blue	yellow	8.0		9.6	blue
Phenolphthalein	colorless	8.2		9.8	red–violet
Thymolphthalein	colorless	9.3		10.5	blue
Alizarin yellow GG	light yellow	10.2		12.1	brownish-yellow
Epsilon blue	orange	11.6		13.0	violet

[a]Adapted from *pH Indicators*, E. Merck and Co.